Macromolecular Engineering

Edited by
Krzysztof Matyjaszewski,
Yves Gnanou,
and Ludwik Leibler

1807–2007 Knowledge for Generations

Each generation has its unique needs and aspirations. When Charles Wiley first opened his small printing shop in lower Manhattan in 1807, it was a generation of boundless potential searching for an identity. And we were there, helping to define a new American literary tradition. Over half a century later, in the midst of the Second Industrial Revolution, it was a generation focused on building the future. Once again, we were there, supplying the critical scientific, technical, and engineering knowledge that helped frame the world. Throughout the 20th Century, and into the new millennium, nations began to reach out beyond their own borders and a new international community was born. Wiley was there, expanding its operations around the world to enable a global exchange of ideas, opinions, and know-how.

For 200 years, Wiley has been an integral part of each generation's journey, enabling the flow of information and understanding necessary to meet their needs and fulfill their aspirations. Today, bold new technologies are changing the way we live and learn. Wiley will be there, providing you the must-have knowledge you need to imagine new worlds, new possibilities, and new opportunities.

Generations come and go, but you can always count on Wiley to provide you the knowledge you need, when and where you need it!

William J. Pesce
President and Chief Executive Officer

Peter Booth Wiley
Chairman of the Board

Macromolecular Engineering

Precise Synthesis, Materials Properties, Applications

Edited by
Krzysztof Matyjaszewski, Yves Gnanou, and Ludwik Leibler

Volume 2
Elements of Macromolecular Structural Control

WILEY-VCH Verlag GmbH & Co. KGaA

The Editors

Prof. Dr. Krzysztof Matyjaszewski
Carnegie Mellon University
Department of Chemistry
4400 Fifth Ave
Pittsburgh, PA 15213
USA

Prof. Dr. Yves Gnanou
Laboratoire de Chimie des Polymères Organiques
16, ave Pey-Berland
33607 Pessac
France

Prof. Dr. Ludwik Leibler
UMR 167 CNRS-ESPCI
École Supérieure de Physique
et Chimie Industrielles
10 rue Vauquelin
75231 Paris Cedex 05
France

Library of Congress Card No.: applied for

British Library Cataloguing-in-Publication Data
A catalogue record for this book is available from the British Library.

Bibliographic information published by the Deutsche Nationalbibliothek
The Deutsche Nationalbibliothek lists this publication in the Deutsche Nationalbibliografie; detailed bibliographic data are available in the Internet at http://dnb.d-nb.de.

Composition K+V Fotosatz GmbH, Beerfelden
Printing betz-druck GmbH, Darmstadt
Bookbinding Litges & Dopf GmbH, Heppenheim
Cover Grafik-Design Schulz, Fußgönheim
Wiley Bicentennial Logo Richard J. Pacifico

Printed in the Federal Republic of Germany
Printed on acid-free paper

ISBN 978-3-527-31446-1

Contents

Macromolecular Engineering. Precise Synthesis, Materials Properties, Applications.
Edited by K. Matyjaszewski, Y. Gnanou, and L. Leibler

ISBN: 978-3-527-31446-1

Volume 2
Elements of Macromolecular Structural Control

Volume 3
Structure-Property Correlation and Characterization Techniques

Volume 4
Applications

Preface

Macromolecular Engineering: From Precise Macromolecular Synthesis to Macroscopic Materials Properties and Applications aims to provide a broad overview of recent developments in precision macromolecular synthesis and in the design and applications of complex polymeric assemblies of controlled sizes, morphologies and properties. The contents of this interdisciplinary book are organized in four volumes so as to capture and chronicle best, on the one hand, the rapid advances made in the control of polymerization processes and the design of macromolecular architectures (Volumes I and II) and, on the other, the noteworthy progress witnessed in the processing methods – including self-assembly and formulation – to generate new practical applications (Volumes III and IV).

Each chapter in this book is a well-documented and yet concise contribution written by noted experts and authorities in their field. We are extremely grateful to all of them for taking time to share their knowledge and popularize it in a way understandable to a broad readership. We are also indebted to all the reviewers whose comments and remarks helped us very much in our editing work. Finally, Wiley-VCH deserves our sincere acknowledgements for striving to keep the entire project on time.

We expect that specialist readers will find *Macromolecular Engineering: From Precise Macromolecular Synthesis to Macroscopic Materials Properties and Applications* an indispensable book to update their knowledge, and non-specialists will use it as a valuable companion to stay informed about the newest trends in polymer and materials science.

November 2006
Pittsburgh, USA
Bordeaux, France
Paris, France

Krzysztof Matyjaszewski
Yves Gnanou
Ludwik Leibler

Macromolecular Engineering. Precise Synthesis, Materials Properties, Applications.
Edited by K. Matyjaszewski, Y. Gnanou, and L. Leibler

ISBN: 978-3-527-31446-1

List of Contributors

Volker Abetz
GKSS Research Centre Geesthacht GmbH
Institute of Polymer Research
Max-Planck-Straße 1
21502 Geesthacht
Germany

Basudam Adhikari
Indian Institute of Technology
Materials Science Centre
Polymer Division
Kharagpur 721302
India

Takuzo Aida
University of Tokyo
School of Engineering
Department of Chemistry and Biotechnology
7-3-1 Hongo, Bunkyo-ku
Tokyo 113-8656
Japan

Michaël Alexandre
Materia Nova Research Centre asbl
Parc Initialis
1 avenue Nicolas Copernic
7000 Mons
Belgium

Xavier Allonas
University of Haute Alsace, ENSCMu
Department of General Photochemistry, UMR 7525 CNRS
3 Alfred Werner street
68093 Mulhouse Cedex
France

Denis Andrienko
Max Planck Institute for Polymer Research
Ackermannweg 10
55128 Mainz
Germany

René Androsch
Martin Luther University Halle-Wittenberg
Institute of Materials Science
06099 Halle
Germany

Markus Antonietti
Max Planck Institute of Colloids and Interfaces
Colloid Department
Research Campus Golm
14424 Potsdam
Germany

Macromolecular Engineering. Precise Synthesis, Materials Properties, Applications.
Edited by K. Matyjaszewski, Y. Gnanou, and L. Leibler

ISBN: 978-3-527-31446-1

Clemens Auschra
CIBA Specialty Chemicals, Inc.
Research and Development
Coating Effects
Schwarzwaldallee 215
4002 Basel
Switzerland

Ramakrishnan Ayothi
Cornell University
Department of Materials Science
and Engineering
Bard Hall
Ithaca, NY 14853-1501
USA

Vincent Ball
Institut National de la Santé
et de la Recherche Médicale
INSERM Unité 595
11 rue Humann
67085 Strasbourg Cedex
France
and
Université Louis Pasteur
Faculté de Chirurgie Dentaire
1 place de l'Hôpital
67000 Strasbourg
France

Nitash P. Balsara
University of California
Department of Chemical Engineering
and Materials Sciences
and Environmental Energy
Technologies Divisions
Lawrence Berkeley National
Laboratory
Berkeley, CA 94720
USA

Nadia Benkirane-Jessel
Institut National de la Santé
et de la Recherche Médicale
INSERM Unité 595
11 rue Humann
67085 Strasbourg Cedex
France
and
Université Louis Pasteur
Faculté de Chirurgie Dentaire
1 place de l'Hôpital
67000 Strasbourg
France

Christopher W. Bielawski
The University of Texas at Austin
Department of Chemistry
and Biochemistry
Austin, TX 78712
USA

Hans G. Börner
Max Planck Institute of Colloids
and Interfaces
Colloid Department
Research Campus Golm
14424 Potsdam
Germany

Redouane Borsali
Université Bordeaux 1
CNRS, ENSCPB
Laboratoire de Chimie des Polymères
Organiques
16 avenue Pey-Berland
33607 Pessac
France

Bernard Boutevin
Ingénierie et Architectures Macromoléculaires
Institut Gerhardt, UMR 5253
Ecole Nationale Supérieure Chimie de Montpellier
8 rue de l'Ecole Normale
34296 Montpellier
France

Cyrille Boyer
Ingénierie et Architectures Macromoléculaires
Institut Gerhardt, UMR 5253
Ecole Nationale Supérieure Chimie de Montpellier
8 rue de l'Ecole Normale
34296 Montpellier
France

Daniel Bratton
Cornell University
Department of Materials Science and Engineering
Bard Hall
Ithaca, NY 14853-1501
USA

Wade A. Braunecker
Carnegie Mellon University
Department of Chemistry
4400 Fifth Avenue
Pittsburgh, PA 15213
USA

Cyril Brochon
Laboratoire d'Ingénierie des Polymères pour les Hautes Technologies
UMR 7165 CNRS
Ecole Européenne Chimie Polymères Matériaux
Université Louis Pasteur
25 rue Becquerel
67087 Strasbourg
France

Manoj B. Charati
University of Delaware
Department of Materials Science and Engineering
201 DuPont Hall
and
Delaware Biotechnology Institute
15 Innovation Way
Newark, DE 19716
USA

Bernadette Charleux
Université Pierre et Marie Curie
Laboratoire de Chimie des Polymères
4, Place Jussieu, Tour 44, 1er étage
75252 Paris Cedex 5
France

Yong Mei Chen
Hokkaido University
Section of Biological Sciences
Faculty of Science
Laboratory of soft & wet matter
North 10, West 8
060-0810 Sapporo
Japan

Zhiyun Chen
Washington University in Saint Louis
Center for Materials Innovation and
Department of Chemistry
One Brookings Drive
St. Louis, MO 63130-4899
USA

Chong Cheng
Washington University in Saint Louis
Center for Materials Innovation and
Department of Chemistry
One Brookings Drive
St. Louis, MO 63130-4899
USA

Eric Cloutet
Université Bordeaux 1
Laboratoire de Chimie des Polymères
Organiques
Unité Mixte de Recherche
(UMR 5629) CNRS
ENSCPB
16 avenue Pey-Berland
33607 Pessac Cedex
France

Geoffrey W. Coates
Cornell University
Department of Chemistry
and Chemical Biology
Baker Laboratory
Ithaca, NY 14853
USA

Helmut Cölfen
Max Planck Institute of Colloids
and Interfaces
Colloid Chemistry
Research Campus Golm
Am Mühlenberg 1
14476 Potsdam-Golm
Germany

Henri Cramail
Université Bordeaux 1
Laboratoire de Chimie des Polymères
Organiques
Unité Mixte de Recherche
(UMR 5629) CNRS
ENSCPB
16 avenue Pey-Berland
33607 Pessac Cedex
France

Costantino Creton
Laboratoire PPMD
ESPCI
10 rue Vauquelin
75231 Paris
France

Ghislain David
Ingénierie et Architectures
Macromoléculaires
Institut Gerhardt, UMR 5253
Ecole Nationale Supérieure Chimie
de Montpellier
8 rue de l'Ecole Normale
34296 Montpellier
France

Priyadarsi De
University of Massachusetts Lowell
Polymer Science Program
Department of Chemistry
One University Avenue
Lowell, MA 01854
USA

Gero Decher
Institut Charles Sadron
(C.N.R.S. UPR 022)
6 rue Boussingault
67083 Strasbourg Cedex
France
and
Université Louis Pasteur
Faculté de Chimie
1 rue Blaise Pascal
67008 Strasbourg Cedex
France

Alain Deffieux
Université Bordeaux 1
CNRS, ENSCPB
Laboratoire de Chimie des Polymères Organiques
16 avenue Pey-Berland
33607 Pessac
France

Timothy J. Deming
University of California, Los Angeles
Department of Bioengineering
420 Westwood Plaza
7523 Boelter Hall
Los Angeles, CA 90095
USA

G. Dubois
IBM Almaden Research Center
650 Harry Road
San Jose, CA 95120
USA

Philippe Dubois
Université de Mons-Hainaut
Matériaux Polymères et Composites
Place du Parc 20
7000 Mons
Belgium

Andrzej Duda
Center of Molecular and
Macromolecular Studies
Polish Academy of Sciences
Department of Polymer Chemistry
Sienkiewicza 112
90-363 Łodz
Poland

Karel Dušek
Academy of Sciences
of the Czech Republic
Institute of Macromolecular
Chemistry
Heyrovského nám. 2
162 06 Praha
Czech Republic

Miroslava Dušková-Smrčková
Academy of Sciences
of the Czech Republic
Institute of Macromolecular
Chemistry
Heyrovského nám. 2
162 06 Praha
Czech Republic

Todd Emrick
University of Massachusetts Amherst
Department of Polymer Science
and Engineering
120 Governors Drive
Amherst, MA 01003
USA

Tom A. P. Engels
Eindhoven University of Technology
Department of Mechanical
Engineering
P.O. Box 513
5600 MB Eindhoven
The Netherlands

Robin S. Farmer
University of Delaware
Department of Materials Science
and Engineering
201 DuPont Hall
and
Delaware Biotechnology Institute
15 Innovation Way
Newark, DE 19716
USA

Rudolf Faust
University of Massachusetts Lowell
Polymer Science Program
Department of Chemistry
One University Avenue
Lowell, MA 01854
USA

Nelson Felix
Cornell University
Department of Materials Science
and Engineering
Bard Hall
Ithaca, NY 14853-1501
USA

Michel Fontanille
Université Bordeaux 1
Laboratoire de Chimie des Polymères
Organiques
ENSCPB
16 avenue Pey-Berland
33607 Pessac
France

Jean Pierre Fouassier
University of Haute Alsace,
ENSCMu
Department of General
Photochemistry, UMR 7525 CNRS
3 Alfred Werner street
68093 Mulhouse Cedex
France

Takeshi Fukuda
Kyoto University
Institute for Chemical Research
Uji, Kyoto 611-0011
Japan

François Ganachaud
Ecole Nationale Supérieure de Chimie
de Montpellier
Laboratoire de Chimie
Macromoléculaire
8 rue de l'Ecole Normale
34296 Montpellier Cedex 5
France

David S. Germack
Washington University in Saint Louis
Center for Materials Innovation and
Department of Chemistry
One Brookings Drive
St. Louis, MO 63130-4899
USA

Yves Gnanou
Université Bordeaux 1
Laboratoire de Chimie des Polymères
Organiques
ENSCPB
16 avenue Pey-Berland
33607 Pessac
France

Jian Ping Gong
Hokkaido University
Section of Biological Sciences
Faculty of Science
Laboratory of soft & wet matter
North 10, West 8
060-0810 Sapporo
Japan

Padma Gopalan
University of Wisconsin – Madison
Department of Materials Science and Engineering
1117 Engineering Research Building
1500 Engineering Drive
Madison, WI 53706
USA

Leon E. Govaert
Eindhoven University of Technology
Department of Mechanical Engineering
P.O. Box 513
5600 MB Eindhoven
The Netherlands

Stefanie M. Gravano
University of California, Davis
Department of Chemistry
One Shields Avenue
Davis, CA 95616-5295
USA

Andrew C. Grimsdale
Nanyang Technological University
School of Materials Science and Engineering
50 Nanyang Avenue
Singapore 639798

Nikos Hadjichristidis
University of Athens
Department of Chemistry
Panepistimiopolis Zografou
15771 Athens
Greece

Georges Hadziioannou
Laboratoire d'Ingénierie des Polymères pour les Hautes Technologies
UMR 7165 CNRS
Ecole Européenne Chimie Polymères Matériaux
Université Louis Pasteur
25 rue Becquerel
67087 Strasbourg
France

Dale L. Handlin, Jr.
Kraton Polymers
700 Milam
North Tower
Houston, TX 77002
USA

James L. Hedrick
IBM Almaden Research Center
650 Harry Road
San Jose, CA 95120
USA

Thomas Heiser
Université Louis Pasteur
Institut d'Electronique du Solide et des Systèmes
UMR 7163, CNRS
23 rue du Loess
67087 Strasbourg
France

Valérie Héroguez
Université Bordeaux 1
Laboratoire de Chimie des Polymères Organiques
ENSCPB
16 avenue Pey-Berland
33607 Pessac
France

Richard Hoogenboom
Eindhoven University of Technology
and Dutch Polymer Institute (DPI)
Laboratory of Macromolecular
Chemistry and Nanoscience
P.O. Box 513
5600 MB Eindhoven
The Netherlands

Jeffrey A. Hubbell
Ecole Polytechnique Fédérale
de Lausanne
Institute of Bioengineering
1015 Lausanne
Switzerland

Hermis Iatrou
University of Athens
Department of Chemistry
Panepistimiopolis Zografou
15771 Athens
Greece

Olli Ikkala
Helsinki University of Technology
Department of Engineering Physics
and Mathematics
and Center for New Materials
P.O. Box 2200
02015 Hut
Espoo
Finland

Woo-Dong Jang
The University of Tokyo
School of Engineering
Department of Chemistry
and Biotechnology
7-3-1 Hongo, Bunkyo-ku
Tokyo 113-8656
Japan

Sreeja Jayant
Rutgers, The State University
of New Jersey
Department of Pharmaceutics
160 Frelinghuysen Road
Piscataway, NJ 08854-8020
USA

Robert Jerome
University of Liège
Center for Education and Research
on Macromolecules (CERM)
Sart-Tilman, B6a
4000 Liège
Belgium

Youngjong Kang
Massachusetts Institute of Technology
Department of Materials Science
and Engineering and
Institute for Soldier Nanotechnologies
Cambridge, MA 02139
USA

Jayant J. Khandare
Rutgers, The State University
of New Jersey
Department of Pharmaceutics
160 Frelinghuysen Road
Piscataway, NJ 08854-8020
USA

Kristi L. Kiick
University of Delaware
Department of Materials Science
and Engineering
201 DuPont Hall
and
Delaware Biotechnology Institute
15 Innovation Way
Newark, DE 19716
USA

Tatsuki Kitayama
Osaka University
Department of Chemistry
Graduate School of Engineering
Toyonaka, Osaka 560-8531
Japan

Harm-Anton Klok
Ecole Polytechnique Fédérale
de Lausanne (EPFL)
Institut des Matériaux
Laboratoire des Polymères
STI – IMX – LP, MXD 112
(Bâtiment MXD), Station 12
1015 Lausanne
Switzerland

Bert Klumperman
Eindhoven University of Technology
Laboratory of Polymer Chemistry
P.O. Box 513
5600 MB Eindhoven
The Netherlands

Shiro Kobayashi
Kyoto Institute of Technology
R&D Center for Bio-based Materials
Matsugasaki, Sakyo-ku
Kyoto 606-8585
Japan

Kurt Kremer
Max Planck Institute for Polymer
Research
Ackermannweg 10
55128 Mainz
Germany

Vahik Krikorian
Massachusetts Institute of Technology
Department of Materials Science
and Engineering and
Institute for Soldier Nanotechnologies
Cambridge, MA 02139
USA

Przemyslaw Kubisa
Center of Molecular and
Macromolecular Studies
Polish Academy of Sciences
Department of Polymer Chemistry
Sienkiewicza 112
90-363 Łodz
Poland

Alok Kumar Sen
Indian Institute of Technology
Materials Science Centre
Polymer Division
Kharagpur 721302
India

Jacques Lalevée
University of Haute Alsace, ENSCMu
Department of General
Photochemistry, UMR 7525 CNRS
3 Alfred Werner street
68093 Mulhouse Cedex
France

Katharina Landfester
University of Ulm
Department of Organic Chemistry III
– Macromolecular Chemistry
and Organic Materials
Albert-Einstein-Allee 11
89081 Ulm
Germany

Philippe Lavalle
Institut National de la Santé
et de la Recherche Médicale
INSERM Unité 595
11 rue Humann
67085 Strasbourg Cedex
France
and
Université Louis Pasteur
Faculté de Chirurgie Dentaire
1 place de l'Hôpital
67000 Strasbourg
France

Nicolas Leclerc
Université Louis Pasteur
Laboratoire d'Ingénierie des Polymères pour les Hautes Technologies
UMR 7165 CNRS
Ecole Européenne Chimie Polymères Matériaux
25 rue Becquerel
67087 Strasbourg
France

Young Moo Lee
Hanyang University
School of Chemical Engineering
College of Engineering
Seoul 133-791
Korea

Ludwik Leibler
Matière Molle et Chimie
UMR 167 CNRS
ESPCI
10 rue Vauquelin
75005 Paris
France

G. B. W. L. Ligthart
DSM Campus Geleen
Performance Materials
PO Box 18
6160 MD Geleen
The Netherlands

Wei-Shi Li
ERATO-SORST Nanospace Project
Japan Science and Technology Agency (JST)
National Museum of Emerging Science and Innovation
2-41 Aomi, Koto-ku
Tokyo 135-0064
Japan

Pierre Lutz
Institut Charles Sadron
6 rue Boussingault
67083 Strasbourg Cedex
France

Ian Manners
University of Bristol
Department of Chemistry
Cantock's Close
Bristol BS8 1TS
UK

Andrew F. Mason
IBM Almaden Research Center
650 Harry Road
San Jose, CA 95120
USA

Toshio Masuda
Kyoto University
Department of Polymer Chemistry
Graduate School of Engineering
Katsura Campus
Kyoto 615-8510
Japan

Krzysztof Matyjaszewski
Carnegie Mellon University
Department of Chemistry
4400 Fifth Avenue
Pittsburgh, PA 15213
USA

Thomas C. B. McLeish
University of Leeds
IRC in Polymer Science and Technology
Polymers and Complex Fluids
Department of Physics and Astronomy
Leeds LS2 9JT
UK

Michael A. R. Meier
Eindhoven University of Technology
and Dutch Polymer Institute (DPI)
Laboratory of Macromolecular
Chemistry and Nanoscience
P.O. Box 513
5600 MB Eindhoven
The Netherlands

E. W. Meijer
Eindhoven University of Technology
Laboratory of Macromolecular
and Organic Chemistry
P.O. Box 513
5600 MB Eindhoven
The Netherlands

Han E. H. Meijer
Eindhoven University of Technology
Department of Mechanical
Engineering
P.O. Box 513
5600 MB Eindhoven
The Netherlands

R. D. Miller
IBM Almaden Research Center
650 Harry Road
San Jose, CA 95120
USA

Tamara Minko
Rutgers, The State University
of New Jersey
Department of Pharmaceutics
160 Frelinghuysen Road
Piscataway, NJ 08854-8020
USA

Martin Moller
RWTH Aachen
Institut für Technische
und Makromolekulare Chemie
Pauwelsstraße 8
52056 Aachen
Germany

Hideharu Mori
Yamagata University
Faculty of Engineering
Department of Polymer Science
and Engineering
4-3-16, Jonan
Yonezawa 992-8510
Japan

Klaus Müllen
Max Planck Institute
for Polymer Research
Ackermannweg 10
55128 Mainz
Germany

Axel H. E. Müller
University of Bayreuth
Macromolecular Chemistry II
95440 Bayreuth
Germany

Anna Musyanovych
University of Ulm
Department of Organic Chemistry III
– Macromolecular Chemistry
and Organic Materials
Albert-Einstein-Allee 11
89081 Ulm
Germany

Kazukiyo Nagai
Meiji University
Department of Applied Chemistry
1-1-1 Higashi-mita, Tama-ku
Kawasaki 214-8571
Japan

Christopher K. Ober
Cornell University
Department of Materials Science
and Engineering
310 Bard Hall
Ithaca, NY 14853-1501
USA

Joëlle Ogier
Institut National de la Santé
et de la Recherche Médicale
INSERM Unité 595
11 rue Humann
67085 Strasbourg Cedex
France
and
Université Louis Pasteur
Faculté de Chirurgie Dentaire
1 place de l'Hôpital
67000 Strasbourg
France

Masashi Ohmae
Kyoto University
Department of Materials Chemistry
Graduate School of Engineering
Katsura, Nishikyo-ku
Kyoto 615-8510
Japan

Kohji Ohno
Kyoto University
Institute for Chemical Research
Uji, Kyoto 611-0011
Japan

Masami Okamoto
Advanced Polymeric Materials
Engineering
Graduate School of Engineering
Toyota Technological Institute
2-12-1 Hisakata, Tempaku
Nagoya 468-8511
Japan

Joost A. Opsteen
Radboud University Nijmegen
Institute for Molecules and Materials
Toernooiveld 1
6525 ED Nijmegen
The Netherlands

Yoshihito Osada
Hokkaido University
Section of Biological Sciences
Faculty of Science
Laboratory of soft & wet matter
North 10, West 8
Sapporo 060-0810
Japan

Timothy E. Patten
University of California, Davis
Department of Chemistry
One Shields Avenue
Davis, CA 95616-5295
USA

Stanislaw Penczek
Centre of Molecular
and Macromolecular Studies
Polish Academy of Sciences
Department of Polymer Chemistry
Sienkiewicza 112
90-363 Łodz
Poland

Catherine Picart
Université de Montpellier II
Laboratoire de Dynamique
des Interactions Membranaires
Normales et Pathologiques
(C.N.R.S. UMR 5235)
Place Eugène Bataillon
34095 Montpellier Cedex
France

Frank Pirrung
CIBA Specialty Chemicals, Inc.
Research and Development
Coating Effects
Schwarzwaldallee 215
4002 Basel
Switzerland

Marinos Pitsikalis
University of Athens
Department of Chemistry
Panepistimiopolis Zografou
15771 Athens
Greece

Jeffrey Pyun
University of Arizona
Department of Chemistry
1306 E. University Boulevard
Tucson, AZ 85721
USA

Damien Quémener
Université Bordeaux 1
Laboratoire de Chimie des Polymères
Organiques
ENSCPB
16 avenue Pey-Berland
33607 Pessac
France

Karunakaran Radhakrishnan
University of Akron
Institute of Polymer Science
170, University Avenue
Akron, OH 44325
USA

Wolfgang Radke
Deutsches Kunststoff-Institut
Darmstadt
Schlossgartenstraße 6
65289 Darmstadt
Germany

David A. Rider
University of Toronto
Department of Chemistry
80 St. George Street
M5S 3H6 Toronto, Ontario
Canada

Megan L. Ruegg
University of California
Department of Chemical Engineering
Berkeley, CA 94720
USA

Pierre Schaaf
Institut Charles Sadron
(C.N.R.S. UPR 022)
6 rue Boussingault
67083 Strasbourg Cedex
France
and
Ecole Européenne de Chimie
Polymères et Materiaux
25 rue Bequerel
67087 Strasbourg Cedex 2
France

Oren A. Scherman
University of Cambridge
Department of Chemistry
Lensfield Road
Cambridge CB2 1EW
UK

Helmut Schlaad
Max Planck Institute of Colloids
and Interfaces
Colloid Department
Research Campus Golm
14424 Potsdam
Germany

Ulrich S. Schubert
Eindhoven University of Technology
and Dutch Polymer Institute (DPI)
Laboratory of Macromolecular
Chemistry and Nanoscience
P.O. Box 513
5600 MB Eindhoven
The Netherlands

Bernard Senger
Institut National de la Santé
et de la Recherche Médicale
INSERM Unité 595
11 rue Humann
67085 Strasbourg Cedex
France
and
Université Louis Pasteur
Faculté de Chirurgie Dentaire
1 place de l'Hôpital
67000 Strasbourg
France

Sergei S. Sheiko
University of North Carolina
at Chapel Hill
Department of Chemistry
Chapel Hill, NC 27599-3290
USA

Rint P. Sijbesma
Eindhoven University of Technology
Laboratory of Macromolecular
and Organic Chemistry
P.O. Box 513
5600 MB Eindhoven
The Netherlands

Peter F. W. Simon
GKSS Research Centre
Geesthacht GmbH
Institute of Polymer Research
Max-Planck-Straße
21502 Geesthacht
Germany

Stanislaw Slomkowski
Center of Molecular and
Macromolecular Studies
Polish Academy of Sciences
Department of Polymer Chemistry
Sienkiewicza 112
90-363 Łodz
Poland

Hans Wolfgang Spiess
Max-Planck-Institute for Polymer
Research
Spectroscopy
P.O. Box 3148
55021 Mainz
Germany

Richard J. Spontak
North Carolina State University
Departments of Chemical and
Biomolecular Engineering and
Materials Science and Engineering
Raleigh, NC 27695
USA

R. F. T. Stepto
The University of Manchester
School of Materials
Materials Science Centre
Polymer Science
and Technology Group
Grosvenor Street
Manchester M1 7HS
UK

Brent S. Sumerlin
Southern Methodist University
Department of Chemistry
3215 Daniel Avenue
Dallas, TX 75240-0314
USA

Yeshayahu Talmon
Technion – Israel Institute
of Technology
Department of Chemical Engineering
32000 Haifa
Israel

M. Atilla Tasdelen
Istanbul University
Department of Chemistry
Maslak
Istanbul 34469
Turkey

Daniel Taton
Université Bordeaux 1
Laboratoire de Chimie des Polymères
Organiques
LCPO CNRS
ENSCPB
16 avenue Pey-Berland
33607 Pessac
France

Gerrit ten Brinke
University of Groningen
Laboratory of Polymer Chemistry
Materials Science Centre
Nijenborgh 4
9747 AG Groningen
The Netherlands

Edwin L. Thomas
Massachusetts Institute of Technology
Department of Materials Science
and Engineering
and
Institute for Soldier Nanotechnologies
Cambridge, MA 02139
USA

Scott Trenor
Kraton Polymers
Houston, TX 77002
USA

Constantinos Tsitsilianis
University of Patras
and FORTH/ICEHT
Department of Chemical Engineering
Karatheodori 1
26504 Patras
Greece

Yoshinobu Tsujii
Kyoto University
Institute for Chemical Research
Uji, Kyoto 611-0011
Japan

Jan C. M. van Hest
Radboud University Nijmegen
Institute for Molecules and Materials
Toernooiveld 1
6525 ED Nijmegen
The Netherlands

Brooke A. van Horn
Washington University in Saint Louis
Center for Materials Innovation and
Department of Chemistry
One Brookings Drive
St. Louis, MO 63130-4899
USA

Shrinivas Venkataraman
Washington University in Saint Louis
Center for Materials Innovation and
Department of Chemistry
One Brookings Drive
St. Louis, MO 63130-4899
USA

Jean-Claude Voegel
Institut National de la Santé
et de la Recherche Médicale
INSERM Unité 595
11 rue Humann
67085 Strasbourg Cedex
France
and
Université Louis Pasteur
Faculté de Chirurgie Dentaire
1 place de l'Hôpital
67000 Strasbourg
France

C. Grant Willson
The University of Texas at Austin
Departments of Chemistry
and Biochemistry
and Chemical Engineering
Austin, TX 78712
USA

Karen L. Wooley
Washington University in Saint Louis
Center for Materials Innovation and
Department of Chemistry
One Brookings Drive
St. Louis, MO 63130-4899
USA

Kathryn Wright
Kraton Polymers
700 Milam
North Tower
Houston, TX 77002
USA

Bernhard Wunderlich
University of Tennessee
Department of Chemistry
200 Baltusrol Road
Knoxville, TN 37992-3707
USA

Yusuf Yagci
Istanbul University
Department of Chemistry
Maslak
Istanbul 34469
Turkey

Tsutomu Yokozawa
Kanagawa University
Department of Material
and Life Chemistry
Rokkakubashi, Kanagawa-ku
Yokohama 221-8686
Japan

1
Tacticity

Tatsuki Kitayama

One of the important aspects of polymer chemistry has been the precise control of polymerization reactions, which leads to well-defined polymer structures such as molecular weight (MW) and its distribution (MWD), structure of end-groups, stereoregularity, sequence of monomer units, branch structure and topology. This chapter deals with the stereoisomerism in the structure of polymers as a consequence of the polymerization reaction, which often strongly affects the properties of the polymer. Thus the control of the stereochemical structure of a polymer is one of the important steps toward precise control of polymer properties, which is required for advanced polymeric materials such as functional and specialty polymers.

The need to consider stereoisomerism in the chain polymerization of vinyl monomers was recognized early by Staudinger et al. [1]. With just one earlier finding of stereoregular poly(vinyl ether) by Schildknecht et al. [2] in 1947, the field of stereospecific polymerization actually came into existence when Ziegler et al. [3] and Natta et al. [4] developed new polymerization systems which exhibited unique stereoregulating powers in olefin polymerization. For the precise control of the structures of polymer molecule, a combination of these two types of polymerizations is desirable, that is, stereospecific living polymerization [5]. Isotactic (*it*-) living polymerization was first reported in the copolymerization of aldehydes with $(C_2H_5)_2AlNPh_2$ in 1965. Although the initiator efficiency was very low [6]. The term "stereospecific living polymerization" was proposed by Soum and Fontanille in 1980 [7] regarding the isotactic living polymerization of 2-vinylpyridine in benzene with an organomagnesium compound. Later, a syndiotactic (*st*-) living polymerization was also reported for 2-isopropenylpyridine in tetrahydrofuran (THF) with alkyllithium [8].

In the field of stereospecific polymerization, structural analysis of a polymer with respect to stereochemical features is particularly important. This chapter, therefore, first deals with analytical approaches for such a purpose and some basic concepts of tacticity. Then several specific tactic controls of polymerizations are discussed with a focus on vinyl polymerization. Although stereospecific polymerizations of α-olefins are very important historically and industrially, the examples employed are not intended to cover this field in a comprehensive manner.

Macromolecular Engineering. Precise Synthesis, Materials Properties, Applications.
Edited by K. Matyjaszewski, Y. Gnanou, and L. Leibler

ISBN: 978-3-527-31446-1

1.1
Definition of Tacticity [9, 10]

In polymers of vinyl monomers $CH_2{=}CHX$ or vinylidene monomers $CH_2{=}CXY$, the main-chain carbons having substituent group(s), X or X and Y, are termed "pseudoasymmetric" since, if the chain ends are disregarded, such carbons do not have the four different substituents necessary to qualify for being truly asymmetric. Nevertheless, successive alignments of these stereoisomeric sites along the chain imply regularity or the irregularity in the relative configurations or handedness. The simplest regular arrangements along a chain are the isotactic structure, in which all the substituents are located on the same side of the zigzag plane representing the chain stretched out in an all-trans conformation. The structure is often represented by a "rotated Fischer projection" in which the main chain skeleton is represented by a horizontal line (Scheme 1.1). Another regular arrangement is the syndiotactic structure, in which the groups alternate from side to side and thus the configurations of the neighboring units are opposite (Scheme 1.2).

The smallest unit representing the relative configuration of the consecutive monomeric units, as seen in Scheme 1.3, is termed a "diad" (or "dyad" in older references). For a vinyl polymer, two types of diads should be considered, which are designated as "meso" (*m*) and "racemo" (*r*). Using these notations, a se-

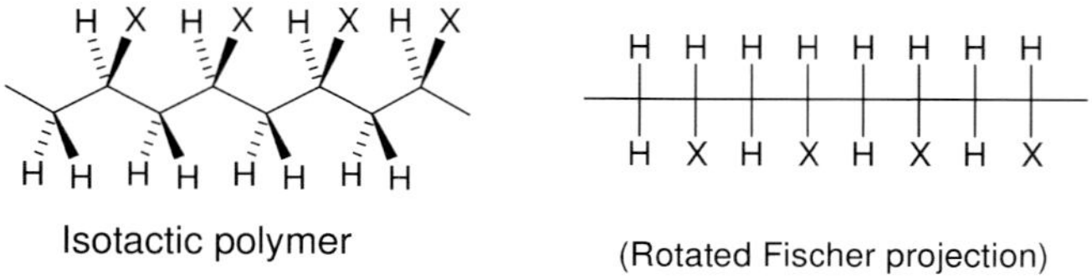

Scheme 1.1

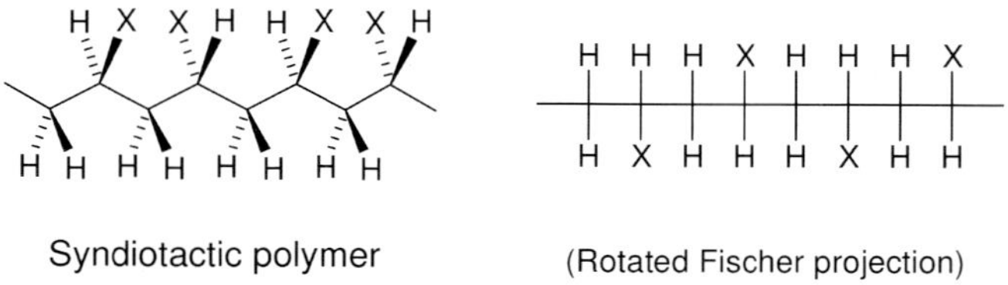

Scheme 1.2

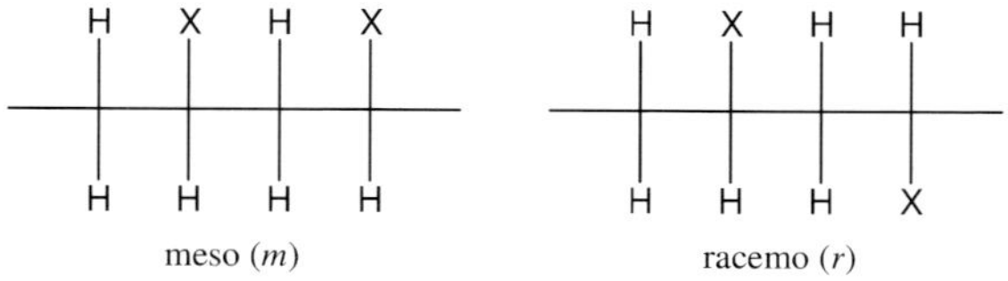

Scheme 1.3

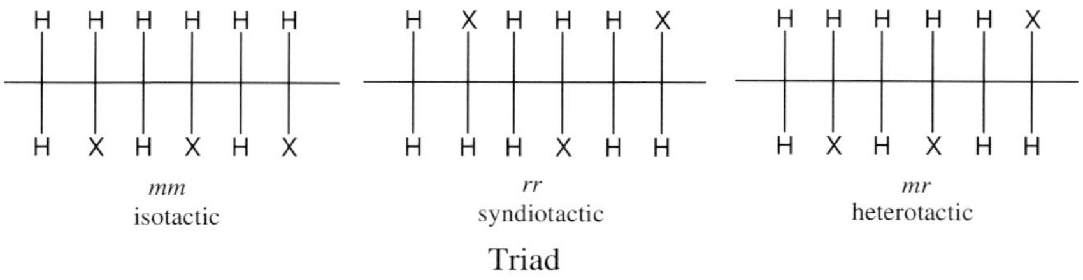

Scheme 1.4

quence in an isotactic polymer can be represented as *–mmmmmm–* and that in a syndiotactic polymer as *–rrrrrrr–*. In reality, however, purely isotactic or syndiotactic polymers are rarely obtainable but the extent of the regularity is the matter of concern. "Tacticity" is the term used for defining such stereochemical features of polymers. The IUPAC Gold Book lists the term tacticity as "The orderliness of the succession of configurational repeating units in the main chain of a regular macromolecule, a regular oligomer molecule, a regular block or a regular chain".

By extending the notation of *m* and *r*, one can define relative configurations for the longer monomeric units along the chain as triad, tetrad, pentad and so on (*n*-ad in general). Scheme 1.4 shows three possible triads, represented by *mm*, *rr* and *mr* (*rm* triad is the same as *mr* triad in vinyl polymers and therefore not shown), which are also named isotactic, syndiotactic and heterotactic triads, respectively. Similarly, for tetrad, the following six distinguishable sequences are possible: *mmm, mmr, rmr, rrr, rrm, mrm*, and for pentad: *mmmm, mmmr, rmmr, mmrm, mmrr, rmrm, rmrr, mrrm, mrrr, rrrr.*

1.2 Methods of Stereochemical Assignments

NMR spectroscopy is the only analytical means that provides quantitative data on tacticity. Tacticity determination of a vinyl polymer by ^{1}H NMR spectroscopy was first achieved for poly(methyl methacrylate) (PMMA) by Bovey and Tiers [11] and Nishioka et al. [12] independently in 1960. The assignments were made based on the consideration of the magnetic equivalency of the main-chain methylene protons in meso and racemo diads. The meso diad has no symmetry axis and H_a and H_b are non-equivalent (Scheme 1.5 a). Thus H_a and H_b have different chemical shifts and exhibit AB quartet signals due to geminal spin coupling. Figure 1.1 a shows the ^{1}H NMR spectrum of an isotactic PMMA comprising almost exclusively of *m* diads, measured in nitrobenzene-d_5 at 110 °C, in which H_a and H_b signals are observed at 1.81 and 2.44 ppm as an AB quartet with a coupling constant of 14.6 Hz. Figure 1.1 b is the spectrum of a syndiotactic PMMA, showing a singlet methylene proton signal at 2.10 ppm, which indicates that the methylene protons in the racemo diad are magnetically equivalent

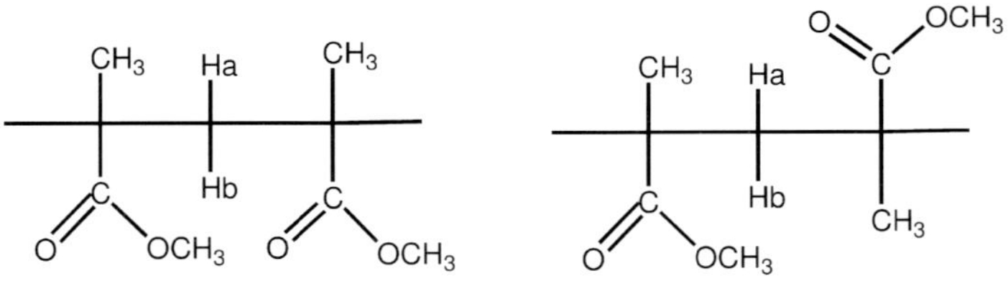

(a) *meso (m)* (b) *racemo (r)*

Scheme 1.5

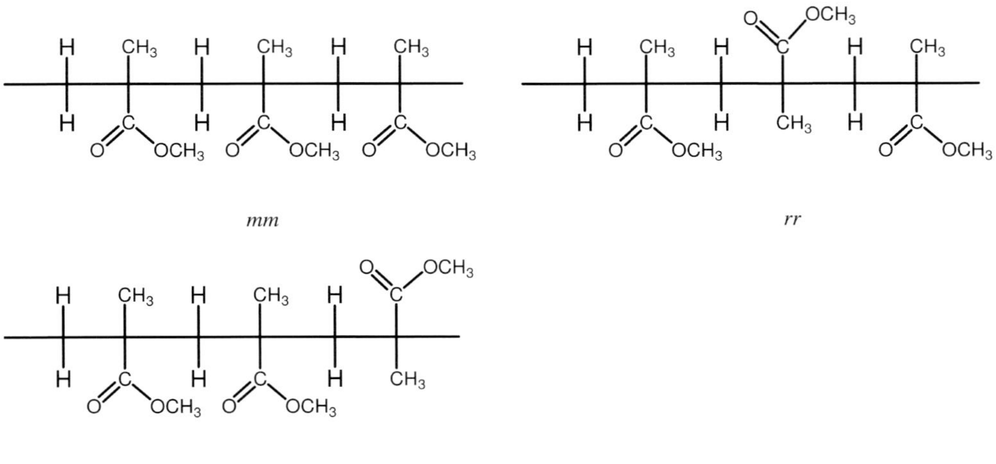

mr

Scheme 1.6

(Scheme 1.5 b). These ^{1}H NMR spectroscopic results are the absolute measure of the stereochemical configuration of the vinyl polymer.

The α-methyl proton signals in these two spectra have different chemical shifts, isotactic PMMA at 1.46 ppm (Fig. 1.1 a) and syndiotactic PMMA at 1.23 ppm (Fig. 1.1 b). This indicates that the α-methyl resonance is sensitive to configuration. The smallest configurational sequence reflected in the shift of the α-methyl resonance is the triad (Scheme 1.6). The heterotactic triad signal is scarcely observed in Fig. 1 a and b. Figure 1.1 c shows the spectrum of PMMA obtained by radical polymerization, which is comprised of both *m* and *r* diads and, as a consequence, *mm*, *rr* and *mr* triads.

Three signals observed in Fig. 1.1 c (1.1–1.5 ppm) are therefore assigned to *mm*, *mr* and *rr* triads, respectively, from lower magnetic field. Closer inspection of the α-CH_3 signals reveals that each signal shows further splittings due to longer stereochemical sequences, i.e. pentads.

When a polymer with relatively low molecular weight is to be analyzed, disturbance of the NMR signals due to end-groups should be considered. Isotactic PMMA prepared with t-C_4H_9MgBr has the structure shown in Scheme 1.7.

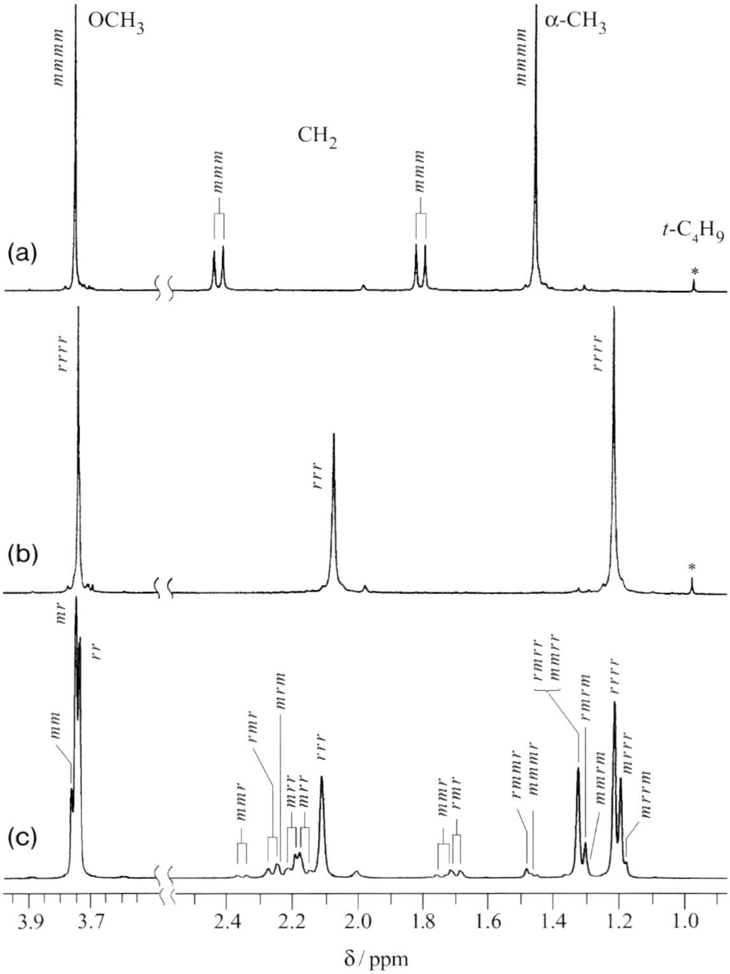

Fig. 1.1 500-MHz ^{1}H NMR spectra of (a) isotactic, (b) syndiotactic and (c) atactic PMMAs measured in nitrobenzene-d_5 at 110 °C. Polymer (a) was prepared using t-C_4H_9MgBr in toluene at −78 °C, polymer (b) was derived from the poly(trimethylsilyl methacrylate) prepared using t-C_4H_9Li–bis(2,6-di-*tert*-butylphenoxy)methylaluminum in toluene at −95 °C and polymer (c) was prepared using AIBN at 60 °C.

The t-C_4H_9 proton signal overlaps the *rr* triad signal when measured in chloroform-*d* but is observed separately when measured in nitrobenzene-d_5. Therefore, in order to obtain correct tacticity data, it is advisable to take the NMR spectrum in nitrobenzene-d_5. An example of such a spectrum is shown in Fig. 1.2. The signals from the first three and the last three monomeric units at

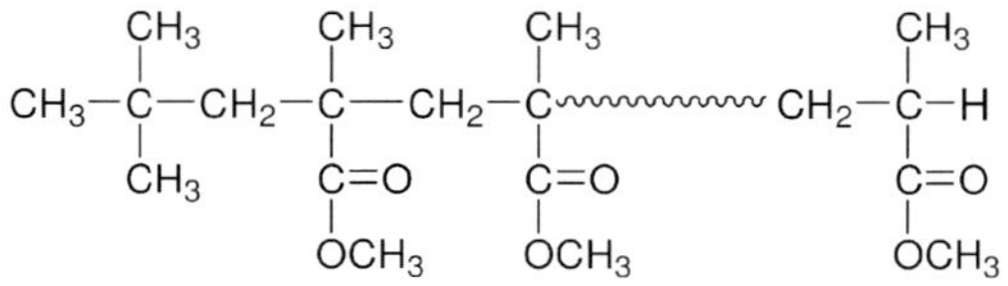

Scheme 1.7

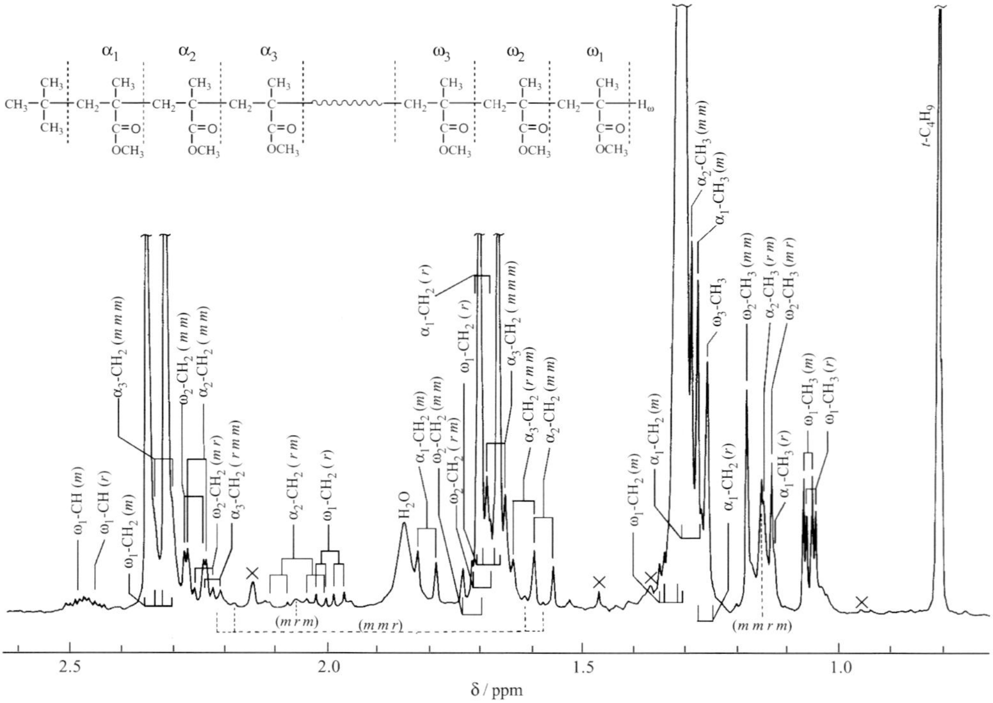

Fig. 1.2 ^{1}H NMR spectrum of isotactic PMMA prepared using t-C_4H_9MgBr in toluene at −78 °C [21]. Signals due to the end-groups and monomeric units at and near the α- and ω-ends are indicated according to the numbering system shown above.

both ends also overlap with the in-chain α-CH_3 signals, as shown in the figure. The assignments shown in the figure were made by ^{1}H COSY. In order to determine the exact triad tacticity, these overlapped signals should be taken into consideration [13]. The isotacticity of the in-chain units, when corrected for the overlapped signals due to the end units, is independent of the molecular weight of the polymer. Without correction, the isotacticity apparently increases with molecular weight and would lead to misunderstanding of the polymerization reaction [13].

1.3
Sequence Statistics

The analysis of tacticity gives information on the stereochemistry of propagation reactions. The simplest case is represented by Bernoullian statistics, where a single probability P*m* can describe the generation of the tactic polymer chain. The probability of an *r* diad is 1–P*m*. It can be readily seen that the probabilities of forming *mm*, *mr* and *rr* triads are given by

$$[mm] = \mathrm{P}m^2 \tag{1}$$

$$[mr] = 2\mathrm{P}m(1 - \mathrm{P}m) \tag{2}$$

$$[rr] = (1 - \mathrm{P}m)^2 \tag{3}$$

The probability of an *mr* triad is given double statistical weighing because both directions *mr* and *rm* must be counted.

When the probability of forming an *m* or *r* diad depends on the previous diad, the first-order Markov sequence is generated by propagating steps in which the mode of addition of the approaching monomer is influenced by whether the growing chain-end is *m* or *r*. We now have four kinds of parameters (conditional probability), P*m*/*m*, P*r*/*m*, P*r*/*r* and P*m*/*r*, for example, P*m*/*m* is the probability of forming an *m* diad after the growing chain-end with an *m* diad. Since the following relations should hold, the number of independent parameters is two:

$$\mathrm{P}m/m + \mathrm{P}m/r = 1 \tag{4}$$

$$\mathrm{P}r/r + \mathrm{P}r/m = 1 \tag{5}$$

These parameters can be derived from triad tacticity as shown in Eqs. (6) and (7):

$$\mathrm{P}m/r = [mr]/(2[mm] + [mr]) \tag{6}$$

$$\mathrm{P}r/m = [mr]/(2[rr] + [mr]) \tag{7}$$

In order to verify whether the tacticity distribution obeys first-order Markovian statistics, the tacticity data of higher order than a triad are required and observed fractions and calculated values should be compared. For example, an *mmrm* pentad fraction can be calculated from triad data and the parameters derived therefrom:

$$[mmrm] = [mm]\mathrm{P}m/r\mathrm{P}r/m + [mr]\mathrm{P}r/m\mathrm{P}m/m \tag{8}$$

If a deviation exists between the observed and calculated values, there is a possibility of second-order Markovian statistics, in which four independent pa-

Table 1.1 Pentad sequence distribution of isotactic-rich PMMA obtained with Li-*i*PrIB in the presence of Me_3SiOLi in toluene at 0 °C.

Pentad (C=O, ^{13}C NMR)	Observed (%) [a)]	Bernoullian statistics (%) [b)]	1st-order Markovian statistics (%) [c)]
mmmm	80.26	74.32	80.13
mmmr	8.14	11.45	8.07
rmmr	0.31	0.44	0.20
mmrm+rmrm	5.69	12.33	5.54
mmrr+rmrr	3.38	0.95	3.36
mrrm	1.18	0.44	1.05
mrrr	0.86	0.07	1.27
rrrr	0.18	0.00	0.39
Error [d)]	–	2.52	0.15

a) *mm*:*mr*:*rr*=88.4:8.9:2.7 (observed value from ^{1}H NMR).
b) P*m*=0.9285.
c) P*m*/*r*=0.0479, P*r*/*m*=0.6223.
d) Σ|population(obsd.) – population(calcd.)|/8.

rameters should be taken into consideration: P*mm*/*r* (=1–P*mm*/*m*), P*mr*/*m* (=1–P*mr*/*r*), P*rm*/*r* (=1–P*rm*/*m*) and P*rr*/*m* (=1–P*rr*/*r*). In order to estimate these parameters, at least tetrad tacticity data are needed.

Table 1.1 [14] shows the pentad tacticity distribution estimated from the carbonyl carbon NMR signals of isotactic-rich PMMA obtained with isopropyl *α*-lithioisobutyrate (Li-iPrIB) in the presence of Me_3SiOLi in toluene at 0 °C. The data are analyzed based on Bernoullian and first-order Markovian statistics. The residual errors from the theoretical values are much smaller for the first-order Markovian analysis, indicating that the stereochemical process of this polymerization obeys Markovian statistics.

When the temperature dependences of the statistical parameters are available, the differences in activation enthalpy and entropy between meso and racemo additions can be derived by plotting ln(P*m*/P*r*) against 1/*T* (Eq. 9) for Bernoullian statistics:

$$\ln(\mathrm{P}m/\mathrm{P}r) = -\Delta(\Delta H^{\ddagger})/RT + \Delta(\Delta S^{\ddagger})/R \quad (9)$$

where $\Delta(\Delta H^{\ddagger}) = \Delta H^{\ddagger}(m) - \Delta H^{\ddagger}(r)$ and $\Delta(\Delta S^{\ddagger}) = \Delta S^{\ddagger}(m) - \Delta S^{\ddagger}(r)$.

For first-order Markovian statistics:

$$\ln(\mathrm{P}m/m/\mathrm{P}m/r) = -\Delta(\Delta H^{\ddagger})/RT + \Delta(\Delta S^{\ddagger})/R \quad (10)$$

where $\Delta(\Delta H^{\ddagger}) = \Delta H^{\ddagger}(m/m) - \Delta H^{\ddagger}(m/r)$ and $\Delta(\Delta S^{\ddagger}) = \Delta S^{\ddagger}(m/m) - \Delta S^{\neq}(m/r)$.

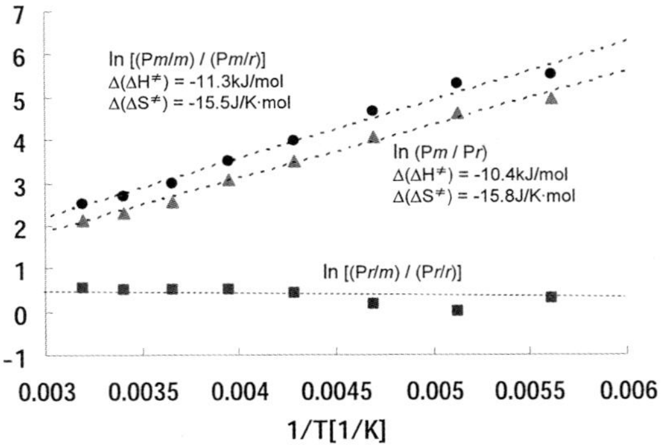

Fig. 1.3 Arrhenius-type plots of ratios of probabilities of stereochemical processes in the polymerization of MMA with Li-iPrIB in the presence of Me_3SiOLi in toluene.

Figure 1.3 [14] illustrates the plots for the isotactic polymerization of MMA with Li-iPrIB in the presence of Me_3SiOLi in toluene, in the temperature range –95 to 40 °C. The linear increase in $\ln(Pm/m/Pm/r)$ with increasing $1/T$ indicates the higher *m*-selectivity of *m*-ended anions at lower temperatures, which contributes high isotacticity to the polymer formed at low temperature, i.e. mm=98.5% at –78 °C. On the other hand, insensitivity and low values of $\ln(Pr/m/Pr/r)$ mean that the *r*-ended polymer anions exhibit low stereoselectivity irrespective of temperature.

To demonstrate the need for higher order statistics, consider a polymer having an *m* diad fraction of 0.5. If the stereochemical sequence distribution obeys Bernoullian statistics, the polymer may have the following random sequence:

$$- m\ r\ r\ r\ m\ m\ m\ r\ m\ r\ r\ m\ m\ m\ r\ r\ m\ r\ m\ m\ m\ r\ r\ m\ r\ r\ r\ m\ m\ m\ r -$$

If the distribution obeys first-order Markovian statistics with $Pm/r=1$ and $Pr/m=1$, the polymer should have a regular sequence (heterotactic) as follows:

$$- m\ r\ m\ r\ m\ r\ m\ r\ m\ r\ m\ r\ m\ r\ m\ r\ m\ r\ m\ r\ m\ r\ m\ r\ m\ r\ m\ r\ m\ r -$$

Both the chains comprise equal numbers of *m* and *r* diads but the latter is a stereoregular polymer and the former is a non-stereoregular polymer. It is therefore obvious that the analysis of longer configurational sequences is important to establish a higher level of stereoregularity.

Figure 1.4 [15] shows methylene and carbonyl carbon signals of a heterotactic poly(allyl methacrylate) with an *mr* triad content of 95.8%. In the carbonyl carbon resonance, the *mr*-centered peaks are most abundant, reflecting the high heterotacticity. In contrast, the methylene signals are comprised of *m*-centered and *r*-centered peak groups with almost the same intensity. If one derives only the diad tacticity from the methylene signals, one may erroneously conclude that the polymer is non-stereoregular. However, once the triad tacticity has been obtained, it is soon realized that the polymer is highly stereoregular. This example clearly demonstrates the importance of the longer sequence analysis.

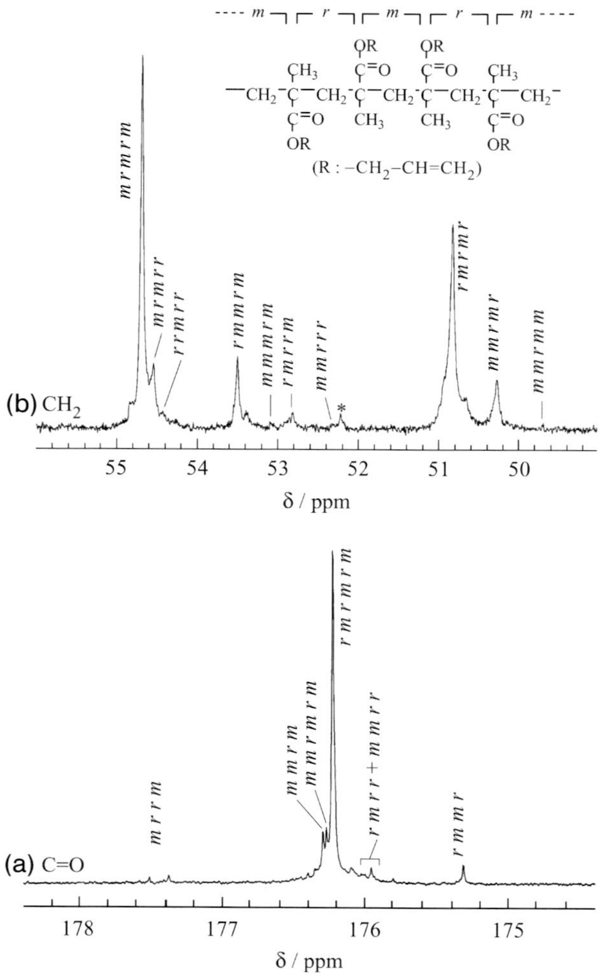

Fig. 1.4 125-MHz NMR signals of the methylene carbon of heterotactic poly(allyl methacrylate) measured in $CDCl_3$ at 55 °C (*mr* triad content 95.8%).

In coordination polymerization, such as the polymerization of α-olefins with Ziegler catalysts, an enantiomer-selective site model has been employed for the analysis of configurational sequence [16]. In this model, a parameter σ, the probability of forming a *d* (or *l*) unit at a *d*- (or *l*-) selective site, is used to define tacticity:

$$[mm] = 1 - 3\sigma(1 - \sigma) \tag{11}$$

$$[mr] = 2\sigma(1 - \sigma) \tag{12}$$

$$[rr] = \sigma(1 - \sigma) \tag{13}$$

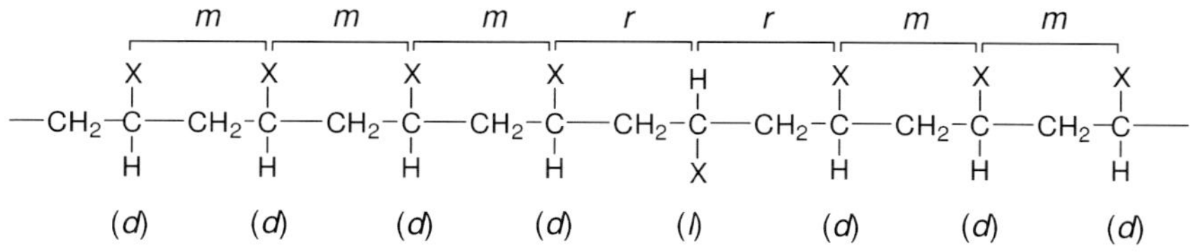

Scheme 1.8

Hence the *mr* triad fraction should be twice as large as the *rr* triad.

Stereochemical defects in an isotactic polymer formed according to this model can be depicted as shown in Scheme 1.8, in which a defect unit with the opposite configuration (*l*) is introduced within the sequence comprising units with *d* configuration. That is, the sequence does not contain an isolated *r* diad. From the pentad level analysis, the following relation should be maintained among the pentads involving *r* diads:

$$[mmmr] = [mmrr] = 2[mrrm] \tag{14}$$

When the stereosequence of a highly isotactic polymer is determined by the relative configuration as represented by a probability such as Pm, the defect is depicted as shown in Scheme 1.9. In this case, the polymer scarcely contains an *rr* triad if the isotacticity is sufficiently high. Hence, in the analysis of highly isotactic polymers, the quantitative determination of the minor sequences is important to distinguish the polymerization mechanisms. An example of this type of isotactic polymer is a PMMA prepared with t-C_4H_9MgBr in toluene at low temperature. Figure 1.5 a displays the carbonyl carbon NMR signals of the isotactic PMMA, in which signals due to *mmrm* and *mmmr* of equal intensity are observed in addition to the strong *mmmm* pentad signal [17].

As the tacticity comes close to 100%, the evaluation of the precision of the tacticity data becomes difficult since the minor defect signals are too small to be detected with high precision. Hence the comparison of two tacticity data reported independently sometimes needs careful inspection. To avoid ambiguity, direct comparison of spectral data taken under identical conditions using the same instrument is advisable. Figure 1.5 [18] shows a carbonyl region ^{13}C NMR spectrum of PMMA with a *mm* content of 98.5% obtained with isopropyl α-lithioisobutyrate (Li-iPrIB)–$LiOSiMe_3$ and of 96.9% obtained with t-BuMgBr. The difference in the triad value is only 1.6%, but the *mmmr* and *mmrm* pentad signals due to the stereochemical defects are evidently smaller in the former than in the latter.

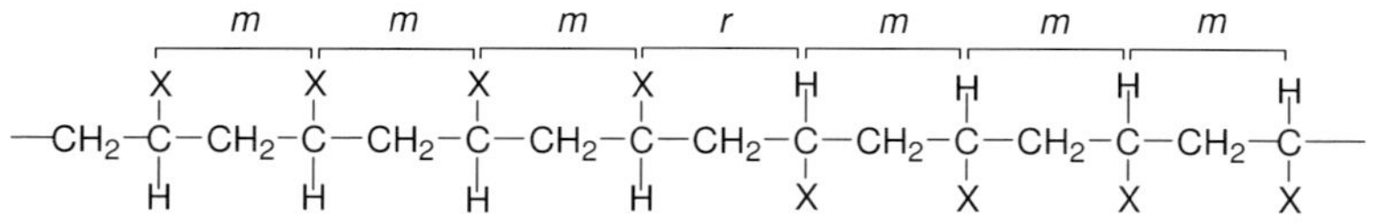

Scheme 1.9

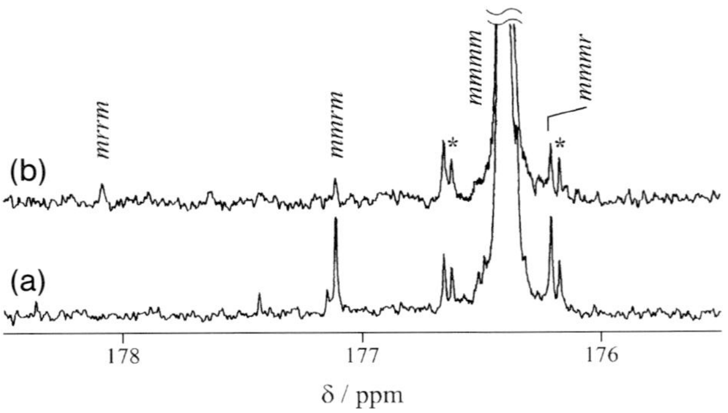

Fig. 1.5 Carbonyl region of 125-MHz ^{13}C NMR spectra of isotactic PMMAs prepared with (a) *t*-BuMgBr (*mm*=96.9%) and (b) Li-iPrIB–Me_3SiOLi (*mm*=98.5%) at –78 °C. The peaks with asterisks are the satellites of the *mmmm* pentad signal due to ^{13}C–^{13}C coupling ($CDCl_3$, 55 °C).

End-group analysis of polymers by NMR spectroscopy often provides important information for the understanding of polymerization mechanism [19]. When a polymer is formed through a living polymerization process, in particular, stereochemical structures near the initiating and terminating chain-ends can be related to the stereospecificity of initiating and propagating species, respectively.

Heterotactic poly(ethyl methacrylate)s [poly(EMA)s] can be prepared with *tert*-butyllithium–bis(2,6-di-*tert*-butylphenoxy)methylaluminum in toluene at –78 °C. The end-group in ^{13}C NMR signals due to the initiating chain-end (*tert*-butyl) and the terminating end (methine) of the poly(EMA) with relatively low degree of polymerization (*DP*) were assigned in terms of triad configurations near the initiating and terminating chain-ends (Fig. 1.6) [20]. The analysis for the initiating chain-end revealed that the dimer anion preferentially undergoes racemo addition to form a trimer anion with a racemo (*r*) diad (*t*-Bu-M$^-$ → *t*-Bu-*r*-*M*$^-$) and that the trimer anion with an *r* diad favors meso (*m*) addition (*t*-Bu-*r*-*M*$^-$ → *t*-Bu-*r*-*m*-*M*$^-$). Even though the dead polymer was analyzed, from the analysis of the terminating chain-end at the triad level, one can estimate the relative abundance of the living polymer having an *r* diad at the chain end (∿∿ *rM*$^-$) and that with *m* diad (∿∿ *rM*$^-$) from the sum of integral intensities, [∿∿ *rr*] + [∿∿ *rm*] and [∿∿ *mr*] + [∿∿ *mm*]. In this particular case, ∿∿ *rM*$^-$ exists more abundantly than ∿∿ *rM*$^-$, suggesting higher stability of the *r*-ended anion. The terminal diad fraction reflects the stereospecificity of the termination reaction with alcohol, indicating the inherently *r*-specific nature of the propagating species.

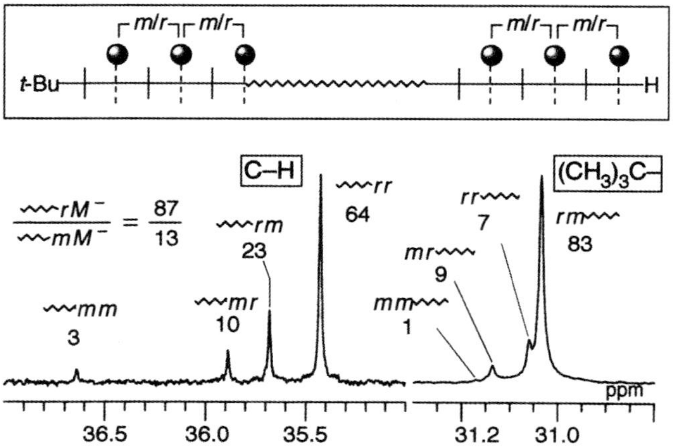

Fig. 1.6 ^{13}C NMR signals of the chain-end groups of heterotactic poly(ethyl methacrylate) prepared using *tert*-butyllithium –bis(2,6-di-*tert*-butylphenoxy)methylaluminum in toluene at −78 °C.

1.4
Isotactic–Stereospecific Polymerizations

1.4.1
Methacrylates

Methacrylate is one of the most extensively studied classes of vinyl monomers with regard to stereoregularity of the polymers obtained. The stereoregularity is a function of monomer structure, initiator, solvent, temperature, etc. Proper selection of the conditions allows the preparation of a wide variety of stereoregular polymethacrylates [22, 23]. However, most of the polymerizations cannot be fully controlled owing to the side-reactions including the reaction of the initiator with the carbonyl group of the monomer [22]. Living polymerizations of polar monomers have been developed in the past 20 years [24], but a few of them give stereoregular polymers. In the 1980s, Hatada and coworkers reported the isotactic-specific living polymerization of methacrylate initiated with *tert*-butylmagnesium bromide (t-C_4H_9MgBr–$MgBr_2$) [25, 26].

it-PMMA is usually prepared using anionic initiators such as alkyllithium and Grignard reagents in non-polar solvents. However, most of the polymerization reactions cannot be fully controlled owing to side-reactions and multiplicity of active species which make the MWDs of the polymers broad [22]. It has been revealed for some polymerization systems that the addition of the initiator to the C=O bond of MMA (leading to the formation of alkyl isopropenyl ketone and CH_3OMgX) occurring in the initial stage of polymerization is the major side-reaction that causes the complexity in the propagation reaction process. To

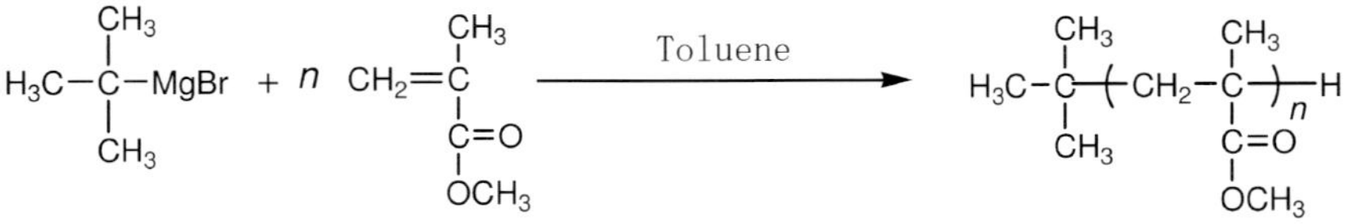

Scheme 1.10

prevent such complexity, we utilized a sterically bulky Grignard reagent, $t\text{-}C_4H_9MgBr$, prepared in diethyl ether, as an initiator for the polymerization of MMA in toluene at low temperature [25, 26]. The polymerization proceeds in a living manner without side-reactions owing to the bulkiness of the initiator and gives a highly isotactic polymer with a narrow MWD.

The number-average molecular weights (M_ns) of the polymers increased proportionally to the yields of polymer and agreed well with the values expected from the amounts of the monomer consumed and of the initiator used. NMR analysis of the resulting polymer revealed that it had a structure similar to that shown in Scheme 1.10, with a $t\text{-}C_4H_9$ group at one end and methine proton at the other end, and the isotacticity was more than 96% in triad. The ether solution of $t\text{-}C_4H_9MgBr$ contains excess amount of $MgBr_2$ produced by a Wurtz-type coupling reaction during the preparation. Hence the initiator may be better described as a binary system, $t\text{-}C_4H_9MgBr\text{–}MgBr_2$.

In the area of polymerization of olefins such as propene, single-site polymerization catalysts called "metallocenes" have greatly developed polymerization control. The relationship between catalyst symmetry and stereospecificity of polymerization has been well established and utilized for rational catalyst design in order to synthesize highly stereoregular polymers [27]. Such transition metal complexes also induced the highly isotactic specific polymerization of methyl methacrylate (MMA). Soga and coworkers reported that a C_2-symmetric zirconocene-based initiating system produced highly isotactic PMMA [28]. An ester-enolate type of such a zirconocene complex formed isotactic PMMA with a fairly narrow MWD at ambient temperature [29]. Highly isotactic specific polymerization ($mm > 95\%$) was also achieved by using half-zirconocene [30], samarium [31] and ytterbium [32] complexes (Table 1.2).

Jérôme and coworkers reported the isotactic-specific polymerization of methacrylate in toluene at moderate temperatures (0 °C) in the presence of a lithium silanolate, which was presumed to be formed *in situ* by the reaction of hexamethylcyclotrisiloxane (D3) and *sec*-butyllithium (*s*-BuLi) with an initial *s*-BuLi:D3 molar ratio of 2.8 (Scheme 1.11) [33, 34].

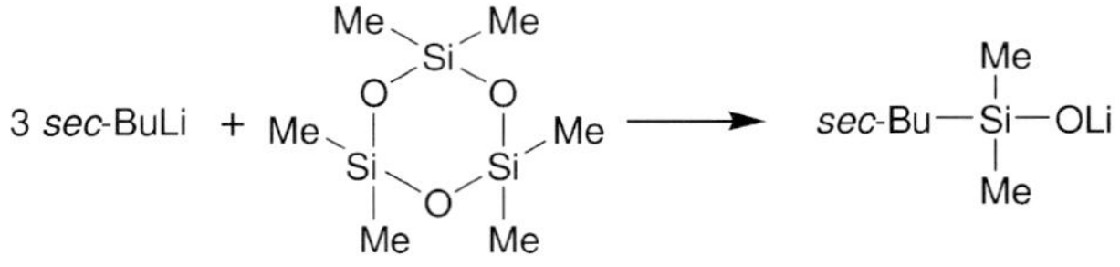

Scheme 1.11

Table 1.2 Examples of highly isotactic-specific MMA polymerization

		Ref.
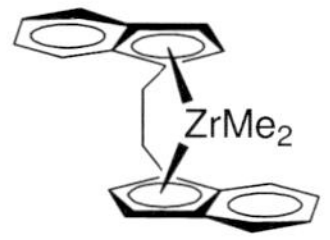	$-R_2Zn-Ph_3CB(C_6F_5)_4$, toluene, 0 °C, *mm* up to 98%, $M_w/M_n > 1.25$	28
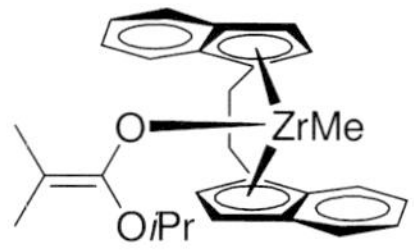	$-B(C_6F_5)_3$, toluene, r.t., *mm* up to 96.7, $M_w/M_n > 1.03$	29
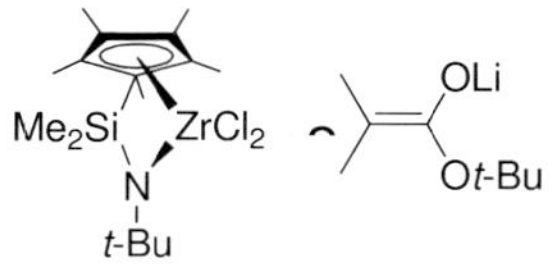	$-[(Et_2O)_2H]\ [B(3,5\text{-}(CF_3)_2C_6H_3)_4]$, −60 to −40 °C, *mm* up to 95.5%, $M_w/M_n > 1.10$	30
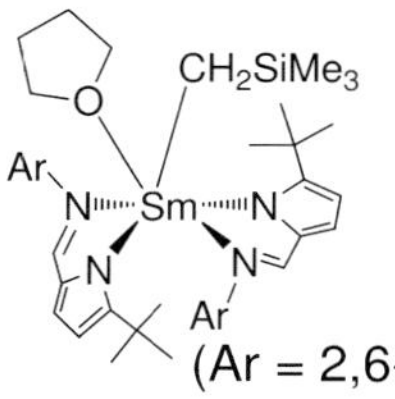 (Ar = 2,6-*i*Pr$_2$C$_6$H$_3$)	Toluene, 0–23 °C, *mm* up to 97.8%, $M_w/M_n > 1.24$	31
$(Me_3Si)_3C-Yb-C(SiMe_3)_3$	Toluene, −78 to 0 °C, *mm* up to 97%, $M_w/M_n > 1.1$	32

More recently, better control of tacticity was realized by using combined anionic initiators comprised of organolithium compounds and a large excess of lithium trimethylsilanolate ($LiOSiMe_3$), a ligated anionic initiator system modified from that of Jérôme and coworkers. Among the organolithiums examined, isopropyl α-lithioisobutyrate (Li-iPrIB), a monomeric anion model of the propagating species of methacrylate polymerization, provided the best results. The new system afforded highly isotactic PMMA (Table 1.3). The isotacticity increased almost linearly with decreasing polymerization temperature and reached *mm* = 98.9% at −95 °C.

The new isotactic-specific initiator was also effective for the polymerization of primary alkyl methacrylate but not for the branched alkyl esters (Table 1.4).

Table 1.3 Polymerization of MMA with Li-iPrIB–Me_3SiOLi in toluene at various temperatures for 1 h [a].

Temperature (°C)	Yield (%)	$M_n/10^3$		M_w/M_n [b]	Tacticity (%) [c]		
		SEC [b]	Calc.		*mm*	*mr*	*rr*
40	85.1	36.5	34.0	1.32	82.6	13.5	3.9
20	100	46.9	40.0	1.19	85.0	11.5	3.5
0	100	41.4	40.0	1.09	88.4	8.9	2.7
−20	100	55.8	40.0	1.10	92.6	5.7	1.7
−40	100	45.3	40.0	1.13	95.2	3.6	1.2
−60	100	50.8	40.0	1.17	97.3	1.9	0.8
−78	100	59.5	40.0	1.23	98.5	1.0	0.5
−95 [c]	82.4	62.9	33.0	1.39	98.9	0.8	0.3

a) Toluene 5 mL, MMA 10 mmol, R_3SiOLi : Li-iPrIB = 25, $[MMA]_0$: $[Li\text{-}iPrIB]_0$ = 400.
b) Determined by SEC (PMMA standards).
c) Polymerization for 2 h.

Table 1.4 Polymerization of alkyl methacrylates with Li-iPrIB–Me_3SiOLi in toluene. [a]

Monomer	Temperature (°C)	Yield (%)	$M_n/10^3$		M_w/M_n [b]	Tacticity (%) [c]		
			SEC [b]	Calc.		*mm*	*mr*	*rr*
MMA	−78	100	59.5	40.0	1.23	98.5	1.0	0.5
	0	100	41.4	40.0	1.09	88.4	8.9	2.7
EMA	−78	100	84.2	45.7	1.13	98.1	1.4	0.5
	0	100	60.0	45.7	1.19	90.8	7.3	1.9
t-BuMA	−78	~0	–	–	–	–	–	–
	0	4.1	20.4	2.7	1.55	77.6	19.1	3.3

a) Toluene 5 mL, monomer 10 mmol, R_3SiOLi : Li-iPrIB = 25, $[monomer]_0$: $[Li\text{-}iPrIB]_0$ = 400. The initiator solution was prepared at ambient temperature.
b) Determined by SEC (PMMA standards).
c) Determined by ^1H NMR for MMA and by ^{13}C NMR for EMA or *t*-BuMA ($CDCl_3$, 55 °C).

1.4.2
Propylene

In 1983, Kaminsky et al. reported that a zirconocene complex, such as Cp_2ZrCl_2 (**3**), combined with methyl aluminoxane (MAO) induced rapid propylene polymerization [36]. The stereoregularity of the polymer obtained was atactic. However, the homogeneous catalysts, so-called single-site catalysts, have attracted

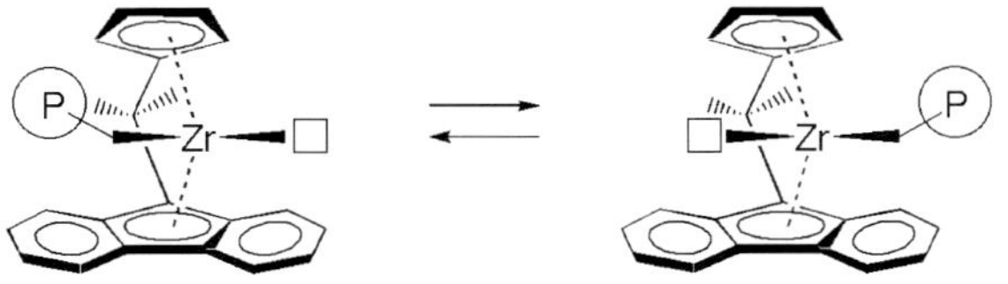

Scheme 1.12

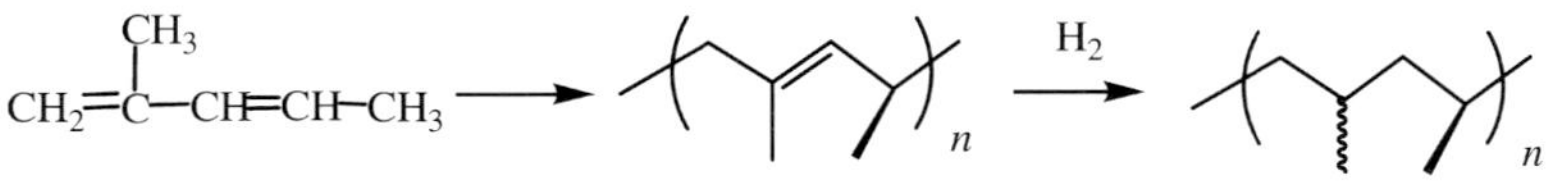

1 2

Scheme 1.13

much attention, because the stereospecificity was expected to be controlled by designing the ligand structure. In fact, polymers with a wide range of stereoregularities have been prepared with properly designed catalysts [37–42].

Two active sites of C_2-symmetric *ansa*-metallocenes (see Table 1.2) are both chiral and equivalent (homotopic) and enantioselective for the same monomer enantioface, leading to isotactic-specific polymerization. In contrast, C_1-symmetric metallocenes produce syndiotactic polymers, as reported by Ewen et al. in 1988 [43]. In contrast, isotactic polymers could be prepared C_1-symmetric metallocenes by utilizing the site epimerization. The introduction of a bulky substituent such as a *tert*-butyl group at the Cp ligand fragment resulted in the formation of isotactic polymers [44], whereas hemiisotactic polymers were obtained by substitution with a methyl group [41, 45] (Scheme 1.12).

The first example of hemiisotactic polypropylene to be reported (**2**) was derived from isotactic 1,4-*trans*-poly(2-methylpenta-1,3-diene) (**1**) via non-stereoselective hydrogenation (Scheme 1.13) [46].

1.5 Syndiotactic-specific Polymerization

1.5.1 α-Olefins

If the C_s symmetry of a catalyst is a sufficient condition for syndiotactic-specific polymerization, all C_s-symmetric catalysts should afford syndiotactic polymers. However, a modified zirconocene complex such as **7** afforded atactic polymers (Table 1.5). Furthermore, the stereoselectivity at the initiating reaction was very low, even when **8** was used as an initiator component [47]. Hence it was assumed that the face of the coordinating monomer was discriminated not by the direct interaction of methyl group of the incoming propylene with the bulkier li-

Table 1.5 Syndiotactic-specific polymerization of propylene with C_s symmetric metallocenes.

Catalyst	Co-catalyst	Temperature (°C)	Syndiotacticity (%)	Ref.
3	MAO	0	43.1 (*r*)	37
4	MAO	0	32.4 (*r*)	37
5	MAO	0	58.4 (*r*)	37
6	MAO	0	61.1 (*r*)	37
7	MAO	60	14 (*rrrr*)	45
8	MAO	25	86 (*rrrr*)	43
b	MAO	65	Hemiisotactic	45
c	MAO	40	78 (*mmmm*)	44
9	MAO	20	91.0 (*rrrr*)	48
10	MAO	40	76.6 (*rrrr*)	48
11	MAO	40	83.0 (*rrrr*)	48
12	MAO	60	72 (*rrrr*)	45
13	MAO	40	94 (*rrrr*)	49
14	MAO	0	91.7 (*rrrr*)	50
15	MAO	0	>99 (*r*)	50
16	MAO	0	95.9 (*rrrr*)	51
17	MAO	0	98.9 (*rrrr*)	51

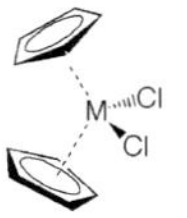

M = Zr (**3**)
=Hf (**4**)

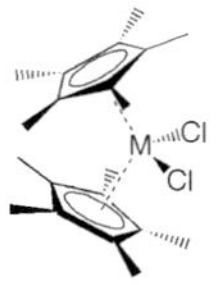

M = Zr (**5**)
=Hf (**6**)

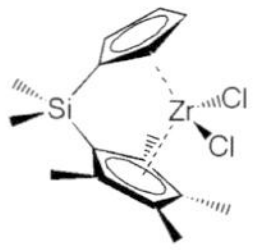

7

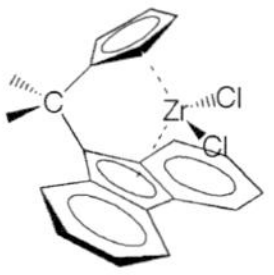

8

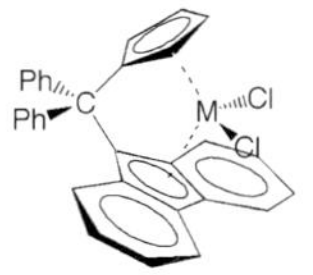

M = Zr (**9**)
=Hf (**10**)

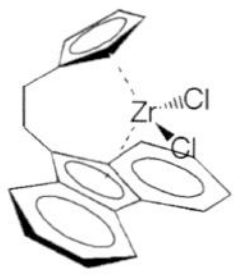

11

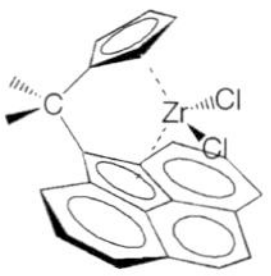

12

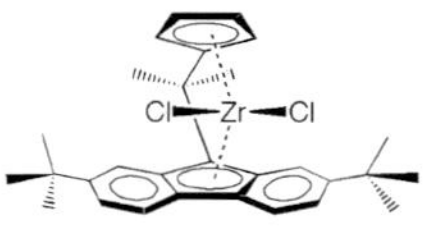

13

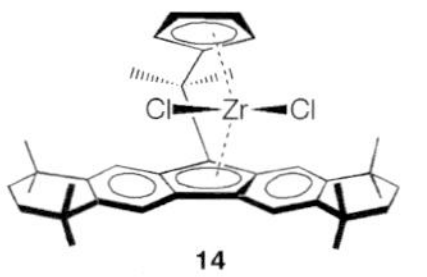

14

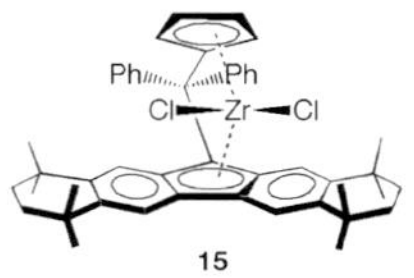

15

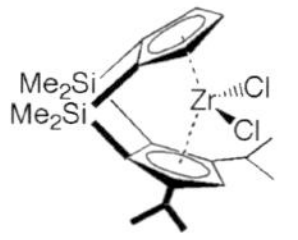

16

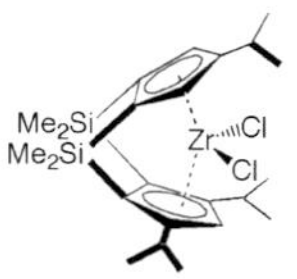

17

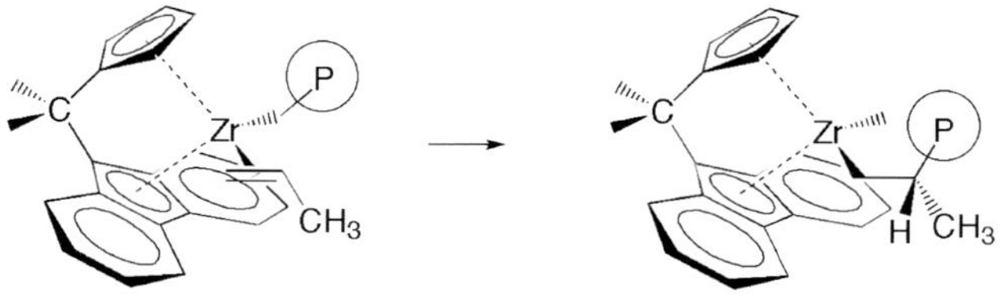

Scheme 1.14

gand fragment such as a fluorenyl group but by the counterbalanced steric effect by the growing polymer chain that is crowded out by the bulkier ligand fragment (Scheme 1.14).

The syndiotactic-specificity was improved by modifying the C_s-symmetric metallocene complex with constrained geometry.

The catalyst systems involving **15** and **17** and MAO exhibit particularly high syndiotactic specificity: **15**, *rrrr* pentad > 99%; **17**, *rrrr* = 98.9%.

1.5.2 Methacrylates and Related Polar Monomers

1.5.2.1 Syndiotactic-specific Polymerization by Anionic Polymerization

Highly syndiotactic living polymerization of MMA was realized by the use of combinations of t-C_4H_9Li and R_3Al in toluene at –78 °C [52, 53]. The polymerization by t-C_4H_9Li alone gives an isotactic-rich polymer with broad MWD and the $\overline{M}_n$ of the polymer is much larger than the calculated value. Addition of an R_3Al such as $(n$-$C_4H_9)_3Al$ decreases the isotacticity accompanied by an increase in syndiotacticity. With an increase in Al : Li ratio the syndiotacticity increases and the MWD becomes narrower and highly syndiotactic PMMAs with narrow MWD are formed at of Al : Li ratios ≥3. The syndiotacticities were around 90% and the $\overline{M}_n$ values were close to the values calculated from the amounts of t-C_4H_9Li and the monomer consumed, t-C_4H_9Li, and R_3Al does not form an ate complex if they are mixed at the polymerization temperature (–78 °C), as evidenced by ^1H NMR analysis of the mixture [16]. The *st*-PMMA obtained contains a *tert*-butyl group at the initiating chain-end and a methine hydrogen at the terminating chain-end, but no alkyl groups originating from the aluminum component. The end-group structure indicates the initiation from t-C_4H_9Li and rules out the possibility of initiation from an ate complex such as t-$C_4H_9R_3Al$–Li^+. The aluminum component is assumed to coordinate to the propagating chain-end to stabilize it and to alter the stereospecificity from isotactic to syndiotactic and also to coordinate with the monomer to prevent it from side-reactions and to enhance its reactivity. The monomer activation by the aluminum compounds is assumed from the observation of the increase in polymer yield with an increase in Al : Li ratio [53], although the kinetics were found not to obey a simple "monomer-activation mechanism" [54]. The use of $(C_6F_5)_3Al$, in place of R_3Al,

Table 1.6 Polymerization of alkyl methacrylate with t-C_4H_9Li–(n-$C_4H_9)_3Al$ in toluene at –78 °C for 24 h[a)] [15, 16].

Alkyl	Yield (%)	$\overline{M}_n$[b)] ($\overline{DP}$) (obs.)	$\overline{M}_n$ ($\overline{DP}$) (calc.)	$\overline{M}_w/\overline{M}_n$[c)]	Tacticity[c)] (%)		
					mm	*mr*	*rr*
C_2H_5–	100	6050 (53.0)	5760 (50.5)	1.07	2	8	90
i-C_3H_7–	92	7200 (56.2)	5940 (46.3)	1.14	3	5	92
n-C_4H_9–	76	7020 (49.4)	5450 (38.3)	1.07	1	7	92
i-C_4H_9–	91	8690 (61.1)	6530 (45.9)	1.06	2	5	93
t-C_4H_9–	87	7140 (50.2)	6260 (44.0)	1.64	10	53	57

a) Monomer 10 mmol, t-C_4H_9Li 0.2 mmol, (n-$C_4H_9)_3Al$ 0.6 mmol, toluene 10 mL.
b) Determined by VPO.
c) Determined by SEC.
d) Determined by ^{13}C NMR.

was also effective in improving the syndiotactic-specificity (rr=95% at –78 °C) [55].

Polymerizations of ethyl, isopropyl, n-butyl and isobutyl methacrylates proceeded in a living manner and gave highly syndiotactic polymers with narrow MWD. Poly(*tert*-butyl methacrylate) [poly(t-BuMA)] obtained with the initiators were less syndiotactic (Table 1.6).

As mentioned above, excess amounts of R_3Al over t-C_4H_9Li are necessary for complete *st*-polymerization to occur. This may indicate that R_3Al in this polymerization coordinates with the propagating species to stabilize it and with the monomer to prevent it from side-reactions (carbonyl attack) and also to activate it. It is well known that less bulky R_3Al tends to form a dimeric aggregate. This implies the possibility of strong interaction between R_3Al at the chain-end and the excess R_3Al that may coordinate to the penultimate monomer unit to fix the chain-end conformation favorable for *st*-placement.

A bulky methacrylate, trimethylsilyl methacrylate (TMSMA), afforded highly syndiotactic polymers when polymerized with t-C_4H_9Li–$MeAl(ODBP)_2$ in toluene at low temperatures [56]. The polymer obtained at –95 °C was 98% syndiotactic in triad. These polymerizations are very useful for preparing various kinds of polymethacrylates since the resulting *st*-polymers can be easily converted into other polymethacrylates through the poly(methacrylic acid)s.

Bulky aluminum phenoxides induced syndiotactic specificity in anionic polymerizations of not only *tert*-butyl acrylate but also less bulky acrylates such as ethyl and n-butyl acrylate (Table 1.7) [57, 58]. Furthermore, polymerization of less bulky acrylates with $EtAl(ODBP)_2$–t-C_4H_9Li afforded syndiotactic polymers with narrow MWDs (M_w/M_n=1.07).

Several examples of stereocontrol of polymerization of N,N-disubstituted acrylamides have been reported. Anionic polymerization of N,N-dimethylacrylamide with Cs^+ as a countercation afforded syndiotactic polymers [65], whereas that

Table 1.7 Polymerization of acrylates with *t*-BuLi–aluminum phenoxide in toluene[a)].

Aluminum phenoxide	M	$[M]_0$	Temperature (°C)	Time (h)	Yield (%)	M_n $\times 10^{-3}$	M_w/M_n	Tacticity (%)		Ref.
								m	*r*	
18	EA	1.0	−60	24	69	3.9	1.77	47	53	10
18	*n*BA	1.0	−60	0.5	71	5.0	1.49	48	52	10
19[b)]	EA	0.5	−60	0.5	100	3.2	1.10	24	76	10
19	EA	0.5	−60	0.5	100	7.2	1.07	24	76	10
19	EA	1.0	−60	0.5	100	6.0	1.09	32	68	10
19	*n*BA	1.0	−60	0.5	96	7.6	1.14	37	63	10
20	*t*BA	1.0	0	24	>99	22.7	2.53	35	65	9
20	*t*BA	0.5	0	24	90	15.1	3.34	33	67	9

a) $[M]_0:[t\text{-BuLi}]_0=50$, $[Al]_0:[Li]_0=5$.
b) $[M]_0:[t\text{-BuLi}]_0=25$.

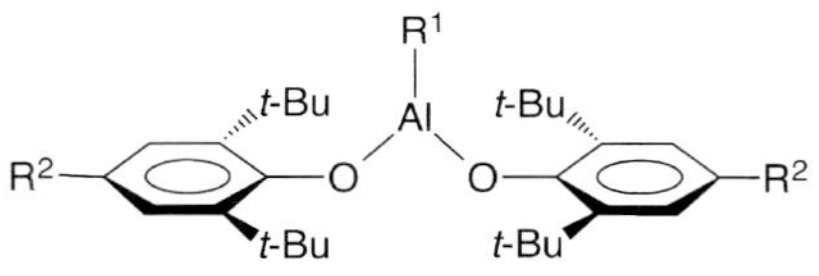

18: R^1 = Me, R^2 = H
19: R^1 = Et, R^2 = H
20: R^1 = Me, R^2 = Me

with Li^+ as a countercation provided isotactic polymers [65–68]. The addition of $ZnEt_2$ [66] or Et_3B [68] changed the stereospecificity so that syndiotactic polymers were obtained even with Li^+ as a countercation. Interestingly, anionic polymerization of *N,N*-diethylacrylamide with K^+ as a countercation in the presence of $ZnEt_2$ or Et_3B afforded heterotactic instead of isotactic or syndiotactic polymers [42(3–3)–45(3–3)]. Syndiotactic polymer (r=75%) was also prepared by anionic polymerization of *N*-methoxymethyl-*N*-isopropylacrylamide (NIPAAm) with Li^+ and $ZnEt_2$ [69]. The polymer obtained was converted to syndiotactic poly(NIPAAm) by an acidic deprotection (Table 1.8).

1.5.2.2 **Syndiotactic-specific Polymerization by Group Transfer Polymerization**

Group-transfer polymerization (GTP) of MMA with $ZnBr_2$ or *i*-Bu_2AlCl produced PMMA containing approximately twice as many syndiotactic as heterotactic sequences, whereas that with a nucleophilic catalyst yielded PMMA the stereoregularity of which was similar to that of PMMA obtained by conventional free radical polymerization [30]. GTP of MMA with $B(C_6F_5)_3$ and $(C_2H_5)_3SiI$ was more efficient in producing syndiotactic polymer, and syndiotactic PMMA (rr=88%) with narrow MWD (M_w/M_n=1.12) was obtained in CH_2Cl_2 at −40 °C [31]. The use of $MeAl(ODBP)_2$ in place of $B(C_6F_5)_3$ resulted in the formation of

Table 1.8 Polymerization of NIPAAm with Lewis bases in toluene [a)]

Lewis base	$[M]_0$	Temperature (°C)	Time (h)	Yield (%)	M_n $\times 10^{-4}$	M_w/M_n	Tacticity (%)		Ref.
							m	*r*	
None	1.0	0	24	>99	6.96	1.9	45	55	13
None	1.0	–80	24	18	1.11	3.5	44	56	13
21	1.0	0	24	98	0.91	1.6	37	63	13
21	1.0	–80	24	86	1.25	1.6	36	64	13
22	0.5	0	24	>99	1.02	2.0	39	61	12
22	0.5	–80	24	>99	0.57	2.1	57	43	12
23	0.5	0	24	>99	0.98	1.9	38	62	15
23	0.5	–80	24	93	0.86	2.8	43	57	15
24	0.5	0	24	83	1.44	1.6	39	61	16
24	0.5	–80	24	32	1.04	1.9	53	47	16

a) $[M]_0:[t\text{-BuLi}]_0=50$, $[Al]_0:[Li]_0=5$.
b) $[M]_0:[t\text{-BuLi}]_0=25$.

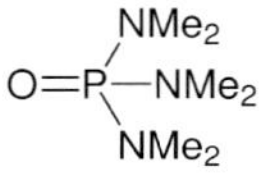

21

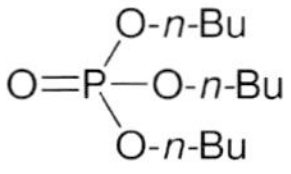

22

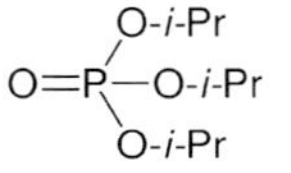

23

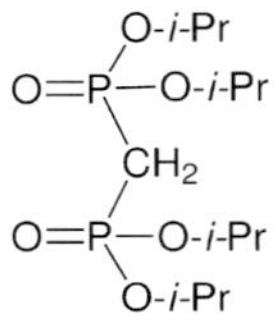

24

syndiotactic-rich PMMA with a bimodal MWD; the higher molecular weight fraction (MW ≈ 106) isolated by SEC fractionation was found to be 97.5% syndiotactic [32]. By appreciating the distinctively different solubility of the highly syndiotactic PMMA, the high-MW fraction was easily separated by solvent fractionation with toluene or acetone.

1.5.2.3 Syndiotactic-specific Polymerization with an Organotransition Metal Initiator

Organolanthanides such as $[SmH(C_5Me_3)_2]_2$ gave highly syndiotactic PMMA (*rr*=95.3%) with narrow MWD ($M_w/M_n=1.05$) [36, 37]. A neutral enolate complex (**25**), formed *in situ* from the initiator and MMA at a 1:2 stoichiometry, was indicated to be responsible for stereospecific propagation.

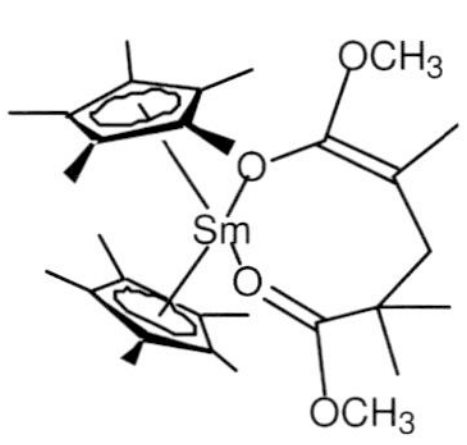

25

Chain-transfer agents such as *tert*-butanethiol were effective for control of the molecular weights of PMMA without decreasing either the polymer yield or the stereoregularity. The use of chain-transfer agents results in a decrease in the amount of initiator fragments in the polymer products, leading to a reduction of coloration of PMMA in use [38]. It was also possible to influence the tacticity of the PMMA obtained through modifications of the lanthanocene ligand environment. Chiral lanthanocenes such as (*R*)-{$Me_2Si(\eta^5$-$C_5Me_4)(\eta^5$-$C_5H_3R^*)$}LuN-$(SiMe_3)_2$ [R* = (–)-menthyl] (**26**) afforded syndiotactic PMMA ($rr \approx 73\%$), whereas chiral lanthanocenes such as (*R*)-{$Me_2Si(\eta^5$-$C_5Me_4)(\eta^5$-$C_5H_3R^*)$}LaN$(SiMe_3)_2$ [R* = (+)-neomenthyl] (**27**) produced isotactic PMMA ($mm \approx 94\%$) [39].

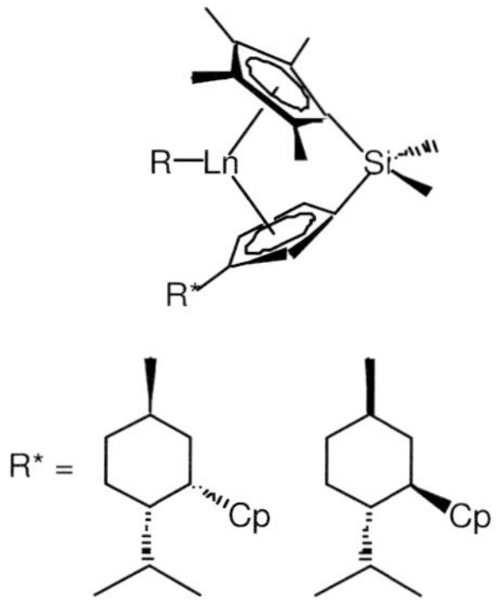

26 **27**

A two-component catalyst system comprised of zirconocene dimethyl (Cp_2ZrMe_2) and a zirconocenium complex such as $[Cp_2ZrMe(THF)]^+[BPh_4]^-$ afforded syndiotactic PMMA ($r = 80\%$) with narrow MWD ($M_w/M_n = 1.4$) at or below room temperature [40, 41]. Furthermore, chiral *ansa*-zirconocenium cations paired with an aluminate anion, such as [*rac*-($Me_2Si(\eta^5$-Ind$)_2$)ZrMe]$^+$ [MeAl-$(C_6F_5)_3]^-$ (**28**), afforded syndiotactic PMMA ($rr = 60.8\%$) [14], whereas the corresponding chiral *ansa*-zerconocenium cations paired with weakly coordinating borate anions, such as [*rac*-($Me_2Si(\eta^5$-Ind$)_2$)ZrMe]$^+$[MeB$(C_6F_5)_3]^-$ (**29**), produced isotactic PMMA ($mm = 92.7\%$) [42–47]. Thus, stereoblock PMMA, isotactic PMMA-*block*-syndiotactic PMMA, was prepared by post-polymerization after the addition of **29** into living polymers derived from isotactic-specific polymerization with **28** to convert cationic zirconocene enolate into enoaluminate [48].

Little attention has been drawn to titanium compounds as initiators of MMA polymerizations compared with zirconium compounds. However, syndiotactic PMMAs could be obtained by using titanium compounds. The combination of $(C_5Me_5)TiMe_3$ and $Ph_3C^+B(C_6F_5)_4^-$ produced syndiotactic PMMA with $rr \approx 70\%$ in the presence of Zn [49]. A catalyst with constrained geometry, [$Me_2Si(C_5$-$Me_4)(t$-BuN)]$TiMe_2$ and $B(C_6F_5)_3$ (**30**), gave PMMA with higher syndiotacticity in a living manner ($rr = 80\%$, $M_w/M_n = 1.09$) [50]. When *n*-butyl methacrylate was used as a monomer, highly syndiotactic polymer was obtained ($rr = 89\%$, $M_w/M_n = 1.12$).

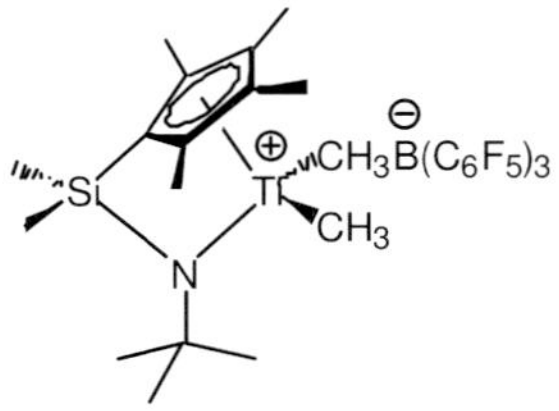

30

1.5.2.4 Preparation of Highly Syndiotactic Polymers by a Radical Mechanism

It is well known that radical polymerization of MMA gives syndiotactic-rich polymers regardless of both polymerization temperature and the kind of solvents, and the syndiotacticity of the polymers obtained increases with decrease in temperature [51–53]. The syndiotactic specificity was enhanced in the presence of a fluoro alcohol. For instance, highly syndiotactic PMMA (rr=93%) was obtained by polymerization in $(CF_3)_3COH$ at –98 °C [53]. Radical polymerization of methacrylic acid was also affected by utilizing a hydrogen-bond interaction with alcohol. Poly(methacrylic acid) [poly(MAD)] with rr=95.0% was obtained at –78 °C in 2-propanol, whereas poly(MAD) with rr=83.8% was obtained at –78 °C in methanol [54].

Highly syndiotactic poly(MAD) (rr=98%) was also prepared by a template polymerization with a macromolecularly porous film of isotactic PMMA, based on stereocomplex formation [55]. These poly(MAD)s were easily converted to PMMAs by reaction with diazomethane.

Similarly to acrylate polymerization, it is difficult to control the stereospecificity of the polymerization of acrylamide derivatives. Atactic polymers were obtained by radical polymerization of acrylamide or *N*-monosubstituted acrylamides regardless of the polymerization conditions such as solvent and temperature. However, it was reported that syndiotactic poly(*N*-isopropylacrylamide) [poly(NIPAAm)] could be obtained by radical polymerization of NIPAAm complexed with Lewis bases, such as hexamethylphosphoramide (**31**) and phosphates (**32**, **33**), through hydrogen-bonding interaction [59–63]. The syndiotacticity reached up to 72% by adding a five-fold amount of HMPA at –60 °C. Interestingly, the stereospecificity of NIPAAm polymerization in the presence of primary alkyl phosphates such as **32** depended strongly on the polymerization temperature: isotactic polymers were obtained at –80 °C, whereas syndiotactic polymers were obtained at –40 to 0 °C. NMR analysis revealed that NIPAAm and **32** formed a 1 : 1 complex at 0 °C and predominantly a 1 : 2 complex at –80 °C, whereas NIPAAm and **31** formed a 1 : 1 complex at both –80 and 0 °C. Hence the stoichiometry of the hydrogen bond-assisted complex plays an important role in the determination of stereospecificity in this polymerization system (Scheme 1.15).

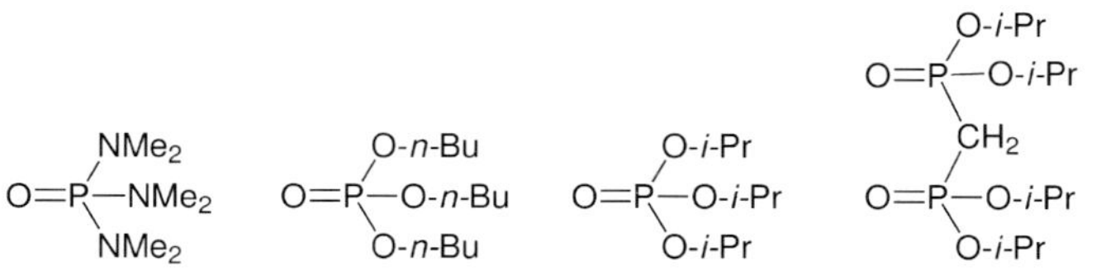

31 **32** **33** **34**

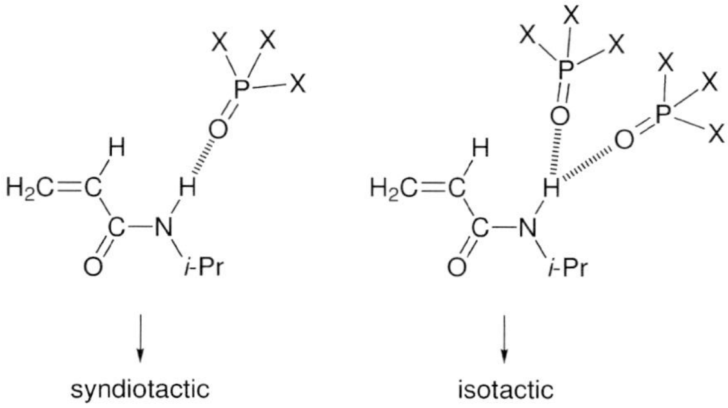

Scheme 1.15

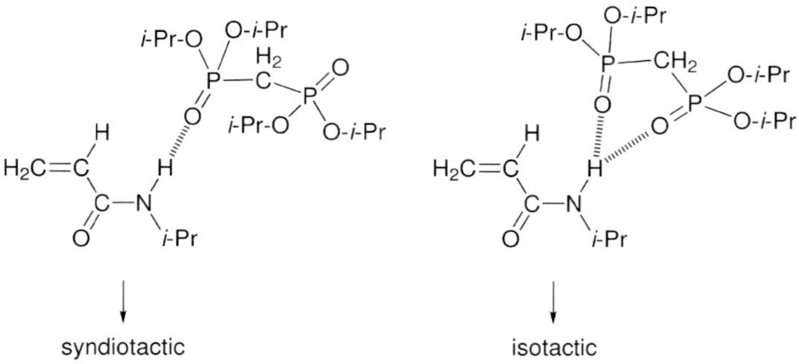

Scheme 1.16

A similar result was observed for NIPAAm polymerization in the presence of bidentate Lewis bases such as tetraisopropyl methylenebisphosphonate (**34**) [64]. Syndiotacticity increased at –40 to 0 °C and isotacticity increased at –80 °C as compared with those in the absence of Lewis bases. Further, NIPAAm and **34** formed a mono-binding complex at 0 °C and a chelate complex at –80 °C. This means that the dependence of stereospecificity in this polymerization depends on the structure of the hydrogen-bond-assisted complex (Scheme 1.16).

Radical polymerization of *N,N*-disubstituted acrylamides provided syndiotactic polymers, whereas those of acrylamide and *N*-monosubstituted acrylamides gave atactic polymers. The stereospecificity depends strongly on the polymerization conditions such as solvent, temperature and substituent {poly(*N,N*-dimethylacrylamide) with $m = 77\%$, toluene, –78 °C, $[M]_0 = 0.1$ M; poly(*N,N*-diphenylacrylamide) with $r = 93\%$, THF, –98 °C, $[M]_0 = 0.22$ M} [70].

1.6
Stereoblock Polymers

A homopolymer with stereochemically different sequences or blocks is termed a stereoblock polymer. Stereoblock PMMA with narrow MWD was first obtained by deriving from *t*-poly(triphenylmethyl methacrylate)-*block-st*-poly(diphenylmethyl methacrylate), that was prepared by sequential block copolymerization of the two monomers in THF at –78 °C. Conversion of the block copolymer to stereoblock PMMA was effected by hydrolysis and subsequent methylation with diazomethane [71]. This example utilizes the drastic difference in stereoregularity of the corresponding polymethacrylates under identical polymerization conditions in addition to the living nature.

Direct synthesis of stereoblock PMMA was effected by polymerizing MMA with *it*-PMMA anion, prepared using *t*-C_4H_9MgBr, in the presence of R_3Al. The stereoblock PMMA, isotactic PMMA-*block*-syndiotactic PMMA, was prepared by adding the second monomer and an excess amount of R_3Al to the isotactic living anion after the consumption of the first monomer [19]. In particular, $(CH_3)_3Al$ was effective for transformation of the stereospecificity of the propagating species while maintaining the living nature (first block, $mm = 96\%$; second block, $rr = 84\%$; $M_w/M_n = 1.39$) (Scheme 1.17).

Bulky aluminum bisphenoxides exhibit an odd stereoregulating power. Polymerization of primary alkyl methacrylates such as MMA and ethyl methacrylate (EMA) with the combination of bis(2,6-di-*tert*-butylphenoxy)methylaluminum $[MeAl(ODBP)_2]$ (**18**) and *t*-C_4H_9Li in toluene at –78 °C afforded syndiotactic polymers in low polymer yields at an Al:Li ratio of 1, whereas highly heterotactic polymers were quantitatively obtained at an Al:Li ratio higher than 2 [20–22].

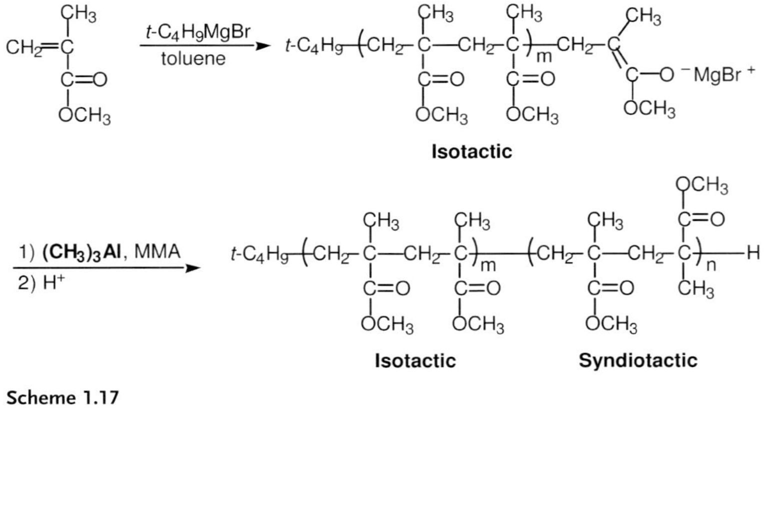

Scheme 1.17

In the polymerization of primary alkyl methacrylates with $MeAl(ODBP)_2$–t-C_4H_9Li (Al : Li > 2), syndiotactic polymers with *rr* contents over 80% were obtained just by increasing the polymerization temperature to –40 °C. Furthermore, highly syndiotactic polymers ($rr \approx 92\%$) were obtained quantitatively by adding bis(2,6-di-*tert*-butylphenoxy)ethylaluminum [$EtAl(ODBP)_2$] (**19**) instead of $MeAl(ODBP)_2$ even at –78 °C [23].

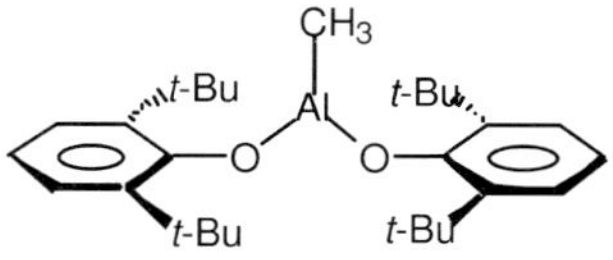

18

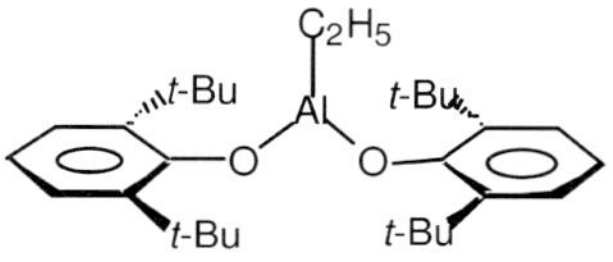

19

The use of tertiary alkyl methacrylates as monomers in polymerization with t-C_4H_9Li–$MeAl(ODBP)_2$ also gave highly syndiotactic polymers regardless of the polymerization temperature [21]. In particular, poly(trimethylsilyl methacrylate) prepared at –95 °C is 98% syndiotactic [24, 25]. The combination of (2,6-di-*tert*-butyl-4-methylphenoxy)diisobutylaluminum and t-C_4H_9Li gave syndiotactic PMMA (*rr*=75%) with narrow MWD ($M_w/M_n = 1.09$) even at high temperatures such as 0 °C [26], and the combination of bis(2,6-di-*tert*-butyl-4-methylphenoxy)-methylaluminum and enamines gave syndiotactic PMMA (*rr*=88%), although enamine alone cannot initiate the polymerization of methacrylates [27]. These results described here indicate that bulky aluminum phenoxides play two important roles, stabilization of propagating anions and activation of monomers, in anionic polymerization of methacrylates [26–29].

Switching stereospecificity during polymerization has also been reported for living radical polymerization of acrylamides (Scheme 1.18).

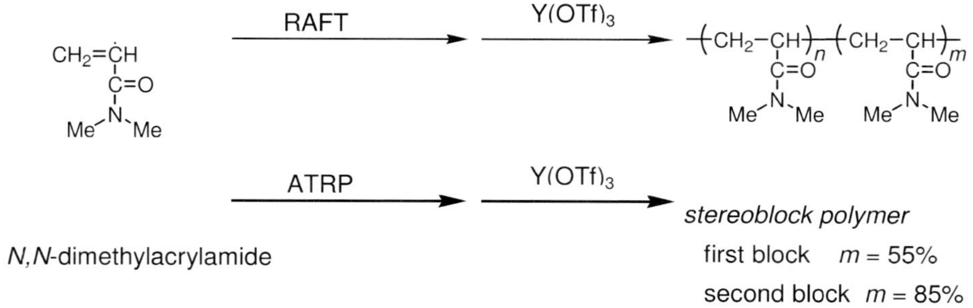

Scheme 1.18

1.7 Heterotactic-specific Polymerization

1.7.1 Preparation of Heterotactic Polymers

The structure of an *ht*-polymer is exemplified by an *ht*-PMMA in Scheme 1.19.

Although the definition of an *ht*-polymer is given for an ideal polymer, polymers containing more than 50% *ht*-triads can be called an *ht*-polymer, since the *mr* triad content can attain at most 50% even when the stereoregulation in the polymerization is completely random [8].

Among stereospecific polymerizations, *ht*-specific polymerization requires higher order stereoregulation, since *m*-addition and *r*-addition have to take place in an alternate manner. One of the rational strategies for the formation of an *ht*-sequence is the cyclopolymerization of divinyl monomers where intra- and intermolecular additions occur alternately. If these two processes exhibit opposite stereoregulations, *m*- and *r*-additions, the polymer obtained should have an *ht*-sequence in the main chain (Scheme 1.20).

Kämmerer et al. studied the cyclopolymerization of 2,2′-methylenebis(4-methyl-2,1-phenylene) dimethacrylate in benzene at 80 °C. Hydrolysis and methylation of the radically prepared polymer gave a PMMA rich in *ht*-triad (*mm*:*mr*:*rr* = 20:52:28) [18]. Although the *mr* content is only slightly higher than 50%, this value is much larger than those for radically prepared polymers of the corresponding monofunctional methacrylate monomers.

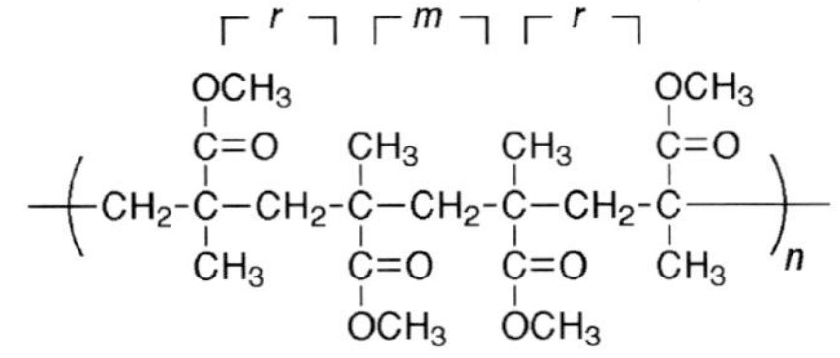

Scheme 1.19

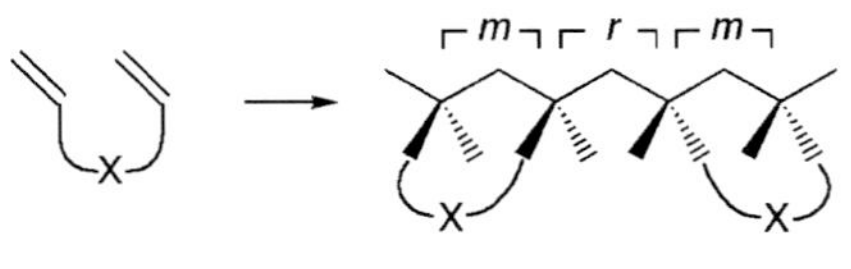

Scheme 1.20

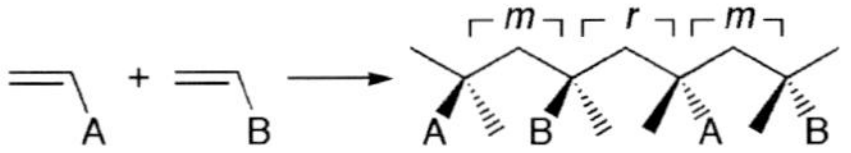

Scheme 1.21

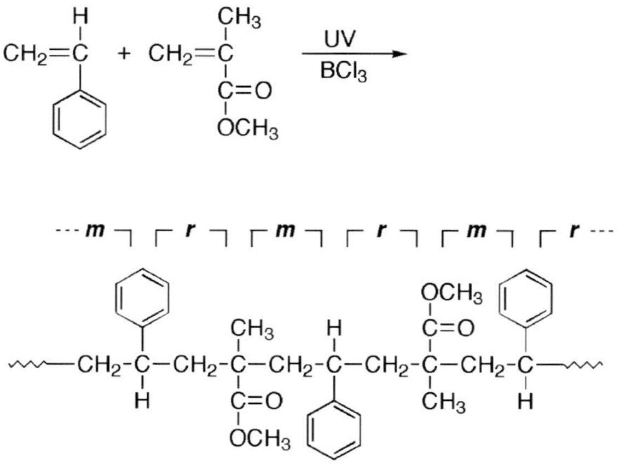

Scheme 1.22

Another possibility of generating an *ht*-sequence is an alternating copolymerization in which stereoselectivities in two cross-propagation processes are opposite (Scheme 1.21).

Gotoh et al. reported the preparation of a highly coheterotactic alternating copolymer of styrene and MMA by a radical copolymerization in the presence of BCl_3 at –95 to –100 °C [19]. The cotacticity of the MMA-centered triad is $mm:mr:rr = 9:89:2$ and that of the styrene-centered triad is $mm:mr:rr = 9:88:3$. The alternating tendency of the copolymerization is essential for this high coheterotacticity and the propagation from MMA radical to styrene was found to be *r*-selective and that from styrene radical to MMA to be *m*-selective (Scheme 1.22) [20].

For the formation of an *ht*-polymer by homopolymerization of vinyl monomers, the diad configuration at the propagating chain-end must affect the stereoselectivity of the propagating reaction. In other words, the chain end with an *m* diad prefers *r*-addition and the chain end with an *r* diad prefers *m*-addition. This obviously requires a higher order stereoregulation than those for *it*- and *st*-polymers. In fact, there have been a limited number of reports on the formation of highly *ht*-homopolymers by vinyl polymerization [15–17, 21–33].

Nozakura et al. reported the preparation of *ht*-poly(triisopropylsilyl vinyl ether) ($mm:mr:rr = 26:69:5$) with $C_2H_5AlCl_2$ in toluene at –78 °C [21]. They proposed that the side-chain ether group in the antepenultimate monomeric unit interacts with a terminal methine hydrogen to form a cyclic transition state, through which the configurations at the penultimate and antepenultimate monomeric units may affect the stereoregulation.

Hatada et al. reported that the polymerization of methacrylates with octylpotassium in THF gave *ht*-rich polymers. The heterotacticity depended on the ester group ($mr = 51–65\%$) and was the highest for the benzyl ester ($mm:mr:rr = 13:65:22$) [22]. The mechanism for the *ht*-propagation in THF with octylpotassium was suggested as a combined effect, i.e. *st*-placement was favored in the polymerization in THF but the bulkiness of the substituent and/or the steric effect of K^+ counterion hindered two successive *st*-additions. Some α-phenyl acrylate derivatives also gave *ht*-rich polymers [22, 23]. A polymer with an *mr* of 67% was obtained by polymerization of phenyl α-(*p*-chlorophenyl)acrylate in toluene with butyllithium [23].

Nakahama et al. reported that the polymerization of *N,N*-diethylacrylamide with 1,1-bis(4′-trimethylsilylphenyl)-3,3-diphenylpropylpotassium–Et_2Zn at 18 °C in THF afforded an *ht*-polymer [24]. The addition of Et_3B instead of Et_2Zn was also effective in *ht*-polymer formation [25].

Okamoto et al. reported the *ht*-specific radical polymerization of vinyl pivalate using fluoro alcohols as solvents [26]. Polymerization with $(n\text{-}C_4H_9)_3B$ in the presence of a small amount of air in $(CF_3)_3COH$ at –40 °C gave an *ht*-polymer with *mr*=61.0%. They claimed that the stereochemical effects observed in the polymerizations may be due to hydrogen-bond interactions between the fluoro alcohol molecules and the ester groups of both the incoming monomer and the monomeric units near the growing chain-end.

In addition to vinyl polymerization, the preparation of *ht*-polymers by a ring-opening polymerization has also been reported. Kasperczyk reported that polymerization of *rac*-lactide with lithium *tert*-butoxide in THF gave predominantly *ht*-poly(lactic acid) (*ht*-PLA) [34]. In addition, Coates et al. reported that *ht*-PLAs could also be prepared by polymerization of *rac*-lactide with achiral zinc alkoxide [35] and by polymerization of *meso*-lactide with *rac*-aluminum alkoxide [36]. Matyjaszewski and coworkers reported that ring-opening polymerization of all-*trans*-1,2,3,4-tetramethyl-1,2,3,4-tetraphenylcyclotetrasilane with silyl cuprates in THF gave a polysilane with an *mr* of 75% [37]. In these cases, the stereochemistry of the monomers is responsible for the heterotactic propagation in addition to the propagation processes.

1.7.2
Heterotactic Living Polymerization and Mechanism of Stereoregulation

As mentioned above, Kitayama et al. reported *ht*-specific living polymerization of certain alkyl methacrylates with a combination of $t\text{-}C_4H_9Li$ and $MeAl(ODBP)_2$ in toluene at low temperatures [15–17]. Table 1.9 gives polymerization results for several methacrylates.

The primary and secondary esters of methacrylic acid gave *ht*-polymers with narrow MWDs. In particular, ethyl, propyl and allyl esters gave highly *ht*-polymers [16, 17]. In the polymerization of ethyl methacrylate (EMA), the $\overline{M}_n$ of the polymer increased linearly with conversion, keeping the MWD narrow. When a fresh feed of EMA was added repeatedly to the system where the monomer was almost consumed, the added monomer was smoothly polymerized and $\overline{M}_n$ of the polymers further increased. These results indicate the living nature of this polymerization system [16, 28].

Table 1.9 Polymerization of alkyl methacrylate with $t\text{-}C_4H_9Li\text{–}MeAl(ODBP)_2$ in toluene at –78 °C or –95 °C [a].

Alkyl	Temperature (°C)	Time (h)	Yield (%)	Tacticity (%) [b]			$\overline{M}_n$ [c]	$\overline{M}_w/\overline{M}_n$ [c]
				mm	*mr*	*rr*		
$-CH_3$	–78	24	>99	11.6	67.8	20.6	8330	1.18
$-CH_2CH_3$	–78	24	>99	7.7	88.6	3.7	7010	1.07
	–95	48	>99	7.0	92.0	1.0	11100	1.07
$-CH_2CH_2CH_3$	–78	24	>99	6.1	91.2	2.7	7650	1.07
	–95	48	>99	5.7	93.3	1.0	7680	1.09
$-CH_2CH{=}CH_2$	–78	24	94	5.8	89.8	4.4	8070	1.06
	–95	48	85	3.1	95.8	1.1	11200	1.08
$-CH_2CH(CH_3)_2$	–78	24	84	12.1	78.4	9.5	6350	1.07
$-CH_2Ph$	–78	48	51	0.5	65.9	33.6	6010	1.08
$-CH(CH_3)_2$	–78	24	50	2.1	69.2	28.7	4730	1.07

a) Monomer 10 mmol, toluene 10 mL, $t\text{-}C_4H_9Li$ 0.2 mmol, $MeAl(ODBP)_2$ 1.0 mmol.
b) Determined by NMR.
c) Determined by SEC.

The Al : Li ratio affects drastically the stereospecificity and yield in this polymerization system. The existence of an excess amount of $MeAl(ODBP)_2$ over $t\text{-}C_4H_9Li$ is essential for the formation of the *ht*-polymer. In fact, an *st*-polymer forms in low yields at an Al/Li ratio of 1, whereas *ht*-polymers are obtained quantitatively at higher Al : Li ratios [16]. Thus, the following concepts are evolved: (1) most of the aluminum phenoxide would coordinate with the propagating species at Al : Li = 1, (2) the coordinating aluminum phenoxide would stabilize the propagating species and change it into an *st*-specific species and (3) *ht*-propagation requires excess $MeAl(ODBP)_2$, which might coordinate with the monomer and activate it. Thus the propagation process involves preferentially the addition of the activated monomer to the less reactive propagating anions. In this process, the steric interaction between the sterically crowded active end (Scheme 1.23 a) and the bulky monomer–$MeAl(ODBP)_2$ complex (Scheme 1.23 b) might be an important factor for *ht*-propagation. Hence $MeAl(ODBP)_2$ in this polymerization plays important roles in the stabilization of the propagating species and in the activation of the monomer, thereby providing not only the livingness but also the unique stereospecificity [29].

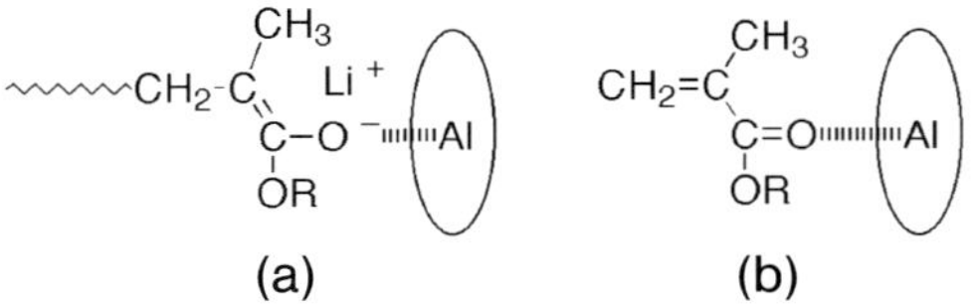

Scheme 1.23

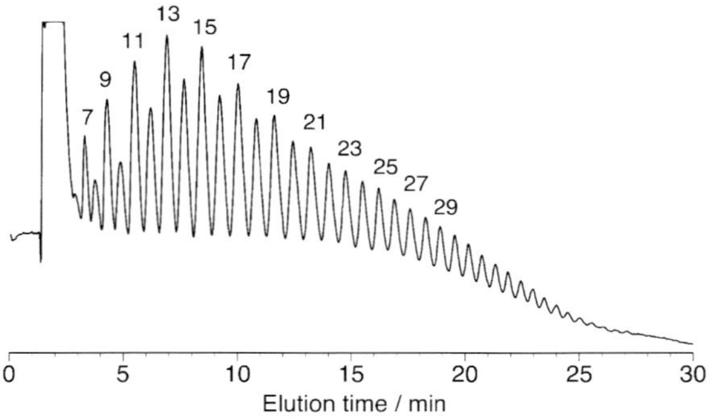

Fig. 1.7 SFC trace of *ht*-oligo(AlMA) ($DP=18.5$, $M_w/M_n=1.22$, $mm:mr:rr=3.1:95.2:1.7$) [31].

The polymerization temperature also affects the stereospecificity in this polymerization system. The stereoregularity of the polymers obtained changed drastically from heterotactic to syndiotactic with increasing polymerization temperature [15, 16].

Supercritical fluid chromatographic (SFC) analysis and matrix-assisted laser desorption/ionization (MALDI-TOF) mass spectrometric analysis of *ht*-oligomers of allyl methacrylate (AlMA) prepared with t-C_4H_9Li–$MeAl(ODBP)_2$ in toluene at −95 °C revealed that the population of the oligomers with respect to *DP* showed even–odd alternation, the oligomer with odd-number *DP* being formed preferentially (Fig. 1.7) [30, 31].

In order to understand the mechanism of stereoregulation in *ht*-specific polymerization, NMR analysis was applied to the *ht*-specific polymerization of EMA. In the polymerization, there exist two kinds of propagating anions, *r*-ended and *m*-ended anions (Scheme 1.24).

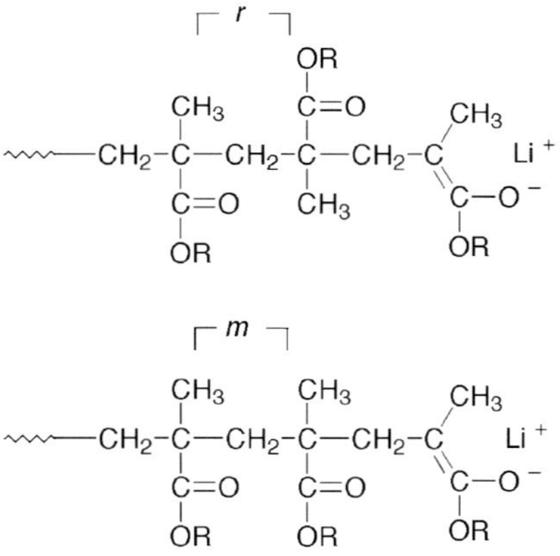

Scheme 1.24

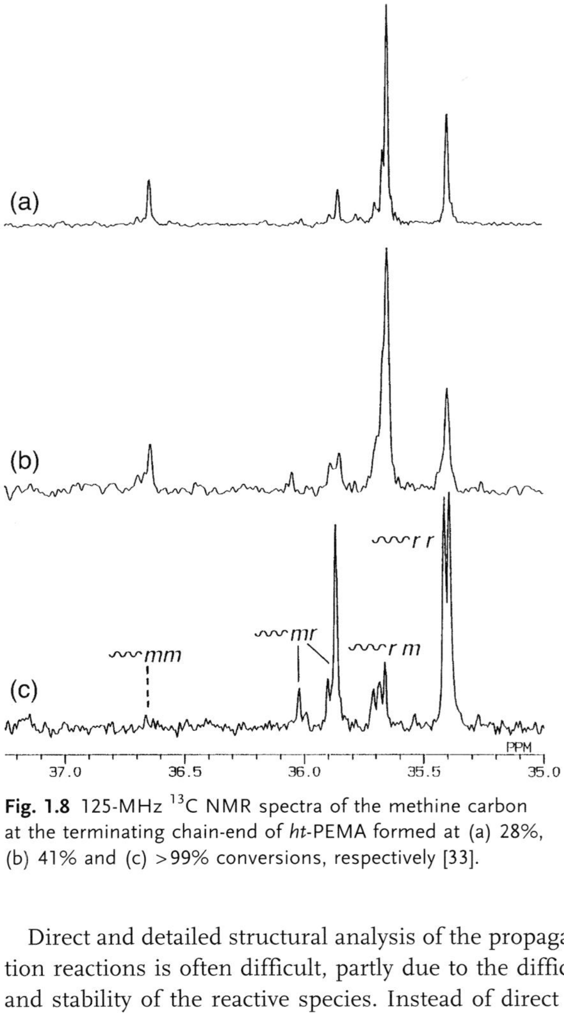

Fig. 1.8 125-MHz ^{13}C NMR spectra of the methine carbon at the terminating chain-end of *ht*-PEMA formed at (a) 28%, (b) 41% and (c) >99% conversions, respectively [33].

Direct and detailed structural analysis of the propagating species in polymerization reactions is often difficult, partly due to the difficulty in sample preparation and stability of the reactive species. Instead of direct observation, NMR analysis of the terminal structure of the unfractionated polymer or oligomer was made first to obtain information on the polymerization mechanism [32, 33]. The results of the ^{13}C NMR analysis of the stereochemical sequences at and near the terminating chain-end of *ht*-PEMA prepared with $t\text{-}C_4H_9Li\text{–}MeAl(ODBP)_2$ are shown in Fig. 1.8 [33].

The signals of the terminal methine carbon showed principally four groups of signals reflecting four possible triad sequences at the chain end, *-mm*, *-mr*, *-rm* and *-rr*, and were assigned as shown in Fig. 1.8c. The assignments were made by comparing the spectra of three types of stereoregular PEMAs; *it-*, *st-* and *ht*-polymers. The ultimate diads were fixed when the polymerization reaction was quenched with methanol and the chain-end anion was protonated, which reflects the stereoselectivity of the protonation reaction. The *r*:*m* ratio of the second diad from the terminal should correspond to the ratio of the *r*-ended and *m*-ended anions existing in the polymerization system before quenching the reaction; that is, the ratio [$\sim$ rM–]:[$\sim$ *m*M–] can be estimated from the terminal triad fractions as

$$[\sim rM-]/[\sim mM-] = ([-rm] + [-rr])/([-mm] + [-mr]) \qquad (15)$$

The analysis of the polymer obtained at 28% conversion (Fig. 1.8a) revealed that the ratio was 78:22, that is, the *r*-ended anion is more abundant than the *m*-ended anion, suggesting higher stability or lower reactivity of the *r*-ended anion [33].

Similar NMR analysis of the initiating chain end of the polymer indicated that the *ht*-propagation starts by *r* placement followed by *m* placement forming an *rm* triad predominantly at the initiating chain end [32]. If the propagation reaction were to proceed in a completely heterotactic manner (alternating placement of *r* and *m* diads) starting with *r* placement and ending mostly with an *r*-ended anion, then the population of odd-numbered oligomers should be predominant, as shown in Scheme 1.25. The *r*-ended anions with $DP=5$, 7, 9 and 11 are dominant species that give odd-numbered oligomers after protonation. Hence the mechanism of even–odd alternation in *DP* distribution could be clearly demonstrated (Fig. 1.9) [30].

In reality, the polymer chains contain some stereochemical defects such as *mm* and *rr* sequences. Such detailed information could be obtained by NMR analyses of fractionated uniform oligomers [31]. It was found that the fractionated uniform oligomers with odd-numbered *DP* have predominantly the structures shown in Scheme 1.25. However, some of the uniform oligomers with even-numbered *DP* have defects of *mm* placement; for example, some of the uniform 8mers were found to have the steric structures of *rmrmmrr* and *rmrmmrm*. The results indicate that the necessary means to increase heterotacticity is to reduce *mm* placements. It is interesting that the spectra of the chain end of the polymer changed with conversion and became more complicated at high conversions (Fig. 1.8b and c). This is due to the increasing irregularity of the chain end tacticity in the later stages of the polymerization. Hence the analysis of the products from the polymerization reaction at different conversions may be useful in understanding how the polymerization reaction proceeds.

1.8 Ditactic Poly(α,β-disubstituted Ethylene)s–Polycrotonates

Stereospecific polymerizations of α,β-disubstituted ethylenes have attracted considerable interest because the stereostructures of the resulting polymers give detailed information about the mechanism of stereoregulation in the polymerization.

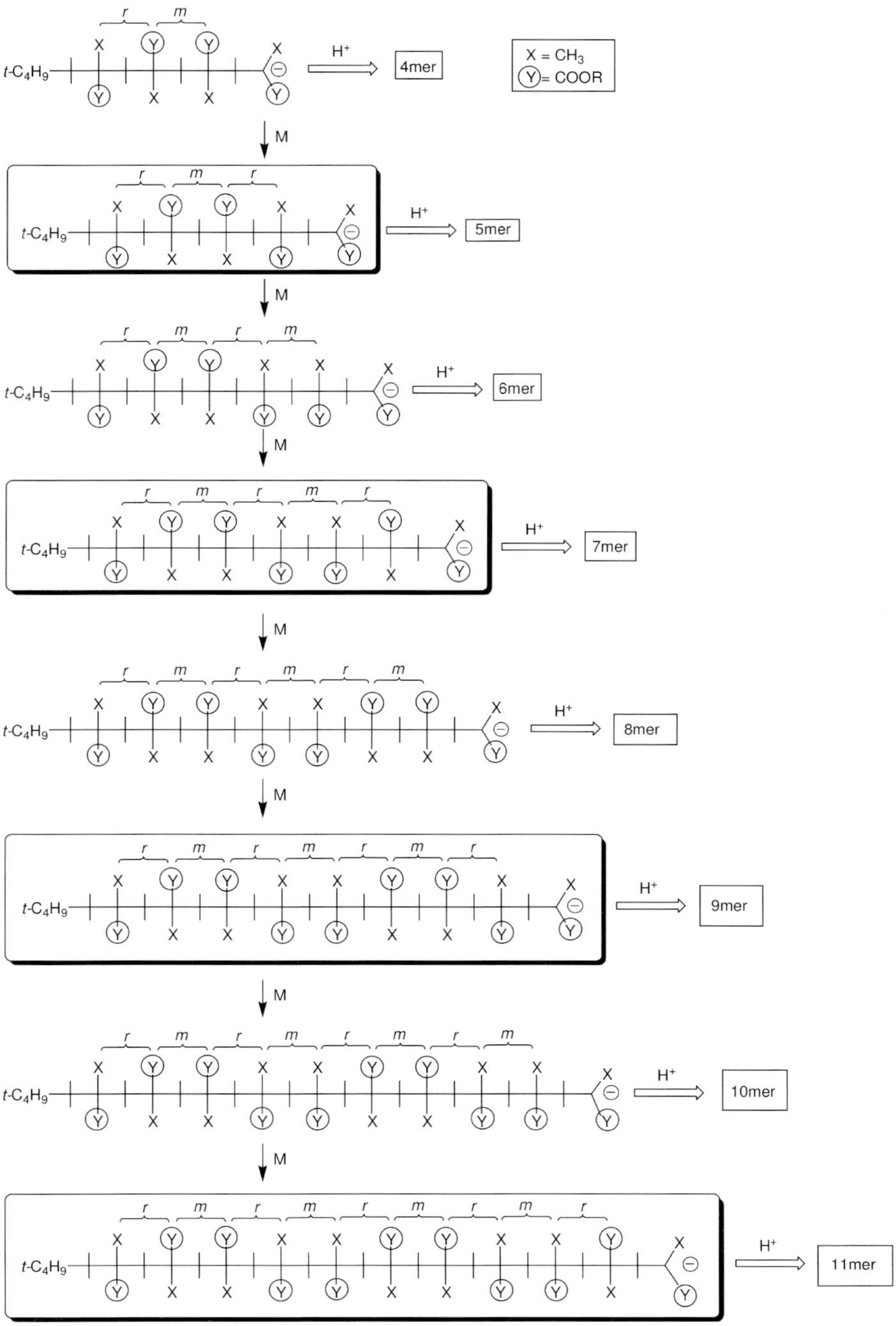

Scheme 1.25

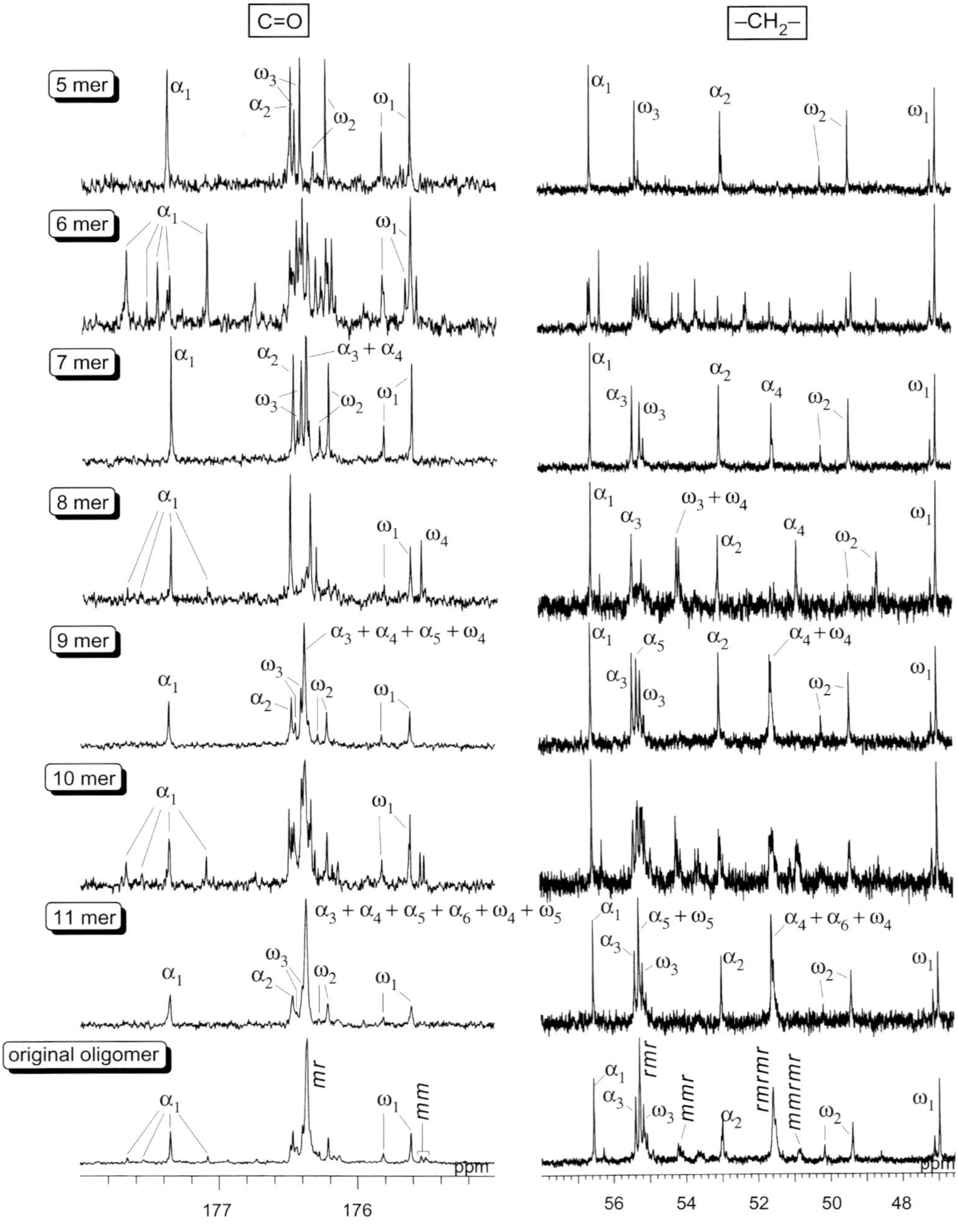

Fig. 1.9 125-MHz ^{13}C NMR spectra of methylene and carbonyl carbons of 5- to 11-mers and the original oligomer, measured in benzene-d_6 at 35 °C.

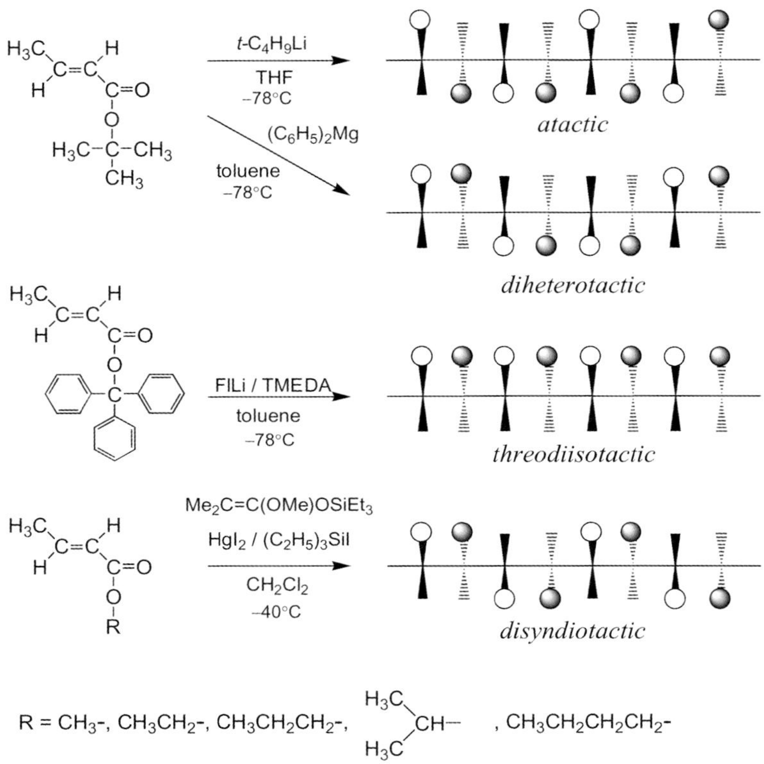

Fig. 1.10 Ditactic polymers of crotonates.

Pure *cis-* or *trans-α,β*-disubstituted ethylenes can give three types of stereoregular polymers: *erythrodiisotactic, threodiisotactic* and *disyndiotactic*. Recently, we have succeeded in preparing *threodiisotactic* [115] and *disyndiotactic* [116–118] polymers and a new type of ditactic polymer, *diheterotactic* [119], from several crotonates, as shown in Fig. 1.10.

tert-Butyl crotonate gave an atactic polymer with t-C_4H_9Li in THF at –78 °C but a diheterotactic polymer with $(C_6H_5)_2Mg$ in toluene at –78 °C. A diheterotactic polymer represents a new type of ditactic polymer which had not previously been obtained. Triphenylmethyl crotonate afforded a threodiisotactic polymer with fluorenyllithium (FlLi) in toluene at –78 °C in the presence of TMEDA. Group transfer polymerization of methyl crotonate gave a disyndiotactic polymer when a ketene silyl acetal, 1-methoxy-1-(triethylsiloxy)-2-methyl-1-propene, HgI_2 and $(C_2H_5)_3SiI$ were used as initiator, catalyst and cocatalyst, respectively.

The polymerizations of *tert*-butyl and triphenylmethyl crotonates proceed in a living manner. Group transfer polymerization of methyl crotonate includes a termination reaction owing to the cyclization reaction at the propagating chain ends. Ethyl crotonate and other bulkier crotonates polymerized in a living manner to give high molecular weight polymers with narrow MWDs.

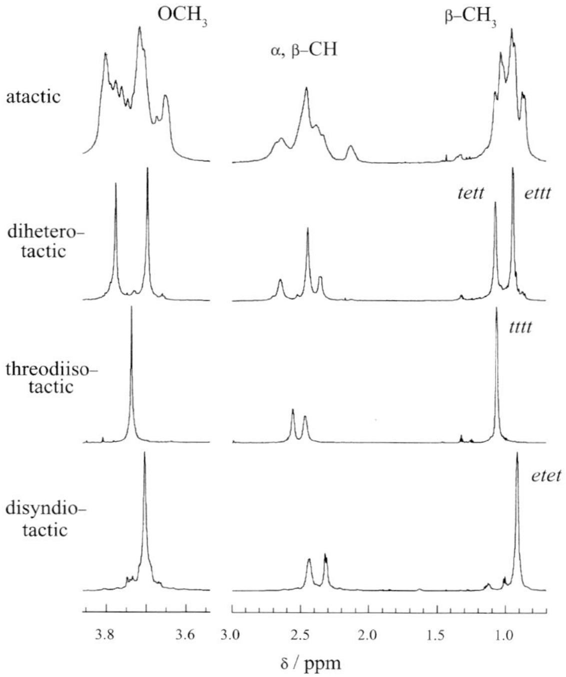

Fig. 1.11 750-MHz ^{1}H NMR spectra of poly(methyl crotonate)s with various stereostructures [$(CF_3)_2CHOH$–C_6D_6 (95 : 5), 55 °C] (*e*, *erythro*; *t*, *threo*).

The polymers of *tert*-butyl and triphenylmethyl crotonates were easily converted into the poly(methyl crotonate)s with corresponding stereoregularities by transesterification. The 750-MHz ^{1}HNMR spectra of the derived poly(methyl crotonate)s are shown in Fig. 1.11. Comparison of the spectra of the three ditactic poly(methyl crotonate)s with that of the atactic polymer clearly indicates that the ditactic polymers, particularly the diheterotactic and threodiisotactic ones, are very highly stereoregular.

Ditacticity assignment of poly(α,β-disubstituted ethylene)s by NMR spectroscopic analysis is usually difficult and requires more elaborate methods such as specific deuteration of one of the vinyl protons. In the cases of the ditactic poly(methyl crotonate)s mentioned above, single-crystal X-ray analyses of the oligo(methyl crotonate)s have been successfully applied. For example, Fig. 1.12 illustrates the results of the X-ray analysis of a single crystal of a disyndiotactic hexamer of methyl crotonate formed in the GTP of this monomer. The results clearly demonstrate that C=C bond opening generates an *erythro* configuration whereas monomer addition forms a *threo* enchainment so that an *etet* (*e*, *erythro*; *t*, *threo*) disyndiotactic sequence is formed.

Stereospecific polymerization of crotonates requires a higher order of stereoregulation than that for methacrylate polymerization since polycrotonate chains

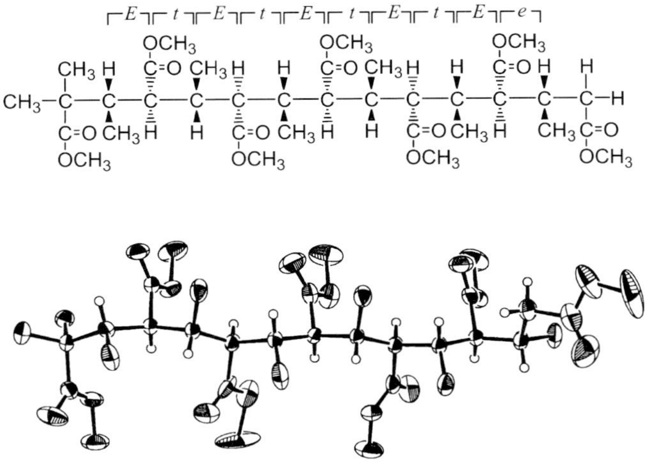

Fig. 1.12 Single-crystal X-ray analysis of pure disyndiotactic hexamer of methyl crotonate isolated from the oligomer mixture formed in the GTP with 1-methoxy-1-(triethylsiloxy)-2-methyl-1-propene, HgI_2 and $(C_2H_5)_3SiI$ in CH_2Cl_2 at 0 °C.

have two types of asymmetric carbons in the constituent repeating units. In particular, in the diheterotactic polymerization of *tert*-butyl crotonate, stereoregulation at the pentad level is required; the configurational repeating units at the pentad level are *ettt* and *tett*.

1.9 Stereoregularity and Uniformity

To demonstrate the effect of tacticity on the polymer properties, polymer samples differing only in their stereochemical purities with exactly the same molecular weight and MWD are most valuable. If uniform polymers with an identical *DP* and different tacticity are available, the evaluation of the tacticity effect should be much simplified and allow straightforward conclusions to be drawn.

SFC has been proved effective for isolating uniform oligomers with relatively high *DP*. By applying SFC as a preparative method to two *it*-PMMA samples, one prepared with *t*-C_4H_9MgBr in toluene at –78 °C (*mm*:*mr*:*rr*=95.9:3.5:0.6) and the other completely isotactic PMMA derived from poly(1-phenyldibenzosuberyl methacrylate) prepared in THF at –78 °C with methyl *α*-lithioisobutyrate as an initiator and methyl iodide as a terminator (Scheme 1.26), a series of uniform *it*-PMMAs with slightly different isotacticities were isolated.

The main-chain tacticity of the latter PMMA is 100% isotactic, whereas only the *meso* fraction at the initiating diad was less than 100% (*m*:*r*=89:11). The crystallization behaviors of the uniform *it*-PMMAs with 100% isotacticity were

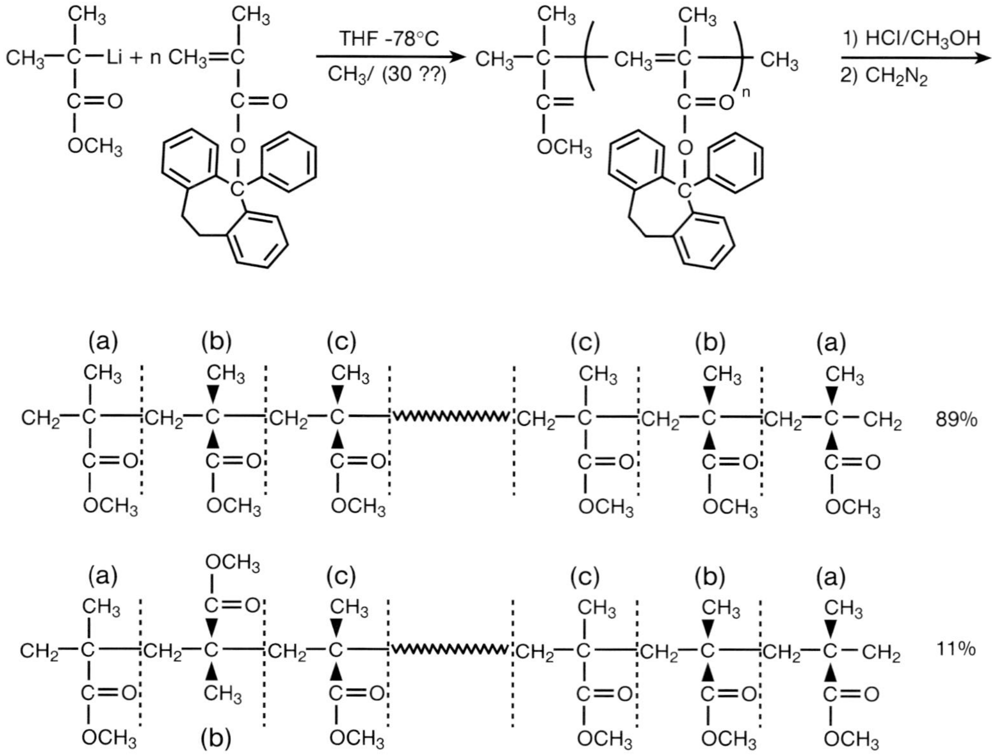

Scheme 1.26

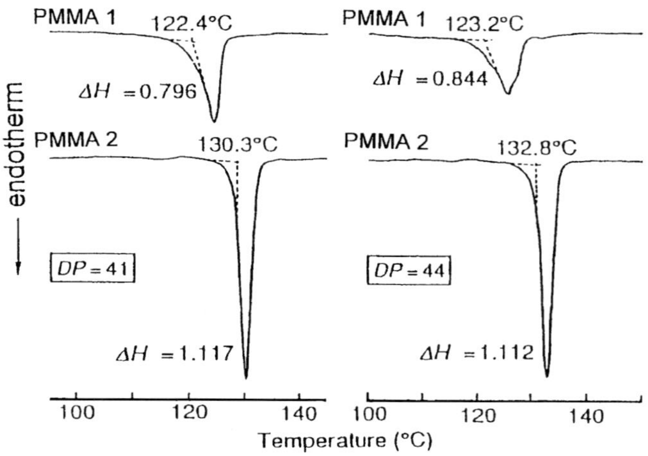

Fig. 1.13 Melting endotherms of uniform 41- and 44-mers of highly *it*-PMMA 1 (*mm* = 96.1%) and the 100% *it*-PMMA 2 (heating rate; 5 °C/min) [122]. ΔH values are given in kcal/mol.

compared with those of uniform *it*-PMMAs with identical *DP*s but with a small number of in-chain stereochemical defects. Figure 1.13 shows melting endotherms of 41- and 44-mers of the two kinds of uniform *it*-PMMAs crystallized from methanol by solvent evaporation. At each *DP*, the uniform *it*-PMMA with 100% isotacticity had higher T_m and larger ΔH^*, clearly indicating that the crystallinity and T_m of uniform *it*-PMMA increase significantly with improvement in tacticity from 96.1 to 100%. The reported ΔH_u (1.20 ± 0.08 kcal/mol) [93] indicates that the crystallinities of the 100% isotactic uniform PMMAs are, at least, almost 90%.

References

1 Staudinger H, Ashdown AA, Brunner M, Bruson HA, Wehrli S, *Helv Chim Acta 12*: 934 (**1929**).
2 Schildknecht CE, Zoss AO, McKinley C, *Ind Eng Chem 39*: 180 (**1947**).
3 Ziegler K, Holzkamp E, Breil H, Martin H, *Angew Chem 67*: 426 (**1955**).
4 Natta G, Pino P, Corradini P, Danusso F, Mantica E, Mazzanti G, Moraglio G, *J Am Chem Soc 77*: 1708 (**1955**).
5 Hatada K, Kitayama T, Stereospecific living polymerization of methacrylate and construction of stereoregular polymer architecture, in *Macromolecular Design of Polymeric Materials,* Hatada K, Kitayama T, Vogl O (Eds). Marcel Dekker, New York, pp. 139–162 (**1997**).
6 Tanaka A, Hozumi Y, Hatada K, *Kobunshi Kagaku 22*: 216 (**1965**).
7 Soum AH, Fontanille M, *Makromol Chem 181*: 799 (**1980**).
8 Soum AH, Tien CF, Hogen-Esch TE, *Makromol Chem Rapid Commun 4*: 243 (**1983**).
9 Bovey FA, in *Chain Structure and Conformation of Macromolecules,* Jelinski LW (Ed.). Academic Press, New York (**1982**).
10 Hatada K, Kitayama T, *NMR Spectroscopy of Polymers – Practical Use in Polymer Chemistry.* Springer, Berlin (**2004**).
11 Bovey FA, Tiers GVD, *J Polym Sci. 44*: 173 (**1960**).
12 Nishioka A, Watanabe H, Abe K, Sono Y, *J Polym Sci, 48*: 241 (**1960**).
13 Hatada K, Ute K, Tanaka K, Imanari M, Fujii N, *Polym J 19*: 425 (**1987**).
14 Kitayama T, Kitaura T, unpublished results.
15 Hirano T, Heterotactic living polymerization of methacrylates, *Dissertation,* Osaka University (**1999**).
16 Bovey FA, *Polymer Conformation and Configuration.* Academic Press, New York, p. 13 (**1969**).
17 Hatada K, Ute K, Tanaka K, Kitayama T, *Polym J 18*: 1037 (**1986**).
18 Kitayama T, Kitaura T, presented at the World Polymer Congress, 41st International Symposium on Macromolecules (**2006**).
19 Hatada K, Kitayama T, Ute K, *Annu Rep NMR Spectrosc 26*: 99 (**1993**).
20 Kitayama T, Hirano T, Hatada K, *Polym J*: *28*, 61 (**1996**).
21 Yuki H, Hatada K, *Adv Polym Sci 31*: 1 (**1979**).
22 Hatada K, Kitayama T, Ute K, *Prog Polym Sci 13*: 189 (**1988**).
23 Webster OW, *Science 251*: 887 (**1991**).
24 Yasuda H, Yamamoto H, Yokota K, Miyake S, Nakamura A, *J Am Chem Soc 114*: 4908 (**1992**); Yasuda H, Yamamoto H, Yamashita M, Yokota K, Nakamura A, Miyake S, Kai Y, Kanehisa N, *Macromolecules 26*: 7134 (**1993**).
25 Hatada K, Ute K, Tanaka K, Kitayama T, Okamoto Y, *Polym J 17*: 977 (**1985**).
26 Hatada K, Ute K, Tanaka K, Okamoto Y, Kitayama T, *Polym J 18*: 1037 (**1986**).
27 Resconi L, Cavallo L, Fait A, Piemontesi F, *Chem Rev 100*: 1253 (**2000**).
28 Deng H, Shiono T, Soga K, *Macromolecules 28*, 3067 (**1995**).

29 Nguyen H, Jarvis AP, Lesley MJG, Kelly WM, Reddy SS, Taylor NJ, Collins S, *Macromolecules 33*, 1508 (**2000**).
30 Frauenrath H, Keul H, Hartwig H, *Macromolecules 34*: 14 (**2001**).
31 Bolig AD, Chen EY-X, *J Am Chem Soc 126*: 489 (**2004**).
32 Cui C, Shafir A, Reeder CL, Arnold J, *Organometallics 22*: 3357 (**2003**).
33 Qi G, Nitto Y, Saiki A, Tomohiro T, Nakayama Y, Yasuda H, *Tetrahedron 59*: 10409 (**2003**).
34 Zundel T, Teyssié P, Jérôme R, *Macromolecules 31*: 2433 (**1998**).
35 Zune C, Zundel T, Dubois P, Teyssié P, Jérôme R, *J Polym Sci, Part A: Polym Chem 37*: 2525 (**1999**).
36 Kaminsky W, Miri M, Sinn H, Woldt R, *Makromol Chem Rapid Commun 4*: 417–421 (**1983**).
37 Resconi L, Abis L, Franciscono G, *Macromolecules 25*: 6814 (**1992**).
38 Resconi L, Piemontesi F, Franciscono G, Abis L, Fiorani T, *J Am Chem Soc 114*: 1025–1032 (**1992**).
39 Ewen JA, *J Am Chem Soc 106*: 6355–6364 (**1984**).
40 Kaminsky W, Külper K, Brintzinger HH, Wild FRWP, *Angew Chem Int Ed Engl 24*: 507 (**1985**).
41 Farina M, Di Silvestro G, Sozzani P, *Macromolecules 26*: 946 (**1993**).
42 Coates GW, Waymouth RM, *Science 267*: 217 (**1995**).
43 Ewen JA, Jones RL, Razavi A, Ferrara JD, *J Am Chem Soc 110*: 6255 (**1988**).
44 Razavi A, Bellia V, De Brauwer Y, Hortmann K, Peters L, Sirole S, Van Belle S, Thewalt U, *Macromol Chem Phys 205*: 347 (**2004**).
45 Ewen JA, Elder MF, Jones RL, Haspeslagh L, Atwood JL, Bott SG, Robinson K, *Makromol Chem, Makromol Symp 48/49*: 253 (**1991**).
46 Silvestro GD, Sozzani P, Sacare B, Farina M, *Macromolecules 18*: 928 (**1985**).
47 Longo P, Proto A, Grassi A, Ammendola P, *Macromolecules 24*: 4626 (**1991**).
48 Razavi A, Peters L, Nafpliotis L, *J Mol Catal A Chem* 115 (**1997**).
49 Shiomura T, Kohno M, Inoue N, Asanuma T, Sugimoto R, Iwatani T, Uchida O, Kimura S, Harima S, et al., *Macromol Symp* (**1996**).
50 Miller SA, Bercaw JE, *Organometallics 23*: 1777 (**2004**).
51 Herzog TA, Zubris DL, Bercaw JE, *J Am Chem Soc 118*: 11988 (**1996**).
52 Kitayama T, Shinozaki T, Masuda E, Yamamoto M, Hatada K, *Polym Bull 20*: 505 (**1988**).
53 Kitayama T, Shinozaki T, Sakamoto T, Yamamoto M, Hatada K, *Makromol Chem Suppl 15*: 167 (**1989**).
54 Schlaad H, Muller AHE, *Macromol Rapid Commun 16*: 399 (**1995**).
55 Bolig AD, Chen EYX, *J Am Chem Soc 123*: 7943 (**2001**).
56 Kitayama T, He S, Hironaka Y, Hatada K, *Polym J 27*: 314 (**1995**).
57 Liu W, Nakano T, Okamoto Y, *Polymer 41*: 4467 (**2000**).
58 Tabuchi M, Kawauchi T, Kitayama T, Hatada K, *Polymer 43*: 7185 (**2002**).
59 Hirano T, Miki H, Seno M, Sato T, *J Polym Sci, Part A: Polym Chem 42*: 4404 (**2004**).
60 Hirano T, Ishii S, Kitajima H, Seno M, Sato T, *J Polym Sci, Part A: Polym Chem 43*: 50 (**2005**).
61 Hirano T, Miki H, Seno M, Sato T, *Polymer 46*: 3693 (**2005**).
62 Hirano T, Miki H, Seno M, Sato T, *Polymer 46*: 5501 (**2005**).
63 Hirano T, Kitajima H, Ishii S, Seno M, Sato T, *J Polym Sci, Part A: Polym Chem 43*: 3899 (**2005**).
64 Hirano T, Kitajima H, Seno M, Sato T, *Polymer 47*: 539 (**2006**).
65 Nakahama S, Kobayashi M, Ishizone T, Hirao A, Kobayashi M, *Polym Mater Sci Eng 76*: 11 (**1997**).
66 Kobayashi M, Ishizone T, Hirao A, Nakahama S, Kobayashi M, *Polym Mater Sci Eng 76*: 304 (**1997**).
67 Kobayashi M, Okuyama S, Ishizone T, Nakahama S, *Macromolecules 32*: 6466 (**1999**).
68 Kobayashi M, Ishizone T, Nakahama S, *Macromolecules 33*: 4411 (**2000**).
69 Ito M, Ishizone T, *Des Monom Polym 7*: 11 (**2004**).
70 Liu W, Nakano T, Okamoto Y, *Polym J 32*: 771–777 (**2000**).

71 Kitayama T, Zhang Y, Hatada K, *Polym Bull 32*: 439 (**1994**).
72 Kitayama T, Zhang Y, Hatada K, *Polym J 26*: 868 (**1994**).
73 Kitayama T, Hirano T, Hatada K, *Polym J 28*: 61 (**1996**).
74 Kitayama T, Hirano T, Zhang Y, Hatada K, *Macromol Symp 107*: 297 (**1996**).
75 Ute K, Miyatake N, Osugi Y, Hatada K, *Polym J 25*: 1153 (**1993**).
76 Hatada K, Ute K, Miyatake N, *Prog Polym Sci 19*: 1067 (**1994**).
77 Nozakura S, Ishihara S, Inaba Y, Matsumura K, Murahashi S, *J Polym Sci 11*: 1053 (**1973**).
78 Gotoh Y, Yamashita M, Nakamura M, Toshima N, Hirai H, *Chem Lett* 53 (**1991**).
79 Hatada K, Sugino H, Ise H, Kitayama T, Okamoto Y, Yuki H, *Polym J 12*: 55 (**1980**).
80 Hirano T, Yamaguchi H, Kitayama T, Hatada K, *Polym J 30*: 767 (**1998**).
81 Hirano T, Cao J-Z, Kitayama T, Hatada K, *Polym J*, to be submitted.
82 Kitayama T, Hirano T, Hatada K, *Tetrahedron 53*: 15263 (**1997**).
83 Kanetaka S, Miyamoto M, Saegusa T, *Polym Prepr Jpn 39*: 220 (**1990**); *Polym Prepr Jpn Engl Ed* E104.
84 Doherty MA, Hogen-Esch TE, *Makromol Chem 187*: 61 (**1986**).
85 Ute K, Asada T, Nabeshima Y, Hatada K, *Macromolecules 26*: 7086 (**1993**).
86 Ute K, Tarao T, Hatada K, *Polym J 29*: 1904 (**1996**).
87 Ute K, Tarao T, Hongo S, Ohnuma H, Hatada K, Kitayama T, *Polym J 31*: 177 (**1999**).
88 Ute K, Hongo S, Tarao T, Kitayama T, in preparation.
89 Ute K, Asada T, Hatada K, *Macromolecules 29*: 1904 (**1996**).
90 Hatada K, Ute K, Nishiura T, Kashiyama M, Saito T, Takeuchi M, *Polym Bull 23*: 157 (**1990**).
91 Ute K, Miyatake N, Asada T, Hatada K, *Polym Bull 28*: 561 (**1992**).
92 Ute K, Yamasaki Y, Naito M, Miyatake N, Hatada K, *Polym J 27*: 951 (**1995**).
93 Ute K, in *Supercritical Fluid Chromatography with Packed Columns*, Berger C, Anton K (Eds). Marcel Dekker, New York, p. 347 (**1997**).
94 Hatada K, Kitayama T, Ute K, Nishiura T, *Macromol Symp 129*: 89 (**1998**).
95 Ute K, Miyatake N, Hatada K, *Polymer 36*: 1415 (**1995**).
96 Kusy RP, *J Polym Sci, Polym Chem Ed 15*: 1527 (**1996**).
97 Hatada K, Kitayama T, Ute K, Nishiura T, *Macromol Symp 132*: 221 (**1998**).
98 Hatada K, Kitayama T, Ute K, Nishiura T, *Macromol Symp*, in press.
99 Ute K, Niimi R, Matsunaga M, Suzuki T, Hatada K, unpublished results.
100 Hatada K, Ute K, Kashiyama M, Imanari M, *Polym J 22*: 218 (**1990**).
101 Ute K, Niimi R, Hongo S, Hatada K, *Polym J 30*: 439 (**1998**).

2
Synthesis of Macromonomers and Telechelic Oligomers by Living Polymerizations

Bernard Boutevin, Cyrille Boyer, Ghislain David, and Pierre Lutz

2.1
Introduction

The definitions given for "telechelic oligomers" and "macromonomers" are not accurate and often lead to some confusion between these two terms in the literature. For instance the term "macromonomer" is often replaced by "semi-telechelic" [1]. Without prescribing any normalization, it is necessary to define these two terms well in order to review correctly the studies made in both areas. To simplify, the definitions will be based on the functionality. Hence, a functionality of 2 on one chain-end will relate to macromonomers. These comprise molecules bearing either a double bond or two polycondensable groups at the same chain-end, see Scheme 2.1. On the other hand, a functionality of 1 at the chain-end will relate to telechelic compounds (Scheme 2.1). These include diols, diamines, diacids etc. Of course they also include diolefin compounds that usually lead to gels or networks.

It is also necessary to specify the particular case of macromolecules bearing a well-identified G functionality at one chain-end and a thermally reactive group at the other chain-end. These groups can be nitroxides, iodine atom, xanthate etc. and are commonly used in living radical polymerization. These compounds may be classified as monofunctional oligomers. According to the above definitions, macromonomers can be considered as precursors of graft copolymers whereas telechelic oligomers will lead to multi-block copolymers.

A great variety of methods to synthesize macromonomers have been developed over the years. These are essentially based on controlled radical polymerization processes, well suited for the polymerization of a vide variety of monomers. This is not the case for ionic polymerization which is applicable only to a restricted number of monomers. In spite of this, ionic polymerization methods are still of interest for the synthesis of macromonomers. Without being exhaustive, the first part of the present chapter emphasizes some typical examples of synthesis of well-defined macromonomers via ionic polymerization, especially those providing access to macromonomers not accessible by other polymeriza-

Macromolecular Engineering. Precise Synthesis, Materials Properties, Applications.
Edited by K. Matyjaszewski, Y. Gnanou, and L. Leibler

ISBN: 978-3-527-31446-1

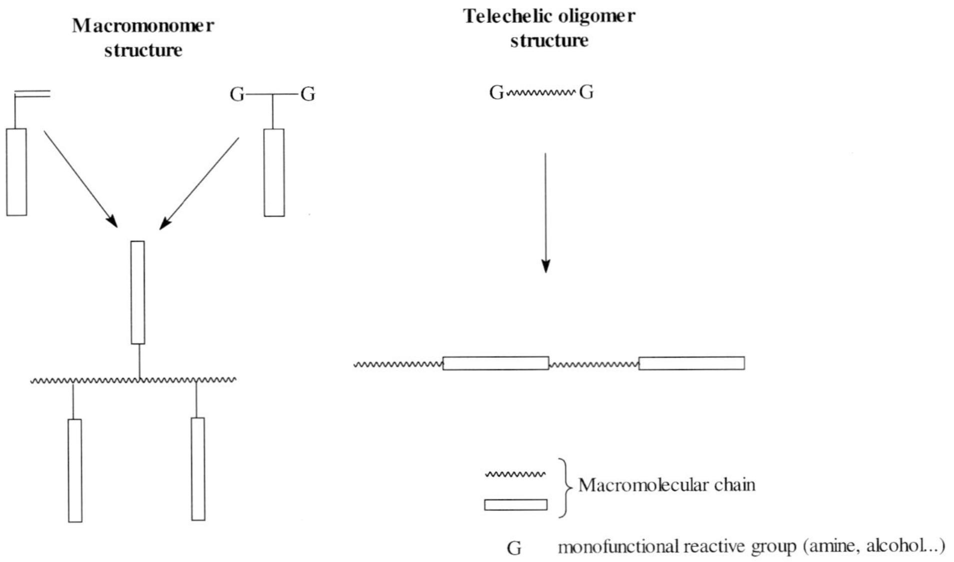

Scheme 2.1 Schematic view of the telechelic and macromonomer structures.

tion processes such as poly(ethylene oxide) (PEO) macromonomers and polydiene macromonomers of controlled microstructure. Pioneering work in the domain of the synthesis of macromonomers by anionic polymerization was carried out by Milkovitch [2].

The major part of this chapter will be devoted to the preparation of both telechelic and macromonomers by controlled radical techniques. In the 90s, the syntheses of telechelic oligomers and macromonomers were reviewed, for instance by Boutevin [3] and Rempp [4], respectively. These surveys deal with the use of conventional radical polymerizations, such as telomerization or dead-end polymerization (DEP), to reach the telechelic or macromonomer structure. DEP is often preferred to reach a telechelic structure as many combinations between functional initiators and vinyl monomers are possible, enabling a quantitative recombination termination reaction.

The telomerization process, based on conventional radical polymerization, enables the synthesis of macromonomer structures, bearing either a polymerizable double bond situated in the ω-position or two condensable groups situated at the same chain-end. However, in 1994 the living radical polymerizations (LRP) officially appeared, allowing a major evolution of both telechelic oligomers and macromonomers. Indeed, LRP techniques provided new end-groups such as nitroxides, dithioesters, xanthates or even halogens, which could be easily modified.

The scope of this chapter is to consider the use of the living techniques (i.e. ionic and radical ones) and especially the different strategies found in the literature in order to reach the targeted telechelic or macromonomer structure.

2.2
Ionic Polymerization

Anionic polymerizations – when carried out under proper conditions, in aprotic solvents, with an efficient metalorganic initiator – were the first examples of living ionic polymerizations [5]. The cationic polymerization of some heterocyclics, vinyl ethers or alkenes follows a similar scheme [6, 7]. The large potential offered has been used in a great variety of instances for the synthesis of macromolecules with complex macromolecular architectures [8] and, among these, macromonomers [9–11]. The reaction scheme of living polymerizations comprises initiation and propagation, with no spontaneous termination or transfer reactions. Thus, the number of active sites remains constant, and is equal to the number of initiator molecules introduced at the onset of the reaction. As a consequence, the number average degree of polymerization is determined by the molar ratio of monomer converted to initiator used. Furthermore, the molar mass distribution within a sample prepared under such conditions is narrow (Poisson distribution).

Another consequence of termination-free processes is that the active sites remain located at the chain-end once the polymerization reaction is over. There is then the opportunity to functionalize polymers at their chain-end, upon induced deactivation. If, instead, another appropriate monomer is added to the solution of living po1ymer, a block copolymer results.

Such processes have been used for many syntheses of macromonomers in the last twenty years. Two approaches essentially have been considered, either the use of an initiator that contains a polymerizable group which is not affected by the polymerization reaction, or the capping of the living chain-end with an electrophile containing a polymerizable group or a living cationic polymer with a nucleophile able to polymerize. The synthesis of macromonomers via ionic polymerization and their homo-or copolymerization have been extensively reviewed in several publications [4, 10–12]. Some typical examples will be given in the following section.

2.2.1
Macromonomer Synthesis by Ionic Initiation

A few examples of macromonomers generated from unsaturated anionic or cationic initiators are presented below. It should always be checked that the initiation reaction is fast and quantitative, and that no side-reaction involves the double bond.

2.2.1.1 Polystyrene Macromonomers by Ionic Initiation

In the 60s, the synthesis of polystyrene (PS) macromonomers by initiation with vinyllithium or allylithium was reported [13]. These initiators gave, however, low rate constants in the initiation reaction, which led to unsymmetrical broad mo-

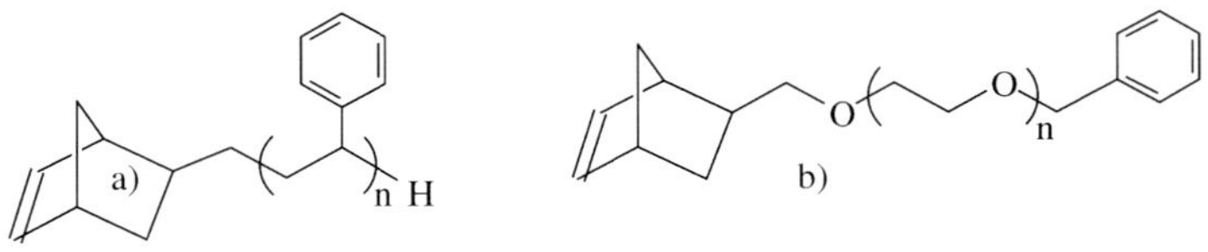

Fig. 2.1 α-Norbornenyl macromonomers obtained by initiation.

lar mass distributions, which was explained by the resonance stabilization of the carbanions. The system was improved by Takano et al. [14] who introduced methylene groups between the vinyl group and the carbanion. They demonstrated the efficiency of 4-pentenyllithium as initiator, and obtained well-defined PS macromonomers exhibiting one olefinic vinyl group per chain-end. This method, however, cannot be applied to the synthesis of α-methacryloyloxy PS macromonomers.

Gnanou et al. introduced an original strategy to design well-defined PS macromonomers with a norbornenyl group in the α position [15]. They started from 5-hydroxymethylbicyclo [2,2,1] hept-2-ene. This compound was chlorinated, then lithiated upon treatment with a large excess of lithium powder in ether. The resulting anionic initiator was used, without further purification, to polymerize styrene. Well-defined α-norbornenyl PS macromonomers (Fig. 2.1 a) were obtained, characterized by sharp molar mass distributions confirming the absence of side-reactions between the growing chain-end and the norbornenyl polymerizable end-group. In addition, in contrast to ω-norbornenyl macromonomers obtained by deactivation reactions, they do not bear an ester function between the unsaturation and the rest of the chain. This strategy was extended to α-norbornenyl polybutadiene macromonomers [16] and to α-norbornenyl PEO macromonomers [17] (Fig. 2.1 b). These macromonomers were homopolymerized under living conditions using a Schrock catalyst. However, much higher homopolymerization degrees were obtained for ω-norbornenyl PS macromonomers. This may be explained by a possible interaction of the catalyst with the remaining double bonds of the former macromonomers.

2.2.1.2 **Poly(ethylene oxide) Macromonomers by Ionic Initiation**

Poly(ethylene oxide) is a water soluble polymer, with outstanding biocompatibility, and its macromonomer is the building block for a large number of structures for various applications. Several approaches have been developed over the years to design PEO macromonomers by initiation as discussed below or by deactivation as discussed later in the chapter. It has to be pointed out that PEO cannot be prepared by free or controlled radical polymerization and that, increasingly, PEO macromonomers are prepared by end-modification of already existing polymers.

Potassium *p*-vinyl or *p*-isopropenyl benzylates belong to the first generation of heterofunctional initiators aimed to design well-defined PEO macromonomers

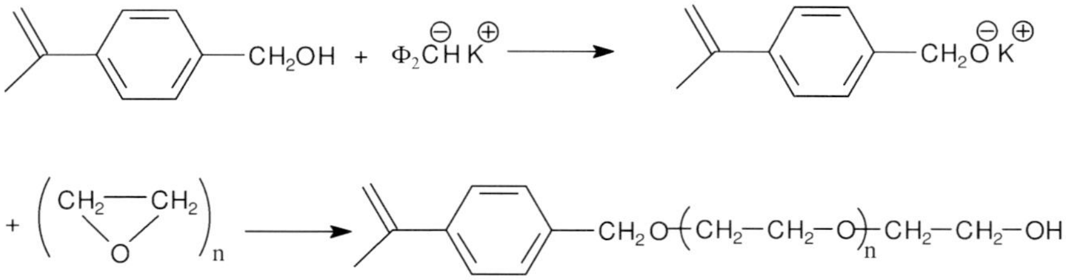

Scheme 2.2 Synthesis of PEO macromonomers by initiation.

via anionic polymerization of oxirane (Scheme 2.2). They have been described in many publications [9, 18].

Briefly, the alcoholate is made with diphenylmethylpotassium acting as the metalating agent that initiates the anionic ring opening polymerization of oxirane. The relatively low nucleophilicity of this compound prevents attack on the double bond under the conditions used. As expected the resulting alcoholate is insoluble in THF. This does not affect the molar mass distribution of the resulting PEOs as the medium becomes homogeneous after addition of a few oxirane units. The molar masses are those expected from the monomer consumed to initiator molar ratio, and the polydispersity is low. The functionalization degree was found to be in agreement with expectations.

Selected functional initiators can also be used to synthesize macromonomers fitted at the chain-end with a polymerizable heterocycle or with two functions, which in the latter case may be involved in polycondensation reactions. PEO macromonomers exhibiting a ring-opening polymerizable 2-oxazoline group at the chain-end, prepared by anionic polymerization of oxirane starting from lithiated 2(-*p*-hydroxyphenyl)-2-oxazoline [19] represent a typical example of the former macromonomers (Fig. 2.2).

Cationic polymerization was also attempted for the synthesis of macromonomers. However, one has to be aware that most cationic polymerizations involve transfer and/or termination reactions, and therefore are not well suited for the synthesis of well-defined macromonomers. The living character of the polymerization could be established only in a limited number of cases (cyclic monomers such as oxolane dioxolane and N-substituted aziridines) or for vinylethers.

The cationic polymerization of oxolane (THF), initiated either by an unsaturated oxocarbenium salt such as methacryloyl hexafluoroantimonate [20], or by *p*-vinyl (or *p*-isopropeny) benzylium hexafluoroantimonate [9] yields poly(THF) chains that are quantitatively fitted with a polymerizable double bond at the

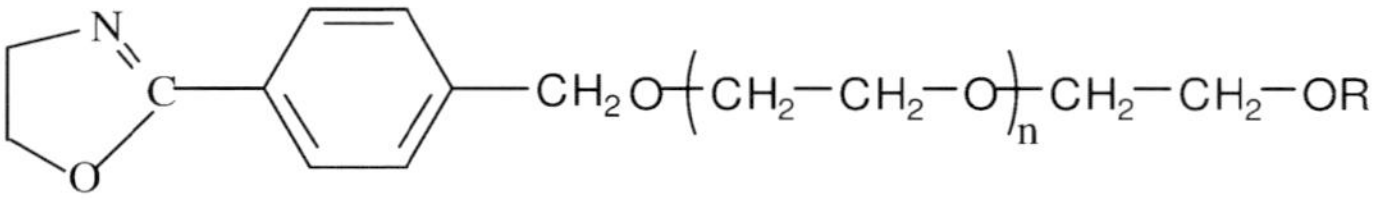

Fig. 2.2 α-2-Oxazoline PEO macromonomer (R=H, CH_3).

chain-end. In the latter case it has to be noted that the presence of the unsaturation in the para position enhances the efficiency compared to unsubstituted benzylium hexafluoroantimonate.

2.2.2
Macromonomers Synthesis by Induced Deactivation

As stated above, unsaturated initiators can be used efficiently to design well-defined macromonomers in only a limited number of cases. Therefore most of the macromonomers based on ionic polymerization were obtained upon deactivation. The deactivation approach involves the choice of an efficient organometallic initiator to allow control of the molar mass of the polymer formed. Once the polymerization is completed, the living sites are deactivated by means of an electrophile, whereupon the double bond is linked at the chain-end. If necessary, the nucleophilicity of the actives sites is lowered to prevent side-reactions affecting the unsaturation.

2.2.2.1 Polystyrene Macromonomers by Ionic Deactivation

Detailed investigations on the synthesis and the characterization of ω- or α,ω-functional PS macromonomers have been performed by several groups [9, 11, 21]. The polymerization of styrene can be carried out in polar or nonpolar aprotic solvents. *Sec*-butylithium and cumylpotassium are efficient initiators for the synthesis of low molar mass samples. To achieve quantitative reaction with an unsaturated electrophile such as 4-vinylbenzyl bromide, 4-vinylbenzyl chloride or allyl bromide, the reaction temperature has to be lowered to $-78\,^{\circ}C$. A higher reaction temperature would give rise to side-reactions involving the double bond. These side-reactions can also be prevented by the intermediate addition of 1,1-diphenylethylene, meant to lower the nucleophilicity of the active sites. This monomer adds, but cannot homopolymerize. The resulting diphenylmethyl anion is resonance stabilized, and therefore less nucleophilic. For ω-methacryloyloxy PS macromonomers, the intermediate addition of oxirane is necessary to lower the reactivity of the chain-end [22].

Hirao et al. [23] investigated the possibility of using less reactive alkyl halide moieties in the reactions with living polymers. They studied the influence of the nature of the halogen and the methylene chain length on the functionalization yield. They deactivated living PS (or polyisoprene) chains with five different (ω-haloalkylstyrenes).

ω-Allyl, ω-undecenyl (and α,ω-undecenyl) polystyrene, polyisoprene or ω-undecenyl poly(styrene-*block*-isoprene) macromonomers well defined in molar mass and functionality, were synthesized via anionic polymerization (Scheme 2.3) recently by Lutz et al. [24, 25].

The PS macromonomers were homo- (or co-polymerized with ethylene) in the presence of coordination catalysts to yield a new type of comb-shaped (or graft) copolymer comprising a polyethylene backbone and polystyrene grafts. It

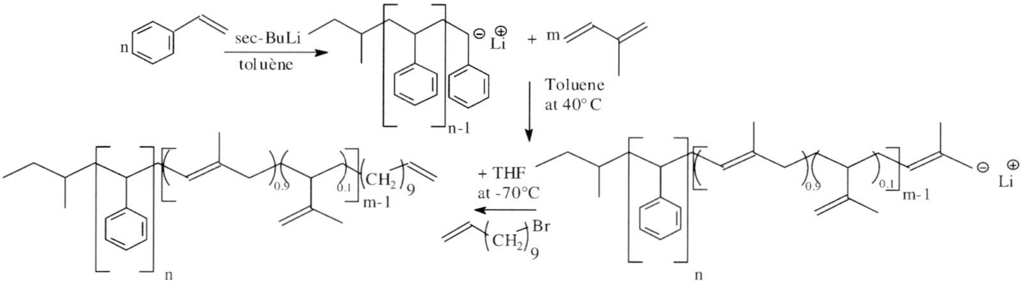

Scheme 2.3 ω-Undecenyl poly(styrene-*block*-isoprene) macromonomer synthesized by induced deactivation.

was shown that the environment of the terminal double bond of the PS macromonomers has an enormous influence on the polymerization behavior. Indeed, an undecenyl end-group is more reactive than an allyl end-group [24].

Anionic polymerization was applied successfully to the synthesis of many other polymers to yield quantitatively well-defined macromonomers fitted at the chain-end with either a styrene type or a methacrylic ester unsaturation. Poly(2-vinylpyridine) macromonomers were prepared by Takaki et al. [26] by polymerizing 2-vinylpyridine with 2-pyridylethylithium in THF at −78 °C followed by deactivation with 4-vinylbenzyl chloride. Senoo et al. [27] and Hadjichristidis et al. [28] synthesized poly(isoprene) and poly(styrene-*block*-isoprene) fitted also with styrene type unsaturations. ω-Living polybutadiene chains were reacted with 4-(chlorodimethylsilyl) styrene to access end functionalization. Similarly to ω-methacryloyloxy PS macromonomers, Schmidt et al. prepared by anionic deactivation ω-methacryloyloxy poly(styrene-*block*-vinyl-2-pyridine) macromonomers. Their homopolymerization lead to polymacromonomers characterized by a high degree of polymerization of the backbone, behaving as cylindrical polymer brushes [29]. Due to the presence of the poly(vinyl-2-pyridine) sequence these polymacromonomers can complex metal ions. Many other examples can be found in the literature.

Gnanou et al. [15–17, 30–32] synthesized well-defined norbornene end-functionalized PS macromonomers by deactivation of living polystyryl lithium chains with 5-carbonyl chloride bicyclo[2,2,1]hept-2-ene. They extended the strategy to ω-norbornenyl polybutadiene macromonomers of controlled microstructure and to ω-norbornenyl poly(styrene-*block*-butadiene) or poly[styrene-*block*-(ethylene oxide)] macromonomers. They polymerized the different macromonomers under living conditions using a Schrock catalyst and examined the influence of the position of the polymerizable unit (α or ω), the molar mass of the macromonomer and the chemical nature or the microstructure of the chain, on the homopolymerization yield and the average polymerization degree.

2.2.2.2 PEO Macromonomers by Deactivation

The alternative way to prepare PEO macromonomers involves deactivation of the alkoxide function of a monofunctional living PEO by means of an unsaturated electrophile. In the example quoted, advantage is taken of the very large electroaffinity difference between the styryl unsaturation and the oxirane ring. Various efficient initiators have been developed to provide access to a good control of the molar mass of the PEO chain. Among these, diphenylmethyl potassium or potassium methoxyethanolate have been extensively used for the preparation of PEO macromonomers. The former is well soluble in THF, efficient, and leads to PEO chains with diphenyl groups in the α-position. This facilitates the characterisation of the samples by ^{1}H NMR, UV spectroscopy or Maldi-Tof MS. However the presence of this bulky hydrophobic entity at the chain-end affects the solubility. Potassium methoxyethanolate is insoluble, and initiation is heterogeneous. Nevertheless, as in the case of *p*-isopropenyl benzylates this has no consequence on the control of PEO molar mass and molar mass distribution.

Once oxirane polymerization is completed, the deactivation of the living PEOs is achieved by means of either methacryloyl choride or *p*-vinyl benzyl bromide to yield the expected PEO macromonomers. Indeed, a wide range of hydrophilic/hydrophobic balance is accessible with anionic polymerization by modifying the initiator type (R), the length of the methylene spacer or the nature of the terminal polymerizable unit (Fig. 2.3). A series of such PEO macromonomers was prepared by Ito et al. [33] and their solution behavior was investigated systematically. It is now well established that PEO macromonomers exhibit an amphiphilic character. They homopolymerize or copolymerize by a conventional free radical process with hydrophobic monomers, unusually rapidly in water, as a result of their organization in micelles [34].

It has to be mentioned that the same authors were able to synthesize well-defined ω-hydroxypoly(ethylene oxide) macromonomers by anionic polymerization starting from the potassium alkoxide of the *tert*-butyldimethylsilyl ether of ethylene glycol [35]. The free radical (co-)polymerization of these macromonomers provided access to a new type of star-like structure in which the outer end of the branches are fitted with hydroxy end groups. The same type of PEO macromonomer fitted at one chain-end with methyl methacrylate units was also (co-)polymerized via ATRP with hydrophobic monomers. The (co-)polymerization of such crystallizable macromonomers with octadecyl(meth)acrylates yielded densely heterografted brush macromolecules with interesting bulk properties and represents a typical examples of copolymerization of macromonomers via ATRP, a process also discussed extensively [36].

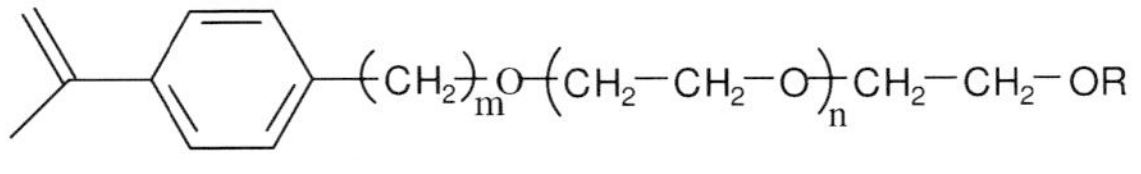

Fig. 2.3 ω-Undecenyl poly(styrene-*block*-isoprene) macromo-

Polyester macromonomers can be obtained from lactones and lactides by anionic ring opening polymerization, using ionic as well as Lewis acid initiators. They have gained increasing interest due to their large domain of applications as a biodegradable material, for controlled drug release, or as building blocks in polyurethanes. However, typical anionic initiators lead to samples with broad molecular distribution because of inter- and intramolecular transesterification and formation of cyclic molecules by back-biting reactions. In contrast, initiators based on (allyl) aluminum alkoxides give rise to polymerization characterized by a living character whereas other alkoxides based on titanium or zirconium yield, besides the main polymer, some macrocyclic oligomers. A detailed discussion of the different initiating systems is outside the scope of the present chapter. Hamaide et al. [37] took advantage of the exchange reaction between aluminum alkoxides and alcohols to design a new catalytic system for the coordinated anionic ring-opening polymerization of ε-caprolactone. These authors were able to design well-defined ω-hydroxy oligo polycaprolactone macromonomers with 2-hydroxyethylmethacrylate or hydroxymethyl styrene as polymerizable end-groups. Various other systems have been designed to initiate the ring-opening polymerization of propiolactone, providing access to well defined ω-methacryloylpolypropiolactone macromonomers.

Cationic deactivation was also applied to the synthesis of macromonomers, especially those of poly(oxolane) or poly(1,3-dioxolane). ω-Methacryloyl poly-THF macromonomers were first obtained by Asami [38]. Franta et al. [39] used a similar strategy to prepare poly(THF) macromonomers with α-methylstyrene polymerizable groups. They deactivated a living poly(THF) with sodium *p*-isopropenylbenzylate whereupon well-defined poly(THF) macromonomers were obtained. It should be noted that the poly(THF) macromonomers obtained by initiation with a *p*-isopropenylbenzylium salt and those arising from deactivation have exactly the same structure (except for the opposite chain-end).

The deactivation methods presented above can be extended to the synthesis of bifunctional macromonomers, carrying a polymerizable unsaturation at each chain-end. A bifunctional initiator is used to start the polymerization process. After complete polymerization, the active sites at both chain-ends are deactivated with an unsaturated deactivator to yield the macromonomer α,ω-bifunctional PS [24], PEO or PEO macromonomers with a central degradable poly(1,3-dioxolane) degradable block [40].

2.2.3
Macromonomers Synthesis by Two-step Procedures

A large variety of macromonomers have been prepared anionically via functional initiators or by deactivation reactions. There are however cases where this method fails to give satisfactory results. Two-step methods have thus been developed: an ω-functional (such as ω-hydroxy or ω-carboxylic) polymer is made first, and the function at the chain-end is reacted subsequently with a suitable unsaturated compound (sometimes according to methods developed in biochemistry) to have the polymerizable unsaturation tied at the chain-end [41].

2.3
Controlled Radical Polymerizations

2.3.1
Atom Transfer Radical Polymerization

Atom transfer radical polymerization (ATRP) is a new method [42–44] allowing the synthesis of telechelic oligomers and macromonomers. The ATRP process can polymerize a wide range of monomers (styrene, acrylate, methacrylate etc.). Furthermore, it allows the incorporation of reactive groups at the chain-end of oligomers [45–48]; these include epoxides or hydroxyl groups. ATRP is a radical process based on the use of a catalytic complex transition metal/ligand. Generally the metal is copper (CuCl or CuBr) but also Fe(II)[50] or Ru(II) [51, 52] may be used in some cases. The ligand (L) [53] is more often a tertiary amine. 1,1,4,7,10,10-hexamethyltriethylenetetramine (HMTETA) and 2,2′-bipyridine are commonly used [54] but Haddleton [55, 56] also showed the good efficiency of *n*-(octyl)-2-pyridylmethanimine (*n*-Oct-L). The catalytic complex is able to establish equilibrium between the dormant species and the radicals. The equilibrium is shifted to the dormant species. The radical concentration is also low during the polymerization, which limits the termination reactions (disproportionation or recombination) and the amount of dead chains.

Like in a redox telomerization, first described by Asscher and Vofsi [57, 58] and developed by Boutevin [59, 60], the catalytic complex extracts the halogen atom from the initiator (R–X) by a redox process. Then a radical is created that propagates to monomer. The growing chain then fixes a halogen atom from the catalytic complex to form a dormant species. The chain is further reactivated when the catalytic complex traps the chain-end halogen atom to be oxidized. The chain is then able to propagate. Some termination and transfer reactions occur but these remain minor.

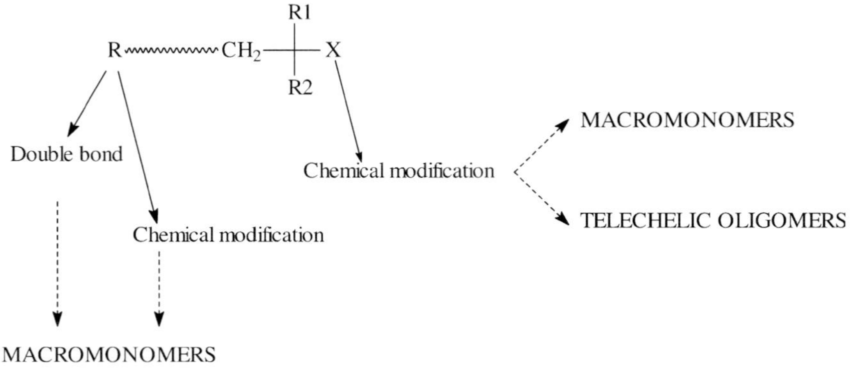

Scheme 2.4 Different routes to obtain macromonomers or telechelic oligomers with a polymerizable double bond.

The particular structure of oligomers obtained by ATRP, i.e. in the presence of both a reactive function and a halogen situated at each extremity of the macromolecular chain, allows one to synthesize macromonomers as well as telechelic oligomers by different routes (Scheme 2.4). According to Scheme 2.4, the resulting oligomers bear an R group (provided by the initiator) in the α position and a halogen atom (provided by the initiator) in the ω position.

The R group, being either a double bond or being chemically modified, only affords the synthesis of macromonomers. The chemical modification of the halogen atom allows one to obtain both macromonomers and telechelics, depending on the technique used to replace the terminal halogen.

2.3.1.1 Synthesis of Macromonomers from the R Group Provided by the Initiator

Initiators with an unsaturated group If the initiators possess a reactive double bond, a chemical modification is not required. For the ATRP process of styrene, Matyjaszewski et al. [61] used initiators bearing an unsaturated group such as vinyl chloroacetate. As VAc was unreactive towards styrene in radical copolymerization, vinyl chloroacetate was able to initiate the ATRP of styrene (Scheme 2.3). The resulting PS macromonomers, with molar masses ranging from 5×10^3 to 15×10^3 g mol^{-1}, were further used in copolymerization reaction with *N*-vinylpyrrolidinone (NVP). The obtained amphiphilic copolymers were used as hydrogels. Zeng et al. [62] extended the use of unsaturated initiator to allyl-type and vinyl-type initiators (Scheme 2.5).

Several ligands were used with allyl-type and vinyl-type initiators, such as HMDETA, PMDETA. Zeng et al. [62] showed that the combination of initiator **1** or initiator **4** gave the best control of the molar mass for the ATRP of 2-(dimethylamino)ethyl methacrylate (DMAEMA). These allylic macromonomers are then able to copolymerize with acrylamide. However, it is well known that trans-

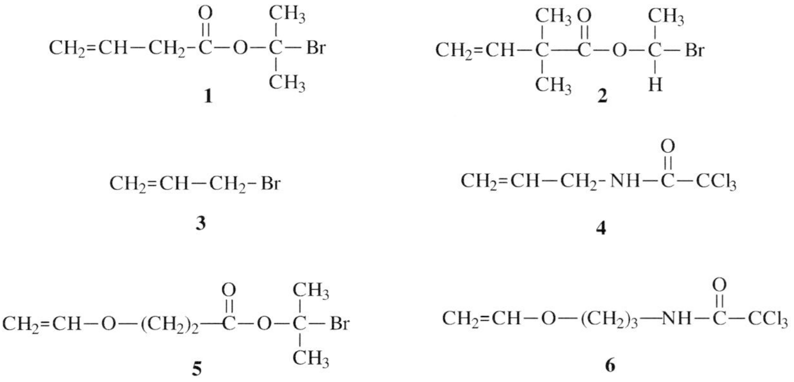

Scheme 2.5 Different allyl- and vinyl-type initiators.

fer reaction often occurs onto the methylene group situated in the α position of the unsaturation which prevents an increase in the molar masses of the copolymers [63–66]. Concerning the vinyl ether-type macromonomers (obtained with compounds **5** and **6**), their copolymerization was studied with several monomers. Unlike styrene and methacrylates, the copolymerization of the vinyl ether is efficient with acrylates. We can also remark that stopping before complete conversion of acrylates prevents any side-reactions of acrylates towards the vinyl chain-end.

Initiators Without an unsaturated group If the initiator used in an ATRP process does not possess any unsaturation, a chemical modification of the functional group is required to introduce the reactive unsaturation. For instance, the functional groups react with a monomer such as glycidyl methacrylate (MAGLY), isocyano ethylmethacrylate (IEM), 1-(isopropenylphenyl)-1,1-dimethylmethyl isocyanate (TMI), or maleic anhydride to lead to macromonomers with either methacrylic, styrenic or maleic groups. However, it is necessary to eliminate the chain-end halogen atom to avoid any side-reaction. In the further copolymerization, several techniques may overcome this problem. Moreover, Mueller et al. [67] employed a new strategy to eliminate the chain-end halogen atom, based on a transfer reaction onto the ligand. Indeed, they investigated the ATRP of acrylate monomer in the presence of 2-hydroxyethyl 2-bromoisobutyrate as a functionalized initiator and also with a large excess of PMDETA ligand (relative to CuBr). These special conditions allow a hydrogen transfer from the ligand onto the ω position (Scheme 2.6). The unsaturation is then obtained by reaction of the terminal group with methacryloyl chloride.

2.3.1.2 Synthesis of Macromonomers and Telechelics by Chemical Modification of the Terminal Halogen

The chain-end halogen atom can be replaced by using various methods. However, the literature shows that most of them were mainly applied to the synthesis of telechelic oligomers. We will also distinguish those essentially leading to telechelic oligomers and those leading both to telechelic and macromonomers.

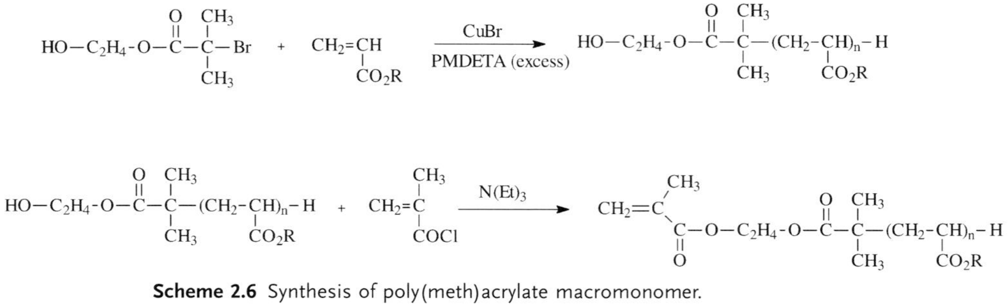

Scheme 2.6 Synthesis of poly(meth)acrylate macromonomer.

Chemical modifications leading to both macromonomers and telechelics

Radical addition reactions (ATRA) 1,2-Epoxy-5-hexene and allyl alcohol [68] are examples of monomers not polymerizable by ATRP, as mentioned in the previous section. The main reason for this is that with the catalytic systems used in ATRP, the activation process is too slow because the radical formed is not stabilized by resonance or by electronic effects. However, when these monomers were added at the end of the polymerization reaction of acrylates [69–71] or methacrylates [72], the radicals of the polyacrylate chain-end were able to add to these monomers and the deactivation provided halogen-terminated polymers. These radical addition reactions can occur due to the sufficient rate constants of poly(methyl acrylate) (Scheme 2.7).

This polymer was previously obtained by ATRP with 95% conversion, using an excess of 1,2-epoxy-5-hexene (25-fold excess towards the end groups). At the same time, Cu(0) (0.5 eq. towards CuBr) was added in order to reduce the amount of Cu(II) in the reaction mixture. Less Cu(II) in the reaction mixture results in a faster radical reaction. However, too high Cu(I) or too low Cu(II) concentrations can result in bimolecular termination reactions and incomplete functionalization.

Other less reactive monomers have been incorporated to the chain-ends of oligomers, including divinyl benzene for MMA [73] and maleic anhydride for styrene [74] (Scheme 2.8) and methacrylates [75].

Use of quencher agents A one-pot synthesis of telechelic polyacrylates with unsaturated end-groups has been developed by Bielawski [1]. Atom transfer radical polymerization of methyl or *n*-butyl acrylate was initiated with either ethyl *α*-bromomethylacrylate or methyl dichloroacetate, as a monofunctional or difunctional initiator, respectively, and mediated with various Cu/amine complexes. Addition of an ethyl 2-bromomethylacrylate excess was found not only to immediately quench the polymerization, but also to insert 2-carbethoxyallyl moieties

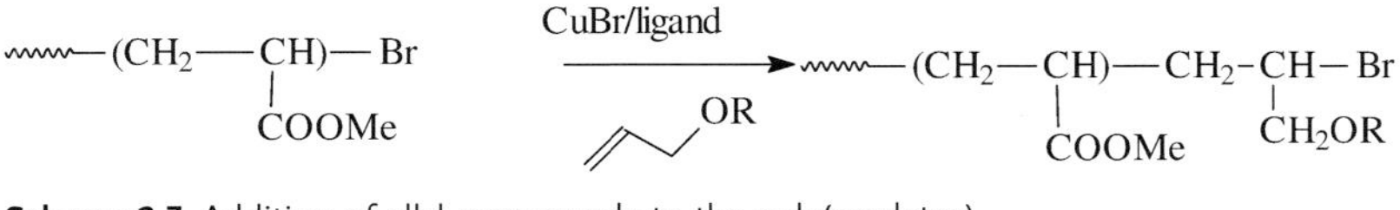

Scheme 2.7 Addition of allyl compounds to the poly(acrylates).

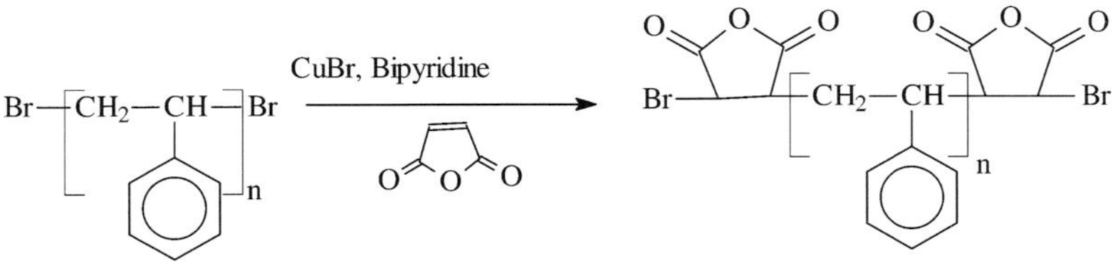

Scheme 2.8 Addition of maleic anhydride to the polystyrene.

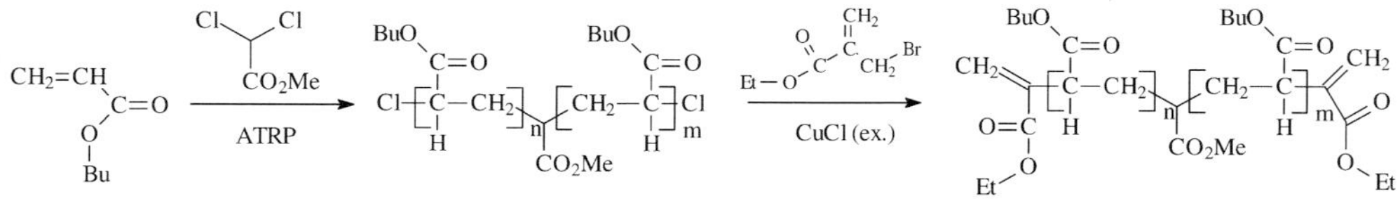

Scheme 2.9 ATRP of *n*-butyl acrylate initiated with a commercially available difunctional initiator and chemical modification of the chain-end by addition of ethyl-2-bromomethylacrylate excess.

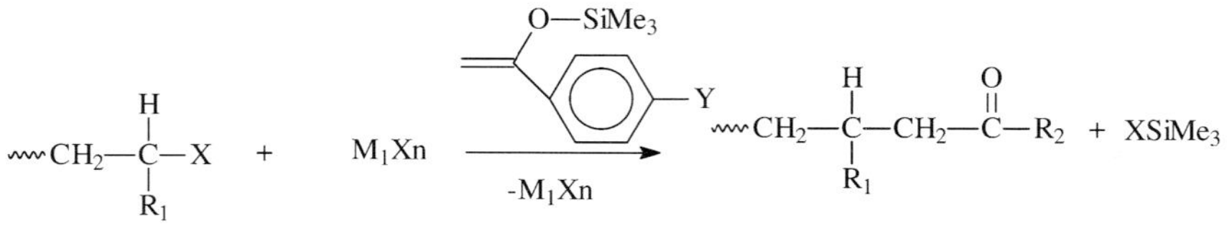

with $R_1 = CO_2CH_3$ and Y = H, OCH_3, Cl, Br.

Scheme 2.10 Reaction of end capping (silyl enol ether) agents onto poly(methacrylates).

at the polymer chain-ends, using the addition–fragmentation process. Thus, the synthesis of telechelic polyacrylate with unsatured end-groups has been accomplished, with a very good functionality ($f \approx 2$) (Scheme 2.9).

Silyl enol ethers such as *p*-methoxy-*α*-(trimethylsilyloxy)styrenes [76], or *iso*-propenoxytrimethylsilane [76] are efficient quenchers in the living radical polymerization of MMA using the $RuCl_2(PPh_3)_3$ complex (Scheme 2.10). They convert the C–X, X being a halogen atom, into a C–C bond with a ketone group. As shown in Scheme 2.8, the silyl compound-mediated quenching reaction probably proceeds via the addition of the growing radical into the C=C double bond of the quencher, followed by elimination of the silyl group and the chlorine that originated from the terminal polymer. In the case of *p*-substituted-*α*-(trimethylsilyloxy)styrenes [76], the quenching is selective and quantitative. Thus, the quenching proceeds faster with electron-donating susbtituents (Y: $OCH_3 > H > F > Cl$) on phenyl groups. The reaction is favored with these silyl enol ethers by the presence of the *α*-phenyl group, which stabilizes the radical by the electron-donating effect of the aromatic group after the addition of the quencher double bond. In contrast, silyl enol ethers with an R-alkyl group (R-silyloxy vinyl ethers) have been proved to be less efficient, indicating that the stability of the resulting silyloxyl radical is the critical factor for the design of good quenchers.

Nucleophilic substitutions The halogen end group can be transformed into other functionalities by means of standard organic procedures such as a nucleophilic displacement reaction.

$$\text{Polystyrene-}N_3 \xrightarrow[\text{Et}_2\text{O, reflux}]{\text{LiAlH}_4} \text{Polystyrene-}NH_2$$

$$\text{Polystyrene-}N_3 \xrightarrow[\text{THF, room temperature}]{\text{PPh}_3} \text{Polystyrene-N=PPh}_3 \xrightarrow[\text{THF, room temperature}]{\text{H}_2\text{O}} \text{Polystyrene-}NH_2$$

Scheme 2.11 Reduction of an azide group.

For instance, the halogen end-group (chlorine and bromine) of oligomers obtained by ATRP can be substituted by azide groups. Thus, poly(styrenes), poly(acrylates) and poly(methacrylates) with bromine end groups were reacted with sodium azide in solvents such as DMF or DMSO, which promoted nucleophilic substitution reactions. The azide group can be reduced in the presence of lithium aluminum hydride or triphenyl phosphine and converted into an amino end-groups (Scheme 2.11). We can note that the azide reduction using lithium aluminum hydride, cannot be applied to azide-terminated poly(meth)acrylate, because the reduction of the ester functionalities will take place.

Coessens and Matyjaszewski [68–71] showed that the direct displacement of a halogen by hydroxide anion is followed by expected side-reactions such as elimination. However, they described the nucleophilic substitution of the halogen end-group by the primary amine, i.e. 2-aminoethanol, to introduce other functionalities. The primary amine, i.e. 2-aminoethanol, gives good and selective nucleophilic substitution of the bromine end-groups of polystyrene oligomers [77]. However, these reactions are not efficient towards poly(meth)acrylates. Indeed, the 2-aminoethanol undergoes degradation of ester functions (Scheme 2.12). We can remark that this reaction does not take place for $H_2N(CH_2)_xOH$, with $x>2$.

The halogen functional polymer can react with a thiol by nucleophilic reaction, resulting in a polymeric thioether and a hydrogen halide. The latter is trapped by a basic additive, preventing a reverse reaction. Snijder et al. [78] used this technique to modify the end-group of poly(*n*-butyl acrylate) into hydroxy-functional polymer in the presence of 2-mercaptoethanol and 1,4-diazabicyclo-[2,2,2]octane (DABCO) (Scheme 2.13).

We can note that all these above techniques of nucleophilic substitution lead mainly to telechelic oligomers. However, Muehlebach et al. [79] developed an original method that consists in replacing the terminal bromine atom by a methacrylate function, using 1,8-diazabicyclo[5.4.0]undec-7-ene (DBU) with acrylic acid or methacrylic acid in ethyl acetate at room temperature. These macromonomers obtained are copolymerized with *N,N'*-dimethylethylamino-methacrylate (DMAEMA) to confirm their reactivities.

In a similar study, Ivan et al. and other groups [43, 80–82] suggested the synthesis of macromonomers by reaction of oligomers obtained by ATRP with allyl trimethylsilane (ATMS), followed by addition with a Lewis acid ($TiCl_4$) without any monomer. In these conditions, the terminal halogen is replaced by a carbocation and $Ti_2Cl_9^-$. The obtained carbocation will lead directly to the allylic dou-

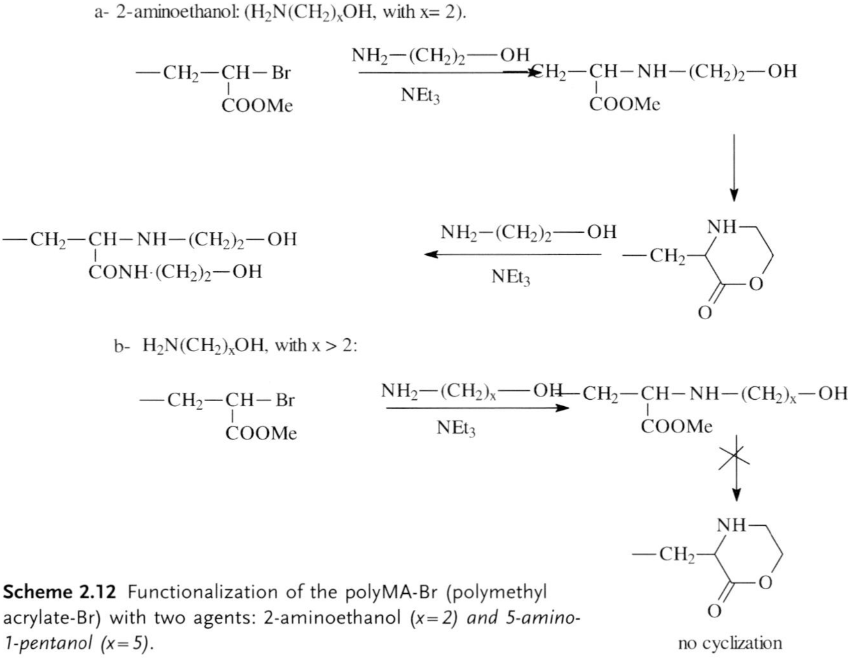

Scheme 2.12 Functionalization of the polyMA-Br (polymethyl acrylate-Br) with two agents: 2-aminoethanol ($x = 2$) *and 5-amino-1-pentanol* ($x = 5$).

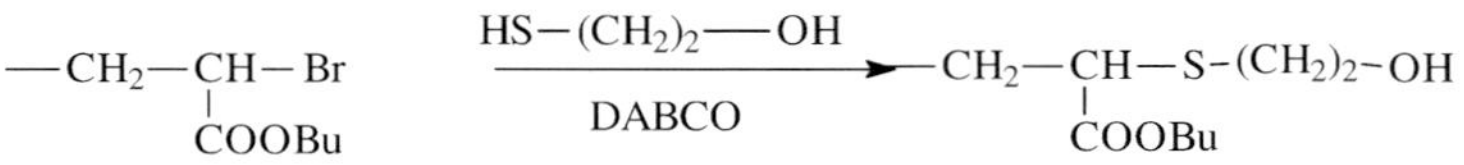

Scheme 2.13 Functionalization of the poly *n*-BuA-Br (polybutyl acrylate-Br) with 2-mercaptoethanol in the presence of DABCO.

ble bond. The macromonomer functionality is almost 1 and corresponds to that of the initial halogen.

Norman et al. [83] synthesized PMMA oligomers by ATRP. The methacrylic double bond of the resulting macromonomer was obtained directly by elimination of the terminal halogen by catalytic chain transfer agents, such as 5,10,15,20-tetraphenyl-21*H*,23*H*-porphine cobalt(II) (Co(tpp)) and bis-[(2,3-butanedione dioximato) (2-) *O*:*O*′] tetrafluorodiborato (2-) *N*,*N*′,*N*″*N*‴cobalt(II), (CoBF) near the end of an atom-transfer polymerization. Low molar mass, narrow polydispersity PMMA polymers prepared by ATRP have been converted in high yield (85%) to the ω-unsaturated PMMA species by the *in situ* addition of Co(tpp) to an ATRP reaction mixture.

Chemical modifications leading only to telechelics To the best of our knowledge, techniques to replace the terminal halogen were only applied to the synthesis of telechelic oligomers.

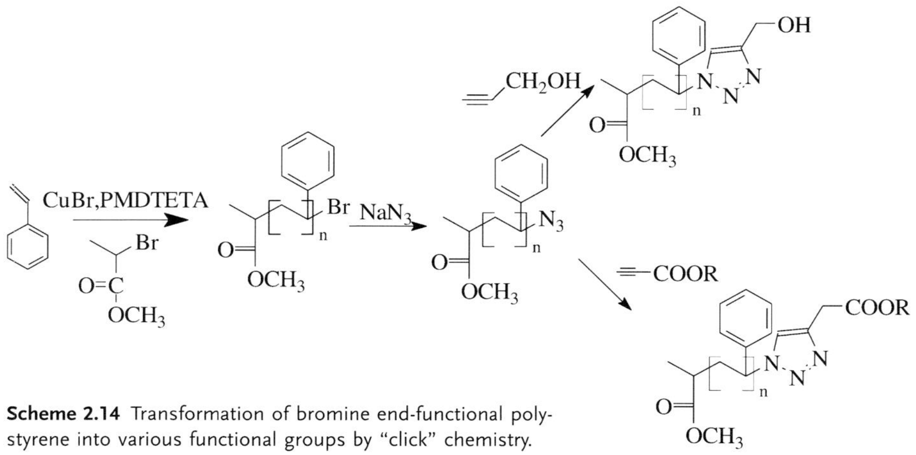

Scheme 2.14 Transformation of bromine end-functional polystyrene into various functional groups by "click" chemistry.

Functionalization by "click" chemistry Recently, Sharpless et al. [84, 85] popularized the 1,3-dipolar cycloaddition of azides and terminal alkynes, catalyzed by copper(I) in organic synthesis. This process was proven to be very practical, because it can be performed in several solvents (polar, nonpolar, protic etc.) and in the presence of different functions. These cycloadditions were classified as "click" reactions, defined by Sharpless. Lutz et al. [86, 87] used this technique to functionalize oligomers of polystyrene ($M_n = 2700\ \mathrm{g\ mol^{-1}}$). The synthesis was performed in THF in the presence of a CuBr and 4,4′-di-(5-nonyl)-2,2′bipyridine (dNbipy) complex. The choice of this ligand is very important, because it can accelerate the catalysis of cycloaddition [88]. This technique can therefore be considered as a "universal" method and allows a quantitative transformation of the polystyrene chain-end into the desired function [86, 87] (Scheme 2.14).

Radical coupling This reaction is based on the Wurtz radical coupling [89, 90]. Three groups [91–95] developed this radical coupling, also called atom transfer radical coupling (ATRC), at the same time.

This reaction takes place in the presence of a transition metal such as copper, iron etc. and consists in coupling α-halogen oligomers, previously synthesized by an ATRP process (Scheme 2.15).

This reaction was first performed on molecules able to model the chain-end of oligomers. For instance, Otazaghine et al. [93] performed the radical coupling of 1-bromoethylbenzene at 65 °C in anisole, in the presence of Cu(0) and HMTETA, with a quantitative yield. They then applied the same experimental conditions to α-halogen oligomers of polystyrene. In a similar way, α-halogen oligomers of acrylates [91], previously synthesized by ATRP, were used in ATRC. However the yield of the radical coupling was lower than that of styrene (60%). A similar result was observed for α-fluoroacrylate monomers [92]. Concerning the ATRC of α-halogen oligomethacrylates, the coupling does not occur, due to steric effects.

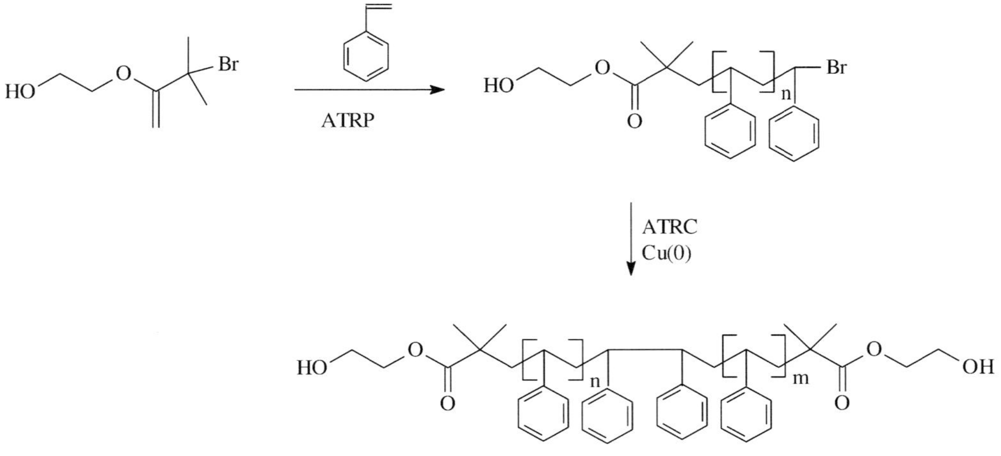

Scheme 2.15 Coupling process based on ATRC.

2.3.2
Nitroxide-mediated Polymerization

The use of nitroxides as mediating radicals has been revealed to be highly successful in living free radical polymerization and has received considerable attention for more than ten years [96–101]. Much attention has been devoted to understanding the mechanism and kinetic details for nitroxide-mediated radical polymerization (NMRP) [102–104]. NMRP was first viable for styrene and its substituted derivatives [105] but was extended to acrylates, acrylamides and some other vinyl monomers [106]. Hawker et al. [100, 107] perfectly reviewed the overall mechanism of NMRP as well as the different nitroxides and their reactivity. The macromolecules obtained by an NMRP process exhibit the following general structure: a functional group provided by the initiator at a chain-end (ω position) and an aminoxyl function at the other chain-end (α position). A macromonomer structure may be reached by modifying one of the two positions, whereas telechelic oligomers can only be obtained by modifying the aminoxyl function.

2.3.2.1 Modification of the α Position

To obtain a macromonomer starting from the α position requires a chemical modification of the function provided by the initiator used during the NMRP process. For instance, Hawker et al. [107] replaced the benzoyl group, provided by the benzoyl peroxide initiator, by a hydroxyl group. This latter is then able to react with acryloyl chloride to obtain the reactive double bond of the macromonomer. Similarly, Hawker et al. [108] synthesized original polystyrene macromonomers with two amine groups situated at the α position. These amine

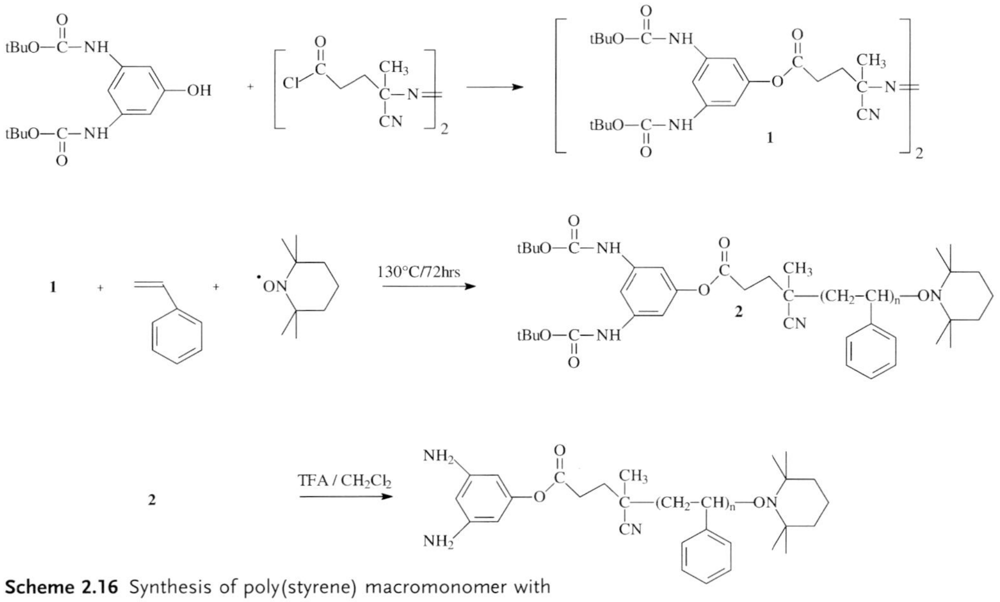

Scheme 2.16 Synthesis of poly(styrene) macromonomer with two condensable amine groups.

groups can further react through a condensation reaction. This macromonomer was synthesized in the presence of a peculiar diazoic initiator **1** synthesized in a first step, according to Scheme 2.16.

In most cases, macromonomers obtained by chemical modification of the *α* position exhibit two polycondensable groups at one end of the macromolecular chain and an aminoxyl function at the other end. The remaining aminoxyl group will however undergo a weak thermal stability to the macromonomer.

2.3.2.2 **Modification of the *ω* Position**

The modification of the *ω* position from the resulting oligomers allows one to reach both a telechelic and a macromonomer structure. For instance, Kuckling et al. [109] polymerized 2-vinylpyridine in the presence of hydroxy-Tempo. The macromonomer structure was obtained by reacting the hydroxyl group, situated at the *ω* position, with acryloyl chloride. The acrylated poly(vinylpyridine) was then copolymerized with *N*-isopropylacrylamide (NIPAAm) in the presence of *N,N′*-methylenebisacrylamide to obtain graft copolymer gels, which were found to be temperature and pH dependent.

Other teams worked on the functionalization of the aminoxyl group to reach a macromonomer structure. For instance, the method of Holdcroft et al. [110] is original for the synthesis of a novel series of poly(sodium styrenesulfonate) (PSSNa) macromonomers (compound **3** of Scheme 2.17) based on stable free radical polymerization in the presence of Tempo.

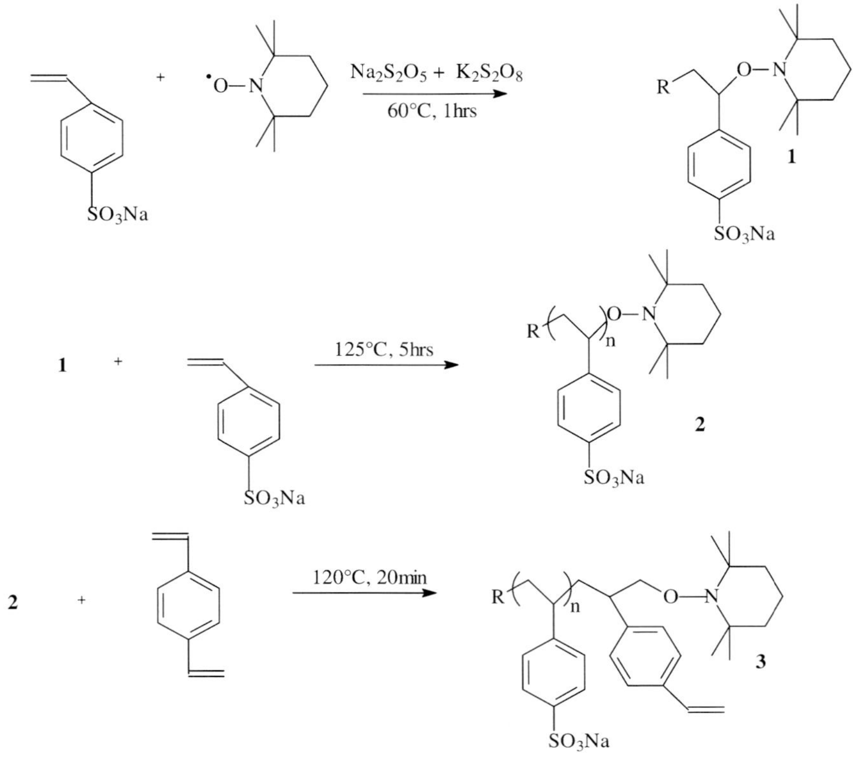

R= SO_4^-

Scheme 2.17 Synthesis of poly(sodium styrenesulfonate) macromonomer.

Unlike for the α position, a modification of the aminoxyl group results in a macromonomer structure showing a reactive double bond at the extremity of the macromolecular chain.

However, we can remark that most of the works realized on the chemical modification of the aminoxyl group are devoted to the synthesis of telechelic oligomers. Indeed, several techniques allow one to replace the aminoxyl group by a further polycondensable group. Rizzardo et al. [96] developed a technique allowing the replacement of the terminal-aminoxyl by a hydroxyl function. To this aim, they reacted the terminal-aminoxyl-containing oligomer with acetic acid catalyzed by zinc. Pradel et al. [111–113] achieved the synthesis of hydroxy-telechelic polybutadiene by applying Rizzardo's methodology to α-hydroxyl,ω-aminoxyl polybutadiene at 80 °C [113]. After 2 h, they obtained a quantitative reduction of the aminoxyl functions, evidenced by ^{1}H NMR. The average hydroxy-functionality of oligobutadiene was 2.06 (Scheme 2.18). Interestingly, our team has used an original method for coupling oligobutadiene initiated by H_2O_2 and terminated by TEMPO [114]. This method is based on the continuous elimina-

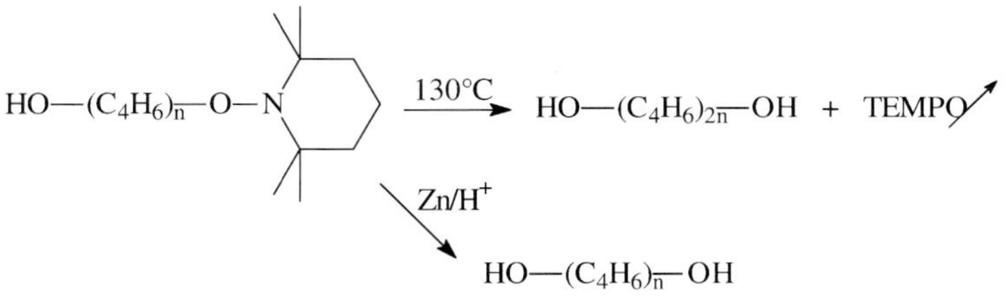

Scheme 2.18 Synthesis of hydroxy-telechelic poly(butadiene).

tion of the TEMPO unit by sublimation, allowing displacement of the reaction equilibrium by simple thermal means (Scheme 2.18).

Harth et al. [115] recently developed a new methodology to replace the terminal-aminoxyl based on the addition of one single maleic anhydride unit, considering that addition of a second unit is disfavored. To mimic this approach, α-hydrido alkoxyamine was reacted with 2 equivalents of *N*-phenyl maleimide, leading to addition of one unit. Upon heating, the corresponding product eliminated the terminal-aminoxyl to give the substituted maleimide derivative with more than 90% yield when conducted in DMF. By using the incorporation methodology of maleimide unit, Lohmeijer et al. [116] were able to synthesize terpyridine-telechelic polystyrene (Scheme 2.19). They indeed utilized a terpyridine-functionalized maleimide that replaced the nitroxide chain-end of polystyrene. The obtained polymer would be of great value to prepare ABA metallo-supramolecular triblock copolymers. This "construction" is based on the use of a metal complex, serving as a supramolecular linker between blocks. Terpyridine ruthenium was proved to be an efficient linker [117].

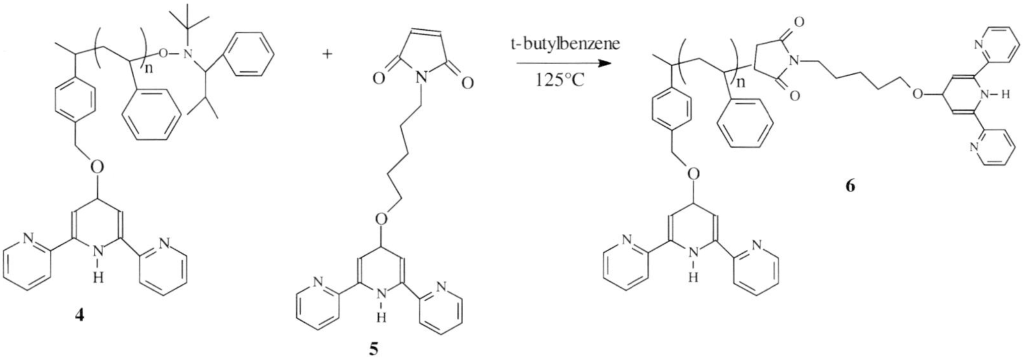

Scheme 2.19 Synthesis of terpyridine-telechelic polystyrene.

2.3.3 Addition–Fragmentation Processes

Two different addition–fragmentation processes have to be distinguished. The first one is addition–fragmentation transfer (AFT) that involves the use of a specific transfer agent. The AFT technique is generally coupled to catalytic chain transfer (CCT) to produce new chain transfer agents (CTAs) easily. Such an AFT/CCT technique essentially enables the synthesis of macromonomers bearing a reactive double bond at the chain-end. The second addition–fragmentation process consists of the reversible addition–fragmentation chain transfer (RAFT). This process is a degenerative transfer reaction that involves the use of xanthate species and mainly leads to a telechelic structure. In this section, we will depict both AFT/CCT and RAFT processes leading to macromonomers and telechelics, respectively.

2.3.3.1 Addition–Fragmentation Transfer and Catalytic Chain Transfer

The mechanism of the AFT is rather complex [118, 119] as it involves different steps: not only addition of the macro-radical onto the chain transfer agent (CTA) followed by a subsequent β-fragmentation, but also an intramolecular substitution on a peroxidic bond may occur, depending on the CTA structure. The overall mechanism of addition–fragmentation is often more complicated [120] and was studied especially in terms of driving forces in free radical AFT [121]. Colombani and Chaumont presented a general review [122], in which they mainly focused on recent developments made in the large area of addition–fragmentation. Two distinct sites of the CTAs are involved in the addition step and the fragmentation step [120]. Thus, it is theoretically possible to design each site separately, in order to: (i) control the reactivity of the CTA, i.e. the chain transfer constant value, which is mainly influenced by the nature of the addition site and (ii) control the nature of the α-functional group, which is mainly influenced by the evolution of the fragmentation site. For the design of this latter site, the reactions involved in the fragmentation process generally deal with two classical reactions studied in organic chemistry: the β-scission reaction and the intramolecular homolytic substitution called SHi. The general form [122] for CTAs involved in addition–fragmentation is CX=C(Y)–W–G (Scheme 2.20). However, CTAs can potentially be separated into three distinct types A, B and C as shown in Scheme 2.20, leading to three different kinds of macromonomers.

We can however remark that only the C-type CTA leads to efficient macromonomers in further reactive graft copolymerization. Indeed, concerning the A-type and B-type CTAs the unsaturations present a poor reactivity, due to steric effects.

The preparation of such CTAs involves a non-negligible part of organic chemistry. The catalytic chain transfer (CCT) may give an answer to this drawback. The CCT technique is based upon the fact that certain cobalt complexes Co(II) such as cobaloximes catalyze the chain transfer to monomer reaction. The

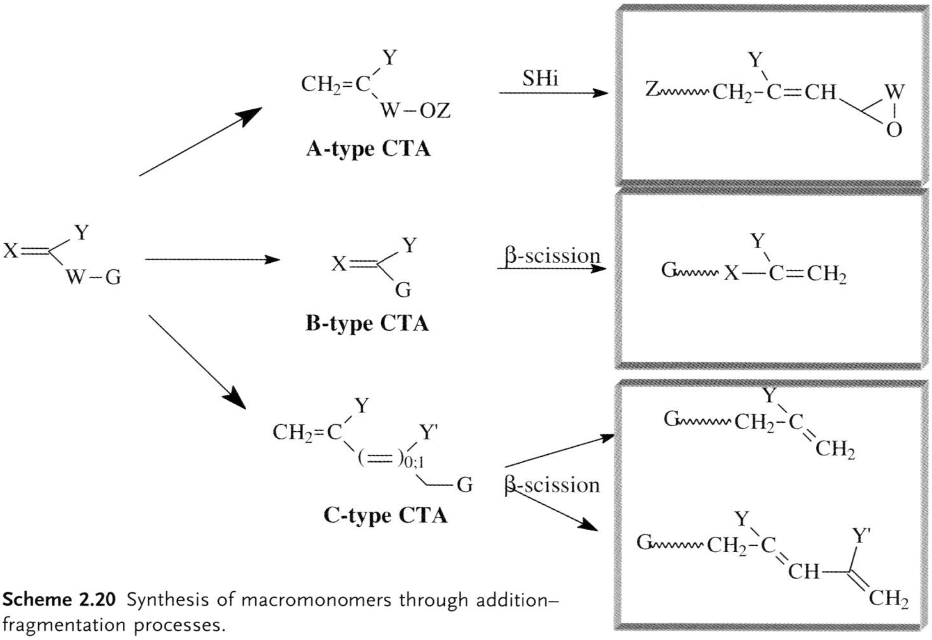

Scheme 2.20 Synthesis of macromonomers through addition–fragmentation processes.

mechanism is believed to consist of two consecutive steps [123, 124]: first a growing polymeric radical R_n undergoes a hydrogen transfer reaction with the Co(II) LCo complex to form a polymer (or an oligomer) with a terminal double bond $P_n^=$ and the corresponding Co(III) hydride LCoH. Then Co(III) hydride LCoH reacts with a monomer to produce both Co(II) and a monomer radical. The mechanism of CCT is perfectly described by Gridnev et al. [125–127] who also proved that propagation in the presence of cobalt catalyst occurs by a free-radical mechanism and not by a coordination mechanism. Davis et al. [128] also supported the mechanism of CCT by the use of Maldi-Tof analyses. The efficiency of CCT for making good CTAs was mainly proved on methyl methacrylate [129] (Scheme 2.21).

These methacrylate-based CTAs are easily synthesized by a CCT process and have been involved in an AFT process to reach the corresponding methacrylate macromonomers. Such methacrylate-based macromonomers have been involved as precursors in the synthesis of graft copolymers [130]. However, Moad et al. [131] also characterized the use of such macromonomers in the synthesis of block copolymers. They explained that for macromonomers based on methacrylic monomers **1** (Scheme 2.22), fragmentation of the adduct **2** (formed by addition of the methacrylate monomer to the methacrylate macromonomer) always dominates over reaction with monomer. This fragmentation leads to block copolymers and graft copolymerization does not occur.

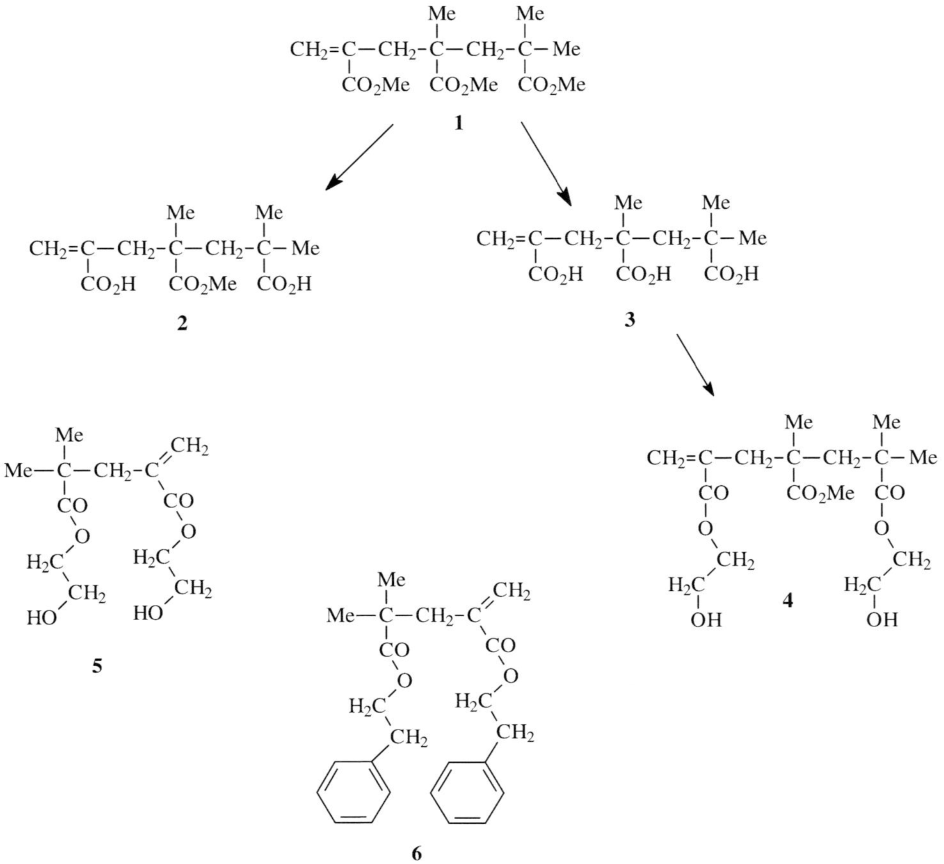

Scheme 2.21 Several CTAs synthesized by catalytic chain transfer.

2.3.3.2 Reversible Addition–Fragmentation Transfer

Reversible addition–fragmentation chain transfer (RAFT) [132, 133] is a versatile technique to produce polymer architectures, such as telechelic ones. Unlike the AFT process, RAFT is based on a radically induced degenerative transfer reaction, first reported by Zard's group [134], between a thiocarbonyl-thio-containing compound and a propagating radical. The mechanism of RAFT, proposed by Rizzardo [135], consists of many complex equilibrium steps and involves a rapid exchange of the radical among all the growing polymeric chains via addition–fragmentation reaction with the chain transfer agent. The mechanism of the RAFT process is very complex and was recently deeply investigated (in terms of kinetic parameters) by several authors [136, 137]. RAFT, allowing for predictable molar mass with low polydispersities, is applicable to a wide range of vinyl monomers [138–140], some of them being not always polymerizable by other

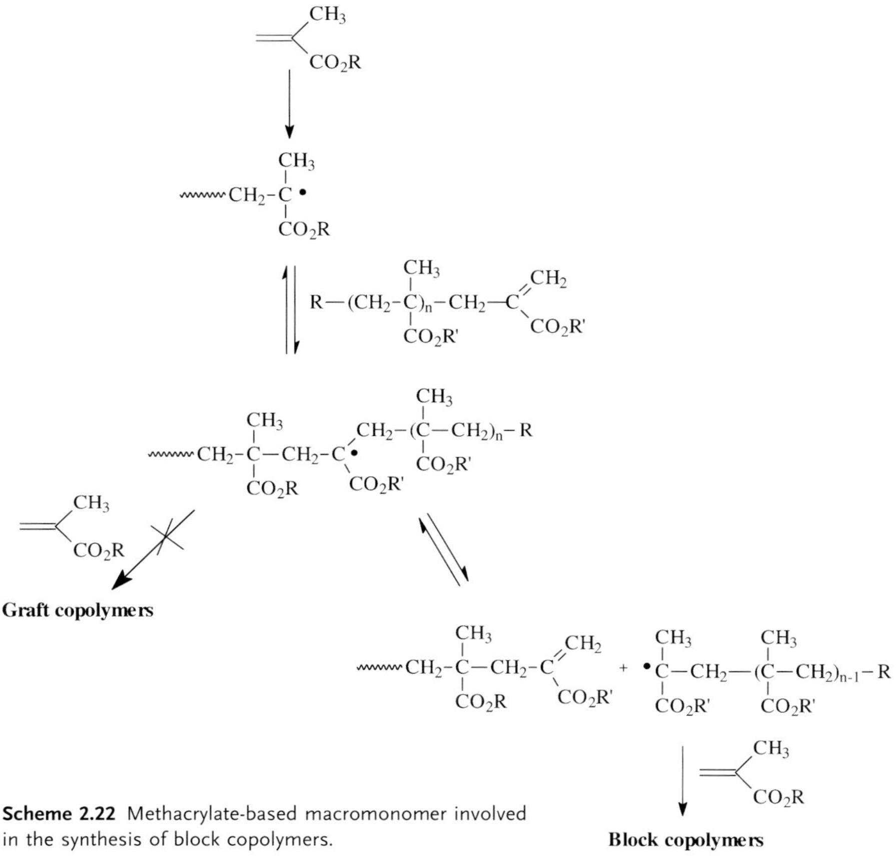

Scheme 2.22 Methacrylate-based macromonomer involved in the synthesis of block copolymers.

LRPs (i.e. vinyl acetate [141] or monomers bearing protonated acid groups). Moreover, RAFT is employed in many polymerization processes such as bulk, solution, suspension, emulsion and miniemulsion [142–144].

To achieve the chain-end functionality in the polymer by RAFT polymerization it is necessary either to adjust the structure of the transfer agent or to combine it with a modification of the terminal dithioester. Scheme 2.23 summarizes two different pathways for getting the telechelic structure. In the first pathway, the transfer agent is a trithioester compound, bearing two leaving groups R^1. The telechelic structure is obtained directly with an expected trithioester group at the middle of the polymer. The second pathway considers a dithioester as transfer agent, bearing a leaving group R^1 at one end and a non-living group at the other end. After RAFT polymerization onto the monomer M_1, the polymer contains the chain-end R^1 group but also the chain-end thioester. The bifunctionality is then obtained by chemically modifying the chain-end thioester into a chain-end R^1 group.

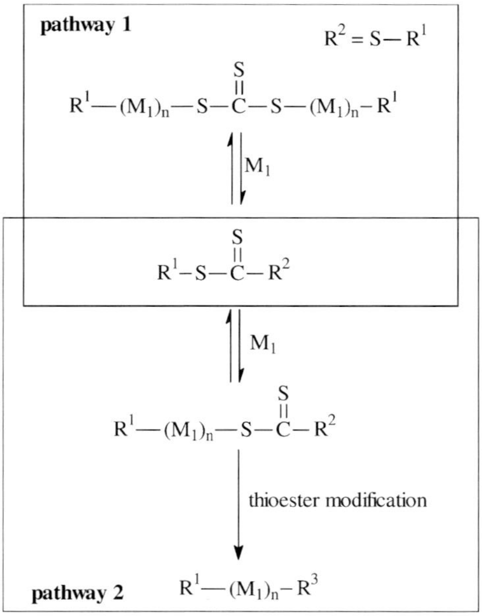

Scheme 2.23 Synthesis of telechelic polymers by the RAFT mechanism (R^1 and R^3 being functional groups).

Use of a trithioester transfer agent Although only one step is necessary to get the desired telechelic structure, this pathway is much less developed than the chemical modification of the terminal dithioester function. This may be attributed to a non-easy synthesis of such trithioester transfer agents [145, 146] and also to the poor stability of this group located at the center of the molecule. As an example, Baussard et al. [147] synthesized a new trithioester sodium *S*-benzyl-*S'*-2-sulfonatoethyl trithiocarbonate. This trithiocarbonate was employed as transfer agent for the RAFT polymerization of vinylbenzyltrimethylammoniumchloride (VBAC) in an aqueous medium. Although good control of the polymerization occurred, the benzyl end-group is not suitable for polyaddition reaction. Some trithioester compounds however have a suitable end-group for further polycondensation reactions of the obtained telechelic oligomers. Liu et al. [148–151] performed the synthesis of hydroxy-telechelic polystyrene or polymethyl acrylate by direct RAFT polymerization of styrene and MA respectively with *S,S'*-bis(2-hydroxyethyl-2'-butyrate)trithiocarbonate (BHEBT):

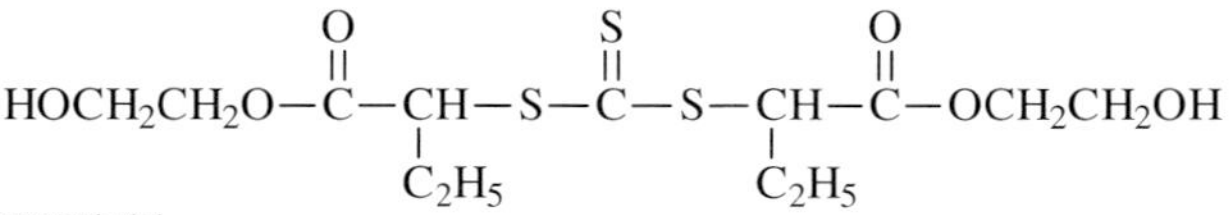

BHEBT

BHEBT was proved to be a highly efficient transfer agent towards styrene and MA by plotting M_n against monomer conversion. Polydispersity indexes were found to be less than 1.2. They demonstrated that the trithiocarbonate group was in the middle of the polymer chain because of the similar fragmentation reactivity of the two leaving groups, 2-hydroxylethyl-2′-butyrate. Finally the telechelic structure was proved for both styrene and methyl acrylate by means of ^{1}H NMR.

More recently, Convertine et al. [152] demonstrated the use of *S*,*S*′-bis(*αα*′-dimethylacetic acid) trithiocarbonate for the RAFT polymerization of both acrylamide (AM) and *N*,*N*-dimethylacrylamide (DMA) in aqueous media at room temperature. They showed that RAFT polymerizations were conducted to high conversion with a living character. The dicarboxyl functionality was evidenced.

This first pathway using a trithioester transfer agent afforded functionality close to 2. However, the final oligomer contains a trithioester group in the middle of the chain that is highly labile. A further polycondensation, which requires high temperature, with such oligomers, obtained by this technique, seems not to be favored.

Thioester modification The second pathway consists of at least a two-step reaction: the first step is the RAFT polymerization in the presence of a dithioester transfer agent, whereas the second step is the removal of the thioester terminal. The first step occurs with RAFT agents bearing only one polycondensable function. Such transfer agents are numerous [153, 154] compared to their trithioester homologues, even if the syntheses are usually costly and require multi-step reactions. We can remark that only a few workers have been interested in removing the thioester end-group to reach the bifunctionality. The removal of the terminal thioester group is indeed more complicated than undertaking a RAFT polymerization because it involves chemical modification followed by a purification of the new difunctional oligomer.

As an example, Lima et al. [155–157] performed the RAFT polymerization of MMA in the presence of (4-cyano-1-hydroxylpent-4-yl) dithiobenzoate (RAFT-ACP) chain transfer agent. Monohydroxy oligomethylmethacrylates were obtained. To get the telechelic structure, aminolysis of monofunctional PMMA with 1-hexylamine was undertaken, leading to an *α*-OH,*ω*-SH-PMMA. A hydroxyl group can replace the thiol terminal by Michael addition with hydroxyl ethyl acrylate (HEA) [157] (Scheme 2.24).

Another example concerns the work of Perrier et al. [158] who first proposed to remove the terminal thioester group (after RAFT polymerization) and second to recover the chain transfer agent. To achieve the bifunctionality and recovery of the CTA, the monofunctional oligomer is placed in solution with a large amount of initiator ([Polymer]:[Ini] = 1 : 20). The radical formed from the initiator will react with the reactive C=S bond of the terminal thioester. By using an excess of initiator radical, fragmentation will occur to release the new leaving thioester group, directly replaced by a radical from the excess of initiator (Scheme 2.25).

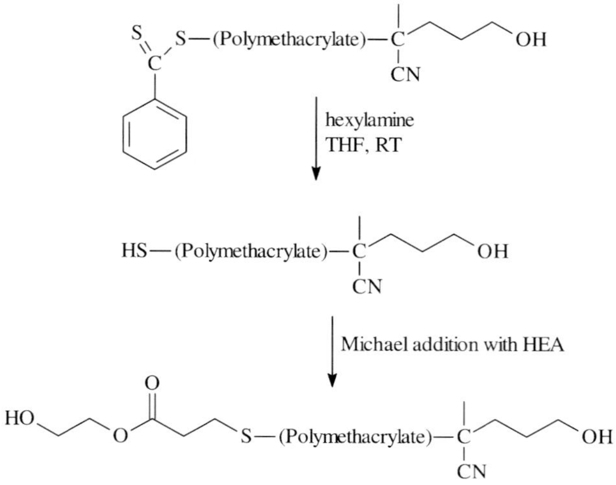

Scheme 2.24 Synthesis of hydroxy-telechelic PMMA.

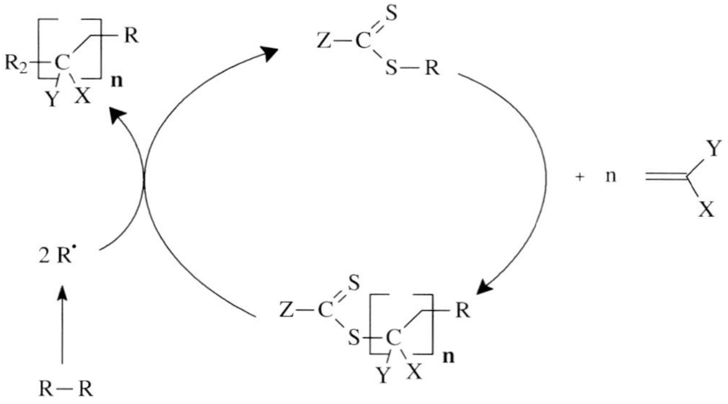

Scheme 2.25 Reaction cycle to obtain telechelic polymers and to recover the CTA.

Scheme 2.25 shows that it is necessary to choose the right initiator, i.e. one bearing a further condensable function. This function will correspond to the second end-group in the polymer. For instance, Perrier et al. have undertaken the RAFT polymerization of methyl acrylate with a dithioester bearing a carboxyl function. The monofunctional oligoacrylate was reacted with an excess of 4,4′ azobiscyanovaleric acid (ACVA). It is noteworthy that carboxy-telechelic oligomethyl acrylate was obtained in a one-step reaction. Due to the structure of ACVA, the same transfer agent is recovered at the end of the reaction.

2.3.4
Iodine Transfer Polymerization

Like RAFT polymerization, iodine transfer polymerization (ITP) is a degenerative transfer polymerization (DT) also using alkyl halides [159, 160]. ITP was developed in the late 70s by Tatemoto et al. [161–164]. In ITP, a transfer agent RI reacts with a propagating radical to form the dormant polymer chain P–I. The new radical $R^{\bullet}$ can then reinitiate the polymerization. In ITP the concentration of the polymer chains is indeed equal to the sum of the concentrations of the transfer agent and of the consumed initiator. The newly formed polymer chain $P^{\bullet}$ can then propagate or react with the dormant polymer chain P–I or R–I [165].

Several investigations have shown that ITP can produce telechelic oligomers or macromonomers. Indeed, oligomers obtained by ITP show a structure similar to that obtained by ATRP. Oligomers bear an iodine atom at the chain-end that can be replaced by a further reactive group. Oligomers can also be functionalized by reaction with a functional monomer, such as 2-isocyanatoethyl methacrylate, maleic anhydride, etc. to reach a macromonomer structure. Scheme 2.26 describes the different routes to obtain macromonomers and telechelic oligomers by ITP. We can note that in the case of telechelic oligomers, the degenerative transfer process then requires the use of diiodide compounds instead of the iodide compounds, usually used in ITP. In a second step, the iodine group is modified to obtain the telechelic structure.

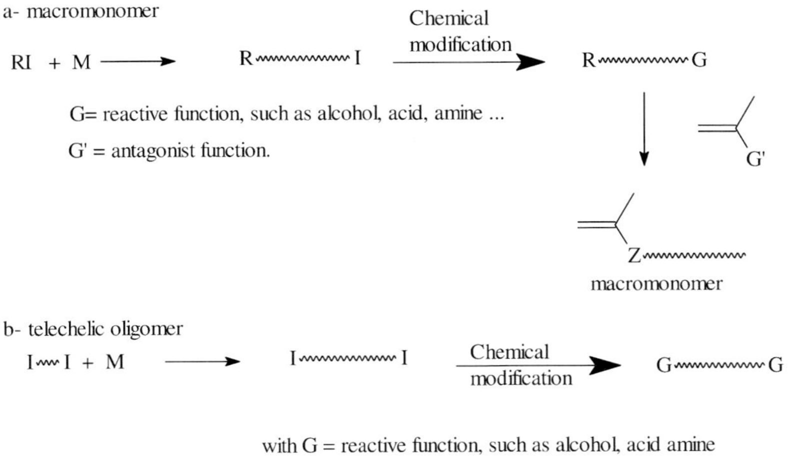

Scheme 2.26 Different routes to obtain macromonomers and telechelic oligomers by iodine transfer polymerization.

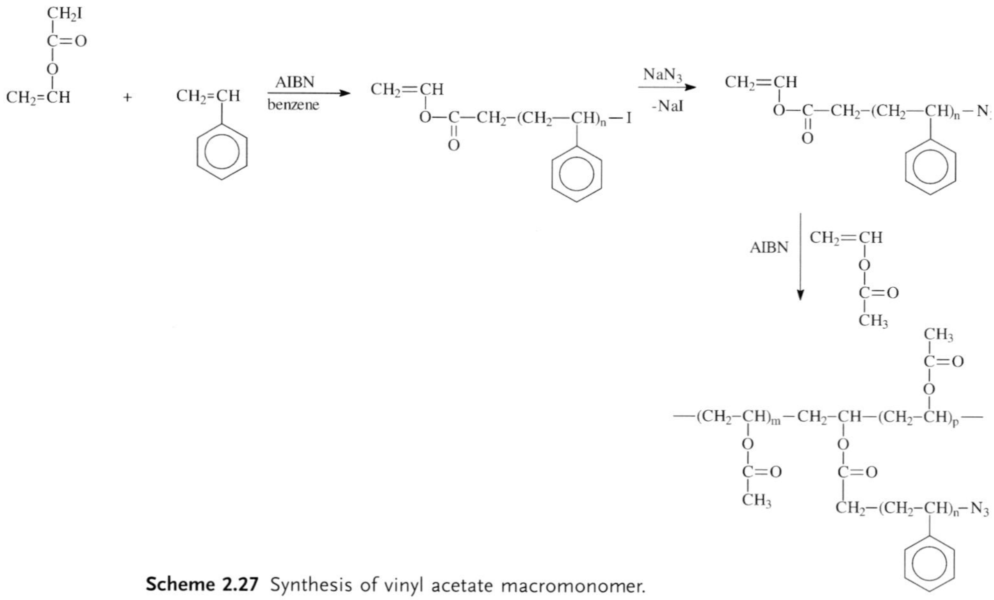

Scheme 2.27 Synthesis of vinyl acetate macromonomer.

2.3.4.1 **Synthesis of Macromonomers by Using a Chain Transfer Agent with a Polymerizable Function**

Teodorescu [166] developed a direct method for obtaining a polymerizable double bond by using the iodovinylacetate, as the CTA. This CTA is used in the presence of styrene. The vinyl acetate function cannot copolymerize during the reaction of ITP (Scheme 2.27).

2.3.4.2 **Synthesis of Macromonomers and Telechelics by Chemical Modification**

Like the ATRP process, some techniques of chemical modifications only afford the synthesis of telechelic oligomers. This is especially the case for direct chemical replacement of the iodine atom by a condensable group and also for radical coupling.

Direct chemical change Only a few studies have used a direct chemical change of the terminal iodide atom into a further condensable function. To our knowledge, the direct chemical change only concerns fluoromonomers, such as vinylidene fluoride (VDF) or tetrafluoroethylene (TFE), which were polymerized through an ITP process in the presence of diiodide transfer agents. We have described these works of direct chemical change in a specific book [167].

As an example, Feiring [168] synthesized fluorinated diamine as follows:

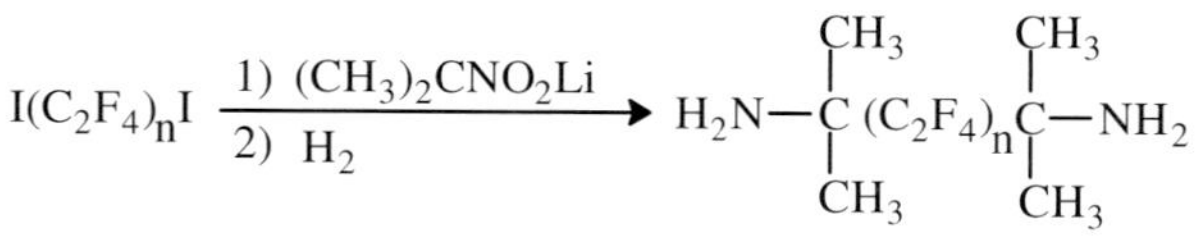

Radical coupling Like the direct chemical change, radical coupling mainly concerns fluorinated monomers. Ameduri et al. [167] summarized the different studies concerning the modification of α,ω-fluoropolymers in a recently published book. For instance, they showed that extensive research [169] has been carried out on the synthesis of diaromatic difunctional compounds linked to fluorinated chains according to the following Ullman coupling reaction:

where G may represent a functional group such as hydroxyl (i.e., bisphenol), carboxylate, isocyanate [170] or nitro (precursor of amine) in the para and meta positions to the fluorinated chain. Similarly, our team has spent much effort in functionalizing α,ω-diiodoperfluoroalkanes into fluorotelechelic compounds. These studies were summarized by Ameduri and Boutevin in a review [171] on the synthesis of fluoropolymers. For instance, our team synthesized α,ω-diols or dienes of perfluoroalkanes [172–175]. These compounds are precursors of hybrid fluorosilicones [176] but also of thermoplastic elastomers by polycondensation with polyimide sequences [177].

Functionalization by radical addition Unlike the previous techniques, it can be seen from the literature that this technique allows the synthesis of both macromonomer and telechelic structures.

Synthesis of macromonomers [178, 179] In the first step, the telomerization is performed with tetrafluoroethylene (TFE) and C_2F_5I. In a second step, the oligomer is modified by addition of allyl alcohol. Macromonomer can then be obtained by reaction of the functional oligomers with methacryloyl chloride. This method allowed the syntheses of fluorinated (meth)acrylates, fluorinated styrene, and fluorinated vinyl ether.

As another example, telomerization of VDF was performed in the presence of CF_3I, followed by chemical modification with allyl alcohol. The hydroxy-terminated PVDF was then reacted with acryloyl chloride to lead to the corresponding acrylate-terminated PVDF (Scheme 2.28).

Synthesis of telechelics Percec et al. [180, 181] demonstrated that the chloroiodomethyl chain-ends of PVC can be replaced by other further condensable functional groups. For instance, the PVC was functionalized by single electron transfer (SET) $Na_2S_2O_4$-catalyzed with 2-allyloxyethanol (Scheme 2.29). After precipitation, the functionalization resulted in α,ω-hydroxy polyvinylchloride in 90% yield. The catalytic effect of $Na_2S_2O_4$ first led to the abstraction of the chain-end iodide atom followed by radical addition of 2-allyloxyethanol. Then H-abstraction onto 2-allyloxyethanol allowed one to get the hydroxyl-telechelic structure (Scheme 2.29).

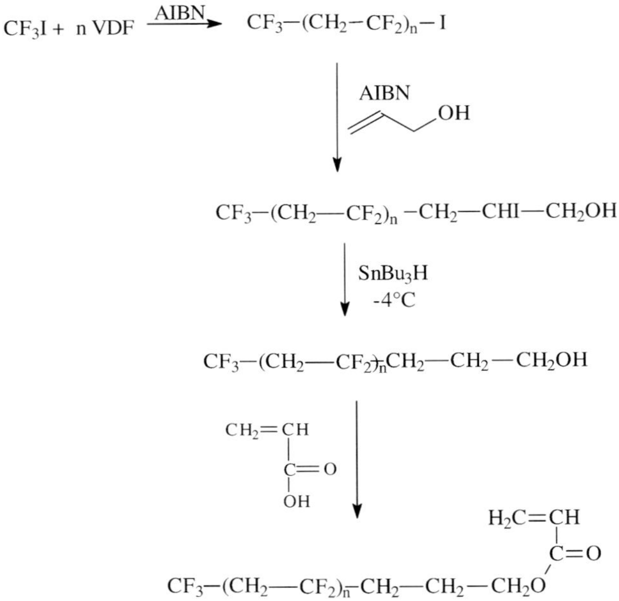

Scheme 2.28 Synthesis of macromonomer of poly(VDF) by iodine transfer polymerization.

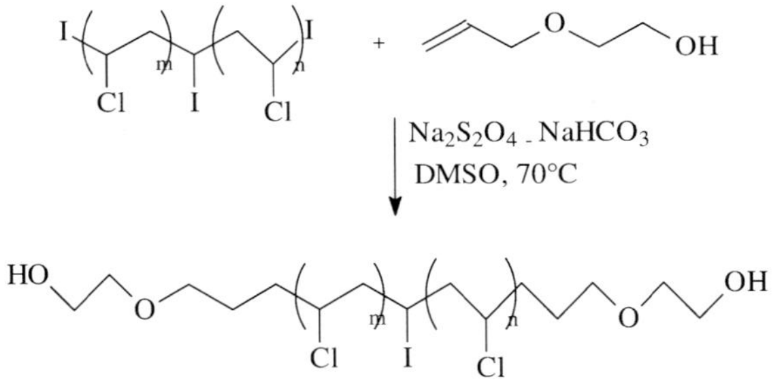

Scheme 2.29 Synthesis of dihydroxy PVC by SET with 2-allyloxyethanol in the presence of $Na_2S_2O_4$.

2.4 Conclusion

This chapter aims chiefly to describe the new designs of macromonomers and telechelic oligomers and especially their syntheses by using controlled radical polymerizations (CRP). Indeed, the development of CRPs in the 90s represents a major breakthrough for the syntheses of macromolecular structures because they afford a very good control of the macromolecular architectures (control of the molar masses and low polydispersity index). Hence, this chapter shows how atom transfer radical polymerization (ATRP), nitroxide-mediated polymerization (NMP), addition–fragmentation (AF) processes and iodine transfer polymerization (ITP) can lead to both macromonomers and telechelic oligomers. For all these living techniques, the obtained oligomers bear a reactive function at the chain-end, i.e. xanthate, bromide, iodine, aminoxyl, etc. The synthesis of telechelic oligomers or macromonomers requires a chemical modification of these reactive functions. The literature offers many possibilities to modify such reactive groups: radical reactions, nucleophilic substitution etc.

To the author's knowledge, most telechelic oligomers and macromonomers are obtained presently by ATRP. This may be explained by the relatively easy replacement of the terminal halogen atom. However, even after chemical modification, CuBr traces remain in the final product, which represents a major drawback for further industrial developments.

Such halogen-terminated polymers are not easily accessible by anionic polymerization. However, as discussed in the first part of the chapter, in the latter years anionic polymerization has still contributed substantially to the synthesis of other functional polymers including macromonomers, especially those which could not be obtained by other polymerization processes. Typical examples such as functional PEOs or polydienes of controlled microstucture (or diblock structures based on polydienes) have been briefly presented. Free (controlled) radical polymerization processes still fail to access polyethers or to control the microstructure of polydienes.

The syntheses of macromonomers and telechelic oligomers by living radical polymerizations are not yet industrially developed. Indeed, unlike conventional radical polymerizations (i.e. telomerization and dead end polymerization), the cost of CRPs still remains very high. However, despite such high cost, new prospects are now opening up for telechelic oligomers and macromonomers obtained by CRPs. This concerns for instance the recent investigations on nanostructuration and especially through non-covalent linkages. The work of Lohmeijer et al. [116] illustrated the synthesis of "metallo-supramolecular copolymers", and that of Leibler et al. [182] linked bi and trifunctional oligomers by hydrogen bonding. For now the innovating works mainly concern (macro)molecules of low molar mass, obtained by polycondensation or radical copolymerization, in which the linkage groups are statistically dispersed in the chain. But these new telechelic oligomers obtained by living radical polymerization may help in building more complex macromolecular structures. On the other hand, the synthesis of new macromonomers aims at obtaining new graft copolymers with controlled architectures. These new types of graft copolymers should provide interesting properties and should find new applications in various areas such as biological applications (transfer of peptides/proteins) or new membranes for fuel cells.

References

1 C.W. Bielawski, J.M. Jethmalani, R.H. Grubbs, *Polymer* 44 (2003) 3721–3726.
2 G.O. Schulz, R. Milkovich, *J. Appl. Polym. Sci.* 27 (1982) 4773–86.
3 B. Boutevin, *Adv. Polym. Sci.* 94 (1990) 69–105.
4 P.F. Rempp, E. Franta, *Adv. Polym. Sci.* 58 (1984) 1–53.
5 M. Szwarc (1968) *Carbanions, Living Polymers, and Electron Transfer Processes.*
6 K. Matyjaszewski, *Makromol. Chem., Macromol. Symp.* 60 (1992) 107–17.
7 M. Sawamoto, T. Higashimura, Makromol. Chem., Macromol. Symp. 60 (1992) 47–56.
8 N. Hadjichristidis, M. Pitsikalis, S. Pispas, H. Iatrou, *Chem. Rev.* 101 (2001) 3747–3792.
9 P. Rempp, P. Lutz, P. Masson, E. Franta, Makromol. Chem., Suppl. 8 (1984) 3–15.
10 K. Ito, *Prog. Polym. Sci.* 23 (1998) 581–620.
11 N. Hadjichristidis, M. Pitsikalis, H. Iatrou, S. Pispas, *Macromol. Rapid Commun.* 24 (2003) 979–1013.
12 K. Ito, S. Kawaguchi, *Adv. Polym. Sci.* 142 (1999) 129–178.
13 R. Waack, M.A. Doran, *J. Org. Chem.* 32 (1967) 3395–3399.
14 A. Takano, T. Furutani, Y. Isono, *Macromolecules* 27 (1994) 7914–716.
15 V. Heroguez, Y. Gnanou, M. Fontanille, *Macromol. Rapid Commun.* 17 (1996) 137–142.
16 V. Heroguez, J.-L. Six, Y. Gnanou, M. Fontanille, *Macromol. Chem. Phys.* 199 (1998) 1405–1412.
17 V. Heroguez, S. Breunig, Y. Gnanou, M. Fontanille, *Macromolecules* 29 (1996) 4459–4464.
18 P. Masson, G. Beinert, E. Franta, P. Rempp, *Polym. Bull.* 7 (1982) 17–22.
19 S. Kobayashi, M. Kaku, T. Mizutani, T. Saegusa, *Polym. Bull.* 9 (1983) 169–173.
20 J.S. Vargas, J. G. Zilliox, P. Rempp, E. Franta, *Polym. Bull.* 3 (1980) 83–89.
21 J.-F. Lahitte, F. Peruch, S. Plentz-Meneghetti, F. Isel, P. J. Lutz, *Macromol. Chem. Phys.* 203 (2002) 2583–2589.
22 Y. Tsukahara, K. Mizuno, A. Segawa, Y. Yamashita, *Macromolecules* 22 (1989) 1546–1552.
23 A. Hirao, M. Hayashi, S. Nakahama, *Macromolecules* 29 (1996) 3353–3358.
24 J.-F. Lahitte, S. Plentz-Meneghetti, F. Peruch, F. Isel, R. Muller, P.J. Lutz, *Polymer* 47 (2006) 1063–1072.
25 F. Peruch, E. Catari, S. Zahraoui, F. Isel, P.J. Lutz, *Macromol. Symp.* 236 (2006) 168–176.
26 M. Takaki, R. Asami, S. Tanaka, H. Hayashi, T.E. Hogen-Esch, *Macromolecules* 19 (1986) 2900–2903.
27 K. Endo, K. Senoo, Y. Takakura, *Eur. Polym. J.* 35 (1999) 1413–1417.
28 S. Christodoulou, H. Iatrou, D.J. Lohse, N. Hadjichristidis, *J. Polym. Sci., Part A: Polym. Chem.* 43 (2005) 4030–4039.
29 R. Djalali, S.-Y. Li, M. Schmidt, *Macromolecules* 35 (2002) 4282–4288.
30 S. Breunig, V. Heroguez, Y. Gnanou, M. Fontanille, *Polym. Preprints* 35 (1994) 526–527.
31 V. Heroguez, Y. Gnanou, M. Fontanille, *Macromolecules* 30 (1997) 4791–4798.
32 V. Heroguez, E. Amedro, D. Grande, M. Fontanille, Y. Gnanou, *Macromolecules* 33 (2000) 7241–7248.
33 K. Ito, K. Tanaka, H. Tanaka, G. Imai, S. Kawaguchi, S. Itsuno, *Macromolecules* 24 (1991) 2348–2354.
34 M. Maniruzzaman, S. Kawaguchi, K. Ito, *Macromolecules* 33 (2000) 1583–1592.
35 K. Ito, K. Hashimura, S. Itsuno, E. Yamada, *Macromolecules* 24 (1991) 3977–3981.
36 D. Neugebauer, M. Theis, T. Pakula, G. Wegner, K. Matyjaszewski, *Macromolecules* 39 (2006) 584–593.
37 K. Tortosa, C. Miola, T. Hamaide, *J. Appl. Polym. Sci.* 65 (1997) 2357–2372.
38 M. Takaki, R. Asami, T. Kuwabara, *Polym. Bull.* 7 (1982) 521–525.
39 J. Sierra-Vargas, P. Masson, G. Beinert, P. Rempp, E. Franta, Polym. Bull. 7 (1982) 277–282.
40 K. Naraghi, N. Sahli, M. Belbachir, E. Franta, P.J. Lutz, *Polym. Int.* 51 (2002) 912–922.

41 Y. Gnanou, P. Rempp, *Makromol. Chem.* 188 (1987) 2111–2119.

42 M. Kamigaito, T. Ando, M. Sawamoto, *Chem. Rev.* 101 (2001) 3689–3745.

43 K. Matyjaszewski, J. Xia, *Chem. Rev.* 101 (2001) 2921–2990.

44 J.-S. Wang, K. Matyjaszewski, *J. Am. Chem. Soc.* 117 (1995) 5614–5615.

45 X. Zhang, K. Matyjaszewski, *Macromolecules* 32 (1999) 7349–7353.

46 T. Ando, M. Kamigaito, M. Sawamoto, *Macromolecules* 31 (1998) 6708–6711.

47 Y. Nakagawa, K. Matyjaszewski, *Polym. J. (Tokyo)* 30 (1998) 138–141.

48 Y. Nakagawa, S. G. Gaynor, K. Matyjaszewski, *Polym. Preprints* 37 (1996) 577–578.

49 V. B. Sadhu, J. Pionteck, D. Voigt, H. Kober, D. Fischer, B. Voit, *Macromol. Chem. Phys.* 205 (2004) 2356–2365.

50 T. Ando, M. Kamigaito, M. Sawamoto, *Macromolecules* 30 (1997) 4507–4510.

51 T. Ando, M. Kato, M. Kamigaito, M. Sawamoto, *Macromolecules* 29 (1996) 1070–1072.

52 M. Kato, M. Kamigaito, M. Sawamoto, T. Higashimura, *Macromolecules* 28 (1995) 1721–1723.

53 C. Granel, P. Dubois, R. Jerome, P. Teyssie, *Macromolecules* 29 (1996) 8576–8582.

54 J. Xia, K. Matyjaszewski, *Macromolecules* 30 (1997) 7697–7700.

55 D. M. Haddleton, S. Perrier, S. A. F. Bon, *Macromolecules* 33 (2000) 8246–8251.

56 D. M. Haddleton, M. C. Crossman, B. H. Dana, D. J. Duncalf, A. M. Heming, D. Kukulj, A. J. Shooter, *Macromolecules* 32 (1999) 2110–2119.

57 M. Asscher, E. Levy, H. Rosin, D. Vofsi, *Ind. Eng. Chem. Prod. Res. Devel.* 2 (1963) 121–126.

58 M. Asscher, E. Levy, D. Vofsi, *Atti Congr. Int. Mater. Plastiche* 12 (1960) 151–152.

59 M. Destarac, B. Boutevin, *Curr. Trends Polym. Sci.* 4 (1999) 201–223.

60 B. Boutevin, *J. Polym. Sci., Part A: Polym. Chem.* 38 (2000) 3235–3243.

61 K. Matyjaszewski, K. L. Beers, A. Kern, S. G. Gaynor, *J. Polym. Sci., Part A: Polym. Chem.* 36 (1998) 823–830.

62 F. Zeng, Y. Shen, S. Zhu, R. Pelton, *Macromolecules* 33 (2000) 1628–1635.

63 P. D. Bartlett, F. A. Tate, *J. Am. Chem. Soc.* 75 (1953) 91–95.

64 P. D. Bartlett, K. Nozaki, *J. Am. Chem. Soc.* 68 (1946) 1495–1504.

65 P. D. Bartlett, R. Altschul, *J. Am. Chem. Soc.* 67 (1945) 816–822.

66 P. D. Bartlett, R. Altschul, *J. Am. Chem. Soc.* 67 (1945) 812–816.

67 F. Schoen, M. Hartenstein, A. H. E. Mueller, *Macromolecules* 34 (2001) 5394–5397.

68 V. Coessens, K. Matyjaszewski, *Macromol. Rapid Commun.* 20 (1999) 127–134.

69 V. Coessens, K. Matyjaszewski, *J. Macromol. Sci., Pure Appl. Chem.* A36 (1999) 667–679.

70 A. K. Shim, V. Coessens, T. Pintauer, S. Gaynor, K. Matyjaszewski, *Polym. Preprints* 40 (1999) 456–457.

71 A. K. Shim, V. Coessens, T. Pintauer, K. Matyjaszewski, Book of Abstracts, 218th ACS National Meeting, New Orleans, Aug. 22–26 (1999) POLY-514.

72 H. Keul, A. Neumann, B. Reining, H. Hocker, *Macromol. Symp.* 161 (2000) 63–72.

73 S. A. F. Bon, S. R. Morsley, C. Waterson, D. M. Haddleton, *Macromolecules* 33 (2000) 5819–5824.

74 E. G. Koulouri, J. K. Kallitsis, G. Hadziioannou, *Macromolecules* 32 (1999) 6242–6248.

75 S. A. F. Bon, A. G. Steward, D. M. Haddleton, *J. Polym. Sci., Part A: Polym. Chem.* 38 (2000) 2678–2686.

76 H. Fukui, M. Sawamoto, T. Higashimura, *Macromolecules* 26 (1993) 7315–7321.

77 K. Matyjaszewski, Y. Nakagawa, S. G. Gaynor, *Macromol. Rapid Commun.* 18 (1997) 1057–1066.

78 A. Snijder, B. Klumperman, R. Van Der Linde, *J. Polym. Sci., Part A: Polym. Chem.* 40 (2002) 2350–2359.

79 A. Muehlebach, *PMSE Preprints* 90 (2004) 180.

80 U. Schulze, T. Fonagy, H. Komber, G. Pompe, J. Pionteck, B. Ivan, *Macromolecules* 36 (2003) 4719–4726.

81 B. Ivan, T. Fonagy, *Polym. Preprints* 40 (1999) 356–357.

82 J. R. Isasi, L. Mandelkern, M. J. Galante, R. G. Alamo, *J. Polym. Sci., Part B: Polym. Phys.* 37 (1999) 323–334.

83 J. Norman, S. C. Moratti, A. T. Slark, D. J. Irvine, A. T. Jackson, *Macromolecules* 35 (2002) 8954–8961.

84 H.C. Kolb, M.G. Finn, K.B. Sharpless, *Angew. Chem., Int. Ed.* 40 (2001) 2004–2021.

85 V. Rostovtsev Vsevolod, G. Green Luke, V.V. Fokin, K.B. Sharpless, *Angew. Chem. Int. Ed. Engl.* 41 (2002) 2596–2599.

86 J.-F. Lutz, H.G. Boerner, K. Weichenhan, *Polym. Preprints* 46 (2005) 486–487.

87 J.-F. Lutz, H.G. Boerner, K. Weichenhan, *Macromol. Rapid Commun.* 26 (2005) 514–518.

88 W.G. Lewis, F.G. Magallon, V.V. Fokin, M.G. Finn, *J. Am. Chem. Soc.* 126 (2004) 9152–9153.

89 Y.K. Kim, O.R. Pierce, *J. Org. Chem.* 33 (1968) 442–443.

90 J.P. Critchley, V.C.R. McLoughlin, J. Thrower, I.M. White, *Br. Polym. J.* 2 (1970) 288–294.

91 B. Otazaghine, C. Boyer, J.-J. Robin, B. Boutevin, *J. Polym. Sci., Part A: Polym. Chem.* 43 (2005) 2377–2394.

92 B. Otazaghine, B. Boutevin, *Macromol. Chem. Phys.* 205 (2004) 2002–2011.

93 B. Otazaghine, G. David, B. Boutevin, J. J. Robin, K. Matyjaszewski, *Macromol. Chem. Phys.* 205 (2004) 154–164.

94 T. Sarbu, K.-Y. Lin, J. Spanswick, R.R. Gil, D.J. Siegwart, K. Matyjaszewski, *Macromolecules* 37 (2004) 9694–9700.

95 S. Yurteri, I. Cianga, Y. Yagci, *Macromol. Chem. Phys.* 204 (2003) 1771–1783.

96 D.H. Solomon, E. Rizzardo, P. Cacioli, (Commonwealth Scientific and Industrial Research Organization, Australia), *Eur. Patent 135 280,* 1985.

97 M.K. Georges, R.P.N. Veregin, P.M. Kazmaier, G.K. Hamer, M. Saban, *Macromolecules* 27 (1994) 7228–7229.

98 M.K. Georges, R.P.N. Veregin, P.M. Kazmaier, G.K. Hamer, *Macromolecules* 26 (1993) 2987–2988.

99 K. Matyjaszewski, S. Gaynor, D. Greszta, D. Mardare, T. Shigemoto, *J. Phys. Org. Chem.* 8 (1995) 306–315.

100 C.J. Hawker, A.W. Bosman, E. Harth, *Chem. Rev.* 101 (2001) 3661–3688.

101 C.J. Hawker, *J. Am. Chem. Soc.* 116 (1994) 11185–11186.

102 A. Goto, T. Fukuda, *Macromolecules* 30 (1997) 4272–4277.

103 T. Fukuda, Y. Tsujii, T. Miyamoto, *Polym. Preprints* 38 (1997) 723–724.

104 T. Fukuda, T. Terauchi, A. Goto, K. Ohno, Y. Tsujii, T. Miyamoto, S. Kobatake, B. Yamada, *Macromolecules* 29 (1996) 6393–6398.

105 D. Benoit, E. Harth, P. Fox, R.M. Waymouth, C. J. Hawker, *Macromolecules* 33 (2000) 363–370.

106 D. Benoit, V. Chaplinski, R. Braslau, C. J. Hawker, *J. Am. Chem. Soc.* 121 (1999) 3904–3920.

107 C.J. Hawker, D. Mecerreyes, E. Elce, J. Dao, J.L. Hedrick, I. Barakat, P. Dubois, R. Jerome, I. Volksen, *Macromol. Chem. Phys.* 198 (1997) 155–166.

108 C.J. Hawker, J.L. Hedrick, *Macromolecules* 28 (1995) 2993–2995.

109 D. Kuckling, S. Wohlrab, *Polymer* 43 (2001) 1533–1536.

110 J. Ding, C. Chuy, S. Holdcroft, *Macromolecules* 35 (2002) 1348–1355.

111 J.-L. Pradel, B. Ameduri, B. Boutevin, *Macromol. Chem. Phys.* 200 (1999) 2304–2308.

112 J.-L. Pradel, B. Ameduri, B. Boutevin, P. Lacroix-Desmazes, *Polymer Preprints* 40 (1999) 382–383.

113 J.L. Pradel, B. Boutevin, B. Ameduri, *J. Polym. Sci., Part A: Polym. Chem.* 38 (2000) 3293–3302.

114 B. Boutevin, M. Cerf, J.-L. Pradel, (Elf Atochem S.A., Fr.), *Wo Patent 9 746 593,* 1997.

115 E. Harth, C.J. Hawker, W. Fan, R.M. Waymouth, *Macromolecules* 34 (2001) 3856–3862.

116 B.G.G. Lohmeijer, U.S. Schubert, *J. Polym. Sci., Part A: Polym. Chem.* 42 (2004) 4016–4027.

117 B.G.G. Lohmeijer, U.S. Schubert, *Angew. Chem., Int. Ed.* 41 (2002) 3825–3829.

118 B. Yamada, S. Kobatake, *Prog. Polym. Sci.* 19 (1994) 1089–1131.

119 D. Colombani, I. Beliard, P. Chaumont, *J. Polym. Sci., Part A: Polym. Chem.* 34 (1996) 893–902.

120 D. Colombani, P. Chaumont, *Acta Polym.* 49 (1998) 225–231.

121 D. Colombani, *Prog. Polym. Sci.* 24 (1999) 425–480.

122 D. Colombani, P. Chaumont, Prog. Polym. Sci. 21 (1996) 439–503.

123 N. S. Enikolopov, G. V. Korolev, A. P. Marchenko, G. V. Ponomarev, B. R. Smirnov, V. I. Titov, (Institute of Chemical Physics, Chernogolovka, USSR; Siberian Institute of Petroleum Chemistry), *Su Patent 664434*, 1980.

124 L. V. Karmilova, G. V. Ponomarev, B. R. Smirnov, I. M. Bel'govskii, *Uspek. Khim.* 53 (1984) 223–235.

125 A. A. Gridnev, S. D. Ittel, *Chem. Rev.* 101 (2001) 3611–3659.

126 A. A. Gridnev, W. J. Simonsick, Jr., S. D. Ittel, *J. Polym. Sci., Part A: Polym. Chem.* 38 (2000) 1911–1918.

127 A. Gridnev, *J. Polym. Sci., Part A: Polym. Chem.* 38 (2000) 1753–1766.

128 C. Barner-Kowollik, T. P. Davis, M. H. Stenzel, *Polymer* 45 (2004) 7791–7805.

129 A. F. Burczyk, K. F. O'Driscoll, G. L. Rempel, *J. Polym. Sci., Polym. Chem.* 22 (1984) 3255–3262.

130 G. F. Meijs, E. Rizzardo, *J. Macromol. Sci., Rev. Macromol. Chem. Phys.* C30 (1990) 305–377.

131 J. Krstina, C. L. Moad, G. Moad, E. Rizzardo, C. T. Berge, M. Fryd, *Macromol. Symp.* 111 (1996) 13–23.

132 J. Chiefari, Y. K. Chong, F. Ercole, J. Krstina, J. Jeffery, T. P. T. Le, R. T. A. Mayadunne, G. F. Meijs, C. L. Moad, G. Moad, E. Rizzardo, S. H. Thang, *Macromolecules* 31 (1998) 5559–5562.

133 M. Destarac, D. Charmot, X. Franck, S. Z. Zard, *Macromol. Rapid Commun.* 21 (2000) 1035–1039.

134 P. Delduc, C. Tailhan, S. Z. Zard, *J. Chem. Soc., Chem. Commun.* (1988) 308–310.

135 J. Chiefari, J. Jeffery, R. T. A. Mayadunne, G. Moad, E. Rizzardo, S. H. Thang, *ACS Symp. Ser.* 768 (2000) 297–312.

136 M. J. Monteiro, J. Polym. Sci., Part A: Polym. Chem. 43 (2005) 3189–3204.

137 C. Barner-Kowollik, J. F. Quinn, D. R. Morsley, T. P. Davis, *J. Polym. Sci., Part A: Polym. Chem.* 39 (2001) 1353–1365.

138 M. Stenzel-Rosenbaum, T. P. Davis, V. Chen, A. G. Fane, *J. Polym. Sci., Part A: Polym. Chem.* 39 (2001) 2777–2783.

139 M. J. Monteiro, J. de Barbeyrac, *Macromolecules* 34 (2001) 4416–4423.

140 B. S. Sumerlin, M. S. Donovan, Y. Mitsukami, A. B. Lowe, C. L. McCormick, *Macromolecules* 34 (2001) 6561–6564.

141 M. Destarac, K. Matyjaszewski, E. Silverman, B. Ameduri, B. Boutevin, *Macromolecules* 33 (2000) 4613–4615.

142 H. de Brouwer, J. G. Tsavalas, F. J. Schork, M. J. Monteiro, *Macromolecules* 33 (2000) 9239–9246.

143 W. Smulders, M. J. Monteiro, *Macromolecules* 37 (2004) 4474–4483.

144 S. W. Prescott, M. J. Ballard, E. Rizzardo, R. G. Gilbert, *Macromolecules* 35 (2002) 5417–5425.

145 A. Postma, T. P. Davis, G. Moad, M. S. O'Shea, *Macromolecules* 38 (2005) 5371–5374.

146 A. Theis, A. Feldermann, N. Charton, M. H. Stenzel, T. P. Davis, C. Barner-Kowollik, *Macromolecules* 38 (2005) 2595–2605.

147 J.-F. Baussard, J.-L. Habib-Jiwan, A. Laschewsky, M. Mertoglu, J. Storsberg, *Polymer* 45 (2004) 3615–3626.

148 J. Liu, C.-Y. Hong, C.-Y. Pan, *Polymer* 45 (2004) 4413–4421.

149 Z. Liu, J. Ebdon, S. Rimmer, *Reactive Functional Polym.* 58 (2004) 213–224.

150 P. Liu, H. Ding, J. Liu, X. Yi, *Eur. Polym. J.* 38 (2002) 1783–1789.

151 R. C. W. Liu, F. Segui, T. Viitala, F. M. Winnik, *PMSE Preprints* 90 (2004) 105–106.

152 A. J. Convertine, B. S. Lokitz, A. B. Lowe, C. W. Scales, L. J. Myrick, C. L. McCormick, *Macromol. Rapid Commun.* 26 (2005) 791–795.

153 K. M. Lewandowski, D. D. Fansler, M. S. Wendland, S. M. Heilmann, B. N. Gaddam, (3M Innovative Properties Company, USA), *US Patent 6762257*, 2004.

154 K. M. Lewandowski, D. D. Fansler, M. S. Wendland, B. N. Gaddam, S. M. Heilmann, (3M Innovative Properties Company, USA), *US Patent 6753391*, 2004.

155 V. Lima, J. Brokken-Zijp, B. Klumperman, G. van Benthem-van Duuren, R. van der Linde, *Polym. Preprints* 44 (2003) 812–813.

156 V. G. R. Lima, J. Brokken, B. Klumperman, G. van Benthem-van Duuren, R. van der Linde, Abstracts of Papers, 225th ACS National Meeting, New

Orleans, LA, United States, March 23–27, 2003 (2003) POLY-047.
157 V. Lima, X. Jiang, J. Brokken-Zijp, P.J. Schoenmakers, B. Klumperman, R. Van Der Linde, *J. Polym. Sci., Part A: Polym. Chem.* 43 (2005) 959–973.
158 S. Perrier, P. Takolpuckdee, C.A. Mars, *Macromolecules* 38 (2005) 2033–2036.
159 S.G. Gaynor, J.S. Wang, K. Matyjaszewski, *Polym. Preprints* 36 (1995) 467–468.
160 S.G. Gaynor, J.-S. Wang, K. Matyjaszewski, *Macromolecules* 28 (1995) 8051–8056.
161 M. Tatemoto, T. Suzuki, M. Tomoda, Y. Furukawa, Y. Ueta, (Daikin Kogyo Co., Ltd., Japan), *Patent 2815187*, 1978.
162 M. Tatemoto, T. Nakagawa, (Daikin Industries, Ltd., Japan), *Jp. Patent 61049327*, 1986.
163 M. Tatemoto, Y. Yutani, K. Fujiwara, (Daikin Industries, Ltd., Japan), *Eur. Patent 272698*, 1988.
164 M. Tatemoto, *Kobunshi Ronbunshu* 49 (1992) 765–83.
165 K. Matyjaszewski, D. Greszta, D. Mardare. 1994, Dept. Chemistry, Carnegie-Mellon Univ., Pittsburgh, PA, USA. FIELD URL: p. 27 pp.
166 M. Teodorescu, *Eur. Polym. J.* 37 (2001) 1417–1422.
167 B. Ameduri, B. Boutevin *Well-Architectured Fluoropolymers: Synthesis, Properties and Applications*, Elsevier, Amsterdam, 2004.
168 A.E. Feiring, *J. Macromol. Sci., Pure Appl. Chem.* A31 (1994) 1657–1673.
169 V.C.R. McLoughlin, J. Thrower, *Tetrahedron* 25 (1969) 5921–5940.
170 K. Baum, *Synth. Fluorine Chem.* (1992) 381–393.
171 B. Ameduri, B. Boutevin, G. Kostov, *Prog. Polym. Sci.* 26 (2001) 105–187.
172 D. Boulahia, A. Manseri, B. Ameduri, B. Boutevin, G. Caporiccio, *J. Fluorine Chem.* 94 (1999) 175–182.
173 A. Manseri, B. Ameduri, B. Boutevin, M. Kotora, M. Hajek, G. Caporiccio, *J. Fluorine Chem.* 73 (1995) 151–158.
174 D. Lahiouhel, B. Ameduri, B. Boutevin, *J. Fluorine Chem.* 107 (2001) 81–88.
175 B. Ameduri, B. Boutevin, F. Guida-Pietrasanta, A. Manseri, A. Ratsimihety, G. Caporiccio, *J. Polym. Sci., Part A: Polym. Chem.* 34 (1996) 3077–3090.
176 B. Ameduri, B. Boutevin, G. Caporiccio, F. Guida-Pietrasanta, A. Manseri, A. Ratsimihety, *Fluoropolymers* 1 (1999) 67–80.
177 B. Ameduri, G. Colomines, A. Rousseau, B. Boutevin, S. Andre, X. Andrieu, Fluorine in Coatings V, Conference Papers, 5th, Orlando, FL, United States, Jan. 21–22, 2003 (2003) Paper19/M, Paper19/1-Paper19/22.
178 F. Montefusco, R. Bongiovanni, A. Priola, B. Ameduri, *Macromolecules* 37 (2004) 9804–9813.
179 F. Montefusco, R. Bongiovanni, A. Priola, B. Ameduri, Fluorine in Coatings V, Conference Papers, 5th, Orlando, FL, United States, Jan. 21–22, 2003 (2003) Paper25/S, Paper25/1-Paper25/23.
180 V. Percec, A.V. Popov, E. Ramirez-Castillo, O. Weichold, *J. Polym. Sci., Part A: Polym. Chem.* 41 (2003) 3283–3299.
181 V. Percec, A.V. Popov, *J. Polym. Sci., Part A: Polym. Chem.* 43 (2005) 1255–1260.
182 L. Leibler, *Prog. Polym. Sci.* 30 (2005) 898–914.

3
Statistical, Alternating and Gradient Copolymers

Bert Klumperman

3.1
Introduction

The synthesis of statistical and alternating copolymers is a mature technology. Even in the early 20th century, copolymers were synthesized on a routine basis. At that stage, the fundamental understanding of the processes governing the chain growth was still very basic. Pioneers in this field such as Mayo and Lewis contributed greatly to the understanding of radical copolymerization [1]. Not until the end of the 20th century were discoveries made that revealed some general mismatches between experimental data and predictions according to the Mayo–Lewis description of copolymerization [2]. In this review, a historical overview of the development of copolymerization models will be presented. The validity of the models in describing radical copolymerization will be assessed on the basis of current knowledge of the field.

During the 1980s, a revival of interest in radical polymerization was initiated by the discovery of techniques that allow control over chain growth [3–6]. A family of so-called *controlled/living radical polymerization* (LRP) techniques was developed. Their significance for copolymerization is that it became possible to superimpose a gradient in copolymer composition on each individual chain. Since all polymer chains grow simultaneously, a change in instantaneous copolymer composition will be reflected in the composition of a copolymer chain along its backbone. Some theoretical and experimental aspects of gradient copolymers will be discussed in this review.

3.2
Copolymerization Models

In the early days of radical copolymerization, models were soon developed to describe the features of this process. Initially, these were relatively simple models where the reactivity of chain ends was assumed to depend only on the na-

Macromolecular Engineering. Precise Synthesis, Materials Properties, Applications.
Edited by K. Matyjaszewski, Y. Gnanou, and L. Leibler

ISBN: 978-3-527-31446-1

ture of the terminal monomer unit in the growing chain (Mayo–Lewis model or terminal model) [1]. This model by definition leads to first-order Markov chains.

Fairly soon after the introduction of the terminal model, it was realized that there were comonomer pairs that did not obey this model. This led to several other models that usually are extensions of the terminal model. In other words, the terminal model is usually a special case within each of these newly developed models. Examples of such models are the penultimate unit model [7] (where the last but one monomer unit is taken into account) and the complex participation model [8] (where comonomer complexes are assumed to take part in the propagation step after the addition of individual monomer units).

In the first part of this review, the two most common copolymerization models will be discussed. Most importantly, this is done in order to show the models with consistent nomenclature.

3.2.1
Terminal Model

As indicated above, the simplest model to describe radical copolymerization is the terminal model (TM) [1], in which the rates of the propagation reactions are governed by the nature of the propagating radical and that of the adding monomer. This leads to four propagation reactions:

$$P_1^\bullet + M_1 \xrightarrow{k_{11}} P_1^\bullet \tag{1}$$

$$P_1^\bullet + M_2 \xrightarrow{k_{12}} P_2^\bullet \tag{2}$$

$$P_2^\bullet + M_1 \xrightarrow{k_{21}} P_1^\bullet \tag{3}$$

$$P_2^\bullet + M_2 \xrightarrow{k_{22}} P_2^\bullet \tag{4}$$

Each of the four propagation reactions has its own rate constant (k_{ij}), where subscript i is indicative of the nature of the propagating radical chain end and subscript j denotes the nature of the adding monomer. In the description of copolymerization rate and copolymer composition as a function of comonomer feed composition, it is common practice to use so-called reactivity ratios, defined as

$$r_i = \frac{k_{ii}}{k_{ij}} \tag{5}$$

where i and j are 1 or 2 and $i \neq j$.

When describing copolymer composition, but also monomer sequence distribution, it is convenient to use conditional probabilities. These conditional probabilities are defined as the chance that a certain event takes place, out of all possibilities at a certain stage [9]. In case of the TM, an example of a conditional probability is the chance that monomer 2 will add to a monomer 1 chain end radical (p_{12}). The two relevant conditional probabilities are defined as

$$p_{12} = \frac{1}{1 + r_1 q} \tag{6}$$

$$p_{21} = \frac{1}{1 + \frac{r_2}{q}} \tag{7}$$

where $q = f_1/f_2$, the comonomer ratio, and by definition, $p_{12} + p_{11} = p_{21} + p_{22} = 1$.

Copolymer composition and monomer sequence distribution are defined as functions of the above-mentioned conditional probabilities as follows:

$$\frac{F_1}{F_2} = \frac{p_{21}}{p_{12}} \tag{8}$$

$$F_{111} = (1 - p_{12})^2 \tag{9}$$

$$F_{112} + F_{211} = 2p_{12}(1 - p_{12}) \tag{10}$$

$$F_{212} = p_{12}^2 \tag{11}$$

where F_1 and F_2 are the monomer fractions in the copolymer and F_{111}, F_{112}, F_{211} and F_{212} are the mole fractions of the monomer 1 centered triads relative to the total amount of monomer 1 centered triads. Monomer 2 centered triads are obviously obtained by simply inverting the indices.

A useful description of the average propagation rate constant as a function of comonomer feed composition is given in a review by Fukuda et al. [10]:

$$\langle k_p \rangle = \frac{r_1 f_1^2 + 2 f_1 f_2 + r_2 f_2^2}{r_1 f_1 / k_{11} + r_2 f_2 / k_{22}} \ [\mathrm{L\ mol^{-1}\ s^{-1}}] \tag{12}$$

The derivation of Eq. (12) is based on the observation that the average propagation rate constant of a copolymerization is the weighted average over the four different propagation reactions. This leads to the mathematical expression

$$\langle k_p \rangle = k_{11} p_1 f_1 + k_{12} p_1 f_2 + k_{21} p_2 f_1 + k_{22} p_2 f_2 \tag{13}$$

where p_i is the fraction of monomer i chain end radicals, with $\sum p_i = 1$, and f_j is the fraction of monomer j in the comonomer mixture, with $\sum f_i = 1$. The main problem in this derivation is that the magnitude of p_i is unknown. In order to overcome this dilemma, a steady-state assumption is applied. This assumption entails that the instantaneous value of p_i does not vary significantly. In other words, the rate at which chain end radical 1 reacts with monomer 2 equals the rate at which chain end radical 2 reacts with monomer 1. The mathematical form of this steady-state assumption is

$$k_{12}p_1f_2 = k_{21}p_2f_1 \tag{14}$$

Substitution of Eq. (14) in Eq. (13), inclusion of the reactivity ratios and some mathematical rewriting leads to Eq. (12).

3.2.2
Penultimate Unit Model

As indicated in the Introduction, the penultimate unit model (PUM) was developed as an extension of the TM. In the PUM, the chain end reactivity is assumed to be dependent on both the terminal and the penultimate unit in a growing polymer chain [7]. Long ago it was clear that some comonomer pairs (such as styrene–acrylonitrile) cannot be described by the TM [11]. Although not completely illogical, there was at that time little theoretical evidence that the PUM is a physically meaningful model. Only in the 1990s, work by Heuts et al. led to the theoretical insight that a penultimate unit effect is a realistic explanation for deviations from the TM [12]. In *ab initio* quantum theoretical calculations, they showed that the transition state of the propagation reaction can be represented by three hindered rotors. It is then observed that one of the rotors is significantly influenced by the penultimate unit. On the basis of their calculations, they predict a factor two as a realistic magnitude of the penultimate unit effect.

The PUM is represented by eight different propagation reactions, each with its own rate constant. The indexes of the rate constants are composed of the penultimate unit and the terminal unit in a propagating chain end and the adding monomer, as first, second and third index, respectively.

$$P_{11}^{\bullet} + M_1 \xrightarrow{k_{111}} P_{11}^{\bullet} \tag{15}$$

$$P_{11}^{\bullet} + M_2 \xrightarrow{k_{112}} P_{12}^{\bullet} \tag{16}$$

$$P_{12}^{\bullet} + M_1 \xrightarrow{k_{121}} P_{21}^{\bullet} \tag{17}$$

$$P_{12}^{\bullet} + M_2 \xrightarrow{k_{122}} P_{22}^{\bullet} \tag{18}$$

$$P_{21}^{\bullet} + M_1 \xrightarrow{k_{211}} P_{11}^{\bullet} \tag{19}$$

$$P_{21}^{\bullet} + M_2 \xrightarrow{k_{212}} P_{12}^{\bullet} \tag{20}$$

$$P_{22}^{\bullet} + M_1 \xrightarrow{k_{221}} P_{21}^{\bullet} \tag{21}$$

$$P_{22}^{\bullet} + M_2 \xrightarrow{k_{222}} P_{22}^{\bullet} \tag{22}$$

Note that it is easy to comprehend that the TM is a special case of the PUM, namely, if $k_{1ij}=k_{2ij}$ for all i and j, no penultimate unit effect is observed.

Similarly to the TM, reactivity ratios are defined also for the case of the PUM. Instead of two reactivity ratios as is the case for the TM, four reactivity ratios are defined for the PUM. In the literature, two ways of representing these reactivity ratios are encountered: on one hand r_i and r_i' and on the other r_{ii} and r_{ji}. Preference is given to the latter notation, since it provides a more transparent comparison with the notation of the TM. The definition of the reactivity ratios is

$$r_{ij} = \frac{k_{ijj}}{k_{ijk}} \tag{23}$$

where i, j and $k=1$ or 2 and $j \neq k$.

Similarly to the description of the TM, also in the case of the PUM conditional probabilities are used to describe copolymer composition and copolymerization rate conveniently as a function of comonomer feed composition. The conditional probabilities in this case are

$$p_{112} = \frac{1}{1 + r_{11}q} \tag{24}$$

$$p_{211} = \frac{r_{21}q}{1 + r_{21}q} \tag{25}$$

$$p_{221} = \frac{1}{r + \frac{r_{22}}{q}} \tag{26}$$

$$p_{122} = \frac{\frac{r_{12}}{q}}{1 + \frac{r_{12}}{q}} \tag{27}$$

Also in the case of the PUM, the copolymer composition and monomer sequence distribution are readily written as functions of these conditional probabilities. The relevant expressions are

$$\frac{F_1}{F_2} = \frac{1 + \frac{p_{211}}{p_{112}}}{1 + \frac{p_{122}}{p_{221}}} \tag{28}$$

$$F_{111} = \frac{p_{211}(1 - p_{112})}{p_{112} + p_{211}} \tag{29}$$

$$F_{112} + F_{211} = \frac{2p_{211}p_{211}}{p_{112} + p_{211}} \tag{30}$$

$$F_{212} = \frac{p_{112}(1 - p_{211})}{p_{112} + p_{211}} \tag{31}$$

In the late 1980s, it was shown that even in cases where the TM is perfectly well able to describe copolymer composition and monomer sequence distribution as a function of comonomer feed composition, it often fails in describing the average rate constant of propagation [13]. Mainly through the work of Fukuda's group it became clear that it is often necessary to invoke the penultimate unit effect for describing the average rate constant of propagation. In order to do so, they introduced the so-called implicit and explicit penultimate unit effect (IPUE and EPUE) [2].

The EPUE is encountered when $r_{11} \neq r_{21}$, and/or $r_{22} \neq r_{12}$, which basically means that the full PUM as shown in Eqs. (15)–(22) is operational. This turned out to be rather exceptional. The prime example is probably the copolymerization of styrene and acrylonitrile [11].

In cases where the EPUE is not necessary to describe the copolymer composition and monomer sequence distribution, there may still be an IPUE. The IPUE is characterized by $r_{11} = r_{21} = r_1$ and $r_{22} = r_{12} = r_2$, furthermore by $k_{111} \neq k_{211}$, and/or $k_{222} \neq k_{122}$. It is important to note that polymerizations that exhibit an IPUE are perfectly well described by the TM when it comes to copolymer composition and monomer sequence distribution. However, the TM fails dramatically in some cases when it comes to the description of the average propagation rate constant as a function of comonomer feed ratio.

In order to describe the IPUE mathematically, Fukuda and coworkers defined two additional reactivity ratios in the PUM, $s_1 = k_{211}/k_{111}$ and $s_2 = k_{122}/k_{222}$. When this terminology is used to describe the average propagation rate constant as a function of the monomer feed composition, the following expressions apply:

$$\langle k_p \rangle = \frac{\bar{r}_1 f_1^2 + 2f_1 f_2 + \bar{r}_2 f_2^2}{(\bar{r}_1 f_1/\bar{k}_{11}) + (\bar{r}_2 f_2/\bar{k}_{22})} \tag{32}$$

$$\bar{r}_1 = r_{21}\frac{r_{11}f_1 + f_2}{r_{21}f_1 + f_2} \tag{33}$$

$$\bar{k}_{11} = k_{111}\frac{r_{11}f_1 + f_2}{r_{11}f_1 + (f_2/s_1)} \tag{34}$$

Equivalent expressions to those shown in Eqs. (33) and (34) hold for the parameters with index 2.

3.2.3
Other Copolymerization Models

Various other models have been developed to account for the deviations from the TM. Most of them include the fact that electron-rich and electron-poor monomers have a tendency to form charge-transfer complexes. These complexes are then assumed to participate in the propagation reaction in some way or other. Typical examples of such models are:
- the complex participation model (CPM) [8]
- the complex dissociation model (CDM) [14]
- the comp-pen model [15].

Especially the CPM is used relatively frequently to describe the copolymerization of electron donor and electron acceptor monomers. One of the examples of such copolymerizations is that of styrene (STY) with maleic anhydride (MAnh). Although even very recently the STY/MAnh copolymerization was mentioned as an example of a system that obeys the CPM [16], this was proven wrong already in the mid-1990s. A combined data set of copolymer composition and average propagation rate constant versus comonomer feed composition was fitted using the TM, the PUM and the CPM [17]. It was clearly shown that the PUM is the only model that properly describes both data sets with one set of parameters. Figure 3.1 a and b show examples of the fits, albeit at two different temperature intervals. The parameters were linked using Arrhenius parameters for all the individual rate coefficients. Figure 1 a and b are based on previously published graphs [17].

To the best of the author's knowledge, the study on STY–MAnh is the only occasion where the CPM was critically tested on a combined data set. The fact that it fails to describe the system may create some questions regarding the physical meaning of the CPM. STY–MAnh is often cited as one of the prime examples of the CPM. Additional research on other alleged examples of the CPM would be necessary to answer the more general question about its physical meaning.

3.2.4
Model Discrimination

From the previous sections, it is clear that for the proper description of a copolymerization system it is necessary to know which copolymerization model is governing the reaction. There have been extensive studies on model discrimination for various comonomer pairs. Initially these studies were aimed solely at copolymer composition and monomer sequence distributions as a function of comonomer feed composition [18]. The basic strategy was to devise carefully a set of experiments that would allow the experimentalist to draw conclusions on the basis of fitting the candidate models to the experimental data. This approach was increasingly perfected and probably the best example was carried

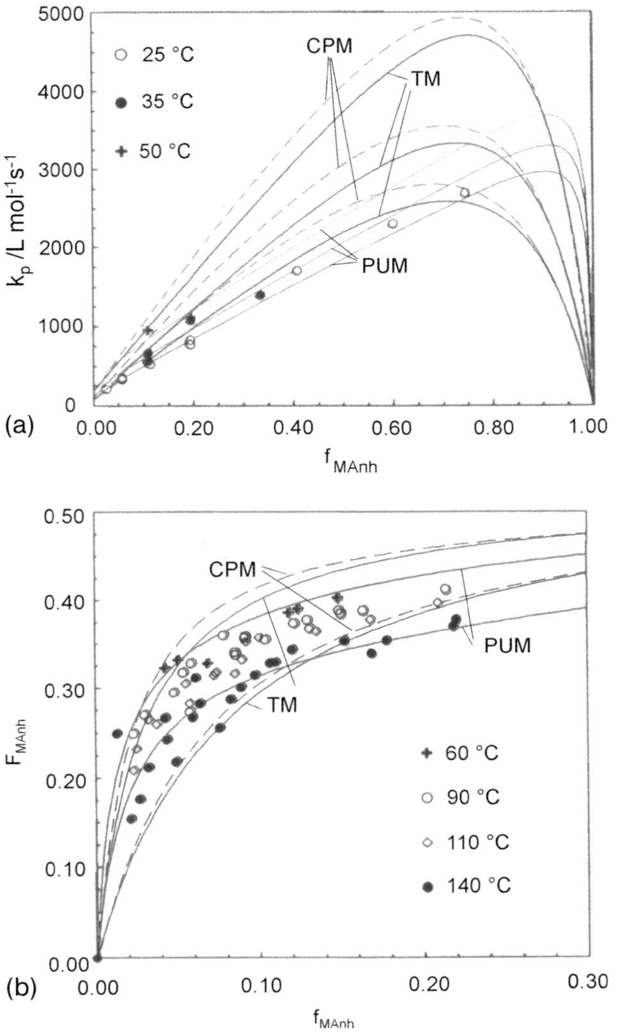

Fig. 3.1 (a) Mean propagation rate constant of the copolymerization of styrene and maleic anhydride (MAnh) as a function of the fraction MAnh in the feed. (b) Copolymer composition (F_{MAnh}) as a function of the fraction MAnh in the feed (f_{MAnh}) for the copolymerization of styrene and MAnh. The curves are best fits to the data in (a) and (b) simultaneously, for the terminal model (TM), penultimate unit model (PUM) and complex participation model (CPM).

out in the group of Duever [19]. Their approach basically distinguished between experiments needed to carry out model discrimination and experiments needed to obtain optimal values for reactivity ratios. It all comes down to an advanced exercise of sensitivity analysis, i.e. the highest sensitivity for making the correct choice between two copolymerization models does not necessarily coincide with that for obtaining the most accurate reactivity ratios. One of the main conclusions from model discrimination studies in general is that data sets often lack the accuracy to perform the required discrimination.

At a later stage, it was discovered that much higher sensitivity can be obtained if the ratio of propagating chain end radicals can be determined as a function of comonomer feed ratio. Especially with the introduction of the IPUE (Section 3.2.2), this propagating radical ratio is indispensable for model discrimination. However, this immediately introduces an experimental complication. The only viable way to monitor propagating radicals directly is by electron spin resonance (ESR) spectroscopy. Unfortunately, the typical radical concentrations in a conventional radical (co)polymerization are too low to perform this measurement with the required accuracy. This problem is partially overcome by the introduction of a trapping technique to capture the propagating radicals with a suitable compound and analyze the resulting product. An example of such an approach was published by Keleman and Klumperman [20]. They initiated a copolymerization by an alkoxyamine that contained a ^{15}N label. The stability of the alkoxyamine was such that it provided a steady flux of radicals, fairly similar to the situation with a conventional radical initiator. The released nitroxide with the ^{15}N label would trap the propagating radical, where the stability of the resulting macromolecular alkoxyamine was such that this trapping was effectively irreversible under the chosen experimental conditions. Reaction conditions and reactant concentrations were chosen such that the molar mass of the trapped radicals were large enough to translate the results to "the long-chain limit". ^{15}N NMR was subsequently used to measure the ratio of the two comonomer units in the trapped chains. The potential pitfall with this approach is a difference in trapping rate between the two propagating radicals, and in the publication on this approach, only upper limits of the radical ratio are given [20]. However, in the investigated system (styrene–methyl acrylate), this was sufficient to carry out the desired model discrimination.

It is clear that, in the future, advanced experiments and high-accuracy analytical equipment will further increase the ability to carry out the model discrimination for radical copolymerization.

3.3
Statistical Copolymers

The traditional linear copolymers can be divided into three categories, random copolymers, alternating copolymers and statistical copolymers. The first two categories can be seen as extremes of statistical copolymers. The determining factor for placement in one of the three categories is the product of the reactivity ratios (r_1r_2). If the product $r_1r_2=1$, a random copolymer is obtained. This essentially means that, in a 50/50 copolymer, there is a completely random sequence of the comonomers in the chain. An alternating copolymer is obtained if both comonomers do not homopolymerize and therefore $r_1r_2=0$. If one of the two reactivity ratios equals zero, but the other does not, the system usually has a strong tendency to alternate, but non-alternating sequences (i.e. longer block lengths of the monomer that is able to homopolymerize) are also possible.

In all other cases in terms of the r_1r_2 product, a statistical copolymer is obtained. The magnitude of the r_1r_2 product determines the proclivity of the copolymer to exhibit a tendency to alternate or a tendency to form blocky structures. In essence, the closer the product is to zero, the greater is the tendency to form alternating copolymers. Conversely, the larger the product is, the longer is the average block length of one of the comonomers. In practice, this only holds true up to a certain maximum product of the reactivity ratios. If the product is too large, cross-propagation will virtually be absent and two almost independent homopolymerizations will take place. The copolymerization of *N*-vinylpyrrolidone with *N*-vinylcaprolactam shows a tendency towards this behavior.

3.3.1 Homogeneous Versus Heterogeneous Copolymers

The introduction to statistical copolymers as given above implicitly speaks about instantaneous copolymer composition, i.e. the composition at zero conversion or the composition over a negligible conversion interval. If a copolymerization

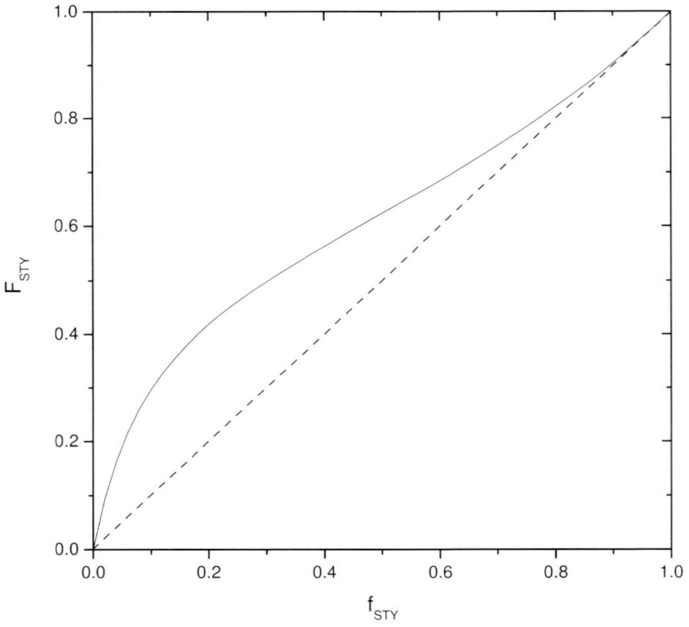

Fig. 3.2 Copolymer composition (F_{STY}) as a function of comonomer feed composition (f_{STY}) for the copolymerization of styrene (STY) and butyl acrylate (BA) with reactivity ratios r_{STY}=0.95 and r_{BA}=0.18.

is carried out under non-azeotropic conditions, i.e. the comonomer ratio built into the copolymer instantaneously deviates from the comonomer ratio in the comonomer feed, a phenomenon called *composition drift* will occur. One of the two comonomers will be built in at a larger rate so that the comonomer mixture is depleted of that comonomer. Essentially, the comonomer fraction in the residual monomer will follow the well-known *F–f* curve, which is a diagram that shows copolymer composition as a function of comonomer feed composition. An example of such a diagram is shown in Fig. 3.2 for the copolymerization of styrene and butyl acrylate. Equations (6)–(8) are used to construct the graph according to the TM with reactivity ratios set to $r_{STY}=0.95$ and $r_{BA}=0.18$. If a polymerization were to be carried out in batch, starting at a fraction of styrene in the comonomer feed of $f_{STY}=0.50$, the initially formed copolymer would contain a fraction of styrene of approximately $F_{STY}=0.62$. This obviously would lead to a depletion of styrene from the mixture. Eventually, all the styrene would be consumed and the residual butyl acrylate would homopolymerize. Hence the resulting material would be a mixture of STY–MA copolymer and MA homopolymer.

Due to the nature of radical polymerization, i.e. due to the short time of growth of individual polymer chains, there will be a heterogeneous mixture of copolymer chains with an approximate *chemical composition distribution (CCD)* as shown in Fig. 3.3. This figure is constructed on the basis of a simulation and lacks the statistical broadening that occurs in a real copolymerization. At 100% monomer conversion, obviously the overall copolymer composition is identical

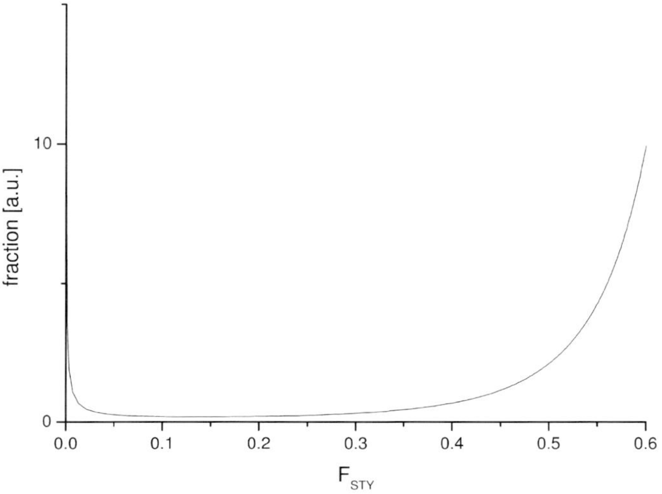

Fig. 3.3 Simulated chemical composition distribution (CCD) of a copolymer of styrene and butyl acrylate at 100% monomer conversion, according to the *F–f* curve in Fig. 3.2, with an initial comonomer feed composition containing $f_{STY}=0.50$ (batch reaction conditions).

with the initial comonomer feed composition. However, as seen from Fig. 3.3, the distribution is strongly bimodal. In terms of material properties, it will have a major effect that there is this bimodal CCD. It is generally accepted that polymers hardly ever mix homogeneously. That also holds true for copolymers of strongly different composition. The phase separation that will take place in the material with the CCD of Fig. 3.3 will strongly influence mechanical properties such as Young's modulus and elongation at break. It is beyond the scope of this contribution, but the particular morphology of this material will basically determine whether it possesses good or bad mechanical properties.

In order to synthesize a homogeneous copolymer, i.e. a copolymer with a narrow CCD, technological solutions need to be sought. The general strategies are semi-continuous operation or continuous operation in a stirred tank reactor. The semi-continuous operation relies on the fact that the most reactive monomer in the copolymerization, i.e. styrene in the example described above, is continuously fed at a rate that ensures a constant comonomer ratio in the reaction mixture. This procedure requires either very accurate monitoring tools and process control or a very accurate predictive model to control the feed rate. The degree of success relies to a large extent on the tendency to undergo composition drift. If copolymer composition and monomer feed composition are strongly different, it is generally very hard to control the feed in such a way that a narrow CCD is obtained. In that case it is usually better to apply continuous operation in a continuous stirred tank reactor (CSTR). In that case, there is a continuous feed into the reactor of both comonomers and potentially a solvent and a continuous removal of copolymer solution. Upon doing this, a steady state will be established after about 3–4 times the mean residence time in the CSTR. The copolymer composition lies between the low conversion (instantaneous) copolymer composition that one reads from an *F*–*f* plot as shown in Fig. 3.2 and the comonomer feed ratio as fed to the reactor. The higher the conversion at steady state, the closer the copolymer composition will be to the comonomer feed to the reactor. The copolymer composition is dependent on the steady-state conversion in the reactor. It can be calculated from an iterative process starting with a mass balance for both comonomers over the reactor. Some examples of the resulting curves for the STY–BA copolymerization in the example above are shown in Fig. 3.4, which indicates how different feed compositions can be used to obtain a homogeneous copolymer with identical composition, by establishing the appropriate steady-state conversion in the CSTR. As indicated, the curves are obtained via an iterative process. It is dependent on the software used to carry out the simulation whether the iteration converges to a stable solution throughout the whole conversion regime. In the present case, a spreadsheet was used to carry out the iterations and instability occurred above ~60% monomer conversion.

The whole discussion about homogeneous versus heterogeneous copolymers overlooked the polymerization medium. Like any radical polymerization, copolymerizations can also be carried out in bulk, solution, suspension, emulsion, etc. Especially some of the technological means to produce homogeneous copoly-

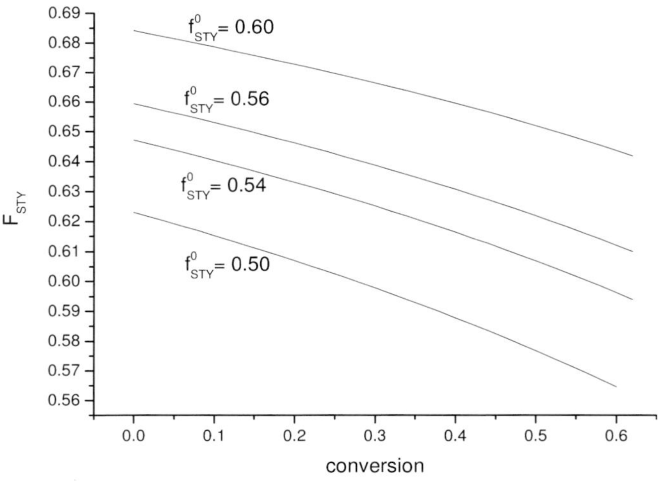

Fig. 3.4 Calculated copolymer compositions as a function of (steady-state) conversion for styrene–butyl acrylate copolymerizations carried out in a CSTR using various feed compositions. Reactivity ratios are $r_{STY}=0.95$ and $r_{BA}=0.18$.

mers will not be easily applicable in heterogeneous reaction media, such as suspension or emulsion polymerization. Emulsion polymerization in a CSTR is generally not desired due to the intrinsically incomplete conversion. Typically, one would like to run emulsion polymerization to full conversion in order to prevent the presence of residual monomer in the latex. Full conversion is fundamentally incompatible with a continuous process in a CSTR. However, in some cases, the use of emulsion polymerization allows the use of a different strategy. This strategy consists of a change in the monomer to water ratio. If the comonomers with different reactivity also possess different water solubility, the apparent azeotropic conditions, i.e. conditions where copolymer composition and comonomer feed composition are identical, may be changed. Without going into too much detail in this contribution, if the most reactive comonomer is also the most water soluble, it will "hide" in the aqueous phase and the apparent azeotropic point may be adjusted. The interested reader is referred to earlier work from the group of German and van Herk (e.g. [21]).

3.3.2
Reactivity Ratios

It has already been pointed out that reactivity ratios are the crucial parameters in describing copolymer composition versus monomer feed composition. Knowledge of these parameters is therefore of great importance if one wishes to control the synthesis of a copolymer. Already in the early days of copolymerization studies, experimental determination of reactivity ratios and their theoretical prediction were investigated. In this contribution, some of the important aspects will be reviewed.

3.3.2.1 Experimental Determination

The principle of the experimental determination of reactivity ratios is independent of the copolymerization model. In this chapter, the procedures will be described for the TM, since its simplicity does not complicate the descriptions unnecessarily. The basic principle for the determination of reactivity ratios involves an experimental data set of e.g. copolymer composition versus monomer feed composition data and a subsequent data treatment to extract the reactivity ratios that best describe the experimental data. Before the period of general access to computers, a significant number of methods were developed to extract reactivity ratios from experimental data. These methods were all named after their inventors, e.g. Kelen–Tüdös [22], Ito–Yamashita [23], Fineman–Ross [24]. In all of these methods, the experimental data are mathematically manipulated in such a way that the graphical representation and a simple linear least-squares data fit provide easy access to the reactivity ratios. One of the main drawbacks and concerns with these methods is that the error structure of the experimental data is distorted. The result of that is a risk of arriving at the wrong values of reactivity ratios, but certainly great difficulty in expressing the accuracy of the reactivity ratios, i.e. the confidence interval. Of the linearized methods, the Kelen–Tüdös method is certainly the most reliable, but with the easy access to personal computers it is no longer necessary to use linearized methods at all. Instead, statistically sound techniques can be used such as the nonlinear least-squares method [25] or the error-in-variables method [26]. In these methods, the original equations are used without transformation and reactivity ratios are varied to obtain the lowest sum of squares, i.e. the curve that shows the smallest deviation from the experimental data points. A very elegant way of doing this was published by van Herk [25]. In his method, he mapped out the sum of squares in a selected reactivity ratio range. This results in a surface graph, where the minimum represents the best values of the reactivity ratios (r_1, r_2). By showing the intersection of this surface graph with a horizontal plane, representations of the joint confidence interval (JCI) can be constructed.

There are two ways to improve the accuracy of reactivity ratios. The first is to carry out experiments at the optimal comonomer feed composition. The intuitive approach is to carry out experiments at compositions that are equally distributed over the entire composition range. For obtaining a first impression of the value of the reactivity ratios, this approach is very well suited. However, once initial estimates of reactivity ratios are available, experiments can be carried out at compositions where the sensitivity towards changes in reactivity ratios is maximal. Tidwell and Mortimer derived expressions for these comonomer feed compositions [27]. They did this exercise for the TM and derived the expressions

$$f_{21} \approx \frac{r_1}{2 + r_1} \tag{35}$$

$$f_{22} \approx \frac{2}{2 + r_2} \tag{36}$$

where f_{21} and f_{22} are the fractions of monomer 2 in the reaction mixture that are most suitable for the accurate determination of reactivity ratios. The same procedure can be applied to different copolymerization models, e.g. PUM, and different types of data sets, e.g. triad distributions as a function of monomer feed composition. Both will lead to more complex mathematical derivations that are difficult, if not impossible, to solve analytically. Fortunately, it is fairly straightforward to solve the equations numerically. The principal equation that needs to be solved is the maximization of the modulus of the determinant D:

$$D = \begin{vmatrix} \dfrac{\partial G(f_{21}; r_1; r_2)}{\partial r_1} & \dfrac{\partial G(f_{21}; r_1; r_2)}{\partial r_2} \\ \dfrac{\partial G(f_{22}; r_1; r_2)}{\partial r_1} & \dfrac{\partial G(f_{22}; r_1; r_2)}{\partial r_2} \end{vmatrix} \tag{37}$$

The second method to improve the accuracy of the reactivity ratios is the use of other types of data. If the monomer sequence distribution is measured as a function of comonomer feed ratio, the accuracy will usually be larger than in the case of copolymer composition versus comonomer feed ratio. One of the frequently used methods is the measurement of so-called triad fractions. For example, in a copolymer of styrene (S) and butyl acrylate (B), the styrene-centered triads are SSS, BSS, SSB and BSB. Their fractions can often be measured by ^{13}C NMR spectroscopy. The assignments in the NMR spectra are often difficult to make and quantitative ^{13}C NMR measurements can be tedious. However, after going to the trouble of doing this, accurate reactivity ratios can be obtained. Also here, as indicated above, the method of Tidwell and Mortimer can be applied to maximize the accuracy even further.

3.3.2.2 **Theoretical Predictions**

Throughout the history of radical copolymerization, attempts have been made to predict the value of reactivity ratios. One of the earliest attempts was the so-called *Q–e* scheme [28]. The basic principle of this method is that the general reactivity of a propagating radical $\sim M_i^\bullet$ is represented by B_i, whereas the general reactivity of a monomer M_j is represented by Q_j. The polarities of monomer and radical are assumed to be identical and are represented by e_i and e_j, respectively, in the example above. The rate constant for the addition of monomer M_2 to radical $\sim M_1^\bullet$ is the written as

$$k_{12} = B_1 Q_2 \exp(-e_1 e_2) \tag{38}$$

This then results in reactivity ratios as follows:

$$r_1 = (Q_1/Q_2)\exp[-e_1(e_1 - e_2)] \qquad (39)$$

$$r_2 = (Q_2/Q_2)\exp[-e_2(e_2 - e_1)] \qquad (40)$$

Although this theoretical prediction only provides rough estimates of reactivity ratios, it is certainly a method that has been used numerous times. In part this is due to the tables of Q and e values listed in the *Polymer Handbook* [29], which makes it easy to do a quick calculation for a comonomer pair. The Q–e scheme is a semi-empirical relationship. A similar methodology was used in later stages by Bamford and Jenkins [30] for an improved prediction method that they called "patterns of reactivity" and which was further optimized by Jenkins [31]. However, with increasing computer power, it is now possible to calculate rate constants from quantum mechanics or by using density functional theory (DFT) calculations. It is expected that these advanced calculation techniques will be used increasingly in predictions of rate constants and therefore also in predictions of reactivity ratios.

3.4
Alternating Copolymers

As indicated above, alternating copolymers can be seen as a special case of statistical copolymers. If the product of the reactivity ratios equals zero, the synthesis of an alternating copolymer is possible. The monomer for which the reactivity ratio equals zero will always be situated between two residues of the comonomer. If then the mole fraction of both monomers equals 0.5, the polymer must be alternating. An example of such a system is the copolymerization of styrene and maleic anhydride. Maleic anhydride does not homopropagate. However, the cross-propagation to styrene occurs at a high rate. Furthermore, although styrene is obviously able to homopropagate, the cross-propagation with maleic anhydride is much faster. Hence there is a large tendency towards alternation and when working at higher fractions of maleic anhydride in the feed, a truly alternating copolymer is formed.

The most peculiar examples of alternating copolymerizations are those systems where neither of the two comonomers is able to homopolymerize. A typical example of this class of copolymerizations is that between 1,2-diphenylethylene (stilbene) and maleic anhydride. Stilbene is an electron-rich monomer that does not homopolymerize for steric reasons. Maleic anhydride, on the other hand, is an electron acceptor monomer that only oligomerizes under severe conditions. It turns out that the cross-propagation between stilbene and MAnh is an efficient reaction. This results in the formation of an alternating copolymer regardless of the ratio of the comonomers in the feed.

The copolymerization mechanism that causes the formation of alternating copolymers is still a matter of debate. The detectable presence of charge-transfer complexes in the reaction mixture has often been used as a proof or at least a strong indication that a complex participation model is operational [32]. Part of that proof involves a study of the effect of temperature on the copolymerization. When temperature is raised, the concentration of charge-transfer complex decreases, until above a certain temperature it completely disappears. At the same time, the tendency towards alternation decreases with increasing temperature [33]. The alleged proof of the complex participation model links these two observations by saying that the tendency towards alternation decreases due to a decreased concentration of the charge-transfer complex. In Section 3.2.2 it was mentioned that the copolymerization of STY and MAnh is still often mistakenly mentioned as an example of a system that obeys the complex participation model. The copolymerization can only properly be described by the PUM with the restriction of no homopolymerization of MAnh. The fact that the tendency towards alternation decreases with increasing temperature is simply a matter of the temperature dependence of the reactivity ratios. This phenomenon was reported already in the 1960s by O'Driscoll, who concluded that copolymerizations have a tendency towards random incorporation of the comonomers with increasing temperature [34].

Apart from the comonomer pairs that have a spontaneous tendency towards alternation, there have been several studies on the addition of Lewis acids to copolymerizations in order to increase the alternating tendency [35]. Especially the combination of styrene and acrylates is susceptible to this technique. Upon addition of a Lewis acid, these copolymerizations turn from a statistical into an alternating copolymerization [36, 37]. In more recent years, the same methodology was combined with controlled/living radical polymerization. Matyjaszewski and coworkers used RAFT-mediated polymerization to obtain well-defined alternating copolymers of styrene and methyl methacrylate in the presence of a Lewis acid [38]. Also comonomer pairs that form alternating copolymers spontaneously have been combined with controlled/living radical polymerization. Alternating styrene–maleic anhydride copolymers have been synthesized via nitroxide-mediated polymerization [39] and via RAFT-mediated polymerization [40].

3.5 Solvent Effects

In the basic theory of radical copolymerization, reactivity ratios are assumed to be independent of monomer conversion, polymer chain length, rate of polymerization, dilution, etc. In this list, solvent is also often mentioned. However, it is commonly known that, in many cases, the solvent can have a strong influence on experimentally determined reactivity ratios. In the case of two comonomers that differ significantly in polarity, such as styrene and acrylamide, it makes a large difference whether the polymerization is conducted in a polar or in a less polar solvent. Many attempts to explain this solvent dependence have been published [41]. One of the most remarkable findings in the area of solvent effects was published by Harwood [9]. It turns out that copolymers with identical compositions that have been synthesized in different solvents possess the same monomer sequence distribution. This may sound trivial at first glance, but it needs to be remembered that the experimentally determined reactivity ratios for these copolymerizations vary significantly. Harwood introduced a concept that he named the bootstrap effect. The basic thought behind the bootstrap effect is that a growing polymer chain influences its own environment. The local comonomer ratio in the vicinity of a growing chain may deviate from the overall comonomer ratio as charged to the reaction vessel. Klumperman and coworkers derived a set of equations that mathematically describe the bootstrap effect and proved for a few examples that solvent effects can be adequately described by this theory [42, 43].

At a somewhat earlier stage, Semchikov et al. presented a chain-length dependence of copolymer composition [44]. This means that a copolymer sample, when fractionated according to molar mass, shows a molar mass dependence of its copolymer composition, i.e. the low molar mass chains have a copolymer composition that deviates from the higher molar mass chains. This seems to be reasonably in line with the bootstrap effect concept. As indicated above, a growing chain can influence its own environment. However, before it can do so, it needs to grow a few monomer units. These first few monomer units can be influenced strongly by the nature of the primary radicals. It is known that primary radicals can have a significant selectivity towards one of the two comonomers. Some of the experiments described by Semchikov et al. were repeated in the group of Klumperman, where the analyses were carried out using more advanced analytical tools [45]. It turned out that the molar mass dependence of the chemical composition could not be reproduced. The most likely explanation for the opposing results is that the method of fractionation employed by Semchikov et al. results in non-orthogonal separations and thus in a misjudgment of the shape of the molar mass–chemical composition distribution (MMCCD).

3.6 Gradient Copolymers

Everything discussed so far is based on conventional radical copolymerization. This means that initiation of new chains and termination of growing chains take place continuously. The effect of composition drift as described above is a broad CCD, i.e. chains of varying composition coexist in a sample after a non-azeotropic batch polymerization. Here, the focus will be on copolymers that are made via controlled/living radical polymerization.

3.6.1
Controlled/Living Radical Copolymerization versus Conventional Radical Copolymerization

The main difference between conventional radical polymerization and controlled/living radical polymerization is the period over which a single polymer chain grows. As indicated, for a conventional polymerization this is a matter of seconds, compared with a total duration of the polymerization reaction of hours in most cases. For controlled/living radical polymerization, the growth of a polymer chain takes place in the course of the entire polymerization reaction. The chains start to grow immediately or shortly after the start of the reaction. They will be dormant for the majority of the reaction time, but every now and again they will be activated and add one or a few monomer units before they are reversibly deactivated again. This means that, in a copolymerization, the history of instantaneous copolymer compositions as a function of monomer conversion is projected along the backbone of each individual polymer chain. In other words, rather than a broad CCD as in conventional radical copolymerization, all the copolymer chains are now identical within the inevitable statistical spread. However, the composition drift that caused the broad CCD in the conventional radical copolymerization now causes a gradient in composition along the copolymer backbone. The difference between conventional radical copolymerization and controlled/living radical copolymerization is illustrated in Fig. 3.5. Figure 3.5 a shows the simulated CCD of a copolymer synthesized via conventional radical copolymerization of styrene and butyl acrylate. In Fig. 3.5 b, the simulated copolymer composition as a function of the position in the chain (0 equals the start of the chain and 1 is the end of the chain) is shown for a styrene–butyl acrylate copolymer synthesized via controlled/living radical copolymerization under conditions identical with those in Fig. 3.5 a. It is of interest that, so far in this chapter, no reference has been made to a specific type of controlled/living radical polymerization. What would be the difference among the three prevailing techniques, i.e. nitroxide-mediated polymerization (NMP), atom transfer radical polymerization (ATRP) and reversible addition–fragmentation chain transfer (RAFT)-mediated polymerization? Especially for the case of ATRP, there have been reports on reactivity ratios deviating from values for conventional radical polymerization [46]. At some point this fed the discussion on the "free radical nature" of ATRP. Now the general opinion about controlled/living radical polymerization is that the propagating species is a radical with characteristics identical with those in a conventional radical polymerization [47]. The remaining question then is why these small but significant deviations between reactivity ratios are observed in comparisons between controlled/living radical copolymerization and conventional radical copolymerization. The most likely explanation is that effects of the initiation step are much more pronounced in controlled/living radical copolymerization than in conventional radical copolymerization [48]. This will be detailed in the next section.

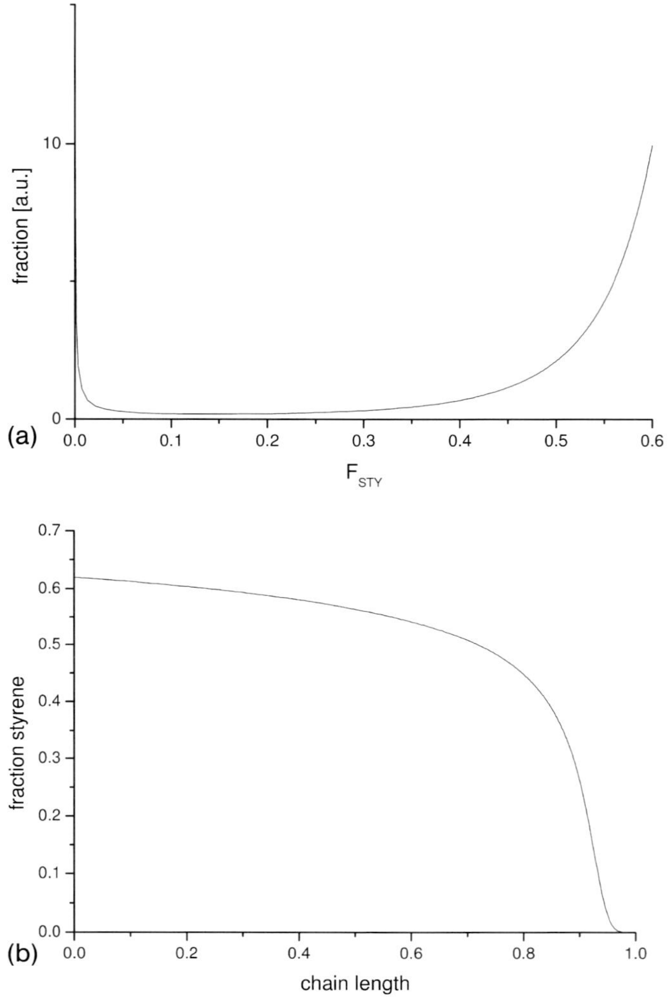

Fig. 3.5 (a) Simulated chemical composition distribution (CCD) of a copolymer of styrene and butyl acrylate, according to the *F*–*f* curve in Fig. 3.2, with an initial comonomer feed composition containing $f_{STY}=0.50$. Reactivity ratios are $r_{STY}=0.95$ and $r_{BA}=0.18$. (b) Calculated copolymer composition as a function of position in the copolymer chain for a controlled/living radical copolymerization of styrene and butyl acrylate with a starting feed composition of $f_{STY}=0.50$. Reactivity ratios are $r_{STY}=0.95$ and $r_{BA}=0.18$.

3.6.2
The Early Stages of Gradient Copolymerization

The typical method for the experimental determination of reactivity ratios (in conventional radical copolymerization) involves the synthesis and composition analysis of low-conversion copolymers at different feed compositions. The general feature of conventional radical polymerization that high molar mass polymer is formed early in the reaction is an implicit, but important, requirement for this method to work. For reactivity ratios to be effective in describing copolymer composition versus comonomer feed composition, the effect of initiation, chain transfer and termination must be negligible. This is exactly the first point where a direct translation of experimental procedures to the controlled/living radical copolymerization case might fail. A low-conversion controlled/living radical polymerization experiment by definition means a low molar mass situation. The selectivity of the first monomer addition determines the initial ratio of adducts. In an efficient LRP system all these low molar mass oligomers will reside in the dormant state. It was shown for atom transfer radical copolymerization that the ratio of the two different dormant chain ends (differing in their terminal monomer unit) can deviate from the expected value for the comonomer ratio and the relevant reactivity ratios [48]. It may take up to 15–20% monomer conversion before the ratio of the two dormant chains reaches a sort of equilibrium value that is in accord with the radical ratio in the equivalent conventional radical copolymerization. Although it has not been thoroughly investigated yet, the indications are that the situation is comparable for nitroxide-mediated copolymerization and for reversible addition–fragmentation chain transfer-mediated copolymerization.

3.6.3
Forced Composition Gradient Copolymers

In order to force a gradient upon a copolymer chain, it is necessary to feed the comonomers in a specific ratio to the reaction vessel. The control strategy is similar to that used for the synthesis of copolymers with a predetermined CCD via a conventional radical copolymerization. In that case there are roughly two methodologies that are sometimes used in combination. The first methodology makes use of a kinetic model that predicts individual rates of comonomer consumption. The feed into the reactor is adjusted to match the required instantaneous copolymer composition. The main drawbacks of this methodology are the assumption that the model is correctly describing the polymerization and the implicit assumption that there are no perturbations of the polymerization that make it deviate from its expected course. The model can be fine-tuned to meet the former requirement as much as possible. However, inhibition phenomena at the start of the reaction and/or retardation phenomena during the reaction are prone to induce deviations from the intended instantaneous comonomer feed ratio in the reactor. Therefore, one might have a preference for the second methodology that relies on *in situ* concentration measurements. The individual comonomer concentrations are measured in real time during the process. Based on these values and a basic model that relates comonomer ratios to the desired copolymer composition, the feed to the reactor is adjusted. Obviously, this method is fully dependent on the response time of the *in situ* measurement in relation to the time-scale of polymerization. Advanced controllers can be designed that minimize the risk of overreaction upon measured comonomer concentrations. As indicated, sometimes the model-based method and the method based on *in situ* concentration measurements are combined. In that case, the model prediction can be used, where the *in situ* measurement is simply used as a check for the absence of anomalies. The whole field of *in situ* measurements and controller development is a science on its own. It is beyond the scope of this review to discuss this in detail.

3.6.4
Analysis of Gradient Copolymers

From the text above, it will be clear that the synthesis of gradient copolymers and that of conventional copolymers with a broad CCD show great resemblance. However, when it comes to the analysis of the polymers, there is a significant difference. The copolymer with the broad CCD consists of a large variety of *different* polymer chains. These chains can be separated on the basis of, e.g., polarity by HPLC techniques, such as gradient polymer elution chromatography (GPEC) [49]. On the other hand, gradient polymers made via LRP techniques (or any other controlled/living polymerization technique) consist of *virtually identical* polymer chains. There is, of course, a statistical variation among the copolymer chains that results from the stochastic process by which the chains are formed. Moreover, chains that happen to terminate in earlier stages of the polymerization process possess a lower molar mass, but also a different chemical composition. That leads to the situation that dormant chains at the end of the polymerization process are more or less identical and cannot be separated on the basis of composition of molar mass. The only handle for tracking the course of the reaction on the polymer product is by carefully picking the terminated chains and analyzing them in terms of molar mass and chemical composition. To the author's knowledge, this exercise has been carried out only once. A styrene–isoprene copolymer was synthesized via living anionic polymerization. Due to the presence of *N,N,N′,N′*-tetramethylethylenediamine (TMEDA), complexation of the lithium counterion takes place and the reactivity of the chain end is modulated. This leads to the spontaneous formation of gradient copolymers. Block copolymers, random copolymers and the above-described gradient copolymers were analyzed by matrix-assisted laser desorption/ionization time-of-flight mass spectrometry (MALDI-TOF-MS) [50]. The acquired spectra were analyzed with a specific software tool, which resulted in the construction of a so-called copolymer fingerprint. The fingerprint is a contour map of the sample that indicates the frequency of occurrence of chains with a specific molar mass

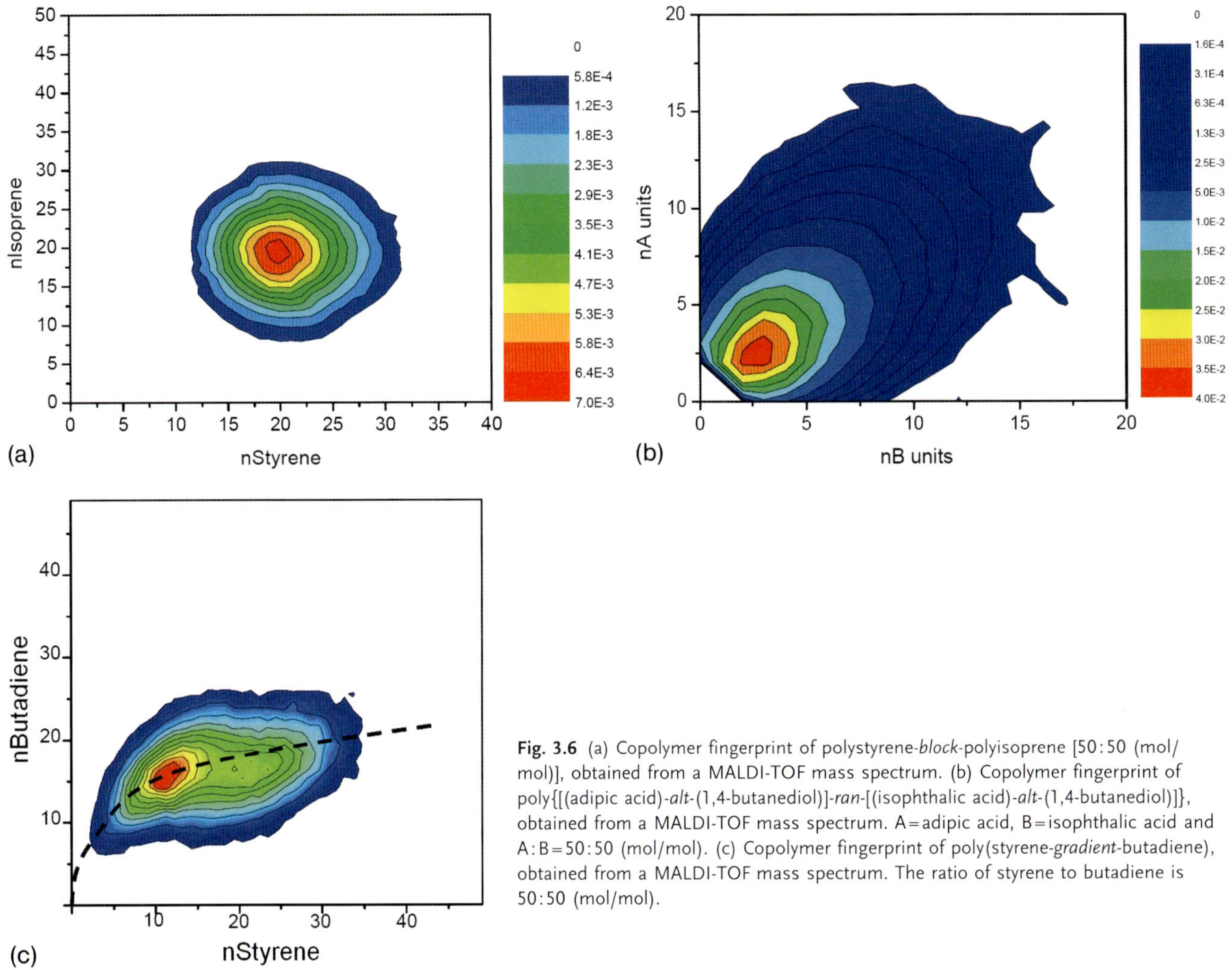

Fig. 3.6 (a) Copolymer fingerprint of polystyrene-*block*-polyisoprene [50:50 (mol/mol)], obtained from a MALDI-TOF mass spectrum. (b) Copolymer fingerprint of poly{[(adipic acid)-*alt*-(1,4-butanediol)]-*ran*-[(isophthalic acid)-*alt*-(1,4-butanediol)]}, obtained from a MALDI-TOF mass spectrum. A = adipic acid, B = isophthalic acid and A:B = 50:50 (mol/mol). (c) Copolymer fingerprint of poly(styrene-*gradient*-butadiene), obtained from a MALDI-TOF mass spectrum. The ratio of styrene to butadiene is 50:50 (mol/mol).

and chemical composition. If the result is an ellipse with axes parallel to the coordinates of the figure, the two comonomers were built in independently. This can only be the case for a block copolymer. An example of such a fingerprint is shown in Fig. 3.6a [51]. If the result is an ellipse with axes that show a certain slope relative to the coordinates, the copolymer is a statistical copolymer. The comonomers were built in according to some type of statistical process. An example of this type of fingerprint is shown in Fig. 3.6b [51]. Finally, if the fingerprint shows nonlinear axes, the copolymer contains a gradient. In that case, the instantaneous composition during the copolymerization has undergone a gradual change. A typical example is shown in Fig. 3.6c [52]. This can be due either to "natural" composition drift or to a forced feed profile as discussed above.

A general strategy for the comprehensive analysis of gradient copolymers is yet to be found. An interesting option that works for the special case where one of the two comonomers is a macromonomer is the use of atomic force microscopy (AFM) [53]. Images of individual copolymer chains on a flat surface allow the detection of the side-chains that result from macromonomer copolymerization. The presence of a gradient along the backbone is then seen as a gradient in the density of the corona around the polymer backbone.

3.7 Properties of Copolymers

In homopolymers, physical properties are to some extent dependent on molar mass. The glass transition temperature (T_g) and various mechanical properties will increase with molar mass and eventually reach some sort of plateau value. The same is true, of course, for a (statistical) copolymer, but in the case of a copolymer there is an additional degree of freedom. The copolymer composition may have a significant influence on macroscopic properties. If the homopolymers of the two monomers in a copolymer have different T_g values, the T_g value of a statistical copolymer will lie between the two homopolymer values. More or less the same holds true for other properties, such as hardness and elasticity modulus. All of this is true if the copolymer has a narrow CCD. If the CCD is broader or even bimodal, phase separation may occur. This will lead to a much more complex situation, where the properties are a function not just of the overall chemical composition, but also of the phase morphology, as was indicated in Section 3.3.1. An additional complication can be introduced if the copolymer chains contain a gradient, as explained in Section 3.6.

Relatively little work has been done on the properties of gradient copolymers. The general idea is that phase boundaries are less sharp. This in general is expected to lead to properties that are between those of random copolymers and block copolymers. Probably the best examples so far are shown in a paper by Matyjaszewski et al. [54]. They point out the different ways of synthesizing gradient copolymers, i.e. via off-azeotropic batch copolymerization and via semicontinuous copolymerization. Then they show with the aid of examples [styrene–*n*-butyl acrylate (STY–BA), styrene–acrylonitrile and methyl methacrylate–butyl acrylate (MMA–BA)] how properties differ among random copolymers, block copolymers and gradient copolymers. From the STY–BA example it becomes clear that the thermal history of a gradient copolymer is extremely important to its thermal properties. The DSC thermograms of quenched and annealed samples differ substantially. In the case of MMA–BA–MMA triblock copolymers, they point to the effect of a sharp transition versus a gradient transition from the central block to the outer blocks. In stress–strain curves it is seen that the presence of a gradient transition leads to a lower stress, but a much higher elongation at break.

Further investigation of the properties of gradient copolymers is necessary to take advantage of the potential of these materials. In addition to the indicated effects on mechanical and thermal properties, it is also expected that optical properties will be affected.

References

1 Mayo, F. R., Lewis, F. M., Inagaki, H., Kubo, K., *J. Am. Chem. Soc.* **1944**, *66*, 1594.
2 Fukuda, T., Ma, Y.-D., Inagaki, H., Kubo, K., *Macromolecules* **1991**, 24, 370.
3 Solomon, D. H., Rizzardo, E., Caciolo, P., *Eur. Patent Appl. 135280*, **1985**; *Chem. Abstr.* **1985**, *102*, 221335q.
4 Wang, J. S., Matyjaszewski, K., *J. Am. Chem. Soc.* **1995**, *117*, 5614.
5 Kato, M., Kamigaito, M., Sawamoto, M., Higashimura, T., *Macromolecules* **1995**, *28*, 1721.
6 Chiefari, J., Chong, Y. K., Ercole, F., Krstina, J., Jeffery, J., Le, T. P. T., Mayadunne, R. T. A., Meijs, G. F., Moad, C. L., Moad, G., Rizzardo, E., Thang, S. H., *Macromolecules* **1998**, *31*, 5559.
7 Fordyce, R. G., Ham, G. E., *J. Am. Chem. Soc.* **1951**, *73*, 1186.
8 Seiner, J. A., Litt, M., *Macromolecules* **1971**, *4*, 308.
9 Harwood, H. J., *Makromol. Chem. Macromol. Symp.* **1987**, *10/11*, 331.
10 Fukuda, T., Kubo, K., Ma, Y.-D., *Prog. Polym. Sci.* **1992**, *17*, 875.
11 Hill, D. J. T., O'Donnell, J. H., O'Sullivan, P. W., *Macromolecules* **1982**, *15*, 960.
12 Heuts, J. P. A., Gilbert, R. G., Radom, L., *Macromolecules* **1995**, *28*, 8771.
13 Davis, T. P., O'Driscoll, K. F., Piton, M. C., Winnik, M. A., *J. Polym. Sci., Part C: Polym. Lett.* **1989**, *27*, 181.
14 Hill, D. J. T., O'Donnell, J. H., O'Sullivan, P. W., *Macromolecules* **1983**, *16*, 1295.
15 Brown, P. G., Fujimori, K., *J. Polym. Sci., Polym. Chem.* **1994**, *32*, 2971.
16 Du, F.-S., Zhu, M.-Q., Guo, H.-Q., Li, Z.-C., Kamachi, M., Kajiwara, A., *Macromolecules* **2002**, *35*, 6739.
17 Klumperman, B., Free radical copolymerization of styrene and maleic anhydride – kinetic studies at low and intermediate conversion. *PhD Thesis*, Eindhoven University of Technology, **1994**.
18 McFarlane, R. C., Reilly, P. M., O'Driscoll, K. F., *J. Polym. Sci., Part C: Polym. Lett.* **1980**, *18*, 81.
19 Burke, A. L., Duever, T. A., Penlidis, A., *J. Polym. Sci., Polym. Chem.* **1996**, *34*, 2665.
20 Kelemen, P., Klumperman, B., *Macromolecules* **2004**, *37*, 9338.
21 Verdurmen-Noël, E. F. J., Monomer partitioning and composition drift in emulsion copolymerization. *PhD Thesis*, Eindhoven University of Technology, **1994**.
22 Kelen, T., Tüdös, F., *J. Macromol. Sci. Chem.* **1975**, *A9*, 1.
23 Ito, K., Yamashita, Y., *J. Polym. Sci. B* **1965**, *3*, 625.
24 Fineman, M., Ross, S. D., *J. Polym. Sci.* **1950**, *5*, 259.
25 van Herk, A. M., *J. Chem. Educ.* **1995**, *72*, 138.

26 Dubé, M., Amin Sanayei, R., Penlidis, A., O'Driscoll, K. F., Reilly, P. M., *J. Polym. Sci., Polym. Chem.* **1991**, *29*, 703.
27 Tidwell, P. W., Mortimer, G. A., *J. Polym. Sci., Polym. Chem.* **1965**, *3*, 369.
28 Alfrey, T., Jr., Price, C. C., *J. Polym. Sci.* **1947**, *2*, 101.
29 Brandrup, J., Immergut, E. H., Grulke, E. A., *Polymer Handbook*, 4th edn. Wiley, New York, **1999**.
30 Bamford, C. H., Jenkins, A. D., *Trans. Faraday Soc.* **1963**, *59*, 530.
31 Jenkins, A. D., *Curr. Org. Chem.* **2002**, *6*, 83.
32 Tsuchida, E., Tomono, T., *Makromol. Chem.* **1971**, *141*, 265.
33 Seymour, R., Garner, D. P., *J. Coat. Technol.* **1976**, *48*, 41.
34 O'Driscoll, K. F., *J. Macromol. Sci. Chem.* **1969**, *A3*, 307.
35 Cowie, J. M. G. (Ed.) *Alternating Copolymers*, Plenum Press, New York, **1985**.
36 Hirai, H., *J. Polym. Sci., Macromol. Rev.* **1976**, *11*, 47.
37 Afchar-Momtaz, J., Polton, A., Tardi, M., Sigwalt, P., *Eur. Polym. J.* **1985**, *21*, 583.
38 Kırcı, B., Lutz, J.-F., Matyjaszewski, K., *Macromolecules* **2002**, *35*, 2448.
39 Benoit, D., Hawker, C. J., Huang, E. E., Lin, Z., Russell, T. P., *Macromolecules* **2000**, *33*, 1505.
40 de Brouwer, H., Schellekens, M. A. J., Klumperman, B., Monteiro, M. J., German, A. L., *J. Polym. Sci., Polym. Chem. Ed.* **2000**, *38*, 3596.
41 Coote, M. L., Davis, T. P., Klumperman, B., Monteiro, M., *J. Macromol. Sci. Rev. Macromol. Chem. Phys.* **1998**, *C38*, 567.
42 Klumperman, B., O'Driscoll, K. F., *Polymer* **1993**, *34*, 1032.
43 Klumperman, B., Kraeger, I. R., *Macromolecules* **1994**, *27*, 1529.
44 Semchikov, Y. D., Smirnova, L. A., Knyazeva, T. E., Bulgakova, S. A., Sherstyanykh, V. I., *Eur. Polym. J.* **1990**, *26*, 883.
45 Chambard, G., Control of monomer sequence distribution. *PhD Thesis*, Eindhoven University of Technology, **2000**.
46 Ziegler, M. J., Matyjaszewski, K., *Macromolecules* **2001**, *34*, 415.
47 Matyjaszewski, K., *Macromolecules* **1998**, *31*, 4710.
48 Chambard, G., Klumperman, B., Brinkhuis, R. H. G., *ACS Symp. Ser.* **2003**, *854*, 180.
49 Braun, D., Kramer, I., Pasch, H., Mori, S., *Macromol. Chem. Phys.* **1999**, *200*, 949.
50 Willemse, R. X. E., Staal, B. B. P., Donkers, E. H. D., van Herk, A. M., *Macromolecules* **2004**, *37*, 5717.
51 Willemse, R. X. E., New insights into free-radical (co)polymerization kinetics. *PhD Thesis*, Eindhoven University of Technology, **2005**.
52 Staal, B. B. P., Characterization of (co)polymers by MALDI-TOF-MS kinetics. *PhD Thesis*, Eindhoven University of Technology, **2005**.
53 Lee, H.-I., Matyjaszewski, K., Yu, S., Sheiko, S. S., *Macromolecules* **2005**, *38*, 8264.
54 Matyjaszewski, K., Ziegler, M. J., Arehart, S. V., Greszta, D., Pakula, T., *J. Phys. Org. Chem.* **2000**, *13*, 775.

4
Multisegmental Block/Graft Copolymers

Constantinos Tsitsilianis

4.1
Introduction

Macromolecular engineering is an integrated chemical process aimed at designing polymeric materials for specific advanced applications. In order to achieve this goal, tailor-made block copolymers with a specific macromolecular architecture, chemical composition/functionality and low molecular polydispersity and heterogeneity have to be synthesized and thoroughly characterized in a first step. The establishment of structure–property relationships for the spontaneously macromolecular self-assemblies in specific environments (e.g. solution, interfaces, bulk) is the following step that will allow a rational retro design of the macromolecular characteristics of copolymers so as finally to obtain nanostructured polymeric materials with tailor-made macroscopic properties suitable for a specific function. For the successful outcome of the above procedure, special synthetic techniques, the so-called controlled/living polymerization methods, have been developed.

The most important achievement of the living/controlled polymerization methods is the synthesis of block copolymers that in most of the cases lead to nanostructured polymeric materials with potential application in nanotechnology [1] and biomedicine [2].

The simplest and most widely studied type of block copolymer is the AB diblock where only two building blocks of different nature are joined together by a single covalent bond. Soon after the establishment of "living" anionic polymerization by Szwarc et al. [3], a new ABA triblock architecture appeared. This polymeric structure led to very interesting polymeric materials, the so-called thermoplastic elastomers, if a relative long elastomer such as polyisoprene is endcapped by shorter glassy polystyrene blocks. From this architecture the concept of *topology* (i.e. the position of the elastic building block in the copolymer) was introduced and designated a key factor for the design of block copolymers with specific macroscopic properties.

Since then, novel macromolecular architectures constituted of more than two building blocks in linear and branched modes, such as ABABA pentablocks,

Macromolecular Engineering. Precise Synthesis, Materials Properties, Applications.
Edited by K. Matyjaszewski, Y. Gnanou, and L. Leibler

ISBN: 978-3-527-31446-1

ABC terpolymers and/or ABCD quarterpolymers, star-shaped and graft copolymers were prepared and studied in solution and in the bulk. All this scientific evolution demonstrated the great importance of macromolecular engineering and motivated polymer chemists to develop new synthetic controlled polymerization methods such as atom transfer radical polymerization (ATRP) [4], nitroxide-mediated [5], group transfer polymerization (GTP) [6], reversible addition fragmentation chain transfer (RAFT) [7] and ring openning metathesis polymerization (ROMP) [8], targeted at creating block copolymers comprising any type of monomer and prepared under the less demanding polymerization conditions.

In this chapter, the synthetic strategies based on controlled/living polymerization techniques that lead to well-defined multisegmental (more than two building blocks) linear ABA and (AB)*n* block copolymers, ABC and ABCBA terpolymers, ABCD quarterpolymers and graft copolymers of various types are demonstrated. Other multisegmental star-shaped block copolymers/terpolymers and even more complicated architectures, such as polymeric brushes and dendrimers, are presented in other chapters. Due to space limitations, only the basic synthetic routes and selected examples can be presented here. Moreover, it is attempted to include new developments on multisegmental copolymers exhibiting sophisticated properties that lead to "smart" polymeric materials.

4.2 Linear Multisegmental Block Copolymers

4.2.1 ABA Triblock Copolymers

The simplest multisegmental block copolymer architecture is the linear ABA triblock one, consisting of a central block B covalently bonded at its both ends by different A blocks. By choosing the appropriate chemical nature of the monomer pair (e.g. soft–hard segments, hydrophobic–hydrophilic) and the right topology (the choice of the central block is crucial), very interesting polymeric materials can be fabricated, e.g. thermoplastic elastomers and physical gels, the properties of which cannot be exhibited by the diblock counterparts. It is obvious that the principles of block copolymerization (for instance the initiation of the second monomer by the active end of the first block, must proceed fast and quantitatively) as established for the diblock copolymers, have to be followed to achieve well-defined multisegmental block copolymers.

Depending on the nature of monomers and irrespective of the chosen polymerization technique, several synthetic strategies have been developed, each exhibiting advantages and limitations.

4.2.1.1 Synthetic Strategies

The most straightforward and widely explored method so far is the use of a difunctional initiator (DI). The middle block A is made first, bearing at both ends active sites capable of initiating the polymerization of the second monomer B, which is added sequentially to the reaction medium after the consumption of the first monomer (Fig. 4.1 a). The advantage of this method is that it can be performed in a one-pot procedure. Moreover, possible contamination of the final product with A homopolymer is negligible and easily removable by selective extraction.

In cases where the outer monomer B cannot polymerize the monomer A and has to constitute the central block, a coupling agent (CA) has to be used. Reaction proceeds through a monofunctional initiator and sequential addition of the monomers A and B. After the consumption of the second monomer, the living diblock copolymer is reacted in stoichiometric amount with a difunctional CA (Fig. 4.1 b). To achieve a high coupling yield, a small excess of the living diblock is required. In this case, residual AB diblock remains in the final product and can be removed by fractionation. The latter is the major drawback of this route.

In a third route, sequential addition of the monomers by using a monofunctional initiator (Fig. 4.1 c) could be followed only if the electron affinity (reactivity) of both monomers is almost the same. In other words, the macroinitiator arising from the monomer A can polymerize the monomer B and *vice versa*. This situation is rare and only a small number of monomer pairs fulfill this requirement. Styrene/dienes in anionic polymerization and different methacrylates in GTP are characteristic examples. The sequential monomer addition (SMA) route allows, on the other hand, the synthesis of asymmetric ABA triblock copolymers (i.e. outer blocks with different degrees of polymerization).

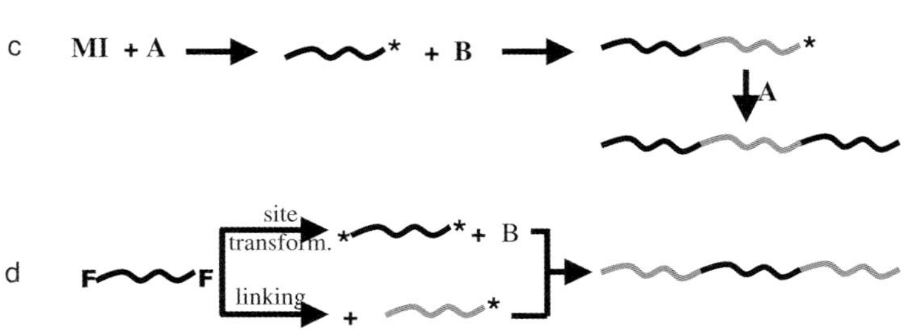

Fig. 4.1 Schematic representation of synthetic strategies towards ABA triblock copolymers: (a) by difunctional initiator (DI); (b) using a coupling agent (CA); (c) sequential monomer addition (SMA); (d) through α,ω-functional polymer precursor (FPP). MI, monofunctional initiator; DI, difunctional initiator; F, functional end-group; *, active site.

Finally, ABA triblock copolymers can be prepared by involving an α,ω-functional polymer precursor (FPP). This procedure is needed when the two monomers cannot be polymerized by the same polymerization technique and have to be carried out in a two-pot reaction (Fig. 4.1 d). Two sub-routes could be distinguished: (a) the generation of new sites suitable for the polymerization of the incoming monomer B, the so-called site transformation technique [9], and (b) the use of the end functions to react with the active sites of a living and/or functionalized B block in a stoichiometric ratio.

4.2.1.2 **Synthesis of ABA by Anionic Polymerization**

Anionic living polymerization (ALP) is the oldest polymerization method suitable for macromolecular engineering and has inspired polymer chemists to develop new controlled polymerization techniques that are carried out with different mechanisms. Anionic polymerization proceeds via organometallic sites (carbanions or oxyanions) through nucleophilic reactions in aprotic media. After the wide use of protective monomers and post-polymerization chemical modifications, the versatility and potential of this method for obtaining novel model block copolymers have been significantly expanded.

Thermoplastic elastomers are one of the most useful achievements afforded by model ABA block copolymers synthesized by ALP. Polystyrene-*b*-polybutadiene-*b*-polystyrene (PS–PBd–PS) and/or polyisoprene (PI) in the middle have been synthesized using difunctional initiators in non-polar media to ensure a high 1,4-diene microstructure necessary for the elastomeric properties (Scheme 4.1) [10]. Diadducts of *sec*-butyllithium (*s*BuLi) with diisopropenylbenzene or 1,3-bis(1-phenyl)benzene (PEB) have been used as DI. The latter is most advantageous since propagation of PEB during the preparation of the initiator is prevented.

Other synthetic routes have also been studied to prepare PS-PI-PS using, for example, $(CH_3)_2SiCl_2$ as a chlorosilane CA [11]. Another type of thermoplastic elastomer constituted of poly(dimethylsiloxane) (PDMS) central block and crystallizable polyethylene (PE) outer blocks has been synthesized by the chlorosilane CA route and hydrogenation of PBd outer blocks of the PBd–PDMS–PBd triblock precursor (Scheme 4.2) [12].

Combination of various methacrylates with low and high T_g building blocks in an ABA architecture has given interesting thermoplastic elastomers exhibit-

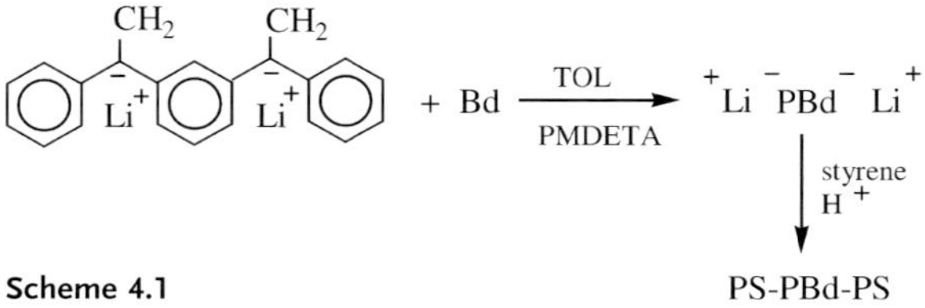

Scheme 4.1

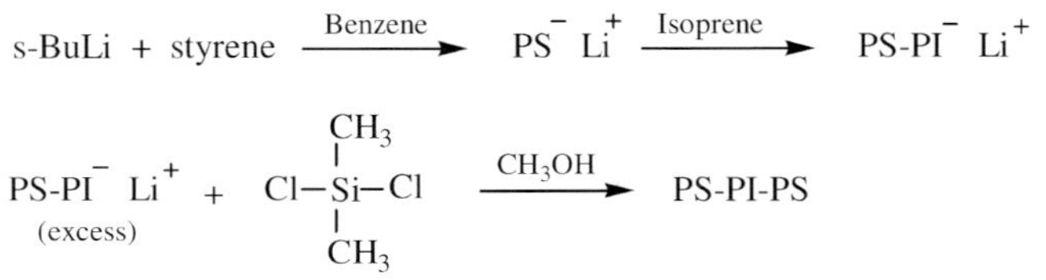

Scheme 4.2

ing better stability against oxidation than the dienic analogues. A representative type is PMMA–PiOcA–PMMA, which was synthesized by the SMA route using sBuLi–diphenylethylene diadduct and two-step addition of *tert*-butyl acrylate (*t*BA) and methyl methacrylate (MMA). The final product, with a low T_g poly-(isooctyl acrylate) (PiOcA) central block, resulted from post-polymerization transalcoholysis at 150 °C in the presence of *p*-toluenesulfonic acid according to Scheme 4.3 [13].

Amphiphilic ABA triblock copolymers, with very interesting rheological properties useful in water-derived formulations, have been prepared by ALP and explored in aqueous media. A well-known example is the so-called telechelic polyelectrolytes that form, through hydrophobic association, reversible networks in water, in a similar manner with phase separation of thermoplastic elastomers in the bulk. Poly(acrylic acid) (PAA) end-capped by PS short blocks was prepared by the CA route using *t*BA and post-polymerization hydrolysis to obtain the desired PAA central block (Scheme 4.4) [14].

The reaction was carried out in THF in the presence of LiCl to ensure control of *t*BA polymerization [15]. Terephthaldicarboxaldehyde (dialdehyde) was used as a CA yielding more than 95% coupling. Due to the weak polyelectrolyte character of the PAA bridging chains, the rheological properties of the formed physical gel can be tuned by adjusting the pH. The same route has been applied for the synthesis of a similar PS–PMAA–PS (MAA = methacrylic acid) telechelic polyelectrolyte using 1,4-di(bromomethyl)benzene as CA [16].

A reverse triblock copolymer with the hydrophobic long PS as the central block and with short poly[5-(*N*,*N*,*N*-diethylmethylammonium)isoprene] (PAI)

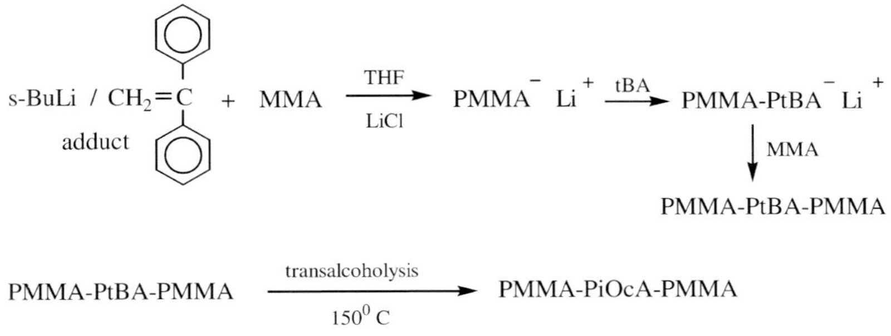

Scheme 4.3

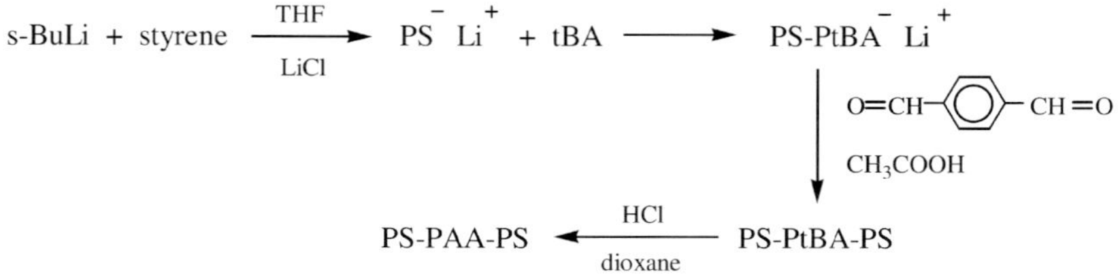

Scheme 4.4

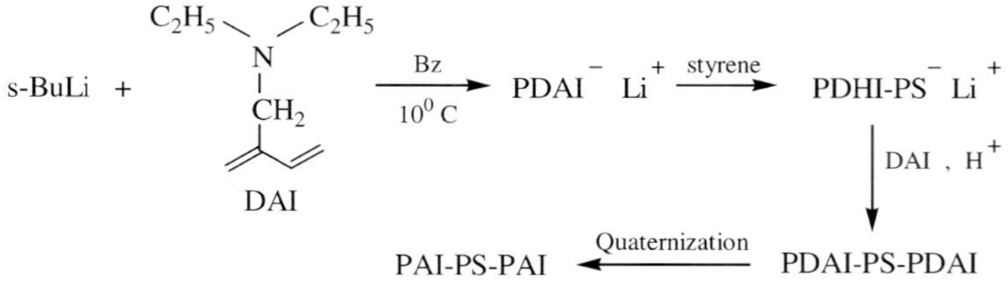

Scheme 4.5

positively charged outer blocks (PAI–PS–PAI) has been synthesized recently following the SMA route (Scheme 4.5).

*s*BuLi initiated the polymerization of 5-(*N,N*-diethylamino)isoprene (DAI) in benzene (Bz) at 10 °C followed by sequential addition of styrene (S) and DAI (equal amount with the first DAI addition). Quaternization of the outer blocks by dimethyl sulfate yielded positively charged PAI blocks [17]. This highly asymmetric amphiphilic ABA triblock copolymer exhibits a unique bowl-shaped micellar morphology upon self-association in THF.

Non-ionic amphiphilic triblocks with poly(ethylene oxide) (PEO) as the hydrophilic part and PS as the hydrophobic part have been synthesized with PEO either in the middle or in the outer block topology [18]. These copolymers have been studied also in the bulk since they were constituted of glassy and crystallizable segments. Recent interest in the case of PS–PEO–PS was to investigate polymer crystallinity in confinement geometries [19].

Recently, an ionizable double hydrophilic triblock copolymer comprising poly(2-vinylpyridine) (P2VP) and PAA was synthesized by the DI route and post-polymerization hydrolysis as described in Scheme 4.6 [20].

The water-soluble PAA–P2VP–PAA triblock copolymer may exist in different states due to protonation–deprotonation equilibrium tuned by adjusting the pH of the solution. At high pH it behaves as an amphiphile, at intermediate pH as a segmented polyampholyte, where positively charged protonated P2VP segments coexist with negatively charged PAA segments, and finally at low pH as a cationic polyelectrolyte capped with AA moieties capable of developing H-bonding. This unique multifunctionality allows it to self-assemble through different interactions (electrostatic or hydrophobic), forming either physical networks (pH 3.5) or compact micelles (pH 8).

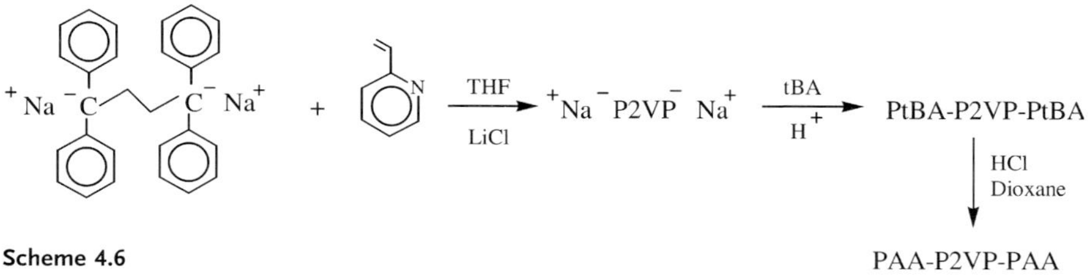

Scheme 4.6

4.2.1.3 **Synthesis of ABA by Group Transfer Polymerization (GTP)**

GTP is a controlled living polymerization method suitable for the preparation of block copolymers composed of meth(acrylic) repeating units. It has been applied on an industrial scale (DuPont) for the fabrication of amphiphilic block copolymer dispersing agents for pigmented water-based inks. According to recent considerations, GTP seems to proceed through a dissociative (anionic) mechanism (proposed by Quirk and Kim [21]) in the presence of nucleophilic anions (catalysts) [22]. One of the advantages of GTP versus classical anionic polymerization apart from the experimental conditions (polymerization at room temperature) consists of the easy block copolymerization by sequential addition of monomers irrespective of the order of monomer addition. Therefore, ABA triblock copolymers can be synthesized by either the DI or SMA route (see Fig. 4.1).

Poly(benzyl methacrylate)-*b*-poly(lauryl methacrylate)-*b*-poly(benzyl methacrylate) (PBzMA–PLMA–PBzMA) was synthesized using 1,5-bis(trimethylsiloxy)-1,5-dimethoxy-2,4-dimethyl-1,4-pentadiene (BDDB) initiator and tetrabutylammonium dibenzoate (TBADB) catalyst following the DI route (Scheme 4.7) [23].

Cationic telechelic polyelectrolytes bearing poly[2-(dimethylamino)ethyl methacrylate] (PDMAEMA) as the amphiphilic central block and PMMA hydrophobic outer blocks were prepared through the same synthetic route and explored in aqueous media as a function of pH [24]. Due to protonation of the tertiary amine groups of the PDMAEMA block at low pH, the copolymer self-associates, forming a stiff physical gel at relative low concentration (1 wt%), functioning as a low-cost, very efficient thickener useful for aqueous formulations.

4.2.1.4 **Synthesis of ABA by Cationic Living Polymerization**

Tailor-made block copolymers constituted of important monomeric units that cannot be polymerized by other methods such as isobutene (IB) and alkyl vinyl

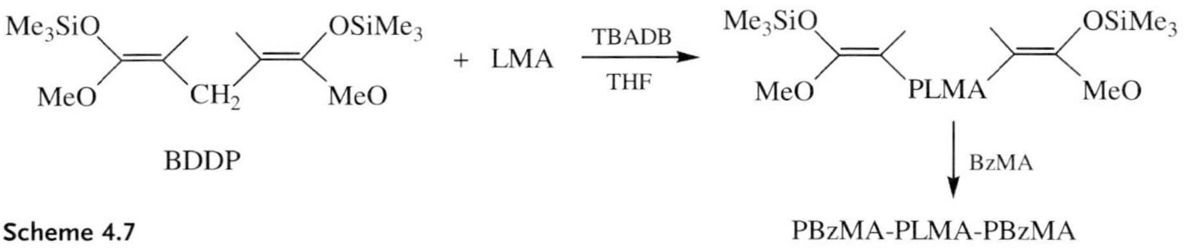

Scheme 4.7

ethers (VE) can be afforded by cationic living polymerization (CLP), thereby broadening the potential of macromolecular engineering [25]. Cationic polymerization proceeds through carbanion and/or oxonium sites in a controlled/living mode if appropriate conditions have been chosen.

Important polymeric materials such as thermoplastic elastomers based on PIB, associative amphiphiles based on PVE and telechelic ionomers have been prepared by CLP. Several ABA triblock copolymers have been synthesized by the DI method. A typical example is the synthesis of poly(*tert*-butylstyrene)-*b*-polyisobutene-*b*-poly(*tert*-butyl styrene) (P*t*BuSt–PIB–P*t*BuSt) that exhibits thermoplastic elastomer properties [26]. *p*-Dicumyl methyl ether (DicumOMe)–$TiCl_4$ was used as a bifunctional initiating system that polymerized sequentially IB and P*t*BuSt in the presence of 2,6-di-*tert*-butylpyridine (D*t*BP) to prevent side-reactions (Scheme 4.8).

Ionomer-like ABA triblocks with good elastomeric properties constituted of poly(tetrahydrofuran) (PTHE) middle block end-capping with polyethylenimine (LPEI) short blocks is another example involving ring-opening oxetane polymerization by CLP. The synthetic strategy involves polymerization of THF by trifluoromethanesulfonic anhydride difunctional initiator followed by the addition of 2-methyl-2-oxazoline in a different solvent and post-polymerization hydrolysis with concentrated HCl. Complexation of the LPEI block ends by $CuCl_2$ led to a physical network structure behaving as a tough thermoplastic elastomer [27].

The CA route has also been used for the preparation of ABA triblocks. PS–PIB–PS synthesis has been conducted, using bis-diphenylethylene as coupling agent, according to Scheme 4.9.

It should be mentioned that IB must be added at ≤95% St conversion to obtain living PS–PIB diblocks with negligible PS homopolymer contamination.

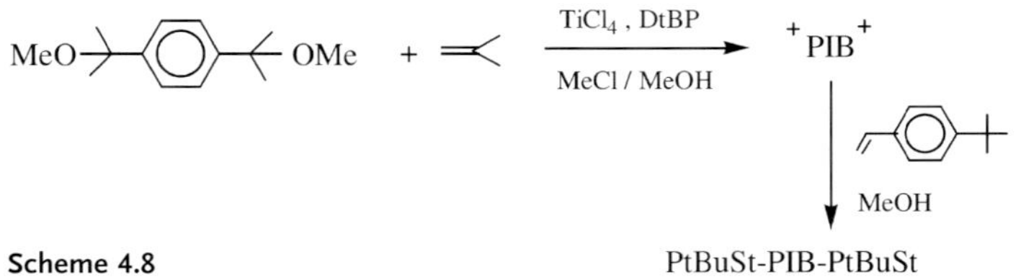

Scheme 4.8

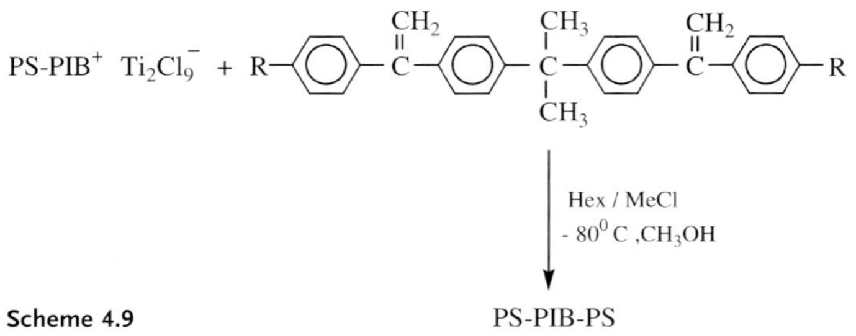

Scheme 4.9

Due to residual PS–PIB diblock in the final product, the mechanical properties were inferior to those of the best triblocks made by the DI route [28]. On the other hand, PαMeSt–PIB–PαMeSt was synthesized successfully since *in situ* coupling of the living diblocks with BDPEP was rapid and nearly quantitative. Note that the DI route for the preparation of this copolymer cannot be followed due to unfavorable crossover from living PIB to αMeSt [29].

4.2.1.5 Synthesis of ABA by Controlled Free Radical Polymerization

After the advent of controlled free radical polymerization techniques, new horizons have been opened towards macromolecular engineering. ATRP, RAFT and 2,2,6,6-tetramethylpiperidinoxy (TEMPO)-mediated approaches are nowadays very promising controlled polymerization mechanisms that can be utilized to prepared tailor-made macromolecules with very important monomers [e.g. *N*-isopropylacrylamide (NIPAAm)] that could not be achieved before.

ABA triblock architectures have been developed using mono- or difunctional initiators and sequential addition of the monomers. Early examples were the synthesis of symmetric PMMA–PBMA–PMMA and PBMA–PMMA–PBMA by the SMA route [30]. Amphiphilic copolymers constituted of PMMA and PDMAEMA that can both be located as the middle block were synthesized through the FPP route (Fig. 4.1c) yielding symmetric triblocks (Scheme 4.10) [31].

The polymerization of MMA was conducted first, using 1,2-bis(2-bromopropionyloxy)ethane as initiator in the presence of CuBr and a bidentate ligand. In the second step, the difunctional Br–PMMA–Br macroinitiator was utilized to polymerize DMAEMA in the presence of CuBr–HMTETA.

Recently, ABA triblocks capable of forming biocompatible responsive physical gels have been synthesized by ATRP. A two-step FPP route was followed using diethyl *meso*-2,5-dibromoadipate (DEDBA) as initiator. In the first step an α,ω-difunctional macroinitiator of 2-methacryloyloxyethylphosphorylcholine (MPC) was prepared in the presence of CuBr–2bpy catalyst. The Br–PMPC–Br obtained after isolation and purification was used in the second step to polymerize NIPAAm using CuBr–Me_4cyclam yielding the triblock PNIPAAm–PMPC–PNIPAAm (Scheme 4.11) [32].

The fact that the reaction was carried out at room-temperature is very advantageous. The phosphorylcholine motif is an important component of cell membranes and polymers based on this component could offer materials for poten-

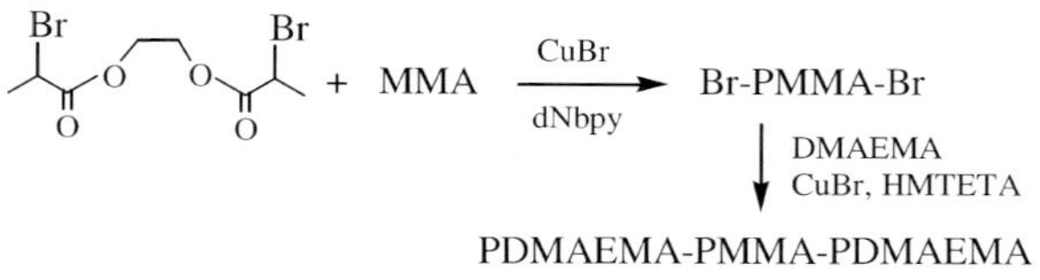

Scheme 4.10

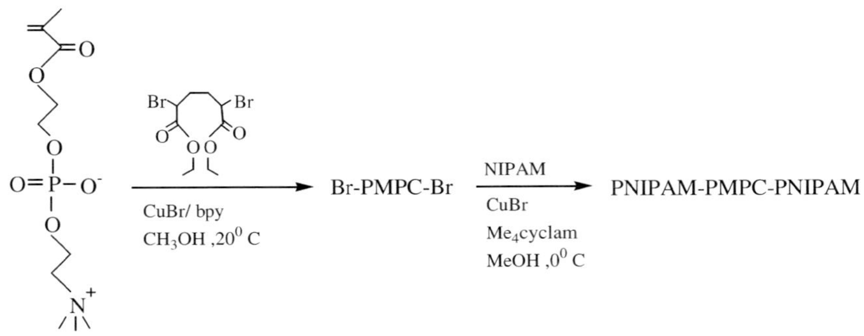

Scheme 4.11

tial applications in biomedicines. The double amphiphilic PNIPAAm–PMPC–PNIPAAm block copolymer exhibits interesting thermosensitive properties due to the PNIPAAm building blocks. At low temperatures the copolymer is molecularly dissolved whereas above the lower critical solution temperature (LCST) of PNIPAAm it is self-assembled forming a physical gel.

Another attempt to incorporate PNIPAAm segments to the ABA architecture has been made recently through the RAFT mechanism [33]. PDMA–PNIPAAm–PDMA (DMA = *N,N*-dimethylacrylamide) copolymers were synthesized at room temperature in water using a novel water-soluble trithiocarbonate RAFT agent. These conditions with ecological impact are very advantageous and promising from a technological point of view. 4,4′-Azobis[2-(imidazolin)-2-yl)propane] dihydrochloride (VA-044) was employed as the radical source and 2-(1-carboxy-1-methylethylsulfanyltrithiocarbonylsulfanyl)-2-methylpropionic acid (CMP) as the RAFT CTA. Thermosensitive reversible micelles are obtained at temperatures above the LCST of PNIPAAm by hydrophobic association, showing that these triblock are promising candidates for drug delivery applications.

Finally, nitroxide-mediated controlled radical polymerization has been used for ABA triblock preparation using the SMA route. Benzoyl peroxide (BPO) as

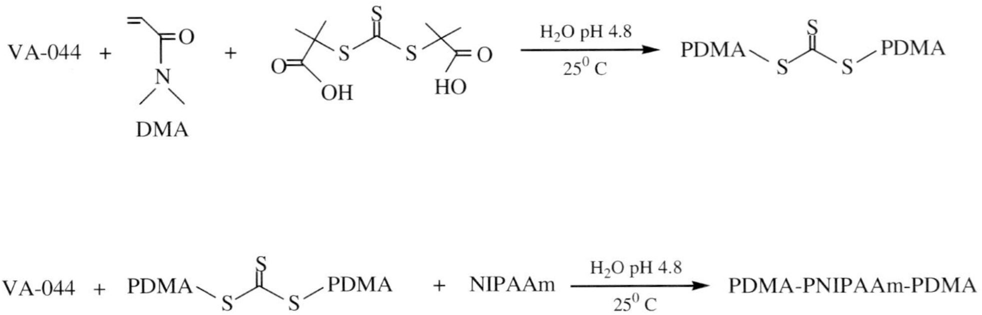

Scheme 4.12

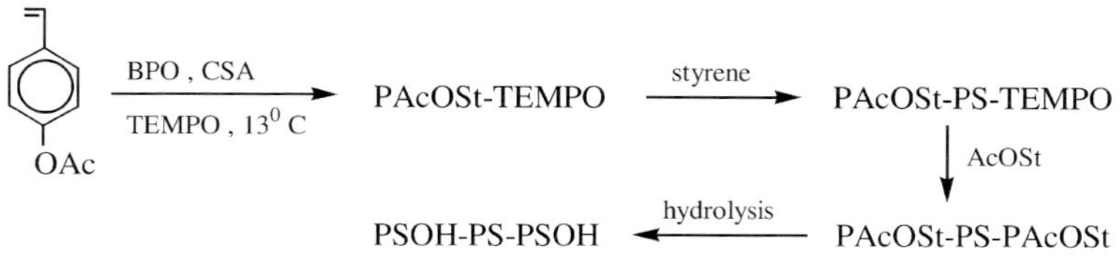

Scheme 4.13

initiator, TEMPO as the nitroxide stabilizer and camphorsulfonic acid as the accelerator were used to polymerize sequentially 4-acetoxystyrene, styrene and again 4-acetoxystyrene. The PAcOSt–PS–PAcOSt obtained can be subjected to hydrolysis to yield the corresponding amphiphilic copolymer (Scheme 4.13) [34].

4.2.1.6 **Synthesis of ABA by Combination of Methods**

In attempts to design block copolymers having specific properties for targeting applications, often the monomers involved cannot be polymerized by the same controlled/living polymerization mechanism. To overcome this difficulty, specific synthetic strategies named site-transformation techniques have been developed. For the case of ABA triblock synthesis, the FPP pathway (Fig. 4.1) is used by transforming the functional end-groups of the A polymer precursor to suitable active sites capable of initiating the polymerization of the monomer B (macroinitiator approach).

α,ω-Polyethylene oxide bearing OH terminal groups and prepared by anionic polymerization can be transformed to ATRP macroinitiator by a prior polymerization reaction according to Scheme 4.14. The conversion of OH groups was performed by reaction with 2-chloropropionyl chloride in methylene dichloride (CH_2Cl_2) in the presence of 4-(dimethylamino)pyridine (DMAP) and triethylamine (TEA). This ATRP macroinitiator was used in a subsequent step for the polymerization of styrene yielding PS–PEO–PS triblock copolymer [35].

Biocompatible and biodegradable amphiphilic triblock copolymers bearing PEO in the middle end-capped at both ends by poly(ε-benzyloxycarbonyl-L-lysine) (PLLZ) were synthesized by anion ring-opening polymerization of the *N*-carboxyanhydride of ε-benzyloxycarbonyl-L-lysine [L-Lys(Z)-NCA] using H_2N–PEO–NH_2 as initiator in DMF [36]. The resulting PLLZ–PEO–PLLZ triblock copolymer exhibits amphiphilic character and forms aggregates with interesting

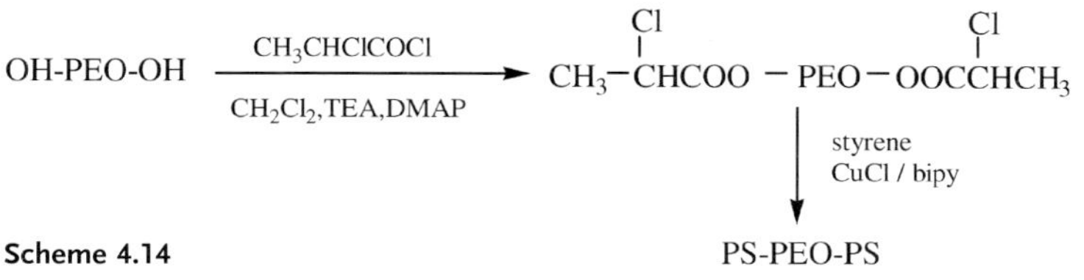

Scheme 4.14

morphologies that is attributed to the ability of the PLLZ segments to form different secondary structures such as α-helix and β-sheet. The above synthetic strategy is a good way to incorporate polypeptide segments on block copolymers with potential applications in biomedicines.

4.2.2 (AB)*n* Linear Multiblock Copolymers

Linear multiblock copolymers comprising more than three A and B sequential segments of the same degree of polymerization have been synthesized following two routes: sequential addition of monomers, which have to exhibit similar reactivity, and condensation polymerization of the α,ω-functionalized A and B building blocks. The former method leads to well-defined $(AB)_n$ multiblock copolymers whereas in the latter the molecular weight distribution and therefore the total number of blocks cannot be controlled.

Representative species of this class of materials, composed of perfectly alternating polystyrene and polyisoprene segments, $(S–I)_n$, were synthesized by living anionic polymerization using *sec*-butyllithium as the initiator and sequential addition of styrene and isoprene in cyclohexane at 60 °C. Exploring the effect of the architecture on their microstructural characteristics and bulk properties, it was observed that increasing n generally promoted reductions in both the lamellar period and upper (styrenic) glass-transition temperature, but noticeable increases in tensile modulus and yield strength. These observed trends are more pronounced in the copolymer series with constant chain length due to the coupled relationship between n and the molecular weight of the blocks [37].

4.2.3 ABC Triblock Terpolymers

Linear triblock terpolymers consisting of three segments of different nature, each conferring on the polymer different functions, are of increasing interest due to their ability to form a rich variety of novel functional nanostructures with potential applications in nanotechnology. The large morphological diversity, observed either in solution or in the bulk by the ABC architecture, is due to the large number of variables involved governing the phase separation/self-assembly of these novel materials, i.e. three interaction parameters in bulk (χ_{AB}, χ_{AC}, χ_{BC}) or six in solution (plus χ_{AS}, χ_{BS}, χ_{CS}) as well as composition (f_A, f_B, f_C) and topology (ABC, ACB, BAC). The synthesis of ABC terpolymers was initiated in the early 1970s using ALP, but in recent years an impressive array of new ABCs have appeared, reflecting the new possibilities of macromolecular engineering after the important development of controlled/living polymerization techniques and the new challenges that the ABCs offer towards potential advanced applications [38].

4.2.3.1 Synthetic Strategies

It is obvious that the synthesis of ABC can be achieved straightforwardly if all the monomers can be polymerized sequentially under the same polymerization mechanism (SMA) (Fig. 4.2 a). However, the need to prepare ABCs with specific segment combinations and topology has led to the development of other synthetic pathways (Fig. 4.2). Therefore, depending on the nature of the monomer (which defines the mechanism) and the topology of the blocks (ABC versus ACB versus BAC), three more strategies can be designed assuming that the order of monomer addition is A+B+C.

The end-function of an ω-functionalized polymer precursor is transformed to an active site capable of initiating the polymerization of monomer B. After the consumption of B the monomer C is added (Fig. 4.2 b). This method, termed the macroinitiator approach, can incorporate two different polymerization mechanisms, for instance OH-terminated PEO synthesized by ALP can be transformed to an ATRP macroinitiator (Scheme 4.14).

If the C monomers are desired to be in the middle segment, an ω-functionalized diblock copolymer, prepared by sequential addition of the monomers B and C, has to be prepared in a first step. Two sub-routes could be followed: either using the diblock precursor as macroinitiator by a site transformation reaction to polymerize monomer A or using the end-function for linking a living polymer A (Fig. 4.2 c). Again, two different polymerization mechanisms can be involved.

Macromonomers can also be used. The double bond can react with an organometallic compound to create an active site that will be used for the sequential polymerization of monomers A and B. This route is an alternative to the macroinitiator approach if monomer A cannot polymerize monomer C.

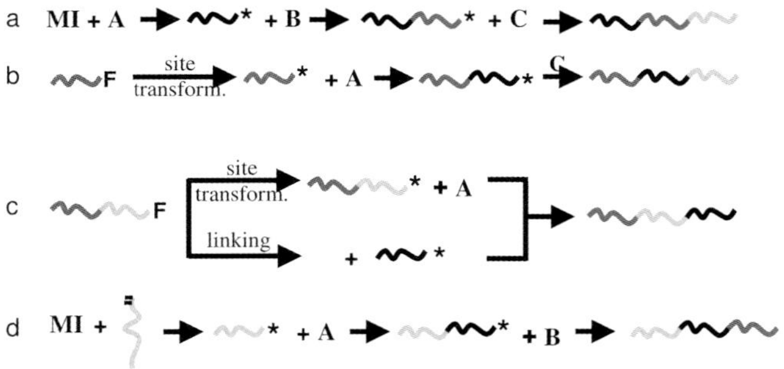

Fig. 4.2 Schematic representation of synthetic strategies towards ABC triblock terpolymers: (a) sequential monomer addition (SMA); (b) by macroinitiator (macro-I); (c) by ω-functional diblock precursor (FDP); (c) through macromonomer (MM). MI, monofunctional initiator; F, functional end-group; *, active site.

4.2.3.2 **Synthesis of ABC by Anionic Polymerization**

A large variety of ABC block terpolymers have been synthesized by ALP using in most cases the SMA pathway and has been described in a recent book [39]. Here only some selective examples concerning the various routes and recent results dealing with important solution properties will be presented.

PS–P2VP–PBd triblock copolymer cannot be synthesized straightforwardly by SMA. This problem was overcome by applying the FDB method in a two-step anionic polymerization procedure. In the first step, an ω-functionalized PS–P2VP–CH_2–Cl diblock was synthesized by SMA using *s*BuLi as initiator and a large excess of *p*-xylene dichloride to terminate the reaction. In such cases where difunctional reagents are used, a large excess is needed to prevent coupling. In the second step, living butadiene capped by 1,1-diphenylethylene (DPE) was deactivated by the PS–P2VP–CH_2–Cl (Scheme 4.15) [40]. DPE served to lower the nucleophilicity of the active sites in order to avoid attack on the pyridine rings.

The same strategy was followed for the preparation of PS–P2VP–PEO but through the macroinitiator sub-route. In the first step, OH ω-functionalized PS–P2VP was synthesized by capping the 2VP active sites with ethylene oxide (EO). Since Li was used as the counteranion (from *s*BuLi initiator), propagation of EO is prevented and excess EO can be used to ensure quantitative functionalization. In the second step, potassium alkoxides were generated by reacting potassium naphthalene with the OH functions of PS–P2VP–OH. This macroinitiator polymerized EO to yield the final product (Scheme 4.16) [41].

This triblock, bearing two different hydrophilic segments, self-assembles in water forming three-layer micelles comprising a PS hydrophobic core and a pH-sensitive P2VP inner layer surrounded by a PEO-soluble corona.

Water-soluble double hydrophilic ABC terpolymers constituted of two ionogenic segments, that can be charged positively (P2VP) or negatively (PAA) depending on pH, and a hydrophobic P*n*BMA were synthesized by SMA and post-polymerization reaction according to Scheme 4.17. The central P*t*BA block was

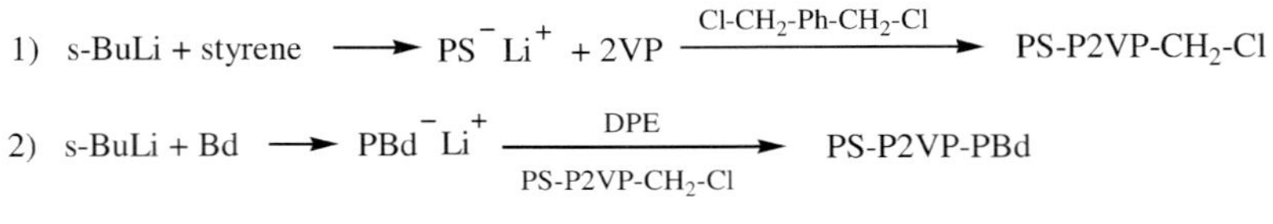

Scheme 4.15

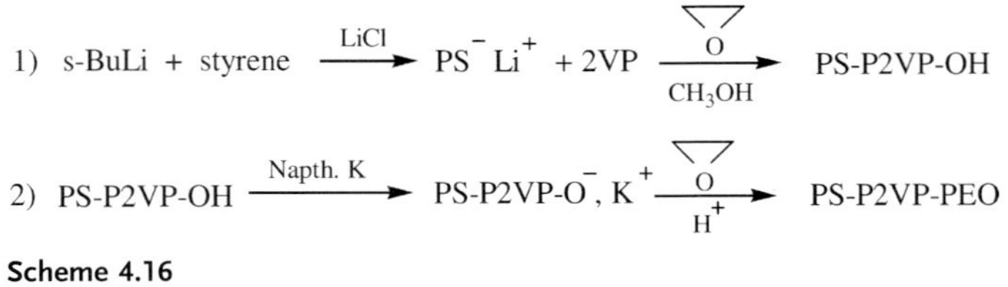

Scheme 4.16

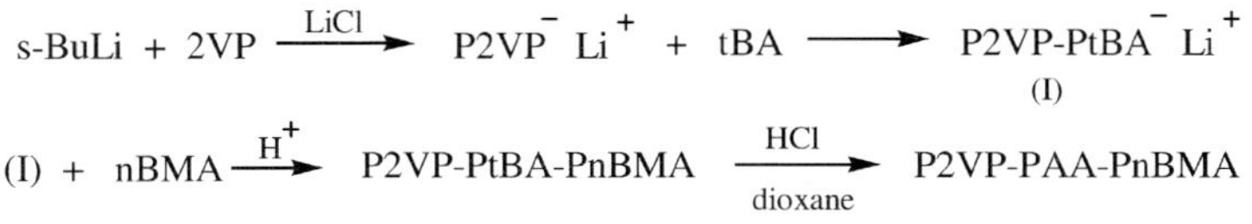

Scheme 4.17

selectively hydrolyzed by HCl in dioxane at 85 °C. ^{1}H NMR inspection showed that the P*n*BMA segment remains intact under the reaction conditions chosen.

A rich polymorphism was observed for the P2VP–PAA–P*n*BMA aqueous solution as a function of pH. Among others it forms three-compartment centrosymmetric micelles at pH 1 and a three-dimensional network at pH 7, functioning as a potential carrier and/or as a gelator, respectively [42].

The topology of the hydrophobic block is critical in the double hydrophilic ABC terpolymers. Using (meth)acrylates, which have very similar reactivity, one can use the same route as above, just changing the sequence of the monomer addition. For instance, the synthesis of P2VP–PMMA–PAA can be accomplished by the SMA route using MMA as the second monomer (through *t*BA precursor). This polymer now forms micelles with the PMMA core and two different chains (P2VP, PAA) in the corona. These "living" self-assemblies, called hetero-arm star-like micelles, are sensitive to pH changes and can further associate in a second level of hierarchy [43].

Another PAA–PS–P4VP ABC triblock bearing segments with similar functionalities as in the previous cases but with a long PS in the middle was synthesized according to the macromonomer pathway. A P*t*BA macromonomer with a styryl unsaturation was first prepared. In the second step, *α*-methylstyryllithium reacted with the macromonomer, generating an active site that initiated the polymerization of styrene and subsequently of 4-vinylpyridine (Scheme 4.18) [44]. Acidic hydrolysis yielded the final double hydrophilic terpolymer, which forms vesicular nanostructures that can interchange the outside and inside soluble chains (P2VP or PAA) by changing the pH of the solution [45].

Sequential addition of monomers was also used to prepare ABC terpolymers with 5-(*N*,*N*-dimethylamino)isoprene, styrene and *tert*-butyl methacrylate seg-

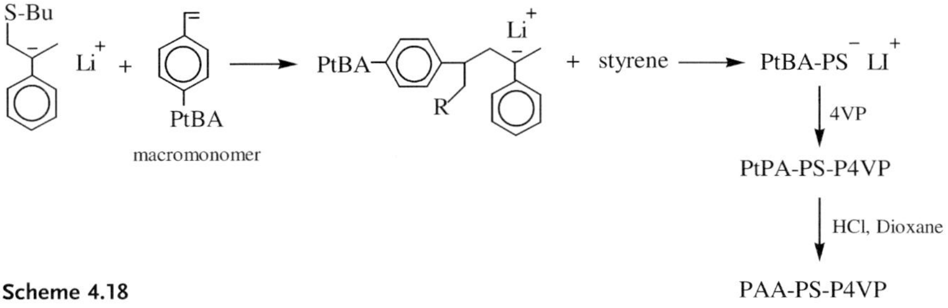

Scheme 4.18

ments using *s*BuLi in THF. Hydrolysis of the third block yielded the final PDMI–PS–PMAA [46].

An interesting monohydrophilic ABC terpolymer with highly incompatible hydrophobic segments was designed and explored in aqueous media. PBd–PS–PEO was prepared in several steps by the FDC route. An HO–PS–PBd *α*-functionalized diblock precursor was prepared first using 3-(*tert*-butyldimethylsiloxy) = 1-proxylithium as initiator. By deprotection of the TBDMS group and reacting the resulted OH function with potassium naphthalene, the diblock was transformed into a macroinitiator suitable for the polymerization of oxirane. Finally, the PBd block was modified to a fluorinated segment PF by reacting the pendant double bonds with *n*-perfluorohexyl iodide. The resulted ABC triblock forms multicompartment micelles in water with well-segregated PS and PF compartments in the core and PEO in the corona [47].

The ABC triblock terpolymer is the simplest, best suited architecture for the stabilization of self-assemblies, which leads to structured nano-objects with potential application in nanotechnology. This can be achieved when one of the three segments affords cross-linkable moieties. A good paradigm is the fabrication of organic nanotubes made by poly(isoprene-*b*-cinnamoylethyl methacrylate-*b*-*tert* butyl acrylate) (PI–PCEMA–P*t*BA) terpolymers that form three-compartment centrosymmetric cylindrical micelles in methanol. These micelles were stabilized by photo-cross-linking of the PCEMA intermediated shell and subsequently they were subjected to ozonolysis, which decomposed the PI core yielding nanotubes. The central block was made from 2-trimethylsiloxyethyl methacrylate [P(HEMA–TMS)], which was modified by deprotection and reaction with cinnamoyl chloride (Scheme 4.20) [48].

Shell-cross-linked micelles are another representative example of nano-objects. PEO–PDMAEMA–PMEMA (MEMA = morpholinoethyl methacrylate) was prepared by the MI approach and SMA of DMAEMA and MEMA. The macroinitiator was generated by reacting OH–PEO with the potassium salt of DMSO [49b]. These terpolymers self-assemble in aqueous media forming three-compartment micelles with a PMEMA hydrophobic core, a PDMAEMA inner layer and a PEO corona. The inner layer was cross-linked by a quaternization reaction of DMAEMA moieties [49a].

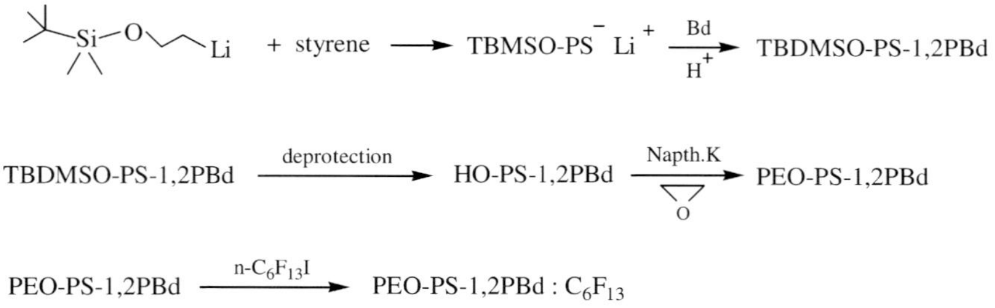

Scheme 4.19

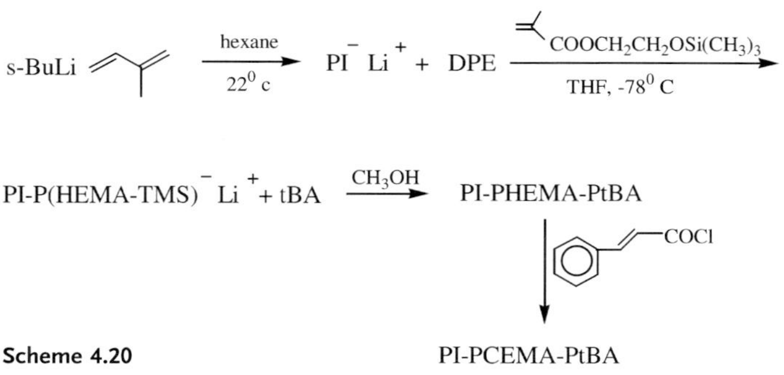

Scheme 4.20

Non-centrosymmetric compartmentalized micellar nanoparticles constituted of a cross-linked PBd core and PS and PMMA hemispheres named Janus micelles were prepared using PS–PBd–PMMA terpolymers synthesized by sequential anionic polymerization and cross-linking of the PBd spherical mesophase in the solid state. After solubilization and alkaline hydrolysis of PMMA to PMAA, amphiphilic Janus micelles were obtained that further self-assembled in water, forming supermicelles [50].

4.2.3.3 **Synthesis of ABC Terpolymers by GTP**

GTP is a versatile polymerization method for preparing ABC terpolymers using the SMA route. An advantage of the method is that the order of addition of the monomers is unimportant since methacrylic types are usually used. Therefore, all the topological isomers ABC, ACB and BAC can be made by just changing the order of monomer addition with the same initiator and conditions. A representative example is the synthesis of all isomer terpolymers with PMMA, PDDMAEMA and THPMA (THPMA = tetrahydropyranyl methacrylate) segments of about equal degree of polymerization. It was shown that the block topology determines the type of micelles formed in a selective solvent (onion or heteroarm star-like) and therefore their size [51].

4.2.3.4 **Synthesis of ABC Terpolymers by Cationic Polymerization**

ABC terpolymers bearing three different vinyl ether monomers have been synthesized by cationic living polymerization using the SMA pathway. A very interesting recent paradigm is the non-ionic triple hydrophilic poly(2-ethoxyethyl vinyl ether)-*b*-poly(2-methoxyethyl vinyl ether)-*b*-poly(2-ethoxy)ethoxyethyl vinyl ether) (PEOVE–PMOVE–PEOEOVE). It was synthesized by using 1-(isobutoxy)ethyl acetate in the presence of $Et_{1.5}AlCl_{1.5}$ in toluene–THF at 0 °C and adding EOVE, MOVE, EOEOVE sequentially.

The unique behavior of this copolymer is due to the different LCSTs of the building blocks, the middle of which exhibits the highest LCST (most hydrophilic block). A multi-stage association occurs in aqueous media upon heating, as the temperature passes through the LCSTs of the different segments. Three reversible sequential states were observed: a molecularly dissolved state, core–shell–corona micelles and a physical gel [52].

4.2.3.5 **Synthesis of ABC Terpolymers by Controlled Free Radical Polymerization**

Controlled free radical polymerization has also been employed to design ABC terpolymers, mainly through the ATRP and more recently the RAFT polymerization mechanisms. P*t*BA–PS–PMAA was synthesized exclusively by ATRP using a three-pot SMA route in which the second and third monomers were polymerized by the macroinitiator prepared in the previous steps. The macroinitiators were purified before use. In the first step, the reaction was performed by the 2-bromopropionate, CuBr–PMDETA initiating system and *t*BA yielding the P*t*BA–Br macroinitiator. In the second step, a P*t*BA–PS–Br diblock macroinitiator was generated which was used in the third step to give the final product according to Scheme 4.21 [53].

The macroinitiator approach is often used to prepare ABC terpolymers with PEO–Br macroinitiator. A nice example of its unique functionality is the preparation of PEO–PDEAEMA–PHEMA (HEMA = 2-hydroxyethyl methacrylate). The synthesis was performed in a one-pot reaction with sequential addition of DEAEMA and HEMA monomers (Scheme 4.22). In a second step, the PHEMA block was converted by esterification, using excess succinic anhydride in pyridine, yielding the final PEO–PDEAEMA–PSEMA. This triple hydrophilic block terpolymer exhibits a rich association behavior as a function of pH. It forms three types of micelles, the corona of which changes its nature from cationic protonated PDEAEMA (low pH) to natural PEO (intermediate pH) and to anionic neutralized PSEMA (high pH), simply by adjusting the solution pH [54].

RAFT polymerization has been employed recently in the synthesis of ABC terpolymers. Poly(vinylbenzyl chloride)-*b*-polystyrene-*b*-poly(pentafluorophenyl 4-vinylbenzyl ether) (PVBCl–PS–PVBFP) was synthesized by a three-step SMA route, using benzyl dithiobenzoate as CTA. The PVBCl block was quaternized by *N*-methylmorpholine yielding the cationic amphiphilic PVBM–PS–PVBFP terpolymer comprising two highly incompatible hydrophobic blocks. The inter-

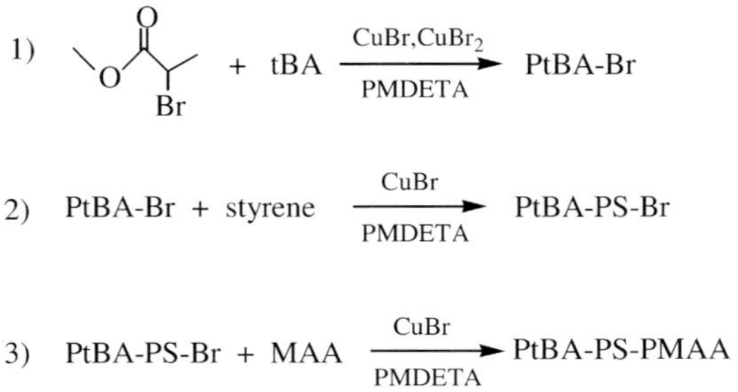

Scheme 4.21

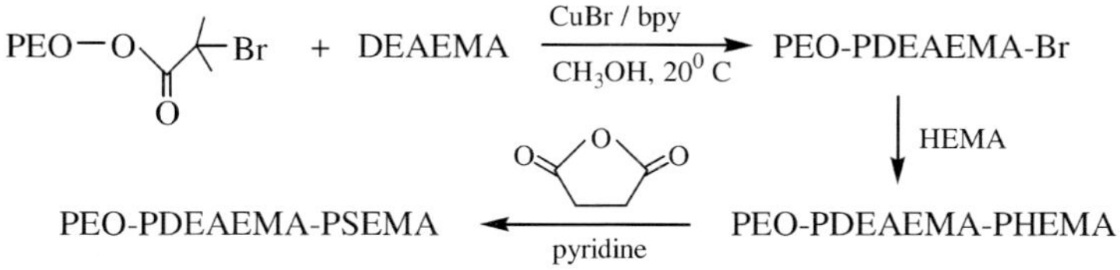

Scheme 4.22

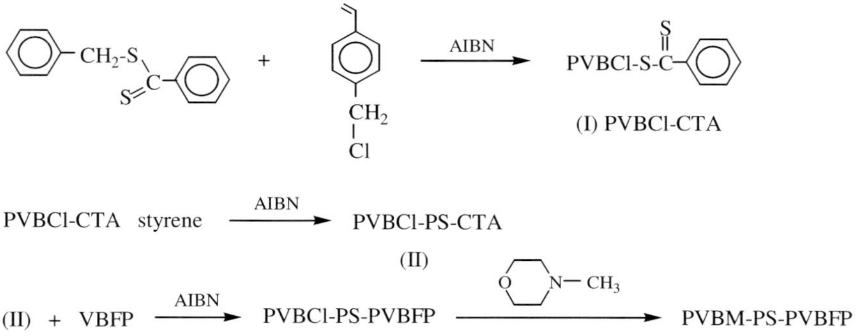

Scheme 4.23

est in this terpolymer arises from its ability to form multicompartment micelles with a novel nanostructure in aqueous media [55].

4.2.3.6 **Synthesis of ABC Terpolymers by Combination of Methods**

Combination of different methods, although they are carried out in a multi-pot procedure and are of high cost since many reagents and solvents are involved, may offer interesting polymeric materials. Nanoporous polystyrene containing hydrophilic pores can be fabricated from a suitable ABC terpolymer bearing polylactide (PLA) etchable segment, poly(*N,N* dimethylacrylamide) as the hydrophilic middle block and polystyrene as the matrix material. This can be achieved by designing the composition of the terpolymer to form a centrosymmetric cylindrical morphology in a continuous matrix. The pores will be formed by selec-

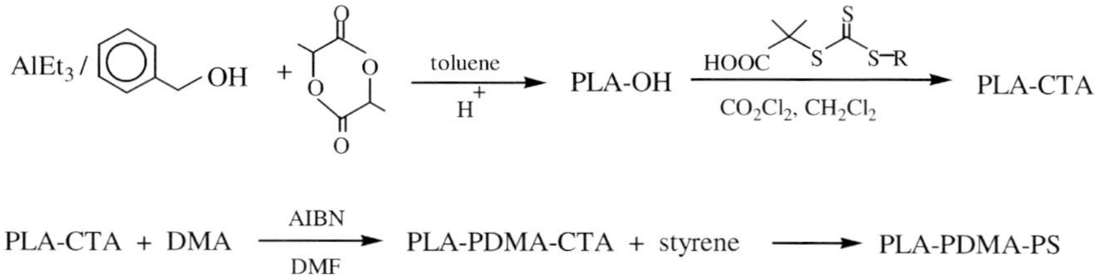

Scheme 4.24

tive degradation of the inner PLA core cylinders. The polymer was prepared by combination of controlled ring-opening and RAFT polymerization according to Scheme 4.24 [56].

Other examples of ABC terpolymer design can be found in a recent review [57]. It is obvious that the combination of the available monomers in three-block macromolecules can give an enormous number of novel block copolymers with unexpected and unique properties, and this will be one of the future trends in polymer and material science.

4.2.4
Synthesis of ABCA Tetra- and ABCBA Penta-block Terpolymers

Linear terpolymers comprising more than three segments that are different in nature (e.g. ABCA and ABCBA) have been designed and investigated in solution and in the bulk.

A very interesting synthetic pathway was applied to prepare PEO–PS–PBd–PEO in a multi-step ALP. First, an *α,ω*-hydroxy functionalized PS-*b*-PBd was synthesized, using the protected initiator 3-*tert*-butyldimethylsiloxy-1-propyllithium and SMA of styrene and butadiene. The reaction was terminated by ethylene oxide. After deprotection with tetra(*n*-butyl)ammonium fluoride (TBAF), both OH ends were converted to potassium alkoxides and used to polymerize ethylene oxide (Scheme 4.25). These ABCA amphiphiles form vesicles or worm-like micelles in water depending on the length of the outer PEO segments [58].

Amphiphilic ABCA terpolymers have been prepared by the SMA route and modified by post-polymerization reactions. PS–PI–PBd–PS synthesized by ALP was first hydrogenated using Ni–Al catalyst and subsequently the PS outer blocks were sulfonated and neutralized with NaOH. A novel aggregate morphology was obtained in water, showing the influence of the ABCA asymmetric architecture in copolymer self-assembly [59].

Symmetric ABCBA terpolymers can be made by using a difunctional initiator. PMMA–PS–PBd–PS–PMMA is a characteristic example prepared by LAP. It was initiated by *t*BuLi–*m*-diisopropenylbenzene (in-DIB) diadduct and sequential addition of butadiene, styrene and MMA, in cyclohexane to ensure a 1,4-microstructure of the central PBd block. The last MMA block was polymerized in the presence of THF at low temperature [60].

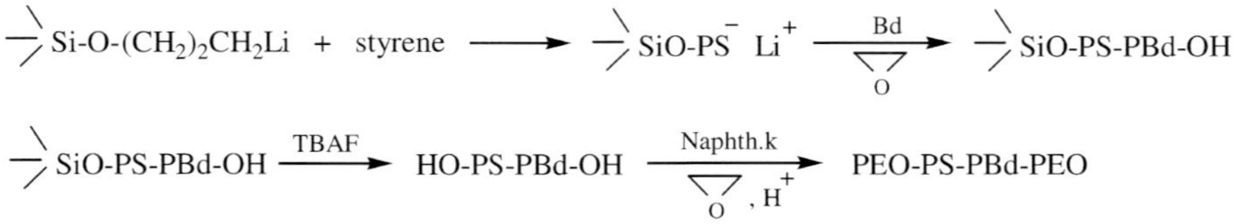

Scheme 4.25

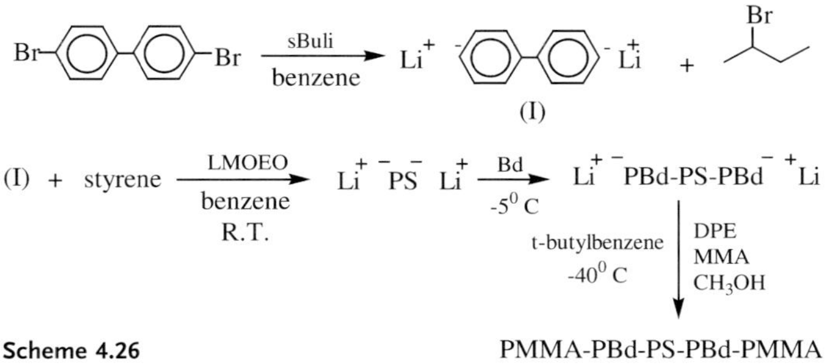

Scheme 4.26

A novel approach to prepare multifunctional initiating systems soluble in apolar organic media susceptible to anionic polymerization was developed recently. For the synthesis of PMMA–PBd–PS–PBd–PMMA, a bis(aryl halide) was metalated by a lithium–halide exchange reaction. The resulting initiator was used to polymerize sequentially styrene, butadiene and MMA (Scheme 4.26). To ensure solubilization of the initiator in benzene, a 4-fold amount (with respect to the initiator) of lithium 2-methoxyethoxide (LMOEO) σ/μ coordinating ligand was added to the reaction medium prior to monomer addition. This novel route was applied also to the preparation of well-defined star-shaped $(ABC)_n$ block terpolymersn block terpolymers“,4,1> [61].

ATRP synthesis has been employed for the preparation of PDEMAEMA–PEO–PPO–PEO–PDEAEMA pentablock terpolymers following the macroinitiator approach. An HO–PEO–PPO–PEO–OH (pluronic) triblock precursor was transformed into a bifunctional ATRP initiator according to the method described in Scheme 4.14 and was used to polymerize DEAEMA. A reversible gelation at around physiological temperatures and pH make this pentablock a potential candidate for use in injectable drug delivery devices [62].

4.2.5 Synthesis of ABCD Quaterpolymers

Four different building blocks incorporated in a single linear macromolecule, termed ABCD quaterpolymers, have appeared recently, increasing our expectations for even higher diversity and functionality towards multifunctional nanostructured polymeric materials.

A novel honeycomb morphology, arising from phase separation of all the incompatible blocks of PS–PI–PDMS–P2VP quaterpolymer, was observed in the bulk. This macromolecule was synthesized by ALP in two steps. SMA was applied in the first step to prepare ω-functionalized PS–PI–PDMS terpolymer precursor capping the living sites of the terpolymer with the heterofunctional linking agent chloromethylphenylethenyldimethylchlorosilane (CMPDMS). In the second step, living P2VP was deactivated on the chloromethylene function of the ABC precursor (Scheme 4.27) [63].

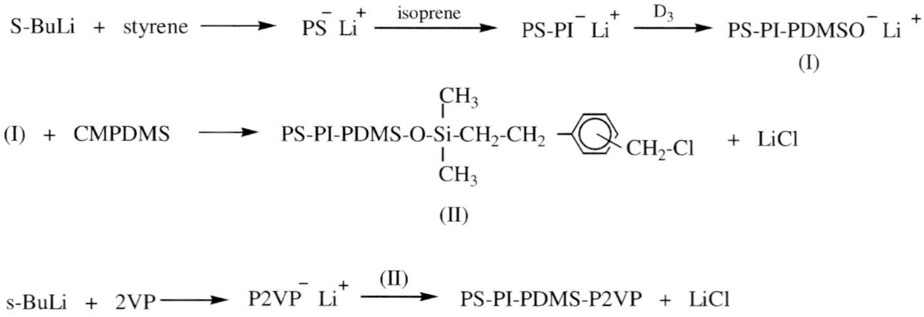

S-BuLi + styrene $\longrightarrow$ PS$^-$Li$^+$ $\xrightarrow{\text{isoprene}}$ PS-PI$^-$Li$^+$ $\xrightarrow{D_3}$ PS-PI-PDMSO$^-$Li$^+$
(I)

(I) + CMPDMS $\longrightarrow$ PS-PI-PDMS-O-Si(CH$_3$)$_2$-CH$_2$-CH$_2$-C$_6$H$_4$-CH$_2$-Cl + LiCl
(II)

s-BuLi + 2VP $\longrightarrow$ P2VP$^-$Li$^+$ $\xrightarrow{\text{(II)}}$ PS-PI-PDMS-P2VP + LiCl

Scheme 4.27

Reactive ABCD quaterpolymers bearing glycidyl methacrylate (GMA) were designed to control the morphology in epoxyamine networks. PS–PBd–PMMA–PGMA was synthesized straightforwardly by anionic polymerization following the SMA route. It was shown that the use of reactive block copolymers of well-defined composition permits the adjustment of the final morphology of block copolymer/thermoset nanocomposites [64].

4.3
Multisegmental Graft Copolymers

Graft copolymers constitute a family of non-linear, branched block copolymers of special interest due to their high segment density and specific architecture. They are comprised of a backbone macromolecular chain bearing a number of side-chains (branches) of different chemical nature. Graft copolymers are characterized by the grafting density, i.e. the ratio of the number of branch points (side-chains) to the number of monomers units of the backbone chain, which is usually relative low (<10%). In the special case that every repeating unit of the backbone bears a side-chain (grafting density 100%), the graft copolymers are referred as molecular brushes and are discussed in Chapter 9.

4.3.1
Synthetic Strategies

Three main synthetic routes have been developed for the preparation of graft copolymers, as depicted in Fig. 4.3. The "grafting on to" method involves the reaction of living macromolecular (B) chains with specific pendant functional groups distributed along the backbone (A) chain (synthesized independently) (Fig. 4.3a and b). In this method, two synthetic pathways can be distinguished: (a) the active chains are deactivated on to the backbone (e.g. anionic sites reacting with electrophilic groups) and (b) the living chains react with unsaturations and therefore the active sites are preserved in the branch points. The latter case

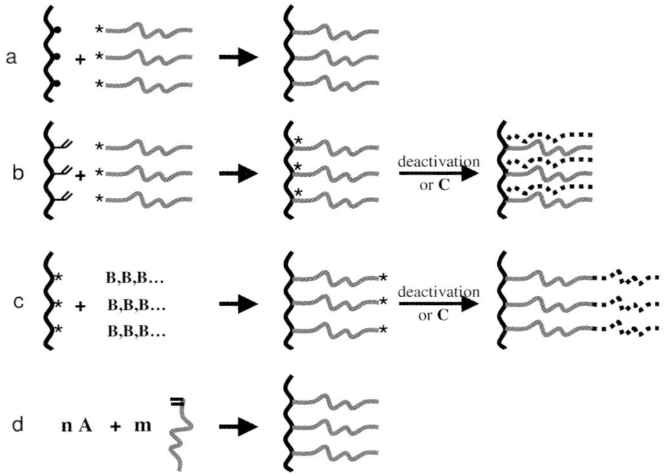

Fig. 4.3 Schematic representation of synthetic strategies towards graft copolymers: (a), (b) "grafting on to"; (c) "grafting from"; (d) "grafting through".

allows further copolymerization by adding a monomer C. The possibility of undesired bridging between backbone chains is the major drawback of this pathway. In general, the "grafting on to" route leads to well-defined graft copolymers with molecular control of the backbone and the side-chains since precise characterization of the side-chains can be performed by sampling out prior grafting.

In the "grafting from" synthetic route, a number of active sites are generated along the backbone (A) chain which are used to initiate the polymerization of monomer B creating the side-chains (Fig. 4.3 c). The significant drawback of this route is the inability to characterize directly the molecular features of the branches and therefore the grafting density. An average molecular weight of the grafts can be calculated from the overall M_w of the copolymer and the number of sites per backbone chain assuming that all the sites have been involved in the polymerization of the incoming monomer.

Macromonomers, i.e. ω-functionalized polymers bearing a polymerizable unsaturation [65], are involved in the "grafting through" synthetic route. A given mixture of monomer A and macromonomer B is copolymerized to yield the graft copolymer (Fig. 4.3 d). This route is suitable for polymerization methods that permit random copolymerization such as the free radical mode. After the establishment of controlled living methods for the free radical polymerization, this method seems to be more promising as one can control the length of the backbone chain and the grafting density from the monomer feed ratio.

4.3.2
A-g-B Graft Copolymers

Graft copolymers were synthesized by anionic polymerization in the early 1960s and have been extensively discussed in previous reviews and books [66]. In this chapter, some recent examples and new developments will be presented.

The most common synthetic strategy is the "grafting on to" route. Amphiphilic PS-g-PMAA polymers with ionizable PMAA branches were prepared and studied in selective media. First a linear PS of known degree of polymerization was partially bromomethylated by trimethylbromosilane and trioxane with the aid of tin tetrabromide catalyst. In the grafting step, 1,1-diphenylhexyllithium (DPHLi) was used to prepare the *tert*-butyl methacrylate branches (P*t*BMA) in THF at low temperature. After monomer consumption, the bromomethylated PS backbone was introduced into the reaction medium and the living P*t*BA ends were deactivated by the bromo electrophiles, yielding the graft copolymer. Hydrolysis of the branches in acidic media give graft copolymers with poly(methacrylic acid) side-chains (Scheme 4.28) [67]. The main conclusion from their association behavior in selective solvents was the low aggregation numbers, attributed to the graft architecture.

A reversible-type graft copolymer with a polyacid as the backbone bearing short PS side-chains was prepared through the "grafting on to" route but using a condensation reaction instead of deactivation as described above. The reaction was carried out in dioxane at 25 °C between amino-functionalized PS (PS–NH_2) and PAA in the presence of 1,3-dicyclohexylcarbodiimide (DCC) as promoter (Scheme 4.29) [68].

The same reaction has been used to prepare the so-called hydrophobically modified polyelectrolytes constituted of PAA grafted by hydrocarbon or fluorocarbon pendant hydrophobes. These amphiphilic polymers have been used as associative thickeners, forming physical gels in aqueous formulations [69].

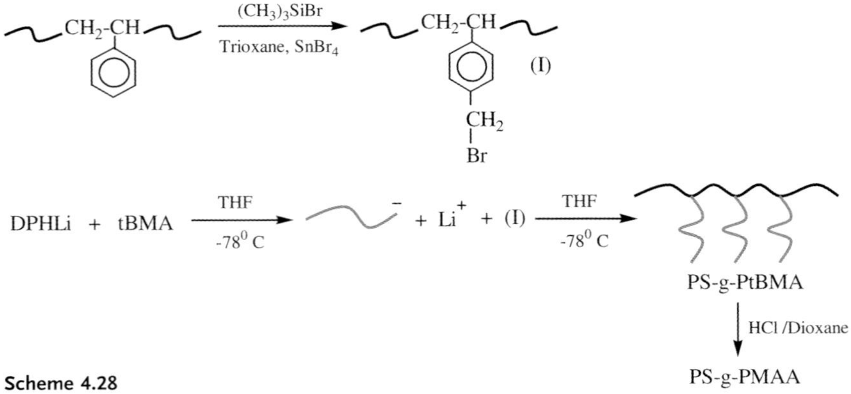

Scheme 4.28

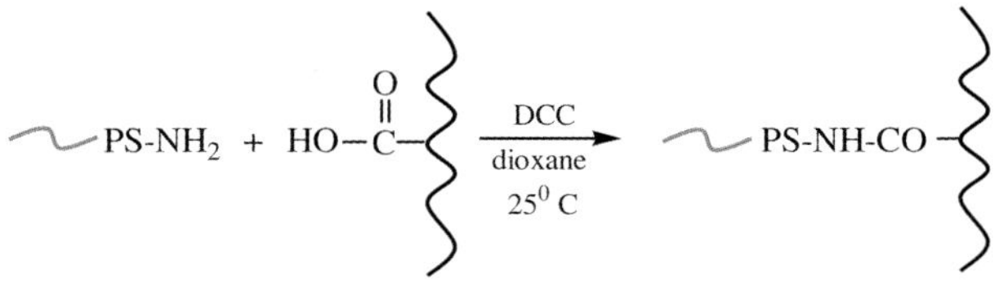

Scheme 4.29

Polyvinylpyridines (P2VP or P4VP) can be used in anionic polymerization as a backbone chain to generate graft copolymers following the "grafting on to" approach. Living PS and/or PI react with the pyridine rings forming P2VP-g-PS copolymers as shown in Scheme 4.30. The anion at the PS end is preserved on nitrogen and is deactivated by addition of CH_3OH [70].

Anionic random copolymerization is a relative rare case since the reactivities of the monomers have to be very similar. However, it has been applied for the preparation of the backbone chain in PMMA-g-PS graft copolymers. 1,1-Diphenylhexyllithium (DPHLi) was used as initiator to copolymerize 1-ethoxyethyl methacrylate (EEMA) and glycidyl methacrylate (GMA). The resulted random copolymer exhibits a narrow molecular weight distribution. In a subsequent step, living PS was reacted with the epoxy pendant groups of P(EEMA-*co*-GMA) and the enolates generated were protonic deactivated. Finally, the EEMA moieties were hydrolyzed to yield the amphiphilic graft (Scheme 4.31). This "grafting on to" method leads to well-defined graft copolymers since it allows control of the grafting density and the length of backbone and branches [71].

A double hydrophilic and stimuli-responsive graft copolymer comprising a PAA backbone and poly(*N*-isopropylacrylamide) (PNIPAAm) side-chains constitutes a very interesting example. Two synthetic methods were applied to prepare PAA-g-PNIPAAm: (a) by reacting α-functionalized H_2N–PNIPAAm on to PAA as in Scheme 4.29 and (b) using the "grafting through" route by conventional radical copolymerization of AA and PNIPAAm macromonomer according to Scheme 4.32 [72].

This graft copolymer represents a characteristic paradigm of a polymer comprising a pH-sensitive backbone (PAA is a weak polyelectrolyte) grafting by temperature-sensitive side-chains due to the LCST of PNIPAAm. It undergoes marked solubility changes in water in response to temperature and/or pH changes, exhibiting potential applications in medicine and biotechnology.

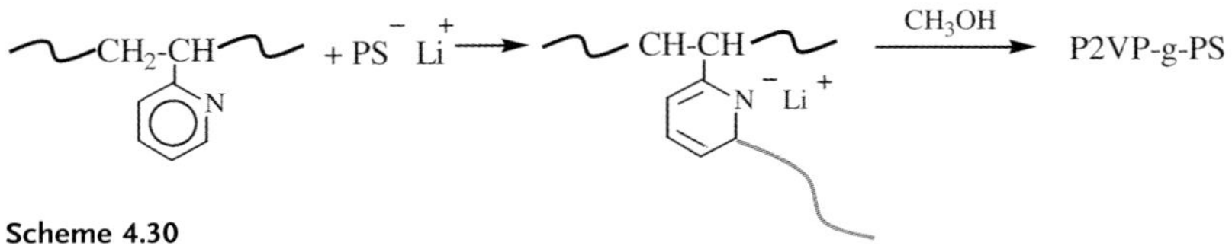

Scheme 4.30

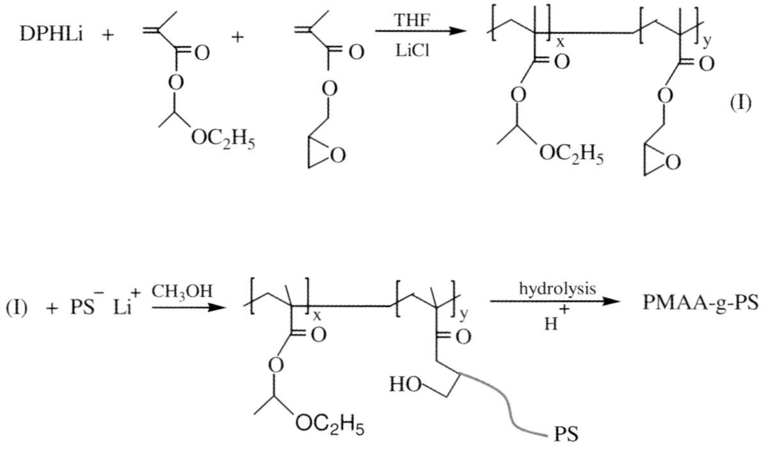

Scheme 4.31

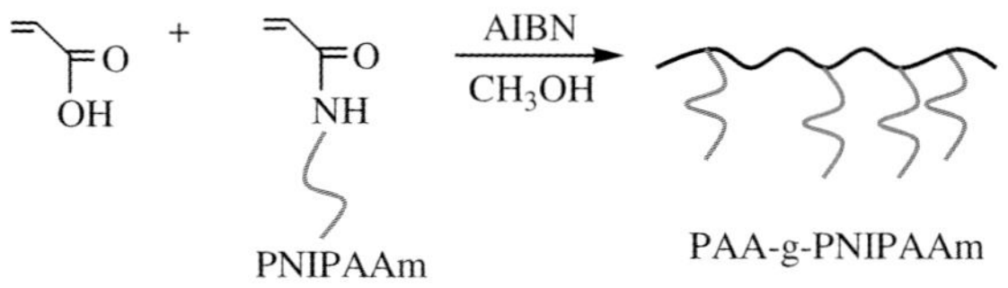

Scheme 4.32

A similar double hydrophilic graft copolymer in which the branches are of *N,N*-dimethylacrylamide (PDMAAm) was accomplished by the same route [73]. Interpolymer complexation through hydrogen bonding between AA and DMAAm moieties governs the aqueous solution properties of this polymer.

In both of the previous cases, the "grafting through" strategy was applied by using the conventional free radical copolymerization of monomer/macromonomer, which fails to control the molecular characteristics of the backbone chain. Controlled free radical polymerization is now used to overcome this problem.

A well-defined P*n*BA-g-PMMA graft copolymer was obtained by the macromonomer approach using the ATRP method [74]. In this investigation, it was indicated that the relative reactivity of the macromonomer is much closer to that of MMA in ATRP than in conventional copolymerization which resulted in a longer time-scale of monomer addition, i.e. seconds for ATRP and milliseconds for the conventional radical approach.

A similar macromonomer approach was used to synthesize PMMA-g-PDMS (PDMS = polydimethylsiloxane) graft copolymers in which again the reactivity ratio of PDMS macromonomer was determined to be much closer to MMA in the ATRP mechanism. The consequence of this fact was that the branches are constantly incorporated into all copolymer chains at a certain rate according to the

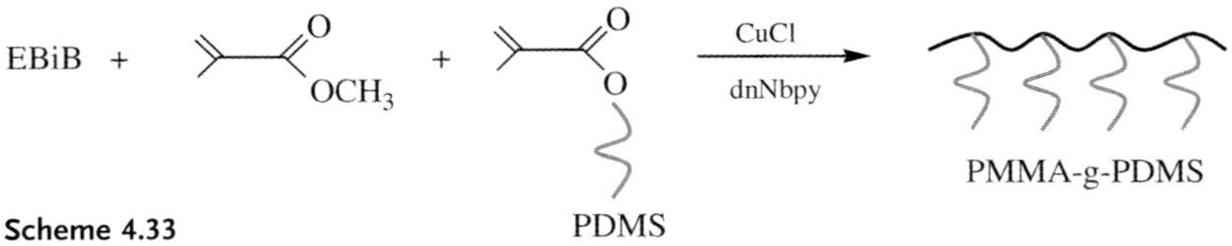

Scheme 4.33

macromonomer/comonomer composition in the feed. Therefore, it was shown that the grafting through macromonomer approach performed by ATRP is the most beneficial and effective method for controlling the graft copolymer structure in terms of main chain length/polydispersity and branch distribution/homogeneity (Scheme 4.33) [75].

As shown later, the effect of the microstructure of graft copolymers on the mechanical properties is significant. Comparing the tensile elongation (ε) of three samples with the same overall M_w and PMMA composition but exhibiting different branch distribution, the following results were observed. The tapered graft (RAFT) had an ε value of 30%, the irregular product (FRP) 120% and that with the regular structure (ATRP) the highest at 280% [76].

Graft copolymers with poly(ε-caprolactone) (PCL) as the backbone and PMMA branches constitute a special case prepared by combination of ring-opening polymerization (ROP) and ATRP. Three synthetic approaches were followed: (a) by the "grafting from" route, (b) by "grafting through" using PMMA macromonomer with a lactone functionality and (c) by a one-step/one-pot route in which ROP and ATRP polymerization procedures occur at the same time. The method relies on the synthesis of γ-(2-bromo-2-methylpropionyl)-ε-caprolactone ATRP (initiator)-functionalized lactone monomer f(ε-CL) which can be copolymerized with ε-CL to build the backbone chain (Scheme 4.34) [77].

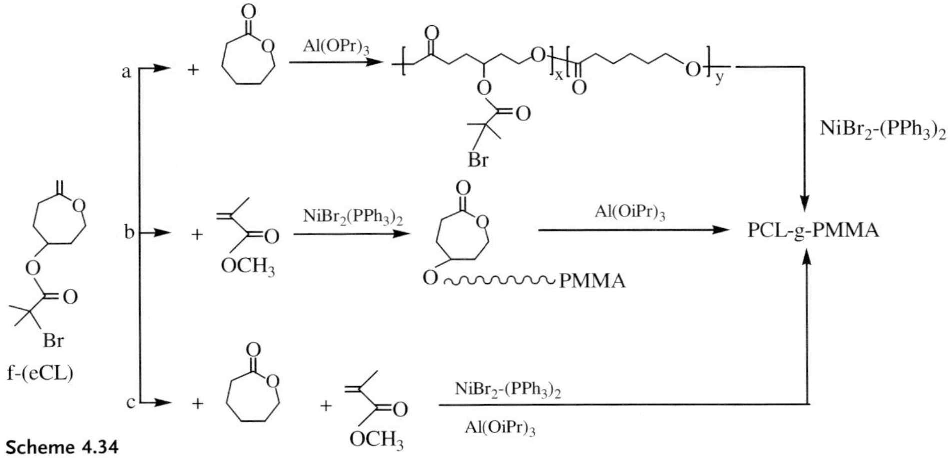

Scheme 4.34

The most pronounced difference was between the one-step/one-pot and the macromonomer routes. The control of molecular weight and polydispersity was clearly compromised in the one-step procedure.

The "grafting from" synthetic route has also been applied in ATRP polymerization to prepare graft copolymers using suitably modified (e.g. brominated monomeric units) macromolecular chains or statistical copolymers as macroinitiators (Fig. 4.3c) [78]. Although this method seems to lead to less well-defined graft copolymers, it permits diblock grafts to be generated. A nice example is the preparation of polyethylene-based graft copolymer. Poly(ethylene-*co*-styrene) was brominated using *N*-bromosuccinimide (NBS) to create benzyl bromide moieties randomly distributed along the chain. These functions were used to initiate methyl methacrylate (MMA) and, in a subsequent step, 2-hydroxyethyl methacrylate (HEMA) (Scheme 4.35) [79].

Heterografted block copolymers, i.e. graft copolymers comprising two kinds of pure branches of different nature, have been synthesized by ATRP following two synthetic strategies. In the first case, a two-step "grafting through" followed by "grafting from" procedure was proposed. In the first step, a graft copolymer precursor is prepared by the macroinitiator approach. Copolymerization of 2-(trimethylsilyloxy)ethyl methacrylate (HEMA–TMS) and PEO macromonomers (PEOMA) was conducted under typical ATRP conditions yielding P[(HEMA–TMS)-*co*-PEOMA)]. This graft copolymer precursor was transformed to ATRP macroinitiator by cleavage of the TMS protecting groups and esterification with 2-bromopropionyl bromide. A new set of poly(*n*-butyl acrylate) (P*n*BA) branches was grown from the active sites along the backbone chain (Scheme 4.36) [80].

Copolymerization of two different macromonomers constitutes an alternative route to prepare heterografted copolymers. This was exemplified by the synthesis of PDMS–PEO heterograft copolymers by an ATRP one-pot procedure [81]. In both cases described above the final products are high densely grafted copo-

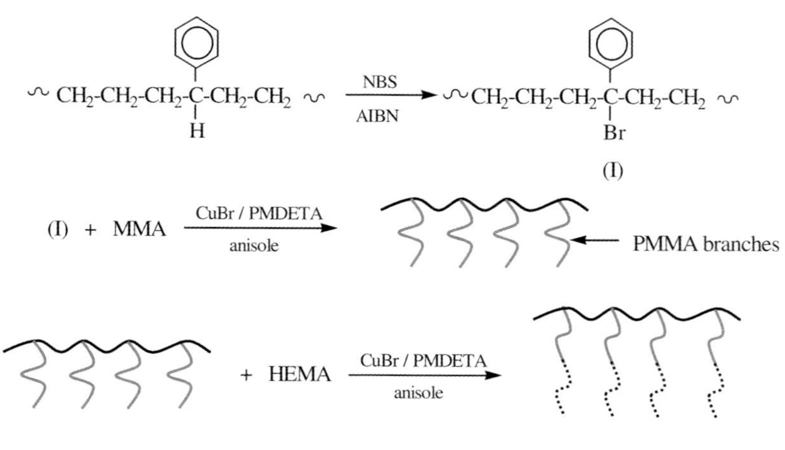

Scheme 4.35 P(E-co-St)-g-(PMMA-b-PHEMA)

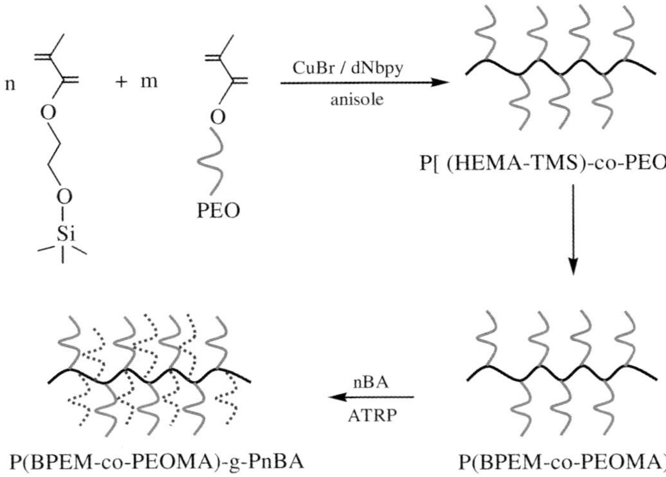

Scheme 4.36

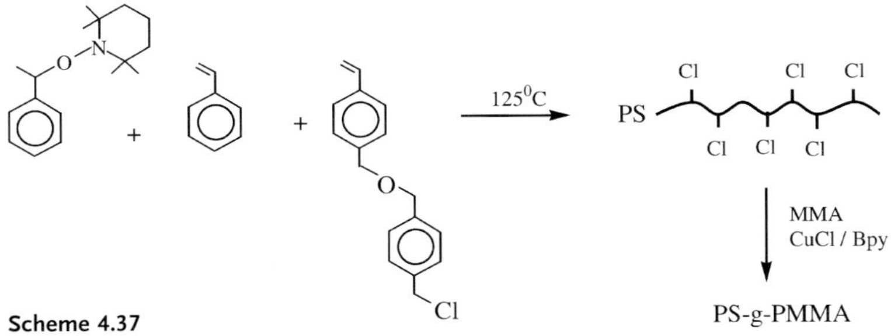

Scheme 4.37

lymers belonging to the category of molecular brushes that exhibit unique properties (see Chapter 9).

Nitroxide-mediated controlled free radical polymerization has been employed to afford graft copolymers. For example, a mixture of styrene and *p*-(4′-chloromethylbenzyloxymethyl)styrene initiated by TEMPO produces a PS backbone chain bearing latent ATRP initiating sites. In a second step, PMMA branches can grow under ATRP standard conditions (Scheme 4.37) [82].

If *p*-chloromethylstyrene is used in the above procedure, then TEMPO-based initiating groups can be generated along the PS backbone by reacting with the sodium salt of the hydroxyl-functionalized alkoxyamine derivative [83].

4.3.3
Model Graft-like Architectures

The establishment of structure–property relationships and testing of theoretical concepts requires precise control of the macromolecular characteristics. This motivated the design of model graft-like block copolymers of specific architectures in which the grafting distance between adjacent grafts and the number of branches per chain and grafting center are well-defined (Figure 4.4) [84].

The synthetic routes towards these model branched macromolecules comprises mainly anionic polymerization under high vacuum and chlorosilane chemistry.

The synthesis of regular PI-g-PS graft copolymers was conducted according to Scheme 4.38 [85]. Excess of $MeSiCl_3$ is needed to prevent dimerization of PSLi. The statistical nature of the final linking (in other words, the mean value of the number of branching points) is not strictly controlled but its average value can be targeted through the Si–Cl/living ends ratio.

Centipede (tetrafunctional) and barbed-wire (hexafunctional) multigrafts were accomplished by using tetrachlorosilane and hexachlorosilane coupling agents, respectively (Schemes 4.39 and 4.40) [86, 87].

It is known that the mechanical properties of styrene/diene-based thermoplastic elastomers (TPE) depend on the morphology they adopt in the solid state after microphase separation. It was shown recently that the PI-g-PS_2 centipede2 centipede“,4,1> with 10 branch points and 22% PS exhibits an exceptional combination of high strength and very high elongation at break as compared with the commercial TPE Kraton (20% PS) and Styroflex (58% PS) [88].

The chlorosilane approach has also been utilized to prepare H-shaped, super H and pom-pom PS/PI block copolymers according to Scheme 4.41 [88–90]. In all these cases fractionation is necessary to isolate the targeted final product. This fact, together with the difficulties of the high vacuum technique and the long reaction times, limits the use of these interesting materials to basic research.

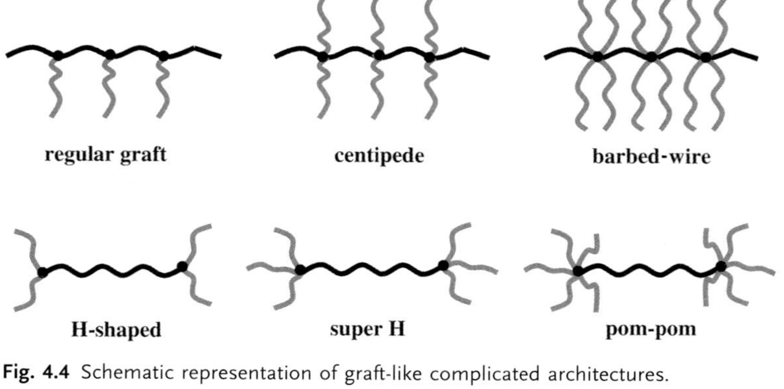

Fig. 4.4 Schematic representation of graft-like complicated architectures.

$$PS^{-}Li^{+} + \text{excess } MeSiCl_3 \longrightarrow (PS)(Me)SiCl_2 + LiCl + MeSiCl_3\uparrow$$

$$Li^{+}\,{}^{-}PI^{-}\,Li^{+} + (PS)(Me)SiCl_2 \xrightarrow{\text{polycondensation}} PI\text{-}g\text{-}PS + LiCl$$

Scheme 4.38

$$2PS^{-}Li^{+} + SiCl_4 \longrightarrow (PS)_2SiCl_2 + 2LiCl$$

$$Li^{+}\,{}^{-}PI^{-}\,Li^{+} + (PS)_2SiCl_2 \xrightarrow{\text{polycondensation}} PI\text{-}g\text{-}(PS)_2 + LiCl$$

Scheme 4.39

$$4PS^{-}Li^{+} + Cl_3Si(CH_2)_6SiCl_3 \longrightarrow Cl\text{-}Si(PS)_2\text{-}(CH_2)_6\text{-}Si(PS)_2\text{-}Cl + 4LiCl$$

(I)

$$Li^{+}\,{}^{-}PI^{-}\,Li^{+} + (I) \xrightarrow{\text{polycondensation}} PI\text{-}g\text{-}(PS)_4$$

Scheme 4.40

$$PS^{+}Li^{-} + MeSiCl_3 \longrightarrow PS_2MeSiCl$$

$$\text{excess } PS_2MeSiCl + {}^{+}Li\,{}^{-}PI^{-}\,Li^{+} \longrightarrow (PS)_2\text{-}PI\text{-}(PS)_2$$

(a)

$${}^{+}Na\,{}^{-}PS^{-}\,Na^{+} + SiCl_4 \text{ (excess)} \longrightarrow Cl_3Si\text{-}PS\text{-}SiCl_3$$

$$Cl_3Si\text{-}PS\text{-}SiCl_3 + PI^{-}Li^{+} \text{ (excess)} \longrightarrow (PI)_3\text{-}PS\text{-}(PI)_3$$

(b)

Scheme 4.41

Recently, amphiphilic H-shaped $(PS)_2$–PEO–$(PS)_2$ block copolymers were synthesized by ATRP using a tetrafunctional PEO macroinitiator [2,2-bis(methylene α-bromopropionate)propionyl-terminated PEO] and studied in selective media. In PS-selective solvents, worm-like structures were observed by atomic force microscopy. In contrast, in PEO-selective solvents, the worm-like aggregates were transformed into spheres coexisting with unimers [91].

Even more complex branched topologies have been proposed through an iterative multi-step synthetic procedure using ALP. A rich variety of such multi-segmental copolymers have been reported in a recent review [92].

4.4 Conclusion

It is beyond doubt that the rapid progress of controlled/living polymerization methods in recent years together with the possibility of using combined synthetic strategies and post-polymerization selective chemical modifications could satisfy the imagination of polymer and material scientists for novel polymeric materials with fascinating properties and targeted applications. Macromolecular architecture is one of the key factors for designing block copolymers. Nowadays it is becoming more complex by incorporating an increasing number of building blocks, differing in chemical nature (e.g. ABA to ABC to ABCD,...), and increasing topological configurations (linear versus branched). This chemical diversity and multifunctionality of multisegmental block copolymers generates new challenges in polymer physics in terms of phase behavior, self-assembly and nanostructured morphology, that determines the structure–property relationships and in turn the potential applications. Therefore, the development of novel multifunctional multisegmental block copolymers with controlled macromolecular characteristics will continue to grow rapidly as it constitutes one of the main trends in polymer science.

Acknowledgment

The contribution of Mrs. Ourania Kouli in dealing with the artwork of the Schemes is greatly appreciated.

References

1 Park, C., Yoon, J., Thomas, E. L. *Polymer* **2003**, *44*, 6725.

2 Satchi-Fainaro, R., Duncan, R., Eds. Polymer Therapeutics I and II. Special Issues. *Adv. Polym. Sci.* **2006**, *192, 193*.

3 Szwarc, M., Levy, M., Milcovick, R. *J. Am. Chem. Soc.* **1956**, *78*, 2656.

4 (a) Wang, J.-S., Matyjaszewski, K. *J. Am. Chem. Soc.* **1995**, *171*, 5614; (b) Kato, M., Kamigaito, M., Sawamoto, M., Higashimura, T. *Macromolecules* **1995**, *28*, 1721.

5 Moad, G., Rizzardo, E., Solomon, D. H. *Macromolecules* **1982**, *15*, 909.

6 Webster, O. W., Hertler, W. R., Sogah, D. V., Farnham, W. B., Rajanbabn, T. V. *J. Am. Chem. Soc.* **1983**, *105*, 5706.

7 Chiefari, J., Chong, Y. K., Ercole, F., Krstina, J., Jeffery, J., Le, T. P. T., Mayadunne, R. T. A., Meijs, G. F., Moad, C. L., Moad, G., Rizzardo, E., Thang, S. H. *Macromolecules* **1998**, *31*, 5559.

8 Nguyen, S. T., Johnson, L. K., Grubbs, R. H., Ziller, J. W. *J. Am. Chem. Soc.* **1992**, *114*, 3975.

9 Burgess, E.J., Cunliffe, A.V., Richards, D.H., Sherrington, D.C. *J. Polym. Sci., Polym. Lett. Ed.* **1976**, *14*, 471.

10 (a) Quirk, R.P., Ma, J.-J. *Polym. Int.* **1991**, *24*, 197; (b) Lo, G.Y.-S., Otterbacher, E.M., Gatzeke, A.L., Tung, L.H. *Macromolecules* **1994**, *27*, 2233.

11 (a) Morton, M., Fetters, L.J. *Macromol. Rev.* **1967**, *2*, 71; (b) Morton, M. *Anionic Polymerization: Principles and Practice.* Academic Press, New York, **1983**.

12 Hahn, S.F., Vosejpka, P.C. *Polym. Prepr.* **1999**, *40*, 970.

13 Tong, J.D., Leclerc, P., Doneux, C., Predes, Li., Lazzaroni, R., Jérôme, R. *Polymer* **2000**, *41*, 4617.

14 Tsitsilianis, C., Iliopoulos, I., Ducouret, G. *Macromolecules* **2000**, *33*, 2936.

15 Fayt, R., Forte, R., Jacobs, C., Jérôme, R., Ouhadi, T., Teyssié, P., Varshney, S.K. *Macromolecules* **1987**, *20*, 1442.

16 Zaroslov, Y.D., Fytas, G., Pitsikalis, M., Hadjichristidis, N., Philippova, O.E., Knokhlov, A.R. *Macromol. Chem. Phys.* **2005**, *206*, 173.

17 Riegel, I.C., Eisenberg, A., Petzhold, C.L., Samios, D. *Langmuir* **2002**, *18*, 3358.

18 Xie, H.-Q., Xie, D. *Prog. Polym. Sci.* **1999**, *24*, 275.

19 Zhu, L., Zhu, L., Mimnaugh, B.R., Ge, Q., Quirk, R.P., Cheng, S.Z.D., Thomas, E.L., Lotz, B., Hsiao, B.S., Yeh, F., Liu, L. *Polymer* **2001**, *42*, 9121.

20 Sfika, V., Tsitsilianis, C. *Macromolecules* **2003**, *36*, 4983.

21 Quirk, R.P., Kim, J.-S. *J. Phys. Org. Chem.* **1995**, *8*, 242.

22 Webster, O.W. *Adv. Polym. Sci.* **2004**, *167*, 1.

23 Purcell, A., Armes, S.P., Billingham, N.C. *Polym. Prepr.* **1997**, *38*, 502.

24 Gotzamanis, G., Tsitsilianis, C., Hadjiyannakou, S.C., Patrickios, C.S., Lupitskyy, R., Minko, S. *Macromolecules* **2006**, *39*, 678.

25 Sawamoto, M. *Prog. Polym. Sci.* **1991**, *16*, 111.

26 Kennedy, J.P., Meguriya, N., Keszler, B. *Macromolecules* **1991**, *24*, 6572.

27 Wang, Y., Goethals, E.J. *Macromolecules* **2000**, *33*, 808.

28 Gao, X., Faust, R. *Macromolecules* **1999**, *32*, 5487.

29 Kwon, Y., Faust, R. *Adv. Polym. Sci.* **2004**, *167*, 107.

30 Kotani, Y., Kato, M., Kamigaito, M., Sawamoto, M. *Macromolecules* **1996**, *29*, 6979.

31 Matyjaszewski, K., Acar, M.H., Beers, K.L., Coca, K.A., Scott, D., Gaynor, G., Miller, P.J., Paik, H.-J., Shipp, D.A., Teodorescu, M., Xia, J., Zhang, X. *Polym. Prepr.* **1999**, *40(2)*, 966.

32 Li, C., Tang, Y., Armes, S.P., Morris, C.J., Rose, S.F., Lloyd, A.W., Lewis, A.L. *Biomacromolecules* **2005**, *6*, 994.

33 Convertine, A.J., Lokitz, B.S., Vasileva, Y., Myrick, L.J., Scales, C.W., Lowe, A.B., McCormick, C.L. *Macromolecules* **2006**, *39*, 1724.

34 Listigovers, N.A., Georges, M.K., Honeyman, C.H. *Polym. Prepr.* **1997**, *38*, 410.

35 Jankova, K., Chem, X., Kops, J., Batsberg, W. *Macromolecules* **1998**, *31*, 538.

36 Yary, Z., Yuan, J., Cheng, S. *Eur. Polym. J.* **2005**, *41*, 267.

37 Spontak, R.J., Smith, S.D. *J. Polym. Sci. Part B: Polym. Phys.* **2001**, *39*, 947.

38 (a) Bates, F.S., Fredrickson, G.H. *Phys. Today* **1999**, *52*, 32; (b) Lodges, T.P. *Macromol. Chem. Phys.* **2003**, *204*, 265.

39 Hatzichristidis, N., Pispas, S., Floudas, G. *Block Copolymers: Synthetic Strategies, Physical Properties and Applications.* Wiley-Interscience, New York, **2003**.

40 Watanabe, H., Shimura, T., Kotaka, T., Tirrel, M. *Macromolecules* **1993**, *26*, 6338.

41 Gohy, J., Willet, N., Varshney, S., Zhange, J., Jérôme, R. *Angew. Chem.* **2001**, *113*, 3314.

42 Katsampas, I., Roiter, Y., Minko, S., Tsitsilianis, C. *Macromol. Rapid Commun.* **2005**, *26*, 1371.

43 Sfika, V., Tsitsilianis, C., Kiriy, A., Gorodyska, G., Stamm, M. *Macromolecules* **2004**, *37*, 9951.

44 Liu, F., Eisenberg, A. *Angew. Chem. Int. Ed.* **2003**, *42*, 1404.

45 Liu, F., Eisenberg, A. *J. Am. Chem. Soc.* **2003**, *125*, 15059.

46 Bieringer, R., Abetz, V., Müller, A.H.E. *Eur. Phys. J.* **2001**, *5*, 5.

47 Zhon, Z., Li, Z., Ren, Y., Hillmyer, M., Lodge, T. *J. Am. Chem. Soc.* **2003**, *125*, 10182.

48 Stewart, S., Liu, G. *Angew. Chem. Int. Ed.* **2000**, *39*, 340.

49 (a) Butun, V., Wany, X.-S., de Paz Banez, M.V., Robinson, K.L., Billingham, N.C., Armes, S.P. *Macromolecules* **2000**, *33*, 1; (b) de Paz Banez, M.V., Robinson, K.L., Armes, S.P. *Polymer* **2001**, *42*, 29.

50 Erhardt, R., Boker, A., Zettl, H., Kaya, H., Pyckhout-Hintzen, W., Krausch, G., Abetz, V., Müller, A.H.E. *Macromolecules* **2001**, *34*, 1069.

51 Patrickios, C., Lowe, A., Armes, S.P., Billingham, N. *J. Polym. Sci. Part A: Polym. Chem.* **1998**, *36*, 617.

52 Subihara, S., Kanaoka, S., Aoshima, S. *J. Polym. Sci. Part A: Polym. Chem.* **2004**, *42*, 2601.

53 Davis, K.A., Matyjaszewski, K. *Macromolecules* **2001**, *34*, 2101.

54 Cai, Y., Armes, S.P. *Macromolecules* **2004**, *37*, 7116.

55 Kubowicz, S., Baussaard, J.-F., Lutz, J.-F., Thunemann, A.F., von Berlepsch, H., Laschewsky, A. *Angew. Chem. Int. Ed.* **2005**, *44*, 5262.

56 Rzayev, J., Hillmyer, M.A. *Macromolecules* **2005**, *38*, 3.

57 Hadjichristidis, N., Iatrou, H., Pitsikalis, M., Pispas, S., Avgeropoulos, A. *Prog. Polym. Sci.* **2005**, *30*, 725.

58 Brannan, A.K., Bates, F.S. *Macromolecules* **2004**, *37*, 8816.

59 Gomez, E.D., Rappl, T.J., Agarwal, V., Bose, A., Schmutz, M., Marques, C.M., Balsara, N.P. *Macromolecules* **2004**, *38*, 3567.

60 Yu, J.M., Yu, Y., Dudois, P., Teyssié, P., Jérôme, R. *Polymer* **1997**, *38*, 3091.

61 Matmour, R., Lebreton, A., Tsitsilianis, C., Kallitsis, I., Héroguez, V., Gnanou, Y. *Angew. Chem. Int. Ed.* **2005**, *44*, 284.

62 Daterman, M.D., Cox, J.P., Swifert, S., Thiyagarajan, P., Mallapragada, S.K. *Polymer* **2005**, *46*, 6933.

63 Takahashi, K., Hasegawa, H., Hashimoto, T., Bellas, Y., Iatrou, H., Hatzichristidis, N. *Macromolecules* **2002**, *35*, 4859.

64 Rebizant, V., Abetz, V., Tournilhac, F., Court, F., Leibler, L. *Macromolecules* **2003**, *36*, 9889.

65 Rempp, P., Franta, E. *Adv. Polym. Sci.* **1984**, *58*, 1.

66 (a) Rempp, P., Franta, E., Herz, J.-E. *Adv. Polym. Sci.* **1988**, *86*, 147; (b) Kennedy, J.P., Ivan, B. *Designed Polymers by Carbocationic Macromolecular Engineering: Theory and Practice.* Hanser, Munich, **1992**.

67 Pitsikalis, M., Woodward, J., Mays, J.W., Hadjichristidis, N. *Macromolecules* **1977**, *30*, 5384.

68 Ma, Y., Cao, T., Webber, S.E. *Macromolecules* **1998**, *31*, 1773.

69 (a) Winnik, M.A., Yekta, A. *Curr. Opin. Colloid Interface Sci.* **1997**, *2*, 424; (b) Berret, J.-F., Calvet, D., Collet, A., Viguier, M. *Curr. Opin. Colloid Interface Sci.* **2003**, *8*, 296.

70 Watanabe, H., Amemiya, T., Shimura, T., Kotaka, T. *Macromolecules* **1994**, *27*, 2336.

71 Zhang, H., Ruckenstein, E. *Macromolecules* **2000**, *33*, 814.

72 Chen, G., Hoffman, A.S. *Nature* **1995**, *373*, 49.

73 Shibanuma, T., Aoki, T., Sanui, K., Ogata, N., Kikuchi, A., Sakurai, Y., Okano, T. *Macromolecules* **2000**, *33*, 444.

74 Roos, S.G., Müller, A.H.E., Matyjaszewski, K. *Macromolecules* **1999**, *32*, 8331.

75 Shinoda, H., Miller, P.J., Matyjaszewski, K. *Macromolecules* **2004**, *34*, 3186.

76 Shinoda, H., Matyjaszewski, K., Okrasa, K., Mierzwa, M., Pakula, T. *Macromolecules* **2003**, *36*, 4772.

77 Mecerreyes, D., Atthoff, B., Boduch, K.A., Trollsas, M., Hedrick, J.L. *Macromolecules* **1999**, *32*, 5175.

78 Gaynor, S.G., Matyjaszewsky, K. *ACS Symp. Series* **1998**, *685*, 396.

79 Liu, S., Sen, A. *Macromolecules* **2001**, *34*, 1529.

80 Neugebauer, D., Zhang, Y., Pakula, T., Matyjaszewski, K. *Polymer* **2003**, *44*, 6863.

81 Neugebauer, D., Zhang, Y., Pakula, T., Matyjaszewski, K. *Macromolecules* **2005**, *38*, 8687.

82 Grubbs, R.B., Hawker, C.J., Dao, J., Fréchet, J.M.J. *Angew. Chem. Int. Ed. Engl.* **1997**, *36*, 270.

83 Hawker, C.J., Bosman, A.W., Harth, E. *Chem. Rev.* **2001**, *101*, 3661.

84 (a) Hadjichristidis, N., Pitsikalis, M., Pispas, S., Iatrou, H. *Chem. Rev.* **2001**, *101*, 3747; (b) Hadjichristidis, N. *Eur. Phys. J.* **E 2003**, *10*, 83.
85 Uhrig, D., Mays, J.W. *Macromolecules* **2002**, *35*, 7182.
86 Iatrou, D., Mays, J.W. *Macromolecules* **1998**, *31*, 6697.
87 Weidish, R., Gido, S.P., Uhrig, D., Iatrou, H., Mays, J.W., Hadjichristidis, N. *Macromolecules* **2001**, *34*, 6333.
88 Gido, S.P., Lee, C., Pochan, D.J., Pispas, S., Mays, J.W., Hadjichristidis, N. *Macromolecules* **1996**, *29*, 7022.
89 Iatrou, H., Avgeropoulos, A., Hadjichristidis, N. *Macromolecules* **1994**, *27*, 6232.
90 Velis, G., Hadjichristidis, N. *Macromolecules* **1999**, *32*, 534.
91 Cong, Y., Li, B., Han, Y., Li, Y., Pan, C. *Macromolecules* **2005**, *38*, 9836.
92 Hirao, A., Hayashi, M., Loykulnant, S., Sugiyama, K., Ryu, S.W., Haraguchi, N., Matsuo, A., Higashihara, T. *Prog. Polym. Sci.* **2005**, *30*, 111.

5
Controlled Synthesis and Properties of Cyclic Polymers

Alain Deffieux and Redouane Borsali

Macrocyclic polymers and other model macromolecules have been a fascinating curiosity for theoreticians and chemists. The preparation of well-defined cyclic polymers and the study of their intrinsic properties are still a challenge in polymer science, although the interest in such chain architectures began more than 50 years ago with the discovery of cyclic DNA [1–3] in living cells and the first theoretical calculations of the cyclization effect of polymer chains on their bulk and solution properties [4–6]. The absence of chain ends and consequently the topological restriction imposed by the cyclic architecture result in a variety of molecular characteristics and physical properties that significantly distinguish them from their linear counterparts. For example, it has been shown that cyclic polymers have smaller hydrodynamic volumes and radii and are less viscous that the linear analogues. Their glass transition temperatures were shown to be almost independent of their molar mass, even for very low oligomers. Other specific properties such as melt diffusion and viscoelastic behaviors that should also be strongly influenced by the cyclic chain topology have been only marginally explored due to the limited availability of the corresponding high molar mass polymers. How they are influenced by the ring topology remains a question.

The preparation and study of cyclic polymers was first concentrated on macromolecular systems exhibiting ring–linear chain equilibria. They correspond to polymer chains bearing reactive functions on their backbone, such as polydimethylsiloxanes [7], polyethers [8, 9] and polyesters [10]. The synthesis of cyclic polymers by end-to-end coupling of chains free of reactive functions was first proposed by Casassa [6]. After several attempts involving different strategies, this was achieved experimentally in 1980 [11–14]. By the end-to-end ring closure of living difunctional polystyrene chains by α,α'-dichloro-*p*-xylene under very dilute conditions.

Since that time, the methods of preparation of ring polymers have multiplied while their efficiency was significantly improved thanks to important achievements in the control of polymerization reactions and the progress in analytical and preparative fractionation and characterization techniques.

Macromolecular Engineering. Precise Synthesis, Materials Properties, Applications.
Edited by K. Matyjaszewski, Y. Gnanou, and L. Leibler

ISBN: 978-3-527-31446-1

Many textbooks and reviews on the synthesis and study of cyclic polymers have been published during the last 50 years including those by Semlyen [15–17], Rempp [18], Keul [19], Deffieux [20], Lutz [21], Roovers [22], Hogen-Esch [23] and their coauthors. We will focus here on some important basic characteristics of cyclic polymers and on recent advances achieved during the last few years. The first part of the chapter deals with the main methods of preparation of cyclic macromolecules and the second presents some specific properties and characteristics of ring polymer and copolymer systems.

5.1
General Methods of Synthesis of Cyclic Macromolecules

Cyclic polymers can be recovered, as a more or less important fraction, in polymer systems having ring–chain equilibration or they can be synthesized on purpose using different synthetic strategies such as ring–chain expansion or ring closure of a linear chain by end-to-end coupling. The advantages and limitations of these different approaches will first be examined and illustrated for the preparation of various cyclic homopolymers. The preparation of cyclic block copolymers and more complex cyclic and multi-cyclic architectures (tadpoles, height–shape, etc.) will be described in a separate section.

5.1.1
Cyclics from Linear–Ring Chain Equilibrates

The concurrent formation of linear and cyclic molecules is a general characteristic of polymer systems containing reactive functions in their main backbone chain. As schematized in Eq. (1), this process may be represented as a propagation/depropagation equilibrium, each elementary reaction being characterized by its own rate constants, k_{pM} and k_{dpM}.

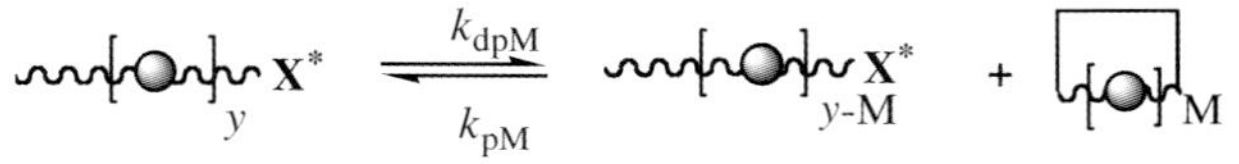

(1)

However, as illustrated in Scheme 5.1, ring–chain equilibria proceed through a more complex series of elementary reactions, which can be related to different back-biting/cyclization pathways.

Thus rings of different size are produced by various intramolecular attacks of the active chain-end –**X*** on any of the functions located on the polymer backbone. Each of these elementary reactions is characterized by a specific rate constant $k_{bb1,\ldots x}$, with $k_{bbM}=k_{dpM}$ when the cyclic product is identical with the monomer. This unimolecular process is in competition with the reverse bimolecular reaction, i.e. the ring-opening polymerization of the cyclics formed characterized by a propagation rate constant, $k_{p1,\ldots x}$, as presented in Eq. (1′).

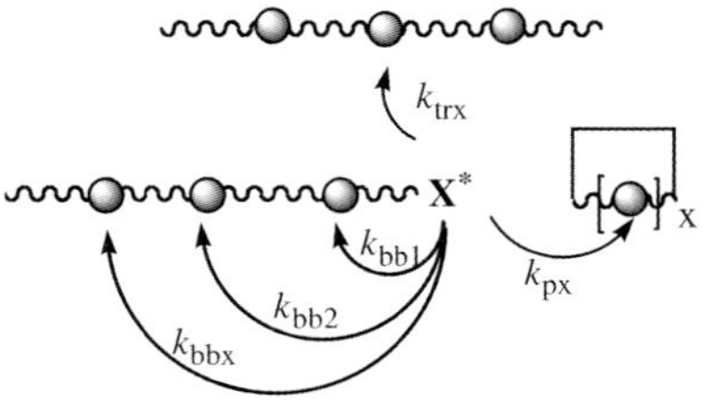

Scheme 5.1

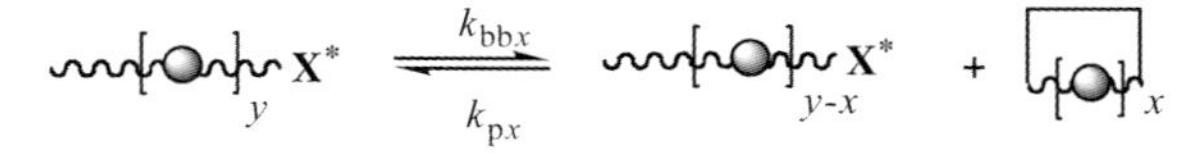

(1′)

The possibility of attack of a function x of another linear chain (k_{trx}) as illustrated in Scheme 5.1 has also been considered. This process, however, only results in linear chain to linear chain transfer (chain scrambling) and does not affect the linear–cyclic equilibrium.

Once equilibration between linear and cyclics of different sizes has been reached, the polymerization system is at thermodynamic equilibrium. At this stage, the molar cyclization equilibrium constant Ke_x of an individual cyclic x-mer is given by Eq. (2).

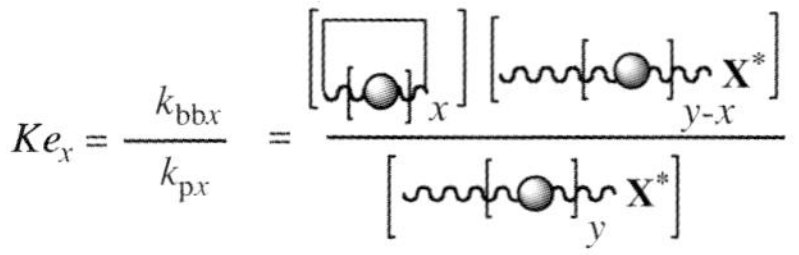

(2)

Considering the most probable Flory distribution of chains in the linear polymer fraction, Ke_x can be expressed by

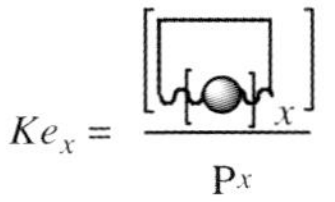

(3)

where P is the probability of reaction of the function at position x. Since P, which can be obtained experimentally, is often close to unity, Ke_x values for macrocyclics that are not too large can be approximated to

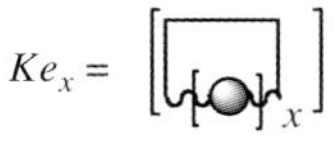

(4)

Considering that the polymer obeys Gaussian statistics the probability of cyclization at equilibrium of an x-mer is given by

$$Ke_x = \left(\frac{3}{2\pi\langle r_x^2\rangle}\right)^{\frac{3}{2}} \frac{1}{2\,N_A\,x} \tag{5}$$

where $\langle r^2\rangle$ is the mean square distance between the active chain end and the x-mer function and N_A is Avogadro's number.

For such idealized systems, the concentration of a given x-mer macrocycle at equilibrium can be finally approximated to

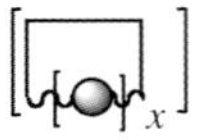

$$= Ax^{-\frac{5}{2}} \tag{6}$$

As indicated by Eq. (6), with some exceptions, rings produced by a back-biting mechanism are generally of small or medium size, typically ranging from dimers to $DP_n=20$ and their large majority have a $\overline{DP}_n$ of less than 10.

Ring–linear chain equilibration reactions are observed in polymer systems obtained by ring-opening polymerization of heterocyclic monomers (lactones, lactides, cyclic ethers, siloxanes, etc.), in polycondensation reactions (polyesters, polyamides, etc.), in ring-opening metathesis of cyclic alkenes (polyalkenamers) and in depolymerization of polymers bearing double bonds on their backbone such as polydienes.

Cationic and anionic polymerizations of heterocyclic monomers provide many examples in which the concurrent formation of cyclics is observed during the polymerization reaction. As illustrated in Scheme 5.1, in these systems active species face at least three types of different pathways; towards the monomer and the polymer (intramolecular) or other linear chains (intermolecular). When the functions involved are those present in the polymer chains, chain scrambling and macrocycle formation take place, as observed in particular in cationic polymerization of cyclic ethers, acetals, esters, amides, siloxanes and so forth.

The relative contribution of the monomer and polymer reactive groups, which are competing sites for the active species, depends on their relative reactivity and concentration. Very different situations can be found, some of them not allowing linear–cyclic equilibrium to be reached. Depending on the monomer size and ring strain, the propagation may be fast or slow compared with the back-biting reaction. When k_p is much higher than k_{bb}, the monomer can be converted under kinetic control into a purely linear polymer. Only afterwards, if the active species are still present, cyclic oligomers are formed. The thermodynamic equilibrium conditions can be reached only after a very long or even non-realistic time. This is the situation observed for cationic polymerization of THF [24]. Conversely, for systems in which k_p is close to k_{bb}, both linear and macrocycles will form almost concomitantly, directly approaching the thermodynamic cyclic-linear equilibrium. In some cases, however, the absence of ring strain in some of the cyclics formed by back-biting may prevent them from repolymerization, allowing them to accumulate in the system. This is exemplified in the cationic polymerization of ethylene oxide [25] by the almost quantitative formation of the six-membered 1,4-dioxane ring.

Systems that are rather closely approaching idealized linear–ring chain equilibria are polymerizations of hexamethylcyclotetrasiloxane [26–28] and of 1,3-dioxolane [29].

The preparation and isolation of cyclic poly(dimethylsiloxane) (PDMS) has been extensively studied over the past 40 years with special attention to the study of high molar mass cyclic PDMS. Dodgson and Semlyen [30], following the pioneer work of Brown and Slusarczuk [26], prepared cyclic PDMS fractions of narrow polydispersity with molar masses up to 30 000 g mol^{-1}. These samples were used to investigate the specific properties of cyclic chains and compared both with linear ones and the prediction of theories [15, 31].

Indeed, cyclic polymers formed in linear–ring step and chain polymerizations are often undesirable side-products present in linear high molar mass polymers which negatively affect the physical and mechanical properties of polymeric materials and experimental conditions have to be adjusted to limit their formation. In commercial-grade high molar mass linear polyesters and polycarbonates, for example, they still represent about 0.5–2% by weight of the total polymer.

In some specific cases, however, the lack of chain entanglement and low melt viscosities of cyclic oligomers were advantageously used, such as in the bulk preparation of high molar mass linear polymers or in composite fabrication. A special chemistry has even been developed for the preparation of large quantities of cyclic polycarbonate and polyester oligomers [31–33].

In conclusion, cyclic macromolecules are readily available from systems exhibiting linear cyclic equilibration. However, these macrocycles are generally obtained in mixtures with linear chains and are predominantly composed of low or medium molar mass macromolecules and exhibit a broad molar mass distribution, the relative amount of macrocycles formed decreasing with molar mass increase.

5.1.2
Cyclics Obtained by Ring Expansion Polymerization

Only a few reports have described the formation of large rings by direct insertion of a monomer into a reactive cyclic precursor. Although very attractive, this approach requires polymerization processes in which chain transfer and exchange reactions between polymerization-active species are absent. This is illustrated by the attempted synthesis of cyclic polyisobutene by cationic polymerization of isobutene using cyclic γ-tolyl-γ-valerolactone as initiator in the presence of BCl_3 [34]. Despite the cyclic structure of the chain precursor, rapid loss of the initial macrocyclic structure of the growing chain, due to intermolecular exchange reactions between propagating species during the polymerization, was observed [35].

One of the most demonstrative example concerns the ring expansion of a cyclic carbene–ruthenium complex by ring-opening metathesis polymerization

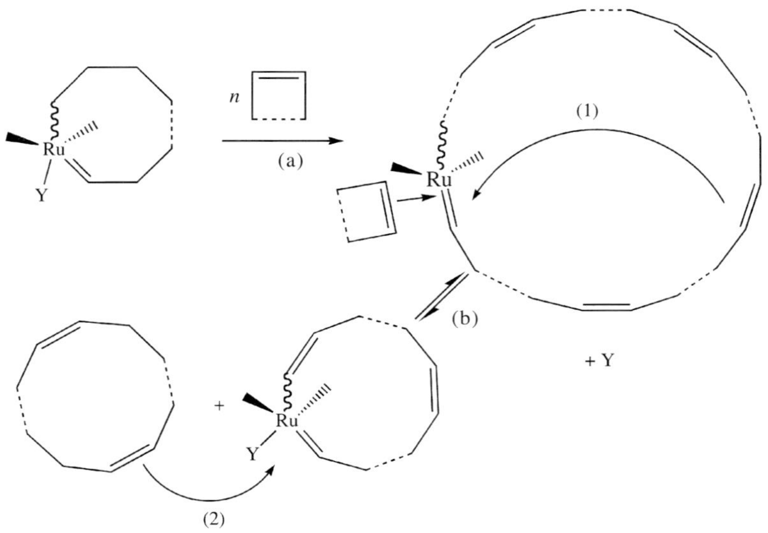

Scheme 5.2 Mechanism of ring expansion polymerization of cyclic alkenes by ring-opening metathesis.

[36–38]. In this coordinated polymerization, the cyclic alkene coordinates on to the ruthenium center before insertion into the cyclic carbene ring, which grows of one monomer unit (Scheme 5.2). In the first stage of the polymerization (a), due to the high monomer reactivity, the macrocycle rapidly expands, yielding a very high molar mass cyclic polyalkenamer containing a metal–carbene ring closing bond. At high monomer conversion, intramolecular back-biting reactions (1) begin to predominate, yielding chain splitting of the large cyclic carbene rings and the formation of smaller sized rings (b), most of them containing only carbon–carbon bonds, while the reverse bimolecular cyclic-to-cyclic chain coupling (2) results in the fusion of two cyclics into a larger one. Interestingly, these different reactions do not affect the cyclic nature of the resulting polymers but increase their molar mass distribution, which finally reaches a polydispersity close to two. This is only the presence of a linear alkene or polyalkene contaminant that will disrupt the selectivity of the ring expansion–contraction process and yield linear-chain side-products [39].

It is worth mentioning that another ring expansion metathesis strategy has recently been described by the same group [40, 41] for the preparation of low molar mass cyclic oligomers and co-oligomers. It involves successive ring opening of a cyclic alkene, cross-reaction with an acyclic diene and end-to-end ring closing. However, so far this process remains limited to low molar mass oligomers.

5.1.3
Size-controlled Cyclic Polymers by End-to-End Ring Closure

The most appropriate methods for the synthesis of cyclic polymers of controlled size and narrow polydispersity are based on the end-to-end chain coupling of α,ω-difunctional linear chains in highly dilute reaction conditions. This method offers a series of significant advantages: it may be used both for polymers having in-chain reactive functions and for systems with no labile linkages in their backbone, such as saturated carbon–carbon bond polymers. The absence of potentially reactive end-standing functions in the latter category may avoid undesirable complications. Since the cyclic is directly obtained from the linear parent by ring closing and the size of the linking unit is negligible, this approach provides both linear and cyclic structures of same molar mass. Moreover, the use of living polymerization techniques for the preparation of the linear precursors allows control of the molar mass and a narrow molar mass distribution. This is of special importance for a direct comparison of the properties of linear and cyclic chains.

The synthesis of cyclic polymers by end-to-end ring closure refers to a process in which the intramolecular cyclization and the intermolecular coupling yielding chain extension can occur simultaneously. As for ring–chain equilibrates the probability of ring formation depends on both the molar concentration and the size of the linear difunctional precursor. For a chain that obeys Gaussian statistics, the probability of cyclization, P_{cycl}, is determined by the probability for its α-end and ω-end functions being in the immediate vicinity of each other, within the same reaction cross-section, τ:

$$P_{\text{cycl.}} = \frac{c}{M} N_{\text{A}} V\tau \left(\frac{3}{2\pi \langle r_{x2} \rangle} \right)^{\frac{3}{2}} \tag{7}$$

where c is the concentration, M the molar mass, N_{A} Avogadro's number, $\langle r^2 \rangle$ the mean square end-to-end distance of the linear precursor and V the total volume of the solution.

On the other hand, the probability of intermolecular reaction between an α-end and an ω-end of two distinct chains, P_{inter}, can be determined in the same way from the probability of these two end-functions being in the same volume τ:

$$P_{\text{inter}} = \frac{1}{3} \frac{c^2}{M^2} N_{\text{A2}} V\tau \tag{8}$$

The concentration at which both reactions occurs at the same extent, c_{eq} ($P_{\text{cycl}} = P_{\text{inter}}$), can be obtained from Eqs. (7) and (8):

$$c_{eq} = \frac{M}{\left(\left\langle r_x^2 \right\rangle\right)^{\frac{3}{2}}} \frac{1}{N_A} V\tau \left(\frac{3}{2\pi}\right)^{\frac{3}{2}} \tag{9}$$

To reach a high cyclization yield, the reaction therefore has to be performed far below this concentration.

5.1.3.1 Bimolecular End-to-End Cyclization

Direct coupling of α,ω-polymer dianions In this method, first proposed by Casassa [6], the polymer to be cyclized has two identical end-functional groups and the ring closure requires the use of a bifunctional coupling agent, as illustrated in Scheme 5.3. Living polymerization reactions are generally preferred since they yield linear precursors with a narrow molar mass distribution and controlled chain-end functionality.

Cyclic polystyrene Practical realization was achieved in 1980 with the preparation of cyclic polystyrene from a linear -difunctional polymer obtained by sodium or potassium naphthalene initiation and the reaction of the living polymer with an equimolar amount of *α,α′*-dichloro-*p*-xylene or *α,α′*-dibromo-*p*-xylene under high dilution [11–14, 42]. The bimolecular reaction between the difunctional polymer and the difunctional coupling agent proceeds in two steps. It yields first the formation of an α,ω-heterodifunctional intermediate from which intramolecular cyclization can proceed (Scheme 5.4). In addition to the expected ring closure reaction, the formation of undesirable high molar mass polycondensates, either linear or cyclic, generally takes place. Indeed, even when operating at very low end-group concentration, typically 10^{-4}–10^{-6} M, the reported experimental cyclization yields are often lower than expected from theoretical calculation using Eqs. (7)–(9). Among possible explanations, the exact bifunctionality during the ring closing reaction has been stressed by Hogen-Esch in a recent review on macrocyclization of vinyl polymers [23]. An efficient cyclization implies first the precise synthesis and the use of a linear precursor perfectly α,ω-difunctional. At very low concentration the carbanionic living ends are exposed

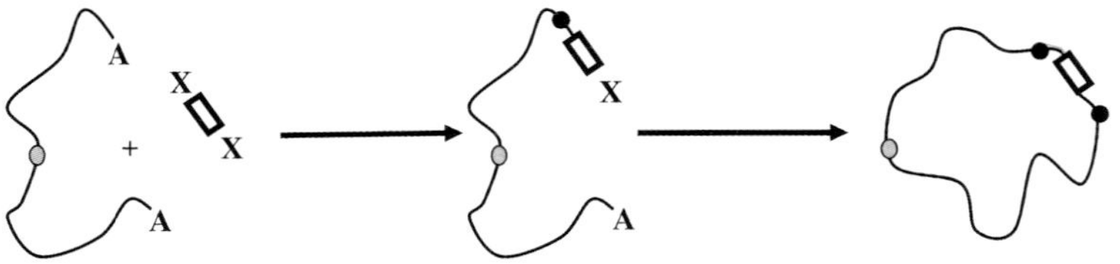

Scheme 5.3 Cyclization of an α,ω-difunctional polymer in the presence of a coupling agent under high dilution and at the stoichiometry between the two reactive functions.

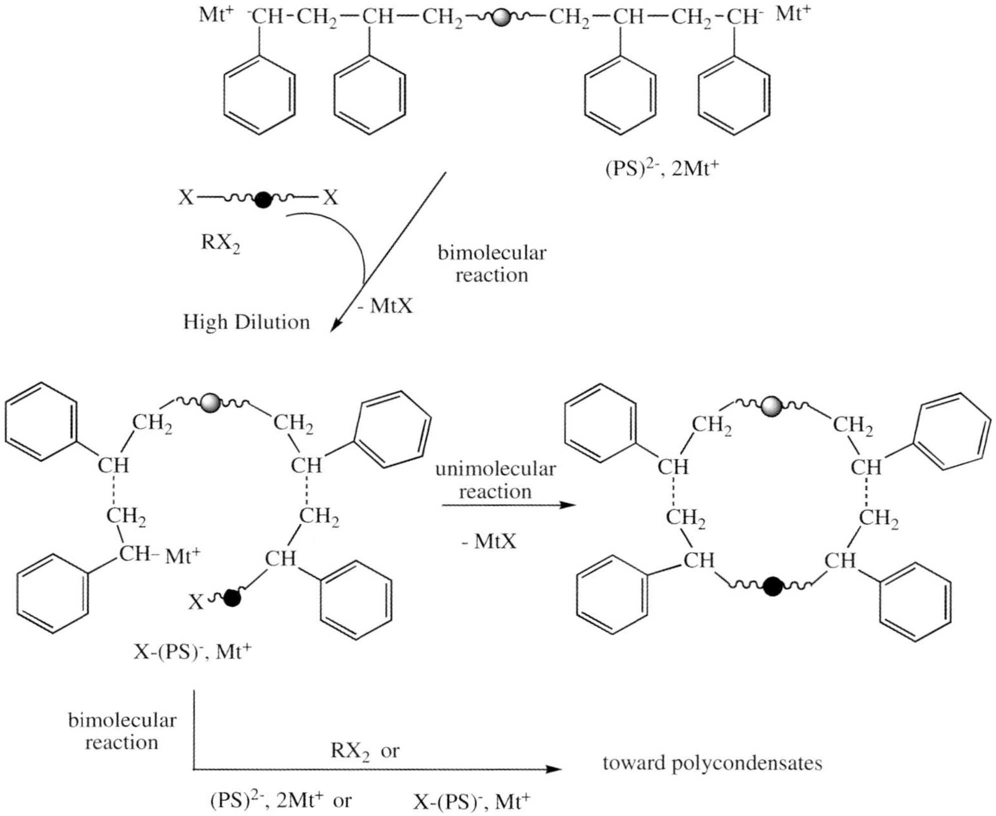

Scheme 5.4

to side-reactions, such as protonation by the solvent, especially in THF, or by a residual impurity [43, 44]. Furthermore, the reaction of carbanions with dihalides carries the risk of halide exchange and/or halide elimination reactions [45, 46]. The stoichiometry of the reactive groups is also an important factor in limiting coupling to cyclization or formation of polycondensates. This means that the living polymer and the coupling agent must be kept at the same low concentration during the ring closure. They should be added incrementally to the reactor. The coupling reaction can be monitored in the case of polystyrene by following the discoloration of the reddish carbanion.

A majority of groups who have subsequently prepared cyclic polymers following this strategy have proposed more or less important experimental modifications to improve the cyclization efficiency. Typical bifunctional initiators, coupling agents and reaction conditions used for the preparation of cyclic polystyrene are listed in Table 5.1. Most of the reactions were conducted in THF at very low temperature (–70 °C). Generally, the first characterization step is a size-exclusion chromato-

Table 5.1 Bifunctional initiators, coupling agents and reaction conditions used for the preparation of cyclic polystyrene by end-to-end bimolecular cyclization.

Synthesis of linear α,ω-precursor			End-to-end coupling		Ref.
Initiating system	Solvent	Polymer	Coupling agent	Solvent	
K naphthalenide	THF	$(PS)^{2-}$, $2K^+$	α,α'-Dichloro-p-xylene	THF	11, 12
K naphthalenide	THF–benzene	$(PS)^{2-}$, $2K^+$	α,α'-Dibromo-*p*-xylene	Benzene–THF	13, 42
Na naphthalenide	THF	$(PS)^{2-}$, $2Na^+$	α,α'-Dibromo-*p*-xylene	THF	14, 48
Na naphthalenide	THF–benzene	$(PS)^{2-}$, $2Na^+$	Dimethyldichlorosilane	Cyclohexane or THF	47
Li naphthalenide		$(PS)^{2-}$, $2Li^+$	1,3-Bis(1-phenylethenyl)benzene	THF or cyclohexane	49
Li naphthalenide	THF	$(PS)^{2-}$, $2Li^+$	α,α'-Dibromo-*p*-xylene, dibromomethane	THF	50
2,7-Dimethyl-3,6-diphenyloctane dilithium salt	THF	$(PS)^{2-}$, $2Li^+$	α,α'-Dibromo-*p*-xylene, dibromomethane	THF	51, 52
K naphthalenide	THF–benzene	PS-$(DPE)_2$	K naphthalenide	THF	53

graphic (SEC) analysis of the crude product, which allows the linear and cyclic polymers to be separated, thanks to the lower hydrodynamic volume of the ring chain. Fractional precipitation can also be applied since cyclic polymers have a θ temperature lower than the linear precursors [42, 47].

Cyclic poly(α-methylstyrene) A cyclization procedure similar to that for cyclic polystyrene was used for this monomer. It involves first the synthesis of poly(α-methylstyrene) (PαMS) by Li naphthalide initiated polymerization of αMS followed by end-to-end coupling with 1,4-bis(bromomethyl)benzene (DBX) under high dilution. The cyclization is carried out by simultaneous addition of THF solutions of the PαMS dianion and DBX at –100 °C. A sample of the matching linear PαMS is obtained by protonation of the PαMS dianion. The purified cyclics are obtained by fractional precipitation [54, 55]. Cyclization yields were in the range 25–75%, depending on the targeted molar mass, which ranges from 2500 to about 30 000 g mol^{-1}.

Cyclic polydienes α,ω-Polybutadiene (62% of 1,2-units) was prepared with potassium naphthalene in THF–hexane and cyclized in cyclohexane at 0 °C [56]. Cyclization was achieved by slowly adding the living polymer solution to a large volume of solvent with concurrent distillation of dichlorodimethylsilane as coupling agent. Madani et al. [57] prepared cyclic polyisoprene from a dilithium initiator in hydrocarbon medium and used dichlorodimethylsilane as coupling agent. The high cyclization yield observed (between 90 and 80% depending on the precursor molar mass) was explained by the intramolecular association of lithium carbanions under the cyclization conditions. This explanation is supported by the effect of the solvent polarity on the coupling reaction; the yield of cyclic compound falls from 88% in pure hexane to 53% in the presence of 15 vol% of THF [58].

Cyclic poly(2-vinylpyridine) A series of macrocyclic poly(2-vinylpyridine)s were synthesized from two-ended living dilithium polymer precursors by high dilution (about 10^{-5} M) coupling in THF with 1,2- and 1,4-bis(bromomethyl)benzene [59–62]. Reported cyclization yields ranged from 70 to 40% with increasing precursor molar mass. After fractionation, SEC showed that the macrocycles contained less than 5% of linear precursor and that the hydrodynamic size of the macrocycles was substantially (approximately 30%) less than that of the linear precursor. Both the linear and cyclic compounds were then converted into the corresponding polyelectrolytes by reaction with alkyl bromide [61].

Cyclic poly(2-vinylnaphthalene) and poly(9,9-dimethyl-2-vinylfluorene) Well-defined and narrow molecular weight distribution (MWD) macrocyclic poly(2-vinylnaphthalene) (P2VN) [63] with $\overline{DP}_n$ up to 120 and poly(9,9-dimethyl-2-vinylfluorene) (PDMVF) with $\overline{DP}_n = 140$ [64–66] were recently prepared from potassium naphthalide (K-Naph)-initiated polymerization of 2VN or DMVF in THF at $-78\,^\circ$C. End-to-end coupling of the resulting dianions was achieved under high dilution conditions (10^{-6}–10^{-4} M) with 1,4-bis(bromomethyl)benzene or 9,10-bis(chloromethyl)anthracene. These polyaromatic macrocycles were extensively studied by Hogen-Esch for their exceptional fluorescence properties [23].

Cyclic block copolymers In addition to the cyclic homopolymers described previously, various types of cyclic AB block copolymers have been prepared using the same strategy. As described later, one obvious interest is the comparison of the domain formation of cyclic diblock copolymers with their linear homologues in solution and in the bulk. Cyclic polystyrene-*b*-polydimethylsiloxane was prepared in THF by cyclization of the two-ended triblock polydimethylsiloxane-*b*-polystyrene-*b*-polydimethylsiloxane with Li counterion in the presence of dichlorodimethylsilane [67–69]. Cyclization yields estimated by SEC were in the range 70–80% for overall molar masses ranging from 3000 to about 80 000 g mol^{-1}.

The synthesis of polystyrene-*b*-polyvinylpyridine [70, 71] with a molar mass of 10 000 g mol^{-1} was also reported. The block copolymer was synthesized by the lithium naphthalide-initiated sequential polymerization of styrene and 2-vinylpyridine followed by the end-to-end coupling with 1,4-bis(bromomethylbenzene) in THF. The block copolymer was isolated by precipitation–extraction procedures.

Cyclic polystyrene-*b*-butadiene systems were prepared using the dilithium derivative of 1,3-bis(1-phenylethylenyl)benzene for initiating the sequential polymerization of butadiene and styrene in benzene in the presence of excess of lithium alkoxide. The triblock copolymer was then cyclized in cyclohexane under high dilution with dichloromethylsilane [72, 73]. A similar methodology for the synthesis of cyclic polystyrene-*b*-polybutadiene diblock copolymers was applied by Iatrou et al. [74]. This involved the reaction of (1,3-phenylene)bis(3-methyl-1-phenylpentylidene)dilithium initiator with butadiene in the presence of *sec*-BuOLi, followed by polymerization of St (or St-d_8). The cyclization of the resulting α,ω-difunctional triblock copolymer was performed by using bis(dimethylchlorosilyl)ethane, under high dilution conditions. The copolymers were characterized by SEC, membrane osmometry, NMR and UV spectrometry and viscometry and their organization and morphology were compared with those of the parent linear triblock PS (or PS-d_8)-*b*-PBd-*b*-PS and (PS-d_8)-*b*-PBd analogues. Similarly, a polystyrene-*b*-oligo(2-*tert*-butylbutadiene)-*b*-polystyrene triblock copolymer was prepared from lithium naphthalenide and end-functionalized using 2 equiv. of 1-[3-(chloropropyldimethylsilyl)phenyl]-1-phenylethylene yielding α,ω-diphenylethylene-ended polymer [75]. This precursor was cyclized in high dilution conditions by addition of potassium naphthalenide. After isolation of the cyclic low molar mass fraction ($\overline{M}_n = 30\,000$ g mol^{-1}), the polymer was ozonized for decomposition of the oligo(2-*tert*-butylbutadiene) sequences to reform a linear chain. This confirmed that the fractionated product was an 86% ring molecule.

Chemical transformation of the anionic ends before ring closing Various organic groups have been introduced through deactivation of linear α,ω-difunctional precursors before the end coupling process in order to generate more stable end-groups. Linear α,ω-dibromopropyl poly(isoprene-*b*-styrene-*b*-isoprene) copolymers were first prepared by coupling the corresponding triblock dianions end-capped with 1,2-diphenylethylene with a large excess of 1,3-dibromopropane. Then end-to-end ring closure of the α,ω-dibromopropyl triblock copolymer was performed by interfacial condensation between the aqueous phase (1,6-diaminohexane) and the organic phase (triblock copolymer). This approach was reported to be very effective and gave a high cyclization yield for a 20 000 g mol^{-1} copolymer [76].

5.1.3.2 **Unimolecular End-to-End Cyclization**

In the bimolecular cyclization process, the ring closure process involves a one-pot, two-step reaction: (a) the bimolecular reaction between one of the polymer ends and one function of the difunctional coupling agent to form an α,ω-heterodifunctional chain and (b) the unimolecular ring closure corresponding to the reaction between the α- and ω-polymer ends. The high dilution required to favor cyclization versus chain extension is not favorable to the quantitative formation of the heterodifunctional polymer intermediate (Scheme 5.3). To overcome

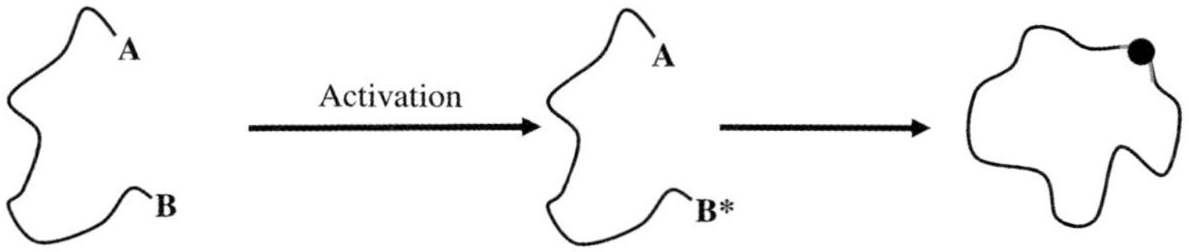

Scheme 5.5 Cyclization of an α,ω-heterodifunctional polymer by selective activation of one end-functional group.

this difficulty, another approach involves the direct synthesis of an α,ω-heterodifunctional linear precursor. The cyclization is then performed in a separate step under high dilution by selective activation of one of the active ends to allow it to react with the other end-function of the same polymer chain, as illustrated in Scheme 5.5.

Cyclic homopolymers by unimolecular end-to-end cyclization The preparation of cyclic polymers from α,ω-heterodifunctional polymers was explored first by Deffieux and coworkers [77–82]. In this approach, a linear polymer precursor is first prepared using a heterodifunctional initiator. Then, the growing chain end is activated at the end of the polymerization to react under dilute conditions with the head polymer function to close the ring. This strategy is outlined in Scheme 5.6 for the preparation of an α-styrenyl poly(2-chloroethyl vinyl ether) chain by cationic polymerization of 2-chloroethyl vinyl ether initiated from a styrenyl-functionalized α-iodoethyl vinyl ether, in the presence of zinc dichloride as catalyst [77]. Under these conditions, the styrenic double bond is unaffected. Cyclization is performed at −10 °C by dropwise addition of the living polymer to a solution of $SnCl_4$ in toluene: $SnCl_4$ activates the α-iodo ether terminus and causes addition to the styrenic double bond. SEC analysis showed 80% cyclization.

Cyclic polystyrenes have been prepared in a similar way [78, 83]. Narrow MWD α,ω-heterodifunctional polystyrene is first prepared anionically starting from lithiopropionaldehyde diethylacetal. After conversion of the styryl anion to diphenylethenyl anion, the polymer is deactivated by p-chloromethylstyrene (Scheme 5.7). Then the acetal group is first converted into α-iodo ether with trimethylsilyl iodide and the cyclization is performed by addition of the linear polymer into a large volume of toluene containing tin tetrachloride or titanium tetrachloride as cationizing agent, thus allowing a fast reaction with the styryl end-group. Cyclization yields of 90% were obtained for PS molar masses of 12 000 g mol^{-1}.

In a modified approach, the active chain was reacted with another functional terminating agent to introduce different specific end-groups, which can be reacted with other type of activation [84, 85] (see Table 5.2). The use of other conventional organic reactions for the cyclization of heterodifunctional linear precursor has also been reported. Macrocyclic polystyrene with $\overline{DP}_n$ 20–35 were prepared from α-carboxyl-ω-amino bifunctional polystyrene through the formation of an amide bond. The linear precursor was synthesized from 3-lithiopro-

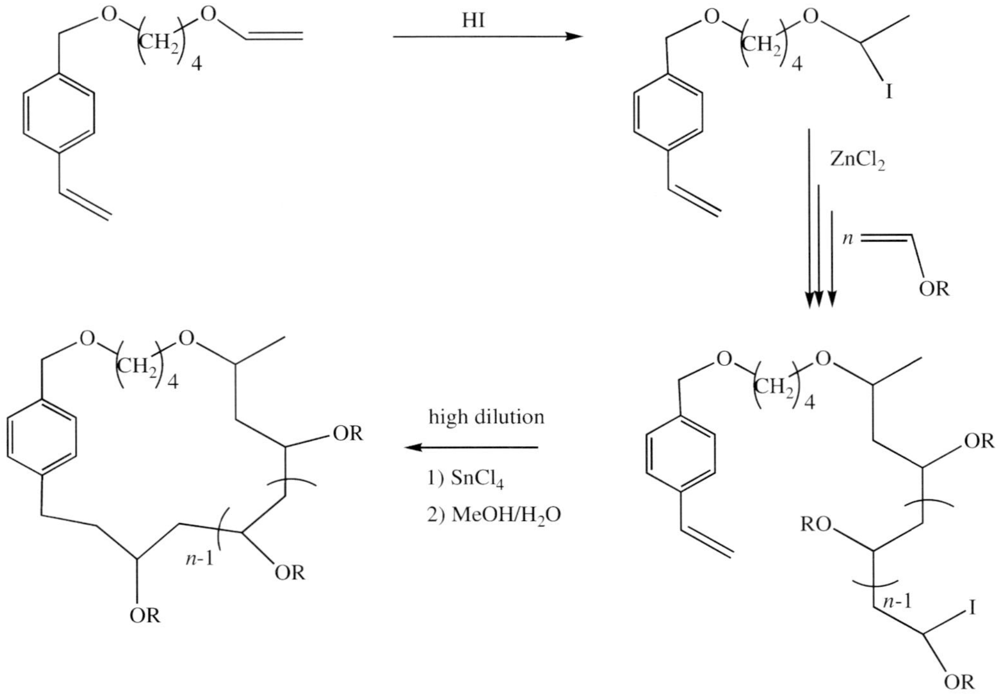

Scheme 5.6

pionaldehyde diethylacetal as the functional initiator and 2,2,5,5-tetramethyl-1-(3-bromopropyl)-1-aza-2,5-disilacyclopentane as the functional terminating agent. Cyclization was achieved after deprotection under high dilution [86].

The direct intramolecular coupling of an α-acetal-ω-bis(hydroxymethyl) hetero-difunctional polystyrene precursor prepared by living anionic polymerization of styrene initiated by 3-lithiopropionaldehyde diethylacetal and terminated by iso-propylidene-2,2-bis(hydroxymethyl)-1-(1-chloroethoxyethoxy)butane has also been reported [84, 85]. Cyclization under high dilution, through acetalization catalyzed by a mild acid, yields crude polymer containing more than 90% of a fraction of cyclic polystyrene ($\overline{DP}_n$ from 13 to more than 300).

Very recently, the high efficiency of click chemistry was used to cyclize α-alkyne-ω-azido polystyrenes ($\overline{M}_n = 2200$ and 4200 g mol^{-1}) prepared by ATRP from an initiator containing an alkyne function and substitution of the bromide terminus in the presence of sodium azide at the end of the polymerization [87]. Both the functionalization and the cyclization appeared to be nearly quantitative, preventing the need for fractionation. So far, however, only the preparation of low molar mass oligomers has been reported. This new technique is expected to be particularly useful owing to the functional group tolerance of ATRP and

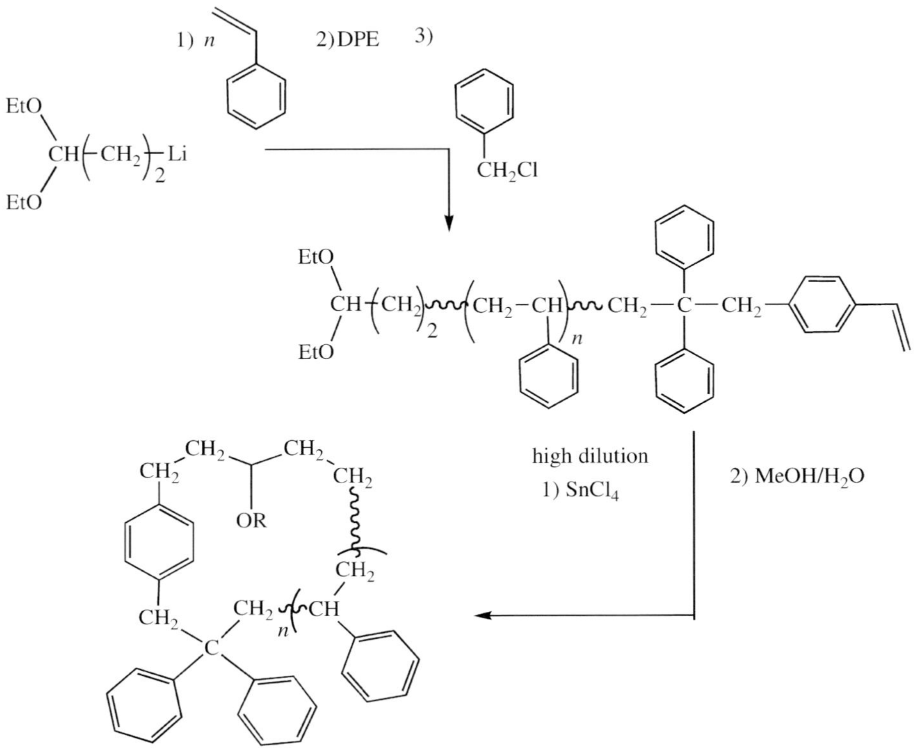

Scheme 5.7

Table 5.2 Functional groups of α,ω-heterodifunctional polymers and activation process for the preparation of cyclic polymers by end-to-end unimolecular cyclization.

Head group	Activation	End-group	Polymer	Ref.
Vinyl ether	HI–$SnCl_4$	Styrenyl	Poly(chloroethyl vinyl ether)	77
Acetal	TMSi–$TiCl_4$	Styrenyl	Polystyrene	78, 80, 88
Acetal	TMSi–$TiCl_4$	Styrenyl	Poly(styrene-*b*-chloroethyl vinyl ether)	82
Acetal	TMSi–$TiCl_4$	Styrenyl	Poly(styrene-*b*-ethylene oxide)	89
Carboxyl	–	Amino	Polystyrene	86
Acetal	HCl–MeOH	Bis(hydroxymethyl)	Polystyrene	84
Acetal	HCl–MeOH	Bis(hydroxymethyl)	Poly(styrene-*b*-isoprene)	85
Amino		Carboxyl	Poly(styrene-*b*-isoprene-*b*-methyl methacrylate)	74, 90
Alkyne	CuBr	Azido	Polystyrene	87

should allow the preparation of cyclic polymers from a broad variety of functional monomers.

In all these systems, the parent linear structures having the same molar mass are easily accessible for direct comparison with the cyclized macromolecules.

Cyclic block copolymers by unimolecular end-to-end cyclization Deffieux et al. made cyclic poly(styrene-*b*-chloroethyl vinyl ether) copolymers from an anionically prepared α-styrenyl-ω-acetal polystyrene [82]. They converted the acetal to an α-iodo ether function in the presence of trimethylsilyl iodide and polymerized the chloroethyl ether monomer cationically. Ring closure catalyzed by $SnCl_4$ yielded block copolymers with molar masses ranging from 4000 to 7500 g mol^{-1}. More recently, they described the synthesis of cyclic poly(styrene-*b*-isoprene) copolymers of various molar masses and composition using the sequential anionic polymerization of styrene and isoprene initiated from 3-lithiopropionaldehyde diethylacetal and terminated by isopropylidene-2,2-bis(hydroxymethyl)-1-(1-chloroethoxyethoxy)butane [85]. Cyclization was catalyzed under high dilution by a mild acid. Cyclic diblocks with $\overline{DP}_n$ of PS and PI ranging from 70 to 300 and from 70 to 100, respectively, were obtained in very high yields (>90%).

The synthesis of a cyclic triblock terpolymer, polystyrene-*b*-polyisoprene-*b*-poly(methyl methacrylate), was achieved by an end-to-end intramolecular amidation reaction of the corresponding linear α,ω-amino acid precursor (S-*b*-I-*b*-MMA) under high dilution conditions [74, 90]. The linear precursor was synthesized by the sequential anionic polymerization of styrene, isoprene and MMA with 2,2,5,5-tetramethyl-1-(3-lithiopropyl)-1-aza-2,5-disilacyclopentane as an initiator and 4-bromo-1,1,1-trimethoxybutane as a terminator. The separation of the unreacted linear polymer from the cyclic terpolymer was facilitated by the transformation of the unreacted species into high molar mass polymers under high concentration conditions. The yield of the final cyclic terpolymer of molar mass about 10 000 g mol^{-1} was estimated by SEC as 90%.

5.1.3.3 Ring Closure of Pre-cyclized Systems

Pre-organization of macromolecular precursors bearing specific ionic end-functions via electrostatic non-covalent interactions is an interesting and original strategy that was recently applied to the preparation of various types of chain architectures. This approach was thoroughly investigated in recent years by Tezuka and coworkers for the preparation of a broad diversity of cyclic polymers [91–98]. The strategy used for the synthesis of a cyclic polymer is illustrated in Scheme 5.8.

In nonpolar solvents, polymers having ionic end-groups, ionomers, tend to form clusters by aggregation through columbic interactions. The aggregation state and dynamics of exchange are significantly influenced by the concentration of the polymer solution. Thus, in diluted media, ionic aggregates tend to dissociate to allow the formation of single molecules. When the ionic functions

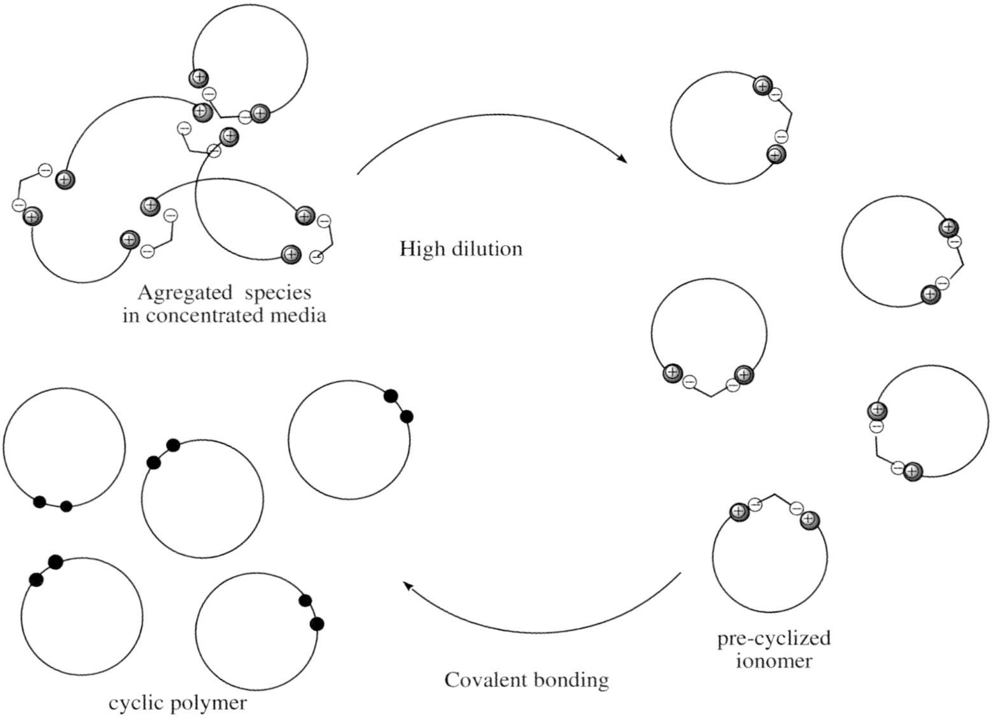

Scheme 5.8

of these telechelic polymers are associated to divalent counterions, they tend to organize thermodynamically into pre-cyclized structures that can be selectively converted to permanently fixed covalent cyclics. This approach was applied first to the preparation of cyclic poly(tetrahydrofuran) starting from a chain precursor having cyclic ammonium salt end groups carrying various dicarboxylate counterions (Scheme 5.9). Upon dilution in an organic solvent to below 1 g L^{-1}, the ionically aggregated polymer precursors completely dissociate into an intrachain assembly system. The subsequent covalent fixation through the ring-opening reaction of the cyclic ammonium salt group by the carboxylate entity produces a covalent bonded cyclic polymer [91, 93, 99]. The cyclization yields reported for the synthesis of poly(THF)s with molar masses ranging from 4000 to 12 000 g mol^{-1}, corresponding to 260–800-membered rings, are higher than 90–95%.

For the synthesis of cyclic polystyrenes, telechelic precursors having quinuclinidium salt groups carrying a difunctional terephthalate counterion were used [99]. This ionically linked pre-cyclized polymer precursor was then subjected to heat treatment in diluted conditions to convert the ionic ends into a stable covalent linkage.

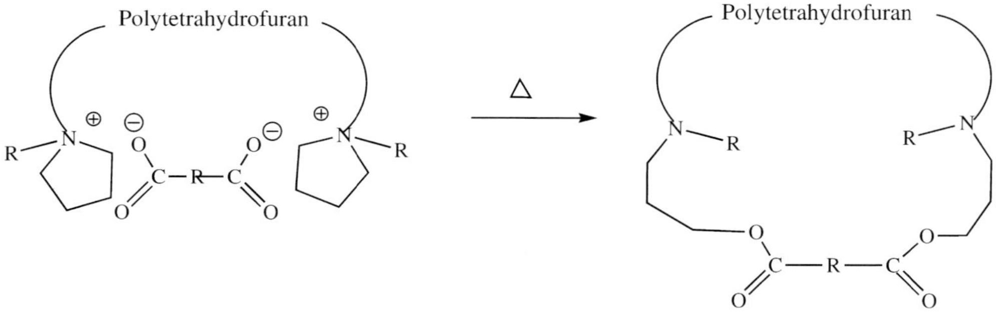

Scheme 5.9

5.1.3.4 **Other Cyclic Architectures**

Various types of polymer architectures based on the combination of cyclic and linear chains or on the association of two or more macrocycles have been synthesized. Some selected examples are illustrated in Scheme 5.10. Their methods of preparation are briefly described in this section.

Hemicyclic polymers or tadpoles Tadpoles are polymers consisting of a macrocyclic part attached to one or more linear chain fragments (Scheme 5.10, structures a, b, c).

A tadpole polybutadiene with two linear polystyrene chains (Scheme 5.9 b) has been prepared by Quirk and Ma [49, 72]: living anionic polystyryllithium chains were first coupled by reacting the living ends with the double diphenyl-

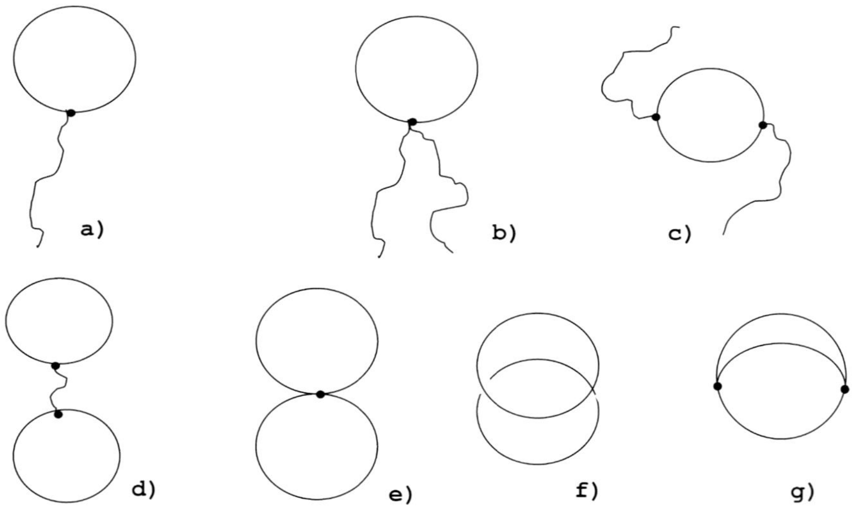

Scheme 5.10

ethylene. Then the resulting polymer bearing two reactive carbanionic centers located in the middle of the chain was used as a bifunctional macroinitiator to initiate the anionic polymerization of butadiene. The two polybutadienyllithium ends were finally coupled under high dilution by dichlorodimethylsilane to form a polybutadiene ring bearing two linear polystyrene branches.

Tadpole polyisoprenes have also been synthesized by reaction of α,ω-dilithio-polyisoprenes with 1,2-(4′-isopropenylphenyl) ethane as linking agent. The ring formed was shown to retain its living character and allowed the synthesis of rings having two arms after addition of isoprene. A bicyclic structure could also be formed after a second cyclization reaction [58].

Deffieux et al. prepared a cyclic poly(chloroethyl vinyl ether) [poly(CEVE)] with a polystyrene tail [100]. Its preparation involves the synthesis of a heterofunctional poly(CEVE)-*b*-PS linear precursor bearing a catiogenic function at the poly(CEVE) chain end and a styrenyl group located at the junction between the poly(CEVE) and PS blocks. The cyclization was then performed in a separate step, by slow addition of the polymer precursor to a large volume of toluene containing $TiCl_4$ as activator for the intramolecular coupling reaction. Tadpole diblock copolymers with molar masses in the range 6000–9000 g mol^{-1} were recovered and characterized.

More recently, Kubo et al. [101] prepared a cyclic polystyrene by end coupling an α-carboxylic-ω-amino linear polystyrene precursor. After transformation of the amide linkage into an amine function by hydrogenation, they used it as an anchoring site to attach an α-carboxylic polystyrene chain.

Tezuka and coworkers, using the electrostatic self-assembly approach, designed a series of well-defined linear poly(THF) precursors bearing pyrrolidinium salts located in the interior of the chain, from which poly(THF) tadpoles constituted of a ring and two branches were prepared [102].

Bicyclic and polycyclic polymers Bicyclic polymer topologies have also been constructed using various approaches. Antonietti [103] prepared difunctional living polystyrene from sodium naphthalene at low temperature. The double cyclization was performed by adding 1,2-bis(dichloromethylsilylethane). The cyclization yield was estimated as around 30% for a polymer of $\overline{M}_n=5600$. Madani et al. [57] reacted α,ω-dilithiopolyisoprenes with $SiCl_4$ as coupling agent to form directly an eight-shaped polymer. Yields of about 70% of cyclic dimers were observed. The high cyclization efficiency reported was attributed to the formation of cyclic aggregates, even at low organolithium concentration, leading to a precyclization of the PS chains as proposed in the electrostatic self-assembly approach. As previously described (Section 5.1.3.4) starting from a cyclized polyisoprene synthesized by reaction of α,ω-dilithiopolyisoprene with 1,2-(4′-isopropenylphenyl)ethane as linking agent, Sigwalt and coworkers used the two carbanionic functions located on the polyisoprene ring to grow two polyisoprene arms that were finally cyclized by addition of dichlorodimethylsilane, yielding a eight-shaped bicyclic polyisoprene [58].

Cyclic polystyrenes having an amide moiety in the main chain were converted by lithium aluminum hydride to macrocyclic amines [101]. The reaction of the macrocyclic amines with glutaric acid gave eight-shaped polystyrenes.

The synthesis of bicyclic polymers has also been approached differently starting from a heterotetrafunctional initiator bearing two acetal end-functions and two styrenyl groups [104]. The two acetal ends were used to polymerize chloroethyl vinyl ether by living cationic polymerization to yield a linear precursor with two acetal ends, the initial styrenyl groups being located in the middle of the poly(CEVE) chain. The cyclization was achieved under high dilution by cationization of the acetal termini by a strong Lewis acid, which allows them to react rapidly with the styrenyl groups. Bicyclic polymers with molar masses ranging from 2500 to 6500 have been prepared, isolated and characterized. Structural characterization of the cyclized poly(CEVE) by SEC and NMR supports the bicyclic structure.

A series of tricyclic poly(CEVE)s with molar masses ranging from 6000 to 15 000 g mol^{-1} was synthesized using the same principles starting from an initiator molecule containing three styrenic and three vinyl ether groups [83, 105]. High cyclization yields were also obtained without the detectable formation of polycondensates resulting from intermolecular reaction. The effect of the cyclization on the apparent molar mass and on the glass transition temperature was examined by GPC and DSC.

Bicyclic and tricyclic poly(THF) were also constructed by using telechelic poly(THF) having *n*-phenylpyrrolidinium salt groups [98, 99, 106]. The triflate counterions were replaced by tetra- and hexacarboxylate counterions. These polymer precursors form in organic media di- or tricyclic aggregates under appropriate dilution conditions. Upon heating the ionic salts are selectively converted into the corresponding stable covalent linkages, thus yielding di- and tricyclic poly(THF). VPO, NMR and SEC analysis of the cyclized polymers confirmed the formation of the expected polymer architectures.

The synthesis of catenated polystyrene-*cat*-polyvinylpyridine with a molar mass of 10000 g mol^{-1} was also reported [107, 108]. Polyvinylpyridine was synthesized by the lithium naphthalide-initiated sequential polymerization of 2-vinylpyridine. The "catenane" copolymer was obtained by end-to-end coupling of the P2VP dianion lithium salt with 1,4-bis(bromomethylbenzene) in THF in the presence of a PS macrocycle (M_p=4500 g mol^{-1}). The catenane was isolated by precipitation–extraction procedures.

The preparation using a similar strategy of polymers with other complex cyclic architecture such as "manacle", cyclic with functional polymerizable groups [96, 99], polycyclic [109, 110] and so forth, has also been reported.

5.2
On the Physics of Cyclic Macromolecules

As stated above, the growing interest reported since the 1940s in cyclic polymers can be attributed both to their presence and specific role in living systems, for instance the DNA ring in the organization of living cells, and to their very unique properties compared with their linear counterparts. They are also of fundamental interest because they provide a simple macromolecular model that can help in elucidating the basic concepts of the physical properties of macromolecular compounds. The static and dynamic properties of cyclic polymers in solution present some differences when compared with the case of linear chains [22, 111]. These differences originate in particular from their thermodynamic behavior, conformational characteristics and evolution in time (dynamics). An example of these facts is the shift in the theta temperature observed in dilute solutions of cyclic polystyrene in cyclohexane, a theta solvent. For instance, it has been shown in dilute solutions of ring polystyrene (r) in cyclohexane using static neutron scattering [112] that the form factor can be represented by the so-called Casassa's function Pr(q) [6]:

$$P_r(q) = \frac{2}{\sqrt{u}}\, e^{-(u/4)} \int_0^{\sqrt{u/2}} dx\, e^{x^2} \tag{10}$$

with $u = q^2 N\sigma^2/6$, where N is the degree of polymerization, σ the length of the statistical segment and q the scattering wavevector, $q = (4\pi/\lambda)\sin(\theta/2)$, where λ is the wavelength of the incident radiation and θ the scattering angle. Neutron scattering experiments made in dilute solutions of linear chains (l) under the same conditions reveal that the form factor satisfies the known Debye function [113]:

$$P_l(q) = \frac{2}{u^2}(e^{-u} + u - 1) \tag{11}$$

The radii of gyration extracted from these data are found to satisfy the relationship

$$Rg_l = \sqrt{2}Rg_r \tag{12}$$

where the subscripts l and r refer to linear and ring polymers, respectively.

As an illustration of such a comparison, the results obtained using neutron scattering on linear and cyclic polystyrene [114] demonstrated clearly that the experimental results and the theoretical predictions are in good agreement. Indeed, the difference in scattering behavior between a cyclic and a linear macromolecule can be easily highlighted by plotting $q^2P(q)$ versus q. Such a representation shows a maximum at intermediate angles observed for the cyclic polymer and is due to the higher segment density of the cyclic macromolecules. At larg-

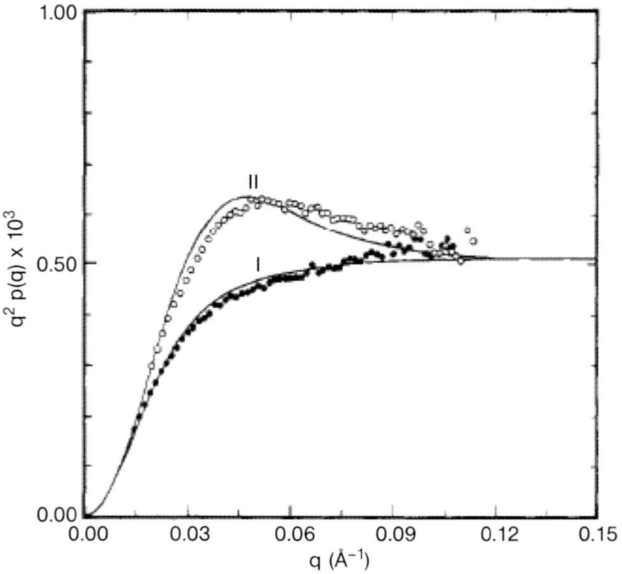

Fig. 5.1 SANS – asymptotic behavior of the particle scattering of (I) a linear polystyrene and (II) a cyclic polystyrene [114].

er angles, the same limit is approached, since then it is no longer possible to observe whether a segment belongs to a cyclic or a linear macromolecule.

Additionally to the radius of gyration and the form factor for linear and cyclic polymers, the hydrodynamic and the intrinsic viscosity are other important parameters that can be easily measured and used to distinguish between the two architectures (ring and linear). Their ratio in a theta solvent is [115, 116]

$$R_{h,r}/R_{,1} = 8/(3\pi) \tag{13}$$

$$[\eta]_r/[\eta]_1 = 0.662 \tag{14}$$

5.2.1
Cyclic Block Copolymer Systems

To extend our understanding of cyclic polymer chains, there has been growing interest recently in extending such architecture (cyclization) to the case of diblock or triblock copolymer systems. The motivations are substantially based on the challenge, in modern colloid chemistry, to find ways to control the size and morphology of particles at the nanometric scale. It is generally recognized that two different polymers will be thermodynamically incompatible with each other provided that specific interactions between the two species do not lead to a negative heat of mixing. Particularly if A and B polymers in an $(A–B)_l$ diblock (or

A–B–A triblock) polymer chain have different cohesive energy densities, they tend to phase separate in the segregation limit $(\chi N) > (N\chi c)$. However, this separation cannot be extended to the macroscopic spatial scale but is limited to molecular dimensions because of the connectivity of A and B. When the two extremities of both species A and B are joined to form a cyclic $(A–B)_r$, it leads to sensitive changes for the static and dynamic scattering properties of the system with respect to those of linear diblock copolymer chain. The aim of this section is to discuss these changes on a molecular scale (single chain) and particularly on the self-assembled structures in solution and in the bulk. Indeed, one of the most interesting and fascinating properties of AB diblock copolymers is their ability to self-assemble into micelles (and other ordered structures such as lamellae and vesicles) if dissolved in a so-called "selective" solvent, i.e. a solvent that is thermodynamically good for one block and poor for the other. Such micelles are characterized by their core–shell structure. In an aqueous environment, the hydrophobic blocks of the copolymer are segregated from the aqueous exterior and form the inner core, whereas the hydrophilic blocks are located in the outer shell or corona. Micelles, microemulsions, vesicles and a huge variety of microphase textures may be used as structural templates for new materials. The increasing number of industrial applications of these systems has attracted great interest in both applied science and basic research. For instance, copolymer micelles have been used as stabilizers in organic reactions and also as nanotemplates for nanotechnology and drug delivery applications.

Most of the published work so far has dealt with *linear block copolymers*. The size and thereby the aggregation number of the micelles are independent of the polymer concentration, but may change with the temperature or the copolymer composition (or volume fraction). They usually have a spherical shape, with a very compact core and a more flexible corona. However, it was also shown in numerous studies that block copolymers can self-assemble into a large variety of objects such as rod-like or worm-like micelles, cylinders, vesicles, hollow spheres and branched tubules. Commercial applications of these different nanostructures call for a high degree of control on the morphology of the resulting materials. As has been demonstrated by Bates, Lodge and coworkers during the past decade [117, 118], several parameters such as temperature, copolymer composition and concentration and sample preparation can be modified and controlled in order to tailor the micellar morphology. The effect of adding salt or changing the pH of the medium on the properties of micelles consisting of polyelectrolyte-type coronas [e.g. polystyrene-*b*-poly(acrylic acid) (PS-*b*-PAA) micelles] has been extensively studied by Eisenberg and coworkers [119].

Other parameters such as the architecture of the block copolymer, namely the cyclization, as described above, have been shown to play a very important role [120–134] in the structure and dynamics of single block copolymer chain properties (disordered phase) and in the micellar morphology (solution) and the nano-organized phases in the bulk state (films), as will be discussed below.

In contrast to linear polymers, cyclic and multi-cyclic polymers present no end-groups and one might expect new architectural properties as compared with their linear homologue polymers.

5.2.1.1 From Disordered to Ordered Phases in Cyclic Diblock Copolymers A–B: Linear vs. Cyclic

To understand better the self-assembly properties in cyclic block copolymer systems, it is important to recall some fundamental aspects in the disordered phase (good solvent for both blocks). Theoretical approaches [120–126] have been developed showing a substantial difference in the elastic and dynamic properties when comparing neutral diblock copolymers having different architectures. This is illustrated, for instance, in Fig. 5.2, showing the scattered intensity as a function of the wavevector at different χFN values (χ is the interaction parameter). For instance, in the case of 50 : 50 (symmetrical system, $N_A = N_B = N/2$), the microphase separation occurs at 10.2 for a linear diblock copolymer and 17.7 for the equivalent cyclic ones. This figure illustrates also the effect of morphology for coil–coil, rod–coil and rod–rod structures.

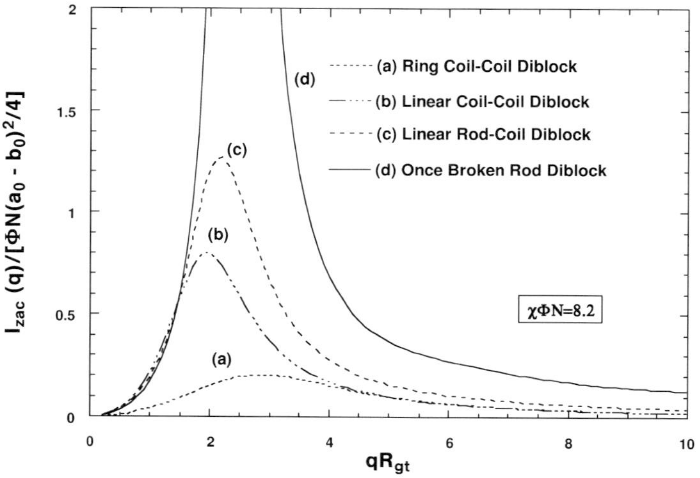

Fig. 5.2 Normalized scattering intensity $I(q)$ versus qR_{gt} for different architectures [120].

So far, no experimental evidence has been clearly established showing such differences that are great indicators for different self-assembled nanostructured phases (solutions and films) in both systems. However, preliminary experiments [127, 128] in the disordered phase on linear and cyclic PS*h*-PS*d* (compatible system where the interaction Flory–Huggins parameter $\chi=0$) and on PS–PDMS–PS (incompatible system where $\chi\neq0$) were carried out using SANS and light scattering, respectively, and showed a very good agreement with the theoretical predictions [120–125].

Self-assemblies in solutions (linear vs. cyclic) Amphiphilic AB diblock copolymers have the ability to self-assemble into micelles (and other ordered structures such as lamellae and vesicles) if dissolved in a so-called "selective" solvent, i.e., a solvent that is thermodynamically good for one block and poor for the other. Such micelles are characterized by their core–shell structure. So far no theoretical models have been developed that predict drastic changes, as will be discussed below, in the self-assembled structure of cyclic block copolymer systems as compared with their linear counterparts in solutions. Very interesting results [129–136] were obtained when comparing linear and cyclic PS–PI, PS–PB or PEO–PPO diblock copolymer systems in selective solvents. Here we repeat some of them obtained in the case of PS–PI, the most studied system, where heptane and decane are selective for the PI (bad solvent for the PS block). These results are illustrated in Fig. 5.3 (DLS experiments), SAXS under shear (Fig. 5.4) and AFM (Fig. 5.5 a and b) and show clearly that micelles made from linear block copolymers are monodisperse and spherical (50 nm in diameter) and those obtained from cyclic block copolymers are giant worm-like micelles (µm in length and 30 nm in cross-section).

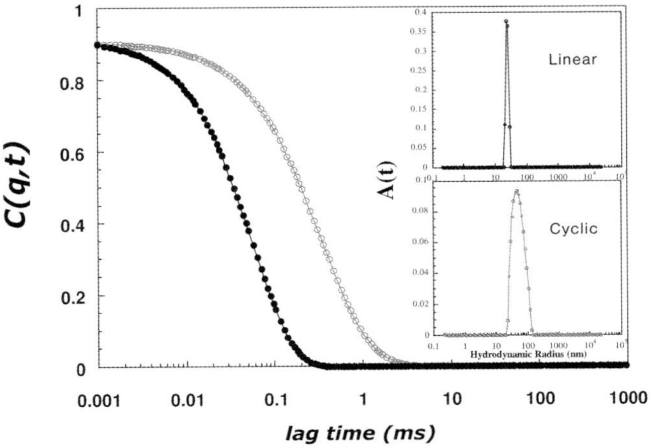

Fig. 5.3 Auto-correlation functions obtained using DLS on linear (bottom curve) and cyclic (top curve) PS_{290}–PI_{110} in heptane (selective solvent for PI) at a concentration of 0.01%.

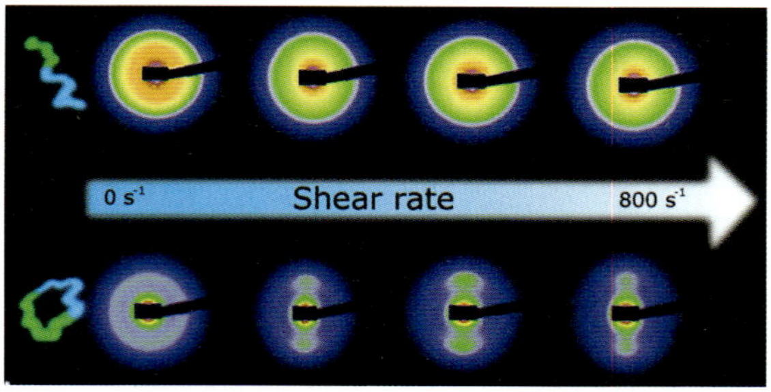

Fig. 5.4 SAXS under shear (ESRF) showing the difference in the response of the applied shear for linear and cyclic PS_{290}–PI_{110} in heptane (selective solvent for PI).

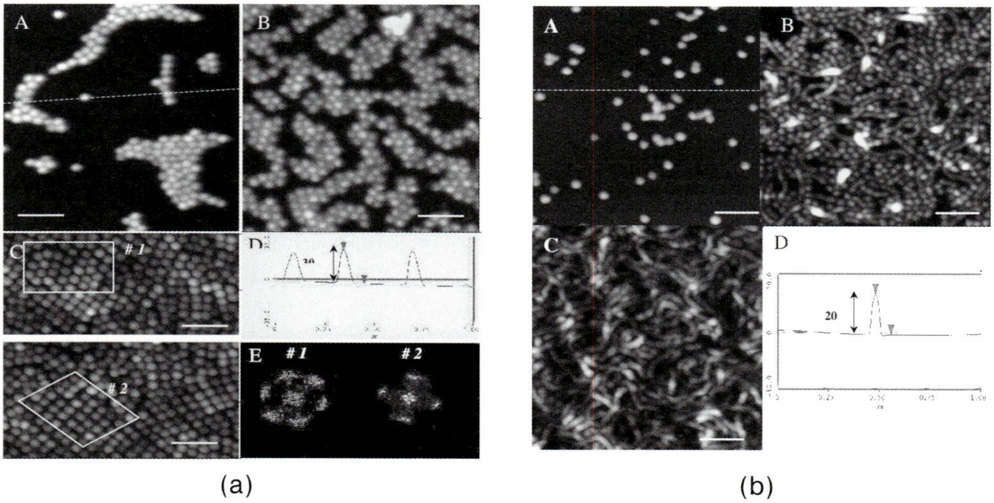

Fig. 5.5 AFM of self-assembled objects obtained from (a) linear (spherical micelles) and (b) cyclic PS_{290}–PI_{110} (giant worm-like micelles) (scale 200 nm).

It is clear through those experiments on cyclic PS–PI that the cyclization of a linear diblock copolymer chain leads to extremely interesting and different morphologies and properties. It is also worth noting that at relatively high concentrations, vesicles are observed in the case of PS–PI cyclic block copolymer in heptane or decane (Fig. 5.6). However, in the same range of concentration only spherical nano-objects (micelles) are observed for the linear PS–PI block copolymer in heptane or decane.

Self-assemblies: effect of additives (homopolymer PS of low molar mass) The morphology of the self-assembled structures could also be monitored by adding a homopolymer PS (low *DP*) to both the linear and cyclic amphiphilic diblock

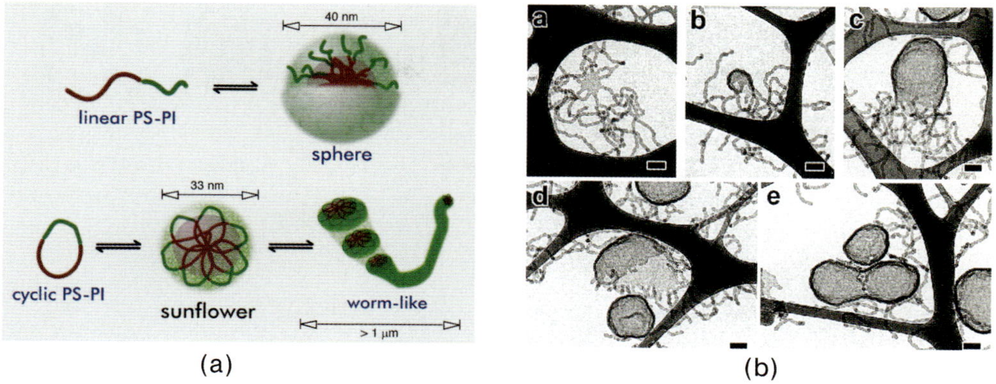

Fig. 5.6 (a) Cartoon showing the difference in the self-assembly in linear and cyclic diblock copolymer PS–PI in heptane (selective solvent for PI). (b) TEM and cryo-TEM: formation of vesicles obtained only in cyclic copolymer systems PS_{290}–PI_{110} (scale 100 nm) at higher concentrations.

copolymers. Such an effect has been already shown in the case of linear and cyclic PS–PI. Addition of low molar mass PS moves the linear diblock copolymer system from spherical micelles to giant worm-like cylinders. Addition of low molar mass PS moves the cyclic case from giant worm-like cylinders to vesicles [136].

5.2.1.2 Nano-organization in Amphiphilic Cyclic Block Copolymer Films

Block copolymer systems, including diblock A–B and triblock A–B–A, constitute a very important class of polymers that lead to fascinating morphologies with a broad spectrum of microstructures/organized morphologies (lamellae, hexagonal cylinders, cubic arrays of spheres, gyroids, etc.) on length scales of 10–100 nm in particular when cast from solutions (in a good solvent for both blocks A and B). The morphologies (shape and size) of the micelles and the bulk organized structures in the ordered phases depend strongly on the total degree of polymerization ($N = N_A + N_B$), the Flory–Huggins parameter χ_{AB} that quantifies the incompatibility between the blocks and the volume fraction of the A block. It is worth recalling that the microphase separation and self-assembly in block copolymer systems are the consequences of unfavorable enthalpy and of a small entropy of mixing.

The phase diagram of linear coil–coil diblock copolymers (e.g. PS–PI) has been extensively studied and is well established (Fig. 5.7, the PS–PI case) [137]. This is far from the case of cyclic block copolymers. However, a theoretical model using the RPA model [120, 138, 139] (Fig. 5.8) and simulations [140, 141] (Fig. 5.9) predicted a transition from the disordered to an ordered phase (microphase separation) at $(\chi N,\ q_m R_{gt}) = (17.8,\ 2.88)$ for a 50:50 diblock copolymer as

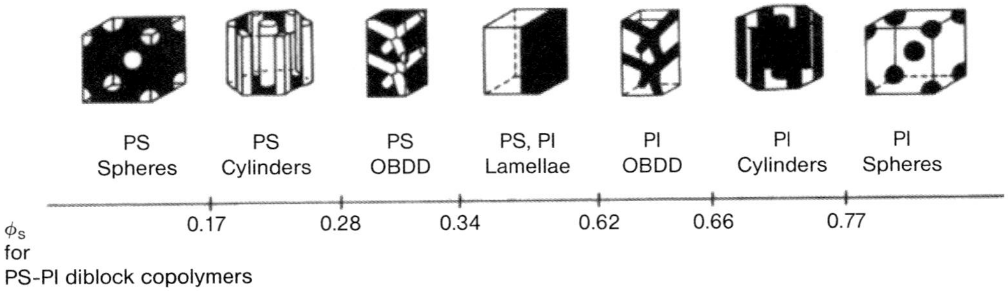

Fig. 5.7 Phase diagram for linear diblock copolymer: PS–PI case [137].

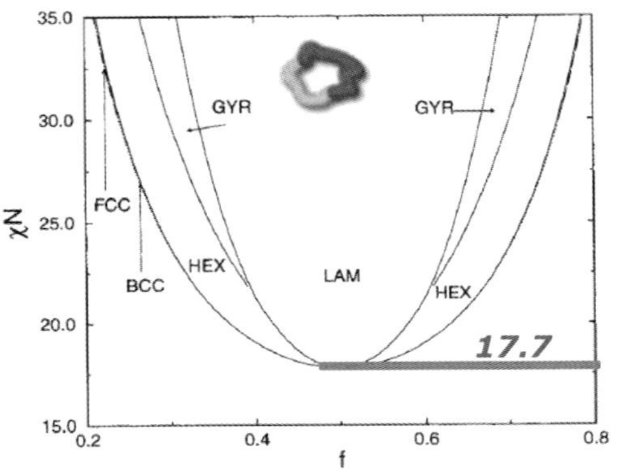

Fig. 5.8 Phase diagram for cyclic block copolymers, from [139]. Note that the microphase separation for a symmetrical linear block copolymer (50:50) occurs at $\chi N = 10.5$; it is at $\chi N = 17.7$ for the cyclic copolymer case.

compared with the classical result for linear diblocks for which $(\chi N, q_{\mathrm{m}} R_{\mathrm{gt}}) = (10.5,\ 1.95)$.

A few experimental studies [142–146] have been carried out on nano-organized films made from cyclic block copolymers. Here we recall some results obtained on linear and cyclic PS–PI films. Using SAXS, the results showed that the linear PS–PI system is microphase separated and organized in a two-dimensional hexagonal packing (reflections 1:3:4:7:9:12: ...) and the morphology adopted by cyclic copolymer chains is a liquid-like micellar phase (single scattering peak) (Fig. 5.10). Both the domain spacing and the temperature effects were found to be very sensitive to the cyclization of the linear diblock copolymer. This shows that the cyclization of a linear block copolymer induces remarkable changes in the morphology of the organized nanostructure.

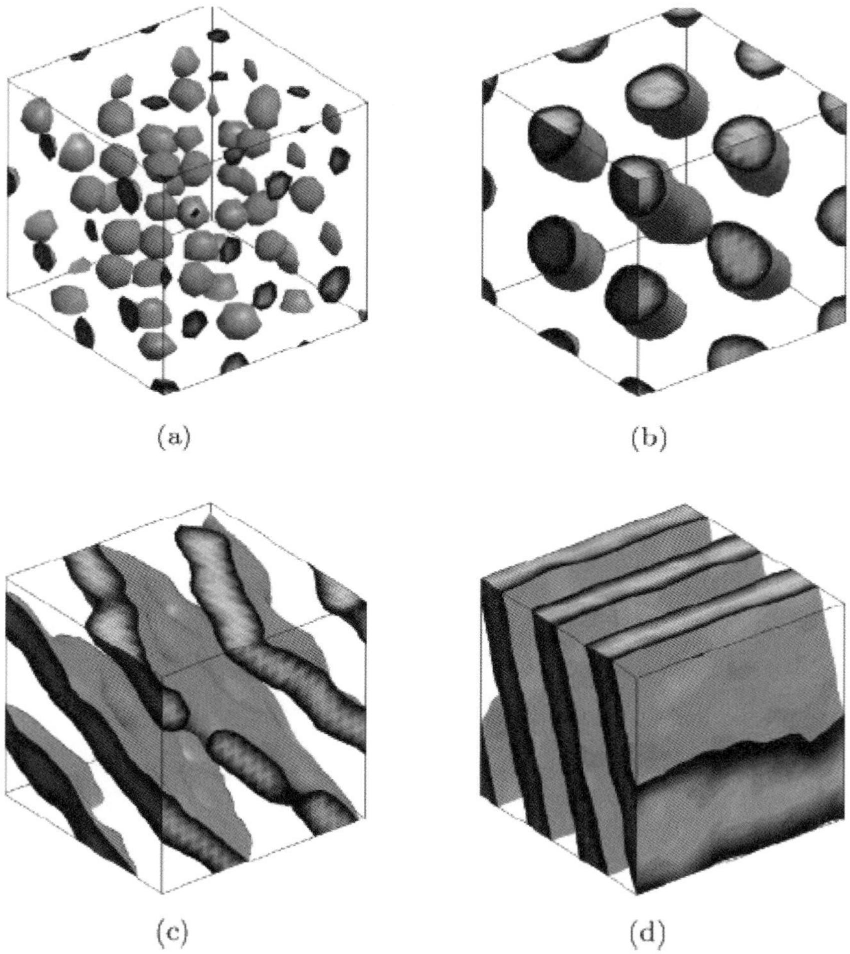

Fig. 5.9 Simulation of meso-structures for cyclic block copolymers with $N = 20$: (a) BCC at $f = 0.1$ and $\chi N = 260$; (b) HEX at $f = 0.3$ and $\chi N = 180$; (c) PL at $f = 0.35$ and $\chi N = 110$; (d) LAM at $f = 0.5$ and $\chi N = 60$ [141].

5.3 Conclusion

In this chapter, we have reviewed the main routes available today for the preparation of cyclic polymers and copolymers. A large number of cyclic structures are now available thanks to the progress achieved in the control of polymerization reactions and to the emergence of new strategies illustrated by the metathesis ring expansion reaction and the electrostatic pre-cyclization method.

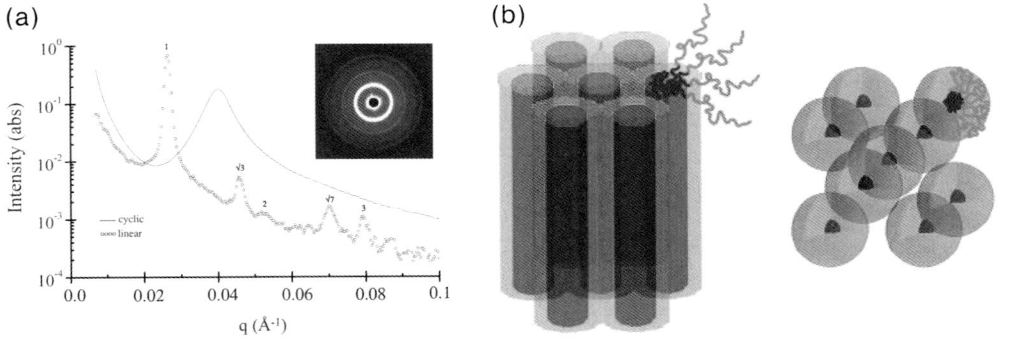

Fig. 5.10 (a) SAXS intensity profiles of linear and cyclic PS_{290}-b-PI_{110} obtained at room temperature (synchrotron, ESRF). (b) Schematic representation of the cylindrical and spherical structures obtained for linear and cyclic PS_{290}-b-PI_{110}, respectively. The red color is for PS and the blue is for PI [146].

It has been clearly shown for low and medium molar mass ring polymers and copolymers that the architectural change from linear to cyclic has dramatic changes on the solution and bulk properties. For cyclic copolymers in the disordered phase (good solvent for both blocks), the main difference remains the size of the single molecule ($R_{gl} = \sqrt{2}R_{gc}$). In a selective solvent (bad for one of the blocks), the self-assembly, depending on the volume fraction, may lead to the formation of "sunflower" to cylindrical worm-like micelles and ultimately to vesicles at relatively high concentrations in the case of cyclic block copolymers. This is not the case for linear block copolymers, where only spherical micelles are formed. In the bulk state, similar conclusions may be drawn where the cyclization is found to play an important role in the phase diagram. Further studies in this domain are still necessary to obtain a more complete view of the properties and behavior of macromolecular systems constituted by ring chains.

It is worth noting that a majority of cyclic and pluricyclic polymers studied are of low or medium molar masses. This can be related to the fact that most of the specific properties of ring molecules, in particular their differences from the corresponding linear polymers, are more pronounced in the low molar mass range. Another reason is the difficulty in preparing in acceptable yield very large cyclic polymers of high purity, free of linear chains. This remains a strong limitation for the investigation of the physical and mechanical properties of cyclic polymers of very high molar mass, the synthesis of which is still an open challenge for polymer chemists.

References

1 R. Dulbecco, M. Vogt, *Proc. Natl. Acad. Sci. USA* **1963**, *50*, 236.

2 V. Vologodskii, in *Cyclic Polymers*, 2nd edn (Ed. J.A. Semlyen), Kluwer, Dordrecht, **2000**, p. 47.

3 F. Jacob, E.L. Wollman, *Symp. Soc. Exp. Biol.* **1958**, *12*, 75.

4 B.H. Zimm, W.H. Stockmayer, *J. Chem. Phys.* **1949**, 1301.

5 V. Bloomfield, B.H. Zimm, *J. Chem. Phys.* **1966**, 315.

6 E.F. Casassa, *J. Polym. Sci., Part A* **1965**, 605.

7 J.A. Semlyen, G.R. Walker, *Polymer* **1969**, *10*, 597.

8 K.J. Ivin, J. Leonard, *Polymer* **1965**, *6*, 621.

9 D. Vofsi, J. Tobolsky, *J. Polym. Sci., Part A* **1965**, 2361.

10 D.R. Cooper, J.A. Semlyen, *Polymer* **1973**, *14*, 185.

11 D. Geiser, H. Höcker, *Polym. Bull.* **1980**, *2*, 591.

12 D. Geiser, H. Höcker, *Macromolecules* **1980**, *13*, 653.

13 G. Hild, A. Kohler, P. Rempp, *Eur. Polym. J.* **1980**, *16*, 525.

14 B. Vollmert, J.X. Huang, *Makromol. Chem. Rapid Commun.* **1980**, *1*, 333.

15 J.A. Semlyen, *Cyclic Polymers*, 1st edn, Elsevier Applied Science, Barking, **1986**.

16 J.A. Semlyen, *Cyclic Polymers*, 2nd edn, Kluwer, Dordrecht, **2000**.

17 J.A. Semlyen, *Large Ring Molecules*, Wiley, Chichester, **1996**.

18 P. Rempp, C. Strazielle, P. Lutz, *Encycl. Polym. Sci. Eng.* **1987**, *9*, 183.

19 H. Keul, H. Höcker, in *Large Ring Molecules* (Ed. J.A. Semlyen), Wiley, Chichester, **1996**, p. 128.

20 A. Deffieux, in *Polymeric Materials Encyclopedia* (Ed. J.C. Salamone), CRC Press, Boca Raton, FL, **1996**, p. 3887.

21 Y. Ederle, K.S. Naraghi, P.J. Lutz, in *Materials Science and Technology* (Eds. H.P. Cahn, R.W. Kramer), Wiley-VCH, Weinheim, **1999**, p. 622.

22 J. Roovers, in *Cyclic Polymers*, 2nd edn (Ed. J.A. Semlyen), Kluwer, Dordrecht, **2000**, p. 347.

23 T.E. Hogen-Esch, *J. Polym. Sci., Part A: Polym. Chem.* **2006**, *44*, 2139.

24 G. Pruckmayr, T.K. Wu, *Macromolecules* **1977**, *10*, 877.

25 D. Worsfold, A.M. Eastham, *J. Am. Chem. Soc.* **1957**, *79*, 900.

26 J.F. Brown, M.J. Slusarczuk, *J. Am. Chem. Soc.* **1957**, *79*, 897.

27 J.A. Semlyen, P.V. Wright, *Polymer* **1965**, *10*, 643.

28 J. Chojnowski, P. Scibiorek, M. Kowalski, *Makromol. Chem.* **1977**, *178*, 1351.

29 J.M. Andrews, J.A. Semlyen, *Polymer* **1972**, *13*, 142.

30 K. Dogson, J.A. Semlyen, *Polymer* **1977**, *18*, 1265.

31 S.J. Clarson, in *Cyclic Polymers*, 2nd edn (Ed. J.A. Semlyen), Kluwer, Dordrecht, **2000**, p. 161.

32 D.J. Brunelle, in *Cyclic Polymers*, 2nd edn (Ed. J.A. Semlyen), Kluwer, Dordrecht, **2000**, p. 185.

33 D.J. Brunelle, E.P. Boden, T.G. Shannon, *J. Am. Chem. Soc.* **1990**, *112*, 2399.

34 A.F. Fehervari, R. Faust, J.P. Kennedy, *Polym. Prepr.* **1987**, *28*, 382.

35 A.F. Fehervari, R. Faust, J.P. Kennedy, *Polym. Bull.* **1990**, *23*, 525.

36 C.W. Bielawski, R.H. Grubbs, *Angew. Chem. Int. Ed.* **2000**, *39*, 2903.

37 C.W. Bielawski, D. Benitez, R.H. Grubbs, *Science* **2002**, *297*, 2041.

38 C.W. Bielawski, D. Benitez, R.H. Grubbs, *J. Am. Chem. Soc.* **2003**, *125*, 8424.

39 C.W. Bielawski, R.H. Grubbs, *PMSE Prepr.* **2003**, *89*, 285.

40 C.W. Lee, T.-L. Choi, R.H. Grubbs, *J. Am. Chem. Soc.* **2002**, *124*, 3224.

41 C.W. Lee, R.H. Grubbs, *J. Org. Chem.* **2001**, *66*, 7155.

42 G. Hild, C. Strazielle, P. Rempp, *Eur. Polym. J.* **1983**, *19*, 721.

43 D. Dong, T.E. Hogen-Esch, S.J. Shaffer, *Polym. Prepr.* **1996**, *37*, 593.

44 D. Dong, T.E. Hogen-Esch, J.S. Shaffer, *Macromol. Chem. Phys.* **1996**, *197*, 3397.

45 W.F. Bailey, J.J. Patricia, *J. Organomet. Chem.* **1988**, *1*, 352.

46 D. Dong, T.E. Hogen-Esch, *Polym. Prepr.* **1996**, *37*, 589.

47 J. Roovers, P. Toporowski, *Macromolecules* **1983**, *16*, 843.
48 B. Vollmert, J.X. Huang, *Makromol. Chem. Rapid Commun.* **1981**, *2*, 467.
49 R.P. Quirk, J. Ma, *Polym. Prepr.* **1988**, *29*, 10.
50 Y. Gan, T.E. Hogen-Esch, *Polym. Prepr.* **1995**, *36*, 406.
51 K. Alberty, E. Tillman, S. Carlotti, S.E. Bradforth, T.E. Hogen-Esch, *Polym. Prepr.* **2001**, *42*, 584.
52 K.A. Alberty, E. Tillman, S. Carlotti, D. King, S.E. Bradforth, T.E. Hogen-Esch, D. Parker, W.J. Feast, *Macromolecules* **2002**, *35*, 3856.
53 D. Cho, K. Masuoka, T. Asari, D. Kawaguchi, A. Takano, Y. Matsushita, K. Koguchi, *Polym. J.* **2005**, *37*, 506.
54 D. Dong, T.E. Hogen-Esch, *Polym. Prepr.* **1996**, *37*, 595.
55 D. Dong, T.E. Hogen-Esch, *e-Polymers [online computer file]* **2001**, no page given.
56 J. Roovers, P.M. Toporowski, *J. Polym. Sci., Part B: Polym. Phys.* **1988**, *26*, 1251.
57 A. Madani, J.-C. Favier, P. Hemery, P. Sigwalt, *Polym. Int.* **1992**, *27*, 353.
58 J.-M. Boutillier, B. Lepoittevin, J.-C. Favier, M. Masure, P. Hémery, P. Sigwalt, *Eur. Polym. J.* **2002**, *38*, 243.
59 T.E. Hogen-Esch, J. Sundararajan, W. Toreki, *Makromol. Chem., Macromol. Symp.* **1991**, *47*, 23.
60 W. Toreki, T.E. Hogen-Esch, G.B. Butler, *Polym. Prepr.* **1987**, *28*, 343.
61 W. Toreki, T.E. Hogen-Esch, G.B. Butler, *Polym. Prepr.* **1988**, *29*, 17.
62 J. Sundararajan, T.E. Hogen-Esch, *Polym. Prepr.* **1991**, *32*, 63.
63 G.G. Nossarev, T.E. Hogen-Esch, *Macromolecules* **2002**, *35*, 1604.
64 R. Chen, T.E. Hogen-Esch, *Abstracts of Papers, 224th ACS National Meeting*, Boston, MA, August 18–22, **2002**.
65 R. Chen, T.E. Hogen-Esch, *Polym. Prepr.* **2003**, *44*, 972.
66 G.G. Nossarev, T.E. Hogen-Esch, *Polym. Prepr.* **2001**, *42*, 590.
67 R. Yin, T.E. Hogen-Esch, *Polym. Prepr.* **1992**, *33*, 239.
68 R. Yin, T.E. Hogen-Esch, *Macromolecules* **1993**, *26.*, 6952.
69 R. Yin, E.J. Amis, T.E. Hogen-Esch, *Macromol. Symp.* **1994**, *85*, 217.
70 Y. Gan, J. Zoller, T.E. Hogen-Esch, *Polym. Prepr.* **1993**, *34*, 69.
71 Y.D. Gan, J. Zoller, R. Yin, T.E. Hogen-Esch, *Macromol. Symp.* **1994**, *77*, 93.
72 J. Ma, *Macromol. Symp.* **1995**, *91*, 41.
73 R.P. Quirk, J. Ma, *Polym. Prepr.* **1988**, *29*, 10.
74 H. Iatrou, N. Hadjichristidis, G. Meier, H. Frielinghaus, M. Monkenbusch, *Macromolecules* **2002**, *35*, 5426.
75 A. Takano, O. Kadoi, K. Hirahara, S. Kawahara, Y. Isono, J. Suzuki, Y. Matsushita, *Macromolecules* **2003**, *36*, 3045.
76 K. Ishizu, A. Ichimura, *Polymer* **1998**, *39*, 6555.
77 M. Schappacher, A. Deffieux, *Makromol. Chem. Rapid Commun.* **1991**, *12*, 447.
78 L. Rique-Lurbet, M. Schappacher, A. Deffieux, *Macromolecules* **1994**, *27*, 6318.
79 A. Deffieux, M. Schappacher, L. Rique-Lurbet, *Polym. Prepr.* **1994**, *35*, 494.
80 A. Deffieux, M. Schappacher, L. Rique-Lurbet, *Macromol. Symp.* **1995**, *95*, 103.
81 A. Deffieux, S. Beinat, M. Schappacher, *Macromol. Symp.* **1997**, *118*, 247.
82 M. Schappacher, C. Billaud, C. Paulo, A. Deffieux, *Macromol. Chem. Phys.* **1999**, *200*, 2377.
83 A. Deffieux, M. Schappacher, *Macromol. Rep.* **1994**, *A31*, 699.
84 M. Schappacher, A. Deffieux, *Macromolecules* **2001**, *34*, 5827.
85 M. Schappacher, A. Deffieux, *Macromol. Chem. Phys.* **2002**, *203*, 2463.
86 M. Kubo, T. Hayashi, H. Kobayashi, K. Tsuboi, T. Itoh, *Macromolecules* **1997**, 2805.
87 A.B. Laurent, S.M. Grayson, *J. Am. Polym. Soc.* **2006**, *128*, 4238.
88 A. Deffieux, M. Schappacher, L. Rique-Lurbet, *Polymer* **1994**, *35*, 4562.
89 S. Cramail, M. Schappacher, A. Deffieux, *Macromol. Chem. Phys.* **2000**, *201*, 2328.
90 D. Pantazis, Schultz, N. Hadjichristidis, *J. Polym. Sci., Part A: Polym. Chem.* **2002**, *40*, 1476.
91 Y. Tezuka, H. Oike, *Macromol. Rapid Commun.* **2001**, *22*, 1017.
92 Y. Tezuka, H. Oike, *J. Am. Chem. Soc.* **2001**, *123*, 11570.

93 H. Oike, T. Mouri, Y. Tezuka, *Macromolecules* **2001**, *34*, 6592.

94 Y. Tezuka, K. Mori, H. Oike, *Macromolecules* **2002**, *35*, 5707.

95 Y. Tezuka, H. Oike, *Macromol. Symp.* **2000**, *161*, 159.

96 Y. Tezuka, H. Oike, *Prog. Polym. Sci.* **2002**, *27*, 1069.

97 Y. Tezuka, *Macromol. Symp.* **2003**, *192*, 217.

98 H. Oike, H. Imaizumi, T. Mouri, Y. Yoshioka, A. Uchibori, Y. Tezuka, *J. Am. Chem. Soc.* **2000**, *122*, 9592.

99 H. Oike, M. Hamada, S. Eguchi, Y. Danda, Y. Tezuka, *Macromolecules* **2001**, *34*, 2776.

100 S. Beinat, M. Schappacher, A. Deffieux, *Macromolecules* **1996**, *29*, 6737.

101 M. Kubo, T. Hayashi, H. Kobayashi, T. Itoh, *Macromolecules* **1998**, *31*, 1053.

102 H. Oike, M. Washizuka, Y. Tezuka, *Macromol. Rapid Commun.* **2001**, *22*, 1128.

103 M. Antonietti, K. J. Fölsch, *Makromol. Chem., Rapid Comm.* **1988**, *9*, 423.

104 M. Schappacher, A. Deffieux, *Macromolecules* **1995**, *28*, 2629.

105 M. Schappacher, A. Deffieux, *Macromolecules* **1992**, *25*, 6744; A. Deffieux, M. Schappacher, L. Rique-Lurbet, *Macromolecular Engineering: Recent Advances. Proceedings of the International Conference on Advanced Polymers via Macromolecular Engineering*, Poughkeepsie, NY, 24–28 June, **1995**, 271.

106 H. Oike, A. Uchibori, A. Tsuchitani, H.-K. Kim, Y. Tezuka, *Macromolecules* **2004**, *37*, 7595.

107 Y. Gan, D. Dong, T. E. Hogen-Esch, *Polym. Prepr.* **1995**, *36*, 408.

108 Y. Gan, D. Dong, T. E. Hogen-Esch, *Macromolecules* **2002**, *35*, 6799.

109 B. Lepoittevin, P. Hemery, *J. Polym. Sci., Part A: Polym. Chem.* **2001**, *39*, 2723.

110 B. Lepoittevin, P. Hémery, *Polym. Adv. Technol.* **2002**, *13*, 771.

111 M. Benmouna, U. Maschke, in *Cyclic Polymers*, 2nd edn (Ed. J. A. Semlyen), Kluwer, Dordrecht, **2000**, Chap 16, p. 741.

112 C. Strazielle, H. Benoît, *Macromolecules;* **1975**, *8*, 203–205.

113 P. Debye, *J. Phys. Chem.;* **1947**; *51*, 18–32.

114 G. Hadziioannou, P. M. Cotts, G. Ten Brinke, C. C. Han, P. Lutz, C. Strazielle, P. Rempp, A. J. Kovacs, *Macromolecules* **1987**, *20*, 493–497.

115 M. Fukatsu, M. Kurata, *J. Chem. Phys.* **1966**, *44*, 4539.

116 V. A. Bloomfield, B. H. Zimm, *J. Chem. Phys.* **1996**, *44*, 315.

117 M. Bohdana, M. Discher, Y.-Y. Won, D. S. Ege, J. C.-M. Lee, F. S. Bates, D. E. Discher, D. A. Hammer, *Science* **1999**, *284*, 1143.

118 J. Bang, S. Jain, Z. Li, T. P. Lodge, J. S. Pedersen, E. Kesselman, Y. Talmon, *Macromolecules* **2006**, *39*, 1199.

119 L. Zhang, K. Yu, A. Eisenberg, *Science* **1996**, *272*, 1777.

120 R. Borsali, R. Pecora, H. Benoît, *Macromolecules*, **2001**, *34*, 4229.

121 R. Borsali, M. Benmouna, *EuroPhys. Lett.* **1993**, *23*, 263.

122 M. Benmouna, R. Borsali, H. Benoît, *J. Phys. II (Paris)* **1993**, *3*, 1401.

123 R. Borsali, M. Benmouna, H. Benoît, *Physica A* **1993**, *201*, 129.

124 M. Benmouna, R. Borsali, *J. Polym. Sci., Polym. Phys. Ed., Part B* **1994**, *32*, 981.

125 R. Borsali, M. Benmouna, *Makromol. Chem. Symp.* **1994**, *79*, 153.

126 C. Vlahos, N. Hadjichristidis, K. M. Kosmas, A. M. Rubio, J. J. Freire, *Macromolecules* **1995**, *28*, 6854.

127 R. Borsali, M. Schappacher, M. de Souza Lima, A. Deffieux, P. Lindner, *Polym. Prepr.* **2002**, *43*, 201.

128 E. J. Amis, D. F. Hodgson, W. Wu, *J. Polym. Sci., Part B: Polym. Phys.* **1993**, *31*, 2049.

129 G.-E. Yu, C. A. Garrett, S.-M. Mai, H. Altinok, D. Attwood, C. Price, C. Booth, *Langmuir* **1998**, *14*, 2278.

130 H. Iatrou, H. Hermis, N. Hadjichristidis, G. Meier, H. Frielinghaus, M. Monkenbusch, *Macromolecules* **2002**, *35*, 5426.

131 E. Minatti, R. Borsali, M. Schappacher, A. Deffieux, T. Narayanan, J.-L. Putaux, *Macromol. Rapid Commun.* **2002**, *23*, 978.

132 E. Minatti, P. Viville, R. Borsali, M. Schappacher, A. Deffieux,

R. Lazzaroni, *Macromolecules* **2003**, *36*, 4125.
133 R. Borsali, E. Minatti, J.-L. Putaux, M. Schappacher, A. Deffieux, P. Viville, R. Lazzaroni, T. Narayanan, *Langmuir* **2003**, *19*, 6.
134 J.-L. Putaux, E. Minatti, C. Lefèbvre, R. Borsali, M. Schappacher, A. Deffieux, *Faraday Discuss.* **2005**, *128*, 163.
135 N. Ouarti, P. Viville, R. Lazzaroni, E. Minatti, M. Schappacher, A. Deffieux, R. Borsali, *Langmuir* **2005**, *21*, 1180.
136 N. Ouarti, P. Viville, R. Lazzaroni, E. Minatti, M. Schappacher, A. Deffieux, J.-L. Putaux, R. Borsali, *Langmuir* **2005**, *21*, 9085.
137 F. S. Bates, G. H. Fredrickson, *Annu. Rev. Phys. Chem.* **1990**, *41*, 525.
138 J. F. Marko, *Macromolecules* **1993**, *26*, 1442.
139 A. N. Morozov, J. G. E. M. Fraaije, *Macromolecules* **2001**, *34*, 1526.
140 W. H. S. S. Jang, *J. Chem. Phys* **1999**, *111*, 1712.
141 H.-J. Qian, Z.-Y. Lu, L.-J. Chen, Z.-S. Li, C.-C. Sun, *Macromolecules* **2005**, *38*, 1395.
142 R. L. Lescanec, D. A. Hajduk, G. Y. Kim, Y. Gan, R. Yin, S. M. Gruner, T. E. Hogen-Esch, E. L. Thomas, *Macromolecules* **1995**, *28*, 3485.
143 Y.-Q. Zhu, S. P. Gido, H. Iatrou, N. Hadjichristidis, J. W. Mays, *Macromolecules* **2003**, *36*, 148.
144 A. Takano, O. Kadoi, K. Hirahara, S. Kawahara, Y. Isono, J. Suzuki, Y. Matsushita, *Macromolecules* **2003**, *36*, 3045.
145 Y. Matsushita, H. Iwata, T. Asari, T. Uchida, G. ten Brinke, A. Takano, *J. Chem. Phys.* **2004**, *121*, 1129.
146 S. Lecommandoux, R. Borsali, M. Schappacher, A. Deffieux, T. Narayanan, C. Rochas, *Macromolecules* **2004**, *37*, 1843.

6
Polymers with Star-related Structures

Nikos Hadjichristidis, Marinos Pitsikalis, and Hermis Iatrou

6.1
Introduction

Star polymers are the simplest branched polymers, consisting of several linear chains linked to a central core [1–3]. The core may be an atom, a small molecule or a macromolecular structure itself. When the core is more than 50% of the weight of the macromolecule, the structure is characterized as a microgel. Due to their high segment density, branching results in more compact structures, compared with the analogous linear forms, and dramatically affects mechanical, viscoelastic and solution properties [1, 4, 5].

The first report concerning the synthesis of star polymers was published in 1948 by Schaefgen and Flory [6], who polymerized, by step-growth reactions, ε-caprolactam with cyclohexanonetetrapropionic or dicyclohexanoneoctacarboxylic acid to obtain star-shaped polyamides with four or eight arms, respectively. In 1962, Morton et al. [7] attempted the synthesis of three- and four-arm polystyrene stars by linking living anionic chains to trichloromethylsilane and tetrachlorosilane, respectively. Although a mixture of stars was obtained in this study, a major breakthrough was achieved showing the validity of living polymerization methods for the synthesis of well-defined star structures. Since this discovery, the great progress that has been made in living polymerization techniques allowed for the synthesis of several types of star polymers, such as star-block copolymers, functionalized, asymmetric and miktoarm stars.

Seminal and recent developments in the synthesis of star structures by the most important polymerization techniques, such as anionic, cationic, controlled radical, ring-opening, ring-opening metathesis and group-transfer methods and their combinations, will be presented in this review. In addition, the synthesis of star polymers by step-growth polycondensation, metal template-assisted and "electrostatic self-assembly and covalent fixation" methods will also be discussed. Until 1990, the average number of papers published on stars ranged between 2 and 23 per year. With the introduction of controlled/living radical polymerization, in the 1990s, this number increased dramatically. Since 2001, the average

Macromolecular Engineering. Precise Synthesis, Materials Properties, Applications.
Edited by K. Matyjaszewski, Y. Gnanou, and L. Leibler

ISBN: 978-3-527-31446-1

number of papers per year has approached 400, thanks not only to controlled/living radical polymerizations but also to combinations of practically all methods.

Considering the enormous number of papers published in the area of star-related polymeric structures, it is impossible to present a complete list of publications here. Therefore, selected examples will be presented.

6.2
General Methods for the Synthesis of Star Polymers

Three general synthetic routes have been developed for the synthesis of star polymers [1, 8], as outlined in Scheme 6.1.

6.2.1
Multifunctional Linking Agents

This method is referred to as the "arm-first" or "arm-in" or convergent approach. It involves the synthesis of living macromolecular chains and their subsequent reaction with a multifunctional linking agent. It is probably the most efficient way to synthesize well-defined star polymers because of the absolute control that can be achieved in all synthetic steps. The functionality of the linking agent determines the number of branches of the star polymer, provided that the linking reaction is quantitative. The living arms can be isolated before linking and

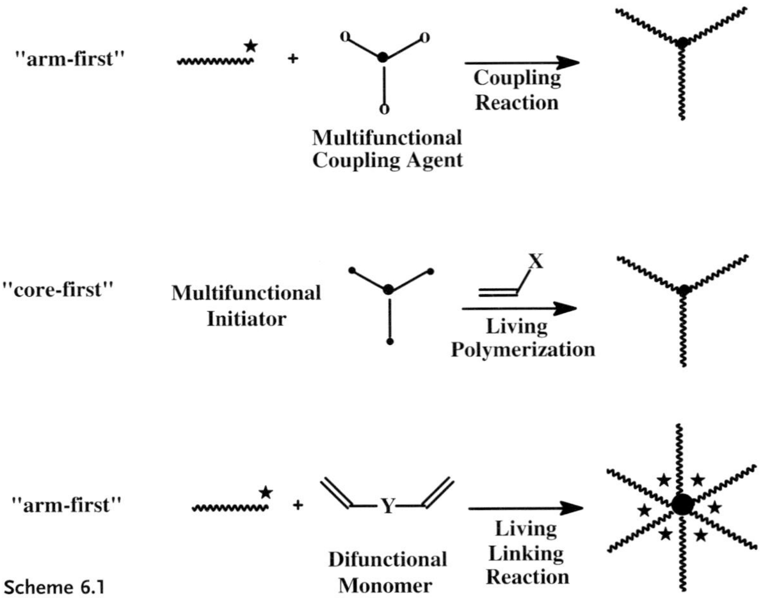

Scheme 6.1

characterized independently along with the final star. Consequently, the functionality of the star can be measured directly and with accuracy. Disadvantages of the method can be considered the long time required for the linking reaction in most cases and the need to perform fractionation in order to obtain the pure star polymer, since a small excess of the living arm is needed to assure complete linking.

6.2.2 Multifunctional Initiators

This method is referred to as the "core-first" or "arm-out" or divergent approach. According to this procedure, multifunctional compounds capable of simultaneously initiating the polymerization of several arms are used. There are several requirements that a multifunctional initiator has to fulfill in order to produce star polymers with uniform arms, low molecular weight distribution and controllable molecular weights. All the initiation sites must be equally reactive and the initiation rate must be higher than the propagation rate. The characterization of the star polymers produced by this method is difficult, since the molecular weight of the arm cannot be measured directly. The number of the arms can be defined indirectly by several methods, such as end-group analysis, determination of the branching parameters, which are the ratios of the mean square radius of gyration, intrinsic viscosity or hydrodynamic radius of the star to the corresponding linear structure with the same molecular weight. Finally, isolation of the arms after cleavage (i.e. hydrolysis), if possible, and subsequent analysis lead to the determination of the functionality.

6.2.3 Difunctional Monomers

In this method, a living polymer precursor is used as initiator for the polymerization of a small amount of a suitable difunctional monomer. Microgel nodules of tightly cross-linked polymer are formed upon the polymerization. These nodules serve as the branch point from which the arms emanate. Although the functionality of the stars can be obtained directly by molecular weight determination of the arms and the star product, it is very difficult to predict and control the number of arms. The number of arms depends mostly on the molar ratio of the difunctional monomer to the living polymer and increases on increasing this ratio. Other parameters influencing the number of branches are the chemical nature, the concentration and the molecular weight of the living polymer chain, the temperature and duration of the reaction, the rate of stirring, etc., and lead to stars with a broad distribution of functionalities. It is obvious that although this method is simple and is applied industrially, it is less suitable for the preparation of well-defined stars.

This method is similar to the "arm-first" approach but the same technique can be applied as a "core-first" approach. In this case, a monofunctional initiator reacts with the difunctional monomer leading to the formation of tightly cross-linked microgel nodules bearing active sites that can be used for the polymerization of a suitable monomer. High-functionality stars can be prepared by this method; however, the disadvantages and restrictions reported previously also apply in this case.

These two procedures can be combined as a single "arm in–arm out" process for the synthesis of more complex structures, such as asymmetric and miktoarm stars of the $A_nA'_n$ and A_nB_n type.

6.3
Star Architectures

Numerous studies have been reported for the synthesis of star homopolymers with different functionalities. However, more complex structures have also appeared in the literature and the most important of them are presented below.

6.3.1
Star-block Copolymers

Star-block copolymers can be envisioned as star polymers in which each arm is a diblock or a triblock copolymer [5, 8] (Scheme 6.2). These structures combine the properties of diblock or triblock copolymers with those of star polymers. The synthesis of star-block copolymers can be achieved by all of the previously mentioned methods using a single living polymerization technique or a combination of techniques.

6.3.2
Functionalized Stars

Functionalized stars (Scheme 6.3) are stars bearing functional groups along the arms (in-chain functionalized stars) or at the end of each arm (end-functionalized star polymers). These molecules can be used as models for the study of

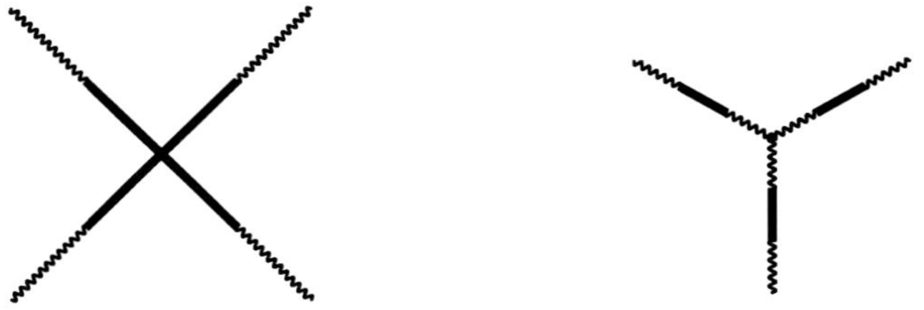

4-arm star diblock copolymer **3-arm star triblock copolymer**

Scheme 6.2

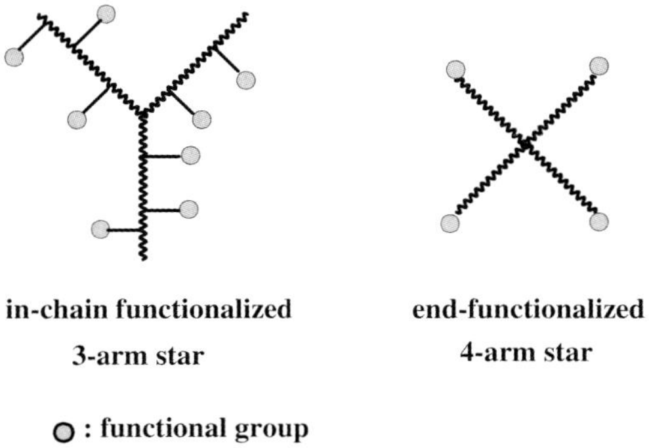

Scheme 6.3

fundamental phenomena in polymer science, such as association, adsorption and chain dynamics.

The synthesis of in-chain functionalized stars can be accomplished by the copolymerization of a non-functionalized with a functionalized monomer followed by a suitable method for the synthesis of star polymers. For the synthesis of end-functionalized stars, two general methods have been developed [9, 10]. The first involves the use of functionalized initiators and is associated with the "arm-first" approach for the synthesis of star polymers. This procedure ensures complete functionalization and therefore can produce well-defined stars, provided that the functionalized initiator is efficient in producing polymers with predicted molecular weights and narrow molecular weight distributions. The second method involves the use of functionalized terminating agents and is associated with the "core-first" approach. It is important to choose a suitable terminating agent, since this reaction may be associated with several side-reactions, leading to incomplete functionalization.

The synthesis of functionalized stars is not always a straightforward procedure, due to the competition or antagonism of functional groups with the active living ends. Therefore, several protective groups are needed to mask reactive functionality, to provide well-defined functionalized stars [10].

6.3.3
Asymmetric Stars

Asymmetric stars consist of a special class of star structures with molecular weight, functional group or topological asymmetry [11] (Scheme 6.4):

- *Molecular weight asymmetry:* In this case, all the arms are chemically identical, but have different molecular weights.
- *Functional group asymmetry:* This is a special class of functionalized stars. The arms are of the same chemical nature and have the same molecular weight, but have different functional groups. Thus, either some of the arms are functionalized or the arms have different functional groups.

Scheme 6.4

- *Topological asymmetry:* The arms of the star are block copolymers having the same molecular weight and the same composition but differ with respect to the polymeric block that is covalently attached to the core of the star.

The asymmetric stars can be synthesized in a manner similar to that for the miktoarm stars reported below.

6.3.4 Miktoarm Stars

Miktoarm stars (from the Greek word μικτός, meaning mixed) are structures containing chemically different arms. The term heteroarm stars (from the Greek word έτερος, meaning other), used previously for the same materials, is not appropriate since it does not convey the concept of a group of dissimilar objects [12].

The synthesis of miktoarm stars has been the subject of intense research over the last decade and several methods have been developed [5, 8, 11–13]. The most common examples of miktoarm stars are the A_2B, A_3B, A_2B_2, A_nB_n ($n>2$) and ABC types, where A, B and C are chemically different chains. Characteristic examples are shown in Scheme 6.5.

Scheme 6.5

6.4 Synthesis of Star Polymers

The syntheses of star polymers discussed below are presented according to the chronological order of the polymerization procedures and not the general synthetic strategies.

6.4.1 Anionic Polymerization

Multifunctional initiators were synthesized by the polymerization of divinylbenzene (DVB) in benzene with butyllithium, at high dilution, to obtain a stable microgel suspension [14–17]. These microgels, which are covered by living anionic sites, were subsequently used to polymerize styrene, isoprene or butadiene. A slight variation was reported, in which the polymerization of DVB was initiated by low molecular weight living poly(*tert*-butylstyryl)lithium chains in order to avoid the solubility problems arising from the strong association of the carbon–lithium groups in non-polar solvents [18, 19].

More recently, the multifunctional initiator approach has been used for the synthesis of PEO stars [20] as depicted in Scheme 6.6. Hyperbranched polyglycerol and polyglycerol modified with short poly(propylene oxide) chains, after reaction with diphenylmethylpotassium (DPMP), were used as multifunctional initiators for the synthesis of PEO stars. The final products were characterized by size-exclusion chromatography (SEC)/low-angle laser light scattering (LALLS), and also by NMR spectroscopy. Rather broad molecular weight distributions were obtained, ranging from 1.4 up to 2.2. The functionalities of these stars were found to vary between 26 and 55.

The same "core-first" approach was adopted for the synthesis of eight-arm PEO stars [21]. *tert*-Butylcalix[8]arene carrying eight hydroxyl groups was used as a multifunctional initiator for the polymerization of ethylene oxide, after transforming the hydroxyl groups to the corresponding alkoxides with diphenylmethylpotassium. The polymerization reaction in tetrahydrofuran (THF) led to well-defined PEO stars.

A hydrocarbon-soluble trifunctional initiator was synthesized by reacting 3 mol of *sec*-butyllithium (*s*-BuLi) with 1 mol of 1,3,5-tris(1-phenylethenyl)benzene (tri-DPE) [22] (Scheme 6.7). It was found that the initiator was efficient for the polymerization of styrene only in the presence of THF in a ratio [THF]/[*s*-BuLi]=20. The polymerization reaction was monitored by UV/VIS spectroscopy. Extreme precision over the stoichiometry of the reaction between *s*-BuLi and tri-DPE is needed and a minimum arm molecular weight around 6×10^3 is required for a successful initiation and synthesis.

The initiator in Scheme 6.7 was also used to produce a three-arm polybutadiene (PBd) star [23]. Although complete monomer consumption was observed by SEC analysis, the star polymer exhibited a bimodal distribution. This was attributed to the strong association effects of the trifunctional initiator in the non-

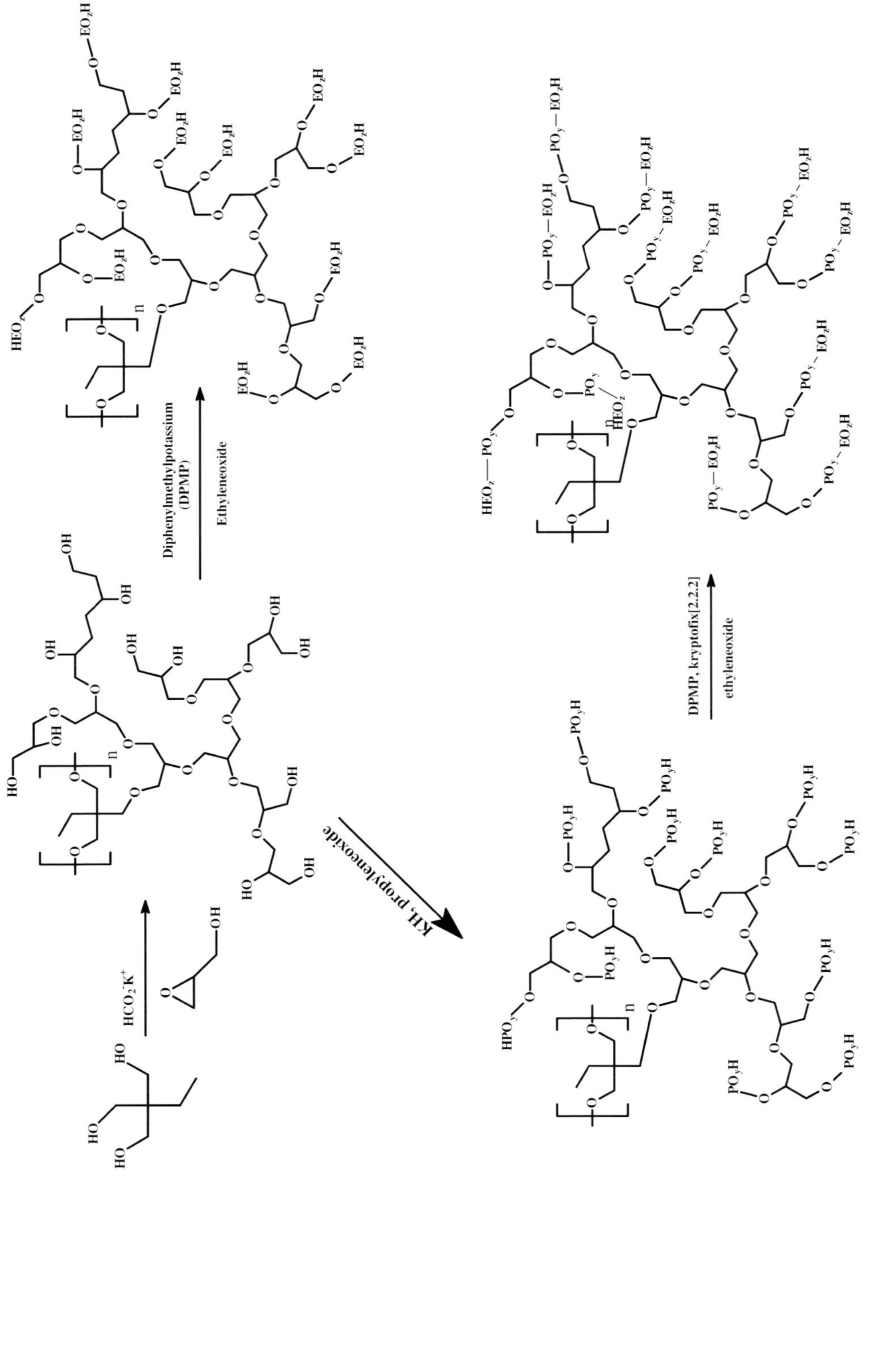
HCO2-K+
Diphenylmethylpotassium (DPMP)
Ethyleneoxide
KH, propyleneoxide
DPMP, kryptofix[2.2.2]
ethyleneoxide

Scheme 6.6

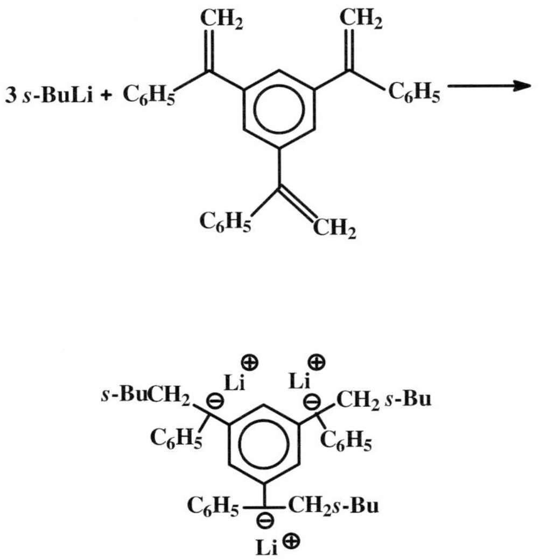

Scheme 6.7

polar solvent benzene. The problem has been overcome when *s*-BuOLi was added to the reaction mixture in a ratio [*s*-BuLi]/[*s*-BuOLi]=2. *s*-BuOLi was shown to be capable of breaking the initiator association without appreciably affecting the microstructure of the PBd chains. Therefore, a well-defined star polymer with low molecular weight distribution was obtained.

Tri- and tetrabromoaryl compounds were lithiated by reaction with *sec*-butyllithium and were then used as multifunctional initiators for the synthesis of monodisperse three- and four-arm PS stars [24, 25], respectively (Scheme 6.8). Tetramethylethylenediamine (TMEDA) was added to improve the solubility of the tetrafunctional initiator in benzene.

In addition to the use of multifunctional initiators, another general and efficient method for the synthesis of star polymers is the linking reaction of the living polymers with a suitable electrophilic reagent. Several linking agents have been used for the synthesis of star polymers [26]. The most efficient are chlorosilanes [27], bromomethyl- and chloromethylbenzene derivatives [28, 29]. However, other linking agents, such as hexafluoropropylene oxide [30], tri(allyloxy)-1,3,5-triazine [31] and tetraphenyl-1,1,4,4-di(allyloxy)triazene-1,4-butane [32] have also been used.

A significant advantage of the chlorosilanes is that the linking reactions with living polymers proceed without any side-reactions. Using appropriate chlorosilanes, model star polymers have been prepared with functionalities ranging from 3 up to 18 [33–40]. The use of carbosilane dendrimers led to the successful preparation of PBd stars having 32, 64 and 128 branches [40–42]. The products were extensively characterized by SEC, membrane osmometry (MO), vapor pressure osmometry (VPO) and light scattering (LS). Low molecular weight dis-

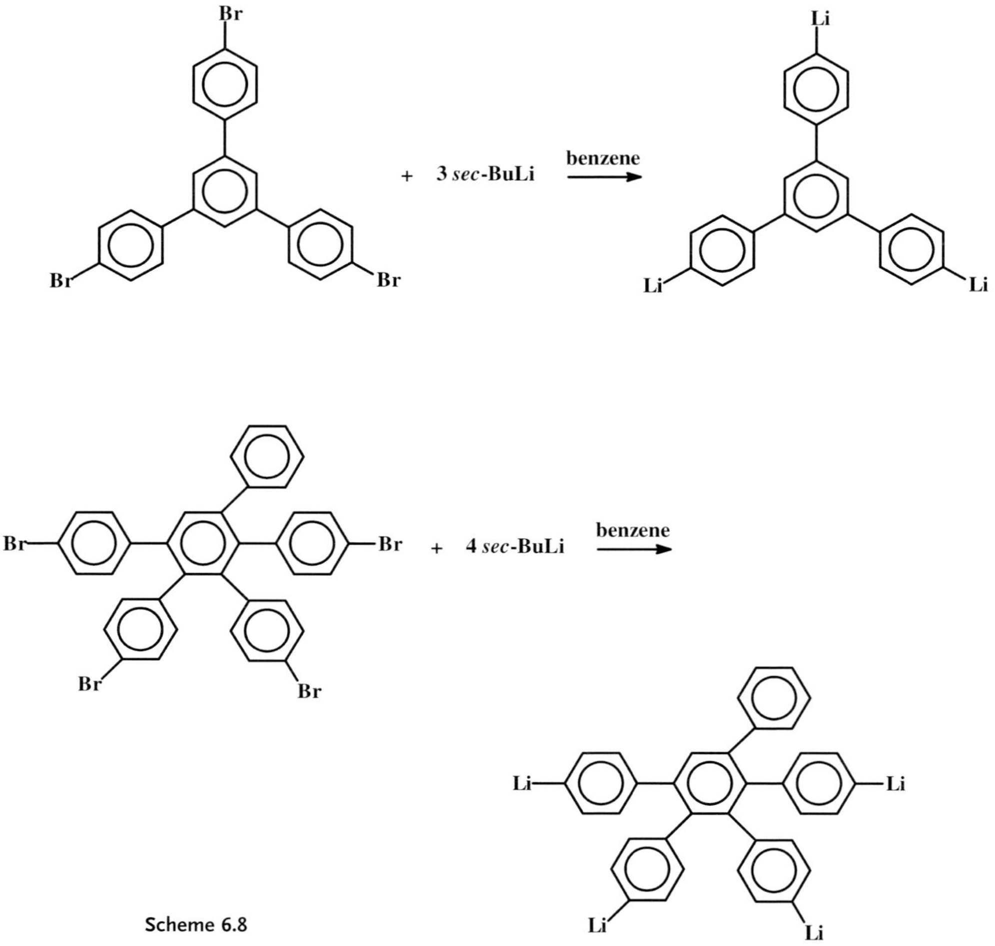

Scheme 6.8

tribution polymers with functionalities close to the theoretical value were obtained in all cases. However, extended periods of time were needed for complete linking reactions and fractionation was required to eliminate both the excess of arm, purposely added, and, in a few cases, stars with lower functionalities.

More recently, the efficiency of the chlorosilane linking was re-evaluated using NMR [43] and matrix-assisted laser desorption/ionization time-of-flight mass spectrometry (MALDI-TOFMS) [44] techniques. Both polyisoprene (PI) and PBd stars with low arm molecular weights ($M \approx 10^3$) and functionalities ranging from 3 to 64 were synthesized and used for this study. It was found that for stars with 16 arms or less, the structural quality with respect to the polydispersity and the functionality agrees very well with the theoretical values. For stars theoretically having 32 and 64 arms, the average functionalities of the

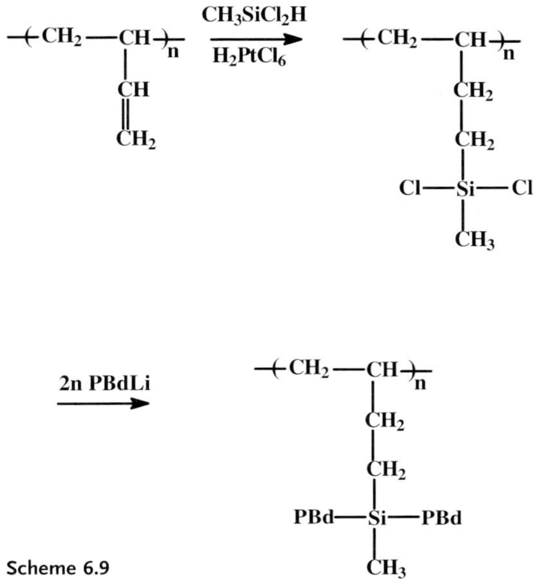

Scheme 6.9

chlorosilane linking agent were found to be 31 and 60, respectively, whereas the number of arms of the stars was 29 and 54, respectively. The linking agents along with the star products exhibited narrow molecular weight distributions. These results clearly demonstrate the steric limitations of this linking reactions.

The use of linking agent methodology was extended to the synthesis of PBd stars having more than 200 arms [45]. Low molecular weight linear and 18-arm star poly(1,2-butadiene) were extensively hydrosilylated with methyldichlorosilane and used as linking agents to prepare multi-arm PBd stars (Scheme 6.9). The synthetic strategy is similar to that used for the preparation of graft polymers, but the low backbone molecular weight makes them behave as star-like structures. This approach led to the synthesis of star polymers with 270 and 200 arms from the linear and the 18-arm star poly(1,2-butadiene), respectively. Although these products present a distribution in the number of arms due to the lack of absolute control over the hydrosilylation reaction, the polydispersity indices were very low.

Low polydispersity star-block copolymers of the type (PS-*b*-PI)$_n$, where $n=4$, 8, 12, 18 [46, 47], have also been synthesized using appropriate chlorosilane linking chemistry. An example is given in Scheme 6.10.

The first asymmetric PS and PBd stars bearing two identical arms and a third one with either half or twice the molecular weight of the other arms were synthesized by selective substitution of the chloride groups of methyltrichlorosilane [48]. The same procedure was later adopted for the synthesis of well-defined miktoarm star copolymer of the A_2B type [49], A being PI and B PS. The synthetic approach involved the reaction of living PS chains with an excess of

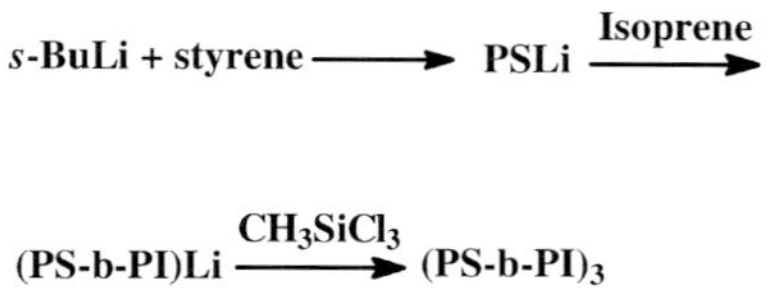

$s\text{-BuLi} + \text{styrene} \longrightarrow \text{PSLi} \xrightarrow{\text{Isoprene}}$

$\text{(PS-b-PI)Li} \xrightarrow{CH_3SiCl_3} \text{(PS-b-PI)}_3$

Scheme 6.10

$PS^{\ominus}Li^{\oplus} + (CH_3)SiCl_3 \text{ (excess)} \longrightarrow PS\text{-}Si(CH_3)Cl_2 + LiCl + (CH_3)SiCl_3\uparrow$

$PS\text{-}Si(CH_3)Cl_2 + PI^{\ominus}Li^{\oplus} \text{ (excess)} \longrightarrow PS\text{-}Si(CH_3)(PI)_2$

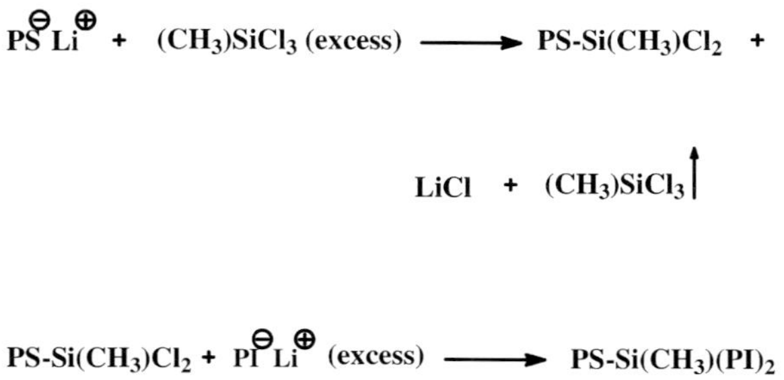

Scheme 6.11

methyltrichlorosilane to produce the monosubstituted macromolecular linking agent. The steric hindrance of the living polystyryllithium (phenyl group) and the excess of the chlorosilane led to the replacement of only one chlorine by PS. After removal of the excess chlorosilane, a slight excess of the living PI chains was added to produce the miktoarm star $PS(PI)_2$. Excess PI was removed by fractionation. The reaction sequence, given in Scheme 6.11, was monitored by SEC and the molecular characterization of the arms and the final product was performed by MO.

This approach was later extended [50] to the synthesis of the A_2B stars, where A and B were all possible combinations of PS, PI and PBd. In this case, a more sophisticated high-vacuum technique was employed to ensure the formation of products with high degrees of chemical and compositional homogeneity, as revealed by SEC, MO, LS, differential refractometry and NMR spectroscopy.

A three-arm (PS)(PI)(PBd) miktoarm star terpolymer has been also reported [51]. The synthetic strategy involved reaction of living PI chains with a large excess of methyltrichlorosilane to produce the dichlorosilane end-capped polyisoprene, followed, after the elimination of the excess silane, by slow stoichiometric addition (titration) of living PS arm. Samples taken during the addition were analyzed by SEC to monitor the progress of the reaction and determine the end-point of the titration. When the formation of the intermediate product $(PS)(PI)Si(CH_3)Cl$ was completed, a small excess of the living PBd chains was added to give the final product. The reaction sequence is outlined in Scheme 6.12.

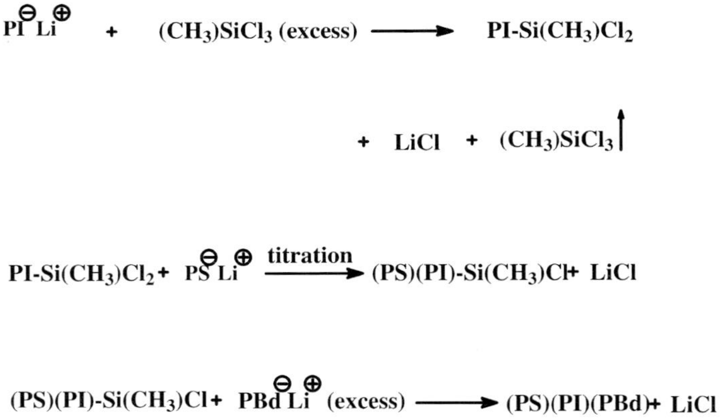

Scheme 6.12

The method was subsequently extended to the synthesis of A_2B_2 miktoarm star copolymers ABCD miktoarm star quarterpolymers [52]. Since then, many studies have been reported on the synthesis of miktoarm stars with a great variety of monomers. Recently, four-miktoarm star quarterpolymers of styrene, isoprene, dimethylsiloxane and 2-vinylpyridine were synthesized [53] using a different synthetic strategy (Scheme 6.13). 4-(Dichloromethylsilyl)diphenylethylene (DCMSDPE), a dual-functionality compound with two SiCl groups and a non-homopolymerizable double bond, was used as the nodule for the incorporation of four incompatible arms. The synthetic approach involved the selective replacement of the two chlorines of DCMSDPE with PI and PDMS by titration with the corresponding living chains, addition of PSLi to the in-chain double bond of the PI–PDMS intermediate and finally polymerization of 2VP. Extensive characterization by SEC/LALLS, membrane osmometry, NMR spectroscopy and DSC confirmed the high degree of molecular and compositional homogeneity and the incompatibility of the arms.

Another important class of linking agents includes chloromethyl- and bromomethylbenzene derivatives [28, 29]. The drawback of these compounds is the occurrence of side-reactions due to lithium–halogen exchange leading to linking agents with higher functionalities [26]. Nevertheless, these compounds are valuable for the synthesis of poly(meth)acrylates and poly(2-vinylpyridine) (P2VP) stars, since they can be used efficiently at –78 °C, where the polymerization of these polar monomers takes place. The chlorosilanes cannot be used because the linking reaction leads either to unstable products [poly(meth)acrylates] or does not occur (P2VP).

The synthesis of three-arm P2VP stars using 1,3,5-tri(chloromethyl)benzene [54], and the synthesis of four-arm poly(*tert*-butyl methacrylate) (P*t*BuMA), poly(methyl methacrylate) (PMMA) and P2VP stars using 1,2,4,5-tetra(bromomethyl)benzene [55] was reported (Scheme 6.14). Combined characterization re-

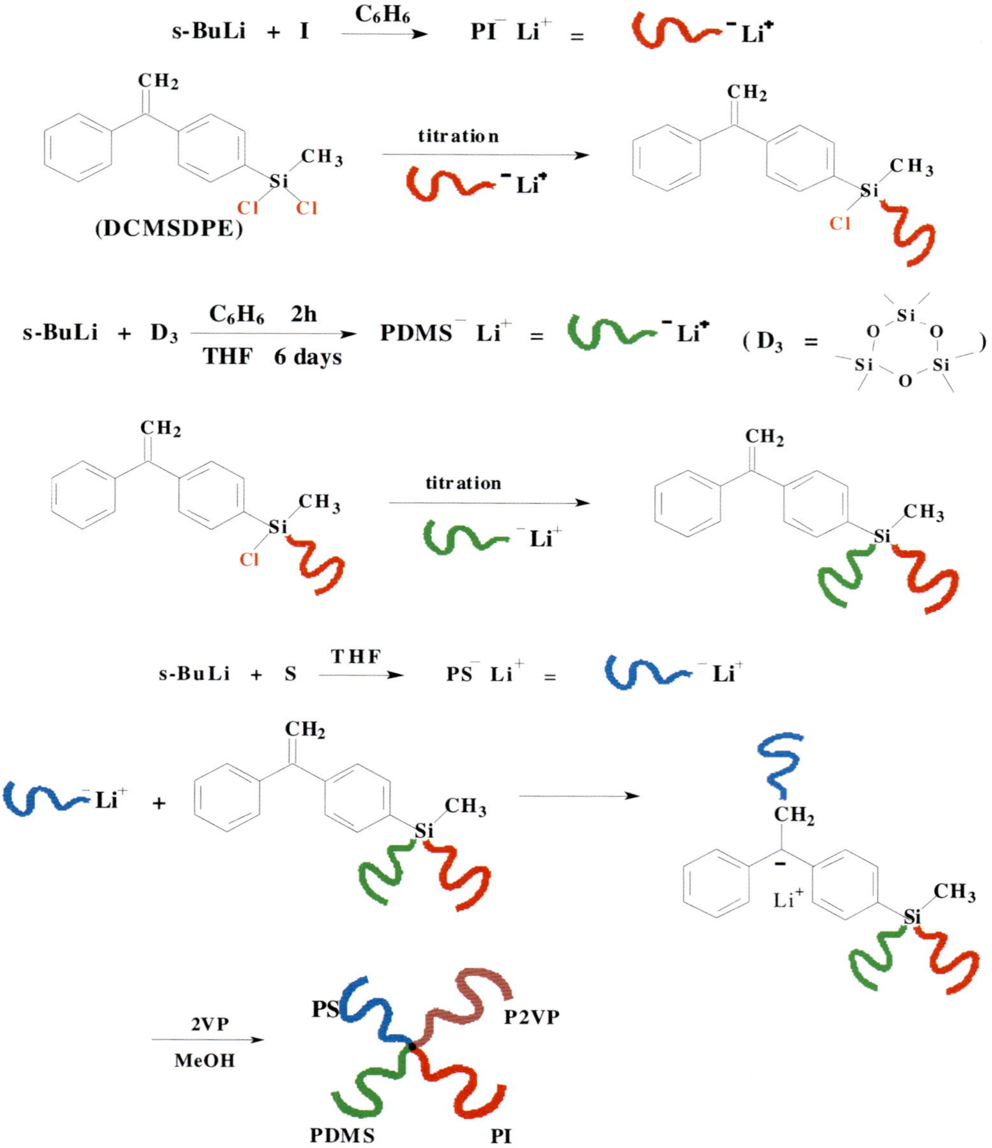

Scheme 6.13

sults revealed that the use of bromo instead of chloro derivatives and low temperatures leads to well-defined products.

Diphenylethylene derivatives form another class of linking agents. 1,3,5-Tris(1-phenylethenyl)benzene was used for the synthesis of a well-defined three-arm PS star [22] (Scheme 6.15), as found from SEC, MO and LS measurements.

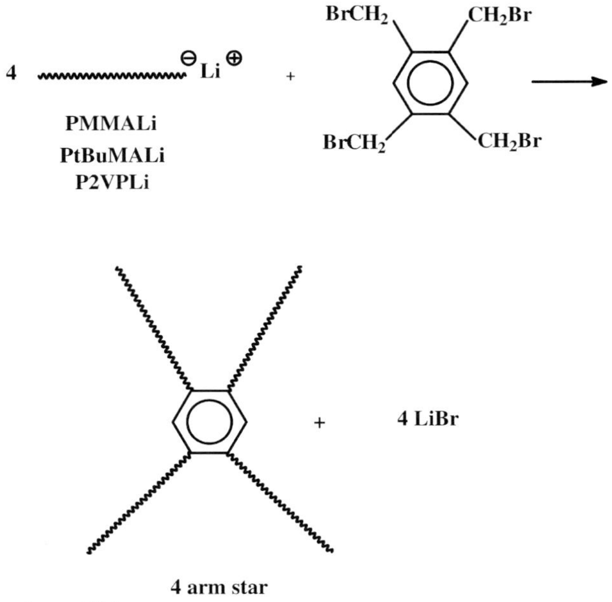

Scheme 6.14

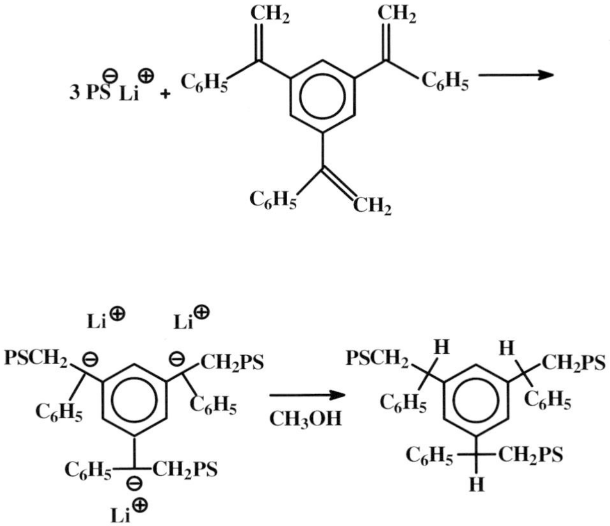

Scheme 6.15

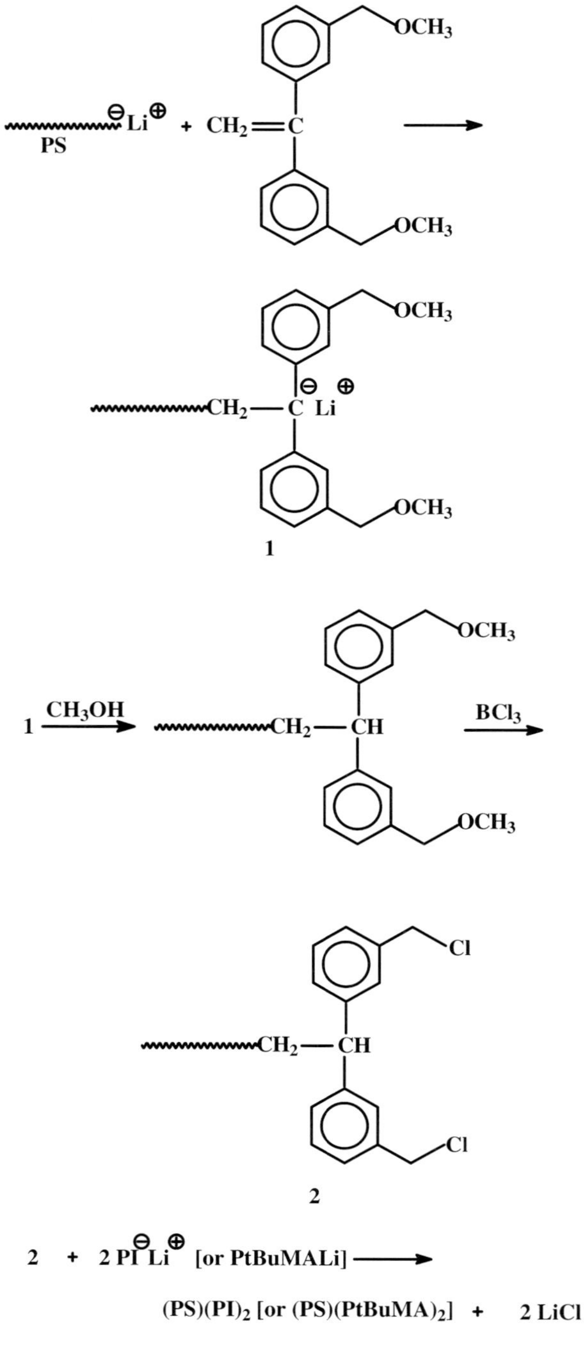
OCH3
Li
PS
CH2=C
OCH3
OCH3
CH2—C Li
OCH3
1
OCH3
1
CH3OH
CH2—CH
BCl3
OCH3
Cl
CH2—CH
Cl
2
2 + 2 PI Li [or PtBuMALi]
(PS)(PI)2 [or (PS)(PtBuMA)2] + 2 LiCl

Scheme 6.16

Although the arm molecular weight used was rather low ($M_n=8.5\times10^3$), there is no steric limitation for the synthesis of three-arm PS stars using this coupling agent. Previous efforts to use methyltrichlorosilane as a linking agent for the synthesis of three-arm PS stars were not successful, due to incomplete coupling (steric hindrance effects) [56].

The use of diphenylethylene derivatives also led to A_2B miktoarm stars [57, 58], where A is PS and B is PI or P*t*BuMA (Scheme 6.16). Living PS chains were reacted with a 1.2-fold excess of 1,1-[bis(3-methoxymethylphenyl)]ethylene in THF at –78 °C for 1 h. Only the monoadduct product was obtained under these conditions. The end methoxymethyl groups were then transformed to chloromethyl moieties by reaction with BCl_3 in CH_2Cl_2 at 0 °C for 2 h. NMR studies showed that the transformation reaction goes to completion. The macromolecular linking agent was carefully purified (repeated precipitation and freeze-drying from benzene solution) and then reacted with living PILi or P*t*BuMALi chains to give the desired products. A small amount (5%) of the dimeric product was observed by SEC analysis. It was proposed that this by-product is obtained by the Li–Cl exchange and/or electron transfer reactions. SEC and NMR methods have been the only methods used for the molecular characterization and low molecular weight stars were employed to facilitate NMR analysis.

The synthesis of a wide variety of regular and asymmetric stars using functionalized 1,1-diphenylethylene derivatives in a similar way has recently been reported in a detailed review [59]. By means of an iterative methodology, 3-, 5-, 9-, 17-arm PS and PMMA star homopolymers along with 33-arm PS stars were synthesized. Extending the same methodology, asymmetric and miktoarm stars

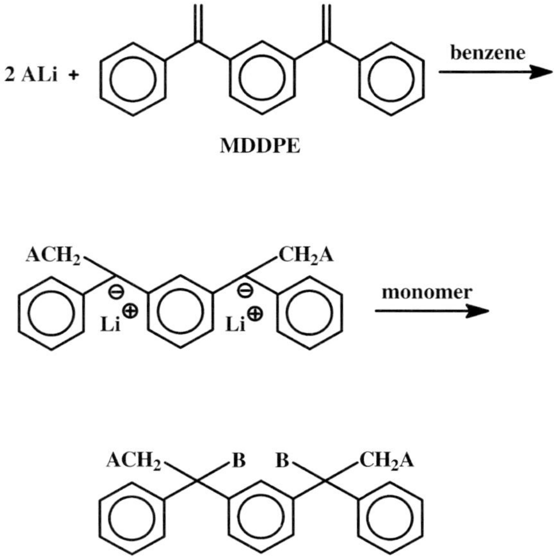

Scheme 6.17

of the type AA'_3, AB_4, AB_8, A_2B_4, A_2B_8, A_2B_{12}, $AA'A''_2$, ABC_2, ABC_4, $A_2B_2C_2D$, $A_4B_4C_4D$, ABC, A_3B, A_2B_2, AB_3, A_2BC, AB_2C, ABC_2, ABCD, where A is PS, B is PI or P*α*MS and C is PI or poly[4-(4-1,2:5,6-di-*O*-isopropylidene-*α*-glucofuranose-3-oxy)butoxy]styrene and D is poly(4-trimethylsilylstyrene), were prepared.

1,3-Bis(1-phenylethenyl)benzene (MDDPE) was used for the synthesis of A_2B_2-type miktoarm stars, where A is PS and B one of PI, PBd or PS-*b*-PBd [60]. The method involves the reaction of the living A arms with MDDPE in a molar ratio of 2:1, leading to the formation of the living dianionic coupled product. The active sites were subsequently used as initiator for the polymerization of another monomer to give the A_2B_2 structure (Scheme 6.17). This approach can be considered as an "in–out" procedure.

Using a similar diphenylethylene derivative, 1,1-(1,2-ethanediyl)bis[4-(1-phenylethenyl)benzene] (EPEB) and an "in–out" procedure, the synthesis of PI_2PMMA_2 miktoarm stars was achieved [61] (Scheme 6.18). A solution of EPEB was slowly added to a solution of the living PI chains, leading to the formation of a living coupled product. LiCl was then introduced and the polymerization of MMA was initiated at –78 °C to give the desired product. Unreacted PI chains formed by

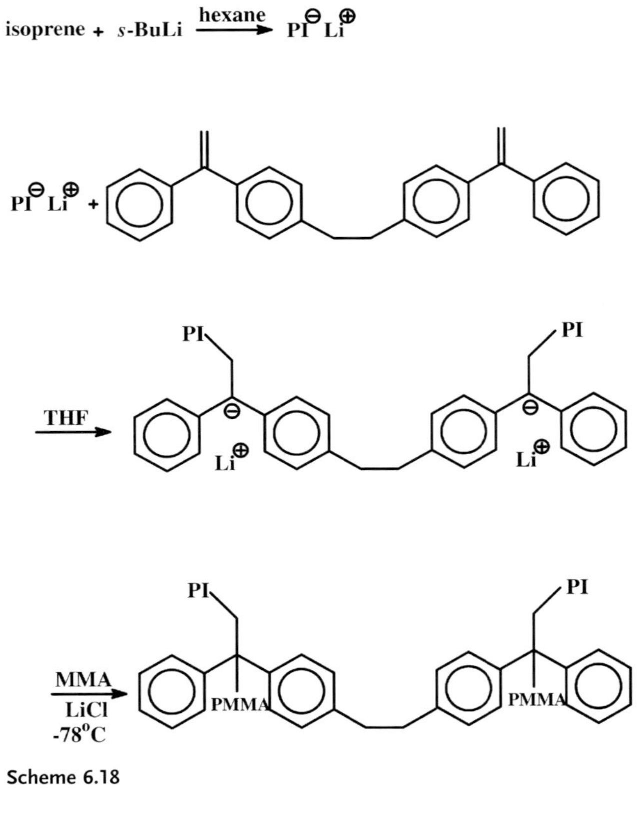

Scheme 6.18

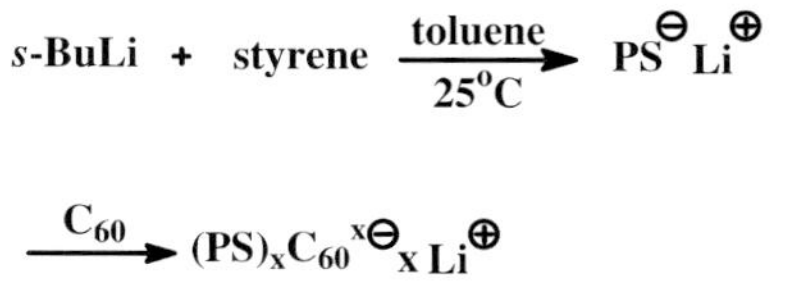

Scheme 6.19

accidental deactivation during the coupling reaction were removed by fractionation. Molecular characterization data showed that well-defined stars were prepared by this method.

Fullerenes have also been used as coupling agents for the preparation of star polymers [62, 63]. In non-polar solvents, such as toluene, it was found that an excess of living PSLi or PILi chains over the C_{60} is needed to prepare a six-arm star by addition of the carbanions to the double bonds of the fullerene (Scheme 6.19). However, when the living PS chains were end-capped with DPE, only a three-arm star was produced, showing that the functionality of the star can be adjusted by changing the steric hindrance of the living chain end. It was also found that the functionality could be controlled by the stoichiometry of the reaction between the living polymers and the C_{60}. However, it was impossible to incorporate selectively one or two chains per C_{60} molecule.

In addition to its use in the synthesis of star homopolymers, C_{60} was also employed for the synthesis of star block copolymers [63]. Living PS-*b*-P2VP diblocks, having short P2VP chains, were prepared by sequential anionic polymerization in THF. The living diblocks were reacted with a suspension of C_{60} in THF, leading to the formation of a three-arm star-block copolymer (Scheme 6.20). SEC analysis showed that the product had a broad molecular weight distribution, indicating that a mixture of stars with different functionalities was obtained.

In the past, several studies using the difunctional monomer DVB for the synthesis of PS [64, 65] and polydiene [66] stars have been reported. It was found that when the [DVB]/[PSLi] ratio was varied from 5.5 to 30, rather narrow molecular weight distribution PS stars were obtained, with the corresponding functionality between 13 and 39. For polydiene stars, when the [DVB]/[PDLi] ratio varied from 5 to 6.5, the functionality of the star varied between 9 to 13. For higher ratios, broad distributions were observed due to the large distribution of the stars' functionalities prepared by this method.

More recently, PMMA stars were prepared using the difunctional monomer ethylene glycol dimethacrylate (EGDM) [67]. EGDM was reacted with isotactic living PMMA chains obtained using *tert*-butylmagnesium bromide as initiator in the presence of 1,8-diazabicyclo[5.4.0]undec-7-ene [68]. A star polymer with a number of arms estimated between 20 to 30 was synthesized. SEC connected with LS and viscometry detectors was used to characterize the sample. A similar reaction involving syndiotactic living PMMA chains, obtained with the initiator system *tert*-BuLi–R_3Al, failed to give star polymers. However, by replacing

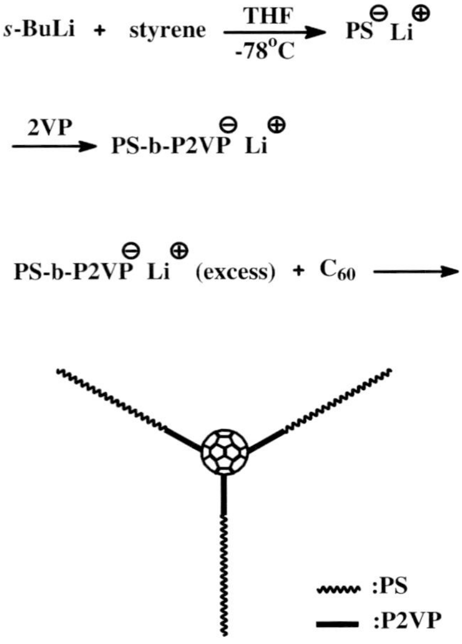

Scheme 6.20

EGDM with butane-1,4-diol dimethacrylate, the synthesis of a PMMA star with 50–120 arms was successful.

The difunctional monomer methodology was also used for the synthesis of miktoarm stars of the type A_nB_n, where A is PS and B is P*t*BuMA, P*t*BuA, PEO, P2VP or PEMA [18, 19, 69–73]. Special care was given to the synthesis of amphiphilic stars carrying both hydrophobic and either cationic or anionic branches. The polymerization of the styrene was initiated with *s*-BuLi, except in the case of the PS_nPEO_n stars, where cumylpotassium was used. After the formation of the living PS star, SEC analysis revealed that a considerable proportion (as high as 15%) of the PS chains was not incorporated into the star structure, mainly because of accidental deactivation. When the second monomer was a (meth)acrylate, the active sites were capped before the polymerization with one unit of DPE to reduce their nucleophilicity. The final stars generally had *n* values between 4 and 20.

6.4.2
Cationic Polymerization

Well-defined eight-arm polyisobutene (PIB) star homopolymers were synthesized using the *tert*-hydroxy and *tert*-methoxy derivatives of the octafunctional initiator 5,11,17,23,29,35,41,47-octaacetyl-49,50,51,52,53,54,55,56-octamethoxycalix[8]arene (Scheme 6.21) [74]. The polymerization was performed in two steps in the same reactor. In the first step, the initiator along with BCl_3 and 25% of the IB monomer were added in CH_3Cl at –80 °C, followed by the second step in

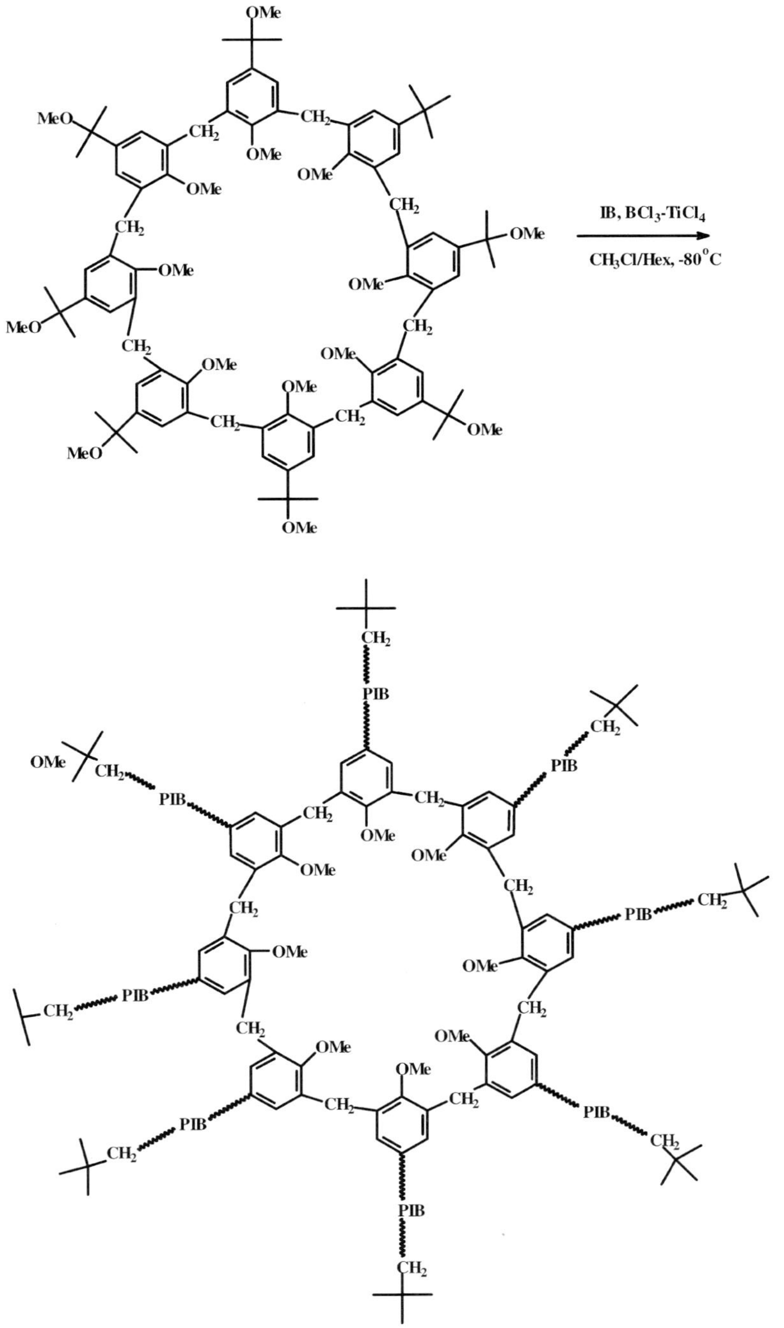

Scheme 6.21

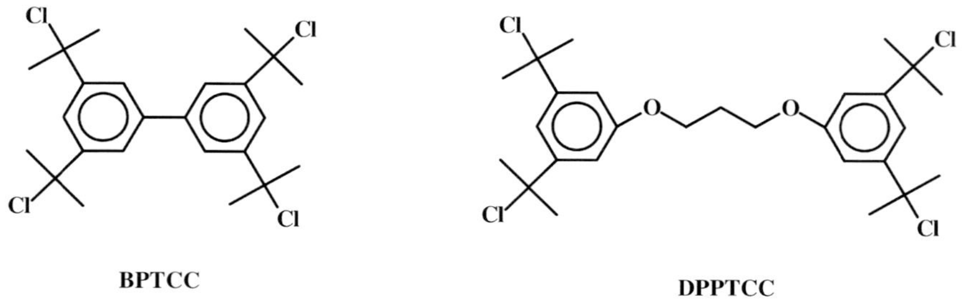

Scheme 6.22

which hexane, $TiCl_4$ and the rest of the monomer were added. SEC with on-line RI, UV and LALLS detectors was used for the determination of the absolute molecular weights and compositions of the stars. The molecular weights obtained were close to those expected theoretically.

A trifunctional initiator was employed to synthesize a (PIB-*b*-PS)$_3$ star-block copolymer using ring-substituted tricumyl chloride as initiator and $TiCl_4$ as co-initiator via the sequential monomer addition methodology [75]. The polymerization was conducted at –80 °C, in the presence of pyridine (electron donor) and 2,6-di-*tert*-butylpyridine (proton trap) in a 60:40 (v/v) methylcyclohexane–methyl chloride solvent mixture. The copolymer synthesized exhibited a rather low polydispersity index (1.17) and the mechanical, thermal and morphological properties were examined extensively.

3,3′,5,5′-Tetrakis(2-chloro-2-propyl)biphenyl (BPTCC) and 1,3-bis[3,5-bis(2-chloro-2-propyl)phenoxy]propane (DPPTCC) were synthesized (Scheme 6.22) and employed as terafunctional initiators for the preparation of four-arm PIB stars [76]. BPTCC led to the synthesis of well-defined products, whereas DPPTCC gave a mixture of two-, three- and four-arm stars. This behavior was attributed to the ether linkages connecting the aliphatic chain to the aromatic rings. These linkages contribute significant electron density to the aromatic rings, making them susceptible to Friedel-Crafts cycloaddition reactions after the addition of one isobutene unit to the initiating cation.

By using tri- and tetrafunctional silylenol ethers as coupling agents (Scheme 6.23), three- and four-arm poly(isobutyl vinyl ether) (PIBVE) star homopolymers have been synthesized [77]. The living arms were synthesized at –15 °C, using an HCl–$ZnCl_2$ initiating system in methylene chloride. It was shown that the coupling of relatively short chains of living PIBVE (DP $\approx$ 10) occurred nearly quantitatively, to give the multi-armed polymers in high yield (>95%), but the yield decreased slightly (85–89%) with a longer living chain (DP $\approx$ 50). The molecular weight distribution of the final stars was low.

Miktoarm star copolymers of the A_2B_2 type, where A is PIB and B is poly(MeVE), were prepared [78]. The synthetic strategy involved the reaction of 2,2-bis[4-(1-phenylethenyl)phenyl]propane (BDPEP) or 2,2-bis[4-(1-tolylethenyl)phenyl]propane (BDTEP) with living PIB, resulting in a dicationic in-chain initiator.

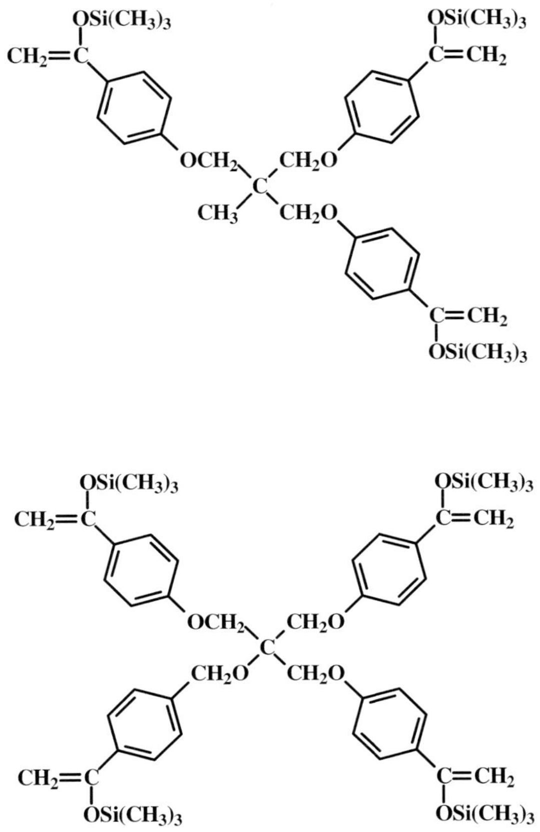

Scheme 6.23

This initiator was used for the polymerization of methyl vinyl ether (MVE) to give the $(PIB)_2(PMVE)_2$ miktoarm copolymer. Purification of the crude A_2B_2 copolymer was performed on a silica gel column and the purity of the resulting star was 93%.

Amphiphilic four-arm star block copolymers of α-methylstyrene (α-MeS) and 2-hydroxyethyl vinyl ether were prepared using tetrafunctional silylenol ether as the coupling agent, as depicted in Scheme 6.20 [79]. Initially α-MeS was polymerized using the HCl adduct of 2-chloroethyl vinyl ether in conjunction with $SnBr_4$ as initiating system in CH_2Cl_2 at $-78\,^{\circ}C$. When the polymerization reached 95% yield, 2-[(*tert*-butyldimethylsilyl)oxy]ethyl vinyl ether was added and the polymerization was continued to reach 85% conversion, followed by addition of the linking agent and *N*-ethylpiperidine to enhance the coupling efficiency. The star copolymer was separated from side-products by preparative gel permeation chromatography. The 2-hydroxyethyl pendant groups were obtained after deprotection using tetra-*n*-butylammonium fluoride, at room temperature, providing a new four-arm amphiphilic block copolymer with hard hydrophobic segments of poly(α-MeS) in the exterior of the star.

A novel methodology for the synthesis of complex non-linear polymeric architectures was proposed, based on the self-assembly principle, through coulombic interactions to preorganize linear polymer precursors and subsequent conversion into covalently linked permanent structures [80]. THF was polymerized cationically using methyl triflate as initiator and was end-capped by reaction with *N*-phenylpyrrolidine to provide the corresponding salt. An efficient ion-exchange reaction of telechelic polytetrahydrofuran (polyTHF) with an *N*-phenylpyrrolidinium end-group, carrying tricarboxylate as the counterion, was performed by repeated precipitation into an aqueous solution containing an excess of sodium tricarboxylate salt. These carboxylate salts are stable under ambient conditions and can be isolated. The subsequent heat treatment of the ionically assembled polymer precursor caused selective ring opening to provide the covalently linked three-arm star polymer in high yield [81] (Scheme 6.24).

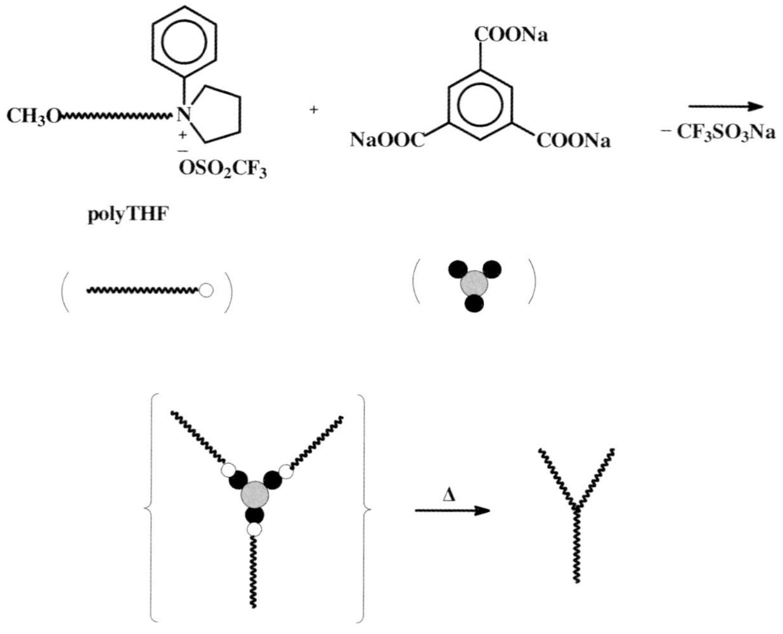

Scheme 6.24

6.4.3 Controlled Radical Polymerization

6.4.3.1 Nitroxide-mediated Polymerization

Three-arm PS-d7 stars have been synthesized by using a trifunctional unimolecular initiator that contains three initiating styrene-2,2,6,6-tetramethylpiperidinyloxy (TEMPO) groups [82] (Scheme 6.25). Bulk polymerization was conducted at 130 °C for 72 h and no detectable amount of cross-linked or insoluble material was observed in the final product. SEC analysis revealed a low polydispersity index (1.2) for the star. Degradation of the polystyrene star by hydrolysis of the central ester links and characterization of the arms showed that the molecular weight was in agreement with the theoretically expected values and the polydispersity indices were lower than those of the corresponding stars (1.10–1.15). The results demonstrate that each of the initiating sites in the trifunctional initiator is active and the individual PS arms grow at approximately the same rate with little or no star–star reaction.

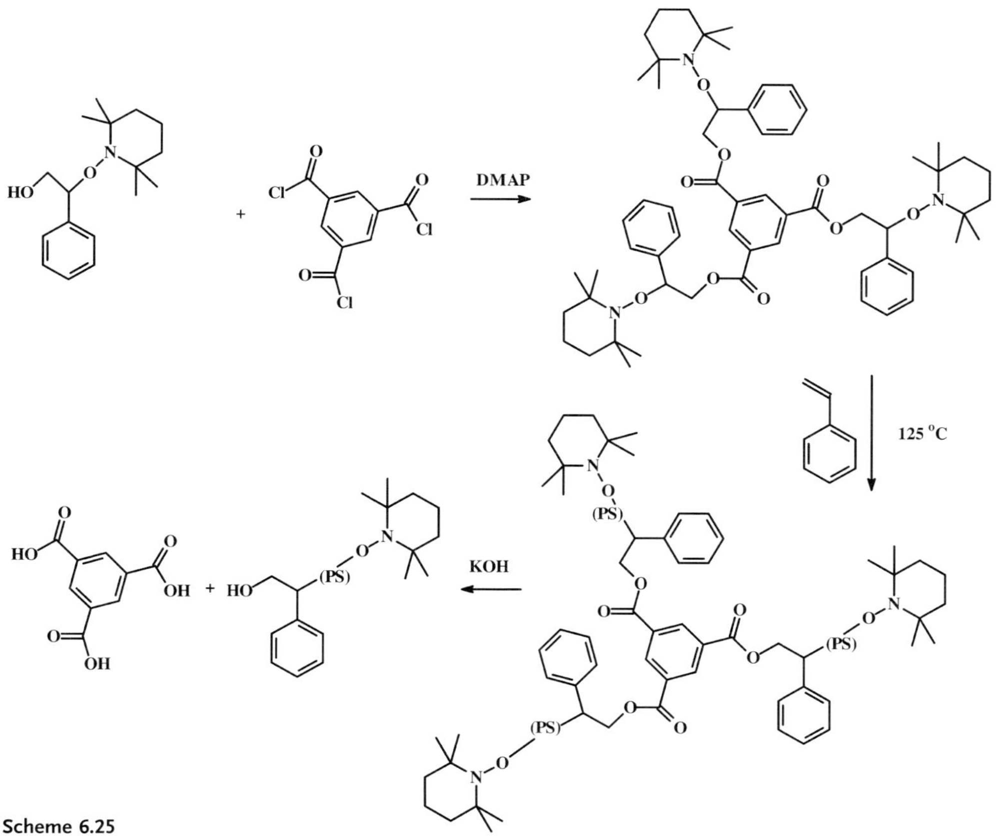

Scheme 6.25

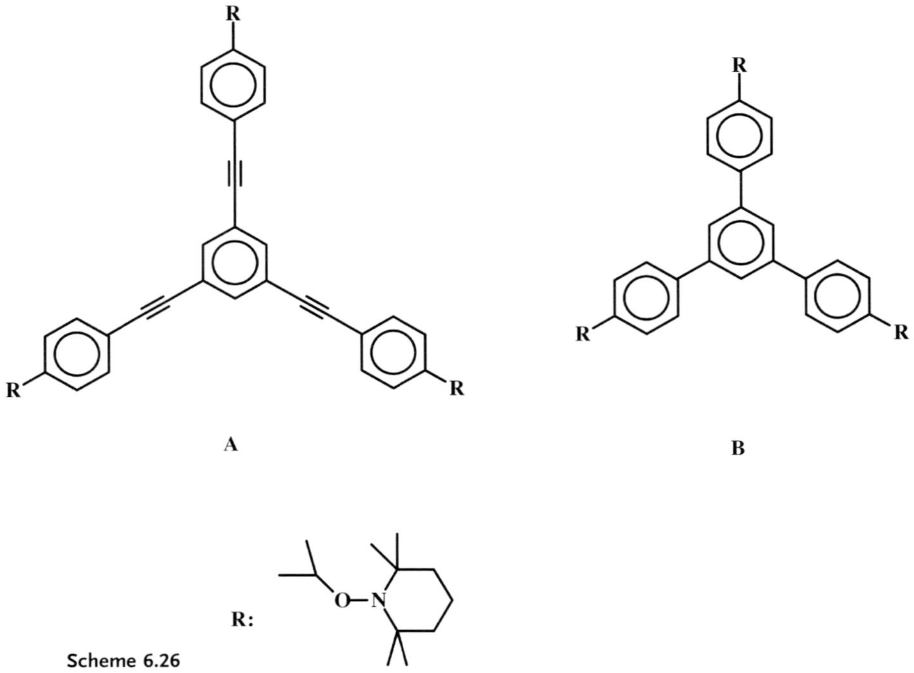

Scheme 6.26

1,3,5-Tris(alkoxyaminophenylethynyl)benzene (A) and 1,3,5-tris(alkoxyamino-phenyl)benzene (B) were synthesized and employed as trifunctional initiators for the polymerization of styrene [83] (Scheme 6.26). Initiator A gave star polymers with a broad molecular weight distribution at high conversion, due to the interference of the diphenylethynyl triple bond during the polymerization. On the other hand, initiator B afforded stars with polydispersity indices between 1.20 and 1.40.

More recently, 12-arm PS star homopolymers as well as 12-arm PS-P4VP star-block copolymers were synthesized, using dendritic dodecafunctional macroinitiators [84], as shown in Scheme 6.27. The polymerizations were carried out in bulk at 120 °C and the synthesized stars exhibited rather low polydispersity indices (1.06–1.26). To evaluate the livingness of the TEMPO-mediated radical polymerization of styrene, hydrolysis of the ester central bonds and subsequent SEC measurements were performed. The analysis revealed that the resulting four-arm star polymers exhibited lower polydispersity indices than the parent 12-arm star polymers, implying that the functionality polydispersity is higher than the molecular weight polydispersity.

It was shown that β-hydrogen-containing nitroxides promote the controlled polymerization of not only styrenic monomers, but also alkyl acrylates and dienes. Taking this into account, a novel trifunctional alkoxyamine (Scheme 6.28, **1**) based on *N-tert*-butyl-1-diethylphosphono-2,2-dimethylpropyl nitroxide

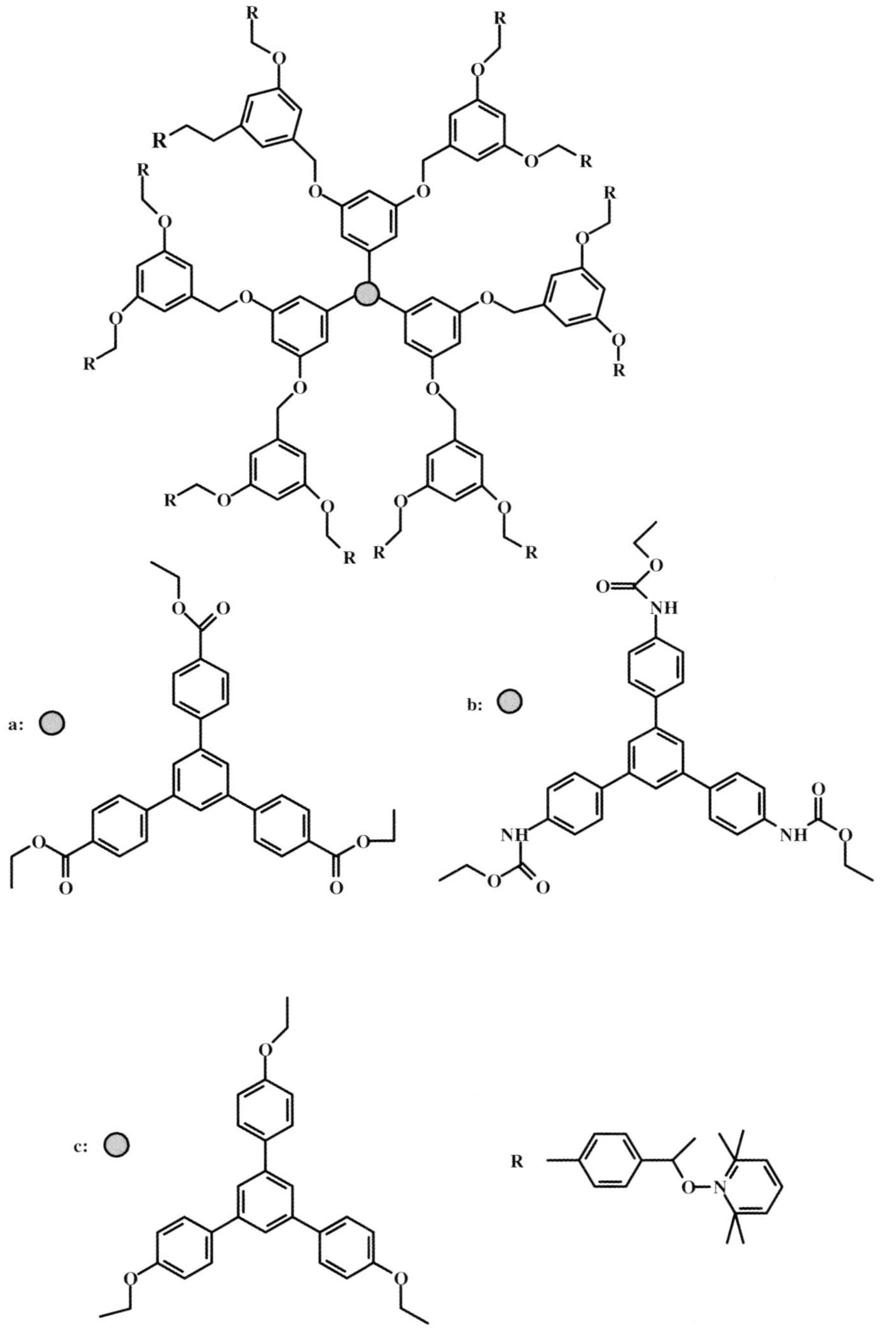

Scheme 6.27

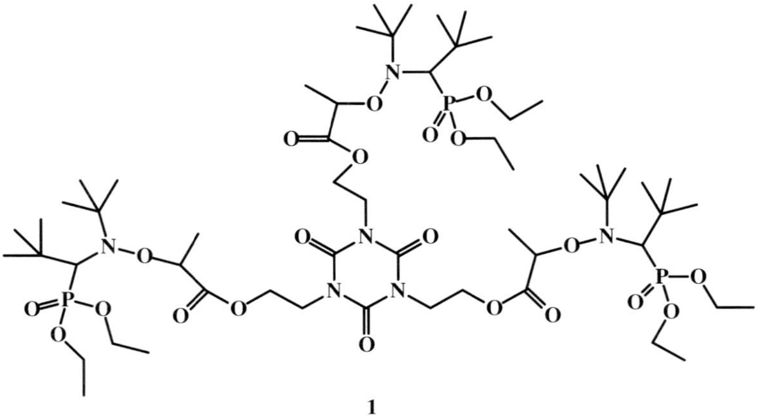

1

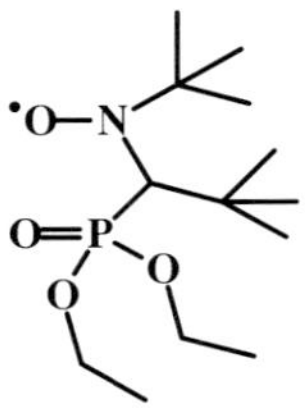

Scheme 6.28 2

(Scheme 6.28, **2**) was developed for the synthesis of three-arm PS and poly(*n*-butyl acrylate) (P*n*BuA) stars along with (P*n*BuA-*b*-PS)$_3$ star-block copolymers [85].

Employment of the difunctional monomer methodology afforded star-block copolymers with a great variety of chemically different chains in terms of molecular weight and composition [86]. The synthetic strategy involved the preparation of TEMPO-terminated linear chains and subsequent coupling with the difunctional monomer divinylbenzene or a bis(maleimide) derivative.

6.4.3.2 **Atom Transfer Radical Polymerization (ATRP)**

Numerous reports have been published regarding the synthesis of star polymers using multifunctional initiators capable of initiating the ATRP of certain monomers, mainly styrene, methacrylates and acrylates [87–96]. The living character of the growing chain ends provides the possibility for the synthesis of star-block copolymers.

One of the first works reported on the synthesis of stars utilizing ATRP was the preparation of three-arm PMMA star homopolymers with a trifunctional di-

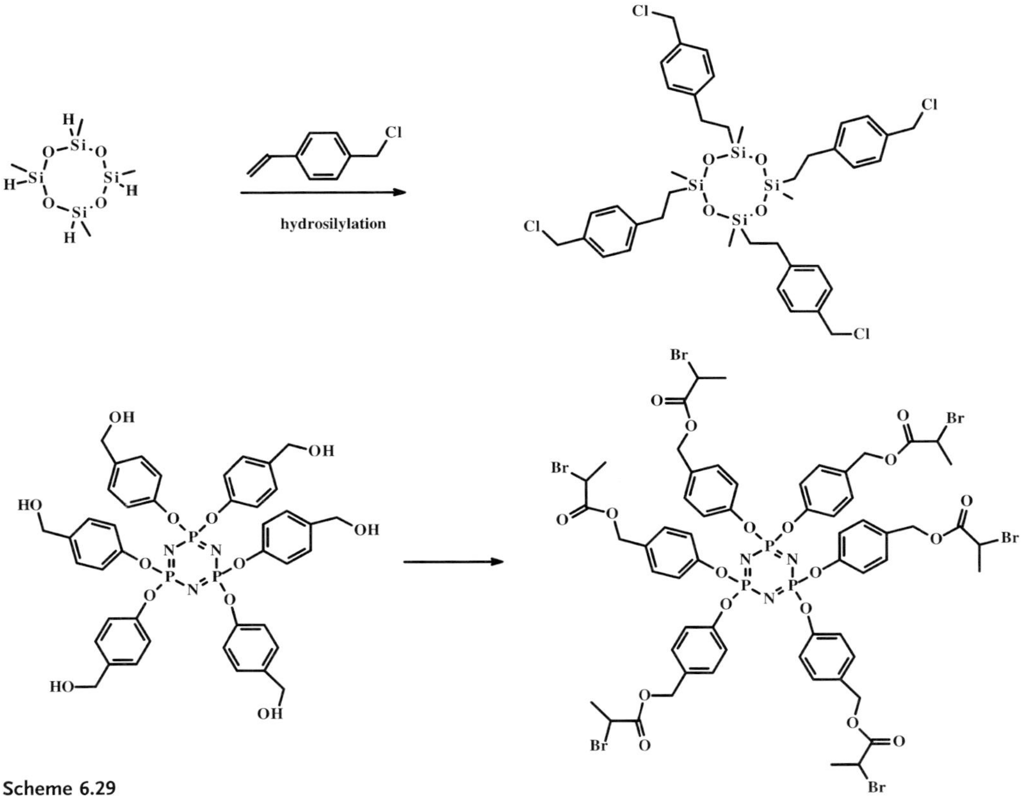

Scheme 6.29

chloroacetate initiator, using $RuCl_2(PPh_3)_3$ in the presence of either $Al(O\textit{i}Pr)_3$ or $Al(acac)_3$ (acac = acetylacetonate) as catalyst [97]. Both aromatic and aliphatic analogues of the initiators were examined. The stars were characterized by SEC and NMR spectroscopy. The polydispersities obtained were rather low (1.2–1.3). It was found that when $Al(O\textit{i}Pr)_3$ was used, deviation from the theoretical molecular weight was observed. Reactions in the presence of $Al(acac)_3$ avoided this problem. NMR studies determined that the more basic $Al(O\textit{i}Pr)_3$ promoted a transesterification reaction with the initiator, which altered the actual monomer to initiator ratio and poisoned the catalyst.

Tetra- and hexafunctional initiators shown in Scheme 6.29 have been used for the polymerization of styrene and acrylates [98]. The resulting stars had low polydispersity indices and the molecular weights obtained by SEC with a LALLS detector and viscometry showed a good correlation with the theoretical values. In addition, a six-armed star-block copolymer composed of a PMMA core and poly(isobornyl acrylate) shell was synthesized in this work.

Octafunctional 2-bromopropionate-modified calixarenes have also been used as initiators for the ATRP polymerization of styrene [99]. It was found that when the polymer yield was lower than 20%, there was good agreement between the measured and the theoretically expected molecular weights. Above this conversion, high molecular weight shoulders were observed by on-line light scattering detection, which the authors attributed to coupling between the stars. By performing the polymerization under high dilution conditions and cessation of the polymerization at low conversion, stars with molecular weights as high as $M_n = 34.0 \times 10^5$ and low polydispersity indices were formed.

More recently, multifunctional ATRP initiators with four, six and eleven 2-bromoisobutyrate groups were prepared for the synthesis of star polymers of styrene and methyl, *n*-butyl and *n*-hexyl acrylates [100]. The polymerization of styrene was performed at 110 °C, employing the catalytic system CuBr–bipyridine. The molecular weights of the resulting stars were up to 51×10^3, the polydispersity indices were lower than 1.1 and the conversions lower than 32%. Hydrolysis of the star PS at the central ester groups led to linear chains with the expected M_n values. Star poly(methyl acrylates) have also been synthesized by performing the polymerization in the bulk, at 60–80 °C with the 1,1,4,7,7-pentamethyldiethylene triamine ligand at higher conversions (up to 82%). Stars of poly(*n*-butyl acrylate) and poly(*n*-hexyl acrylate) with controlled characteristics have been prepared as well under the same experimental conditions.

The difunctional monomer methodology has been used for the synthesis of PS [101] and poly(*tert*-butyl acrylate) star homopolymers [102]. The difunctional monomers used were divinylbenzene, 1,4-butanediol diacrylate and ethylene glycol dimethacrylate. Several factors were investigated for the formation of nearly monodisperse stars, including the choice of exchanging halogen, the solvent, the addition of copper(II) species, the ratio of the coupling agent to the macroinitiator and the reaction time. The highest efficiency (~ 95%) was obtained with a 10–15-fold excess of the difunctional monomer over chain ends.

Living PMMA chains, produced by the $RuCl_2(PPh_3)_3$-catalyzed polymerization of MMA, were reacted with the difunctional monomer bisphenol A dimethacrylate (BPDMA) to afford the corresponding PMMA star polymers adopting the "arm-first" approach [103]. The functionality of the products ranged from 4 to 63. The yield of the linking reaction was found to depend on the concentration and the degree of polymerization of the living arms as well as on the molar ratio of BPDMA over the living chains.

The convergent approach was also utilized for the synthesis of PEO stars. Mono-2-bromoisobutyryl PEO ester was used as macroinitiator for the ATRP of DVB, leading to the synthesis of the desired star polymers. Subsequent addition of styrene led to the synthesis of PEO_nPS_m miktoarm stars through the polymerization of styrene from the initiating sites located at the core of the PEO stars [104] ("in–out" approach). The styrene conversion did not exceed 10% to avoid star–star coupling reactions and, therefore, the production of gels.

DVB and 2-methacryloyloxyethyl disulfide were used as difunctional monomers for the synthesis of $PnBuA_nPS_m$ and $PMMA_nPnBuA_m$ stars, respectively, where n and m range between 4 and 60 [105]. In the last case, with a suitable reducing agent, it was possible to degrade the core of the star structure and to

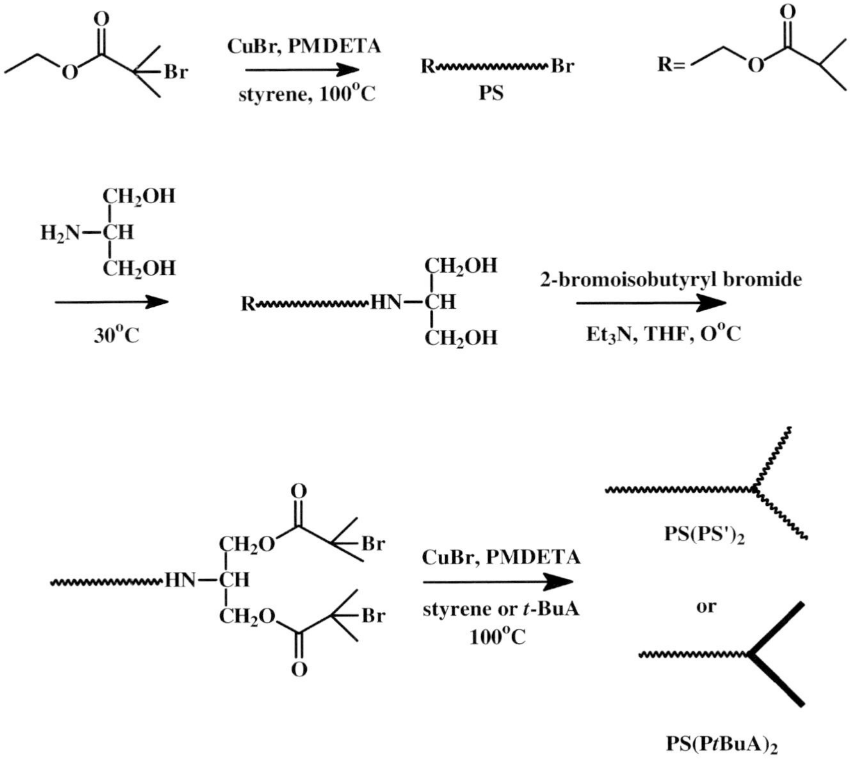

PMDETA: N,N,N',N'',N''-pentamethyldiethylenetriamine

Scheme 6.30

analyze the arms. The interstar and intrastar arm–arm coupling was observed. The initiating efficiency of the alkyl bromide sites in the core of the star polymers was determined after cleavage of the degradable stars and the corresponding miktoarm stars. It was found that only 19% of the initiation sites were active for the polymerization of the second monomer.

Asymmetric and miktoarm stars were prepared by ATRP and suitable chemical modification of the end-groups [106, 107], as shown in Scheme 6.30. ω-Bromopolystyrene was obtained using ethyl 2-bromoisobutyrate as initiator in the presence of the catalyst CuBr and the ligand pentamethyldiethylenetriamine. The end-bromine was reacted with 2-amino-1,3-propanediol to afford PS chains bearing two hydroxyl groups at the same chain end. The OH groups were subsequently transformed to Br after reaction with 2-bromoisobutyryl bromide. The new initiating sites were then available either for the polymerization of styrene, leading to the synthesis of asymmetric $PS(PS')_2$ stars, or the polymerization of *t*BuA, to afford miktoarm $PS(PnBuA)_2$ stars.

6.4.3.3 Reversible Addition Fragmentation Chain Transfer Polymerization (RAFT)

The "arm-first" and "core-first" methodologies have been both employed for the synthesis of star polymers by RAFT polymerization [108–111]. The control of the procedure is achieved by the direction of fragmentation of the RAFT thiocarbonyl agent during the polymerization process. This is influenced by the leaving ability of the groups attached to the two sulfur atoms of the trithiocarbonate moiety. For example, methyl groups are poor leaving radicals relative to benzyl leaving radicals. Thermodynamic, electronic and steric parameters affect the stability of the radicals. A "core-first" approach is displayed in Scheme 6.31. The

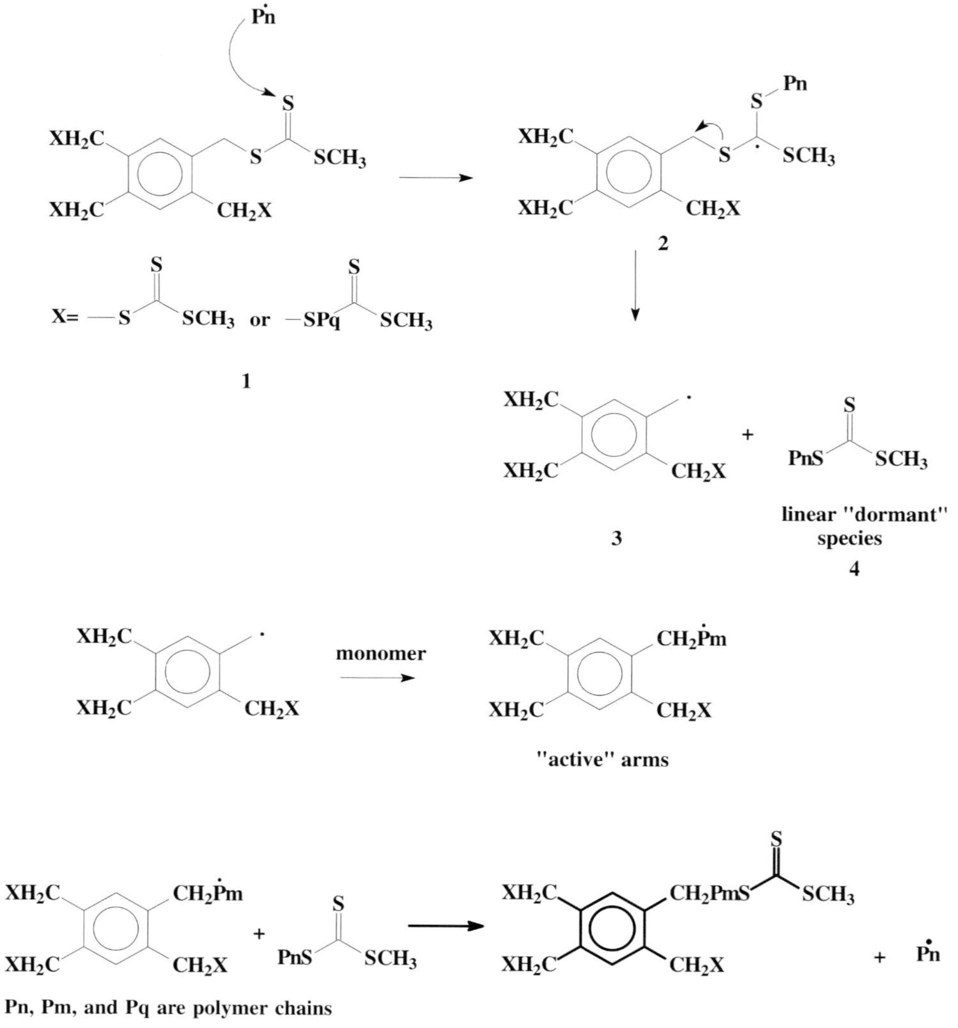

Scheme 6.31

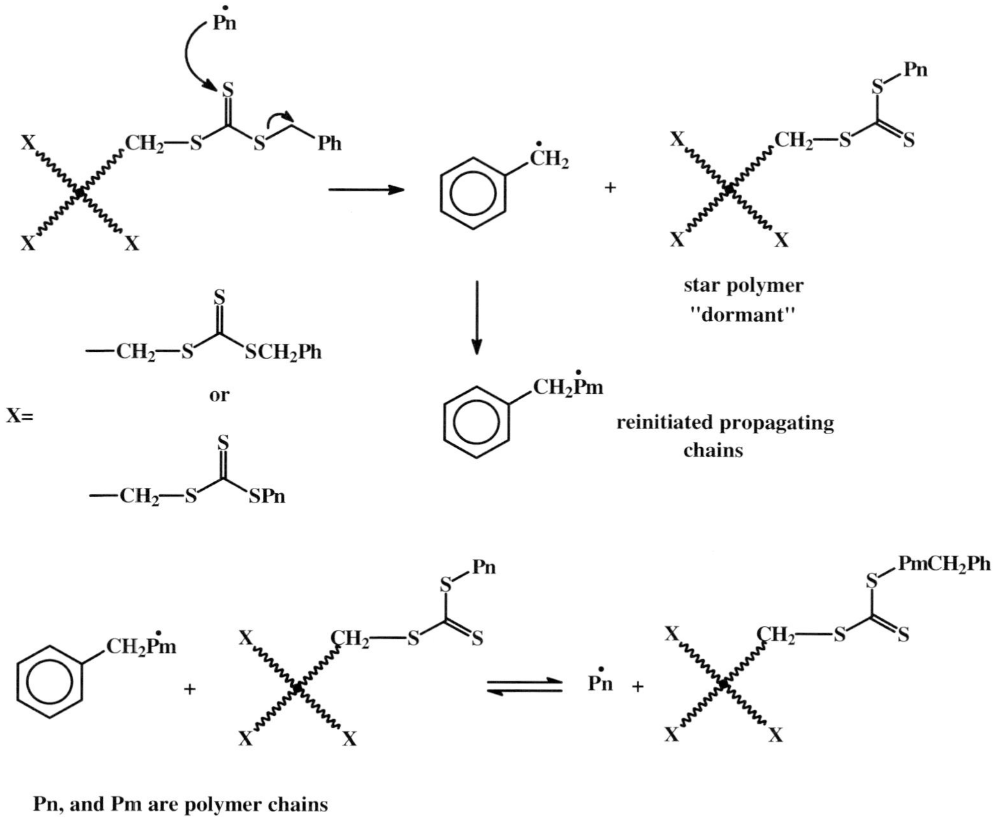

Scheme 6.32

methyl groups of the RAFT agent, **1**, are weak leaving groups. The fragmentation of the resulting intermediate radical, **2**, formed by addition at the sulfur of the thiocarbonyl group, leads to the formation of linear dormant species, **4**, and active radicals, **3**. Propagation of the living arms occurs through radicals **3**, affording a four-arm star polymer. PS stars have been prepared by this method [112].

The "arm-first" approach is presented in Scheme 6.32 using the RAFT agent **1**. In this case, the fragmentation results in the formation of benzyl radicals, which are able to reinitiate polymerization of linear chains. The arms of the star polymer are dormant and the growth of the arms always occurs away from the core. Using this methodology, four-arm PS and poly(methyl acrylate) stars were prepared [112].

The synthesis of dendritic multifunctional RAFT agent carrying 6 or 12 external 3-benzylsulfanylthiocarbonylsulfanylpropionic acid groups was reported [113]. These agents have been used to prepare P*n*BuA and PS stars according to

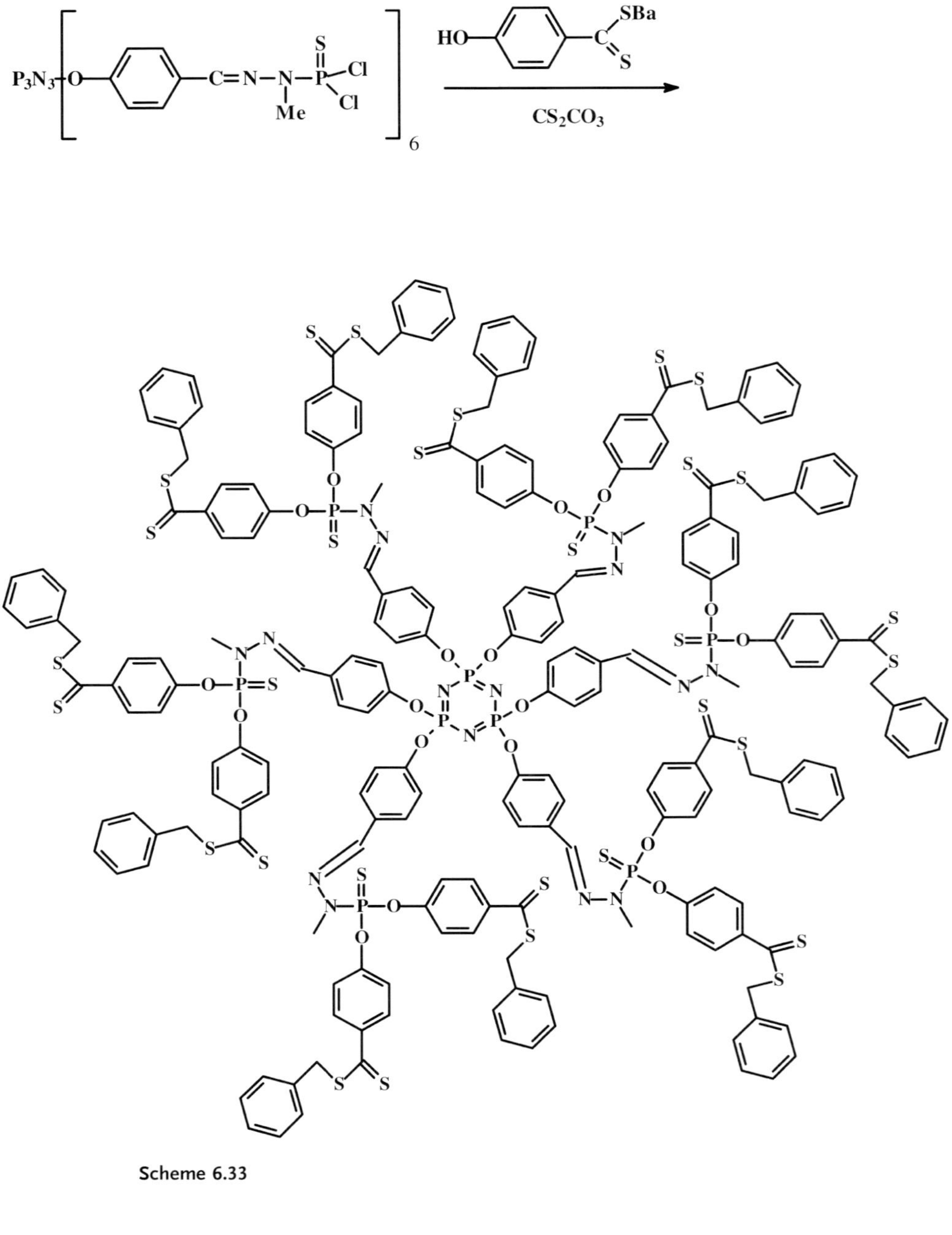

Scheme 6.33

the "arm-first" method. The polymerization was performed in bulk at 60 °C, using AIBN as thermal initiator. The stars were poorly characterized by SEC based on linear PS standards. The molecular weights of the synthesized P*n*BuA and PS stars were lower than 1.60×10^5 and 7.0×10^5 g mol^{-1}, respectively, while the polydispersity indices were rather high, ranging from 1.1 to 1.5. The star-shaped structure of the synthesized polymers was confirmed through the cleavage of the arms from the core and characterization.

Twelve-arm PS star homopolymers have been also synthesized [114] by using a dendritic compound with 12 terminal benzyl dithiobenzoate groups in bulk, at 110 °C in the presence of AIBN (Scheme 6.33). The reaction was monitored by SEC/UV. It was found that conversions higher than 69% produced a bimodal distribution. The high molecular weight peak corresponded to the star molecule while the other peak was attributed to terminated linear chains formed by irreversible terminations. After fractionation, the stars exhibited rather low polydispersity indices (<1.2).

The difunctional monomer methodology has been used for the synthesis of PS star homopolymers along with PS-*b*-poly(*N*-isopropylacrylamide) star-block copolymers [115]. The linear macro RAFT PS agent was synthesized in bulk, at 110 °C, from benzyl dithiobenzoate and AIBN. The synthesis of the diblock PS-*b*-poly(*N*-isopropylacrylamide) copolymer was performed by bulk sequential polymerization under the same experimental conditions. DVB was used as the difunctional monomer. The stars were extensively characterized by NMR and IR spectroscopy, as well as by SEC and DLS. The DVB : PS molar ratio, along with the duration of the polymerization were found to influence the yield, molecular weight and polydispersity index of the stars. In most cases, a significant amount of living linear macroinitiator remained unreacted and was separated from the star by preparative SEC.

6.4.4 Ring-opening Polymerization

Four-arm hydroxyl-terminated poly(ε-caprolactone) stars were synthesized by the ring-opening polymerization (ROP) of ε-caprolactone, with pentaerythritol as the initiator [116] (Scheme 6.34). The terminal hydroxyl groups of the star were then reacted with an α-carboxyl-functionalized PEO, to afford the four-arm PCL-*b*-PEO star-block copolymer. The PCL star homopolymer precursors, along with the final star-blocks, were characterized by NMR spectroscopy and SEC. The star precursors exhibited higher molecular weights than those expected theoretically, their yield was lower than 88% with fairly high polydispersities (1.42–1.56). The polydispersity indices of the star-block copolymers, were lower (1.16–1.36) than these of the star precursors the linking efficiency was high (~95%).

Two series of four-arm star polypeptides of γ-benzyl-L-glutamate *N*-carboxyanhydride (NCA) and ε-benzyloxycarbonyl-L-lysine NCA, were prepared via ROP [117]. A tetraamino-substituted perylene fluorophore was employed as the initiator. The polymerization was performed under dry inert atmosphere conditions

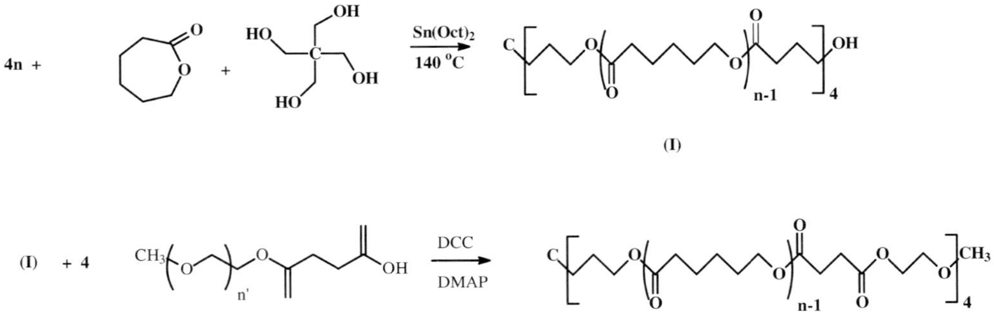

Scheme 6.34

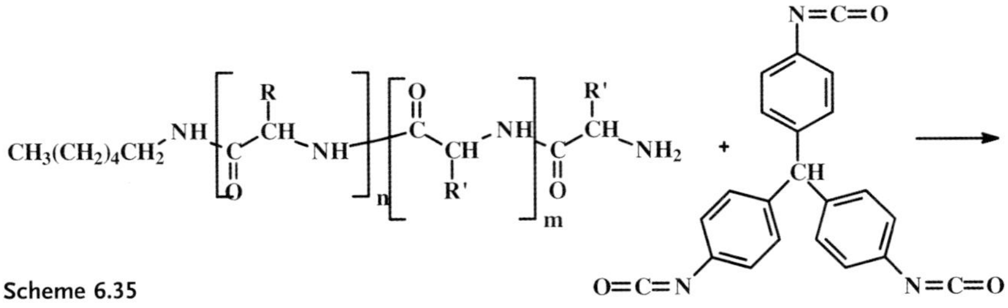

Scheme 6.35

and the star homopolymers produced were characterized by SEC and NMR spectroscopy. The polydispersity indices ranged from 1.22 to 4.96. The removal of the α-amino acid side-chain protecting groups resulted in water-soluble perylene-functionalized star polypeptides that displayed strong fluorescence in aqueous solutions. Additionally, the pH-dependent change of the conformation of the arms from random coil to α-helix was investigated by circular dichroism.

More recently, three-arm homo- and star-block copolypeptides of poly(γ-benzyl-L-glutamate) (PBLG) and poly(ε-benzyloxycarbonyl-L-lysine) (PZLL) in all possible combinations, i.e. $(PBLG)_3$, $(PZLL)_3$, $(PBLG\text{-}b\text{-}PZLL)_3$ and $(PZLL\text{-}b\text{-}PBLG)_3$, were synthesized [118]. The synthetic approach involved the preparation of the corresponding living arms, followed by linking of those arms with triphenylmethane 4,4′,4″-triisocyanate at room temperature (Scheme 6.35). The polymerization of the monomers, γ-benzyl-L-glutamate NCA and ε-benzyloxycarbonyl-L-lysine NCA, was performed using high-vacuum techniques and *n*-hexylamine as the initiator. This method leads to high molecular weight, well-defined living polypeptides in ~100% yield and low polydispersity. The characterization results (membrane osmometry, SEC/LALLS, NMR) revealed that the stars exhibited a high degree of molecular and compositional homogeneity.

6.4.5 Group-transfer Polymerization (GTP)

The group transfer polymerization method, a catalytic Michael addition reaction involving the addition of a silyl ketene acetal to α,β-unsaturated carbonyl compounds in the presence of a suitable catalyst, was initially reported by Webster et al. [119]. This method is best suited for the polymerization of acrylates and especially for methacrylates [120–122]. It is a living polymerization technique and can be applied from room temperature up to 80 °C.

1,3,5-Tris(bromomethyl)benzene was employed as a linking agent for the coupling of living poly(methylmethacrylate) chains leading to the formation of three-arm stars [123], as shown in Scheme 6.36. A re-examination of this reaction revealed that the linking reaction was not quantitative, perhaps because the reaction of the $-CH_2Br$ group with the living chain end occurred in a stepwise manner. Thus, the reactivity of the remaining $-CH_2Br$ groups was reduced after each substitution, as a result of steric hindrance, leading to lower conversions.

The multifunctional initiator method was employed for the synthesis of three- and four-arm stars. Trimethylolpropane triacrylate was converted to a silyl enol ether capable of initiating the polymerization of ethyl acrylate to form the corresponding three-arm star [121] (Scheme 6.37).

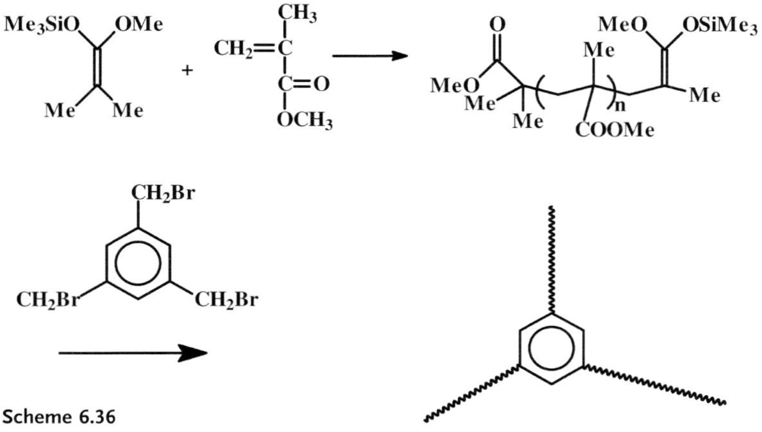

Scheme 6.36

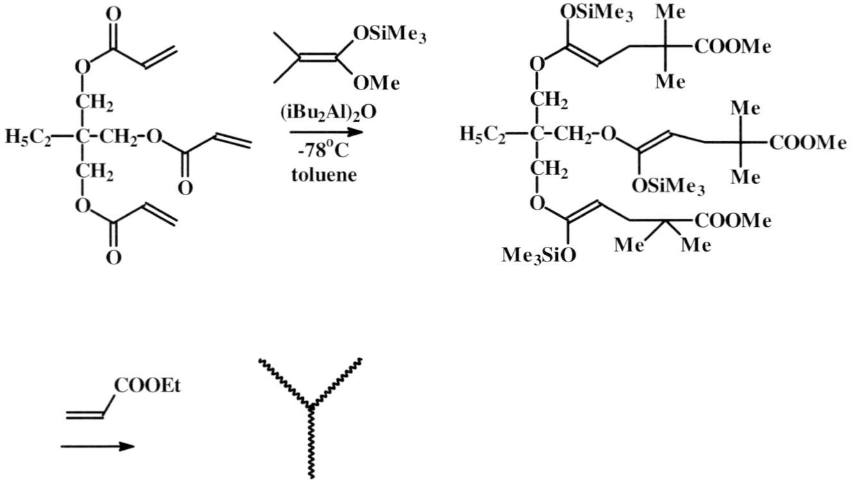

Scheme 6.37

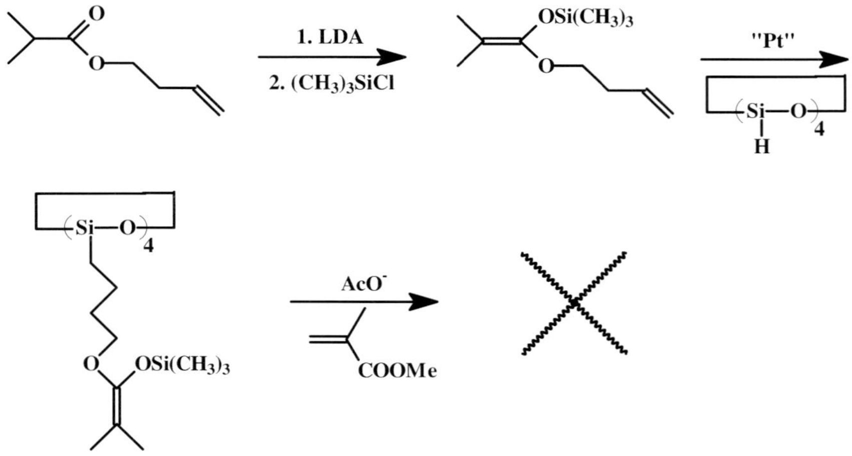

Scheme 6.38

1,3,5,7-Tetramethylcyclotetrasiloxane was converted to a tetrafunctional GTP initiator through the Pt-catalyzed hydrosilylation reaction with 1-[(but-3-en-1-yl)-oxy]-1-(trimethylsiloxy)-2-methyl-1-propene (Scheme 6.38). PMMA stars with four arms and 20–150 MMA repeat units were obtained [124]. The star topology was confirmed by cleavage of the cyclic siloxane cores with triflic acid in the presence of an excess of hexamethyldisiloxane and product analysis.

The most important method for the synthesis of star polymers by GTP is through the use of a difunctional monomer either as an "arm-first" or as a

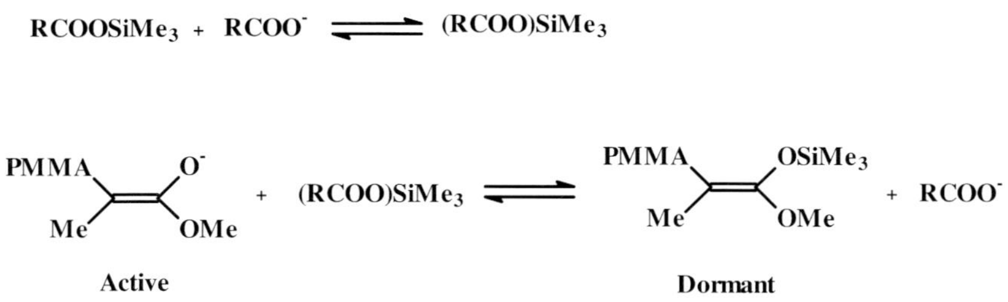

Scheme 6.39

"core-first" approach. Usually ethylene glycol dimethacrylate (EGDM) is employed as the difunctional monomer. However, other multifunctional monomers, such as tetramethylene glycol dimethacrylate, trimethylol propane trimethacrylate and 1,4-butene dimethacrylate, have been used.

In a detailed study of the synthesis of PMMA stars using EGDM as the difunctional monomer according to the "arm-first" approach, the functionality of the stars was shown to be controlled by the concentration of the living arm, the ratio of the concentrations of the living arms over EGDM and the dilution of the star polymer core [125].

The best experimental conditions for minimizing termination reactions of the living polymer chains prior the linking reaction with the difunctional monomer were investigated in another study [126]. The use of 0.2% mol of catalyst (based on initiator) and minimization of the time between arm formation and core formation, by addition of EGDM in about 5 min and maintaining high monomer concentrations, ensured a minimum content of unattached arms in the final product. Certain compounds, such as silyl esters, have been reported to improve the conversion and reduce the amount of free arms [127]. It was proposed that a complex is formed between the silyl ester and the carboxylate anion, ensuring a low concentration of the active species in equilibrium with the dormant species, as shown in Scheme 6.39. These compounds are called enhancing agents and work better when carboxylates are used as nucleophilic catalysts.

Functionalized stars bearing in-chain in addition to end-hydroxyl groups were prepared by copolymerizing methyl methacrylate, ethyl methacrylate and trimethylsilyloxyethyl methacrylate with a functionalized initiator carrying a hydroxyl group protected as a trimethylsilyloxy group. The living polymers formed the star structure through the reaction with EGDM followed by the hydrolysis of the trimethylsilyl masking groups under mild acidic conditions to afford the desired functionalized star [120, 128] (Scheme 6.40).

The cross-linked core of the star structure, formed through the use of a difunctional monomer, is living. The active centers of the core may be deactivated by reaction with a proton source. Alternatively, they can be used to reinitiate the polymerization of another quantity of monomer, thereby offering the possibility for the formation of asymmetric or miktoarm stars. Further addition of difunctional monomer leads to the formation of networks [129–131].

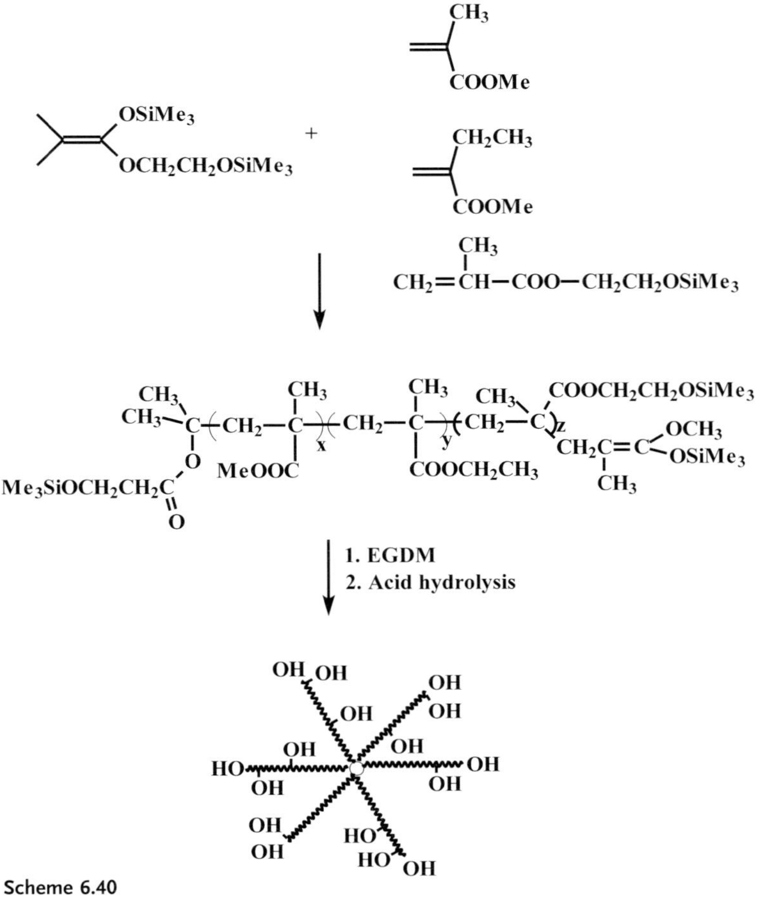

Scheme 6.40

6.4.6 Ring-opening Metathesis Polymerization (ROMP)

Metathesis chemistry has evolved as one of the most rapidly growing methods for the synthesis of well-defined polymeric materials [132–135]. This progress has been triggered by the discovery of the Schrock and Grubbs catalytic systems for the ring-opening metathesis polymerization (ROMP) of cycloolefins, leading to the synthesis of model polymers. Despite the fact that ROMP does not provide the means to synthesize a rich variety of complex macromolecular architectures, several attempts to synthesize star polymers have been reported.

The norbornadiene dimer *exo,trans,exo*-pentacyclo[8.2.1.$1^{4,7}$.$0^{2,9}$.$0^{3,8}$]tetradeca-5,11-diene (**1** in Scheme 6.41) was employed as a difunctional monomer for the synthesis of star polynorbornene using the "arm-first" approach. Norbornene was polymerized to the corresponding living polymer using M(CHR)(NAr)(O-*t*-Bu)$_2$ (M = W or Mo; NAr = *N*-2,6-C_6H_3*i*Pr_2) as catalyst. The living polymer was

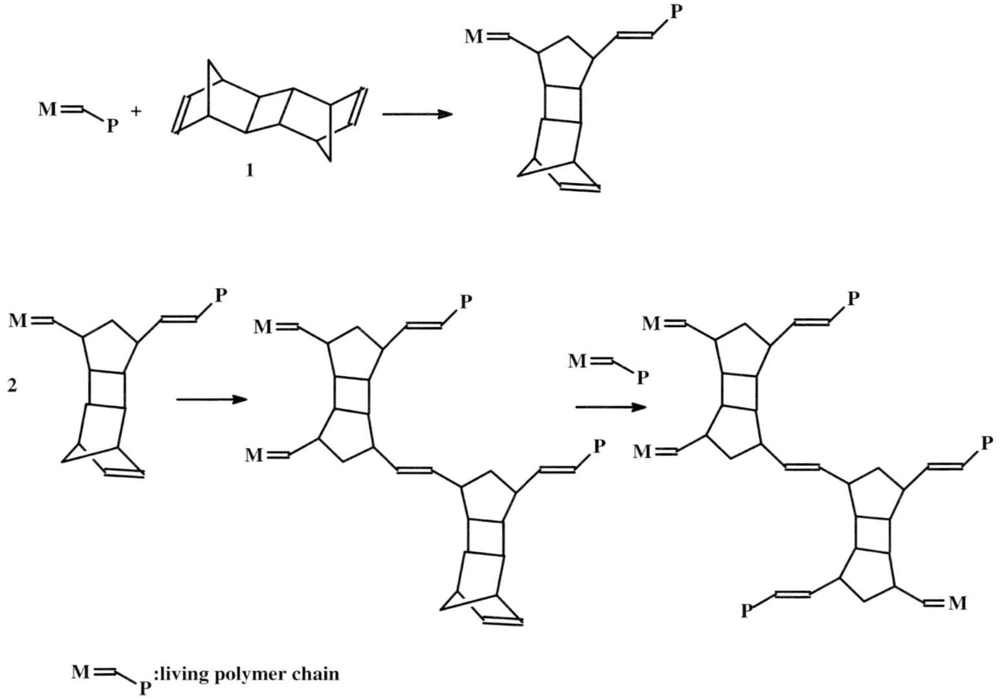

Scheme 6.41

then reacted with the difunctional monomer leading to the synthesis of star polymers (Scheme 6.41). The linear living polymer was completely consumed, but the molecular weight distribution of the stars was broader than that of the corresponding arms, probably due to the formation of products characterized by chemical heterogeneity, i.e. a distribution in the number of arms. These stars are still living and bear active centers at the core. New arms can grow from these living sites, giving rise to the formation of asymmetric stars. However, products with bimodal distributions were obtained, due to the slower initiation at the core of the star and the faster propagation as the reacting alkylidene centers move away from the sterically crowded core [136].

The "core-first" approach was also adopted for the synthesis of star polymers. According to this procedure, the catalyst was reacted with the difunctional monomer to give the multifunctional initiator, followed by the addition of norbornene [136]. Unfortunately, a multimodal product was obtained, revealing the presence of linear chains, dimers and various star structures.

Using the more successful "arm-first" approach, in-chain functionalized stars were prepared [136]. 2,3-Dicarbomethoxynorbornadiene was polymerized with the same catalytic systems followed by the addition of the difunctional monomer leading to the desired functionalized star polymers. SEC revealed the pres-

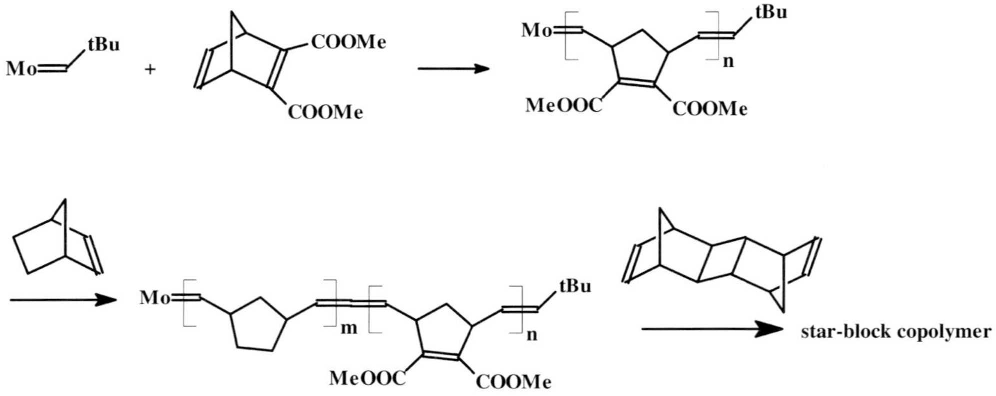

Scheme 6.42

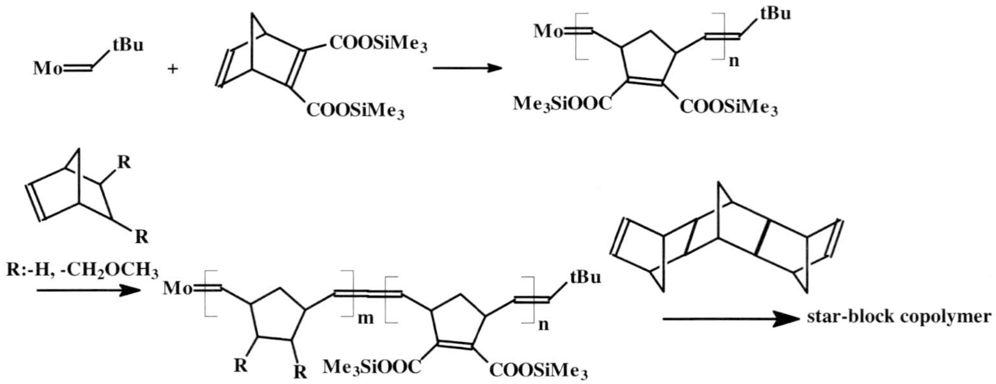

Scheme 6.43

ence of 10–20% of unreacted homopolymers. Using the same procedure, star-block copolymers can also be prepared. 2,3-Dicarbomethoxynorbornadiene was initially polymerized followed by the addition of norbornene for the synthesis of the living diblock copolymers. Reaction of these living diblocks with the difunctional monomer afforded the star-block copolymers with complete consumption of the living arms (Scheme 6.42).

Another difunctional monomer, *endo,cis,endo*-hexacyclo[10.2.1.$1^{3,10}$.$1^{5,8}$.$0^{2,11}$.$0^{4,9}$]-heptadeca-6,13-diene, was also employed for the synthesis of star polymers [137]. Using the "arm-first" approach, star polymers of norbornene, 5,6-bis(methoxymethyl)norbornene (DMNBE) and 5,6-bis(dicarbotrimethylsilyloxy)norbornene (TMSNBE) were obtained. Amphiphilic star-block copolymers were prepared by reacting living block copolymers, from the sequential polymerization of TMSNBE and DMNBE or norbornene, with the same difunctional monomer followed by the

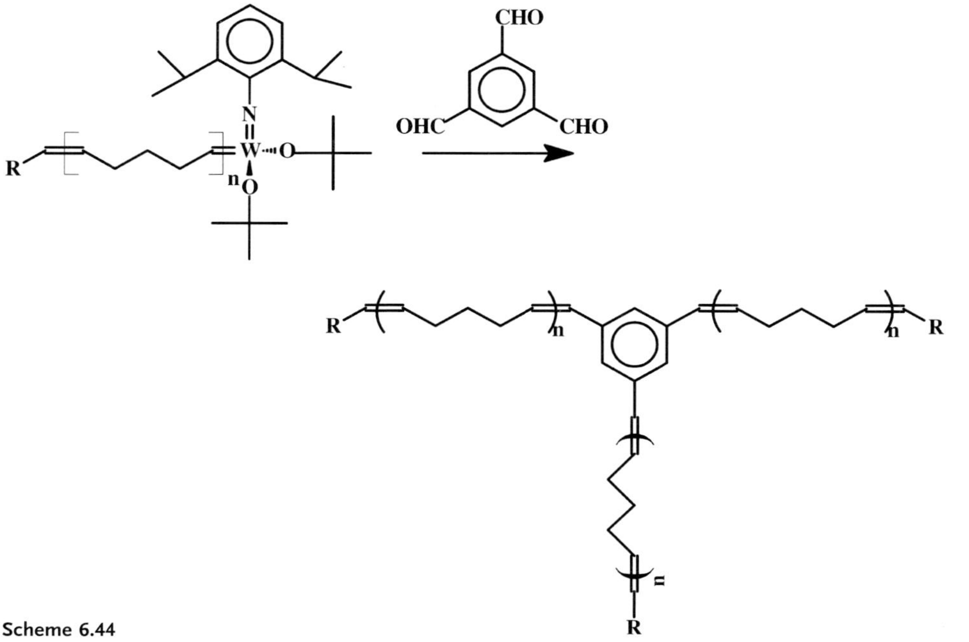

Scheme 6.44

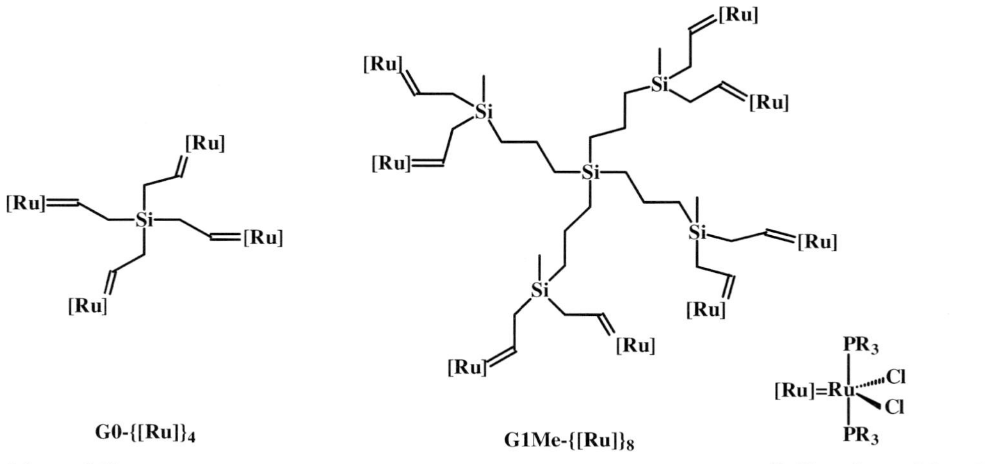

Scheme 6.45

R=Phenyl or cyclohexyl

hydrolysis of the trimethylsilyl groups, as shown in Scheme 6.43. SEC analysis revealed that linear and lightly branched structures were also present along with the desired stars.

Star polymers were also prepared by ROMP through the reaction of living linear chains with suitable linking agents. Cyclopentene was polymerized at –45 °C using W(CH*t*Bu)-(NAr)(O*t*Bu)$_2$ and the living chains were coupled with 1,3,5-benzenetricarboxaldehyde for the synthesis of three arm stars [138] (Scheme 6.44). SEC analysis showed that the final star structure had a rather broad molecular weight distribution. This was attributed to the purity of the linking agent, which was contaminated with the difunctional aldehyde, up to 15%. Subsequent hydrogenation of the poly(1-pentenylene) stars produced the corresponding polyethylene stars.

The multifunctional initiator approach was employed for the synthesis of stars by ROMP. Carbosilane dendrimers bearing ruthenium complexes on their surface (Scheme 6.45) were prepared and utilized for the synthesis of four- and eight-arm polynorbornene stars [139]. End-group analysis by NMR revealed that well-defined structures were obtained.

Dendrimers having 4, 8 or 16 –N(CH$_2$PCy$_2$)$_2$ groups at the periphery of the structure were used as precursors for the synthesis of the corresponding ruthe-

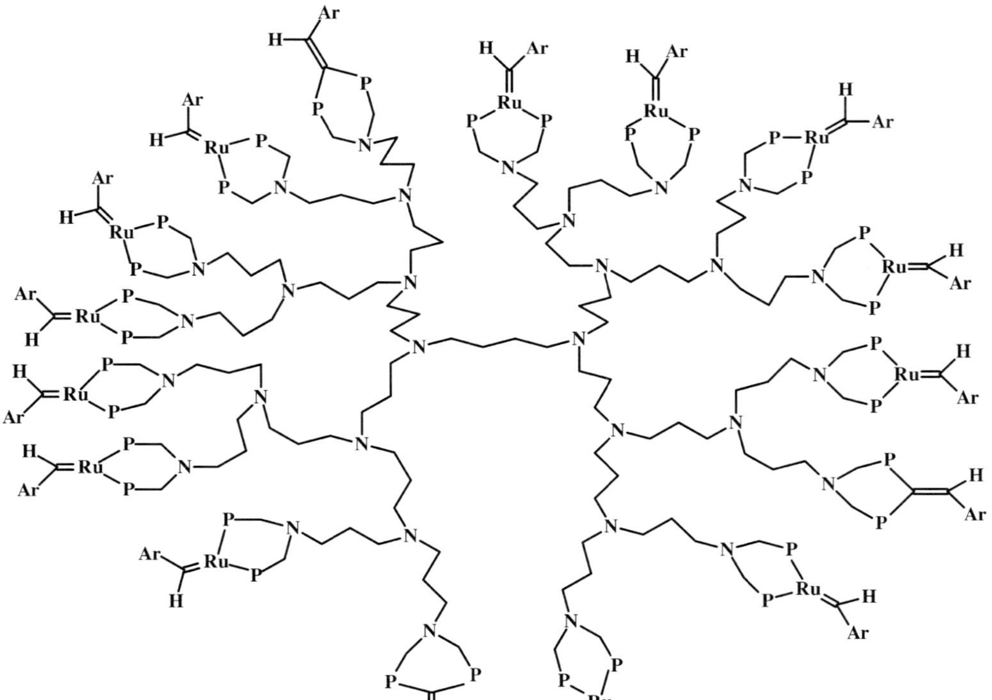

Scheme 6.46

nium–benzylidene complexes [140] (Scheme 6.46). These structures can be used as multifunctional initiators for the synthesis of polynorbornene stars with 4, 8 or 16 arms. The final products were characterized by broad molecular weight distributions.

6.4.7
Step-growth Polycondensation

Step-growth polymerization has been also used for the synthesis of star polymers utilizing the general methodologies previously reported. Linking chemistry was employed for the synthesis of polyimide stars. Imide oligomers by the reaction of 3,3′,4,4′-benzophenonetetracarboxylic dianhydride and 4,4′-diaminodiphenyl ether served as the arms for the synthesis of three-arm stars after reaction with 1,3,5-tris(4-aminophenoxy)benzene, the linking agent [141]. The excess anhydride-ended oligoimide arm was deactivated by primary arylamines (Scheme 6.47).

Polycondensation reactions are very important industrial processes for the manufacture of engineering plastics. However, this method does not offer control over the molecular weight or the molecular weight distribution. The loss of any control is attributed to the step-growth character by which the polymerization proceeds [142]. The procedure is initiated by the reaction of the monomers and is propagated by the reactions between all types of oligomers and between the oligomers and monomers. In contrast, the addition polymerization reactions proceed exclusively through the reaction of the growing chains with monomer. In this case, propagation reactions between monomers and between growing chains are not observed. With this behavior in mind, a new procedure, "chain-growth polycondensation", was proposed [143, 144]. According to this method, the polycondensation reaction proceeds in a chain-growth manner. To achieve this behavior, the monomers react with each other during the initiation process and then, during propagation, react exclusively with the growing chain ends. This is accomplished when the polymer end group becomes more reactive than the monomer by stimulation of bond-forming between the monomer and the chain end-group (Scheme 6.48). Well-defined polycondensation polymers with narrow molecular weight distributions can be prepared.

A characteristic example of the synthesis of aromatic polyamides is given in Scheme 6.49. Well-defined products were obtained by the polycondensation of 4-(alkylamino)benzoate using phenyl 4-nitrobenzoate as initiator in the presence of *N*-triethylsilyl-*N*-octylaniline, as a base, in combination with CsF and 18-crown-6. The proton of the amino group of monomer **1** is abstracted by the base, leading to the formation of the aminyl anion **1**′. This anionic center behaves as an active nucleophilic site and, furthermore, deactivates the phenyl ester moiety of **1**′ by its strong electron-donating ability. Thus the monomers do not react with each other. The anion **1**′ reacts with initiator **2** bearing an electron-withdrawing group, due to the much higher reactivity of the phenyl ester moiety of **2**′ than that of **1**′. The obtained amide **3**′ has a weak amide electron-withdrawing group and the phenyl ester moiety of **3** is more reactive than that

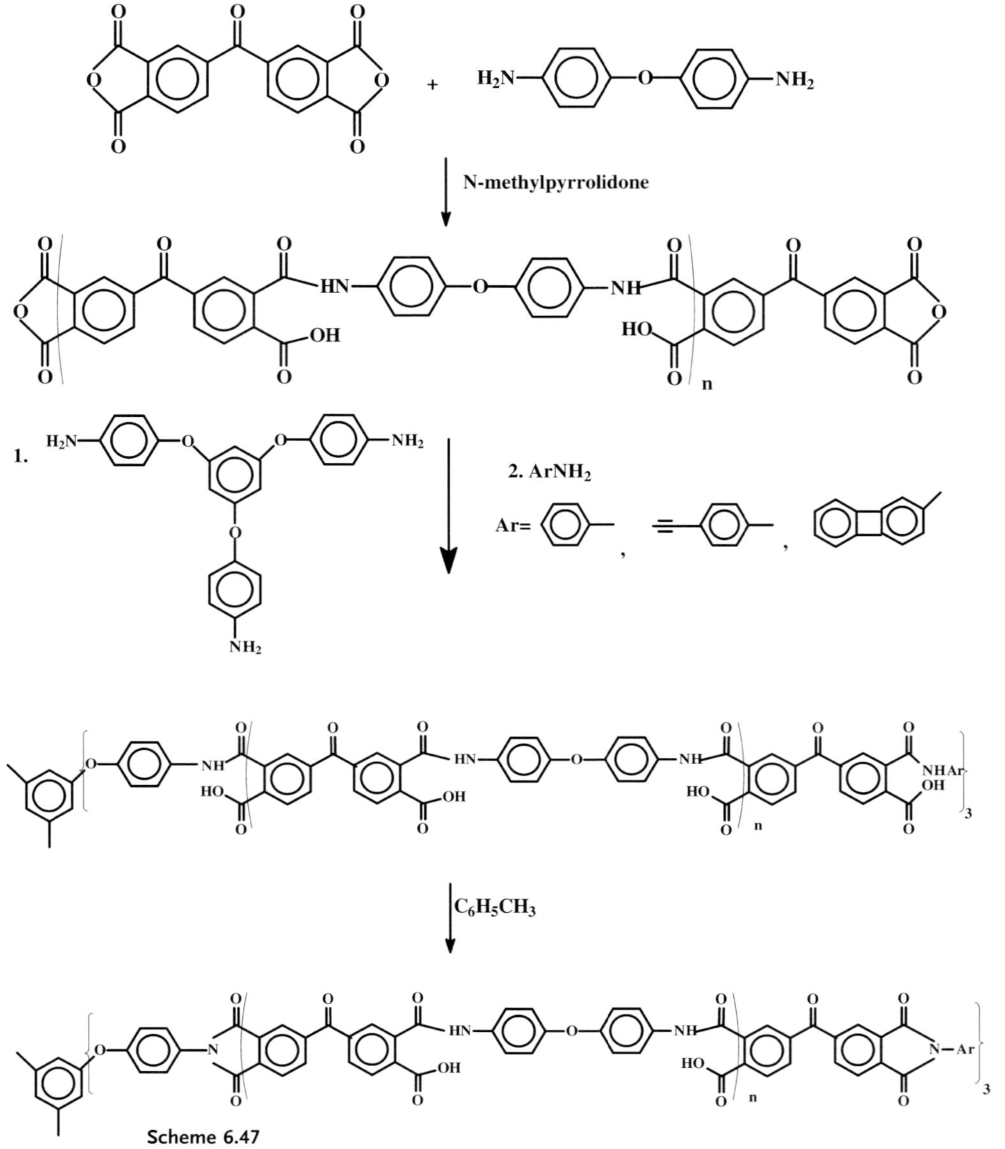

Scheme 6.47

of **1**′. Therefore, the monomer reacts exclusively with the phenyl ester moiety of **3**. Consequently, the polymerization propagates in a chain-growth manner.

Using this methodology and a trifunctional phenyl ester initiator, three-arm star polyamides and star-block copolymers of different polyamides were prepared [145] (Scheme 6.50).

Despite these recent advances, step-growth polymerization does not offer many possibilities for the synthesis of complex star architectures and the molec-

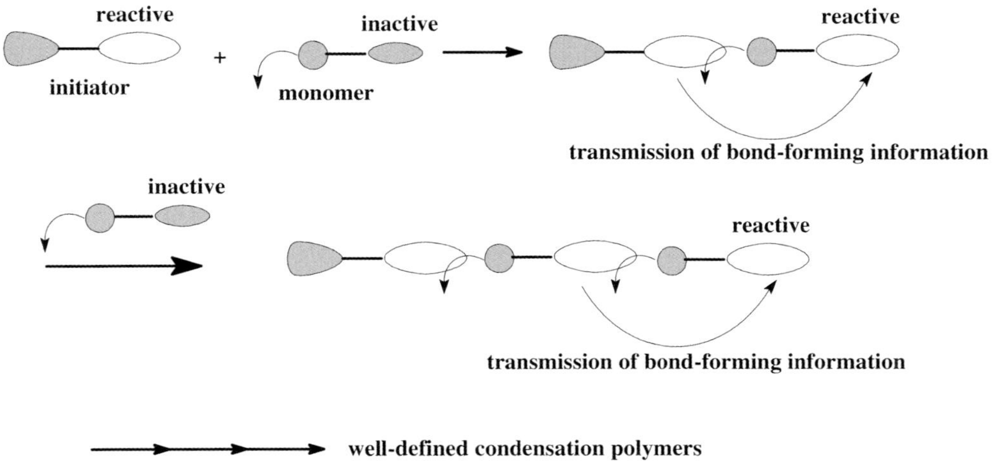

Scheme 6.48

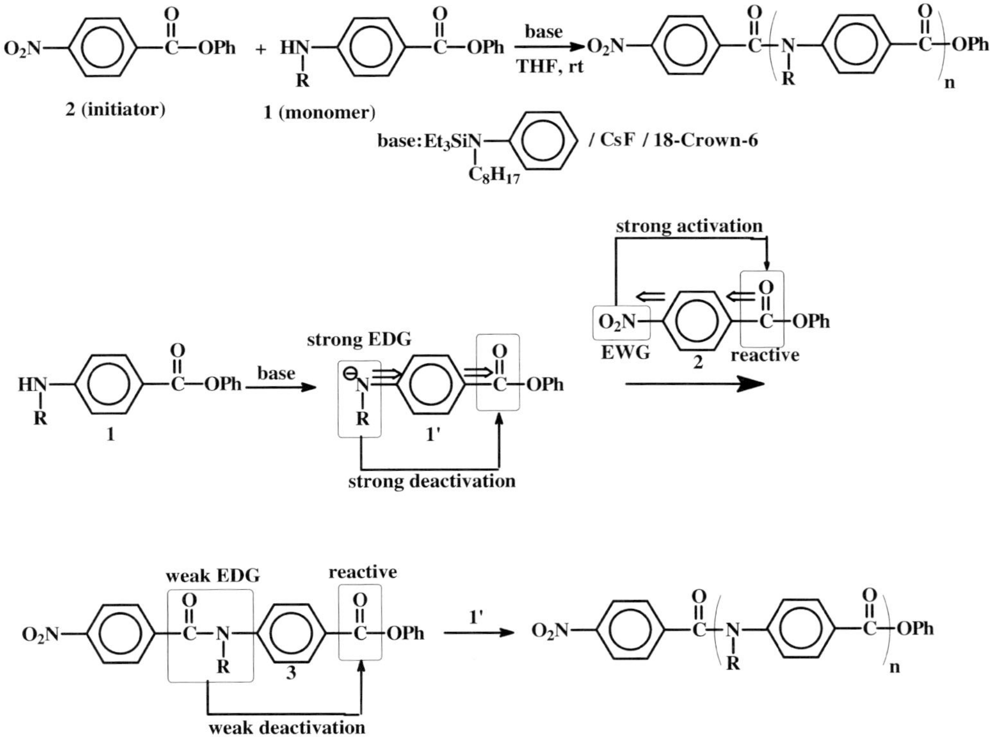

(EDG: electron donating group; EWG: electron withdrawing group)

Scheme 6.49

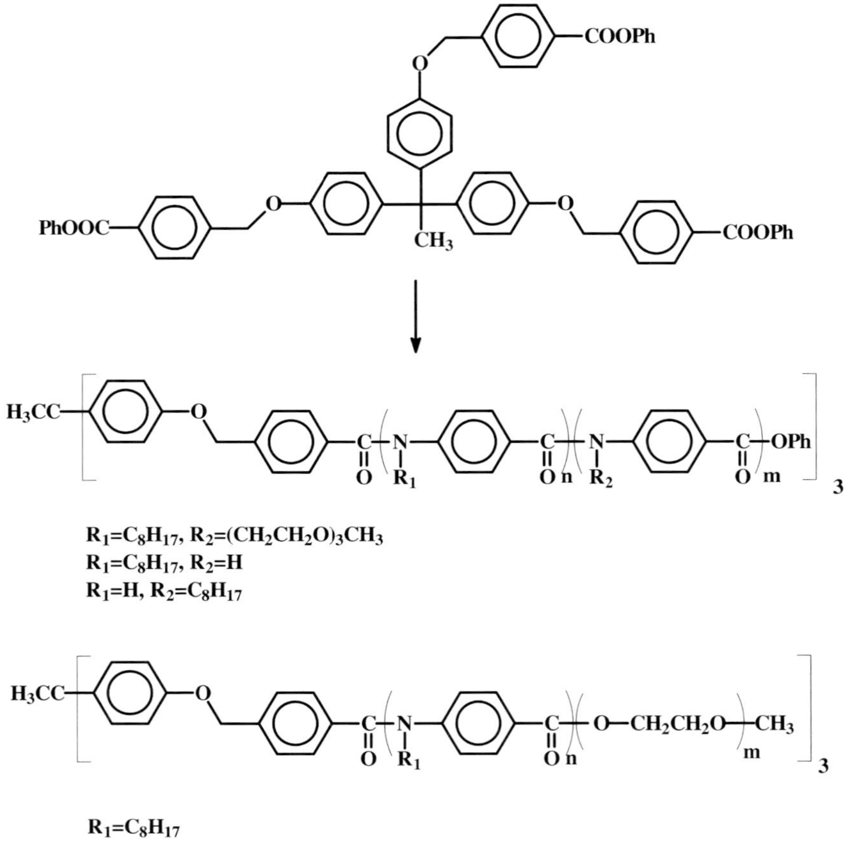

Scheme 6.50

ular characteristics cannot be controlled to the same degree as in the case of the living polymerization methods.

6.4.8
Metal Template-assisted Star Polymer Synthesis

Metal complexes with macromolecular ligands can be prepared with considerable control over the metal binding site and the polymeric environment. Consequently, metal ions serve as highly tunable structural motifs for the assembly of materials leading to the formation of star polymers.

Star structures can be performed by the combination of coordination chemistry with living or controlled polymerization via metalloinitiation, metallotermination, coupling or macroligand chelation approaches [146–160](Scheme 6.51).

Metal complexes bearing initiator functionalities in their ligand periphery may act as multifunctional metalloinitiators according to the "core-first"

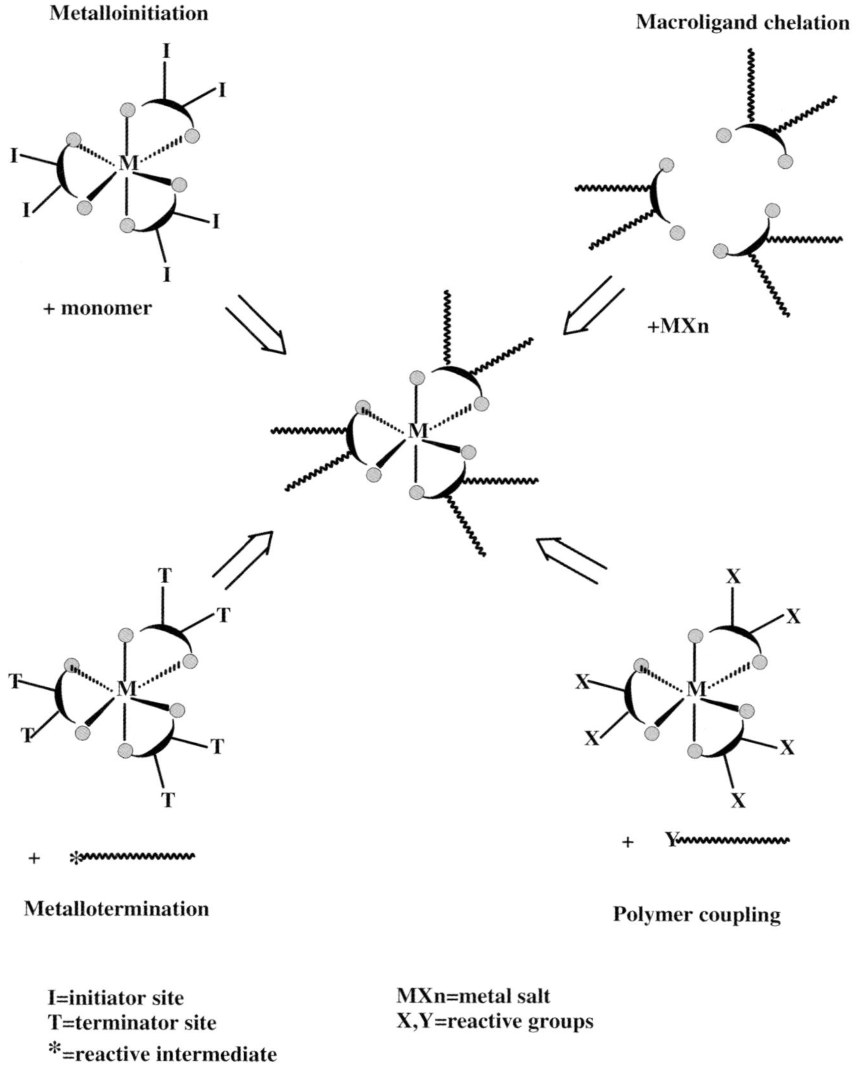

Scheme 6.51

approach. This method is less sensitive to steric restrictions and the products can be easily purified from the unreacted monomer by precipitation. Limitations of the method include the difficulties concerning the synthesis of the metalloinitiator, which often bears sensitive functionalized ligands, and the compatibility problems of the metal complex with the polymerization reaction conditions. Furthermore, the decomposition of the complex has to be avoided during

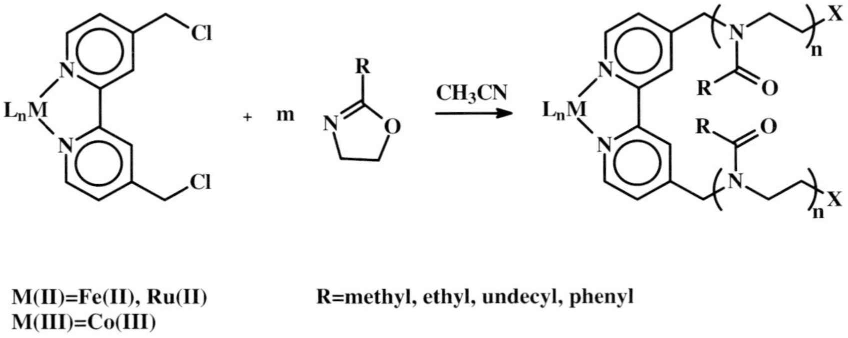

Scheme 6.52

the polymerization. So far, inert Ru tris(bipyridine) [146, 149] and labile Fe complexes [147, 148] have been widely used.

The "arm-first" approach was also adopted by linking properly functionalized macromolecular ligands to the metal center and by termination or coupling of living polymers with suitable metal complexes, bearing functional groups. Steric constraints play an important role in this case. There is usually an upper molecular weight above which the star synthesis is not efficient.

In order to employ the metalloinitiator or divergent method, metal complexes bearing functionalities appropriate for initiating different types of polymerization reactions are needed. Halomethyl-substituted metal tris(bipyridine) complexes [M(II) = Fe, Ru or Zn; M(III) = Co] were found to be efficient initiators for the cationic polymerization of a range of 2-R-2-oxazoline monomers (R = methyl, ethyl, phenyl and undecyl) [161], as shown in Scheme 6.52.

Labile core Fe polyoxazoline star polymers may be cleaved with an aqueous base to remove the metal and to generate bipyridine-centered materials [147, 148]. Specifically, cleavage of star-block copolymers with poly[(2-ethyl-2-oxazoline)-*b*-(2-undecyl-2-oxazoline)] (AB) blocks affords bipyridine-centered BAB triblock copolymers [161, 162] (Scheme 6.53).

Star polymers with four and six arms were prepared by this method [161]. For Ru complexes, an upper molecular weight of 25 000 was obtained and the final products were characterized by relatively narrow molecular weight distributions. For the corresponding hexafunctional Fe complexes, stars with molecular weights up to 75 000 and even higher can be prepared. However, in this case the molecular weight distributions were broader.

The same halomethyl-substituted bipyridine-centered complexes have been utilized as initiation sites for the ATRP. Four- and six-arm polystyrene stars were prepared by this method [146]. Pronounced star–star coupling was obtained at high monomer conversion.

Ru complexes with α-halo ester-functionalized ligands have been used as initiators for the polymerization of acrylates in the presence of $Ni(PPh_3)_2Br_2$ as the

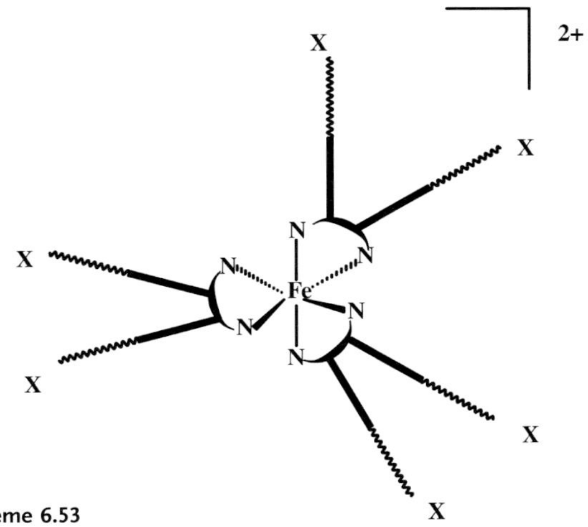

Scheme 6.53

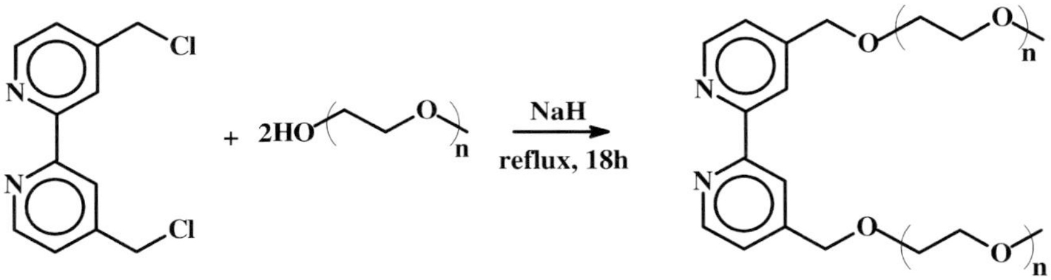

Scheme 6.54

catalyst [163]. Hexafunctional Ru-centered complexes were also used as initiators for the synthesis of six-arm poly(methyl methacrylate) stars.

Hydroxyl-substituted bipyridine Ru-centered complexes have been also used to promote the ring opening polymerization of ε-caprolactone and lactide in the presence of stannous octoate, $Sn(Oct)_2$ [161].

The "arm-first" methodology for the synthesis of star polymers requires the synthesis of proper macromolecular ligands. Bipyridine-functionalized polymers have frequently been employed for this purpose. ATRP techniques have been utilized for the preparation of bipyridine-centered or end-functionalized polystyrenes and poly(methyl methacrylates) [161, 164]. However, other techniques have also been employed. The ROP of ε-caprolactone and lactide was initiated by hydroxymethylbipyridine initiators in the presence of tin catalysts [165]. Coupling reactions of poly(ethylene oxide) chains to functionalized bipyridines have been also carried out [166] (Scheme 6.54).

Star polymers have been synthesized by linking the macroligands with either labile or inert metals. For example, Fe(II) tris(bipyridine) complexes can be formed by the reaction of the suitable bipyridine macroligands with ferrous am-

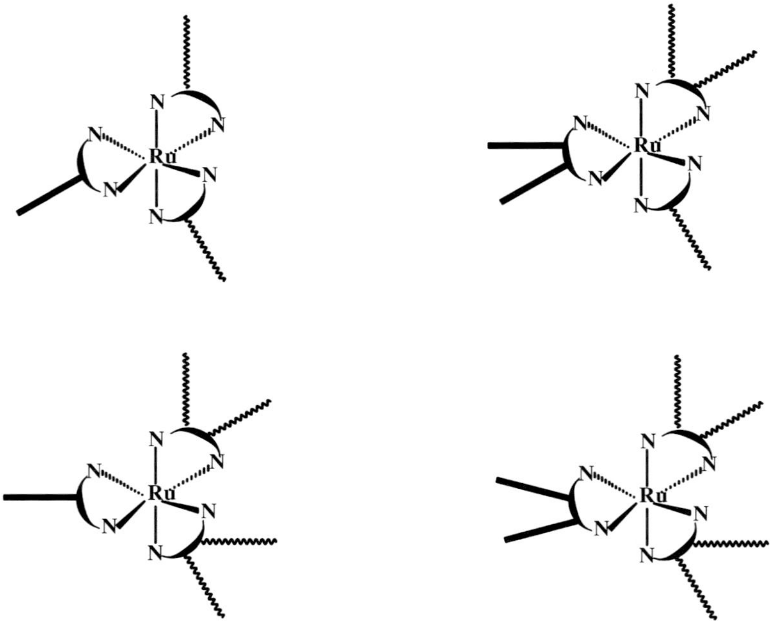

Scheme 6.55

monium sulfate in a common good solvent for the metal ion and the macroligand [167]. Various stable macromolecular star architectures were generated by chelation of macroligands to Ru(II), as depicted in Scheme 6.55. Symmetric PS stars with up to six arms have been synthesized with this procedure. Miktoarm stars of the type $(PS)_2(PMMA)_2$ and $(PS)_2(OE)_2$, where OE is oligoethylene $(C_{23}H_{47})$, were also prepared [161].

6.4.9
Combination of Different Polymerization Techniques

Each polymerization technique is characterized by certain advantages and limitations. Therefore, a combination of several polymerization methods provides the means for the synthesis of complex architectures by the incorporation, into the same structure, of different types of monomeric units that cannot be polymerized by the same technique. This strategy is synthetically very challenging, owing to the incompatibility which often characterizes the various polymerization methods and has been adopted for the synthesis of complex star structures, such as star-block and miktoarm star copolymers [168–172]. Characteristic examples will be given below.

Miktoarm stars of the type A_2B_2, where A is PS and B is PEO, were prepared by the combination of anionic and ATRP methods using the "core-first" procedure. Pentaerythritol was employed as a multifunctional initiator. Two of the hy-

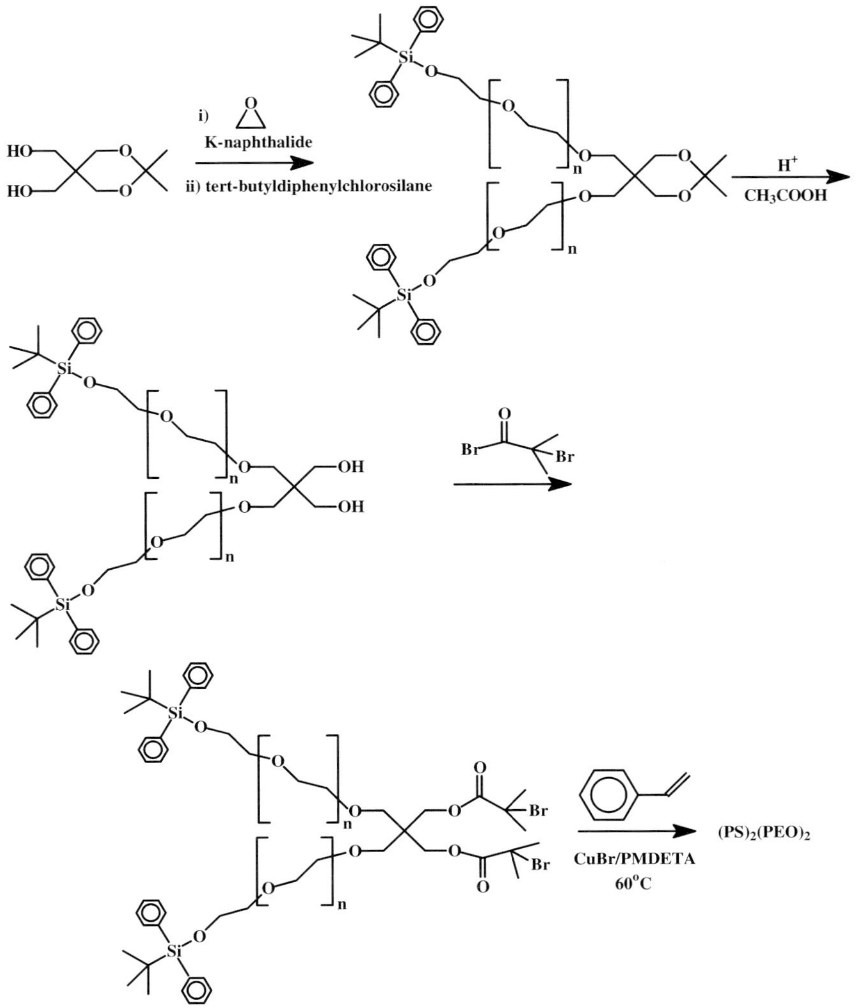

Scheme 6.56

droxyl groups were protected as acetals by reaction with acetone. The remaining hydroxyl groups were activated by potassium naphthalenide, in order to polymerize ethylene oxide. The living chains were terminated with *tert*-butyldiphenylchlorosilane to introduce specific end-groups at the PEO arms. The remaining hydroxyl groups at the core were deprotected under mild acidic conditions and were reacted with 2-bromoisobutyryl bromide to provide the initiation sites for the polymerization of styrene by ATRP, leading to the synthesis of the desired $(PS)_2(PEO)_2$ miktoarm stars [173] (Scheme 6.56). Different functional end-groups were attached by transformation of the bromide or *tert*-butyldiphenylsilyl end-groups of the PS and PEO arms, respectively.

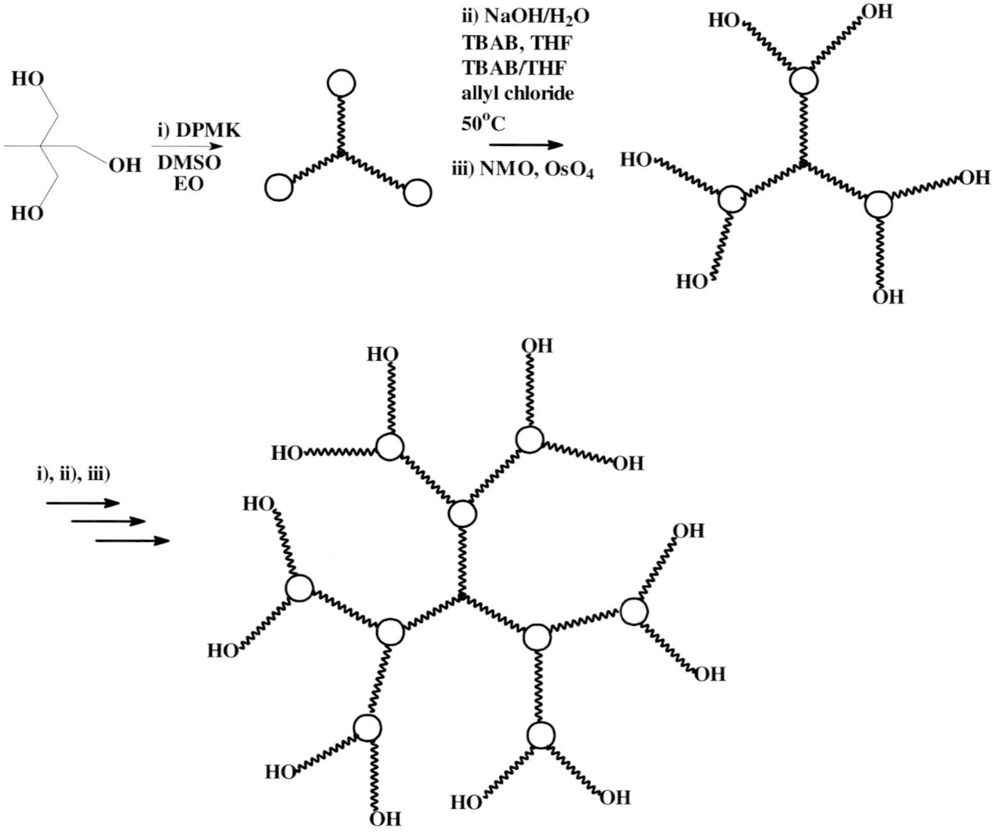

Scheme 6.57

An iterative divergent approach combining anionic polymerization of ethylene oxide from multi-hydroxylated precursors and branching reactions of PEO chain ends was performed for the synthesis of dendrimer-like PEOs [174]. 1,1,1-Tris(hydroxymethyl)ethane was employed as a trifunctional initiator for the synthesis of three-arm PEO stars (Scheme 6.57). The three hydroxyls were deprotonated using diphenylmethylpotassium in dimethyl sulfoxide solutions to avoid association effects of the growing chains. The end-hydroxyl groups of the three-arm PEO stars were derivatized into twice as many hydroxyl groups according to the following procedure. Allyl chloride was reacted with the end-hydroxyl groups in a water–THF mixture, in the presence of tetrabutylammonium bromide (TBAB) as a phase-transfer catalyst. The end-allyl groups were then reacted with OsO_4 in the presence of *N*-methylmorpholine-*N*-oxide (NMO), to introduce two hydroxyl groups per allyl moiety. Repetition of these synthetic steps allows for the synthesis of dendrimer-like polymers of different generations.

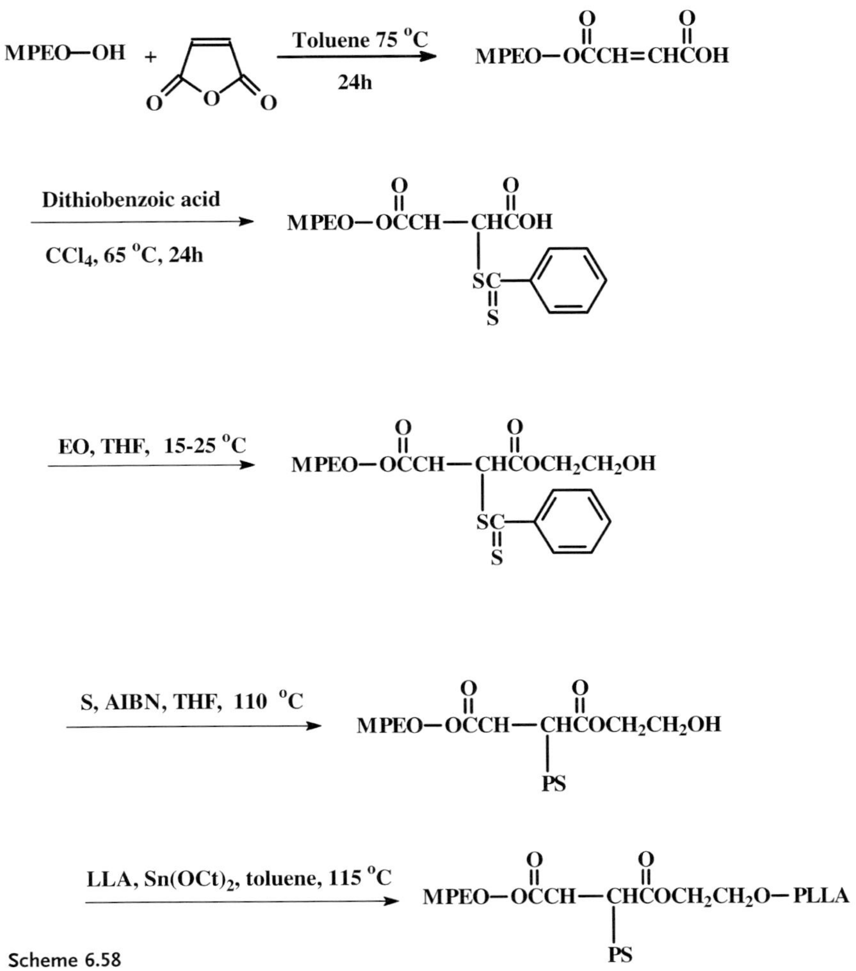

Scheme 6.58

A combination of RAFT and ROP afforded [poly(ethylene oxide) methyl ether](polystyrene)[poly(L-lactide)] [(MPEO)(PS)(PLLA)] three-miktoarm star terpolymers [175]. The synthetic approach involved the reaction of the ω-functionalized OH group of the poly(ethylene oxide) methyl ether with maleic anhydride under conditions, where only one hydroxyl group could be esterified. The double bond of the maleic group was then reacted with dithiobenzoic acid, resulting in a dithiobenzoic-terminated MPEO. The second carboxyl group of the maleic anhydride was then reacted with ethylene oxide, leading to the corresponding ester with a free OH group. The dithiobenzoic group of the MPEO was used for the RAFT polymerization of styrene in THF at 110 °C with AIBN as the initiator. Finally, the OH group attached at the junction point of the diblock copolymer served as the initiating site for the ROP of L-lactide, in the

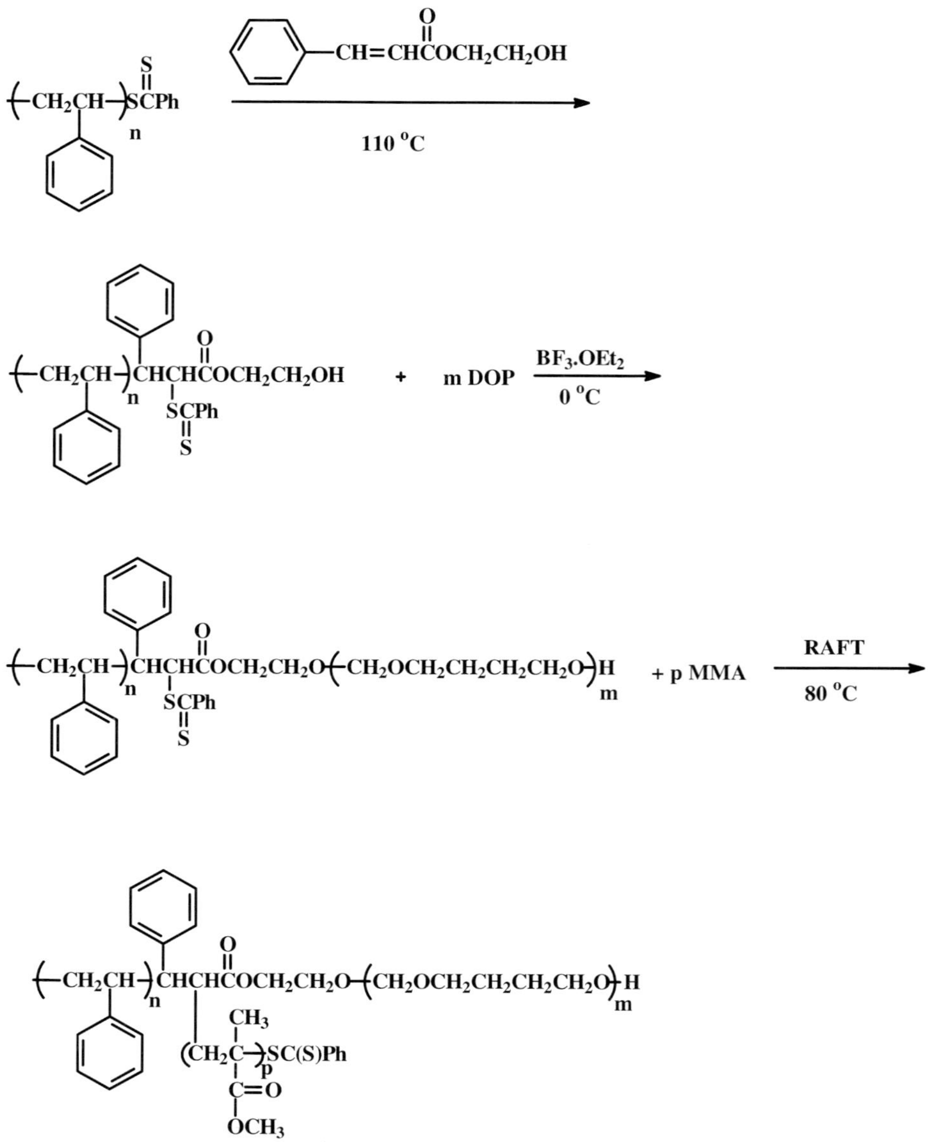

Scheme 6.59

presence of $Sn(Oct)_2$ in toluene at 115 °C (Scheme 6.58). The polydispersity indices of the intermediate along with the final products were between 1.05 and 1.07, indicating a high degree of molecular and compositional homogeneity.

A combination of RAFT and cationic ROP was employed for the synthesis of a series of [poly(methyl methacrylate)][poly(1,3-dioxepane)](polystyrene) three-

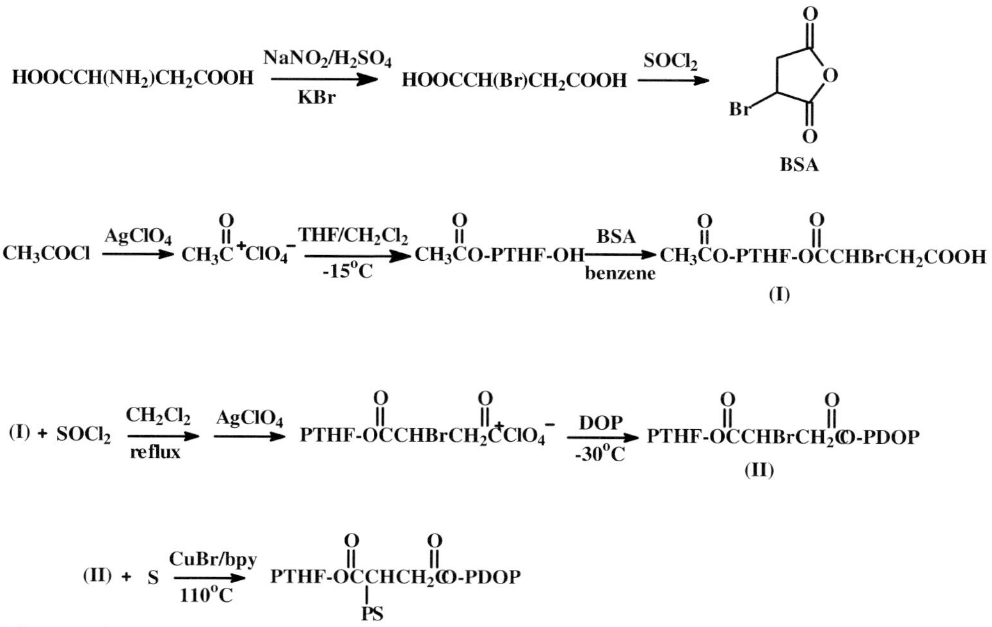

Scheme 6.60

miktoarm star terpolymers [176]. The synthetic approach involved the synthesis of PS functionalized with a dithiobenzoate group, using RAFT polymerization and subsequent reaction of this group with hydroxyethylene cinnamate in THF (Scheme 6.59). The hydroxyl group served as the initiating site for the cationic ROP of 1,3-dioxepane in the presence of triflic acid. Finally, the diblock copolymer with the dithiobenzoate group situated between the two blocks was used for the reversible addition–fragmentation transfer polymerization of methyl methacrylate. The miktoarm star terpolymers were characterized by NMR spectroscopy and SEC.

(Polytetrahydrofuran)[poly(1,3-dioxepane)](polystyrene) three-miktoarm stars, (PTHF)(PDOP)(PS), were synthesized via a combination of cationic ROP (CROP) and ATRP [177]. The star terpolymers were synthesized using two different functional groups, carboxylic acid and CHBr, which can be initiating sites of the two different polymerizations. These two groups were introduced at one end of the polytetrahydrofuran chain, through the reaction of OH-functionalized PTHF with 2-bromosuccinic anhydride. The –COOH group was transformed to –COCl, which with $AgClO_4$ served to form the second arm by CROP. Finally, the remaining Br initiated the ATRP of styrene (Scheme 6.60). The intermediate diblocks along with the final terpolymers were characterized by NMR and SEC. The relatively large polydispersity indices of the terpolymers obtained (1.49–1.54) indicate that the final terpolymers contain some amount of unreacted diblock and PTHF precursors.

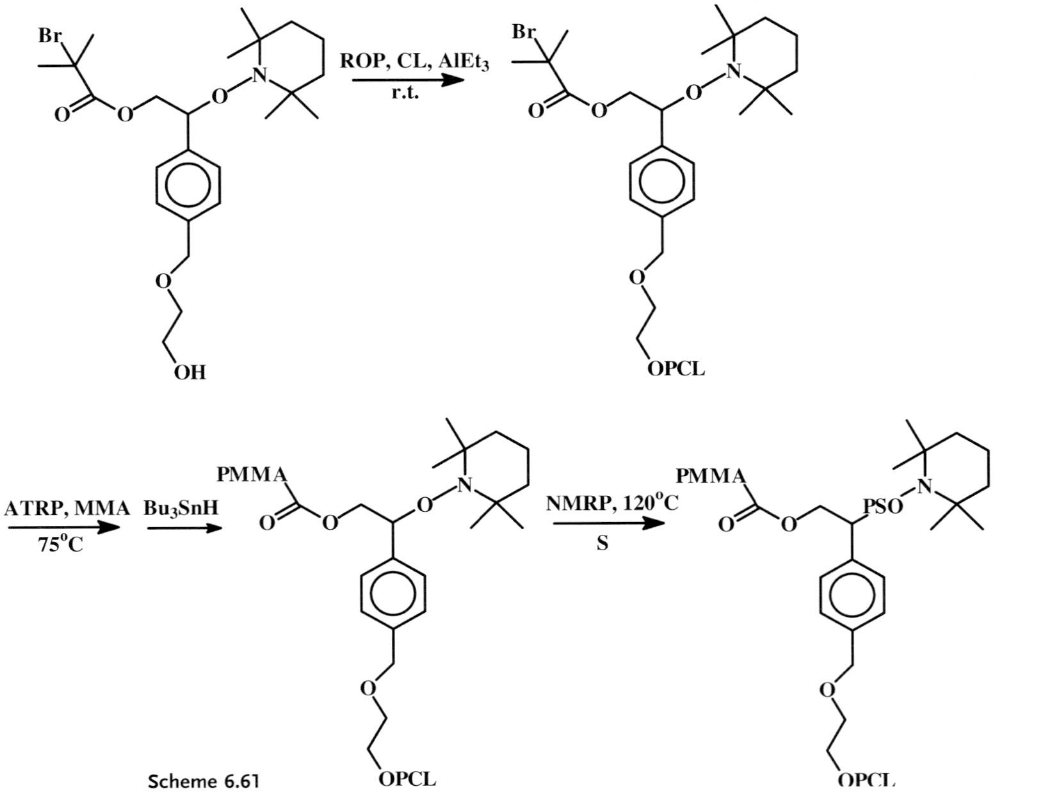

Scheme 6.61

(PS)(PMMA)(PCL) stars were synthesized with a trifunctional initiator bearing a hydroxyl group (ROP), a $-CH_2Br$ (ATRP) initiator and a nitroxide group (NMRP) [178] (Scheme 6.61). SEC characterization and kinetic studies confirmed the structure of the synthesized ABC stars.

Miktoarm stars of the AB_2C_2 type, where A is PS, B is poly(*tert*-butyl acrylate) (PtBA) and C is PMMA, were prepared with the trifunctional initiator 2-phenyl-2-[(2,2,6,6-tetramethyl)-1-piperidinyloxy]ethyl 2,2-bis[methyl(2-bromopropionato)]-propionate [179] (Scheme 6.62). A combination of NMRP and ATRP polymerization techniques and a three-step reaction sequence were employed. In the first step, PS macroinitiator with dual ω-bromo functionality was obtained by NMRP of styrene in bulk at 125 °C. This precursor was subsequently used as the macroinitiator for the ATRP of *tert*-butyl acrylate in the presence of CuBr and pentamethyldiethylenetriamine at 80 °C, to produce the miktoarm star of the $(PS)(PtBA)_2$. This star was the macroinitiator for the subsequent polymerization of MMA, giving the $(PS)(PtBA)_2(PMMA)_2$ miktoarms stars.

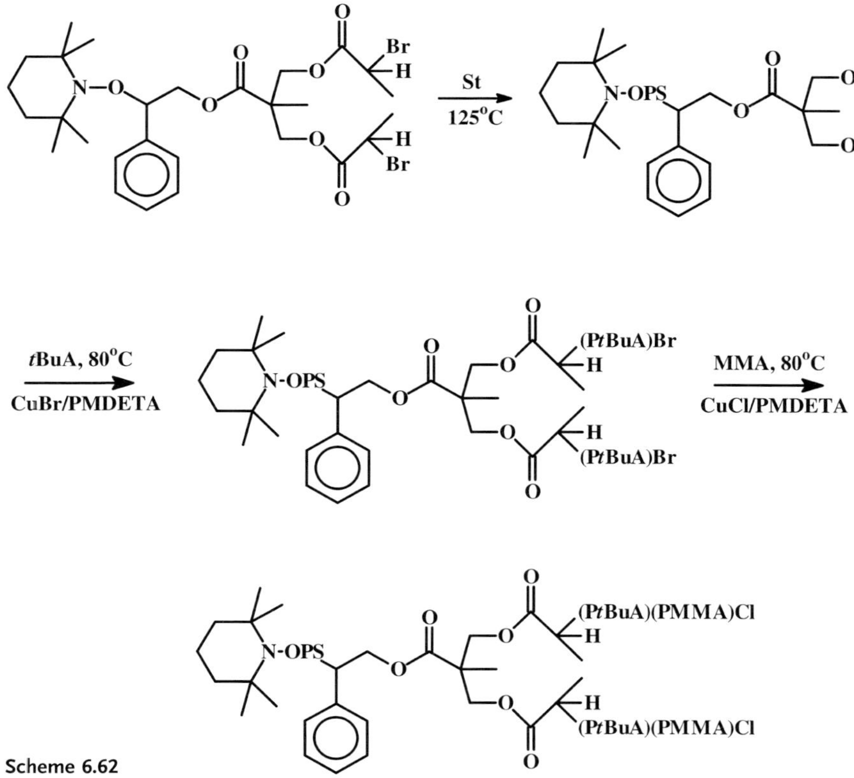

Scheme 6.62

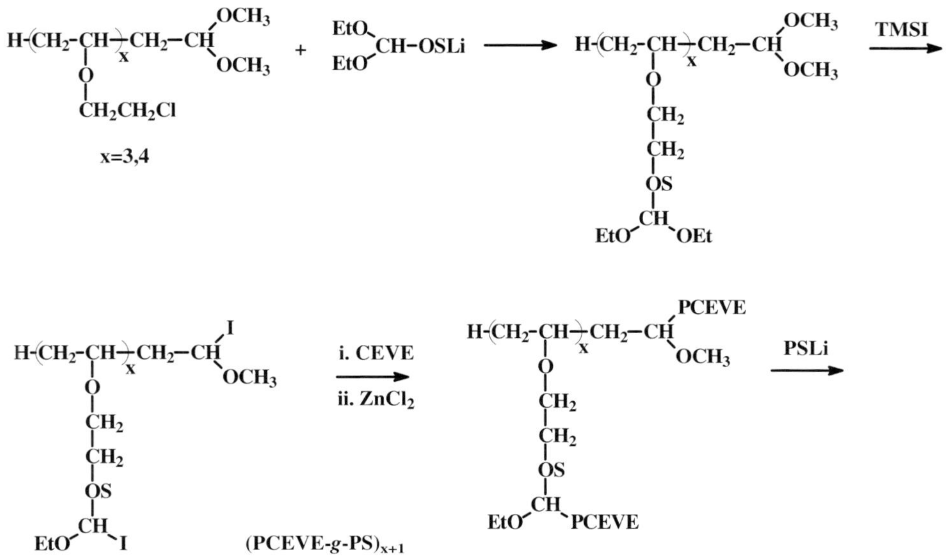

Scheme 6.63

6.5 Conclusions

Past and recent polymerization procedures, and their combinations, serve as a powerful tool for the synthesis of novel, well-defined star polymers with different architectures and chemistry. The study of these polymers, which are the simplest branched structures, in bulk or solution, continuously broadens our knowledge and understanding of the properties of more complex structures and their behavior.

References

1 Mishra, M.K., Kobayashi, S. (Eds) *Star and Hyperbranched Polymers.* Marcel Dekker, New York, **1999**.
2 Hatada, K., Kitayama, T., Vogl, O. (Eds) *Macromolecular Design of Polymeric Materials.* Marcel Dekker, New York, **1997**.
3 Schultz, J.L., Wilks, E.S. *J. Chem. Inf. Comput. Sci.* **1998**, *38*, 85.
4 Roovers, J. *Encyclopedia of Polymer Science and Technology*, Mark, H.F., Bikales, N.M., Overberger, C.G., Menges, G. (Eds) Vol. 2, 2nd ed, p. 478, John Wiley & Sons, Inc., **1985**.
5 Pitsikalis, M., Pispas, S., Mays, J.W., Hadjichristidis, N. *Adv. Polym. Sci.* **1998**, *135*, 1.
6 Schaefgen, J.R., Flory, P.J. *J. Am. Chem. Soc.* **1948**, *70*, 2709.
7 Morton, M., Helminiak, T.E., Gadkary, S.D., Bueche, F. *J. Polym. Sci.* **1962**, *57*, 471.
8 Hadjichristidis, N., Pitsikalis, M., Pispas, S., Iatrou, H. *Chem. Rev.* **2001**, *101*, 3747.
9 Hadjichristidis, N., Pispas, S., Pitsikalis, M. *Progr. Polym. Sci.* **1999**, *24*, 875.
10 Patil, A.O., Schulz, D.N., Novak, B.M. (Eds) *Functional Polymers. Modern Synthetic Methods and Novel Structures.* ACS Symposium Series, Vol. 704. American Chemical Society, Washington, DC, **1998**.
11 Hadjichristidis, N., Pispas, S., Pitsikalis, M., Iatrou, H., Vlahos, C. *Adv. Polym. Sci.* **1999**, *142*, 71.
12 Hadjichristidis, N. *J. Polym. Sci. Polym. Chem. Ed.* **1999**, *37*, 857.
13 Hadjichristidis, N., Iatrou, H., Pitsikalis, M., Pispas, S., Avgeropoulos, A. *Prog. Polym. Sci.* **2005**, *30*, 725.
14 Eschwey, H., Hallensleben, M.L., Burchard, W. *Makromol. Chem.* **1973**, *173*, 235.
15 Burchard, W., Eschwey, H. *Polymer* **1975**, *16*, 180.
16 Lutz, P., Rempp, P. *Makromol. Chem.* **1988**, *189*, 1051.
17 Tsitsilianis, C., Lutz, P., Graff, S., Lamps, J.-P., Rempp, P. *Macromolecules* **1991**, *24*, 5897.
18 Okay, O., Funke, W. *Makromol. Chem. Rapid Commun.* **1990**, *11*, 583.
19 Funke, W., Okay, O. *Macromolecules* **1991**, *24*, 2623.
20 Knischka, R., Lutz, P.J., Sunder, A., Mulhaupt, R., Frey, H. *Macromolecules* **2000**, *33*, 315.
21 Taton, D., Saule, M., Logan, J., Duran, R., Hou, S., Chaikof, E., Gnanou, Y. *J. Polym. Sci. Polym. Chem. Ed.* **2003**, *41*, 1669.
22 Quirk, R.P., Tsai, Y. *Macromolecules* **1998**, *31*, 8016.
23 Quirk, R.P., Yoo, T., Lee, Y., Kim, J., Lee, B. *Adv. Polym. Sci.* **2000**, *153*, 67.
24 Lebreton, A., Kallitsis, J.K., Héroguez, V., Gnanou, Y. *Macromol. Symp.* **2004**, *215*, 41.
25 Matmour, R., Tsitsilianis, C., Kallitsis, I., Héroguez, V., Gnanou, Y. *Angew. Chem. Int. Ed.* **2005**, *44*, 284.
26 Hsieh, H.L., Quirk, R.P. *Anionic Polymerization. Principles and Practical Applications.* Marcel Dekker, New York, **1996**.

27 Morton, M., Helminiak, T. E., Gadkary, S. D., Bueche, F. *J. Polym. Sci.* **1962**, *57*, 471.
28 Orofino, T., Wenger, F. *J. Phys. Chem.* **1963**, *67*, 566.
29 Mayer, R. *Polymer* **1974**, *15*, 137.
30 Turner, S., Blevins, R. *Macromolecules* **1990**, *23*, 1856.
31 Otsu, T., Matsumoto, A., Yoshiaka, M. *Indian J. Technol.* **1993**, *31*, 172.
32 Quirk, R., Lee, B. *Polym. Int.* **1992**, *27*, 359.
33 Zelinski, R. P., Wofford, C. F. *J. Polym. Sci. Part A* **1965**, *3*, 93.
34 Roovers, J., Bywater, S. *Macromolecules* **1972**, *5*, 384.
35 Roovers, J., Bywater, S. *Macromolecules* **1974**, *7*, 443.
36 Hadjichristidis, N., Roovers, J. *J. Polym. Sci. Polym. Phys. Ed.* **1974**, *12*, 2521.
37 Hadjichristidis, N., Guyot, A., Fetters, L. J. *Macromolecules* **1978**, *11*, 668.
38 Hadjichristidis, N., Fetters, L. J. *Macromolecules* **1980**, *13*, 191.
39 Roovers, J., Hadjichristidis, N., Fetters, L. J. *Macromolecules* **1983**, *16*, 214.
40 Zhou, L.-L., Roovers, J. *Macromolecules* **1993**, *26*, 963.
41 Zhou, L.-L., Hadjichristidis, N., Toporowski, P. M., Roovers, J. *Rubber Chem. Technol.* **1992**, *65*, 303.
42 Roovers, J., Zhou, L.-L., Toporowski, P. M., van der Zwan, M., Iatrou, H., Hadjichristidis, N. *Macromolecules* **1993**, *26*, 4324.
43 Pitsikalis, M., Hadjichristidis, N., Di Silvestro, G., Sozzani, P. *Macromol. Chem. Phys.* **1995**, *196*, 2767.
44 Allgaier, J., Martin, K., Rader, H. J., Mullen, K. *Macromolecules* **1999**, *32*, 3190.
45 Roovers, J., Toporowski, P., Martin, J. *Macromolecules* **1989**, *22*, 1897.
46 Nguyen, A. B., Hadjichristidis, N., Fetters, L. J. *Macromolecules* **1986**, *19*, 768.
47 Alward, D. B., Kinning, D. J., Thomas, E. L., Fetters, L. J. *Macromolecules* **1986**, *19*, 215.
48 Pennisi, R. W., Fetters, L. J. *Macromolecules* **1988**, *21*, 1094.
49 Mays, J. W. *Polym. Bull.* **1990**, *23*, 247.
50 Iatrou, H., Siakali-Kioulafa, E., Hadjichristidis, N., Roovers, J., Mays, J. W. *J. Polym. Sci. Polym. Phys. Ed.* **1995**, *33*, 1925.
51 Iatrou, H., Hadjichristidis, N. *Macromolecules* **1992**, *25*, 4649.
52 Iatrou, H., Hadjichristidis, N. *Macromolecules* **1993**, *26*, 2479.
53 Mavroudis, A., Hadjichristidis H. *Macromolecules* **2006**, *39*, 535.
54 Hogen-Esch, T. E., Toreki, W. *Polym. Prepr.* **1989**, *30*, 129.
55 Pitsikalis, M., Sioula, S., Pispas, S., Hadjichristidis, N., Cook, D. C., Li, J., Mays, J. W. *J. Polym. Sci. Part A: Polym. Chem.* **1999**, *37*, 4337.
56 Morton, M., Helminiak, T. E., Gadkary, S. D., Bueche, F. *J. Polym. Sci.* **1962**, *57*, 471.
57 Hayashi, M., Negishi, Y., Hirao, A. *Proc. Jpn. Acad., Ser. B* **1999**, *75*, 93.
58 Hayashi, M., Kojima, K., Hirao, A. *Macromolecules* **1999**, *32*, 2425.
59 Hirao, A., Hayashi, M., Loykulnant, S., Sugiyama, K., Ryu, S., Haraguchi, N., Matsuo, A., Higashihara, T. *Prog. Polym. Sci.* **2005**, *30*, 111.
60 Quirk, R. P., Yoo, T., Lee, Y., Kim, J., Lee, B. *Adv. Polym. Sci.* **2000**, *153*, 67.
61 Fernyhough, C. M., Young, R. N., Tack, R. D. *Macromolecules* **1999**, *32*, 5760.
62 Samulski, E. T., Desimone, J. M., Hunt, M. O., Menceloglu, Y., Jarnagin, R. C., York, G. A., Wang, H. *Chem. Mater.* **1992**, *4*, 1153.
63 Ederle, Y., Mathis, C. *Macromolecules* **1997**, *30*, 2546.
64 Mays, J. W., Hadjichristidis, N., Fetters, L. J. *Polymer* **1988**, *29*, 680; Worsfold, D. J., Zilliox, J. G., Rempp, P. *Can. J. Chem.* **1969**, *47*, 3379.
65 Tsitsilianis, C., Graff, S., Rempp, P. *Eur. Polym. J.* **1991**, *27*, 243.
66 Bi, L.-K., Fetters, L. J. *Macromolecules* **1976**, *9*, 732.
67 Efstratiadis, V., Tselikas, Y., Hadjichristidis, N., Li, J., Yunan, W., Mays, J. W. *Polym. Int.* **1994**, *33*, 171.
68 Hatada, K., Kitayama, T., Vogl, O. *Macromolecular Design of Polymeric Materials.* Marcel Dekker, New York, **1997**, p. 154.
69 Rempp, P., Franta, E., Herz, J.-E. *Adv. Polym. Sci.* **1988**, *86*, 145.
70 Tsitsilianis, C., Boulgaris, D. *Macromol. Chem. Phys.* **1997**, *198*, 997.

71 Tsitsilianis, C., Papanagopoulos, D., Lutz, P. *Polymer* **1995**, *36*, 3745.
72 Tsitsilianis, C., Boulgaris, D. *Macromol. Rep.* **1995**, *A32* (Suppl. 5/6), 569.
73 Tsitsilianis, C., Boulgaris, D. *Polymer* **2000**, *41*, 1607.
74 Jacob, S., Majoros, I., Kennedy, J. *Macromolecules* **1996**, *29*, 8631.
75 Storey, F., Chisholm, J., Lee, Y. *Polymer* **1993**, *34*, 4330.
76 Taylor, S., Storey, R. *J. Polym. Sci. Polym. Chem. Ed.* **1999**, *37*, 857.
77 Fukui, H., Sawamoto, M., Higashimura, T. *Macromolecules* **1994**, *27*, 1297.
78 Bae, Y., Faust, R. *Macromolecules* **1998**, *31*, 2480.
79 Fukui, H., Yoshohashi, S., Sawamoto, M., Higashimura, T. *Macromolecules* **1996**, *29*, 1862.
80 Tezuka, Y., Hideaki, O. *Prog. Polym. Sci.* **2002**, *27*, 1069.
81 Oike, H., Imamura, H., Imaizumi, H., Tezuka, Y. *Macromolecules* **1999**, *32*, 4819.
82 Hawker, J. *Angew. Chem. Int. Ed. Engl.* **1995**, *34*, 1456.
83 Miura, Y., Yoshida, Y. *Macromol. Chem. Phys.* **2002**, *203*, 879.
84 Miura, Y., Dote, H. *J. Polym. Sci. Polym. Chem. Ed.* **2005**, *43*, 3689.
85 Robin, S., Guerret, O., Couturiet, J.-L., Gnanou, Y. *Macromolecules* **2002**, *35*, 2481.
86 Bosman, A., Vestberg, R., Heumann, A., Fréchet, J., Hawker, C. *J. Am. Chem. Soc.* **2003**, *125*, 715.
87 Muthukrishnan, S., Plamper, F., Mori, H., Müller, A. *Macromolecules* **2005**, *38*, 10631.
88 Abraham, S., Ha, C., Kim, I. *J. Polym. Sci. Polym. Chem. Ed.* **2005**, *43*, 6367.
89 Wang, X., Zhang, H., Cheng, E., Wang, X., Zhou, Q. *J. Polym. Sci. Polym. Chem. Ed.* **2005**, *43*, 3232.
90 Pitto, V., Voit, B., Lootjens, T., van Benthem, R. *Macromol. Chem. Phys.* **2004**, *205*, 2346.
91 Li, J., Xiao, H., Kim, Y., Lowe, T. *J. Polym. Sci. Polym. Chem. Ed.* **2005**, *43*, 6345.
92 Limer, A., Rullay, A., Miguel, V., Peinado, C., Keely, S., Fitzpatrick, E., Carrington, S., Brayden, D., Haddleton, D. *React. Funct. Polym.* **2006**, *66*, 51.
93 Zhao, Y., Chen, Y., Chen, C., Xi, F. *Polymer* **2005**, *46*, 5808.
94 Karaky, K., Reynaud, S., Billon, L., François, J., Chreim, Y. *J. Polym. Sci. Polym. Chem. Ed.* **2005**, *43*, 5186.
95 Strandman, S., Luostarinen, M., Niemelä, S., Rissanen, K., Tenhu, H. *J. Polym. Sci. Polym. Chem. Ed.* **2004**, *42*, 4189.
96 Strandman, S., Pulkkinen, P., Tenhu, H. *J. Polym. Sci. Polym. Chem. Ed.* **2005**, *43*, 3349.
97 Ueda, J., Matsuyama, M., Kamigaito, M., Sawamoto, M. *Macromolecules* **1998**, *31*, 557.
98 Matyjaszewski, K., Miller, P., Pyun, J., Kickelbick, J., Diamanti, S. *Macromolecules* **1999**, *32*, 6526.
99 Angot, S., Murthy, S., Taton, D., Gnanou, Y. *Macromolecules* **1998**, *31*, 7218.
100 Jankova, K., Bednarek, M., Hvilsted, S. *J. Polym. Sci. Polym. Chem. Ed.* **2005**, *43*, 3748.
101 Xia, J., Zhang, X., Matyjaszewski, K. *Macromolecules* **1999**, *32*, 4482.
102 Zhang, X., Xia, J., Matyjaszewski, K. *Macromolecules* **2000**, *33*, 2340.
103 Baek, K., Kamigaito, M., Sawamoto, M. *J. Polym. Sci. Polym. Chem. Ed.* **2002**, *40*, 2245.
104 Du, J., Chen, Y. *J. Polym. Sci. Polym. Chem. Ed.* **2004**, *42*, 2263.
105 Gao, H., Tsarevsky, N., Matyjaszewski, K. *Macromolecules* **2005**, *38*, 5995.
106 Francis, R., Lepoittevin, B., Taton, D., Gnanou, Y. *Macromolecules* **2002**, *35*, 9001.
107 Babin, J., Leroy, C., Lecommandoux, S., Borsali, R., Gnanou, Y., Taton, D. *Chem. Commun.* **2005**, 1993.
108 Zheng, Q., Pan, C. *Macromolecules* **2005**, *38*, 6841.
109 Darcos, V., Duréault, A., Taton, D., Gnanou, Y., Marchand, P., Caminade, A., Majoral, J., Destarac, M., Leising, F. *Chem. Commun.* **2004**, *35*, 2110.
110 Bernard, J., Favier, A., Zhang, L., Nilasaroya, A., Davis, T., Barner-Kowollik, C., Stenzel, M. *Macromolecules* **2005**, *38*, 5475.
111 Hong, C., You, Y., Pan, C. *J. Polym. Sci. Polym. Chem. Ed.* **2005**, *43*, 6379; Stenzel-Rosenbaum, M., Davis, T.,

Fane, A., Chen, V. *Angew. Chem. Int. Ed.* **2001**, *40*, 3428.
112 Mayadunne, R., Jeffery, J., Moad, G., Rizzardo, E. *Macromolecules* **2003**, *36*, 1505.
113 Hao, X., Nilsson, C., Jesberger, M., Stenzel, M., Malmström, E., Davis, T., Östmark, E., Barner-Kowollik, C. *J. Polym. Sci. Polym. Chem. Ed.* **2004**, *42*, 5877.
114 Darcos, V., Duréault, A., Taton, D., Marchand, P., Caminade, A., Majoral, J., Destarsc, M., Leising, F. *Chem. Commun.* **2004**, 2110.
115 Zheng, G., Pan, C. *Polymer* **2005**, *46*, 2802.
116 Manglio, G., Nese, G., Nuzzo, M., Palumbo, R. *Macromol. Rapid. Commun.* **2004**, *25*, 1139.
117 Klok, H., Hernandez, J., Becker, S., Müllen, K. *J. Polym. Sci. Polym. Chem. Ed.* **2001**, *39*, 1572.
118 Aliferis, A., Iatrou, H., Hadjichristidis, N. *J. Polym. Sci. Polym. Chem. Ed.* **2005**, *43*, 4670.
119 Webster, O.W., Hertler, W.R., Sogah, D.Y., Farnham, W.B., RajanBabu, T.V. *J. Am. Chem. Soc.* **1983**, *105*, 5706.
120 Hertler, W.R. in *Macromolecular Design of Polymeric Materials*, Chapter 7, Hatada, K., Kitayama, T., Vogl, O. (Eds). Marcel Dekker, New York, **1997**, p. 109.
121 Webster, O.W., Anderson, B.C. in *New Methods for Polymer Synthesis*, Chapter 1, Mijs, W.J. (Ed.). Plenum Press, New York, **1992**, p. 1.
122 Webster, O.W. *Adv. Polym. Sci.* **2004**, *167*, 1.
123 Webster, O.W. *Macromol. Chem. Macromol. Symp.* **1990**, *33*, 133.
124 Zhu, Z., Rider, J., Yang, C.Y., Gilmartin, M.E., Wnek, G.E. *Macromolecules* **1992**, *25*, 7330.
125 Burchard, W., Lang, P., Wolfe, M.S., Spinelli, H.J., Page, L. *Macromolecules* **1991**, *24*, 1306.
126 Haddleton, D.M., Crossman, M.C. *Macromol. Chem. Phys.* **1997**, *198*, 871.
127 Sannigrahi, B., Wadgaonkar, P.P., Sehra, J.C., Sivaram, S. *J. Polym. Sci. Polym. Chem. Ed.* **1997**, *35*, 1999.
128 Spinelli, H.J. *US Patent 5036139*, DuPont, **1991**.
129 Georgiou, T.K., Vamvakaki, M., Patrickios, C.S., Yamasaki, E.N., Phylactou, L.A. *Biomacromolecules* **2004**, *5*, 2221.
130 Themistou, E., Patrickios, C.S. *Macromolecules* **2004**, *37*, 6734.
131 Vamvakaki, M., Hadjiyannakou, S.C., Loizidou, E., Patrickios, C.S., Armes, S.P., Billingham, N.C. *Chem. Mater.* **2001**, *13*, 4738.
132 Grubbs, R.H. *Tetrahedron* **2004**, *60*, 7117.
133 Buchmeiser, M.R. *Chem. Rev.* **2000**, *100*, 1565.
134 Feast, W.J., Khosravi, E. in *New Methods of Polymer Synthesis*, Chapter 3, Ebdon, J.R., Eastmond, G.C. (Eds). Chapman and Hall, London, **1995**, p. 69.
135 Ofstead, E.A., Wagener, K.B. In *New Methods for Polymer Synthesis*, Chapter 8, Mijs, W.J. (Ed.). Plenum Press, New York, **1992**, p. 237.
136 Bazan, G.C., Schrock, R.R. *Macromolecules* **1991**, *24*, 817.
137 Saunders, R.S., Cohen, R.E., Wong, S.J., Schrock, R.R. *Macromolecules* **1992**, *25*, 2055.
138 Dounis, P., Feast, W.J. *Polymer* **1996**, *37*, 2547.
139 Beerens, H., Wang, W., Verdonck, L., Verpoort, F. *J. Mol. Catal. A Chem.* **2002**, *190*, 1.
140 Gatard, S., Kahlal, S., Mery, D., Nlate, S., Cloutet, E., Saillard, J.-Y., Astruc, D. *Organometallics* **2004**, *23*, 1313.
141 Takeichi, T., Stille, J.K. *Macromolecules* **1986**, *19*, 2093.
142 Odian, G. *Principles of Polymerization*, 3rd edn. Wiley, New York, **1991**.
143 Mita, I., Stepto, R.F.T., Suter, U.W. *Pure Appl. Chem.* **1994**, *66*, 2483.
144 Yokozawa, T., Yokoyama, A. *Chem. Rec.* **2005**, *5*, 47.
145 Yokozawa, T., Yokoyama, A. *Polym. J.* **2004**, *36*, 65.
146 Collins, J.E., Fraser, C.L. *Macromolecules* **1998**, *31*, 6715.
147 Lamba, J.J.S., Fraser, C.L. *J. Am. Chem. Soc.* **1997**, *119*, 1801.
148 McAlvin, J.E., Fraser, C.L. *Macromolecules* **1999**, *32*, 1341.

149 McAlvin, J. E., Fraser, C. L. *Macromolecules* **1999**, *32*, 6925.
150 Hochwimmer, G., Nuyken, O., Schubert, U. S. *Macromol. Rapid Commun.* **1998**, *19*, 309.
151 Schubert, U. S., Eschbaumer, C., Hochwimmer, G. *Tetrahedron Lett.* **1998**, *39*, 8643.
152 Naka, K., Yaguchi, M., Chujo, Y. *Chem. Mater.* **1999**, *11*, 849.
153 Naka, K., Kobayashi, A., Chujo, Y. *Macromol. Rapid Commun.* **1997**, *32*, 1025.
154 Chujo, Y., Naka, K., Kramer, M., Sada, K., Saegusa, T. *J. Macromol. Sci., Pure Appl. Chem.* **1995**, *32*, 1213.
155 Matsui, H., Murray, R. W. *Inorg. Chem.* **1997**, *36*, 5118.
156 Maness, K. M., Masui, H., Wightman, R. M., Murray, R. W. *J. Am. Chem. Soc.* **1997**, *119*, 3987.
157 Fraser, C. L., Anastasi, N. R., Lamba, J. J. S. *J. Org. Chem.* **1997**, *62*, 9314.
158 Emmenegger, F., Williams, M. E., Murray, R. W. *Inorg. Chem.* **1997**, *36*, 3146.
159 Peters, M. A., Belu, A. M., Linton, R. W., Dupray, L., Meyer, T. J., Desimone, J. M. *J. Am. Chem. Soc.* **1995**, *117*, 3380.
160 Hecht, S., Ihre, H., Fréchet, J. M. J. *J. Am. Chem. Soc.* **1999**, *121*, 9239.
161 Fraser, C. L., Smith, A. P. *J. Polym. Sci. Polym. Chem. Ed.* **2000**, *38*, 4704.
162 McAlvin, J. E., Scott, S. B., Fraser, C. L. *Macromolecules* **2000**, *33*, 6953.
163 Johnson, R. M., Corbin, P. S., Ng, C., Fraser, C. L. *Macromolecules* **2000**, *33*, 7404.
164 Fraser, C. L., Smith, A. P., Wu, X. *J. Am. Chem. Soc.* **2000**, *122*, 9026.
165 Schubert, U. S., Hochwimmer, G. *Polym. Prepr.* **2000**, *41*, 433.
166 Williams, M. E., Masui, H., Long, J. W., Malik, J., Murray, R. W. *J. Am. Chem. Soc.* **1997**, *119*, 1997.
167 Eisenbach, C. D., Göldel, A., Terskan-Reinold, M., Schubert, U. S. *Macromol. Chem. Phys.* **1995**, *196*, 1077.
168 Miura, Y., Narumi, A., Matsuya, S., Satoh, T., Duan, Q., Kaga, H., Kakuchi, T. *J. Polym. Sci. Polym. Chem. Ed.* **2005**, *43*, 4271.
169 Erdogan, T., Ozyurek, Z., Hizal, G., Tunca, U. *J. Polym. Sci. Polym. Chem. Ed.* **2004**, *42*, 2313.
170 Deng, G., Zhang, L., Liu, C., He, L., Chen, Y. *Eur. Polym. J.* **2005**, *41*, 1177.
171 Zhao, Y., Shuai, X., Chen, C., Xi, F. *Chem. Commun.* **2004**, 1608.
172 Zhao, Y., Shuai, X., Chen, C., Xi, F. *Macromolecules* **2004**, *37*, 8854.
173 Gunawidjaja, R., Peleshanko, S., Tsukruk, V. V. *Macromolecules* **2005**, *38*, 8765.
174 Feng, X.-S., Taton, D., Chaikof, E. L., Gnanou, Y., *J. Am. Chem Soc.* **2005**, *127*, 10956.
175 Shi, P., Li, Y., Pan, C. Y. *Eur. Polym. J.* **2004**, *40*, 1283.
176 Li, Y.-G., Wang, Y.-M., Pan, C.-Y. *J. Polym. Sci. Polym. Chem. Ed.* **2003**, *41*, 1243.
177 Feng, X., Pan, C.-Y. *Macromolecules* **2002**, *35*, 2084.
178 He, T., Li, D., Sheng, X., Zhao, B. *Macromolecules* **2004**, *37*, 3128.
179 Celik, C., Hizal, G., Tunca, U. *J. Polym. Sci. Polym. Chem. Ed.* **2003**, *41*, 2542.

7
Linear Versus (Hyper)branched Polymers

Hideharu Mori, Axel H. E. Müller, and Peter F. W. Simon

7.1
Introduction

Depending on their topology, polymers can be classified as linear, cyclic, branched and cross-linked. Linear polymers are macromolecules in which the monomer units are linked in one continuous string. Cyclic polymers are similar, but they have no end-groups. Branched polymers are characterized by the existence of f-functional branch points with $f \geq 3$ (or dendritic groups, D), resulting in more than two end-groups (or terminal units, T). There are three kinds of branched polymers, depending on the number and functionality of branched units, as shown in Fig. 7.1: stars, combs (or cylindrical brushes at very high density of branch points) and randomly branched or arborescent polymers, also called dendritic or – for a high density of branch points – hyperbranched. A special case of arborescent polymers is dendrimers, which are monodisperse and have a completely regular branching structure; these are discussed in another chapter. Combinations of the three topological motifs in Fig. 7.1 can lead to even more complex structures, such as dumbbell (pom-pom) polymers, where two stars are coupled at the ends of one arm each. In addition, there are networks (including microgels), which contain loops, i.e. at least two subchains are connected by at least two branch points (although hyperbranched polymers may form exactly one loop during their synthesis; see below). In fact, some properties (e.g. rheology) of microgels are similar to those of hyperbranched polymers, but they will not be discussed here.

Many properties of branched polymers are different from those of their linear analogues. Randomly branched polymers, i.e. macromolecules having a small number of long branches along the polymer backbone (and also along the sidechains), show strong differences in melt rheology, mechanical behavior and solution properties. For higher degrees of branching, the possibility of forming entanglements is restricted or absent. The recent interest in hyperbranched polymers arises from the fact that they combine some features of dendrimers, e.g. an increasing number of end-groups and a compact structure in solution,

Macromolecular Engineering. Precise Synthesis, Materials Properties, Applications.
Edited by K. Matyjaszewski, Y. Gnanou, and L. Leibler

ISBN: 978-3-527-31446-1

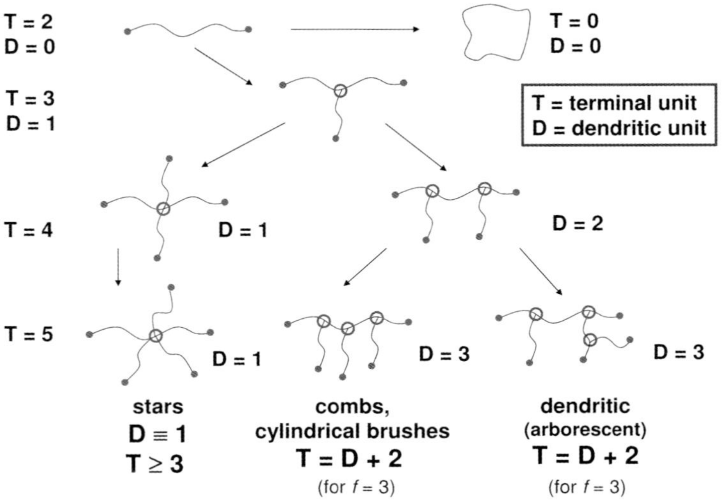

Fig. 7.1 Three classes of branched topologies.

with ease of preparation of linear polymers by means of a one-pot reaction. In addition to their chemical structure, molecular weight and polydispersity, the number and location of the branch points and the chain length of the branches are crucial for the unique properties of (hyper)branched polymers.

This chapter will cover structural control of (hyper)branched polymers using different polymerization techniques and some unique properties of these polymers, in comparison with their linear analogues. The distinction between the terms "hyperbranched", "highly branched" and "randomly branched" is obscure, but it mainly relies on the degree of branching, *DB*, and also on the synthetic procedure leading to branching. However, from a topological point of view, all these terms denote the same class of polymers, namely arborescent polymers. The field of arborescent polymers has been well established with a large variety of synthetic approaches, fundamental studies on the structure and properties of these unique materials and possible applications [1–4].

The degree of branching, *DB*, is the most prominent parameter to describe the topology of a branched pattern. According to the definition first introduced by Hawker et al. [5] and Kim [6] and later modified by Hölter et al. [7] and Yan et al. [8], it reads

$$DB = \frac{D + T - 1}{D + T + L - 1} = \frac{2D}{2D + L} \tag{1}$$

Here, the number or the fraction of the different unit present in the molecule can be used. D represents branched (dendritic) units, whereas L and T represent linear and terminal units. For trifunctional branch points $D = T - 1$. The denominator represents the sum over all units and the focal unit (in linear polymers the initiating group) is not counted. Using this definition, DB is the fraction of branch points relative to a dendrimer, which has $DB=1$.

Alternatively, the fraction of branch points, FB, can be defined as

$$FB = \frac{D}{2D+L} = DB/2 \tag{2}$$

In the case of randomly (long-chain) branched polyolefins, the degree of branching is often expressed as the number of branch points per 1000 monomer units, N_{1000}, or per 1000 carbon atoms, N_{1000C}, which can be correlated with DB or FB (assuming that one monomer unit has two carbons in the backbone) as

$$\begin{aligned} N_{1000} &= 1000FB = 500DB \\ N_{1000C} &= 2000FB = 1000DB \end{aligned} \tag{3}$$

In the case of cocondensation of trifunctional AB_x with bifunctional AB monomers or the copolymerization of AB* inimers with conventional monomers, M, the degree of branching is reduced. The various types of monomers are discussed in detail in the subsequent subsections. This reduction depends on the comonomer ratio, $\gamma=[AB]_0/[AB_x]_0$ or $[M]_0/[AB^*]_0$. For $\gamma \gg 1$, this parameter is equal to the number of linear units between branch points:

$$\gamma \cong \frac{L}{D} \tag{4}$$

which renders for the DB at full conversion solely by using geometrical considerations,

$$DB \cong \frac{2}{2+\gamma} \cong \frac{1}{\gamma} \tag{5}$$

7.2 Synthesis

Statistical branching can be introduced by various approaches. The first is chain transfer to polymer, which is a side-reaction in the polymerization of vinyl monomers, in particular α-olefins and acrylates. This transfer can occur by a hydrogen abstraction from a monomer unit near to the chain end ("back-biting") and leads to short-chain branching (i.e. a comb-like structure) or by intermolecular abstraction of a hydrogen atom of another chain, leading to long-chain branching. Since the degrees of branching are typically very low ($DB<0.01$), the terms N_{1000} and N_{1000C} have been most frequently used. Commercial poly(ethylenimine) (PEI), is a branched polymer prepared by cationic ring-opening polymerization of ethylenimine (aziridine), where intramolecular transfer of a proton occurs from in-chain NH groups to the cationic chain end.

The second approach is the step- or chain-growth polymerization of special monomers of the type AB_x ($x \geq 2$) or AB*, respectively, leading to hyperbranched polymers with *DB* in the range 0.1–0.5. Several strategies for the preparation of hyperbranched polymers are currently employed. The polymerization reactions can be classified into three categories [1, 4]: (1) step-growth polycondensation of AB_x monomers; (2) chain-growth self-condensing vinyl polymerization (SCVP) of AB* initiator-monomers ("inimers"); and (3) chain-growth self-condensing ring-opening polymerization of cyclic inimers. By cocondensation of AB_x monomers with "normal" AB monomers or by copolymerization of AB* inimers with "normal" monomers, M, the degree of branching can be controlled in the range 0.01–0.5. This approach will be described in more detail in the subsections below.

A third approach, called the "two monomer method", the copolycondensation of A_2 with B_3 monomers or the copolymerization of mono- and difunctional vinyl monomers, can lead to hyperbranched polymers, but for statistical reasons these polymers will generally contain a certain number of loops and form microgels. This approach will be included in the subsections below.

A fourth technique involves a series of grafting procedures, where a linear backbone is grafted via a "grafting on to" approach. The resulting comb polymer is subsequently subjected to one or more additional grafting procedures, leading to a highly branched, arborescent polymer [9, 10].

Highly branched polymers are also found in nature, in particular in polysaccharides, e.g. amylopectin and glycogen, which can be regarded as copolymers of 1,4,6-α-glucose (AB_2) and 1,4-α-glucose (AB) units. Even more complex structures are found in proteoglycans and glycoproteins, e.g. aggrecan, mucin or fibrin, which show complex "comb on comb" structures [11].

7.2.1
Step-growth Polymerization

The most common method for the synthesis of hyperbranched polymers is the polycondensation or polyaddition of AB_x monomers, containing one 'A' functional group and two or more 'B' functional groups that are capable of reacting with each other (Scheme 7.1). The reaction involves typical features of a step-growth reaction of multifunctional monomers, resulting in the formation of hyperbranched polymers without gelation. In principle, all polymers contain exactly one A group, the so-called focal group, whereas the number of B groups is given by the degree of polymerization and the functionality of the monomer, $n_B = DP + x - 1$. Many excellent reviews have been published [1, 2, 12–14].

As early as 1952, Flory [15, 16] pointed out that the polycondensation of AB_x-type monomers will result in soluble "highly branched" polymers and calculated the molecular weight distribution (MWD) and its averages using a statistical derivation. Ill-defined branched polycondensates had been known even before then [17, 18]. In 1982, Kricheldorf et al. published the cocondensation of AB and AB_2 monomers to form branched polyesters [19]. However, only after Kim

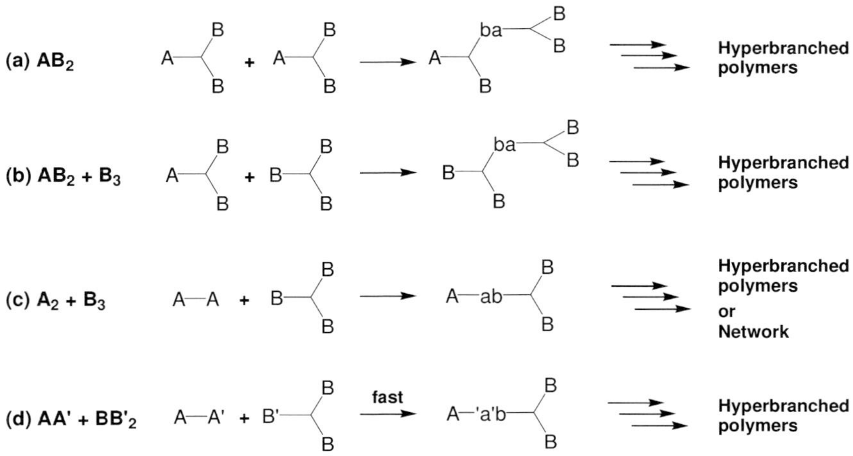

Scheme 7.1 Various step-growth polymerization approaches for hyperbranched polymers. Lower-case letters stand for reacted units.

and Webster [20] had published the synthesis of polyarylenes from AB_2 monomer **1** in 1990 and introduced the term "hyperbranched" did this class of polymers became a topic of intensive research. A multitude of hyperbranched polymers have been synthesized since then via polycondensation of AB_2 monomers, including polyesters, polyethers, poly(ether ketone)s, poly(ether sulfone)s, polyamides, polyurethanes, polyureas, poly(ether imide)s, poly(siloxysilane)s and poly(carbosilane)s. Some representative AB_2 monomers are summarized in Fig. 7.2. For example, various hyperbranched polyesters are formed by esterification reactions of AB_2 monomers having one carboxylic acid derivative and two hydroxyl derivatives (e.g. **2**, **3**) or vice versa.

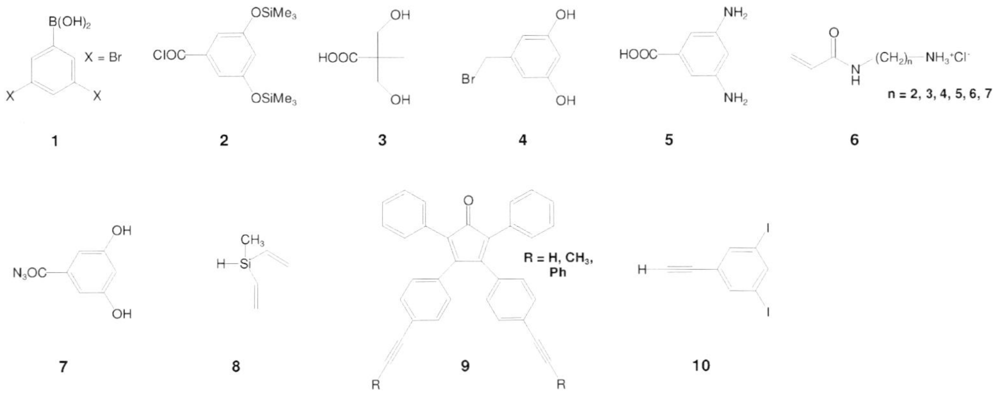

Fig. 7.2 Some AB_2 monomers used for the synthesis of hyperbranched polymers.

In addition to typical polycondensation reactions, other type of reactions, such as addition reactions, can be used for the synthesis of hyperbranched polymers. For example, hyperbranched polyurethane and polycarbosilane are formed by addition reaction mechanisms using classical reactions of isocyanate precursors (**7**) [21] and hydrosilylation (**8**) [22]. Monomer **9** can undergo a Diels–Alder [2+4] cycloaddition reaction [23, 24], resulting in the formation of hyperbranched polyphenylenes. Michael addition reaction of **6** leads to a hyperbranched polymer, which has a structure similar to the PAMAM dendrimers [25].

For a successful synthesis of classical hyperbranched polymers from AB_x-type monomers, side-reactions have to be avoided, since gelation may occur to provide insoluble gels. An intrinsic side-reaction, inhibiting the formation of polymers having high molecular weights, is the so-called "back-biting" reaction, where the only A functional group in the hyperbranched molecule is consumed by reaction with a B unit of the same molecule forming cyclic oligomers. The resulting molecules contain only B groups and when all A groups have been consumed the polymerization ends. Although these polymers contain exactly one loop (and they therefore formally should be regarded as microgels) they are regarded as hyperbranched polymers since one can expect the influence of only one loop per molecule to be negligible for the overall properties of the polymer.

The most common AB_2 approach theoretically leads to $DB=0.5$ [26, 27] and to a very broad molecular weight distribution (MWD) with polydispersity index $M_w/M_n \approx DP_n/2$ [15]. Both values depend on the reactivity of a B group located in a linear unit, –ab(B)–, relative to one in a monomer or terminal unit, AB_2 or $-aB_2$, respectively [27]. In theory, the degree of polymerization should go to infinity at high conversions of the B groups, but in reality it is limited by the cyclization reaction described above. *DB* can be decreased by cocondensation with conventional AB comonomers.

A challenging goal in this field, particularly from the synthetic point of view, is the development of general polymerization methods that increase *DB* and achieve control over the polydispersity. An effective approach for better control over molecular weights and resulting geometrical shape is the addition of a "core" molecule B_f ($f>2$). The polycondensation of dihydroxymethylpropionic acid (**3**) and trimethylolpropane (**12**) is a representative example of so-called "AB_2 and B_3" approach, as shown in Scheme 7.1 b. In this case, a successful addition of the core B_3 monomer leads to an increase in the degree of branching, which is higher than the theoretical limitation. For example, Bharathi and Moore reported a new hyperbranched procedure (one-step AB_2-type polymerization using **10**) which takes place on an insoluble solid support, providing polymers with low polydispersity and controlled molecular weight [28, 29]. Theoretical work by Müller and coworkers [30, 31] and simulation work by Frey and coworkers [32] have shown that copolymerization of a core molecule of the structure B_f (representing a polyfunctional initiator in the case of the SCVP and a polyfunctional core molecule for the AB_x-type polycondensation reaction) can be used to lower the polydispersity considerably, in particular when the AB_x monomers are added so slowly that they only react with the core or the polymer growing from there, but not with each other. In the latter case, the polydispersity index and the degree of branching, respectively, were predicted to be

$$M_w/M_n = 1 + 1/f$$
$$DB = 2/3 \qquad (6)$$

This approach, named "slow monomer addition", was verified experimentally by Frey and coworkers [33].

In addition to the single-monomer approaches reviewed so far, double-monomer methodologies have also been used to obtain hyperbranched polymers [34]. This approach implies the cocondensation of A_2 and B_3 monomers (or B_n, $n \geq 3$, Scheme 7.1c). The potential advantage of the A_2+B_3 polymerization relies on the feasibility of using commercially available monomers to prepare hyperbranched polymers, in contrast to most AB_x monomers. Statistically, however, the condensation of difunctional (A_2) and trifunctional (B_3) compounds will result in gelation when a certain conversion of functional groups is exceeded. Soluble "hyperbranched" polymers could be prepared using the A_2+B_3 method, when the reaction was stopped prior to the gel point, but in most cases one can expect the formation of microgels rather than true hyperbranched polymers. Many parameters, such as the ratio of functionalities, reaction temperature, concentration of monomers and the amount of condensation agents, should be optimized in order to avoid gelation. The slow addition of monomer and employment of special catalysts, inorganic salts and condensation agents are other techniques to prepare hyperbranched polymers without gelation. However, to control the reaction precisely and to obtain hyperbranched polymers exhibiting sufficient high molecular weights are very challenging.

Recently, a more sophisticated double-monomer methodology has been reported, which retains the advantages of the method, such as use of commercially available monomers, but overcomes the essential problem of possible gelation. This approach is based on *in situ* formation of AB_x intermediates from specific monomer pairs, $AA'+BB'_2$, in the initial stage of polymerization, due to the non-equal reactivities of different functional groups. This leads to the formation of hyperbranched polymers without cross-linking reaction. Many commercially available chemicals can be used as reactive monomers for this new double-monomer methodology, as shown in Fig. 7.3. The representative reaction step of this method is shown in Scheme 7.1 d.

When the reactivity of the B′ group in the $B'B_2$ monomer is greater than that of the other two B groups, AB_2 intermediates can be formed *in situ* during the initial reaction. Further self-polycondensation of the AB_2 species formed can provide hyperbranched polymers without gelation. Choosing a suitable monomer pair is crucial for the molecular design of a hyperbranched polymer by this method. Depending on the difference in the reactivities of functional groups, AB_2 intermediates can be formed *in situ* and further reaction of the AB_2 intermediates results in hyperbranched macromolecules, which are essentially different from the conventional "A_2+B_3" method ($A'=A$, $B=B'$). When the AB_2

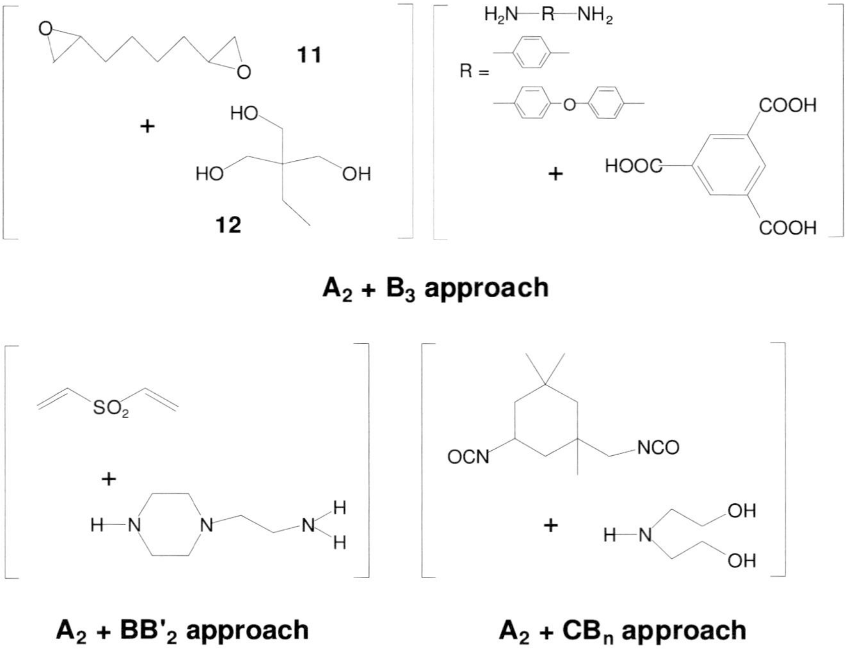

Fig. 7.3 Some monomers used for the double-monomer method.

monomer reacts predominantly with a $B'B_2$ molecule due to the higher reactivity of the B′ group, a molecule containing four B groups should be formed, which can play the role of a core molecule in the preparation of hyperbranched polymers. In this case, the reaction can be regarded as an "AB_2+B_4" polymerization. Due to the presence of a B_4 core unit, the system has the potential to produce hyperbranched macromolecules having a narrower molecular weight distribution. If the polymerization proceeds preferentially through an "AB_2 and B_4" reaction, this synthetic route can be considered as a suitable combination of the advantages of double-monomer method (use of commercially available monomers) and AB_x+B_f approach (polycondensation of AB_2 monomers in the presence of multifunctional core moieties to lower the polydispersity considerably).

7.2.2
Chain-growth Polymerization

Chain-growth approaches use monomers that contain a polymerizable group, i.e. an initiator-monomer ("inimer"), AB*. The approach can be divided into two categories: (i) self-condensing vinyl polymerization (SCVP) and (ii) self-condensing ring-opening polymerization (SCROP) (also named multibranching ring-opening polymerization, MBROP). Both methods are mechanistically similar and we will use the term SCVP for both. Both methods allow the use of vinyl

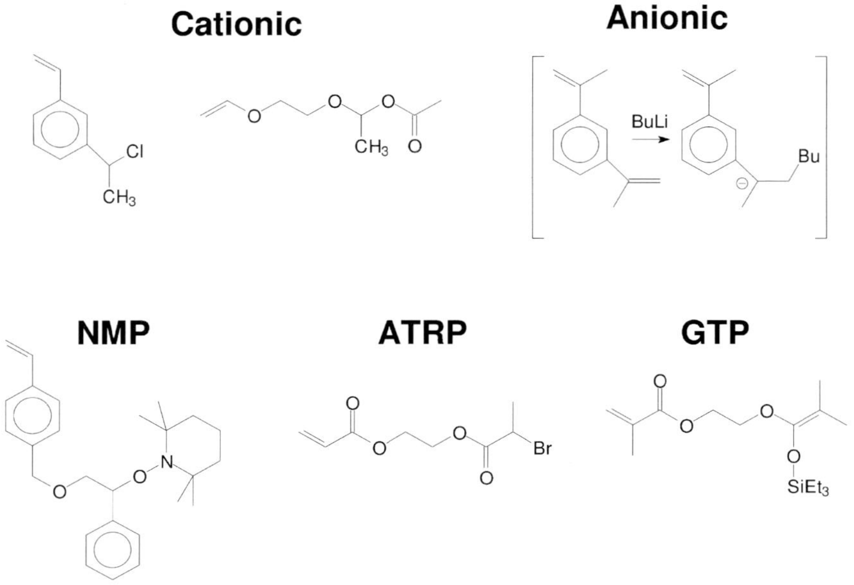

Fig. 7.4 Examples of AB* inimers used for SCVP.

or cyclic monomers for a convenient, one-pot synthesis of hyperbranched polymers [35–42]. Similar to the AB_x approach, $DB \approx 0.5$ can be reached and polydispersities are predicted to be very high, $M_w/M_n \approx DP_n$. Again, significant narrowing is reported due to the back-biting reaction. Similarly to the $A_2 + B_f$ approach, the polymerization can also be initiated by a mono- or multifunctional initiator, leading to better control of MWD and *DB*, especially when the inimer is added slowly [30]. By copolymerizing AB* inimers with conventional monomers, M, the SCVP technique was extended to self-condensing vinyl copolymerization (SCVCP), leading to highly branched copolymers where *DB* is controlled by the comonomer ratio, $\gamma = [M]_0/[AB^*]_0$ [43–48]. Some representative inimers are presented in Fig. 7.4.

By using living anionic, cationic or radical techniques, a variety of hyperbranched polymers such as derivatives of polystyrene, poly(meth)acrylates or poly(vinyl ether)s can be synthesized [49]. Hyperbranched polyamines, polyethers and polyesters can be prepared through SCROP of cyclic monomers [50, 51].

7.2.2.1 Self-condensing Vinyl Polymerization (SCVP) of Inimers

In attempts to synthesize ω-styrylpolyisobutene using *m*-/*p*-(chloromethyl)styrene as an initiator and Al alkyls–H_2O as co-initiators, Kennedy and Frisch [52] found significantly less than 100% of vinyl groups in the product. They concluded that copolymerization of the vinyl group of the "initiator" with isobutene occurred as a "deleterious side-reaction". In a similar experiment, Nuyken et al. [53] observed the formation of soluble polymers with much higher than calculated molecular weights and broad MWD and attributed this to the formation of branched copolymers. On the other hand, copolymerizations of isobutene with *p*-(chloromethyl)styrene or *p*-(chloromethyl)-α-methylstyrene initiated with boron trifluoride or $EtAlCl_2$ were reported earlier [54, 55]. Fréchet et al. recognized the importance of initiator-monomers (later called "inimers" [56, 57]) to synthesize hyperbranched polymers from vinyl monomers and in 1995 they named the process "self-condensing vinyl polymerization" (SCVP) [35].

Inimers have the general structure AB*, where the double bond is designated A and B* is a group capable of being activated to initiate the polymerization of vinyl groups. Schemes 7.2 and 7.3 show the initial steps in SCVP (and identically in SCROP). In order to start the polymerization, the B* group is initially activated and then adds to the A group (double bond) of another inimer, resulting in the formation of the dimer, Ab-A*B*. The asterisk indicates that a structural group can add monomer; it can be either in its active or in its dormant form. Lower-case letters indicate that the site has been consumed and can no longer participate in the polymerization. The resulting dimer has two active sites, A* (propagating) and B* (initiating), for possible chain growth besides the vinyl group. Addition of a third monomer unit at either site results in the formation of two different trimers, which can now grow in three directions. Also, oligomers (e.g. two dimers) or polymers can react with each other, similarly (but mechanistically different) to a polycondensation. Interestingly this process forms two different kinds of linear units: vinyl-type, L_v, and polycondensate type, L_c (Scheme 7.4). The kinetics of SCVP, the molecular weight distribution

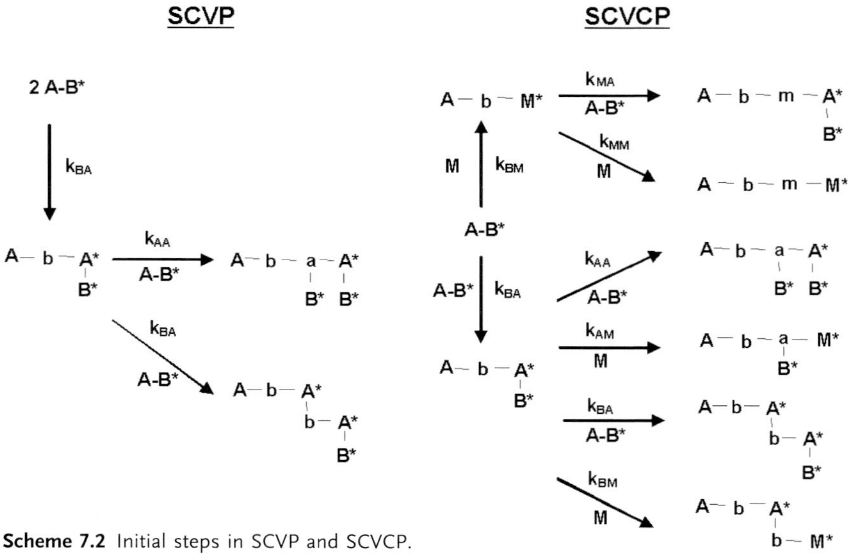

Scheme 7.2 Initial steps in SCVP and SCVCP.

and *DB* that can be obtained in this process were analyzed [8, 56]. Both parameters depend on the reactivity ratio of A* and B* chain ends, $r=k_{AA}/k_{AB}$. The maximum $DB=0.465$ is reached at $r=2.6$. For $r \gg 1$ and $r \ll 1$ two different kinds of linear polymers are formed, one resembling a vinyl polymer and the other a polycondensate.

Some typical AB* inimers used for SCVP are listed in Fig. 7.4. A variety of (meth)acrylate-type inimers have been reported [41, 42, 58–62]. In general, Cu-based atom transfer radical polymerization (ATRP) was employed for SCVP of these acrylate-type inimers having an *α*-bromo ester initiating group, B*, capable of initiating ATRP. For example, the ATRP of an acrylate inimer provided a hyperbranched polymer with $DB=0.49$ [41, 63]. For the preparation of hyperbranched methacrylates, Cu- and Ni-based ATRP was used [64, 65]. Group transfer polymerization (GTP) was employed for a silyl ketene acetal-functional methacrylate [39, 66].

For the SCVP of styrenic inimers, cationic [35, 46], atom-transfer radical [37, 44], nitroxide-mediated radical [36], anionic [38], photo-initiated radical [2, 67, 68], and ruthenium-catalyzed coordinative [69] polymerization systems have been used. Another example involves the cationic polymerization of vinyl ether-

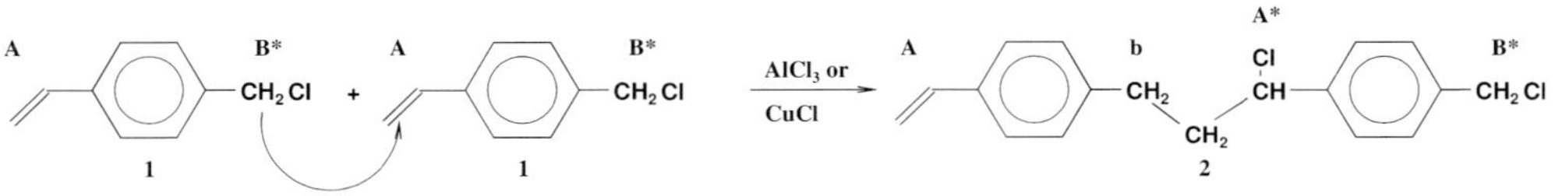

Scheme 7.3 Initiation step in SCVP of *p*-(chloromethyl)styrene.

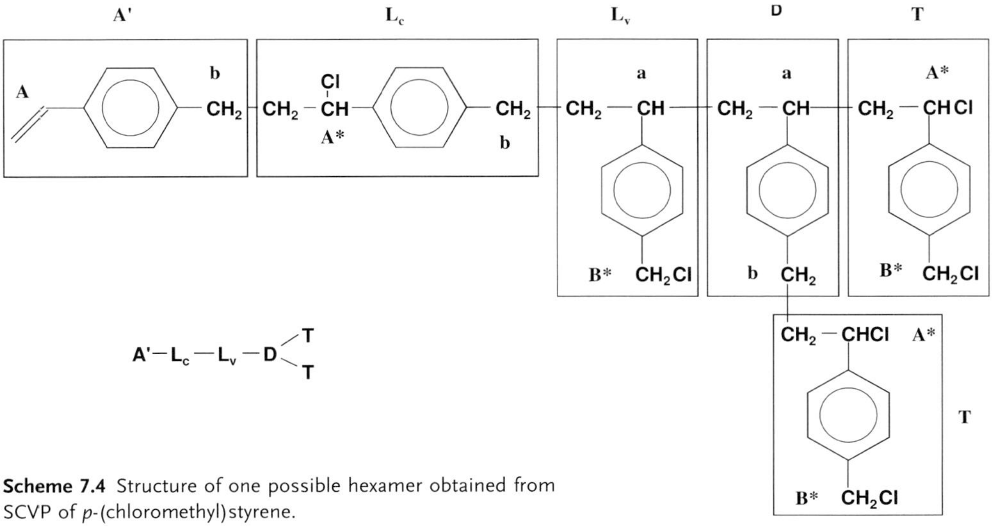

Scheme 7.4 Structure of one possible hexamer obtained from SCVP of *p*-(chloromethyl)styrene.

type inimers [43, 70] and reversible addition–fragmentation chain transfer polymerization (RAFT) [71, 72].

For an ideal SCVP process, living polymerization systems are necessary to avoid cross-linking reactions and gelation caused by chain transfer or recombination reaction. In living cationic polymerization, spontaneous or monomer transfer may occur, leading to more than one vinyl group in some molecules and subsequent cross-linking. In controlled radical polymerization, termination via recombination or disproportionation is to be expected. If recombination occurs intermolecularly, more than one vinyl group per chain can occur, which may lead to gelation. For intramolecular termination (which may be more probable), the number of active sites will decrease and their distribution will be affected. In anionic or group transfer polymerization of (meth)acrylic monomers, it is possible that attack of the anion on the ester groups of its own polymer molecule ("back-biting") will even remove whole branches from the polymer thus decreasing the molecular weight considerably [56].

Similarly to cationic and group transfer polymerization, these systems are based on establishing a rapid dynamic equilibration between a minute amount of growing free radicals and a large majority of dormant species; however, they are more tolerant of functional groups and impurities.

Self-condensing vinyl copolymerization (SCVCP) of AB* inimers with conventional monomers, M, leads to highly branched polymers, allowing for control of MWD and *DB* [43–48]. The copolymerization method is a facile approach to obtain functional branched polymers, because different types of functional groups can be incorporated into a polymer, depending on the chemical nature of the comonomer. In addition, the chain architecture can be modified easily by a suitable choice of the comonomer ratio in the feed. The initial steps of SCVCP are given in Scheme 7.2. Theoretical calculations show that the MWD is narrower than that for polymers obtained by SCVP, i.e. M_w/M_n decreases by $1/\gamma$. Because the number of linear units is higher, the *DB* of the copolymers is lower than that of SCVP homopolymers. For $\gamma \gg 1$, $DB \approx 1/\gamma$ [136]. However, the effect on solution properties such as intrinsic viscosity and radius of gyration is less than proportional, because branched polymers above a limiting molecular weight are self-similar objects (see below). Therefore, SCVCP is an economical approach to obtain highly branched functional polymers, especially when aiming at controlling rheology.

SCVCP can be initiated by two ways (Scheme 7.2): (i) by addition of the active B* group of an AB* inimer to the vinyl group A of another AB* inimer forming a dimer with two active sites, A* and B*, and (ii) by addition of a B* group to the vinyl group of monomer M forming a dimer with one active site, M*. Both the initiating B* group and the newly created propagating centers A* and M* can react with any vinyl group in the system. Hence, we have three different types of active centers, A*, B* and M*, in the dimers, which can react with double bonds A (inimer and macromolecules; each macromolecule contains strictly one double bond) and M (monomer). Kinetics, MWD and *DB* depend on the relative rate constants involved in Scheme 7.2 in a complicated way [193].

Several approaches have been reported for the synthesis of hyperbranched (meth)acrylates via SCVCP. Highly branched poly(methyl methacrylate) was synthesized by GTP of methyl methacrylate with the silyl ketene acetal inimer shown in Fig. 7.4 [45]. Highly branched poly(*tert*-butyl methacrylate) was also obtained by SCVCP of the same inimer with *tert*-butyl methacrylate, which is a precursor of branched poly(methacrylic acid) [73]. A series of functional hyperbranched (meth)acrylates has been synthesized by SCVCP of ATRP-type (meth)-acrylate inimers with *tert*-butyl acrylate [48], methyl acrylate [74], 2-(diethylamino)ethyl methacrylate [75], and glucose-containing (meth)acrylates [76, 77]. Cationic SCVCP of *p*-chloromethylstyrene with isobutene was used to synthesize hyperbranched polyisobutene [57].

An alternative way to obtain highly branched polymers is the use of "macroinimers", i.e. macromonomers carrying an initiating group at the other terminus. Thus, hyperbranched poly(*tert*-butyl acrylate) has been obtained by SCVP via ATRP of the corresponding macroinimer [62]. In fact, the term "macroinimer" had been coined much earlier by Hazer [78]. However, he used conventional radical polymerization, leading to cross-linking and gelation.

7.2.2.2 Self-condensing Vinyl Polymerization of Monomers Containing Chain-transfer or Terminating Groups

Several workers have reported the use of monomers containing a group that is capable of initiation and chain transfer ("iniferter" group), e.g. methacrylates [68] and styrene derivatives [67, 79, 80] with dithiocarbamate groups. Styrene derivatives containing a methylmalodinitrile azo group have also been used [2]. Polymerization of such compounds is based on a dithiocarbamate radical formed by UV radiation and the methylmalodinitrile radical formed by decomposition of an azo group. These fragments act as reversible terminating/transfer agents, but not as initiators, and therefore unfavorable cross-linking reactions can be avoided, even if the polymerizations are non-living processes. Hence these compounds cannot be regarded as inimers, but the mechanism has similarities to SCVP. A styrene derivative, 4-(chlorodimethylsilyl)styrene, which contains a polymerizable vinyl group and a moiety capable of undergoing quantitative S_N2 reactions, was also employed as a monomer-terminator to synthesize dendritic branching by living anionic polymerization [81]. The method requires the slow addition of this monomer. A styrene derivative having a dithio ester group was used as a monomer-transfer agent to prepare randomly branched poly(*N*-isopropylacrylamide)s via a RAFT process, which showed characteristic temperature-responsive properties in aqueous solution [72].

There are also several accounts of preparing branched polymers using copolymerization of a normal monomer with a difunctional monomer. In order to avoid gelation, chain transfer agents were used [82–86]. For example, RAFT polymerization was used to produce hyperbranched poly(methyl methacrylate) via the one-pot copolymerization of methyl methacrylate and ethylene glycol dimethacrylate as a branching agent, mediated by 2-(2-cyanopropyl) dithiobenzoate [82]. Another example is that a cobalt chain transfer catalyst was successfully applied for the free radical polymerization of divinyl monomers, which afforded soluble hyperbranched polymers in one pot by controlling the competition between propagation and chain transfer [87].

7.2.2.3 **Self-condensing Ring-opening Polymerization (SCROP)**

As stated above, SCROP differs from SCVP only in the fact that instead of a vinyl group a heterocyclic group is used as the monomer part of the inimer. All mechanistic considerations of SCVP also hold here.

In 1992, Suzuki et al. [88] reported the palladium-catalyzed SCROP of cyclic carbamates **13** and **14** to produce hyperbranched polyamines. Later, a variety of cyclic ethers and lactones were used for SCROP (Scheme 7.5).

Anionic ring-opening polymerization of glycidol **16**, which has one epoxide and one hydroxyl group, leads to hyperbranched polyglycerol [33]. Partial deprotonation to an initiating multifunctional alkoxide leads to almost simultaneous growth of all chains ends. With slow monomer addition, unfavorable cyclization is suppressed and the molecular weights and polydispersity can be controlled. Copolymerization with propylene oxide allowed control of the polarity of the highly hydrophilic nature of polyglycerol and of the glass transition temperature.

Cationic ring-opening polymerization of the oxetane inimer **17** leads to hyperbranched aliphatic polyethers. SCROP of lactones containing hydroxyl groups as an initiating moiety leads to hydroxy-terminated hyperbranched aliphatic polyesters [89]. Hyperbranched polyesters are also attained via SCROP of the cyclic inimer **18** or by its copolymerization with ε-caprolactone [90]. Initiation and propagation proceed entirely through one type of reactive nucleophile, a primary alcohol. The cyclic monomer **19** has two hydroxymethyl initiating groups, thus forming an AB_2* inimer [91]. Conversely, the inimer **20** has two epoxide groups and can be denoted an A_2B* inimer [92].

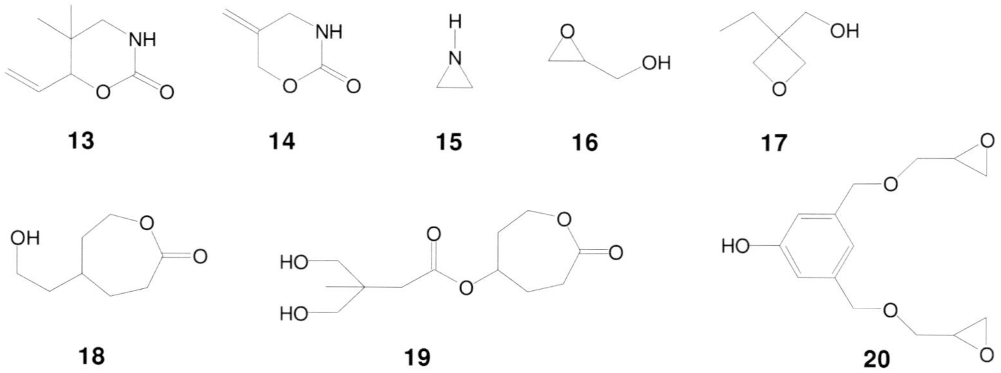

Scheme 7.5 Cyclic AB* monomers used for SCROP.

7.2.2.4 Multiple Grafting Processes

This method is generally based on the combination of ionic polymerization and repeated grafting processes ("graft on graft") [9, 10]. The resulting polymers have also been named "dendrigraft" polymers. Since the grafting reaction is a random process, the resulting polymers belong to the class of arborescent polymers. However, the polydispersities of these materials typically are fairly low ($M_w/M_n \approx 1.1$ in some cases), which are comparable to those observed in dendrimers. Due to the relatively uniform molecular weights, dendrigraft polymers exhibit many of the physical properties characteristic of dendritic molecules on a mesoscopic scale, even though the architecture is not as strictly defined as for dendrimers. On the other hand, multi-step reactions are required to obtain branched architectures, which is different from the synthesis of branched polymers.

The divergent 'grafting on to' method, based on successive coupling reactions of polymer chains with a functionalized substrate polymer, has been most widely applied for the synthesis of dendrigrafts, e.g. dendrigraft polyethylenimines, named Comb-burst® polymers [93]. Arborescent polystyrenes involved the iterative grafting of end-functional polymer chains on to reactive polymer backbones. The method is based on the highly selective coupling of living polystyryllithium on to poly(chloroethyl vinyl ether) chains, which were prepared individually by anionic and cationic living polymerizations [10, 94, 95]. Other examples include arborescent polybutadiene [96], arborescent polystyrene-*graft*-polyisoprene [97] and arborescent polystyrene-*graft*-poly(2-vinylpyridine) [98].

7.3 Effect of Branching on Properties in Dilute Solution

The solution behavior of polymers is experimentally quantified by the intrinsic viscosity, $[\eta]$, the radius of gyration, R_g, the hydrodynamic radius, R_h, and the second virial coefficient, A_2, typically determined by solution viscometry and scattering techniques. The effect of branching may be described using the Flory and Kuhn–Mark–Houwink–Sakurada scaling laws:

$$R_g = K_{LS} M^{\nu} \quad (7)$$

$$[\eta] = K M^{a} \quad (8)$$

To account for a relation between these dependences a draining parameter, Φ, was introduced by Fox and Flory [99, 100]. It describes how deeply a structure is drained by the solvent leading to a relation of the intrinsic viscosity and the radius of gyration:

$$[\eta] = \Phi \frac{R_g^3}{M} \quad (9)$$

Equation (9) is fairly well settled for linear chains of sufficiently high molecular weight, but for branched chains the draining parameter, Φ_{br} is different from that of linear chains, Φ_{lin}. In fact, Φ increases with the number of branch points (see below). As the intrinsic viscosity is strongly governed by polymer–solvent interactions, the draining parameter must be used with caution. This is especially the case for low molar mass hyperbranched polymers where the end-groups influence the solution viscosity much more strongly compared to a linear architecture. For higher molar masses Φ can be assumed to be constant and the polymer exhibits self-similarity [101]. Then, the Mark–Houwink exponent, a, and the Flory exponent, ν, are connected:

$$a = 3\nu - 1 \tag{10}$$

The reciprocal of Flory's scaling exponent, ν, can also be considered as a fractal dimension, $d_f = 1/\nu$ [102–106]. Besides the geometric issue the fractal dimension indicates how much of the mass of an object is accessible at the surface and may be interpreted as the openness of a structure [105, 106].

Branching leads to a contraction of the macromolecule and, for a given molecular weight, M, lowers R_g and $[\eta]$. Compared with its linear analogue, a branched polymer also exhibits lower exponents in Eqs. (7) and (8) and consequently a higher fractal dimension, d_f. Accounting for this effect, Stockmayer and coworkers [107, 108] introduced the contraction factors g and g′:

$$g = \left(\frac{\langle R_g^2 \rangle_{br}}{\langle R_g^2 \rangle_{lin}} \right)_M \tag{11}$$

$$g' = \left(\frac{[\eta]_{br}}{[\eta]_{lin}} \right)_M \tag{12}$$

Combination of Eqs. (11), (14) and (15) leads to [105, 109, 194]

$$g' = \frac{\Phi_{br}}{\Phi_{lin}} g^{\frac{3}{2}} = g^{\varepsilon} \tag{13}$$

Thus, if linear and branched polymers had the same draining parameter, the conversion exponent, ε, should have the fixed value 3/2. However, depending on topology, different values have been found, ranging from 1.5 [108] to 0.5 [109]. This proves that indeed the draining parameter, Φ, increases and hence the conversion parameter, ε, decreases with DB and we can only use Eq. (13) for polymers with the same topology and DB [105].

The Flory and the Mark–Houwink exponents, ν and a, respectively, are most conveniently determined by multi-detector SEC approaches. Universal calibration [110, 111] applying an on-line viscosity detector and also on-line multi-angle light scattering [112, 113] present the possibility of determining absolute molecular weights and solution properties if the molecular weight exceeds about 10^4.

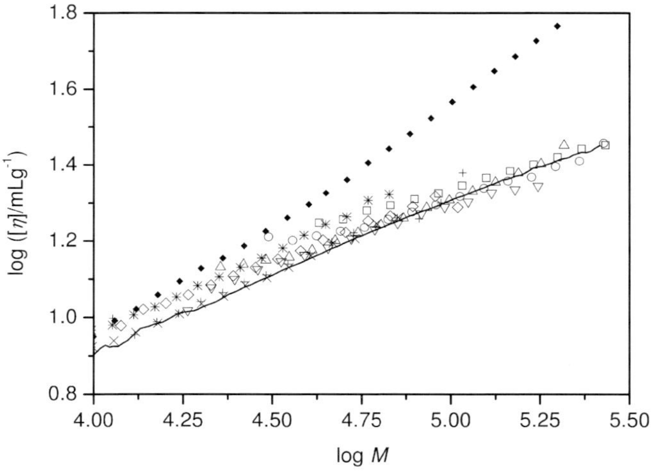

Fig. 7.5 Mark–Houwink plot for fractions of highly branched poly(methyl methacrylate) (□) M_w=121 000; (○) M_w=96 000; (△) M_w=72 000; (▽) M_w=56 000; (◇) M_w=35 000; (+) M_w=33 000; (X) M_w=22 000; (✲) M_w=20 000. (——) Unfractionated feed polymer with DB=0.07; (◆) linear PMMA. Molecular weights determined by universal calibration. Data compiled from [132].

Due to the purely statistical nature of the polymerization process, hyperbranched polymers exhibit a rather high polydispersity with respect to the molecular weight and to the distribution of the branch points in the polymer. At this point, it should be stressed that the latter non-uniformity is not reflected by the degree of branching, because this parameter only deals with the *number* and not the *location* of the branch points in a polymer. Consequently, two macromolecules exhibiting an equal *DB* may differ in their architecture [114, 115]. To gather information on the location of the branch points, graph theory [116, 117] or topological indices [118–131] may be more appropriate. Thus, to explore whether a size-exclusion separation mechanism is still operative, a hyperbranched poly(methyl methacrylate) sample was fractionated by preparative size-exclusion chromatography (SEC) and compared with traces for the unfractionated polymer (feed). The Mark–Houwink plots of the different fractions and the feed polymer – established using the absolute molecular weight obtained by universal calibration and the measured intrinsic viscosities in each slice – are depicted in Fig. 7.5.

Although all Mark–Houwink curves of fractionated samples are nearly identical, the exponents are lower for most fractions ($0.24 \leq a \leq 0.42$) than those determined for the feed polymer (a=0.40). This can be attributed to an artifact due to axial dispersion of these narrowly distributed fractions. The observation that a Mark–Houwink exponent of $a = 0.40 \pm 0.02$ was rendered from the average in-

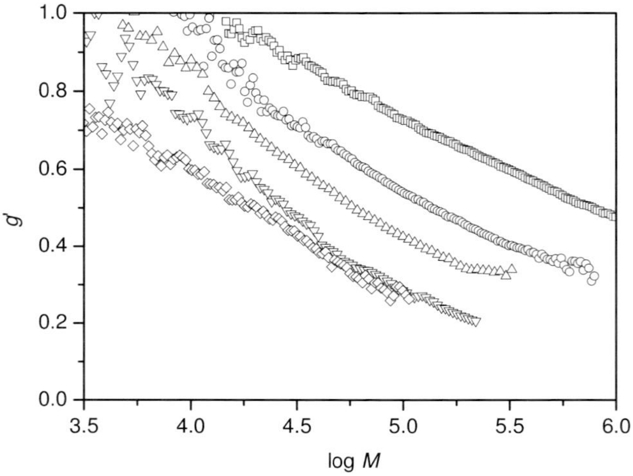

Fig. 7.6 Contraction factors, g', as a function of the molar mass, M, and the comonomer ratio, γ, for highly branched poly(methyl methacrylate). Comonomer ratios: (◇) $\gamma=5.2$; (▽) $\gamma=9.8$; (△) $\gamma=26.0$; (○) $\gamma=46.8$; (□) $\gamma=86.5$. Data compiled from [45].

trinsic viscosities of each fraction and their viscosity-average molecular weight, M_η, corroborated this hypothesis, which was also observed for the R_g–M dependences obtained by SEC–light scattering [112].

The universal calibration method also allows direct comparison with linear analogs to yield the contraction factor g' according to Eq. (12). The contraction factors – depicted in Fig. 7.6 as a function of the comonomer ratio, γ – decrease with increasing molar mass. At a given mass, the samples with a lower γ also exhibit lower g' [45], indicating that the relative density of the hyperbranched polymer increases with the molar mass and the inverse comonomer ratio.

The effect of the degree of branching, DB, on the Mark–Houwink exponent, a, is highlighted for ether imides [133, 134], methacrylates [39, 45] and polyethylene in Fig. 7.7.

It is striking that even a small number of branch points significantly alters the structure, which is reflected in the Mark–Houwink exponent. According to Fig. 7.7, at $\gamma=26$ (corresponding to $DB=0.07$ or alternatively 4% branch points), a is approximately 50% of the value for the linear poly(methyl methacrylate). The results suggest two distinct regimes: For a high degree of branching a varies gradually. In the low DB regime, however, a is strongly affected by variations of the comonomer ratio. For $0.1 \leq DB \leq 0.2$, an intermediate regime may be identified at which major changes from a highly branched and condensed to a more linear and open structure may occur. Moreover, the data in Fig. 7.7 suggest rather sharp than gradual property changes between highly branched and linear polymers and the existence of a critical value independent of the polymer's

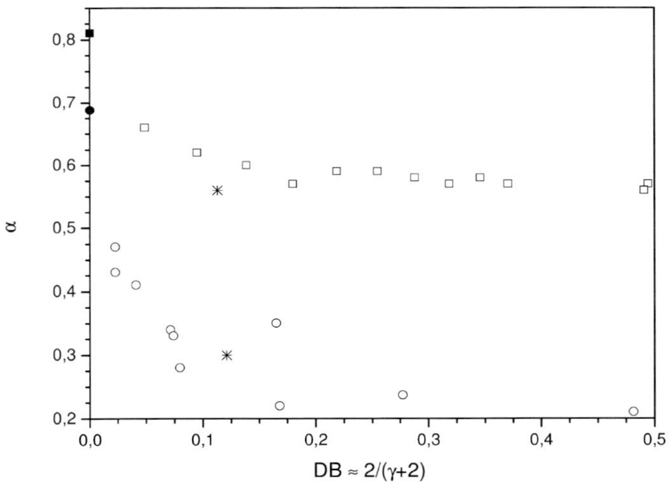

Fig. 7.7 Mark–Houwink exponents, a, as a function of the degree of branching, *DB*. (□, ○, ☆) Branched polymers; (■, ●) linear analogues. (□, ■) Aromatic poly(ether imide) copolymers in 0.05 M NMP solution at 65 °C, data according to [134]; (○, ●) poly(methyl methacrylate) in tetrahydrofuran at 30 °C, data compiled from [39, 45]; (☆) branched polyethylene in tetrahydrofuran at 40 °C [135]. *DB* was determined by NMR in the case of polyethylene or calculated according to Eq. (17) in [136] [poly(methyl methacrylate)] or Eq. (21) in [137] [poly(ether imide)], respectively.

chemistry. This hypothesis is supported by the results of a simulation [138]. Independent of the molecular weights, a transition both of the zero-shear viscosity and the radius of gyration from a perfectly branched to a linear regime occurs at a critical value of approximately 20% branch points (corresponding to $DB \approx 0.32$ or alternatively $\gamma = 4$). Even below that value, a small number of branch points in a linear architecture results in a significant reduction in size and viscosity. To fine-tune the polymer properties and to switch from a linear to a perfectly branched regime by simple variation of the fraction of linear units sounds appealing from the scientific and engineering viewpoint. "Costly" branch points are diluted by "inexpensive" monomers, which makes a copolymerization approach a very economical technique compared with the homopolymerization of AB^*- or AB_2-type monomers.

The results for branched polyethylene shown in Fig. 7.7 deserve further discussion. Although both PEs exhibit a similar degree of branching ($DB = 0.11$ and 0.12; both values calculated form the number of branches per 1000 CH_2 units, N_{1000C}, as determined by ^{13}C NMR spectroscopy), their Mark–Houwink exponents differ significantly. For the sample exhibiting $a = 0.56$ a short-chain branched topology was assigned, whereas the hyperbranched architecture of the other PE-type [87, 135, 139–142] ($a = 0.30$) was also investigated by SEC–light scattering [135, 143].

Although SEC–light scattering has proved to be a powerful tool for structure determination of linear polymers, its use for highly branched systems is aggravated due to small radii in combination with a broad polydispersity. This unfavorable combination may result in a low signal-to-noise ratio, often not permitting the determination of reasonable R_g values. Only if the refractive index increment, dn/dc, of a particular polymer–solvent combination is high enough does this technique lead to meaningful results. For example, branched carbohydrate polymers such as dextran, amylopectin and glycogen have been repeatedly investigated [144–148] using light scattering also in combination with field-flow fractionation [149, 150] as an alternative chromatographic separation technique [101, 151, 152].

For glycogen, the plot $R_g=f(M)$ exhibited two different slopes: for high molar masses, hard sphere behavior was observed with $\nu=0.3$ and $a=0$; for intermediate molar masses the correlation between light scattering and viscosity is lost and the Flory exponent increases to $\nu=0.74$; for low masses the correlation is again retained with $\nu=0.47$. The existence of an intermediate region has been attributed to an overlap of two different components in the distribution. As the radii of gyration are z-averaged, the high molar masses of a two-component mixture are more strongly weighted. In the case of viscosity averaging, the effect of high molar masses is too small to be observed [152].

The difficulties in obtaining meaningful light scattering results also limits the number of polymers for which the contraction factor g and the conversion exponent ε (cf. Eqs. 11 and 13) can be established. For hyperbranched poly[dimethyl 5-(4-hydroxybutoxy)isophthalate], a power-law dependence between the molar mass and the contraction factor was determined:

$$g' = (2.1 \pm 0.2) M_w^{-0.92} \tag{14}$$

and also a relation in accordance with Eq. (13) was established [153, 154]:

$$g' = (0.86 \pm 0.03) g^{0.26 \pm 0.02} \tag{15}$$

The conversion exponent, $\varepsilon=0.26$, is far smaller than the values determined for dextran ($\varepsilon=0.71$) [148], glycogen ($\varepsilon=2.46$ in water and $\varepsilon=2.11$ in 0.5 M sodium hydroxide solution) [152]. The discrepancy was attributed to the strong variance of the Flory–Fox parameter, Φ, on the contraction factor, g, as corroborated by a subsequent small angle neutron scattering study [154]. Smaller values of R_g and A_2 indicated less swelling in deuterated tetrahydrofuran than chloroform used in light scattering. However, the Flory exponent determined by neutron scattering, $\nu=0.4$, agreed very well with the light scattering results and indicted a more uniform density distribution ($d_f=2.5$) than non-randomly branched polymers. The large variations in the experimental values of the conversion parameter, ε, with values even higher than 3/2, indicate that Eq. (13) is not well suited for the comparison of polymers with different DB, in particular when DB changes with the molar mass.

7.4 Effect of Branching on Properties in Semi-dilute Solution

Despite the plethora of results in dilute solution, reports on the behavior of hyperbranched polymers in semi-dilute solutions are rare. Here, inter-particle interaction gains a dominant influence on the system's properties. Branch points will serve as obstacles preventing full interpenetration of segments from different chains. To describe the effects for samples with different topologies, a scaled concentration, X, may be introduced for good solvents ($A_2 > 0$) [155, 156]:

$$X = A_2 M_w c \equiv \frac{c}{c^*} \tag{16}$$

According to Eq. (16), the overlap concentration, c^*, can be calculated from the second virial coefficient, A_2, and the molar mass, M_w, both determined in dilute solution.

When $X > 1$ ($c > c^*$), an additional osmotic pressure ($\partial\pi/\partial c$) is required for the chains to close in. It is determined from the scattering intensity extrapolated to a zero angle, $R_{\theta=0}$, and leads to the dimensionless reduced osmotic modulus $M_w/M_{app}(c)$:

$$\frac{M_w}{M_{app}(c)} \equiv \left(\frac{M_w}{RT}\right)\left(\frac{\partial\pi}{\partial c}\right) = M_w\left(\frac{Kc}{R_{\theta=0}}\right) = 1 + 2A_2 M_w c + 3A_3 M_w c^2 + 4A_4 M_w c^3 + \ldots \tag{17}$$

with A_i the higher osmotic viral coefficients and K the contrast factor in light scattering. Assuming $M_w/M_{app}(c)$ to be a universal function of X, a double-logarithmic plot should lead to typical curves for each topology. These curves are universal as they do not depend on molar mass, concentration and even solvent (provided that good solvent conditions are retained) (cf. Fig. 7.8).

When hard spheres start to overlap at $X > 3$, the increase in $M_w/M_{app}(c)$ diverges [157]. In the case of coils (e.g. of a linear polymer), interpenetration starts and a transient network is formed with the entanglement points – unlike a permanent network – unfixed. Repulsion will be much weaker and no divergence will be observed. Branching points, however, will prevent full interpenetration and alter the behavior.

For the branched carbohydrates amylopectin and glycogen, master plots of the reduced osmotic modulus, M_w/M_{app}, according to Eq. (17) are reproduced in Fig. 7.8.

In Fig. 7.8, the reduced osmotic modulus follows a behavior between those of a flexible chain and a hard sphere. However, beyond a certain concentration, the curves flatten. A similar turnover is also reported for the dependence of the specific viscosity, η_{SP}, on the product of intrinsic viscosity and concentration, $c[\eta]$. For low concentrations, the specific viscosity increases with an exponent in the range 1.3–1.5. In the higher concentrated regime, the less branched amylopectin exhibits a slope of 3.5 whereas the power law behavior of the higher

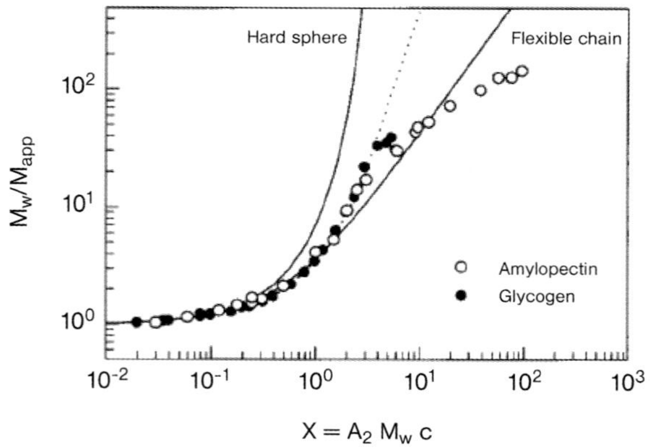

Fig. 7.8 Master plots of the reduced osmotic modulus, $M_w/M_{app}(c)$, versus the scaled concentration, X. Dotted curve: master curve for highly branched systems according Eq. (17). Reprinted with permission from [155]. Copyright 1999 American Chemical Society.

branched glycogen is reminiscent of that of hard spheres (slope 5.1) [155]. At very high concentrations, the behavior deviates from that of hard spheres due to association effects and the interpenetration of short dangling chains on the surface. Moreover, the $c[\eta]$ dependence appears to be independent of molar masses, but is influenced by the degree of branching of the macromolecules [195]. Results in the same direction are also reported for branched poly(ether imide) solutions in NMP. For $DB \geq 0.32$ (corresponding to more than 20% of linear units or $\gamma \geq 4$), the zero shear viscosity of various concentrations showed a slight increase with decreasing DB. For the less branched regime, however, a sharp rise in the zero shear viscosity as a function of the degree of branching is found [158], in analogy with the behavior observed in dilute solution (cf. Fig. 7.11).

7.5
Effect of Branching on Bulk Properties

7.5.1
Crystallinity",4,1>

Generally, the presence of branches decreases the degree of crystallinity by hampering an ordered array of the crystallizable segments. Even if crystallization is observed in branched polymers, a distinct behavior compared with their linear analogues has been reported. Branched low-density polyethylene, for instance, exhibits a rod-like crystallite growth, which is distinct from the spherulitic pattern of linear high-density polyethylene [159, 160]. Analogous effects have been found for poly(ethylene terephthalate), for which a rod-like growth process is suggested – unlike its linear analogue, which crystallizes in a spherulitic manner [161]. It is hypothesized that these distinctions are due to the impeding effect of branch points on the diffusion process during the crystallite growth [160]. For the branched poly(ethylene terephthalate), the effect of linear units on the crystallinity has been investigated in detail [161]. At low degrees of branching ($DB \leq 0.04$), crystallinity is increased; however, it will be lowered again for larger numbers of branch points. Such behavior is reminiscent of blends of poly(ethylene terephthalate) and poly(phenylene sulfide) with a liquid crystalline polyester [162, 163]. Small amounts of the polyester enhance the rate of crystallization by acting as a nucleating agent for the crystallization process.

Crystallinity of branched polymers is known for some natural polymers, e.g. amylopectin [196, 197], which has crystalline domains. In contrast, they are not found in glycogen (with higher *DB*). Many natural polymers (e.g. alginates, which are used as thickeners) do not show crystallization because the length of linear segments is too short [198].

7.5.2
Glass Transition

Studies indicate that the glass transition temperature, T_g, is dependent on the polymer backbone's structure and on the nature and number of end-functional groups. Entanglements are considered to be absent in highly branched systems and therefore do not influence the glass transition temperature [164, 165]. In principle, highly branched and hyperbranched polymers are known to exhibit lower glass transition temperatures than their linear analogues. This effect is often attributed to their end-groups, the number of which increases with the degree of branching. The decrease in glass transition temperatures may not be observed in the case of polar end-groups, as higher polarity leads to an increase in T_g. However, this effect can also be overcompensated by mobility restrictions, especially if the concentration of the end-groups is increased [164, 166–168]. Then the glass transition no longer represents the flexibility and the free volume of the repeating unit and, in consequence, the Flory–Fox relationship [99, 100, 169, 170] cannot be applied. For example, the T_g of the second-pseudo-generation of poly[2,2-bis(hydroxymethyl)propionic acid] with an ethoxylated pentaerythritol core decreases by 58 K if 63% of its hydroxyl end-groups are capped with decanoic acid. The decrease in T_g is possibly due to the decrease in hydrogen bonding, although an increase in the chain-end free volume cannot be completely ruled out [171].

The situation becomes more complicated if the amount of linear units is taken into account: As the distance between the branch points affects the internal mobility of the entire molecule, the presence of linear units will affect the glass transition temperature. The effect of the degree of branching, *DB*, on T_g is shown in Fig. 7.9, where data for copolymers of aromatic poly(ether imide) type [134], poly[(4-ethyleneoxy)benzoate] type [172], of poly[trimethylsilyl-β-(4-acetoxyphenyl)propionate] with poly(3,5-bisacetoxybenzoic acid) [173], and of AB_3-monomer triacetylgallic acid with 3-acetoxybenzoic acid [174] are compiled.

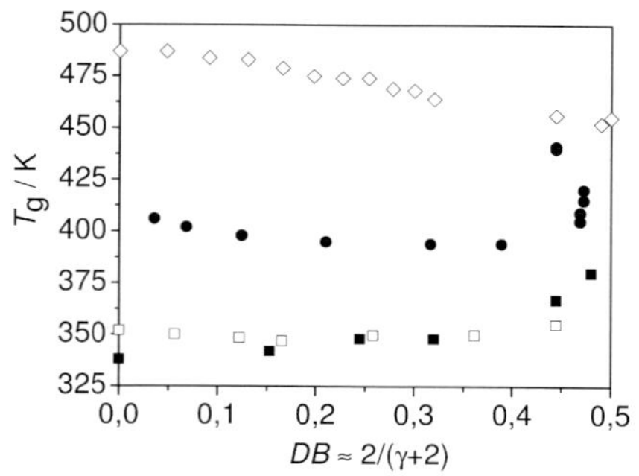

Fig. 7.9 Glass transition temperature, T_g, as a function of the degree of branching, *DB*. (◇) Aromatic poly(ether imide) copolymers according to [134]; (□) poly[(4-ethyleneoxy)-benzoate] copolymers according to [172]; (●) copolymers composed of triacetylgallic acid and 3-acetoxybenzoic acid according to [174]; (!) copolymers of poly[trimethylsilyl-β-(4-acetoxyphenyl)propionate] with poly(3,5-bisacetoxybenzoic acid) according to [173]. *DB* was calculated assuming complete conversion for (◇, □, ■) according to Eq. (21) in [137] and for (●) according to Eq. (33′) in [137], respectively.

In the case of the aromatic poly(ether imide) copolymers a constant decrease in T_g with increasing *DB* can be observed, suggesting a continuous decrease in mobility. For hyperbranched benzoates, the evolution of the glass transition temperature needs further discussion. In the case of benzoates copolymerized with poly(acetoxyphenyl propionate), T_g constantly decreases, whereas for the other samples it passes through a minimum. This may be attributed to variation in the number of end-groups, which leads to a change in the free volume. The effect is most obvious when triacetylgallic acid is used as comonomer; however, the peculiar upturn is due to the maximum *DB* at $\gamma \approx 0.645$ (cf. Eqs. (21) and (33′) in [137]) in the case of AB–AB_3 cocondensations.

Moreover, the results indicate that the glass transition temperature of branched polymers is influenced by a number of factors, with the chemical structure – especially the number of branches and the type of end-groups – being the most prominent. The Flory–Fox relation is generally followed, even for hyperbranched copolymers composed of two AB_2 monomers differing in their flexibility. This leads to the hypothesis that the alteration of the polymer topology affects the T_g to only a minor extent [175]. However, further refinements in theory are necessary to understand all effects and to account for the influence of functional groups and linear units.

7.5.3
Rheology

Unlike linear polymers, the bulk viscosity of highly and hyperbranched polymers is mainly governed by the absence of entanglements [176]. An unentangled linear chain may be described by a bead-spring model, first proposed by Rouse [177, 178]. It postulated that in the terminal flow regime the storage

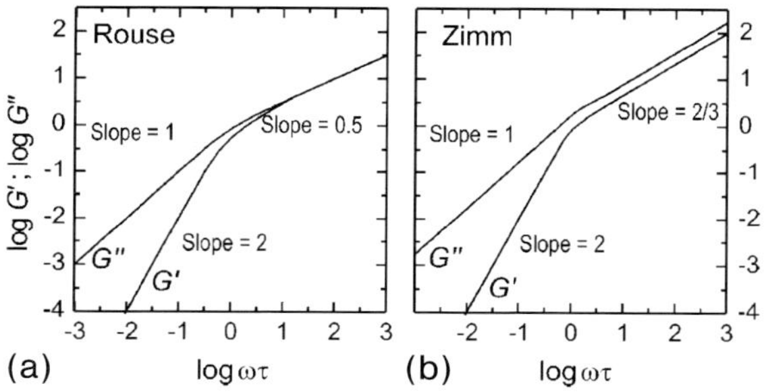

Fig. 7.10 Schematic representation of the Rouse-type (a) and the Zimm-like (b) behavior. Variations of the storage, G', and loss moduli, G'', as a function of the reduced frequency, $\omega\tau$. Representation according to [179].

modulus, G', scales with the square of the applied frequency, ω, and the loss modulus, G'', should directly be proportional to ω. Above the glass transition temperature, both moduli are equal and proportional to $\omega^{1/2}$. In contrast, linear polymers of higher molar masses are entangled and usually are described by the Zimm model [109]. Its predictions are identical with those of the Rouse model in the low-frequency domain only. At higher frequencies, however, both moduli should scale with the frequency with an exponent of 2/3 and differ by a factor of $\sqrt{3}$ (Fig. 7.10). Furthermore, in the reptation regime, the zero shear viscosity is strongly molar mass dependent, exhibiting an exponent of 3.4.

Generally, the melt dynamics of highly branched polymers are rather complex. Neither the Zimm nor the Rouse model (Fig. 7.14) is capable of describing the data quantitatively. Application of the Rouse model to branched polymers predicts the exponent in the high-frequency region to depend on the fractal dimension, d_f [180, 181]:

$$G'' \sim G'' \sim \omega^{\frac{3}{d_f+2}} \tag{18}$$

In the case of branched polyisobutenes, the zero-shear viscosity and length of the rubbery plateau depend on the number of branch points rather than the length of the branches and scales reasonably well with the total molecular weight [182, 183]. For branched polystyrenes of low comonomer ratio, $\gamma=5$, both moduli rise in a nearly parallel fashion, exhibiting scaling exponents of 0.77 for G' and 0.73 for G'', reminiscent of the Zimm model [106]. When $\gamma=200$, the data are reminiscent of Rouse dynamics: The terminal regime is followed by a region in which G' and G'' are nearly equal, exhibiting power-law exponents of 0.47 and 0.53, respectively. Similar behavior has been reported in other highly branched systems, e.g. comb polymers [184], while other gel and hyperbranched systems exhibit a lower exponent of approximately 0.4 [132, 185].

For highly branched polymers with polar end-groups, the elastic melt viscosity decreases strongly with increasing frequency [186–188]. Thus, at low frequen-

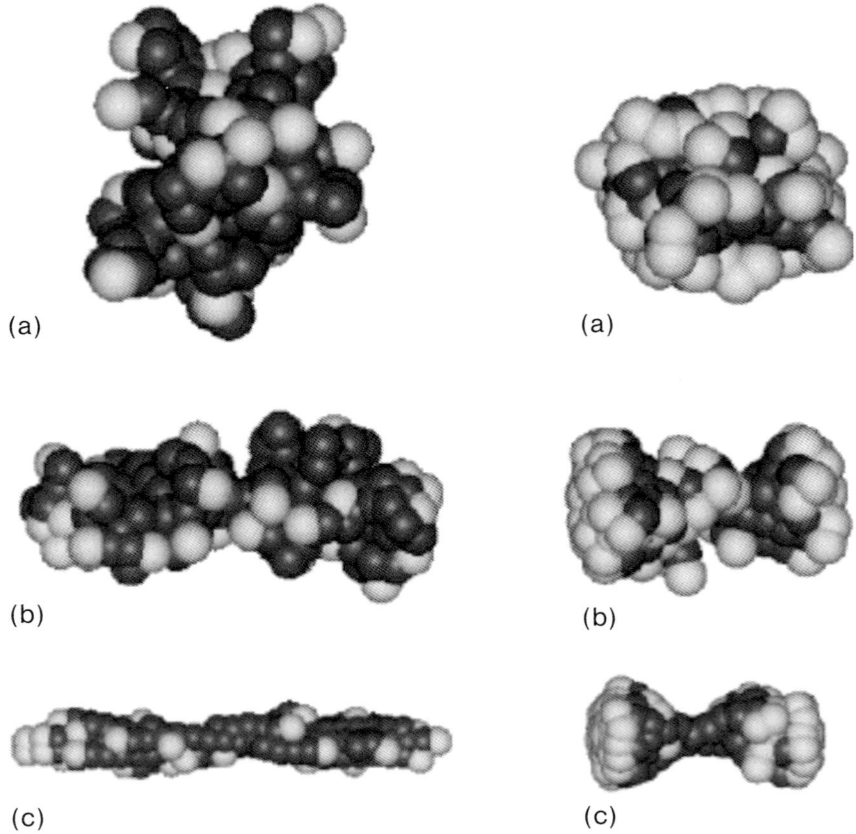

Fig. 7.11 Hyperbranched (left column, $DB=0.4$) and dendritic (right column, $DB=1$) polymers with identical number of repeat units ($N=190$) under different elongational flow rates. Light beads represent terminal monomer units and in the case of the dendrimer the monomers of the last generation. Elongational flow-rates: row (a), before the coil–stretch transition; row (b), within the transition region; and row (c), close to state of maximum extension. Reprinted with permission from [131]. Copyright 2004 American Chemical Society.

cies they might exceed the melt viscosity of even high molar mass but viscous linear polymers. As soon as the polar end-groups are modified with non-polar end-groups, the melt viscosity is drastically reduced, indicating the pronounced influence of polarity on the bulk properties (cf. also the effect on glass transition, see above).

The response to elongational flows of hyperbranched polymers with trifunctional branch points has recently been simulated [131]. For hyperbranched polymers, a sharp transition from a coil to a stretched state at a critical elongational flow-rate, the so-called coil–stretch transition, is sketched in Fig. 7.11. The first row shows the snapshots of a hyperbranched and a dendritic topology before the coil–stretch transition. Both molecules exhibit a rather globular shape, i.e. they orient as a whole along the flow axis without significant deformation. At higher flow-rates near the transition point, the molecules become oriented along the flow axis and their anisotropy increases. Further increase in flow-rate (last row) close to the state of maximum extension leads to a scaling of R_g with the number of the hyperbranched polymer's repeat units at exponents, ν, ranging from 0.44 to 0.55, which are considerably smaller than the $\nu=1.5$ exponent reported for linear polymers.

Reports on the bulk behavior of highly branched polyethylene are rare in the literature. Recently, the rheology and the thermal properties of these types of macromolecules have been investigated. The highly branched samples exhibit typical Newtonian flow behavior [189, 190] and follow the Rouse model with $G' \sim \omega^2$ and $G'' \sim \omega$ dependences in the terminal zone [143, 189, 190]. Short-chain branched polyethylenes on contrary show typical non-Newtonian shear thinning – both in steady shear and in dynamic oscillation measurements. Their master curves also exhibited a behavior very close to the predictions of the Rouse model in the terminal zone. However, at higher frequencies a rubbery plateau was obvious, indicating the elastic character of the polymer melt due to chain entanglements (average entanglement molar mass, $M_E \approx 2000$) [143, 189, 190]. The topological change from a highly branched to a more linear architecture increases the zero-shear viscosity by more than six orders of magnitude.

A major drawback is the broad molecular weight distribution of hyperbranched polymers samples. Thus, highly branched PMMA ($DB=0.074$) was fractionated into fractions with narrow molecular weight distributions [132]. In contrast to the linear sample (Fig. 7.12a), for most of the hyperbranched fractions the dynamic moduli G' and G'' are nearly equal over several decades of frequency (Fig. 7.16b). Two different relaxation processes can be identified in Fig. 7.12: (i) at high frequencies, ω, a segmental relaxation which separates the glassy state of the polymer from the state with local segmental mobility (relaxation time τ_S) and (ii) at low frequencies the chain relaxation separating the rubbery plateau corresponding to the frequency range of internal relaxations of entangled chains from the chain-flow range (relaxation time τ_C). The molecular weight dependence of the normalized relaxation times, τ_C/τ_S, obeys a power law with an exponent of 2.61 – distinctly lower than that typical for linear polymers of 3.4 (Fig. 7.13). Note that normalization is necessary to separate off the effect of end-group mobility. It has been hypothesized that the branched structure causes a compact but highly deformable structure with strongly limited interpenetration of neighboring molecules and the absence of entanglements. This is also corroborated by the complex viscosity, which scales with a frequency with a power law exponent of –0.54, reminiscent of a microgel or near-critical gel [185, 191] – a behavior independently observed for highly branched polystyrenes [192].

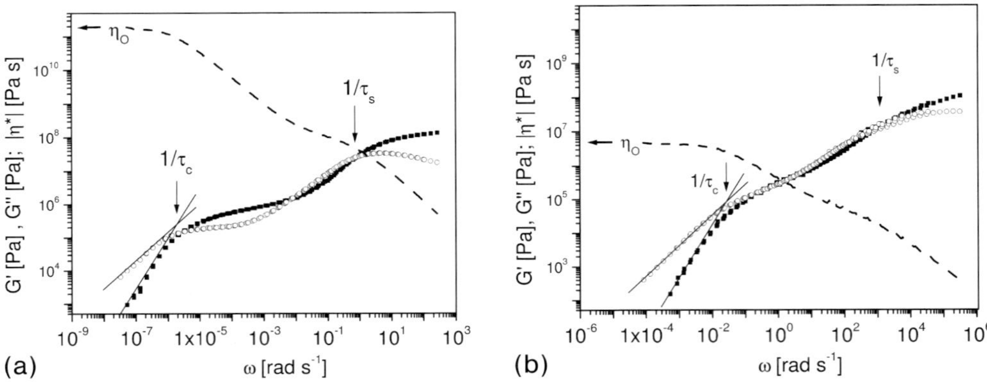

Fig. 7.12 (a) Storage and loss moduli G' (■), G'' (○) and melt viscosity (- - -) of linear PMMA (M_w=94 200) and (b) a fraction of highly branched PMMA (DB=0.074, M_w=96 000) as a function of shear frequency at a reference temperature of 130 °C. Reprinted with permission from [132]. Copyright 2001 American Chemical Society.

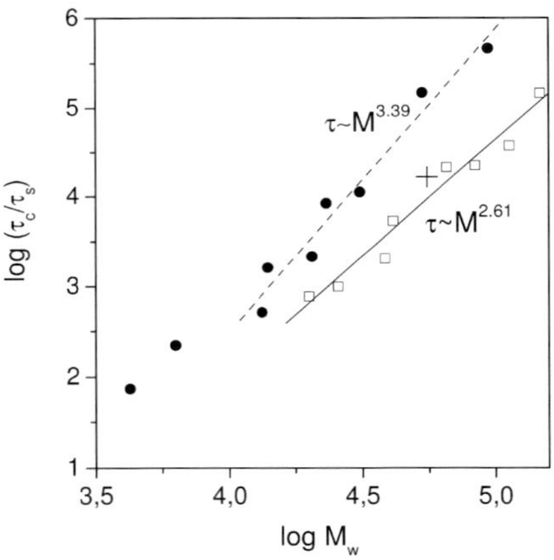

Fig. 7.13 Molecular weight dependences of the normalized chain relaxation time, τ_C/τ_S, for linear PMMA (●), branched PMMA fractions (□) and branched feed polymer (+). Reprinted with permission from [132]. Copyright 2001 American Chemical Society.

7.6 Conclusion

Randomly branched polymers have different solution and bulk properties than the linear analogues. Highly and hyperbranched polymers behave similarly to dendrimers, despite their intermediate degree of branching. The results obtained in dilute solutions and in the melt indicate highly compact molecular conformations with little or no entanglements and a strong influence of physical and rheological properties depending on the chemical nature of the end-groups [171].

Acknowledgement

The authors wish to thank Professor Walther Burchard for many helpful discussions.

References

1 Jikei, M., Kakimoto, M. *Prog. Polym. Sci.* **2001**, *26*, 1233.
2 Voit, B. *J. Polym. Sci., Part A: Polym. Chem.* **2000**, *38*, 2505.
3 Ambade, A.V., Kumar, A. *Prog. Polym. Sci.* **2000**, *25*, 1141.
4 Sunder, A., Heinemann, J., Frey, H. *Chem. Eur. J.* **2000**, *6*, 2499.
5 Hawker, C.J., Lee, R., Fréchet, J.M.J. *J. Am. Chem. Soc.* **1991**, *113*, 4583.
6 Kim, Y.H. *Macromol. Symp.* **1994**, *77*, 21.
7 Hölter, D., Burgath, A., Frey, H. *Acta Polym.* **1997**, *48*, 30.
8 Yan, D., Müller, A.H.E., Matyjaszewski, K. *Macromolecules* **1997**, *30*, 7024.
9 Teertstra, S.J., Gauthier, M. *Prog. Polym. Sci.* **2004**, *29*, 277.
10 Muchtar, Z., Schappacher, M., Deffieux, A. *Macromolecules* **2001**, *34*, 7595.
11 Waigh, T.A., Papagiannopoulos, A., Voice, A., Bansil, R., Unwin, A.P., Dewhurst, C. *Langmuir* **2002**, *18*, 7188.
12 Brenner, A.R., Schmaljohann, D., Voit, B.I., Wolf, D. *Macromol. Symp.* **1997**, *122*, 217.
13 Hult, A., Johansson, M., Malmström, E. *Adv. Polym. Sci.* **1999**, *143*, 1.
14 Malmström, E., Hult, A. *J. Macromol. Sci., Rev. Macromol. Chem. Phys.* **1997**, *C37*, 555.
15 Flory, P.J. *J. Am. Chem. Soc.* **1952**, *74*, 2718.
16 Flory, P.J. *Principles of Polymer Chemistry*, Cornell University Press, Ithaca, NY, **1953**.
17 Hunter, W.H., Woollett, G.H. *J. Am. Chem. Soc.* **1921**, *43*, 135.
18 Jacobson, R.A. *J. Am. Chem. Soc.* **1932**, *54*, 1513.
19 Kricheldorf, H.R., Zang, Q.-Z., Schwarz, G. *Polymer* **1982**, *23*, 1821.
20 Kim, Y.H., Webster, O.W. *J. Am. Chem. Soc.* **1990**, *112*, 4592–4593; Kim, Y.H., Webster, O.W. *Macromolecules* **1992**, *25*, 5561.
21 Kumar, A., Ramakrishnan, S. *J. Polym. Sci., Part A: Polym. Chem.* **1996**, *34*, 839.
22 Drohmann, C., Möller, M., Gorbatsevich, O.B., Muzafarov, A.M. *J. Polym. Sci., Part A: Polym. Chem.* **2000**, *38*, 741.
23 Morgenroth, F., Müllen, K. *Tetrahedron* **1997**, *53*, 15349.
24 Berresheim, A.J., Müller, M., Müllen, K. *Chem. Rev.* **1999**, *99*, 1747.
25 Hobson, L.J., Feast, W.J. *Polymer* **1999**, *40*, 1279.
26 Yan, D., Müller, A.H.E., Matyjaszewski, K. *Macromolecules* **1997**, *30*, 7024.
27 Hölter, D., Frey, H. *Acta Polym.* **1997**, *48*, 298.
28 Bharathi, P., Moore, J.S. *J. Am. Chem. Soc.* **1997**, *119*, 3391.
29 Bharathi, P., Moore, J.S. *Macromolecules* **2000**, *33*, 3212.
30 Radke, W., Litvinenko, G.I., Müller, A.H.E. *Macromolecules* **1998**, *31*, 239.
31 Yan, D., Zhou, Z., Müller, A.H.E. *Macromolecules* **1999**, *32*, 245.
32 Hanselmann, R., Hölter, D., Frey, H. *Macromolecules* **1998**, *31*, 3790.
33 Sunder, A., Hanselmann, R., Frey, H., Mülhaupt, R. *Macromolecules* **1999**, *32*, 4240.
34 Gao, C., Yan, D. *Prog. Polym. Sci.* **2004**, *29*, 183.
35 Fréchet, J.M.J., Henmi, M., Gitsov, I., Aoshima, S., Leduc, M.R., Grubbs, R.B. *Science* **1995**, *269*, 1080.
36 Hawker, C.J., Fréchet, J.M.J., Grubbs, R.B., Dao, J. *J. Am. Chem. Soc.* **1995**, *117*, 10763.
37 Weimer, M.W., Fréchet, J.M.J., Gitsov, I. *J. Polym. Sci., Part A: Polym. Chem.* **1998**, *36*, 955.
38 Baskaran, D. *Macromol. Chem. Phys.* **2001**, *202*, 1569.
39 Simon, P.F.W., Radke, W., Müller, A.H.E. *Makromol. Chem., Rapid Commun.* **1997**, *18*, 865.

40 Matyjaszewski, K., Gaynor, S. G., Kulfan, A., Podwika, M. *Macromolecules* **1997**, *30*, 5192.
41 Matyjaszewski, K., Gaynor, S. G., Müller, A. H. E. *Macromolecules* **1997**, *30*, 7034.
42 Matyjaszewski, K., Gaynor, S. G. *Macromolecules* **1997**, *30*, 7042.
43 Fréchet, J. M. J., Aoshima, S. *Patent WO9614345*, **1996**.
44 Gaynor, S. G., Edelman, S., Matyjaszewski, K. *Macromolecules* **1996**, *29*, 1079.
45 Simon, P. F. W., Müller, A. H. E. *Macromolecules* **2001**, *34*, 6206.
46 Paulo, C., Puskas, J. E. *Macromolecules* **2001**, *34*, 734.
47 Hong, C.-Y., Pan, C.-Y. *Polym. Int.* **2002**, *51*, 785.
48 Mori, H., Chan Seng, D., Lechner, H., Zhang, M., Müller, A. H. E. *Macromolecules* **2002**, *35*, 9270.
49 Mori, H., Müller, A. H. E. *Top. Curr. Chem.* **2003**, *228*, 1.
50 Sunder, A., Mülhaupt, R., Haag, R., Frey, H. *Adv. Mater.* **2000**, *12*, 235.
51 Sunder, A., Bauer, T., Mülhaupt, R., Frey, H. *Macromolecules* **2000**, *33*, 1330.
52 Kennedy, J. P., Frisch, K. C., Jr. *US Patent 4327201*, **1982**.
53 Nuyken, O., Gruber, F., Pask, S. D., Riederer, A., Walter, M. *Makromol. Chem.* **1993**, *194*, 3415.
54 Jones, G. D., Runyon, J. R., Ong, J. *J. Appl. Polym. Sci.* **1961**, *5*, 452.
55 Powers, K. W., Kuntz, I. (Exxon Research and Engineering) *US Patent 4074035*, **1978**.
56 Müller, A. H. E., Yan, D., Wulkow, M. *Macromolecules* **1997**, *30*, 7015.
57 Puskas, J. E., Grasmüller, M. *Macromol. Symp.* **1998**, *132*, 117.
58 Yoo, S. H., Yoon, T. H., Jho, J. Y. *Macromol. Rapid Commun.* **2001**, *22*, 1319.
59 Yoo, S. H., Lee, J. H., Lee, J.-C., Jho, J. Y. *Macromolecules* **2002**, *35*, 1146.
60 Bibiao, J., Yang, Y., Jian, D., Shiyang, F., Rongqi, Z., Jianjun, H., Wenyun, W. *J. Appl. Polym. Sci.* **2002**, *83*, 2114.
61 Hong, C. Y., Pan, C. Y. *Polymer* **2001**, *42*, 9385.
62 Cheng, G., Simon, P. F. W., Hartenstein, M., Müller, A. H. E. *Macromol. Rapid Commun.* **2000**, *21*, 846.
63 Matyjaszewski, K., Gaynor, S. G., Kulfan, A., Podwika, M. *Macromolecules* **1997**, *30*, 5192.
64 Matyjaszewski, K., Pyun, J., Gaynor, S. G. *Macromol. Rapid Commun.* **1998**, *19*, 665.
65 Mori, H., Böker, A., Krausch, G., Müller, A. H. E. *Macromolecules* **2001**, *34*, 6871.
66 Sakamoto, K., Aimiya, T., Kira, M. *Chem. Lett.* **1997**, 1245.
67 Ishizu, K., Ohta, Y., Kawauchi, S. *Macromolecules* **2002**, *35*, 3781.
68 Ishizu, K., Shibuya, T., Mori, A. *Polym. Int.* **2002**, *51*, 4248.
69 Lu, P., Paulasaari, J., Weber, W. P. *Macromolecules* **1996**, *29*, 8583.
70 Zhang, H., Ruckenstein, E. *Polym. Bull. (Berlin)* **1997**, *39*, 399.
71 Rizzardo, E., Chiefari, J., Mayadunne, R., Moad, G., Thang, S. *Macromol. Symp.* **2001**, *174*, 209.
72 Carter, S., Rimmer, S., Sturdy, A., Webb, M. *Macromol. Biosci.* **2005**, *5*, 373.
73 Mori, H., Müller, A. H. E. *Prog. Polym. Sci.* **2003**, *28*, 1403.
74 Hong, C. Y., Pan, C. Y. *Polym. Int.* **2002**, *51*, 785.
75 Mori, H., Walther, A., Andre, X., Lanzendorfer, M. G., Muller, A. H. E. *Macromolecules* **2004**, *37*, 2054.
76 Muthukrishnan, S., Jutz, G., André, X., Mori, H., Müller, A. H. E. *Macromolecules* **2005**, *38*, 9.
77 Muthukrishnan, S., Mori, H., Müller, A. H. E. *Macromolecules* **2005**, *38*, 3108.
78 Hazer, B. *Makromol. Chem.* **1992**, *193*, 1081.
79 Ishizu, K., Mori, A. *Polym. Int.* **2002**, *51*, 50.
80 Ishizu, K., Mori, A., Shibuya, T. *Designed Monom. Polym.* **2002**, *5*, 1.
81 Knauss, D. M., Al-Muallem, H. A., Huang, T., Wu, D. T. *Macromolecules* **2000**, *33*, 3557.
82 Liu, B. L., Kazlauciunas, A., Guthrie, J. T., Perrier, S. *Macromolecules* **2005**, *38*, 2131.
83 Li, Y., Armes, S. P. *Macromolecules* **2005**, *38*, 8155.
84 Isaure, F., Cormack, P. A. G., Sherrington, D. C. *J. Mater. Chem.* **2003**, *13*, 2701.
85 Isaure, F., Cormack, P. A. G., Sherrington, D. C. *Macromolecules* **2004**, *37*, 2096.
86 Guan, Z. *J. Am. Chem. Soc.* **2002**, *124*, 5616.

87 Guan, Z. *Chem. Eur. J.* **2002**, *8*, 3086.

88 Suzuki, M., Ii, A., Saegusa, T. *Macromolecules* **1992**, *25*, 7071.

89 Magnusson, H., Malmström, E., Hult, A. *Macromol. Rapid Commun.* **1999**, *20*, 453.

90 Liu, M., Vladimirov, N., Fréchet, J. M. J. *Macromolecues* **1999**, *32*, 6881.

91 Trollsås, M., Löwenhielm, P., Lee, V. Y., Möller, M., Miller, R. D., Hedrick, J. L. *Macromolecues* **1999**, *32*, 9062.

92 Chang, H.-T., Fréchet, J. M. J. *J. Am. Chem. Soc.* **1999**, *121*, 2313.

93 Tomalia, D. A., Hedstrand, D. M., Ferritto, M. S. *Macromolecules* **1991**, *24*, 1435.

94 Schappacher, M., Billaud, C., Paulo, C., Deffieux, A. *Macromol. Chem. Phys.* **1999**, *200*, 2377.

95 Deffieux, A., Schappacher, M. *Macromolecules* **1999**, *32*, 1797.

96 Hempenius, M. A., Michelberger, W., Möller, M. *Macromolecues* **1997**, *30*, 5602.

97 Kee, R. A., Gauthier, M. *Macromolecules* **1999**, *32*, 6478.

98 Kee, R. A., Gauthier, M. *Macromolecules* **2002**, *35*, 6526.

99 Fox, T. G. J., Flory, J. P. *J. Appl. Phys.* **1950**, *21*, 581.

100 Fox, T. G., Flory, P. J. *J. Polym. Sci.* **1954**, *14*, 315.

101 Hanselmann, R., Burchard, W., Ehrat, M., Widmer, H. M. *Macromolecules* **1996**, *29*, 3277.

102 Isaacson, J., Lubensky, T. C. *J. Phys. Lett.* **1980**, *41*, L469.

103 Muthukumar, M. *J. Chem. Phys.* **1985**, *83*, 3161.

104 de Gennes, P.-G. *Scaling Concepts in Polymer Physics*, 3rd edn, Cornell University Press, Ithaca, NY, **1988**.

105 Burchard, W. *Adv. Polym. Sci.* **1999**, *143*, 113.

106 Dorgan, J. R., Knauss, D. M., Al-Muallem, H. A., Huang, T. Z., Vlassopoulos, D. *Macromolecules* **2003**, *36*, 380.

107 Zimm, B. H., Stockmayer, W. H. *J. Chem. Phys.* **1949**, *17*, 1301.

108 Stockmayer, W. H., Fixman, M. *Ann. N. Y. Acad. Sci.* **1953**, *57*, 334.

109 Zimm, B. H., Kilb, R. W. *J. Polym. Sci.* **1959**, *37*, 19.

110 Grubisic, Z., Rempp, P., Benoît, H. *J. Polym. Sci., Part B* **1967**, *5*, 755.

111 Benoît, H., Grubisic, Z., Rempp, P., Decker, D., Zilliox, J. G. *J. Chem. Phys.* **1966**, *63*, 1507.

112 Wyatt, P. J. *Anal. Chim. Acta* **1993**, *272*, 1.

113 Wyatt, P. J. *J. Chromatogr. A* **1993**, *648*, 27.

114 Hobson, L. J., Feast, W. J. *Chem. Commun.* **1997**, 2067.

115 Maier, G., Zech, C., Voit, B., Komber, H. *Macromol. Chem. Phys.* **1998**, *199*, 2655.

116 Harary, F. *Graph Theory*, Addison-Wesley, Reading, MA, **1969**.

117 Trinajstiæ, N. *Chemical Graph Theory*, 2nd edn, CRC Press, Boca Raton, FL, **1992**.

118 Wiener, H. *J. Am. Chem. Soc.* **1947**, *69*, 17.

119 Wiener, H. *J. Phys. Chem.* **1948**, *52*, 1082.

120 Dobson, G. R., Gordon, M. *J. Chem. Phys.* **1964**, *41*, 2389.

121 Dobson, G. R., Gordon, M. *J. Chem. Phys.* **1965**, *43*, 705.

122 Klein, D. J., Seitz, W. A. In *Chemical Applications of Topology and Graph Theory: a Collection of Papers from a Symposium Held at the University of Georgia, Athens, Georgia, USA, 18–22 April 1983*, King, R. B. (Ed.), Elsevier, Amsterdam, **1983**, p. 430.

123 Mekenyan, O., Dimitrov, S., Bonchev, D. *Eur. Polym. J.* **1983**, *19*, 1185.

124 Bertz, S. H., Herndon, W. C. *ACS Symp. Ser.* **1986**, *306*, 169.

125 Bonchev, D., Mekenyan, O., Kamenska, V. *J. Math. Chem.* **1992**, *11*, 107.

126 Bicerano, J. *Prediction of Polymer Properties*, 2nd edn., Marcel Dekker, New York, **1996**.

127 Bertz, S. H., Sommer, T. J. *Chem. Commun.* **1997**, *24*, 2409.

128 Balaban, T. S., Balaban, A. T., Bonchev, D. *J. Mol. Struct. Theochem* **2001**, *535*, 81.

129 Bonchev, D. *J. Chem. Inf. Comput. Sci.* **2000**, *40*, 934.

130 Bonchev, D., Dekmezian, A. H., Markel, E., Faldi, A. *J. Appl. Polym. Sci.* **2003**, *90*, 2648.

131 Neelov, I. M., Adolf, D. B. *J. Phys. Chem. B* **2004**, *108*, 7627.

132 Simon, P. F. W., Müller, A. H. E., Pakula, T. *Macromolecules* **2001**, *34*, 1677.
133 Thompson, D. S., Markoski, L. J., Moore, J. S., Sendijarevic, I., Lee, A., McHugh, A. J. *Macromolecules* **2000**, *33*, 6412.
134 Markoski, L. J., Moore, J. S., Sendijarevic, I., McHugh, A. J. *Macromolecules* **2001**, *34*, 2695.
135 Cotts, P. M., Guan, Z., McCord, E., McLain, S. *Macromolecules* **2000**, *33*, 6945.
136 Litvinenko, G. I., Simon, P. F. W., Müller, A. H. E. *Macromolecules* **1999**, *32*, 2410.
137 Frey, H., Hölter, D. *Acta Polym.* **1999**, *50*, 67.
138 Lee, A. T., McHugh, A. J. *Macromolecules* **2001**, *34*, 7127.
139 Johnson, L. K., Killian, C. M., Brookhart, M. *J. Am. Chem. Soc.* **1995**, *117*, 6414.
140 Guan, Z., Cotts, P. M., McCord, E. F., McLain, S. J. *Science* **1999**, *283*, 2059.
141 Plentz Meneghetti, S., Kress, J., Lutz, P. J. *Macromol. Chem. Phys.* **2000**, *201*, 1823.
142 Ittel, S. D., Johnson, L. K., Brookhart, M. *Chem. Rev.* **2000**, *100*, 1169.
143 Patil, R., Colby, R. H., Read, D. J., Chen, G., Guan, Z. *Macromolecules* **2005**, *38*, 10571.
144 Senti, F. R., Hellman, N. N., Ludwig, N. H., Babcock, N. H., Tobin, R., Glass, C. A., Lamberts, B. L. *J. Polym. Sci.* **1955**, *17*, 527.
145 Smit, J. A. M., van Dijk, J. A. P. P., Mennen, M. G., Daoud, M. *Macromolecules* **1992**, *25*, 3585.
146 Nordmeier, E. *J. Phys. Chem.* **1993**, *97*, 5770.
147 Ioan, C. E., Aberle, T., Burchard, W. *Macromolecules* **2000**, *33*, 5730.
148 Ioan, C. E., Aberle, T., Burchard, W. *Macromolecules* **2001**, *34*, 3765.
149 Giddings, J. C. *J. Chem. Phys.* **1968**, *49*, 81.
150 Cölfen, H., Antonietti, M. *Adv. Polym. Sci.* **2000**, *150*, 67.
151 Galinsky, G., Burchard, W. *Macromolecules* **1996**, *29*, 1498.
152 Ioan, C. E., Aberle, T., Burchard, W. *Macromolecules* **1999**, *32*, 7444.
153 De Luca, E., Richards, R. W. *J. Polym. Sci., Part B: Polym. Phys.* **2003**, *41*, 1339.
154 De Luca, E., Richards, R. W., Grillo, I., King, S. M. *J. Polym. Sci., Part B: Polym. Phys.* **2003**, *41*, 1352.
155 Ioan, C. E., Aberle, T., Burchard, W. *Macromolecules* **1999**, *32*, 8655.
156 Ioan, C. E., Burchard, W. *Macromolecules* **2001**, *34*, 326.
157 Carnahan, N. F., Starling, K. E. *J. Chem. Phys.* **1969**, *51*, 635.
158 Sendijarevic, I., Liberatore, M. W., McHugh, A. J., Markoski, L. J., Moore, J. S. *J. Rheol.* **2001**, *45*, 1245.
159 Maderek, E., Strobl, G. R. *Colloid Polym. Sci.* **1983**, *261*, 471.
160 Strobl, G. R., Engelke, T., Maderek, E., Urban, G. *Polymer* **1983**, *24*, 1585.
161 Jayakannan, M., Ramakrishnan, S. *J. Appl. Polym. Sci.* **1999**, *74*, 59.
162 Ou, C. F., Lin, C. C. *J. Appl. Polym. Sci.* **1994**, *54*, 1223.
163 Gopakumar, T. G., Ghadage, R. S., Ponrathnam, S., Rajan, C. R., Fradet, A. *Polymer* **1997**, *38*, 2209.
164 Wooley, K. L., Hawker, C. J., Pochan, J. M., Fréchet, J. M. J. *Macromolecules* **1993**, *26*, 1514.
165 Stutz, H. *J. Polym. Sci., Part B: Polym. Phys.* **1995**, *33*, 333.
166 Kim, Y. H., Beckerbauer, R. *Macromolecules* **1994**, *27*, 1968.
167 Malmström, E., Johansson, M., Hult, A. *Macromol. Chem. Phys.* **1996**, *197*, 3199.
168 Brenner, A. R., Voit, B. I., Massa, D. J., Turner, S. R. *Macromol. Symp.* **1996**, *102*, 47–54.
169 Cown, J. M. G. *Eur. Polym. J.* **1975**, *11*, 297.
170 Tadlaoui, K., Pietrasanta, Y., Michel, A., Verney, V. *Polymer* **1991**, *32*, 2234.
171 Luciani, A., Plummer, C. J. G., Nguyen, T., Garamszegi, L., Manson, J. A. E. *J. Polym. Sci., Part A: Polym. Chem.* **2004**, *42*, 1218–1225.
172 Jayakannan, M., Ramakrishnan, S. *J. Polym. Sci., Part A: Polym. Chem.* **2000**, *38*, 261–268.
173 Kricheldorf, H. R., Stukenbrock, T. *Polymer* **1997**, *38*, 3373–3383.
174 Kricheldorf, H. R., Stukenbrock, T. *J. Polym. Sci., Part A: Polym. Chem.* **1998**, *36*, 2347–2357.
175 Behera, G. C., Saha, A., Ramakrishnan, S. *Macromolecules* **2005**, *38*, 7695.
176 McLeish, T. C. B., Milner, S. C. *Adv. Polym. Sci.* **1999**, *143*, 195.

177 Rouse, P. E. *J. Chem. Phys.* **1953**, *21*, 1272.

178 King, D. H., James, D. F. *J. Chem. Phys.* **1983**, *78*, 7473.

179 Ferry, J. D. *Viscoleastic Properties of Polymers*, 3rd edn, Wiley, New York, **1980**.

180 Rubinstein, M., Colby, R. H. *Polymer Physics*, Oxford University Press, New York, **2003**.

181 Lusignan, C. P., Mourey, T. H., Wilson, J. C., Colby, R. H. *Phys. Rev. E* **1995**, *52*, 6271.

182 Robertson, C. G., Roland, C. M., Puskas, J. E. *J. Rheol.* **2002**, *46*, 307.

183 Robertson, C. G., Roland, C. M., Paulo, C., Puskas, J. E. *J. Rheol.* **2001**, *45*, 759.

184 Roovers, J., Graessley, W. W. *Macromolecules* **1981**, *14*, 766.

185 Antonietti, M., Pakula, T., Bremser, W. *Macromolecules* **1995**, *28*, 4227.

186 Schmaljohann, D., Häussler, L., Pötschke, P., Voit, B. I., Loontjens, T. J. A. *Macromol. Chem. Phys.* **2000**, *201*, 49.

187 Böhme, F., Clausnitzer, C., Gruber, F., Grutke, S., Huber, T., Potschke, P., Voit, B. *High Performance Polym.* **2001**, *13*, S21.

188 Suneel, B., Buzza, D. M. A., Groves, D. J., McLeish, T. C. B., Parker, D., Keeney, A. J., Feast, W. J. *Macromolecules* **2002**, *35*, 9605.

189 Ye, Z., Zhu, S. *Macromolecules* **2003**, *36*, 2194.

190 Ye, Z. B., Al Obaidi, F., Zu, S. P. *Macromol. Chem. Phys.* **2004**, *205*, 897.

191 Antonietti, M., Bremser, W., Schmidt, M. *Macromolecules* **1990**, *23*, 3796.

192 Hempenius, M. A., Zoetelief, W. F., Gauthier, M., Möller, M. *Macromolecules* **1998**, *31*, 2299.

193 Litvinenko, G. I., Simon, P. F. W., Müller, A. H. E. *Macromolecules* **2001**, *34*, 2418.

194 Weissmüller, M., Burchard, W. *Acta Polym.* **1997**, *48*, 571.

195 Burchard, W. *Biomacromolecules* **2001**, *3*, 342.

196 Guildot, A., Mercier, Ch. In *The Polysaccharides*, Vol. 3, Aspinall, G. O. (Ed.), Academic Press, New York, **1985**, p. 210.

197 Jenkins, P. J., Cameron, R. E., Donald, A. M., Bras, W., Derbyshire, G. E., Mut, G. R., Ryan, A. *J. Polym. Sci., Polym. Phys. Ed.* **1994**, *32*, 1579; Jenkins, P. J., Donald, A. M. *Int. J. Biol. Macromol.* **1995**, *17*, 315.

198 Skjak-Braek, G., Smidsrod, O., Larson, B. *Int. J. Biol. Macromol.* **1986**, *8*, 330; Skjak-Break, G. *Biochem. Soc. Trans.* **1992**, *20*, 27.

8
From Stars to Microgels

Daniel Taton

8.1
Introduction

Introduction of branching points throughout polymeric structures is a convenient means to adjust the properties of synthetic polymers. Rheological, mechanical, and solution properties of branched polymers, indeed, correlate closely to their type and degree of branching [1]. It is well-documented that branched polymers exhibit distinct properties compared to their linear counterparts, from the idea that the topology of a polymeric assembly dictates its macroscopic behavior. In the past decades, advances in polymer chemistry through the development of "controlled/living" polymerizations (CLP) [2] combined with branching reactions have made it possible to arrange polymer chains in miscellaneous branched architectures (Fig. 8.1), including star-like polymers [3–6], graft copolymers [7], hyperbranched polymers [5, 8], dendrimers [5, 9] dendrimer-like polymers [10], dendrigrafts [10] and even more complex architectures. In this regard, the advent of "controlled/living" radical polymerization (C/LRP) [11] techniques in the mid 90s has provided new and easy routes for macromolecular engineering [12]. To be utilized in specific functions/applications (generally as specialty polymeric additives in formulations), branched polymers must be accessible following relatively simple, low-cost synthetic methodologies relying on environmentally friendly processes, preferably water-borne or solvent-free ones, at moderate to high temperature. Both *star polymers* and *microgels* (Fig. 8.1) belong to the category of branched polymers. They provide an easy access to globular macromolecules and both of them are currently exploited on the industrial scale, which is not yet the case for other branched polymers that have emerged in the literature. For instance, stars and microgels are used in coatings manufacture as rheology modifiers (stars) or as additives for organic binders to decrease the drying time (microgels).

Among all branched architectures, star polymers correspond to the simplest possible arrangement of macromolecular chains in a branched structure, since stars involve only one central branching point per macromolecule. Structurally

Macromolecular Engineering. Precise Synthesis, Materials Properties, Applications.
Edited by K. Matyjaszewski, Y. Gnanou, and L. Leibler

ISBN: 978-3-527-31446-1

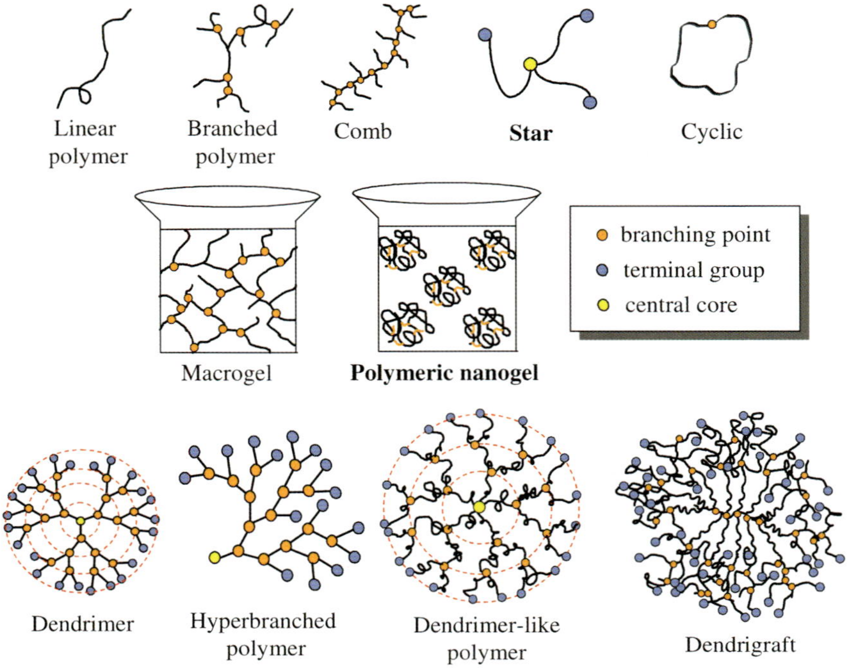

Fig. 8.1 Typical examples of branched polymers.

well-defined stars of precise functionality are quite useful in providing acute insight into how branching affects the overall properties of polymers in solution or in the melt [3, 4]. As for polymeric nanogels/ microgels [13, 14], they can be readily obtained in one-pot processes, including crosslinking copolymerization from different processes (emulsion, mini-emulsion, precipitation or dilute solution), or by inducing intramolecular crosslinking of a pre-formed linear polymer.

This chapter describes the synthetic strategies to star polymers and microgels, from traditional routes to more recent developments. For each category of structures, this chapter is organized according to the polymerization method employed with a special emphasis on the merits and drawbacks of each approach. Due to space limitation, it is not realistic to present an exhaustive list of all star polymer and microgels described so far in the literature. Many recent review articles, book chapters and highlights dealing with general and specific aspects of both stars and microgels are already available. Studies covered in this article are those described by polymer chemists over the two past decades. The objective here is to impart both the synthetic accesses to star polymers and microgels and their relevant characteristics from the structural viewpoint.

8.2 Synthesis of Multiarm Star Polymers

Star polymers are comprised of several polymer chains attached at one end to only one branching point serving as the core. Star-like polymers can be obtained with a well-defined structure and a narrow molar mass distribution from CLP techniques and are therefore well suited for investigations into the structure–property relationships of branched polymers [3, 4]. Three categories of star-shaped polymers can be distinguished (Fig. 8.2), (i) multiarm star polymers or "regular stars" possessing homopolymeric arms, (ii) star-block copolymers exhibiting a core–shell structure, where each arm of the star is a block copolymer and (iii) stars containing chemically different or size different polymeric arms called "miktoarm stars" or "heteroarm stars". The first section of this chapter will only cover the synthetic efforts devoted to the regular star-shaped polymers or multiarm star polymers mentioned above. Chapter 6 in this volume will focus on the two other categories as well as on other star-like polymer architectures.

Six distinct strategies can actually be contemplated for the synthesis of multiarm star polymers (Fig. 8.3). For instance, coupling "living" linear polymeric chains onto a multifunctional comonomer leads to star polymers composed of a microgel-like core, following a convergent ("arm-first") approach (route A). This first method has been the most intensively studied; it has the advantage of allowing the arms to be sampled out before the star formation. Alternatively, one can homopolymerize (or copolymerize) macromonomers and generate polymacromonomers that exhibit a star-type behavior, provided the degree of polymerization of the backbone is low (route B). This macromonomer method can be viewed as an arm-first approach to star polymers where the chain-end of the precursor is a polymerizable group. One can also use a divinylic comonomer to trigger its copolymerization with the macromonomer. In another "arm-first" strategy, one can use a precursor bearing multiple functions complementary to that of the living chains (route C). A fourth strategy involves the polymerization of a monomer "out of" multifunctional initiators following a divergent ("core-first") approach (route D). Next is a very recent approach that may be viewed as a two-step core-first method based on the use of multivinylic crosslinker in CLP, which leads to a microgel-type polymer, followed by chain extension from

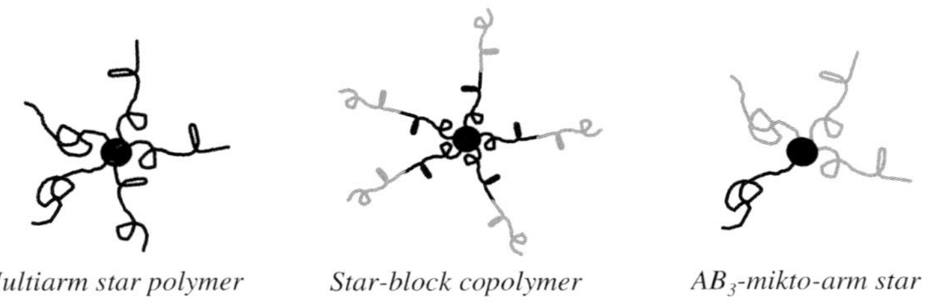

Fig. 8.2 Different categories of star polymers.

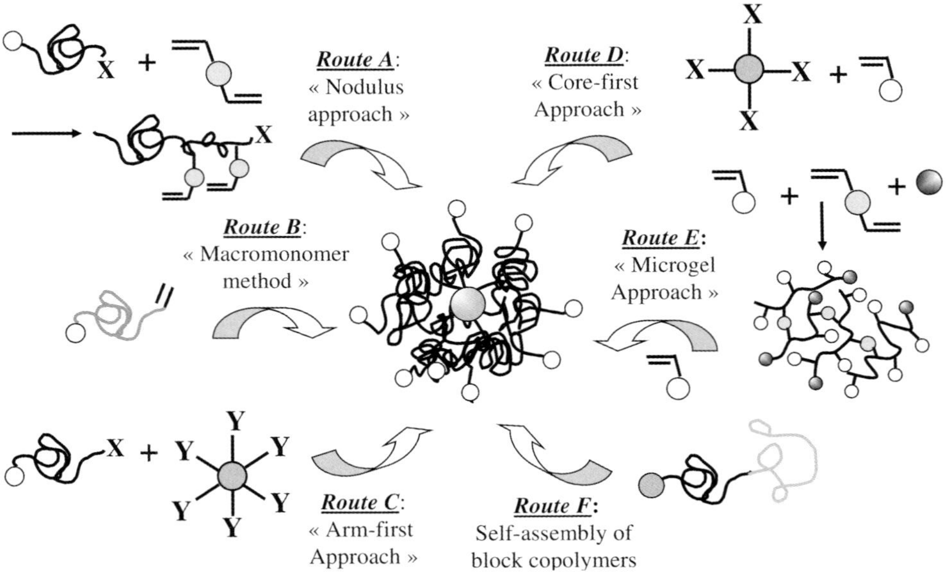

Fig. 8.3 Different synthetic strategies to multiarm star polymers.

the multiple active centers present in the parent polymer (route E). Finally, self-assembly of block copolymers in a selective solvent can be considered as a supramolecular physico-chemical approach to star-like polymeric micelles (route F), but the latter method will not be covered in this chapter.

8.2.1
By the "Nodulus" Approach

Addition of a multivinylic monomer as a crosslinking/coupling agent onto "living" or re-activable linear chains is an arm-first synthetic route to star polymers, sometimes called the "nodulus approach" known since the early 1960s (route A, Fig. 8.3). Figure 8.4 shows the most representative crosslinkers employed in this arm-first strategy. First developed in the context of anionic polymerization [15, 16], such an approach has been recently applied to other CLP techniques including radical ones.

Upon addition of a crosslinker to a solution containing a "living" or a dormant polymer, one can grow a short block carrying pendant double bonds. In route A of Fig. 8.3, the X moiety is either an active center (carbanion, carbocation or carbon-centered radical) or a "pseudo-active center" corresponding to a dormant form of the growing chains capable of being re-activated. Star-like polymers consisting of a microgel-type core (nucleus or nodulus) are thus formed through intermolecular reactions between the remaining "living" precursors and the pendant double bonds. Concomitantly to this mechanism of core for-

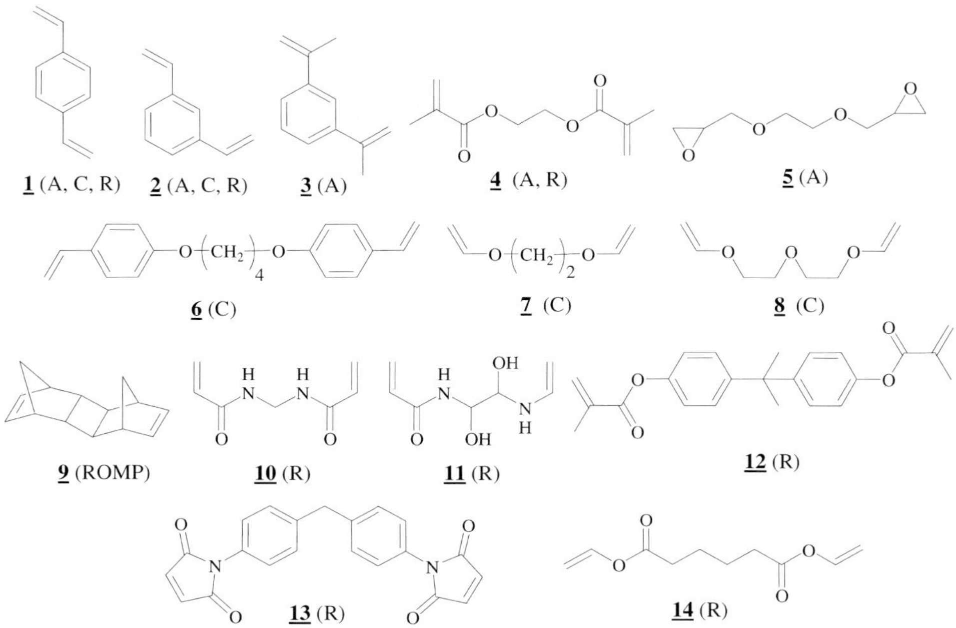

Fig. 8.4 Representative crosslinkers used for star and microgel synthesis (A=anionic; C=cationic; R=radical; ROMP=ring-opening metathesis polymerization).

mation, stars can also be generated through intermolecular reactions involving the divinylic blocks themselves. This eventually leads to the formation of star polymers with a core–shell architecture based on a microgel core and an envelope of linear polymers. Multiple active centers created around the core can further be used as multifunctional initiators for the polymerization of the same or another monomer leading, in this way, to "miktoarm" star polymers (see Chapter 6 in this volume). As illustrated in Fig. 8.3, one can start shaping up α-function-bearing "living" (or dormant) polymers which, after crosslinking, will carry these functional groups at their periphery.

The number of chains attached to the core depends on a number of experimental variables, including the molar ratio (r) of the divinylic compound to the preformed (co)polymer, the nature of the linking agent, the size of the precursor, the chemical nature of the solvent and the concentration of the reaction mixture. In contrast to stars synthesized by the core-first methodology, however, one cannot expect to obtain stars carrying a precise number of arms by this "nodulus approach", one can at best minimize the fluctuation of their functionality. Another drawback of this approach is the presence of substantial linear parent polymers remaining, even with large amounts of crosslinker. Partial conversion of the precursors into star molecules is explained by an incomplete incorporation of the divinylic comonomer, or a loss of the -termini of the "living" or the dormant species, or the build-up of steric hindrance around the core as the coupling reactions proceed. All these features, though established a long time ago for anionically derived star polymers, also apply to stars derived by any CLP methods.

8.2.1.1 By Anionic Means

As already mentioned, the golden rules for efficient star formation by the "nodulus approach" have been established in the context of "living" anionic polymerization and have been a source of inspiration for the synthesis of star polymers by other methodologies. Among multivinylic monomers, 1,4- and 1,3-divinylbenzene (DVB) **1** and **2** (Fig. 8.4) and ethylene glycoldimethacrylate (EGDMA) **3** have been used extensively for arm-first star synthesis grown anionically [17, 18]. For instance, "living" anionic polystyrene (PS) chains can react with a controlled amount of DVB to form a star molecule having a poly(DVB) core. If the reaction is performed under completely dry and inert conditions, the core of the star molecule carries, in principle, a number of anionic sites corresponding to the number of the parent polymer arms. These anionic active species can either be "killed" to generate "regular stars" or serve as initiators for the polymerization of the same or another monomer.

The proportion of the nucleus in regular star polymers represents generally no more than 1–5% of the total mass of the star but it can be increased by using larger amounts of crosslinker [19]. Under specific conditions, up to 30–35 mass% of 1,4-DVB forming the core can be incorporated, and yet soluble objects are obtained (no macrogelation). This demonstrates that the PS chains surrounding the core prevent the intermolecular coupling reactions between the pendant vinyl groups at the surface of the nodulus. Crosslinking reactions occurring upon addition of the divinylic comonomer in batch experiments usually prevent the complete conversion of the polymer arms, as mentioned above. This implies that the separation or analysis of the resulting stars is sometimes troublesome. Moreover, the use of a too large excess of crosslinker and/or prolonged reaction times may produce insoluble macrogel. Bi and Fetters [18] reported, however, that pure stars could be synthesized by slowly adding DVB to polystyryllithium ($PS\text{-}Li^+$) linear precursors with very rapid mixing. In recent contributions, PS stars were synthesized following an incremental procedure in which different loads of DVB were added periodically [20]. The arm number of star polymers proved to increase continuously using this incremental-addition method, preventing gel formation. This method also allowed the authors to dramatically decrease the content of the unreacted $PS\text{-}Li^+$ and finally achieve complete conversion by favoring star–star couplings: "gradient star" polymers were thus obtained.

Works by Funke and Okay [21] and by Solomon et al. [22] have shown the importance of the solvent of polymerization on the kinetics and the overall size and structure of the core–shell star-like systems. For instance, the sequential copolymerization of *tert*-butylstyrene (TBS) and DVB revealed that the use of *n*-heptane or THF as solvent leads to microgels of different size. In THF, both the poly(DVB) core and the poly(TBS) arms are in a good solvent and no precipitation is observed. In contrast, *n*-heptane is a good solvent for the poly(TBS) corona but a rather poor solvent for the poly(DVB) core. Thus, upon addition of DVB to "living" anionic poly(TBS) chains, the block copolymers formed precipitate from the solution and further crosslinking proceeds in a separate phase. For this reason, a better control over the molar mass distribution was noted when THF was employed.

The "arm-first" technique was also applied to synthesize star-like compounds based on a poly(EGDMA) microgel core and poly(*tert*-butylacrylate) (P*t*BA) arms [23].

One can also mention the work by Penczek et al. who combined "arm-first" and "core-first" methods to generate star-like PEOs by using a bis-epoxide **5** (Fig. 8.4) as a crosslinker that was reacted with ω-hydroxy-terminated PEO under anionic conditions. The multi-oxanionic cores thus formed then served to initiate anionic ring opening polymerization of ethylene oxide, leading to highly branched PEOs [24].

Group transfer polymerization (GTP) is a CLP technique of choice for the synthesis of "tailor-made" (meth)acrylate (co)polymers [25]. A silyl ketene acetal is generally employed as the initiator in the presence of either a nucleophilic or an electrophilic catalyst, e.g. tetrabutylammonium fluoride for methacrylate or a Lewis acid for acrylates. This anionic catalyst either cleaves or activates selectively the Si–O bond, which generates enolate-type active species in equilibrium with dormant silyl ketene acetals whereas the electrophilic catalyst activates the monomer. Synthesis of (meth)acrylic regular stars could be achieved by sequential GTP of alkyl (meth)acrylate and EGDMA [25, 26].

8.2.1.2 By Cationic Means

The development of "living" cationic polymerization was achieved in the 1980s [27], which further permitted the preparation of star polymers. Monomers that have been used for this purpose include vinyl ethers [28], styrene [29] and isobutylene [30], the corresponding crosslinkers being DVB and compounds **6**, **7** and **8** in Fig. 8.4. As for the other means, the number of branches of stars was found to depend on the respective proportion of monomer – and also its type – and crosslinker nature. A chemical method to determine the number of branches was proposed by Kennedy et al. for the case of their polyisobutene (PIB) stars obtained from DVB as the crosslinker: using trifluoroperacetic acid, the poly(DVB) core was selectively destroyed without affecting the PIB arms that could be further analyzed.

8.2.1.3 By Ring-opening Metathesis Polymerization (ROMP)

The mechanism of ROMP and its use in macromolecular engineering can be found in Volume 1, Chapter 6. Synthesis of arm-first stars by ROMP was described by Schrock et al. who added a norbornadiene dimer **9**, (Fig. 8.4) to obtain stars based on polynorbornene [31]. Consistent with the mechanism discussed above for the ionic routes, ROMP-derived stars generated by the nodulus approach also exhibit a large fluctuation in their number of branches.

Alternatively, the homo- or copolymerization of macromonomers by ROMP is an efficient pathway to star-shaped polymers if one targets small DPns for the backbone of the polymacromonomers (route C, Fig. 8.3), as established by Gnanou et al. [32]. Indeed, the ROMP of macromonomers based on PS and/or poly(ethylene oxide) (PEO) offers the significant advantage of being quantitative, hence yielding star-like polymers free of any macromonomeric precursors.

8.2.1.4 **By Radical Means**

Conventional radical polymerization The possibility of preparing regular star polymers by "conventional" (uncontrolled) free radical (co)polymerization of macromonomers has been demonstrated in many reports [33]. In most cases, the method consists in polymerizing macromolecular chains fitted with a vinylic (styryl or (meth)acryloyl) group under radical conditions. For instance, stars based on polyisoprene (PI) were prepared using DVB as the crosslinker, an azo-type initiator as a radical source and PI-based macromonomers, the latter being prepared by anionic polymerization of isoprene and post-functionalization with *p*-chloromethylstyrene [34]. In such an approach, the star formation is favored by a proper choice of the solvent capable of inducing the pre-organization of macromonomers in micelle-like structures. Using *n*-heptane, crosslinking reactions the poly(DVB) cores could thus be confined in the core of the micelles. A similar methodology was followed by Ito et al. and by Ishizu et al. to derive star-like PEOs from PEO-based macromonomers [35, 36], but the obtained samples were relatively ill-defined (high fluctuation in functionality and composition) due to the occurrence of chain breaking between propagating chains.

Controlled/living radical polymerization Much substantial research has been devoted in recent years to "controlled/living" radical polymerization (C/LRP) methodologies, which combine the advantages of truly "living" systems for the quality of the polymers formed with the ease inherent in radical processes (Volume 1, Chapter 7). Among C/LRP systems, stable free radical polymerization (SFRP), often referred as to nitroxide-mediated polymerization (NMP) [37], atom transfer radical polymerization (ATRP) [38, 39] and reversible addition–fragmentation chain transfer (RAFT) [40, 41] have been extensively investigated. The same trends as those discussed above for arm-first star polymers obtained by ionic means have been observed when the nodulus approach is applied to C/LRP. For instance, gel formation is generally observed above a critical value of the r molar feed ratio above 15, where r=[divinyl monomer]/[linear precursor]. Also, residual linear precursors often contaminate the star products.

ATRP To synthesize their PS stars, Matyjaszewski et al. used ATRP-derived PS macroinitiators and various divinylic monomers, in the presence of copper-based complexes in anisole at 110 °C [42]. A ratio of 5 to 15 between DVB and the PS macroinitiator was found to be optimal for the formation of stars with good yields. Other experimental parameters including solvent nature, use of Cu(II) as a deactivator of propagating radical chains, and reaction time, proved crucial for efficient star formation. However, the star samples were contaminated with residual linear chains and exhibited rather broad molar mass distribution due to star–star couplings. Following a similar route, the same group reported the synthesis of P*t*BA stars [43]. Through the use of functional initiators from which the linear P*t*BA precursors were grown, various functions (e.g. epoxy, amino, cyano or bromo) could be introduced at the end of each branch of the stars. The same conditions were applied by Ishizu et al. to prepare poly(-acrylic acid) (PAA) stars after hydrolysis of the *tert*-butyl groups of the P*t*BA star precursors; the viscoelastic behavior of the multiarm polyelectrolyte could be further investigated [44]. In a series of reports, the group of Sawamoto thoroughly investigated the scope and limitations of the nodulus synthetic approach to star-like polymers, mostly based on polymethacrylates, using their ruthenium-based complexes as catalysts and using various crosslinkers, including EGDMA and amide-type divinyl compounds (e.g. **10–12**, Fig. 8.4) [45]. They also described that star formation is influenced by the parameters mentioned above. Interestingly, high molar mass stars could be prepared following a one-pot procedure, by adding the divinylic comonomer *in situ*. They have also shown that adding a mixture of the divinylic crosslinker and a monovinylic homologue to the pre-formed linear polymer during the crosslinking process is an easy means to improve the yield of the star formation. In another report, the same team presented the synthesis of star polymers containing ruthenium(II) in the core by the direct encapsulation of the catalysts *in situ* [46]. This could be achieved by copolymerizing the divinyl compound with a specific monomer carrying a phosphine as a ligand for complexing Ru(II). Catalysts embedded in the core of the star polymer were obtained in this way. More recently, star-like PS with arms functionalized with a few hydroxy- or methoxy-ended ethylene oxides were used as organic supports for tridentate bis(imino)pyridinyliron catalyst in the presence of alkylaluminum compounds towards ethylene polymerization, as reported by Cramail et al. [47]. Such star-like organic supports composed of a microgel core afforded polyethylene beads of spherical morphology and high bulk density, with a relatively high catalytic activity.

NMP or SFRP The synthesis of star polymers by SFRP (or NMP) by the “arm-first” approach has also been described in a few reports. The proof of concept was first described by Solomon et al. who employed tetramethyl piperidyl-*N*-1-oxyl (TEMPO) as the persistent counter-radical to mediate the NMP of styrene, followed by crosslinking of the obtained PS precursor with DVB [48]. Surprisingly, Long et al. observed that a ratio of 67 between DVB and a PS precursor also terminated by an alkoxyamine based on TEMPO was needed in order to obtain PS stars efficiently [49]. In contrast, Hadjichristidis et al. employed much smaller ratios (between 3 and 13) to obtain their PS stars [50]. As such stars contained many TEMPO-based alkoxyamines at their core, a large number of arms (up to 600) could be grown outwards ("in–out" approach) by NMP, affording "miktoarm" stars. Further works by Kakuchi et al. described the synthesis of arm-first PS stars functionalized either at the periphery or at the interior of the core with glycoconjugated moieties. This was accomplished through the use of either TEMPO-based glycoconjugated alkoxyamines or glycoconjugated styrenic monomer during the NMP process [51]. As for Hawker et al., they used combinatorial techniques for high throughput synthesis of arm-first star polymers, using unimolecular alkoxyamines based on α-hydrogen-containing nitroxides operating both as initiator and controlling agent for NMP, and either bis-maleimide (**13**, Fig. 8.4) or DVB as the crosslinker. Their aim was to rapidly screen the key experimental parameters pertaining to star polymers synthesis [52]. Functional groups could also be incorporated along the backbone or at the periphery of the stars. The authors proposed an application of such functional stars in catalysis, where stars served as scaffolds for catalytic sites to trigger Heck-type coupling as well as enantioselective addition reactions. In a recent contribution [53], the same group confined specific functions (basic on the one hand, acidic on the other) at the core of two distinct stars. These specific groups were incorporated through the nitroxide-mediated copolymerization of the purposely designed acidic and basic monomers with styrene, before crosslinking with DVB. As a result of this site isolation, sequential catalysis two-step reaction using catalytic species that are normally incompatible with each other was possible.

The "nodulus approach" seems also to be applicable in aqueous dispersed media, as recently reported by Okubo et al. [54]. They have investigated the nitroxide-mediated radical crosslinking copolymerization (RCC) (see Section 8.3) of styrene and DVB using a PS-TEMPO macroinitiator in aqueous miniemulsion and in bulk. They found that the crosslinking density of the microgels was significantly lower in miniemulsion than in the corresponding bulk system.

RAFT Only a few studies of star polymers grown by RAFT from the "nodulus approach" have been reported. A recent work by Taton et al. described the direct synthesis of hydrophilic arm-first stars based on poly(acrylic acid) or polyacrylamide by a RAFT-like process called MADIX (MAcromolecular Design by the Interchange of Xanthates), using a xanthate of general structure $C_2H_5O{-}C(=S)S{-}CH(CH_3)COOCH_3$ as reversible chain transfer agent [55]. Others reported the synthesis of PS stars from dithioethers as RAFT agents, using various solvents to favour the formation of star-like micelles composed of a crosslinked core [56–58]. The same key parameters mentioned above have a dramatic influence on the number of chains attached to the core.

8.2.2
Synthesis of "Arm-first" Stars Using Multifunctional Antagonist Reagent

8.2.2.1 By Anionic Polymerization

Another "arm-first" synthetic approach to star polymers involves the deactivation of linear chains onto a central core containing a precise number of antagonist reactive sites (route B, Fig. 8.3). Representative multifunctional agents employed in such an arm-first star polymer synthesis by anionic polymerization are shown in Fig. 8.5.

Use of chlorosilane reagents For instance, deactivating "living" anionic linear PS or polybutadiene (PB) chains onto chlorosilane derivatives is a very powerful synthetic access to the corresponding star polymers with a precise number of branches and high molar masses. Like in the "nodulus approach", linear chains can be sampled out and independently characterized using this chlorosilane-based method. Also, the coupling reaction can be monitored throughout the polymerization using size exclusion chromatography (SEC) and methods to determine absolute molar masses of starting materials, intermediates and final compounds in order to account for the structure of all species. One main drawback in this approach, however, is the necessity to use an excess of linear chains to obtain star samples with the expected number of arms; hence, a final fractionation step is required to remove unreacted linear precursors. The synthesis of four-armed PS stars by deactivation of living PS anionic chains onto $SiCl_4$ (**15**, Fig. 8.5) was first described by Morton et al. in 1962 [59]. Incomplete coupling was observed despite the use of a large excess of linear chains, which led to a mixture of three- and four-armed PS stars. The same group also showed that, when reacted in stoichiometry, $SiCl_4$ and living PI-Li^+ chains yielded three-armed stars based on PI [60]. In contrast, quantitative coupling was observed when using PB as linear precursor for the synthesis of four-armed PB stars [61]. It turned out that the efficiency of the coupling reaction varies in the following order: PB > PI > PS, which is unexpected if one considers the intrinsic reactivity of the corresponding propagating species. Steric crowding around the core seems to be the main reason for such findings. It was therefore recommended to form a short sequence of PB onto living PS or PI before addition of $SiCl_4$ to obtain well-defined four-armed PI stars [60]. Another possibility is to use chlorosilane derivatives for which the chlorosilyl groups are separated by an alkyl chain. Roovers and Bywater thus prepared PS stars with 4 and 6 arms, respectively [62]. Again, addition of a few isoprene units onto the PS living chains greatly accelerated the star formation. The same linking agents were used by Fetters et al. for the synthesis of six- and four-armed PI stars [63]. Chlorosilane compounds of higher functionality (e.g. **16** and **17**) were designed by the same group [64] and by others [65, 66] and used as linking agents for the synthesis of 8, 12 and 18-armed PI, PB and PS stars. In the latter case, addition of triethylamine as an additive was needed to increase the reactivity of PS-Li^+ chains toward chlorosilane functions in hydrocarbon solvents. Chlorosilane reagents of

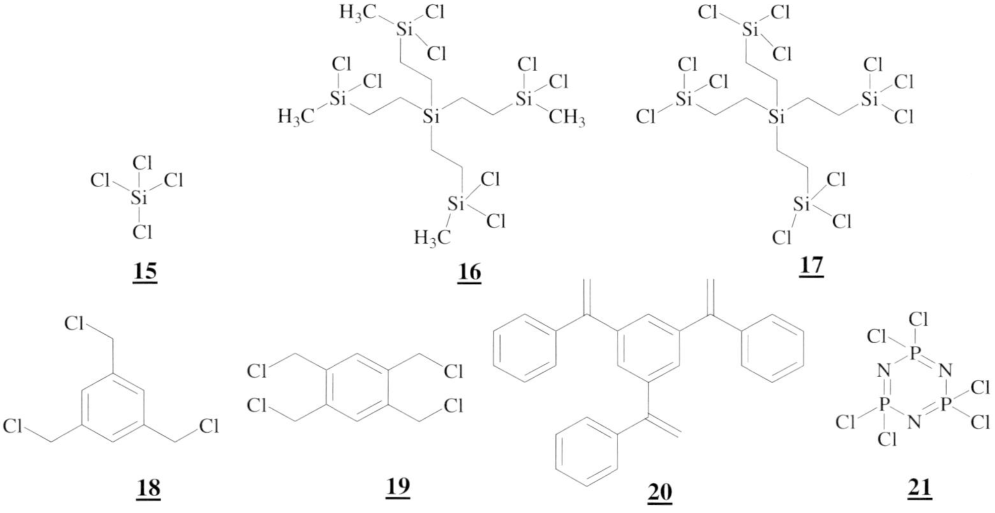

Fig. 8.5 Multifunctional agents for the arm-first approach by anionic polymerization.

even higher functionality (32, 64 and 128) were used by Roovers et al. to derive the corresponding stars based on PB: for this purpose, the authors designed hydrocarbon dendrimers of third and fourth generation carrying peripheral chlorosilane of equal reactivity [67]. This chemistry of chlorosilane was also applied to polymers other than PS, PB and PI: stars based on polydimethylsiloxane or poly(cyclohexadiene) with three and/or four branches were synthesized under appropriate conditions by Ressia et al. [68] and by Mays et al. [69], respectively.

Use of halomethyl benzene reagents Deactivation of carbanionic chains onto chloromethylbenzene functions is less straightforward than using the chlorosilane method described above. Indeed, a side reaction consisting of a halogen–lithium exchange may occur, leading to the formation of -chloro polymer chains and a carbanionic species ($RMetCl_{x-1}$) which can further react with the starting reagent RCl_x, yielding a novel electrophilic reagent of higher functionality. The latter side-product may be responsible for the production of a mixture of stars with different functionalities. For instance, Yen observed the formation of a multimodal distribution of molar masses when employing 1,3,5-trichloromethylbenzene **18** to deactivate $PS\text{-}Li^+$ chains [70]. Similar results were obtained by Allen et al. in their attempts to obtain four-armed stars using, in this case, 1,2,4,5-tetrachloromethylbenzene **19** and a mixture of THF and benzene as solvent [71]. Such a side-reaction can be minimized, however, by first adding the linking agent in the reaction mixture before the linear polymer precursor is introduced [71]. Another solution to avoid the halogen–lithium exchange is to use either potassium as a counter-cation instead of lithium [70, 71] or to perform the reaction in pure THF as a solvent, as reported by different groups [72, 73].

In this case, however, the coupling reaction between PS-Li^+ chains and **18** is restricted to the bis-adduct formation (two-armed PS) [72]. Better results can be obtained with bromomethylbenzene functions and other polymeric precursors such as poly(2-vinylpyridine) (P2VP) or poly(methyl methacrylate) (PMMA), as demonstrated by Hadjichristidis et al. [74] and Hatada et al. [75]: syndiotactic PMMA or P2VP stars with the expected functionality could be obtained in THF.

Use of diphenylethylene derivatives 1,1-Diphenylethylene (DPE) and its functionalized derivatives have also been extensively used as linking agents following the arm-first approach for the synthesis of star-like polymers, including regular stars, miktoarm and asymetric stars. The reader can refer to the recent review by Hirao et al. [76] (see also Chapter 6 in this volume). DPE derivatives do not homopolymerize and can be used as both linking agents and initiators. An illustrative example of the "arm-first" strategy to synthesize three-armed PS stars was described by Quirk et al. [77]: the reaction between 1,3,5- tris(1-phenylethenyl)benzene **20** with a threefold stoichiometric amount of short PS-Li^+ chains proved quantitative. On the other hand, after the living anionic polymer chains are linked into the DPE derivative, the corresponding adduct possesses one or more reactive anionic sites and can then serve for the growth of further chains ("in-out" approach).

Use of other electrophilic reagents Phosphonitrilic chloride trimer ($N_3P_3Cl_6$) also referred to as hexachlorocyclophosphazene **21** proved an efficient hexafunctional reagent for the synthesis of six-armed PS stars [78]. Oxanions from PEO chains grown anionically can also be deactivated onto **21**. The latter reaction is generally quantitative because of the very stable phosphorus–oxygen bonds formed after coupling. For example, six and twelve-armed PEO stars have been prepared in this way [79–82]. Other multifunctional cores including dendrimers [83], or other reagents [84–86] have also been used to derive PEO stars by the arm-first method. A series of original studies based on the arm-first approach was reported by Fraser et al. who first synthesized 2,2′-dipyridyl carrying for instance PS or poly(methyl methacrylate) (PMMA) chains obtained by ATRP, or PEO chains [87]. These two-armed polymeric precursors were further chelated onto a hexadendate ruthenium(II) or iron(II)-based complex to form six-armed star-like polymers containing a metallic core with interesting photophysical properties. Fullerene (C_{60}) has been thoroughly investigated as a linking reagent of carbanionic species by Mathis et al.. For instance, well-defined PS and PI stars with a fullerene core possessing exactly six arms (PS_6C_{60} and PI_6C_{60}) were obtained by grafting PS-Li^+ or PI-Li^+chains onto C_{60} in a nonpolar solvent such as toluene [88]. Interestingly, photophysical, and non-linear optical properties were observed for the corresponding materials.

8.2.2.2 Use of Nucleophilic Reagents

Multifunctional nucleophilic reagents are well-suited to the deactivation of living cationic polymer precursors and generate the corresponding stars. A few examples of such reagents have been described (Fig. 8.6). Living chains based on poly(isobutyl vinyl ether) (PIBVE) can react with both malonate salts (e.g. **22**) and silyl enol ethers **23**. However, the latter species proved more efficient deactivating agents than the former to derive well-defined four-armed PIBVE stars [89]. Higashimura et al. explained these findings by the poorer solubility of malonate salts in organic media in contrast to silyl enol ether functions.

8.2.2.3 Use of Multifunctional RAFT Agents

Recent reports have shown the possibility of deriving multiarm star polymers by RAFT, through the use of multifunctional dithioesters, trithiocarbonates or xanthates [40, 41]. Two different types of multifunctional CTAs can eventually be contemplated for use in a RAFT process: those implying an outward growth of arms from the core and those involving the reaction of linear chains with the functional core, following “core-first” and “arm-first” approaches, respectively, called the R-group and the Z-group approach, in the context of the RAFT process. In the Z-group (convergent) approach [90–93] the core plays the role of the activating group and the chains are grown away from the core and are attached when undergoing transfer reactions. A method to predict and quantify reactions occurring during multiarm polymer synthesis by RAFT was recently proposed by Barner-Kowollik et al. [94]. Addition of an oligomer chain (P_n°) onto one of the dithioester groups results in an intermediate radical that releases an R° group (called the leaving group) upon fragmentation, allowing initiation of a new linear chain. The latter then becomes one of the arms of a given star after addition–fragmentation transfer reaction to its core. In other words, such arm-first stars only contain branches in a dormant form. One main advantage of this Z-group strategy by RAFT is that complications often seen in the core-first star synthesis (see Section 8.2.3), such as star–star and star–linear chain couplings could be ruled out. A potential problem from such an approach, however, is the accessibility to the dithioester groups carried by the core. Indeed, the star core can be shielded by its polymeric arms, its access by linear growing chains being limited as the star arm grows. Consequently, linear chains will experience a higher probability of termination, thus increasing the concentration of dead

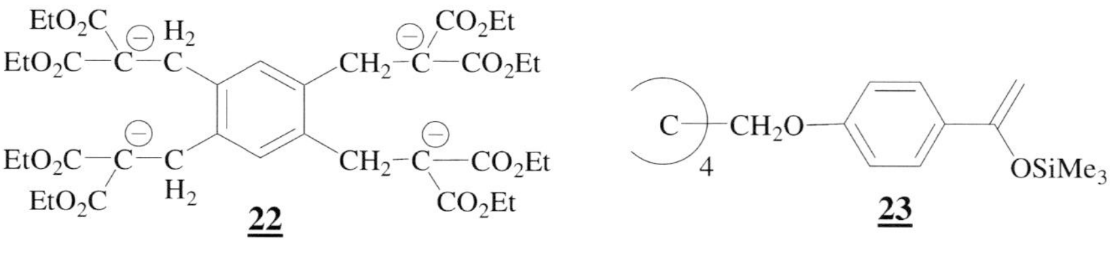

22 **23**

Fig. 8.6 Agents for arm-first star synthesis by cationic polymerization.

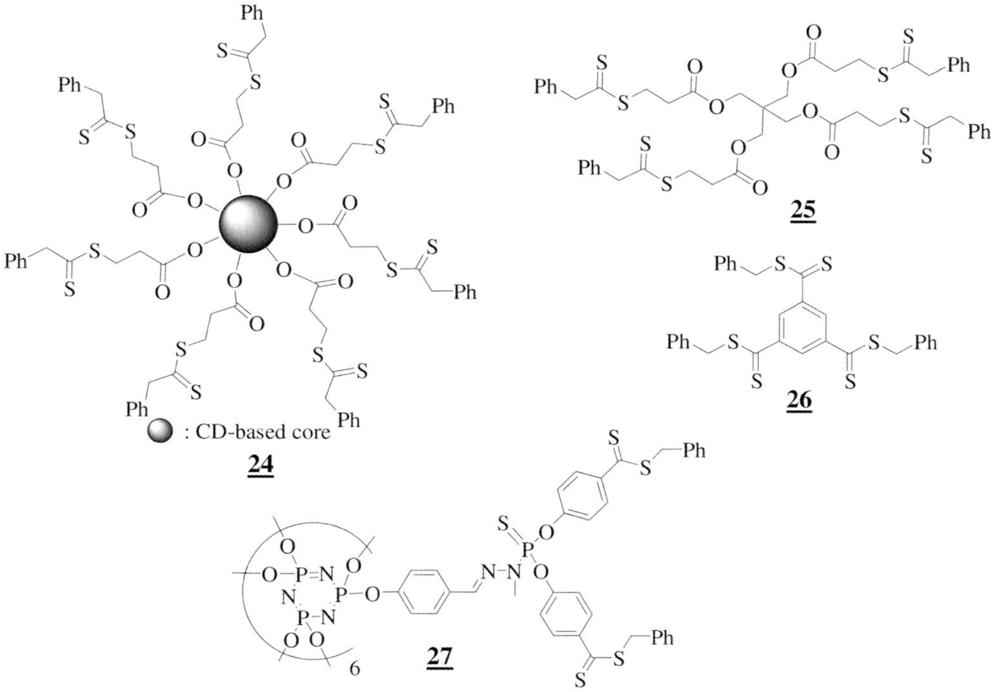

Fig. 8.7 Examples of multifunctional RAFT agent for the Z approach.

polymers that do not contain any thiocarbonylthio group and whose proportion varies with the functionality of the RAFT agent and of the conditions used [92]. Representative multifunctional RAFT agents employed for the Z-approach are displayed in Fig. 8.7. They include hyperbranched polymers, regular dendrimers, β-cyclodextrine derivatives or other small organic molecules.

8.2.2.4 **Use of Multifunctional Agents for Condensation Oligomers**

Well-defined star polymers can be synthesized by coupling very short and monodisperse linear chains (oligomers) prepared by sequential condensation, and not by a real step-growth polymerization which is not "living" in essence. For instance, Bo et al. synthesized four-armed star-shaped oligofluorenes possessing side octyl chains on the fluorene units and containing a porphyrin core [95]. These star oligomers showed blue absorption but emitted in the red by energy transfer. Roncali et al. reported the synthesis of planarized star-shaped oligothiophenes by connecting short-chain oligothiophenes to a central core deriving from a trithiophene [96]. The electrochemical and optical properties of these oligothiophene stars showed enhanced π-electron delocalization as compared to their linear counterparts.

8.2.2.5 The Supramolecular Approach

To synthesize metallo-supramolecular star polymers, Fraser et al. [86, 87, 171, 97] and Schubert et al. [98, 99] developed a coupling strategy that is not based on covalent linkages between pre-formed polymers but on the reversible formation of metal–ligand interactions of polymers that are end-functionalized by bipyridyl or terpyridyl moieties. In this case, prepolymers are coupled by adding ruthenium-based metallic centers. Very recently, Gibson et al. reported the first example of a star polymer based on pseudorotaxane complexation, following a supramolecular approach, i.e. using nonconvalent interactions [100]. The star molecule was formed by self-assembly of a homotritopic tris(crown ether) host and a complementary monotopic paraquat-terminated polystyrene guest based on the bis(*m*-phyenylene)- 32-crown-10/paraquat recognition motif. The α-paraquat functionalized PS chains were synthesized from a purposely designed paraquat-functionalized molecule used as initiator for SFRP of styrene.

8.2.3 Synthesis of Star Polymers by the "Core-first" Approach

The core-first approach has come to maturity after it was shown in the late 90's that stars of precise functionality could be obtained from multiionic initiators. However, the main limitation of the core-first method is the design of suitable multifunctional initiators.

8.2.3.1 By Anionic Polymerization

Divinylic comonomer approach Synthesis of "core-first" stars can be achieved by using some of the crosslinkers described in Fig. 8.4. When such a divinylic monomer is polymerized it first generates multiple active centers that further initiate the polymerization of the second monofunctional monomer. This approach leads to the formation of core–shell star-shaped structures, with a microgel as a core surrounded by polymeric branches: in this context, 1,4-DVB and EGDMA have been the two main crosslinkers employed in anionic polymerization for such core-first star syntheses [13, 14]. For instance, in the anionic polymerization of EGDMA, the latter is consumed without a noticeable increase in molar mass although the concentration of pendant double bonds decreases because of intramolecular cyclization reactions, which dominate during the early stage (more than 50% of the structural units participate in cyclization), and only a minor participation of intermolecular crosslinking. The *microgels* formed (see Section 8.2) possess, in this case, both active centers and reactive double bonds. In a second phase of the process, as the free volume decreases, these microgels react with each other through intermolecular crosslinking, which is evidenced by a build-up of the molar mass of the products formed.

As can be expected, the solvent nature has a dramatic influence on the crosslinking events, on the overall kinetics of the copolymerization, as well as on the orientation of the reaction toward the formation of micro- or macrogels [101].

The behavior of 1,4-DVB during anionic polymerization is distinct from that of EGDMA. Indeed, intramolecular cyclizations are much less prevalent when THF is used as solvent during the anionic polymerization of 1,4-DVB initiated by *n*-BuLi or lithium diisopropylamide used at low concentration [102]. The first stage of the process affords "living" carbanionic polymer chains consisting of pendant vinyl groups left unreacted. This reflects the lower reactivity of the second double bond of DVB once the first one has reacted. This, however, depends on the initial concentrations of both the initiator and the monomer [103, 104]: for instance, a [*n*-BuLi] equal to 2 mol% increases the probability of cyclization, which results in the formation of microgels in the nanometer size range, whereas a concentration higher than 15% leads to insoluble macrogels. Okay and Funke have reported that the domain of the microgel formation can be predicted from kinetic modelling [105]. It is worth pointing out that differences are observed when using 1,3-DVB instead of 1,4-DVB. The anionic polymerization of the former monomer, indeed, is much faster than that of of the latter. This can be rationalized by the fact that the reactivity of the pendant vinylic groups is significantly higher for 1,3-DVB, which favors cyclization and hence the formation of microgels.

1,4-diisopropenylbenzene (DIPB) **3** (Fig. 8.4) is another crosslinker that has been used to obtain microgels and core–shell star polymers by anionic polymerization when sequentially copolymerized with styrene [102, 106]. This approach using a multivinylic monomer for star polymer synthesis is not restricted to styrenics or methacrylates since "core-first" PEO stars could also be obtained in this way [107, 108]. Again, samples with a rather large fluctuation in functionality have been obtained though Lutz et al. reported that a better control can be reached when using poly(DIPB) instead of poly(DVB) as a multicarbanionic core [108].

Precise Multifunctional Initiators for Anionic Polymerization

Vinylic monomers Synthesis of core-first stars from multicarbanionic precursors is complicated by the limited solubility of polylithiated compounds that form insoluble aggregates in most organic solvents. This may explain why the only examples of initiators of precise functionality described by this method are those of Quirk et al. [77] and of Gnanou et al. [109]. Quirk's tricarbanionic precursor **28** (Fig. 8.8) was obtained by addition of *sec*-butyllithium to the DPE derivative **20** (Fig. 8.5). As for Gnanou et al., they reported on a novel approach to the preparation of soluble tri- and tetrafunctional lithium organic compounds **29** by lithium/halogen exchange (polylithiation reaction) of tris- and tetrakis-bromoaryl compounds. In the latter case, the solubility of the polylithium precursors is favored because halogen atoms are carried by separate aryl rings and not by adjacent carbon atoms. Unlike regular halogen/metal exchanges, the polylithiation

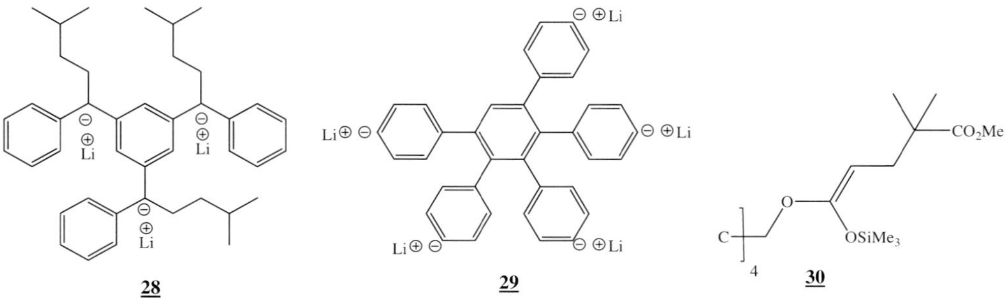

28 29 30

Fig. 8.8 Multifunctional initiators for anionic polymerization.

reaction could be conducted to completion at room temperature in an apolar medium, the polylithiated species formed being unexpectedly soluble in the presence of σ/μ-coordinating ligands. These features could be exploited for the synthesis of the core-first star PS and PB by anionic polymerization.

Stars based on poly(alkyl acrylate) were also derived by GTP from multifunctional precursors possessing three or four silyl ketene acetal functions (e.g. **30**) used in the presence of a Lewis acid catalyst [110]. Similarly, four-armed PMMA stars could be obtained by GTP of methyl methacrylate using a tetrafunctional core based on 1,3,5,7-tetramethylcyclotetrasiloxane containing silyl ketene acetal functions, in the presence of an ammonium salt as catalyst [111].

Heterocyclics Many examples of multioxanionic precursors for anionic ring-opening polymerization (ROP) of heterocyclics (e.g. ethylene oxide, lactones, *N*-carboxyanhydrides) can be found in the literature, in spite of the tendency of alkoxides to aggregate and their lack of solubility (very similar to the case of polycarbanions). The design of PEO chains, aliphatic polyesters as well as synthetic poly(amino acid)s arranged in a star-branched architecture is motivated by the fact that the latter structures may be more efficient or load more active compounds than their linear counterparts in biomedical or pharmaceutical applications. PEO stars could be obtained from precursors containing a precise number of hydroxyl groups in a deprotonated form using polar solvents such as THF and/or dimethylsulfoxide (DMSO). For instance, Roovers et al. used carbosilane dendrimers as multifunctional cores consisting of precisely 4, 8 or 16 hydroxy functions to derive the corresponding PEO stars [112]. Mulhaupt et al. employed hyperbranched polyglycerols serving as multifunctional macroinitiators and carrying a high but uncontrolled number of terminal hydroxyls and obtained by "ring-opening multi-branching polymerization" of glycidol [113]. Alternatively, one can deprotonate only partially the multi-hydroxylated precursor and take advantage of the fast exchange of protons between propagating alkoxides and dormant hydroxylated species, as shown by Rempp et al. for the preparation of 3-armed PEO stars from trimethylolpropane **32** (Fig. 8.9) [107]. This was also demonstrated by Gnanou, Taton et al. for the synthesis of 3- [114],

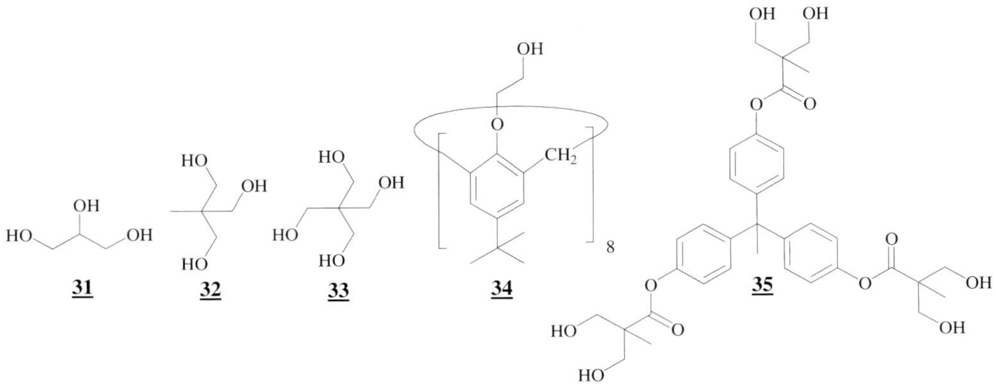

Fig. 8.9 Multifunctional initiators for ROP.

4- [114] and 8-armed [115] PEO stars. In the latter case, an octahydroxylated precursor **34** was designed in a two-step procedure from *tert*-butylcalix[8]arene. The same group took advantage of the exchange reaction to derive dendrimer-like PEOs up to the seventh generation [116]. As mentioned above, there is also a growing interest in the design of biocompatible and biodegradable stars based on aliphatic polyesters. Different groups have contributed to this area through the description of both the synthesis and the analysis of 3- 4-, 5-, 6-, and 13-armed star-shaped polylactide (PLA) or poly(ε-caprolactone) (PCL). For this purpose, regular multihydroxylated precursors such as glycerol **31**,117 trimethylolpropane **32** [118, 119], di(trimethylpropane) [118], pentaerhytritol **33** [117, 118], di(pentaerhytritol) [120], "polyglycerine" possessing 8 or 12 OH groups [121], poly(3-ethyl-3-hydroxymethyloxetane) with an average number of OH groups per macromolecule equal to 13 [118], tetra- and hexahydroxy functionalized perylene chromophores [122], etc. were used, whereas tin(II) alkoxides, diethyl zinc or aluminum-based complexes generally served as catalysts to trigger the ROP of the cyclic esters. A thorough analysis of PLA stars by liquid chromatography under critical conditions revealed structural imperfections due to a lack of control during the synthesis [118].

More complex multifunctional precursors were developed by the Hedrick group who used perfectly defined polyester-based dendrimers of various functionalities such as **35** with the idea of investigating the influence of this parameter on the morphology and the overall properties of stars based on PLA and PCL [123]. Similarly, others used poly(amidoamine) dendrimer to grow PLA chains in multiple directions [124, 125]. In these examples, hybrid stars consisting of a dendritic core (dendritic homopolymers) were synthesized.

Synthetic poly(amino acid)-based stars could be obtained in a divergent fashion by anionic ROP of *N*-carboxyanhydrides from multifunctional amino-containing precursors. Such synthetic polypeptides exhibit an interesting pH-sensitive and reversible *α*-helix to coil transition. For instance, Inoue et al. prepared stars based on poly(benzyl L-glutamate) carrying triethylene glycol monomethyl ether in the side chain by ROP of γ-benzyl-L-glutamate using hexakis(4-benzylamino-1-oxy)cyclotriphosphazene, followed by the displacement of the benzyl groups with triethylene glycol monomethyl ethers [126]. These stars proved efficient in molecular recognition to resolve amino acids such as tryptophan, tyrosine or phenylalanine.

8.2.3.2 Multifunctional Initiators for Cationic Polymerization

As in anionic polymerization, the major difficulty in cationic polymerization is the design of the multifunctional initiators for core-first star synthesis. Examples of such compounds are shown in Fig. 8.10.

Again, one has to distinguish between vinylic monomers (styrene, isobutene, vinyl ethers) and heterocyclic ones (oxazolines and THF essentially). Multifunctional initiators for carbocationic polymerization possessing functions such as secondary and tertiary chlorides that mimic the dormant species of living polymers have been designed (**38**, **37**) for styrene and isobutene, respectively. Precursors containing 3 or 4 fluoroacetate functions (e.g. **36**) activated by AlEt2Cl allowed the group of Higashimura to prepare stars based on poly(isobutylvinylether) (PIBVE) [127]. The well-defined character of these stars was verified by detaching and further analyzing the arms from the core. Gnanou et al. investigated the conditions best suited to derive 6-armed PS stars from a precursor containing 6 phenylethyl chloride initiating groups (**38**) activated with $SnCl_4$ in the presence of NBu_4Cl [128]. Polymerizations of styrene were performed at $-15\,^{\circ}C$ in CH_2Cl_2 and transfer reactions to monomer forming linear chains (from H^+) could be minimized by adding di-*tert*-butylpyridine as a proton trap in the reaction mixture. The ω-chloro groups could be further displaced to introduce fullerene (C_{60}) at the periphery of the PS stars [128]. The group of Kennedy reported the synthesis of PIB stars from tri- (**37**) and tetrafunctional pre-

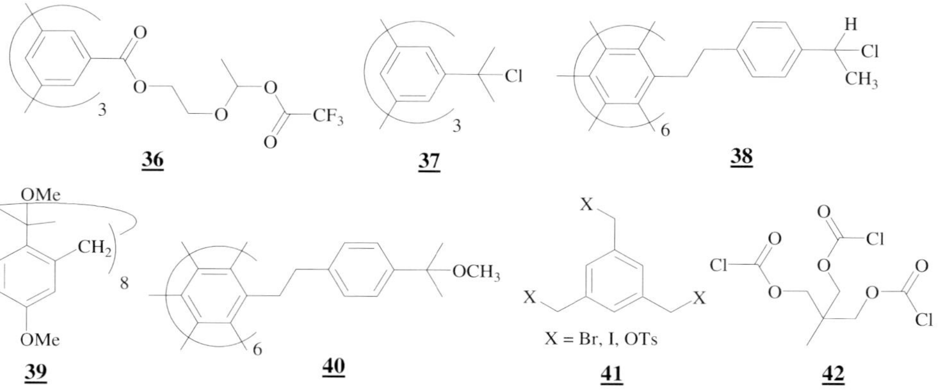

Fig. 8.10 Multifunctional initiators for cationic polymerization.

cursors carrying cumyl chlorides initiating groups, activated by BCl_3 or $TiCl_4$ [129]. The same team further described a calixarene-based derivative **39** and similar conditions to obtain 8-armed PIB stars [130]. Similarly, Gnanou et al. employed the conditions reported by Kennedy et al. to prepare 6-armed PIB stars using **40** [131]. As for Puskas et al., they employed a copolymeric backbone made of styrene and (4-(2-hydroxyisopropyl)styrene) from which 23 PIB on average were grown cationically, the resulting graft copolymer being viewed as star-like macromolecules because of the short size of its backbone [132]. The same group also showed that epoxy-type initiators can trigger the living cationic polymerization of isobutylene [133]. On this basis, they designed a hexakis(epoxysqualene) as initiator and prepared 6-armed PIB stars, though uncomplete initiation was noted.

Numerous heterocyclic compounds can undergo cationic ring-opening polymerization but only a limited number of these monomers offer the possibility of CLP. Among them, 2-alkyl oxazolines are cyclic monomers that can form persistent oxazolinium salts under relatively non-demanding conditions [134]. Examples of star-shaped polyoxazolines have been described in the literature. The well-defined character of polyoxazolines stars mostly depends on the structure of the multifunctional precursor used to initiate the ROP of the oxazoline. For instance, 1,3,5-tris(bromomethylbenzene) **41** used for 2-methyl (or 2-phenyl) oxazoline led to 3-armed stars of relatively broad molar mass distribution, because of a slow initiation [135, 136]. Better results were obtained using compounds such as trichloroformate initiating functions **42** [137] or 2-perbromomethyl-2-oxazoline as initiator [138]. In the latter case, the oxazoline moiety at the core of the star could be further polymerized. Relatively well-defined 4- [139] and 6-armed [140] polyoxazolines could also be prepared following the divergent approach.

8.2.3.3 Multifunctional Precursors for Controlled Radical Polymerization

When using multifunctional precursors in free radical polymerization, conditions have to be found to avoid any of the growing arms undergoing irreversible terminations, which result in the loss of control of the star functionality. The extent of these side-reactions is not only correlated to kinetic parameters such as the equilibrium constant (K_{eq}) between the radical active sites and the dormant species or the ratio between k_p/k_t, where k_p and k_t are the rate constant of propagation and that of termination, but also to the actual concentration of stars in the reaction medium [141]. It was observed, in particular, that the probability for intermolecular coupling is enhanced whenever the concentration of stars crosses their overlapping concentration [C^*].

By SFRP Hawker was among the first to demonstrate the possibility of obtaining well-defined PS stars by the core-first methodology [142]: a TEMPO-based tri-alkoxyamine was designed for NMP of styrene so as to generate a three-armed PS star. The latter was characterized by comparing its molar mass with that of its hydrolyzed arms isolated after cleavage of the central core. Gnanou et

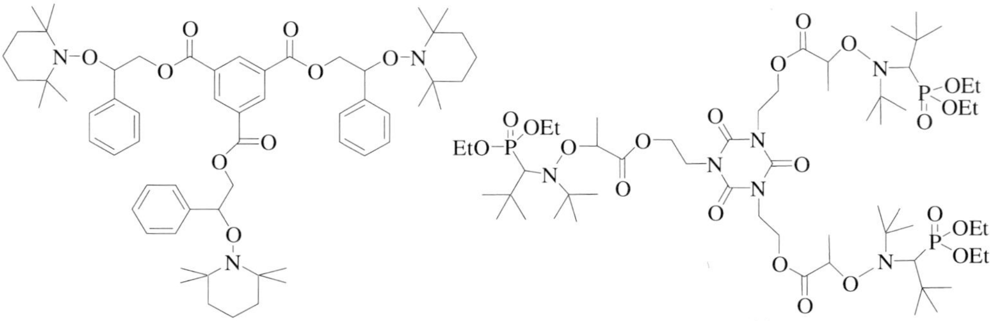

43 **44**

Fig. 8.11 Representative multifunctional alkoxyamines for NMP.

al. also resorted to NMP to prepare three-armed PS and poly(*n*-butylacrylate) (P-nBA) stars of rather high molar mass [143]; they used to this end an original trifunctional alkoxyamine based on the β-hydrogen-containing nitroxide. Others designed TEMPO-based tri-[144–146] or hexafunctional cores [146], a heptafunctional precursor derived from β-cyclodextrin [147] or dendritic precursors possessing up to 12 TEMPO-based alkoxyamines [148] for NMP of styrene so as to obtain the corresponding multiarm star PS. In most examples, the livingness of NMP of styrene was verified by hydrolysis of the arm star polymers and subsequent SEC characterization.

By ATRP Because it is easier to derive multihalides than multialkoxyamines, ATRP has been preferred over NMP for the synthesis of core-first star polymers [149, 150]. First examples were provided in 1998 by Sawamoto et al. who reported the synthesis of three-armed PMMA stars [151] and by Pugh et al. who described the synthesis of three-armed liquid crystalline polyacrylates stars using tris-(bromomethyl)mesitylene **45** (Fig. 8.12) as a trifunctional initiator [152]. In more systematic investigations, Sawamoto et al. [153] and Gnanou et al. [141, 154] first showed that stars with a high number of arms could be obtained using calixarene derivatized initiators (**47** and **48**), with either Ru(II)/PPh3 or Cu(I)/dipyridyl as polymerization catalysts. Many other groups have further resorted to ATRP to derive star polymers by the core-first method, using various families of multifunctional initiators (Fig. 8.12), including inorganic heterocyclics such as cyclotetrasiloxanes [155] or cyclophosphazene **53** [155], activated phenol derivatives (e.g. **50**, **52**) [156], glucose [157, 158], sucrose [159], cyclodextrin [158–161] or other halogenated cores derived from polyol [162, 163], calixarene (e.g. **47**, **48** and **49**) [141, 154, 164] or resorcinarene [165] derivatives with 4 to 8 ATRP-initiating sites, tetrakis(bromomethyl benzene) [166], adamantane-based cores [167], carbosilane dendrimers [168], dendritic polyesters (e.g. **52**) [156] or polyarylether-based dendrimers [169] that were used in conjunction with various activators based on either copper, ruthenium or nickel halides and

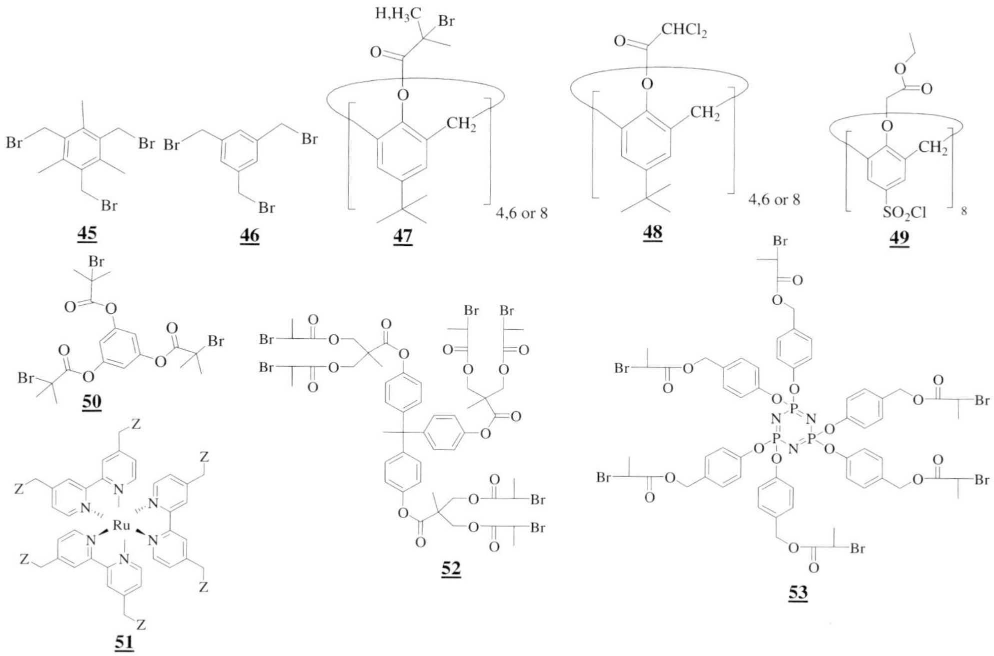

Fig. 8.12 Representative multifunctional initiators for ATRP.

miscellaneous ligands. Stars carrying PS or poly[alkyl (meth)acrylate] arms – carrying functional groups or not – whose number could range from 3 up to 21 were synthesized via this core-first route.

The scope and limitations of the most commonly used ATRP system that is CuBr/dipyridyl catalyst for star synthesis have been investigated by Gnanou et al. [141, 154], PS, PMMA and P*t*BA stars with predictable and high molar masses and polydispersity index close to unity, constituted of precisely 4, 6 and 8 arms were synthesized, starting from calixarene-derived initiators **47**. For instance, octafunctional PS stars exhibiting molar masses as high as 600 000 g mol–1 could be prepared by polymerizing styrene in bulk, discontinuing the polymerization to low conversion, typically below 15–20%, to prevent stars from mutually coupling. In addition, it was found that the lower the functionality of the prepared stars, the higher the conversion above which star–star coupling became detectable. No intermolecular couplings were detected while synthesizing polyacrylate stars, because the ATRP of acrylics is associated with a much lower equilibrium constant (K_{eq}) between dormant and active species. In many of the cases mentioned, the functionality of the multiarm stars obtained was checked by comparing their molar mass with that of their individual arms, the latter being isolated after hydrolysis of the ester functions of the central core.

Recently, Cheng et al. proposed a one-pot synthetic approach to star polymers by ATRP of *N*-[2-(2 bromoisobutyryloxy)ethyl]maleimide (BiBEMI) used as an inimer (INItiator-monoMER) in conjuction with an excess of styrene [170]. It was demonstrated that copolymerization of BiBEMI and styrene first led to the formation of a hyperbranched precursor, which could further continue ATRP of the excess of styrene. Fraser et al. used an approach based on coordination chemistry to design halogeno-2,2′-dipyridyl-based complexes as multifunctional ATRP initiators (e.g. **51**) of MMA and styrene by the core-first route [87, 171]. Multiarm stars incorporating metallic cores were prepared in this way. P*t*BA stars grown by ATRP were also used as precursors for the synthesis of the corresponding poly(sodium acrylate) [158, 172], poly(cesium acrylate) [172] or poly (rubidium acrylate) stars [172]. This was achieved after acidic cleavage of the *tert*-butyl groups of the star precursors followed by treatment with Met^+OH^-, where $Met^+ = Na^+$, Cs^+ and Rb^+, respectively. The aqueous solution behavior of the 4- and 6-armed star polyelectrolytes was investigated by viscometry and by light scattering. The variation of the reduced viscosity (η_{red}) as a function of the polyelectrolyte concentration C_p showed the existence of a maximum in η_{red} vs. C_p and, scattering experiments show the existence of a scattering peak that disappeared with the addition of a simple electrolyte, evidencing the electrostatic character of the interactions [172].

By RAFT As mentioned above in Section 8.2.2.3, multifunctional RAFT agents have been described in the recent literature for the synthesis of star polymers, following either the Z-group or the R-group approach. The group of Rizzardo [173] first attempted the synthesis of four- and six-armed PS stars from RAFT agents containing 4 and 6- thiocarbonylthio groups (**54** and **55**, Fig. 8.13). The use of such RAFT agents implies an outward growth of arms from the core since the homolytic leaving groups generated after the fragmentation step are part of the core of the stars. Davis, Barner-Kowollik et al. later investigated in detail the polymerization of styrene from the same haxafunctional RAFT agent [174]. These authors found that, as for ATRP-derived star polymers, the synthesis of core-first stars by RAFT may be complicated by irreversible terminations between star molecules, especially for poorly reactive radicals such as polystyryl ones. The use of a trithiocarbonate heptafunctional β-cyclodextrin confirmed these observations [175]. In the context of RAFT, star–star couplings seem to depend, however, on the structure of both the activating and the leaving groups and also on the temperature. These parameters have, indeed, a dramatic influence on the overall kinetics of the RAFT polymerization. Overall, a low concentration of propagating radicals can prevent star–star couplings from occurring [176]. On the other hand, more reactive radicals, like those deriving from poly-(acrylic acid) [55], or poly(vinyl acetate) [177] – in the latter case, subsequent methanolysis led to four-armed poly(vinyl alcohol)-based stars – the probability for stars to get coupled can be minimized. In these cases, tri- and tetrafunctional dithiocarbonates (xanthates **57** and **58**) served as RAFT agents. A facile means to synthesize stars by RAFT is to use well-defined poly(styrene-*co*-vinylbenzyl chloride) copolymeric precursors with a short chain length – themselves prepared by RAFT [178]. The chlorobenzyl side-groups can be further deriva-

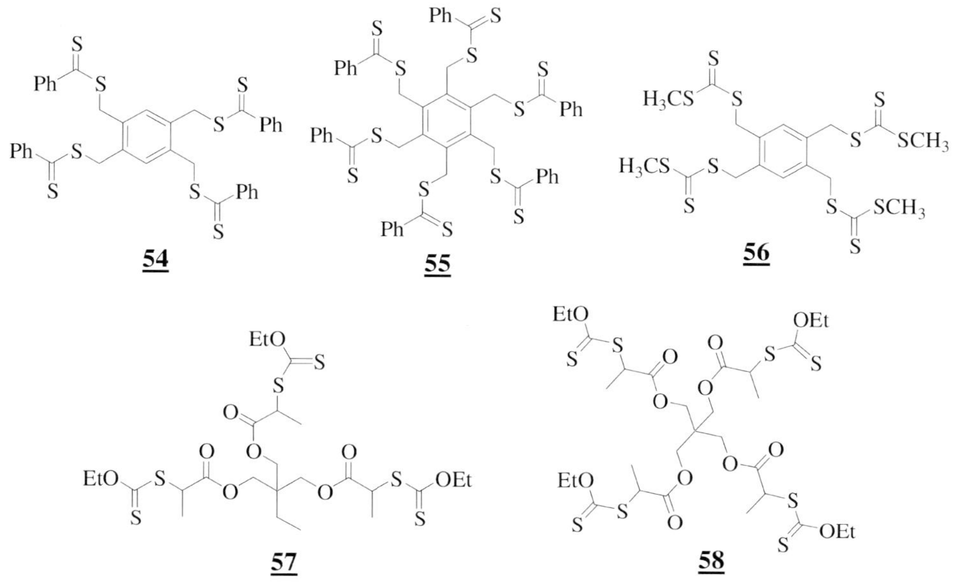

Fig. 8.13 Examples of multifunctional RAFT agent for the R approach.

tized to benzyl dithiobenzoate functions for the growth of PS chains by RAFT, the corresponding graft copolymer adopting a star-like conformation if the backbone is short enough. Regular dendrimers [179] or fractionated hyperbranched polyesters [180] also served as starting materials to obtain multifunctional RAFT agents for the synthesis of temperature-sensitive stars (dendritic homopolymers) based on poly(*N*-isopropylacrylamide), whereas a hexa(dithiobenzoate)-functionalized Ru(II)- tris(bipyridine) was used for the RAFT polymerization of a styrene-functionalized coumarin monomer for the development of star polymers with light-harvesting properties [181].

8.2.3.4 **By ROMP**

Examples of multifunctional initiators used in ROMP for star polymer synthesis have been proposed. Grubbs et al., for example, have described the preparation of a tetrafunctional precursor possessing four metallacyclobutane moieties based on titanium dicyclopentadiene to be used for the ROMP of norbornene [182]. However, the purity of this starting material has been questioned since the resulting polymer compounds did not exhibit the expected functionality of 4. Carbosilane dendrimers functionalized at their periphery with ruthenium carbene have also been designed and used for ROMP of norbornene [183]. Broad distribution of molar masses ($1.6 < PDI > 3.2$) were obtained, however, certainly because of a too slow initiation relative to propagation.

8.2.3.5 By Other Organometallic Routes

Aryl isocyanides can be polymerized in a controlled/living fashion by Pd-Pt μ-ethynediyl complexes, as reported by Onitsuka et al. [184]. This method allowed the synthesis of well-defined polyisocyanides, the latter being capable of adopting helical conformations in solution and exhibiting unique properties. This polymerization system recently proved applicable to multifunctional initiators for the synthesis of star-shaped polyisocyanides [184]. Indeed, multinuclear acetylide complexes possessing three Pd-Pt μ-ethynediyl moities were used as efficient trifunctional initiators for the polymerization of aryl isocyanides. The controlled character of the polymerization was demonstrated after cleavage of the arms from the core. Interestingly, the polymerization of a monomer possessing a chiral ester group led to the formation of a star polymer with three helical chiral arms.

As for 1-alkyne monomers, one can trigger their living polymerization using Rh- or Mo-based catalysts, as shown by Masuda et al. who also reported the synthesis of star-like polymers by polymerization of phenylacetylene and 2-(trifluoromethyl)phenylacetylene, using such catalysts [185].

8.2.3.6 By "Condensative Chain-growth Polymerization"

Polycondensation (or step-growth polymerization) is traditionally not an appropriate method for synthezing well-defined core-first star polymers because it is not a "living/controlled" polymerization process. For instance, Kricheldorf attempted the synthesis of liquid crystalline stars based on polyester by polycondensation, using multifunctional cores that were copolymerized with AB-type monomers or with a mixture of (A2 + B2) monomers [186]. As expected, a relatively poor control over the arm length of the resulting polymers was noted. Recently, however, Yokozawa et al. have developed the so-called "condensative chain-growth polymerization" method, which yields well-defined condensation polymers [187], block copolymers [188] and star-shaped polymers [189]. The chain-growth polymerization of phenyl 4-(octylamino)-benzoate from multifunctional initiator allowed this group to prepare well-defined core-first star-shaped aromatic polyamides [189]. Their initiators consisted of tri- or tetraphenyl ester moieties, such as triphenyl benzene-1,3,5-tricarboxylate, tetraphenyl benzene-1,2,4,5-tetracarboxylate or 1,3,5-tris(4-phenyloxycarbonylbenzyloxy) benzene.

8.3
Microgels/Polymeric Nanogels: Intramolecularly Crosslinked Macromolecules

Microgels are included in the broad category of branched polymers (Fig. 8.1) but there is no clear-cut universal definition for them [190]. A microgel can however be defined as a set of intramolecularly crosslinked macromolecules that are macroscopically soluble with a globular structure of colloidal dimension, typically in the range 10 to 1000 nm [13]. However, the term "microgel" itself may lead to confusion as it refers to macromolecular assemblies whose size would be larger than 1 micron, which is seldom the case for the microgels described in most articles. The term "polymeric nanogels" is certainly more appropriate to account for their mesoscopic dimensions but the term *microgel* is more popular and is generally used to describe all small gel structures [14]. Microgels/polymeric nanogels can thus be viewed as macromolecules of very high molar mass, of limited segmental mobility but with solution and viscometric properties similar to those of linear or branched polymers of low molar masses. Moreover, although their dimensions are compatible with high molar mass linear polymers, their internal structure is very close to that of a macroscopic insoluble network.

The main property of microgels is their ability to swell in an appropriate solvent. The overall dimensions of swollen microgels are often similar to those of single polymers in solution but the presence of internal crosslinks leads to specific properties, such as a defined shape, distinct rheological behavior, better resistance to degradation, and their ability to encapsulate guest molecules. Microgels are of great importance to many industrial disciplines as components of binders for organic coatings [14], carriers of dyes, pharmaceuticals and biochemical compounds [191], and fillers and materials for reinforcing plastics.

Insoluble macroscopic networks (macrogels) that include the important class of "hydrogels" will not be covered here. These have been extensively studied and numerous applications in the field of biomaterials (e.g. contact lenses, drug-delivery systems, wound dressings) utilizing hydrogels have been described [192]. Only chemical microgels formed from true covalent bonds between polymeric chains will be discussed and not microgels generated from weak and reversible interactions. Also included in the category of microgels are systems that are called "shell-crossliked micelles" [193], formed by self-assembly of block copolymers, followed by chemical crosslinking of the core or the shell of the micellar structure (see route F in Fig. 8.3), but these will not be considered here. A few hundreds of reports have been published on the topic of microgels/polymeric nanogels. Needless to say, due to space limitation, this review cannot encompass the exhaustive list of microgels described so far. Below are presented the main synthetic methodologies to obtain microgels/polymeric nanogels and the analytical means to account for the internal structure of this special class of branched polymers.

8.3.1 Synthetic Methodologies of "Microgels/Polymeric Nanogels"

Two main synthetic strategies of formation of microgels have been described [13, 14]. The first method consists in the copolymerization of a monomer in the presence of a crosslinker. The second method utilizes intramolecular crosslinkings of preformed polymers in a monomer-free process, under specific conditions (e.g. ionizing radiation, reaction of functional polymers with a reagent possessing at least two antagonist functional groups).

8.3.1.1 Radical Crosslinking Copolymerization (RCC)

The method generally employed for synthesizing polymeric nanogels is *radical crosslinking copolymerization (RCC)* of a vinylic monomer with a crosslinker, using one of the three following processes: (i) emulsion (or microemulsion) RCC through the compartmentalization of growing chains within micelles in order to prevent macrogelation, (ii) dilute solution RCC that favours intramolecular crosslinking reactions and (iii) precipitation (or dispersion) RCC to induce the precipitation of growing chains in the polymerization solvent. The first reports on microgels date back to the 1930s when Staudinger et al. described the formation of a microgel from the copolymerization of styrene and DVB [194]. From a theoretical viewpoint, Flory and Stockmayer proposed a model to account for the formation of both micro- and macrogels by RCC [195, 196]. In this theory, intramolecular cyclization was neglected. In addition, all vinylic groups were supposed to exhibit the same reactivity toward propagating species. Following these two hypotheses, the crosslinking density of a microgel (or of a macrogel) should be constant throughout the structure and a homogeneous network should be obtained. However, such a model was never experimentally verified. In their recent contribution, Costa et al. well summarized the state-of-art of modelling methods of RCC [197] which include the probabilistic approach [196], the method of moments [198, 199], kinetic models involving different chemical systems and reactors (batch or semi-batch) [197, 200], the Monte Carlo method [201–206] and the statistical crosslinking method [207]. Alternatively, "numerical fractionation" [208] and "real chain-length domain calculation" exploiting the Galerkin finite elements method [209] have been used to describe RCCs.

Actually, intramolecular crosslinkings predominate at the early stages of a regular RCC (>50%), generating highly heterogeneous materials whose branching points are irregularly distributed [13, 14]. In addition, chain growth may imply different types of vinylic groups with distinct reactivity: mono- and divinylic comonomers may exhibit markedly different reactivities from that of pendant vinylic groups formed after addition of the first double bond, due to chemical reasons, steric hindrance, or thermodynamic excluded-volume effects [210]. Average crosslinking density is therefore typical of both micro- and macrogels. The overall structure of the chains resulting from the different modes of synthesis is thus far from perfect and numerous structural defaults can be envisaged: loops, dangling chains, large fluctuation of chain length between branching points, heterogeneous composition from chain-to-chain, etc. [210, 211]. Recent studies have also discussed the possible formation of "local molecular clusters" also called "mechanical network" by reaction of reactive groups on the globular surface of microgels but which progressively disappear as time goes on [212]. Figure 8.14 shows possible events during RCC. The active center at the chain end can thus react with: (a) the double bond of the mono- or (b) the multivinylic monomer, or (c) with pendant double bonds of the same chain, or (d) with double bonds of the same macromolecule but located on other chains than the one carrying the active center, or (e) with a double bond of another macromolecule. As discussed above, cyclization reactions (c) and intramolecular crosslinkings (d) prevail at the early stages of the

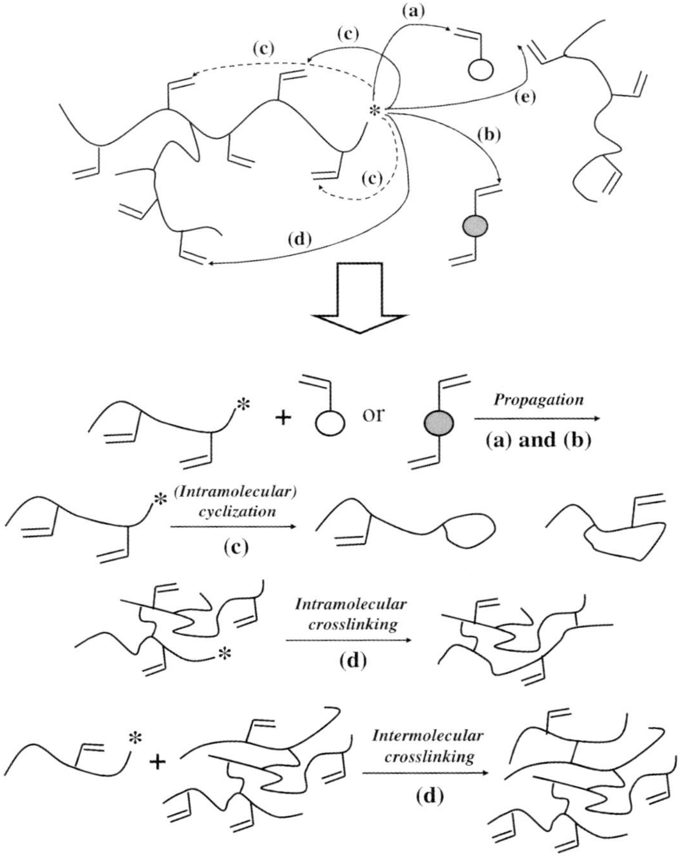

Fig. 8.14 Schematic representation of the crosslinking copolymerization process.

RCC. The pendant double bonds are then less accessible (excluded volume effect) in the internally crosslinked macromolecules formed and their reactivity decreases dramatically, which has been verified by the presence of residual double bonds at high conversion [210].

RCC in bulk RCC in bulk is generally not appropriate for synthesizing microgels because intermolecular crosslinking reactions are mostly favored at high concentrations, which finally leads to the formation of a macroscopic "wall-to-wall" macrogel. Before macrogelation occurrs, however, it is established that bulk RCC gives rise in an initial phase to microgels that subsequently link together to eventually form the final network. This means that microgels may be isolated before RCC has reached the gel point. Besides this, there may be an interest in developing solvent-free RCC when specific applications are targeted. Thus, alternative techniques to the use of conventional initiating systems that decompose upon heating in solution are needed. For instance, photopolymerization, radiation polymerization, initiation from ultrasound or from microwaves can be contemplated for microgel synthesis under mild conditions. These uncommon methods of polymerization can be implemented in environment-friendly solvent-free processes at room temperature; their utilization for micro- and macrogel synthesis has been highlighted in the review article by Ulanski and Rosiak [14].

Simulation of the copolymerization of mixtures including multifunctional monomers confirmed that microgels of structural heterogeneity are formed during the bulk process [213, 214]. From an experimental point of view, various examples of microgel formation before macrogelation in bulk have been reported, e.g. the free-radical photopolymerization of multiacrylates [215, 216], or the free-radical bulk copolymerization of a dimethacrylate derived from bisphenol A with styrene or divinylbenzene [217]. In the latter case, microgels could be visualized by atomic force microscopy.

In a recent report, Morbidelli et al. have investigated the copolymerization of styrene and DVB at very high-temperature (>300 °C) in a steady-state continuous-stirred tank reactor (CSTR) [218]. Large amounts of crosslinker could be employed, which yielded hyperbranched polymers with high crosslink density, i.e. without gelation. These highly branched and soluble polymers can also be viewed as microgels. Under such hard conditions, RCC occurs by a "chain degradation mechanism" involving backbiting followed by β-scission reactions, while bimolecular termination proved negligible. The use of CSTR is thus advantageous with of view of controlling both the occurrence of gelation and the molar mass of the hyperbranched polymers.

RCC in solution When copolymer chains precipitate out of the solvent of polymerization and the comonomers are soluble, the process is named precipitation RCC. In highly dilute RCCs, cyclization reactions are favored in the early stages of the process, i.e. (c) in Fig. 8.14, which results in a decrease in the size of the polymer coil. This is due to the fact that the local concentration of pendant vinylic groups within the same macromolecule is much higher than their overall concentration. As the concentration of microgels increases, reactions between distinct particles are more probable, i.e. (e). Branched species thus formed can undergo further intramolecular crosslinkings (d), leading to globular and compact microgels. However, the number of pendant vinylic groups decreases, which decreases the probability of further intramolecular crosslinkings. The concentrations in crosslinker and in monomer are the key parameters to reach or avoid gelation. Okay has proposed a theoretical study to explain these phenomena [219] and Funke et al. [220], Antonietti et al. [221] and Ishizu et al. [222] have thoroughly investigated the mechanism of RCC in solution – in general DVB is used as the crosslinker. Generally speaking, RCC in solution provides microgels with a less precise size control and with broader size distribution than for microgels formed by emulsion RCC (see next section).

From a practical viewpoint, it is recommended to perform RCC in solution typically below 5% in order to avoid macrogelation. Alternatively, one can discontinue RCC before its completion that is, before the observation of the gel point. However, Graham et al. have described opposite trends, assuming that the critical gelation concentration below which no macrogelation occurs – even at complete conversion – can be as high as 90% [223]. Obviously, the solvent selection is of crucial importance for the the synthesis of microgels in solution at such high solid contents. This particularly applies when synthesizing microgels in a very good solvent where the polymer chains can swell and behave as efficient steric stabilizers, minimizing intermolecular reactions but favoring intramolecular crosslinkings.

A combination of solvents can be used to adjust the solvent quality with regards to the polymer chains. Stöver et al. have thus established the "morphological map" of compounds obtained from the RCC of DVB and either 4-methylstyrene [224] or maleic anhydride [225] in mixtures of methylethylketone (MEK) and heptane. Four types of morphologies have been observed depending on solvent composition, among which, microspheres, microgels, macrogels and "coagulum". Microgels were formed in a mixture corresponding to the limit of solubility of the copolymer.

If original studies on RCC in solution investigated mostly DVB/styrene, numerous examples of RCC involving various divinyl/monovinyl monomers have also been described subsequently [13]: for instance, functionalized/reactive microgels derived from maleic and fumaric acids [226], microgels carrying specific groups that can further bind metal ions [227] and microgels possessing catalytic functional groups for subsequent enantioselective reductions [228]. Fullerene (C_{60}) also served as multivalent crosslinker in RCC with 4-vinylbenzoic acid or 2- and 4-vinylpyridine using *o*-dichlorobenzene as a solvent [229]; the motivation here being to develop functionalized microgels with unique rheological and electroactive properties.

Recently, Sherrington et al. carried out solution RCC of methacrylic-type monomers at relatively high monomer concentrations and succeeded in avoiding macrogelation by employing mercaptans as chain transfer agents (CTAs) [230]. In these examples, the authors claimed that "hyperbranched copolymers" were obtained rather than microgels. However, the latter process is very similar to that used in RCC, except for the presence of a CTA. Indeed, soluble and highly branched copolymers in the sub-micron size range were produced as in the case of the synthesis of hyperbranched (co)polymers [231]. It has to be mentioned that the use of CBr4 as a CTA for RCC in more dilute solution had already been reported by Ishizu et al. [222].

As discussed below, other groups have proposed alternative strategies to avoid macrogelation in RCC. In a series of reports, the group of Sato has developed a method called "initiator-fragment incorporation radical polymerization" (IFIRP) so as to produce hyperbranched polymers/microgels [232]. The key point of the IFIRP method is the use of a high initiator concentration during the free-radical homopolymerization of a multivinylic monomer. A large number of initiator fragments could be incorporated as terminal groups in the resulting branched polymers via initiation and primary radical termination. In these studies, dimethyl 2,2′-azobisisobutyrate (MAIB) was chosen as the radical source while DVB, EGDMA, divinyl adipate **14** (Fig. 8.4) and even polybutadiene were used as crosslinkers. Noteworthy, no monovinylic monomer was added in these IFIRPs so that these experiments were not really RCCs, except when NiPAm was copolymerized with DVB in the presence of MAIB [233].

Rimmer et al. have reported the copolymerization of vinyl acetate with allyl carbonates possessing isopropyl groups, which afforded highly branched poly(vinyl acetate)s (PVAc)s [234]. Indeed, the isopropyl group induces significant transfer reactions with the propagating radical PVAc radical by abstraction of the isopropyl hydrogen, which generates the branching points. Although, compounds described in these works [232–234] are hyperbranched copolymers rather than true microgels [231], they exhibit similar molecular features as those of RCC-derived copolymers.

A few groups have recently exploited the potential of controlled/living radical polymerization (C/LRP) techniques in RCC for the synthesis of better defined and, possibly, more homogeneous microgels. The three main methodologies employed for this purpose were NMP [48, 235], ATRP [236] and RAFT [55, 237, 238]. Figure 8.15 shows the expected differences between conventional and controlled RCCs as the result of differences in chain growth mechanism, which also suggests differences in the crosslinking processes and the spatial distributions of crosslinks arising therefrom [235]. In conventional RCC, polymers of large size are constantly created from the beginning to the end of the copolymerization and corresponding growing radicals exhibit a very short lifetime. As mentioned above, this scenario implies intramolecular cyclization followed by intermolecular couplings,

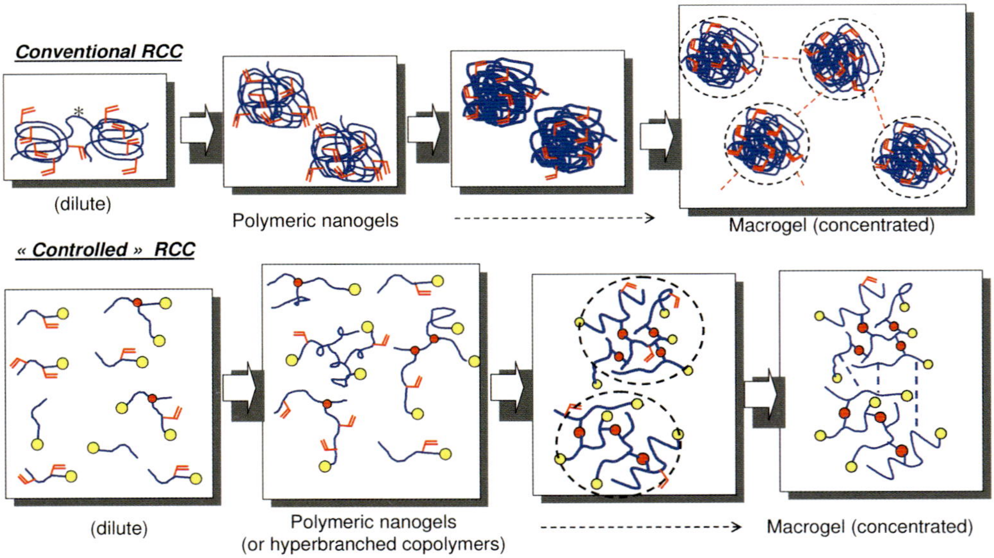

Fig. 8.15 Conventional RCC vs. "controlled" RCC.

yielding heterogeneous microgels containing many defaults. In contrast, nearly all chains are created at the same time in controlled RCC, which generates a high and steady concentration of primary chains from the very beginning of the copolymerization. Therefore, crosslinks are introduced nearly randomly throughout the course of controlled RCC since all chains are subjected to the same effect of the crosslinker. Chain growth in "controlled" RCC is slower than in conventional RCC because the overall concentration of active radicals is reduced and the chains are mostly in their dormant form. Microgels obtained from "controlled" RCC can thus be viewed as "hyperbranched copolymers" [231] and their structures differ from that obtained by conventional RCC. In particular, the crosslinking density from one macromolecule to the other should be the same. In both scenarios, macrogels can be formed from their parent microgels at high solid content. Overall, the presence of the controlling agent offers the following advantages over conventional RCC for the synthesis of microgels: (i) incorporation of higher concentrations in crosslinker while still maintaining soluble nano-objects; in other words, it retards and even prevents macrogelation; (ii) solution RCCs can be run at higher solid contents than the 5% limit above which macrogelation is generally observed in conventional RCC; (iii) branched polymers with an improved homogeneity of their internal structure can be obtained; (iv) can take advantage of the possibility of growing further chains from dormant ones, which involves a covalent attachment of the shell to the core (see route E in Fig. 8.3), thus forming, core–shell star-like structures with a polymeric nanogel core in a divergent manner. Depending on the targeted application, the shell can offer a protection/stabilization of the core by a steric and/or an electrostatic effect, or allow the incorporation of suitable functionalities.

Taton et al. reported the direct synthesis of nanogels based on polyacrylamide and poly(acrylic acid) by RCC in aqueous solutions, with no need to protect the functional group of the monomers, using an *O*-ethylxanthate as a reversible CTA by the MADIX process so as to avoid macrogelation [237]. The molar masses and polydispersities of these branched copolymers could be varied as a function of the concentrations of the starting reagents. Even in the presence of a rather high proportion of crosslinker (up to 15 mol%), no macrogelation was observed provided enough xanthate was utilized. Chain extension from these nanogels afforded multiarm star polymers and a practical pathway to this kind of architecture (route E, Fig. 8.3). A similar approach was followed by Perrier et al. to obtain hydrophobic "hyperbranched copolymers" based on polymethacrylate; in this case RCC was performed in toluene in the presence of a dithioester as a RAFT agent [238]. Microgels possessing glucoside moieties were also synthesized by Kakuchi et al. by TEMPO-mediated RCC of 4-vinylbenzyl glucoside peracetate with DVB in *m*-xylene [239].

Finally, star-like polymers with a core–shell structure where the core is built by adding a crosslinker onto predefined polymer precursors through the arm-first methodology also belong to the category of "controlled" microgels grown by RCC. For information on these types of microgels, see Section 8.2.1 and the references cited therein.

RCC in emulsion RCC in emulsion is a standard method of microgel synthesis. This process is also of great utility for preparing microgels with a core–shell structure [13, 190, 191, 240] or reactive microgels for coating applications [241]. Different morphologies, e.g. microgels with internal pores of different size, are generated as a result of RCC performed under homogeneous (solution) or heterogeneous (dispersion) conditions. The formation of a porous structure has been proved to efficiently enhance the deswelling rate of stimuli-responsive macrogels based on poly(NIPAM) [242, 243]. For this purpose, varying the rate of addition of the comonomers in a semi-continuous process is a convenient means for fine-tuning the structural features of microgels: materials with different morphologies may be eventually obtained when using batch or semi-continuous processes.

Compared to a crosslinker-free emulsion polymerization, RCC in emulsion yields smaller particles of narrower size distribution, which is due to the internal crosslinked structure of the microgels formed [244]. A higher number of polymer particles is obtained in the latter case because of a lower rate of growth of the particles, as the crosslinking proceeds. The pendant vinylic groups present in a given particle may also react with the propagating radical species of a neighboring particle, thus favoring interparticular reactions. Furthermore, there is a higher probability of chain growth within the monomer droplets, which was ascribed to the fact that termination events are slower – thus favoring further propagation in monomer droplets – during RCC in emulsion than during an emulsion crosslinker-free polymerization. Simulations by the Monte Carlo method proved quite useful in providing an insight into the kinetics, mechanism and molar mass distribution of emulsion polymerization of multifunctional monomers [245, 246]. Practically speaking, a wide range of microgels obtained by RCC in emulsion has been reported [14] such as microgels based on lauryl methacrylate/trimethylolpropane trimethacrylate [247], methyl methacrylate/butanediol dimethacrylate [248], *N*-ethylacrylamide/*N,N'*-methylenebisacrylamide [249], *N*-vinylcaprolactam [250] or methyl methacrylate, 2-ethoxyethyl methactylate, ethylene glycol methacrylate, and 2-(tetradecyldimethylammonio) ethyl methacrylate bromide [251]. In most of these examples, conventional surfactants such as sodium dodecylsulfate (SDS), potassium laurate, or alkylated PEG have been used. In contrast, Ishii designed zwitterionic oligomers that were used as an emulsifier for RCC in emulsion and thus obtained functionalized microgel [252]; the corresponding latexes were found to have a better shelf stability than if a SDS was used. An illustrative example of stimuli-resposive microgels has been recently reported by Armes et al.: emulsion RCC of 2-vinylpyridine and DVB in the presence of a cationic surfactant and a monomethoxy-capped PEO methacrylate used as reactive steric stabilizer (macromonomer) afforded monodisperse and stabilized latexes [253]. The mean weight-average particle diameters of these microgels ranged from 370 to 970 nm depending on the initiator, polymeric stabilizer and surfactant concentrations. A further acidic treatment of vinylpyridine units afforded cationic microgels exhibiting very fast swelling capacities.

The particular case of allyl methacrylate is interesting since it has two vinylic groups with a distinct reactivity; the use of this comonomer in microgel synthesis by emulsion was investigated by Matsumoto et al. [254]. Homopolymerization of allymethacrylate, its copolymerization with various comonomers (e.g. diallyl terephtalate, allyl benzoate, methyl methacrylate), as well as its comparative reactivity with vinyl methacrylate have been reported [254].

Above a critical value of surfactant concentration, monomer droplets can be eliminated, which leads to a transparent micellar solution, called a microemulsion, in which RCC allows one to obtain nearly monodisperse microgels. For instance, Antonietti et al. prepared microgels based on PS of various sizes by RCC in a microemulsion [255]. Chemical modifications of the latter structures allowed these authors to derive polyelectrolyte microgels [256]. In most cases, an aqueous continuous liquid phase is used in emulsion RCCs to synthesize dispersed hydrophobic microgels. However, inverse emulsion can be utilized to polymerize hydrophilic monomers by RCC, using hydrophobic organic dispersed media. A typical example is that of RCC of acrylamide, for which the influence of various parameters (solvent nature, type and concentration of surfactant, concentration of reagents, etc.) on the overall properties of the formed polyacrylamide-based microgels has been extensively investigated [257, 258]. RCC in inverse emulsion can also be applied to anionic monomers (e.g. sodium 2-acrylamido-2-methylpropanesulfonate or acrylic acid) or cationic ones (e.g. [2-(methacryloyloxy)ethyl]trimethylammonium chloride) for the synthesis of polyelectrolyte microgels of narrow size distribution [259]. When equimolar amounts of cationic and anionic monomers are used in the presence of a crosslinker for RCC in an inverse emulsion, polyampholyte microgels exhibiting unique swelling and flocculation properties can be obtained [259].

RCC by surfactant-free precipitation As mentioned above, the choice of a proper solvent induces the precipitation (deswelling) of growing chains in precipitation (or dispersion) RCC; the initial mixture of comonomers is soluble in the solvent and the formed polymer precipitates during polymerization. When properly sterically and/or electrostatically stabilized, the corresponding phase-separated particles are monodisperse. Precipitation RCC can be viewed as a special case of RCC in solution but the process also resembles surfactant-free RCC in emulsion [190]. Precipitation polymerization is particularly appropriate for synthesizing thermosensitive aqueous microgels such as poly(NipAm), poly(*N*isopropylmethacrylamide) [260, 261] or poly(*N*-vinyl caprolactam) and their copolymers with other hydrophilic monomers [262, 2637], by carrying out precipitation polymerization above the LCST of the polymer. Polyelectrolyte microgels based on poly(acrylic acid) or poly(methacrylic acid) can also be obtained by precipation RCC [264, 265]. Precipitation RCC in water of vinyl acetate and *N*,*N*-dimethyl-*N*-*N*-divinyl-sebacoyldiamide used as a crosslinker, followed by alkaline methanolysis of the crosslinked poly(vinyl acetate) formed, afforded neutral microgels based on poly(vinyl alcohol) [266].

8.3.1.2 **Anionic Crosslinking Copolymerization**

Such a synthetic approach to microgels where the latter serve as a core to grow core–shell architectures has been described in detail in Section 8.2.2.1 This was exemplified using DVB or EGDMA as crosslinkers.

8.3.1.3 **Step-growth Crosslinking Copolymerization**

A few examples of microgels obtained by step growth polymerization have been described. For instance, Graham et al. reported the step-growth polymerization of α,ω-dihydroxytelechelic PEO and bis-(cyclohexylmethane-4,4-diisocyanate) in various solvent compositions at rather high solid content, using 1,2,6-hexanetriol as crosslinker [267]; microgels based on polyurethanes were thus obtained. In this case, the PEO played the role of a steric stabilizer and prevented macrogel formation, as in the precipitation surfactant-free crosslinking copolymerization described above. Polyorganosiloxane-based microgels have also been synthesized by direct emulsion polycondensation of trimethoxymethylsilane in aqueous media [268, 269]. Further grafting of polystyrene macromonomer chains onto these polyorganosiloxane microgels led to nanospheres with a core–shell structure that were subsequently blended with linear PS chains. No depletion demixing was observed, in sharp contrast to the situation prevailing for standard colloid–polymer mixtures [268].

8.3.1.4 **Crosslinking of Pre-formed Polymers**

An alternative approach to the preparation of microgels is to trigger the crosslinking of linear polymeric precursors at the intramolecular level, following the so-called "monomer-free approach". The starting material, indeed, is a collection of already formed macromolecules corresponding to synthetic or natural polymers. As discussed below, the crosslinking process can be chemically or photochemically induced or achieved by radiation using environmentally friendly conditions.

The chemical route One way to achieve the chemical intramolecular crosslinking of pre-formed polymers is to take advantage of the presence of functional reactive groups present along the polymeric backbone. To avoid the formation of a macroscopic gel, one has to carry out polymerization under dilute conditions, i.e. at a concentration that is generally much lower than the critical overlap concentration (C^*) of the polymer in the solvent. Likewise, the concentration of the crosslinker has obviously a direct impact on the internal crosslinking density. For instance, poly(4-vinyl styrene) can be prepared and further crosslinked intramolecularly using AIBN as the radical source, as reported by Batzilla and Funke [270]. Alternatively, one can induce the crosslinking by photoinitiation [271]. In a recent addition, functionalized polymeric nanogels have been synthesized by intramolecular crosslinking of a RAFT-derived backbone based on PS that also contained amino functional groups further derivatized into styryl moieties. The latter functions served for further crosslinking [272]. In this case, advantage was taken of the controlled radical RAFT polymerization process to induce the formation of well-defined functionalized nanoparticules. Synthesis of hydrophilic polymeric nanogels could be performed in aqueous dilute solutions using an appropriate crosslinker that reacted with the functional groups of the parent polymer. The latter method is also of particular interest for synthesizing microgels based on polysaccharides such as hydroxypropylcellulose [273] or hyaluronic acid [274]; in this case divinylsulfone can be used as crosslinker. Also important in various applications (oil field, water treatment, biomedical applications, etc.), microgels based on poly(vinyl alcohol) could be prepared in aqueous solutions using glutaraldehyde as crosslinker [275, 2760]. The extent of intramolecular reactions and the ratio of aldehyde functions to hydroxyl ones at the critical gel point was found to increase upon decreasing the concentration of PVA [276]. The structure of the PVA hydrogels/microgels could thus be adjusted by varying the concentration of crosslinker or that of the PVA precursor. As for poly(allylamine), it can be intramolecularly crosslinked with 1,4-dimethoxybutane-1,4- diimine dihydrochloride [277–279].

The radiative route Ionizing radiation is another method for synthesizing microgels by intramolecular crosslinking of individual polymer chains. In contrast to the previously described methods, ionization radiation can be carried out in the absence of crosslinkers. In addition, short time-scales (a few microseconds) of an intense pulse of ionizing radiation can be sufficient to trigger intramolecular crosslinking reactions following a radical pathway. The average number of radicals created along the polymer backbone is dependent on the initial conditions of irradiation, which directs the crosslinking towards micro- or macrogel formation: the higher the concentration of radicals present along the same chain, the higher the probability of intramolecular crosslinking. Obviously, there is a competition between inter- and intramolecular reactions as a function of these parameters; both mechanisms occur during the process. A steady-state concentration of radicals (around 10^{-7} mol L^{-1}) can be obtained from low dose rates of continuous irradiation, typically γ-ray irradiation from isotope sources. If the concentration of the polymer chains is higher than the above value, then intermolecular crosslinking reactions are favored. In contrast, microgel formation is achieved as a result of intramolecular crosslinkings using short and intense pulses of radiation from an accelerator generating fast electrons. In this case, the average concentration of radicals is in the range 10^{-4}–10^{-3} mol L^{-1} and the concentration of polymer chains should be kept lower than these values so as to induce the creation of a large enough number of radicals on the same polymeric backbone. The ionizing radiation method has been exemplified for the synthesis of neutral or charged water-soluble microgels such as poly(vinyl alcohol), polyvinylpyrrolidone [280] and poly(vinyl methyl ether) [281, 282] etc.

8.3.2
Hydrophilic Microgels

Water-soluble and stimuli-responsive microgels (aqueous colloidal microgels) represent a special class of branched polymers that have been thoroughly investigated in the two last decades [190–192, 283]. They include pH-sensitive and temperature-sensitive systems such as nanogels based on poly(acrylic acid) and poly(*N*-isopropylacrylamide) -poly(NiPAm)-, the latter polymer being known to undergo a shrinking effect by expelling water above its lower critical solution temperature (LCST) at 32 °C. Poly(NiPAm) microgels were first synthesized by Pelton and Chibante in 1986 [284], and since then they have been the subject of numerous investigations [13, 14, 190–192] Various hydrophilic polymers, including natural polymers such as polysaccharides are also susceptible to the formation of water-soluble microgels. Methods to synthesize pH- as well as temperature-sensitive PNiPAm-based microgels are essentially based on emulsion RCC of NiPAm with a pH-sensitive comonomer such as acrylic acid [285], methacrylic acid [286–288] etc. In this case, the content in pH-sensitive units and the pH of the aqueous solution have a dramatic impact on the LCST of the obtained PNiPAm-based microgels.

8.3.3
Microgels with a Core–Shell Structure

Different types of core–shell systems can be obtained [14, 191]. The core can be constituted of a non-microgel material (e.g. polystyrene latex, silica, etc.) and the shell of a microgel. On the other hand, both the core and the shell can be made of a microgel-like material. In many cases, the preformed core generally serves as a seed for the subsequent growth of the shell by precipitation polymerization. With cores functionalized at their periphery, shells can be either grafted or physically adsorbed through precipitation of the growing oligomers captured by the core. If the crosslinker is consumed faster than the monovinylic comonomer in RCC, a progressive decrease in crosslinking density is induced from the core to the periphery of the particle, which also results in the formation of core–shell microgels [289]. Interestingly, this can be influenced by the polymerization process (batch or semi-continuous conditions). Finally, chain extension from microgels obtained by "controlled" RCC also affords multiarm star polymers with core–shell architectures [237, 238].

"Smart" doubly stimuli-responsive microgels exhibiting tuneable responses to both pH and/or temperature have also been reported. For instance, doubly temperature sensitive microgels [290] or microgels with temperature-sensitive cores and pH-sensitive shells have been described [291, 292]. In addition, Lyon et al. have prepared core–shell microgels where the core was made of crosslinked poly(NiPAm)or poly(NiPAm-co-acrylic acid) and the shells based on either poly(-NiPAm-co-acrylic acid) or poly(NiPAm) [293].

8.3.4 General Methods of Characterization of Microgels

Their solubility permits characterization microgels with standard analytical means used for linear polymers. These include methods of characterization in solution (in the dilute and semi-dilute regimes) such as viscometry, radiation scattering (light, neutrons), size exclusion chromatography, NMR, ultracentrifugation, osmometry, etc. and analytical methods in the solid state such as electron microscopy (TEM) on dried or freeze-dried samples, cryo-TEM and freeze-fracture TEM, as well as atomic force microscopy (AFM) for a visualization of the entire objects. Drying methods may have significant influence on the final structure and properties of aqueous microgels [294]. Interestingly, analytical techniques allowing *in situ* real-time monitoring of RCC in aqueous solutions were developed by Kara and Pekcan for RCC of acrylamide and MBA [295]. The surface charge density of polyelectrolyte microgels could be investigated by determining their zeta potential or by phase analysis light scattering [296]. For instance, size determination of microgels in correlation to their molar mass was monitored using viscosimetry in dilute solutions, through the evolution of the intrinsic viscosity, $[\eta]$. In contrast to the values taken by the a coefficient of the Mark–Houwink–Sakurada law – $[\eta]=KM_v^a$ – for linear and flexible polymers adopting a random coil conformation in a good solvent – which is in the range 0.7–0.8 – values of the exponent obtained for microgels are much lower (for instance, $a=0.25$ for microgels based on PS [297], indicative of their highly branched structure. Changes in volume occurring during the formation of microgels by crosslinking copolymerization of a mixture of comonomers (e.g. microgels based on P*t*BA) [298] or after intramolecular crosslinking of preformed linear polymers [299] could also be followed from $[\eta]$ value measurements.

Aqueous stimuli-responsive microgels have been the subject of detailed investigations of their physico-chemical properties, including their viscometric behavior in aqueous solution. For instance, Richtering and Senff described the influence of temperature on the viscosity of microgels based on poly(NiPAm) [300] while Antonietti et al. reported viscometric data of polyelectrolyte microgels [301].

Rheology is another useful method to gain information about the compactness and stiffness of microgels, their susceptibility to flow-induced deformation, solvent permeability interactions between nanoparticles, etc. The structure–rheology relationship of microgels has been highlighted [302]. Changes in rheological properties of the reaction mixture can also yield important information about the mechanism and kinetics of crosslinking copolymerization [303]. Indeed, the homogeneity/heterogeneity of the internal structure of microgels could be inferred from the evolution of the Newtonian viscosity of a solution with its volume fraction. The viscosity of a colloidal dispersion diverges at high volume fraction, in a way which is related to the compressibility of the particles, and therefore related to the crosslinking density and homogeneity. The evolution of the Newtonian viscosity with the effective volume fraction can, on the other hand, be compared with hard sphere models [304], in order to evaluate the microgel architecture in terms of "hard" versus "soft" colloids.

Static light scattering and dynamic light scattering allow one to determine, respectively, the radius of gyration and the hydrodynamic radius and distributions thereof, whereas data from small-angle neutron scattering (SANS) provide information about the characteristic size of the concentration fluctuations in the internal structure, using appropriate modelling [305]. As a result of the internal structural heterogeneity in microgels, the scattered intensity observed in DLS and SANS can be divided into a static part, which is a result of frozen-in inhomogeneities, and a dynamic part caused by the thermal fluctuations of the network.

8.3.5 Applications of Microgels

It is beyond the scope of this chapter to detail all applications of microgels. A complete description and relevant literature can be found elsewhere [13, 14, 190, 191, 283]. Applications of polymeric nanogels/microgels mainly concern the domain of coatings (paints, varnishes, etc.) where they are used at high solid-contents in formulations, with the noticeable advantage of reducing solvent emission while maintaining the proper viscosity. Highly swellable microgel dispersions obtained with a lower crosslink density can present shear thinning behaviour [306]. Therefore, such nanogels can be used as rheology control agents in high-solid and water-based systems.

Hydrophilic microgels as well as hydrogels are also intensively developed in biotechnology, either in drug and gene delivery or as biomaterials [190–192]. As highly crosslinked microgels are hardly swellable, they provide dense and high solid-content coating formulations with low viscosity. The swellability of pH- or temperature-sensitive hydrophilic microgels is closely linked to their crosslinking density.

Microgels are also currently being investigated for their use as drug delivery systems, in chromatographic separation technology, as catalyst media, etc.

In addition, they can be excellent microreactors for the preparation of different inorganic materials. The reason for this is the network structure of microgels which can be used for the preparation of semiconductor, metal, or magnetic nanoparticles.

8.4 Conclusions

Star polymers and microgels are two categories of branched polymers which share the common feature of being exploited industrially, in particular in the coatings industry. Structurally well-defined star polymers are essentially accessible by controlled/living polymerization involving solution processes and represent good models of branched polymers. As for microgels, they are considered as colloidal macromolecular objects of fixed shape that are usually more readily derived using low-cost synthetic methodologies from environmentally friendly processes, such as direct emulsion, microemulsion or surfactant-free precipitation polymerizations. Microgels are also viewed as a transition from linear macromolecules to larger macroscopic polymeric materials. The first examples of the synthesis of star polymers date back to the late 1940s, whereas a complete and clear description of microgels is more recent and was proposed in the late 1970s. However, synthetic developments, investigation of the properties of microgels and applications of these materials as specialty polymers have been the subject of intensive research in the past two decades. The size of both star polymers and microgels can be varied over a wide range, from nano- to microscale, by choosing the proper polymerization technique/process of polymerization and suitable monomer combinations. In addition, particular functions located either at their periphery and/or within their core can be incorporated in both types of architectures, which allows the adjustment of microgels or star polymers to special applications. Interestingly, "hybrid" core–shell structures that are more complex star-like polymers consisting of a microgel core can be obtained using specific synthetic methods. Of particular interest for use in biomedical or pharmaceutical applications are hydrophilic microgels or core–shell star-like microgels exhibiting stimuli-responsiveness in aqueous solution, i.e. those which respond to a change in their size and/or shape in the environment. Looking at both the patent and the academic literature, it becomes clearer that star polymers and microgels hold great promise as specialty materials in bio- and nanotechnology.

References

1 N. Hadjichristidis, M. Pitsikalis, S. Pispas, H. Iatrou *Chem. Rev.* **2001**, *101*, 3747.

2 (a) O. Webster, *Science* **1991**, *251*, 887; (b) K. Matyjaszewski *J. Phys. Org. Chem.* **1995**, *8*, 197.

3 K. Mishra, S. Kobayashi, *Star and Hyperbranched Polymers*, Marcel Dekker, New York **1999**.

4 G. S. Grest, L. J. Fetters, J. S. Huang, D. Richter *Advances in Chemical Physics*, Vol. XCIV, Wiley & Sons, New York **1996**.

5 K. Inoue *Prog. Polym. Sci.* **2000**, *25*, 453.

6 K. Ishizu, S. Uchida *Prog. Polym. Sci.* **1999**, *24*, 1439.

7 A. Bhattacharya, B. N. Misra *Prog. Polym. Sci.* **2004**, *29*, 7676.

8 For a recent review on hyperbranched polymers see for instance: (a) B. Voit *J. Polym. Sci., Part A: Polym. Chem.* **2000**, *38*, 2505–2525; (b) C. Gao, D. Yan *Prog. Polym. Sci.* **2004**, *29*, 183.

9 For a recent review on dendrimers see for instance: (a) S. M. Grayson, J. M. J. Fréchet *Chem. Rev.* **2001**, *101*, 3819–3867; (b) D. A. Tomalia *Prog. Polym. Sci.* **2005**, *30*, 294.

10 For a review on dendrimers with macromolecular generations see S. J. Teerstra, M. Gauthier *Prog. Polym. Sci.* **2004**, *29*, 277.

11 (a) K. Matyjaszewski *ACS Series 854, Advances in Controlled/Living Radical Polymerization*, K. Matyjaszewski (Ed.), Vol. 854, pp. 2–9, **2003**; (b) *Handbook of Radical Polymerization*, K. Matyjaszewski, T. Davis (Eds.), John Wiley & Sons, New York **2002**.

12 Y. Gnanou, D. Taton *Handbook of Radical Polymerization*, K. Matyjaszewski,

T. Davis (Eds.), John Wiley & Sons, New York **2002**, Vol. 14, p. 775.

13 W. Funke, O. Okay, B. Müller-Joos *Adv. Polym. Sci.* **1998**, *136*, 139.

14 P. Ulanski, J. M. Rosiak *Encyclopedia of Nanoscience and Nanotechnology* **2004**, Vol. X, p. 1.

15 D. J. Worsfold, J. G. Zilliox, P. Rempp *Can. J. Chem.* **1969**, *47*, 3379.

16 H. Eschwey, M. L. Hallensleben, W. Burchard *Makromol. Chem.* **1973**, *173*, 235.

17 A. Kohler, J. G. Zilliox, P. Rempp, J. Pollacek, I. Koessler *Eur. Polym. J.* **1972**, *8*, 627.

18 (a) L. K. Bi, L. J. Fetters *Macromolecules* **1976**, *9*, 732; (b) R. N. Young, L. J. Fetters *Macromolecules* **1978**, *11*, 899.

19 F. A. Taromi, P. Rempp *Makromol. Chem.* **1989**, *190*, 1791.

20 (a) H.-J. Lee, K. Lee, N. Choi *J. Polym. Sci.: Part A: Polym. Chem.* **2005**, *43*, 870; (b) H.-J. Lee, K. Lee, S. N. Lee J. *Polym. Sci.: Part A: Polym. Chem.* **2006**, *44*, 2579.

21 (a) O. Okay, W. Funke, *Makromol. Chem. Rapid Commun.* **1990**, *11*, 583; (b) O. Okay, W. Funke, *Macromolecules* **1990**, *23*, 2623.

22 (a) L. Pille, D. H. Solomon *Macromol. Chem. Phys.* **1994**, *195*, 2477; (b) S. Abrol, D. H. Solomon *Polymer* **1999**, *40*, 6583.

23 D. Held, A. H. E. Muller, *Macromol. Symp.* **2000**, *157*, 225.

24 G. Lapienis, S. Penczek *Macromolecules* **2000**, *33*, 6630.

25 O. W. Webster *Adv. Polym.* Sci. **2004**, *167*, 1.

26 P. Lang, W. Burchard, M. S. Wolfe, H. J. Spinelli, L. Page *Macromolecules* **1991**, *24*, 1306.

27 K. Matyjaszewski, M. Sawamoto *Cationic Polymerizations, Mechanisms, Synthesis, and Applications*, K. Matyjaszewski (Ed.), Marcel Dekker, New York **1996**, Vol. 5, p. 381.

28 S. Kanaoka, M. Sawamoto, T. Higashimura *Macromolecules* **1991**, *24*, 2309.

29 H. Deng, S. Kanaoka, M. Sawamoto, T. Higashimura *Macromolecules* **1996**, *29*, 1772.

30 (a) T. M. Marsalko, I. Majoros, J. P. Kennedy *Macromol. Symp.* **1995**, *95*, 39; (b) J. P. Kennedy, L. R. Ross, J. E. Lackey, O. Nuyken *Polym. Bull.* **1981**, *4*, 67.

31 G. C. Bazan, R. R. Schrock *Macromolecules* **1991**, *24*, 817.

32 V. Héroguez, Y. Gnanou, M. Fontanille *Macromolecules* **1997**, *30*, 4791.

33 See for instance the following review: K. Ito *Prog. Polym. Sci.* **1998**, *23*, 581.

34 K. Ishizu, K. Sunahara *Polymer* **1995**, *36*, 4155.

35 K. Ito, K. Hashimura, S. Itsuno, E. Yamada *Macromolecules* **1991**, *24*, 3977.

36 K. Ishizu, H. Kitano *Macromol. Rapid Commun.* **2000**, *21*, 979.

37 C. J. Hawker, A. W. Bosman, E. Harth *Chem. Rev.* **2001**, *101*,3661.

38 M. Kamigaito, T. Ando, M. Sawamoto *Chem. Rev.* **2001**,101, 3689.

39 K. Matyjaszewski, J. Xia *Chem. Rev.* **2001**, *9*, 2921.

40 G. Moad, E. Rizzardo, S. Thang *Aust. J. Chem.* **2005**, *58*, 379.

41 S. Perrier, P. Takolpuckdee *J. Polym. Sci. A: Polym. Chem.* **2005**, *43*, 5347.

42 J. Xia, X. Zhang, K. Matyjaszewski *Macromolecules* **1999**, *32*, 4482.

43 X. Zhang, J. Xia, K. Matyjaszewski *Macromolecules* **2000**, *33*, 2340.

44 T. Furukawa, K. Ishizu *Macromolecules* **2005**, *38*, 2911.

45 (a) K.-Y. Back, M. Kamagaito, M. Sawamoto *Macromolecules* **2001**, *34*, 215; (b) K.-Y. Back, M. Kamagaito, M. Sawamoto, *Macromolecules* **2001**, *34*, 7629; (c) K.-Y. Baek, M. Kamigaito, M. Sawamoto, *J. Polym. Sci.: Part A: Polym. Chem.* **2002**, *40*, 633; (d) K.-Y. Baek, M. Kamigaito, M. Sawamoto *Macromolecules* **2002**, *35*, 1493; (e) K.-Y. Baek, M. Kamigaito, M. Sawamoto *J. Polym. Sci.: Part A: Polym. Chem.* **2002**, *40*, 2245.

46 T. Terashima, M. Kamigaito, K.-Y. Back, T. Ando, M. Sawamoto *J. Am. Chem. Soc.* **2003**, *125*, 5288.

47 C. Bouilhac, E. Cloutet, H. Cramail, D. Taton, A. Deffieux *Macromol. Rapid Commun.* **2005**, *26*, 1619.

48 (a) S. Abrol, P. A. Kambouris, M. G. Looney, D. H. Solomon *Macromol. Rapid. Commun.* **1997**, *18*, 755; (b) S. Abrol,

M. J. Caulfield, G. G. Qiao, D. H. Solomon *Polymer* **2001**, *42*, 5987.
49 T. E. Long, A. J. Pasquale *J. Polym. Sci.: Part A: Polym. Chem.* **2001**, *39*, 216.
50 T. Tsoukatos, S. Pispas, N. Hadjichristidis *J. Polym. Sci.: Part A: Polym. Chem.* **2001**, *39*, 320.
51 (a) A. Narumi, T. Satoh, H. Kaga, T. Kakuchi *Macromolecules* **2002**, *35*, 699; (b) A. Narumi, S. Yamane, Y. Miura, H. Kaga, T. Satoh, T. Kakuchi *J. Polym. Sci.: Part A: Polym. Chem.* **2005**, *43*, 4373.
52 (a) A. W. Bosman, A. Heumann, G. Klaerner, D. Benoit, J. M. J. Fréchet, C. J. Hawker *J. Am. Chem. Soc.* **2001**, *123*, 6461-6462; (b) A. W. Bosman, R. Vestberg, A. Heumann, J. M. J. Fréchet, C. J. Hawker *J. Am. Chem. Soc.* **2003**, *125*, 715.
53 B. Helms, S. J. Guillaudeu, Y. Xie, M. McMurdo, C. J. Hawker, J. M. J. Fréchet *Angew. Chem. Int. Ed.* **2005**, *44*, 6384.
54 P. B. Zetterlund, M. D. Alalm, H. Minami, M. Okubo *Macromol. Rapid Commun.* **2005**, *26*, 955.
55 D. Taton, J.-F. Baussard, L. Dupayage, Y. Gnanou, Mathias Destarac, C. Mignaud, C. Pitois *ACS Series, Advances in Controlled/Living Radical Polymerization*, K. Matyjaszewski (Ed.), **2006**, Vol. 39, p. 578
56 H. T. Lord, J.-F. Quinn, S. D. Angus, M. R. Whittaker, M. H. Stenzel, T. P. Davis *J. Mater. Chem.* **2003**, *13*, 2819.
57 G. Moad, R. T. A. Mayadunne, E. Rizzardo, M. Skidmore, S. H. Thang *Macromol. Symp.* **2003**, *192*, 1.
58 (a) G. Zheng, C. Pan *Polymer* **2005** *46*, 2802; (b) G. Zheng, Q. Zheng, C. Pan *Macromol. Rapid Commun.* **2006**, *207*, 216; (c) G. Zheng, C. Pan *Macromolecules* **2006** *39*, 95.
59 M. Morton, T. E. Helminiak, S. D. Gadkary, F. Bueche *J. Polym. Sci.* **1962**, *57*, 471.
60 L. J. Fetters, M. Morton *Macromolecules* **1974**, *7*, 552.
61 R. P. Zelinski, C. W. Wofford *J. Polym. Sci. Part A* **1965**, *3*, 93.
62 (a) J. E. L. Roovers, S. Bywater *Macromolecules* **1972**, *5*, 384; (b) J. E. L. Roovers, S. Bywater *Macromolecules* **1974**, *7*, 443.
63 N. Hadjichristidis, J. E. L. Roovers *J. Polym. Sci. Poly. Phys. Ed.* **1974**, *12*, 2521.
64 (a) N. Hadjichristidis, A. Guyot, L. J. Fetters *Macromolecules* **1978**, *11*, 668–672; (b) N. Hadjichristidis, L. J. Fetters *Macromolecules* **1980**, *13*, 191.
65 P. M. Toporowski, J. E. L. Roovers *J. Polym. Sci. Part A: Polym. Chem.* **1986**, *24*, 3009.
66 J. E. L. Roovers, N. Hadjichristidis, L. J. Fetters *Macromolecules* **1983**, *16*, 214.
67 L. L. Zhou, J. Roovers *Macromolecules* **1993**, *26*, 963.
68 J. A. Ressia, M. A. Villar, E. M. Vallés *Macromol. Symp.* **2001**, *168*, 43.
69 K. Hong, Y. Wan, J. W. Mays *Macromolecules* **2001**, *34*, 2482.
70 S. P. S. Yen *Makromol. Chem.* **1965**, *81*, 152.
71 J. T. Altares, D. P. Wyman, V. R. Allen *J. Polym. Sci. Part A.* **1965**, *3*, 4131.
72 W. A. J. Bryce, G. McGibbon, I. G. Meldrum *Polymer* **1970**, *11*, 394.
73 J. C. Meunier. R. Van Leemput *Makromol. Chem.* **1971**, *142*, 1.
74 M. Pitsikalis, S. Sioula, S. Pispas, N. Hadjichristidis, D. C. Cook, L. Jianbo, J. W. Mays *J. Polym. Sci. Part A: Polym. Chem.* **1999**, *37*, 4337.
75 M. Lazzari, T. Kitayama, M. Janco, K. Hatada *Macromolecules* **2001**, *34*, 5734.
76 A. Hirao, M. Hayashi, S. Loykulnant, K. Sugiyama, S. W. Ryu, N. Haraguchi, A. Matsuo, T. Higashihara *Prog. Polym. Sci.* **2005**, *30*, 111.
77 R. P. Quirk, Y. Tsai *Macromolecules* **1998**, *31*, 4372.
78 J. A. Gervasi, A. Gosnell *J. Polym. Sci., Part A-1* **1966**, *4*, 1391.
79 N. N. Reed, K. D. Janda *Org. Lett.* **2000**, *2*, 1311.
80 R. H. Allcock, R. Ravikiran, S. J. M. O'Connor *Macromolecules* **1997**, *30*, 3184.
81 K. Inoue, H. Miyamoto, T. Itaya *J. Polym. Sci. Part A: Polym. Chem.* **1997**, *35*, 1839.
82 S. Hou, D. Taton, M. Saule, J. L. Logan, E. L. Chaikof, Y. Gnanou *Polymer* **2003**, *44*, 5067.
83 M. Liu, K. Kono, J. M. J. Fréchet *J. Polym. Sci.: Part A: Polym. Chem.* **1999**, *37*, 3492.
84 Y. Chen, J. Smid *Langmuir* **1996**, *12*, 2207.
85 H. Liu, A. Jiang, J. Guo, K. E. Uhrich *J. Polym. Sci. Part A: Polym. Phys.* **1999**, *37*, 703.

86 K. Naka, A. Kobayashi, Y. Chujo *Macromol. Rapid Commun.* **1997**, *18*, 1025.

87 C.L. Fraser, A.P. Smith *J. Polym. Sci. Part A: Polym. Chem.* **2000**, *38*, 4704.

88 (a) Y. Ederlé, C. Mathis, *Macromolecules* **1997**, *30*, 2546; (b) J.M. Janot, H. Eddaoudi, P. Seta, Y. Ederlé, C. Mathis *Chem. Phys. Lett.* **1999**, *302*, 103; (c) S. Couris, M. Konstantaki, E. Koudoumas, Y. Ederlé, C. Mathis *Chem. Phys. Lett.* **2001**, *335*, 533; (d) F. Audouin, R. Nuffer, C. Mathis *J. Polym. Sci.: Part A: Polym. Chem.* **2004**, *42*, 4820.

89 See for instance (a) H. Fukui, M. Sawamoto, T. Higashimura *Macromolecules* **1995**, *28*, 3756; (b) H. Fukui, S. Yoshihashi, M. Sawamoto, T. Higashimura *Macromolecules* **1996**, *29*, 1862 and references cited therein.

90 (a) M. Jesberger, L. Barner, M.H. Stenzel, E. Malmström, T.P. Davis, C. Barner-Kowollik *J. Polym. Sci. Part A: Chem.*, **2003**, *41*, 3847; (b) X. Hao, C. Nilson, M. Jesberger, M.H. Stenzel, E. Malmström, T.P. Davis, E. Ostmark, C.J. Barner-Kowollik *J. Polym. Sci. Part A: Chem.*, **2004**, *42*, 5877; (c) J. Bernard, A. Favier, L. Zhang, A. Nilasaroya, T.P. Davis, C. Barner-Kowollik, M.H. Stenzel *Macromolecules* **2005**, *38*, 5475.

91 R.T.A. Mayadunne, J. Jeffery, G. Moad, E. Rizzardo *Macromolecules* **2003**, *36*, 1505–1513.

92 (a) A. Duréault, D. Taton, M. Destarac, F. Leising, Y. Gnanou *Macromolecules* **2004**, *37*, 5513; (b) V. Darcos, A. Duréault, D. Taton, Y. Gnanou, P. Marchand, A.-M. Caminade, J.-P. Majoral, M. Destarac, F. Leising *Chem. Commun.* **2004**, 2110.

93 Q. Zheng, C.-Y. Pan *Macromolecules* **2005**, *38*, 6841.

94 H. Chaffey-Millar, M. Busch, T.P. Davis, M.H. Stenzel, C. Barner-Kowollik *Macromol. Theory Simul.* **2005**, *14*, 143.

95 B. Li, J. Li, Y. Fu, Z. Bo *J. Am. Chem. Soc.* **2004**, *126*, 3430.

96 Y. Nicolas, P. Blanchard, E. Levillain, M. Allain, N. Mercier, J. Roncali *Org. Lett.* **2004**, *6*, 273.

97 R.M. Johnson, C.L. Fraser *Macromolecules* **2004**, *37*, 2718.

98 P.R. Andres, U.S. Schubert *Adv. Mater.* **2004**, *16*, 1043.

99 J.-F.Gohy, B.G.G. Lohmeijer, U.S. Schubert *Chem. Eur. J.* **2003**, *9*, 3472.

100 F. Huang, D.S. Nagvekar, C. Slebodnick, H.W. Gibson *J. Am. Chem. Soc.* **2005**, *127*, 484.

101 W. Straehle, W. Funke *Makromol. Chem.* **1978**, *179*, 2145.

102 J. C. Hiller, W. Funke *Angew. Makromol. Chem.* **1979**, *76/77*, 161.

103 H. Eschwey, M.L. Hallensleben, W. Burchard *Makromol. Chem.* **1973**, *73*, 235; (b) H. Eschwey, W. Burchard *J. Polym. Sci., Polym. Symp.* **1975**, *53*, 1.

104 P. Lutz, P. Rempp *Makromol. Chem.* **1988**, *191*, 1051.

105 O. Okay, W. Funke *Makromol. Chem.* **1990**, *191*, 1565.

106 A. Okamoto, I. Mita *J. Polym. Sci., Polym. Chem. Ed.* **1978**, *16*, 1187.

107 Y. Gnanou, P. Lutz, P. Rempp *Makromol. Chem.* **1988**, *189*, 2885-2892.

108 K.S. Naraghi, S. Plentz Meneghetti, P.J. Lutz *Macromol. Rapid Commun.* **1999**, *20*, 122.

109 R. Matmour, A. Lebreton, C. Tsitsilianis, I. Kallitsis, V. Héroguez, Y. Gnanou *Angew. Chem. Int. Ed.* **2005**, *44*, 284.

110 S.Y. Sogah, W.R. Hertier, O.W. Webster, G.M. Cohen *Macromolecules* **1987**, *20*, 1473.

111 Z. Zhu, J. Rider, C.Y Yang, M.E. Gilmartin, G. E. Wnek *Macromolecules* **1992**, *25*, 7330.

112 B. Comanita, B. Noren, J. Roovers *Macromolecules* **1999**, *32*, 1069.

113 R. Knischka, P.J. Lutz, A. Sunder, R. Mülhaupt, H. Frey *Macromolecules* **2000**, *33*, 315.

114 S. Angot, D. Taton, Y. Gnanou *Macromolecules* **2000**, *33*, 5418.

115 D. Taton, M. Saule, J.L. Logan, R.S. Duran, S. Hou, E.L. Chaikof, Y. Gnanou *J. Polym. Sci. Part A : Polym. Chem.* **2003**, *41*, 1669.

116 X.-S. Feng, D; Taton, E.L. Chaikof, Yves Gnanou *J. Am. Chem.. Soc.* **2005**, *127*, 10956.

117 S.-H.Lee, S.H. Kim, Y.K. Han, Y.H. Kim *J. Polym. Sci. Part A: Polym. Chem.* **2001**, *39*, 973.

118 (a) T. Biela, A. Duda, S. Penczek, K. Rode, H. Pasch *J Polym Sci Part A: Polym. Chem.* **2002**, *40*, 2884; (b) T. Biela, A. Duda, H. Pasch, K. Rode *J. Polym. Sci. Part A : Polym. Chem.* **2005**, *43*, 6116; (c) T. Biela, A. Duda, K. Rode, H. Pasch *Polymer* **2003**, *44*, 1851.

119 A. Finne, A.-C. Albertsson *Biomacromolecules* **2002**, *3*, 684.

120 W. Radke, K. Rode, A.V. Gorshkov, T. Biela *Polymer* **2005**, *46*, 5456.

121 H. Korhonen, A. Helminen, J.V. Seppala *Polymer* **2001**, *42*, 7541.

122 H.-A. Klok, S. Becker, F. Schuch, T. Pakula, K. Muellen *Macromol. Biosci.* **2003**, *3*, 729.

123 See for instance: (a) M. Tröllsas, B. Atthoff, H. Claesson, J.L. Hedrick *J. Polym. Sci. Part A: Polym. Chem.* **2004**, *42*, 1174 and reference therein from the same group; (b) M.S. Hedenquist, H. Yousefi, E. Malmstrom, M. Johanson, A. Hult, U.V. Gedde, M. Tröllsas, J.L. Hedrick *Polymer* **2000**, *41*, 1827.

124 Q. Cai, Y. Zhao, J. Bei, F. Xi, S. Wang *Biomacromolecules* **2003**, *4*, 828.

125 Y.-L. Zhao, Q. Cai, J. Jiang, X. T. Shuai, J.Z. Bei, C.F. Chen, F. Xi *Polymer* **2002**, *43*, 5819.

126 K. Inoue, H. Sakai, S. Ochi, T. Itaya, T. Tanigaki *J. Am. Chem. Soc.* **1994**, 116, 783; (b) K. Inoue, A. Miyahara, T. Itaya *J. Am. Chem. Soc.* **1997**, *119*, 6191.

127 H. Shohi, M. Sawamoto, T. Higashimura *Macromolecules* **1991**, *24*, 4926.

128 E. Cloutet, J.-L. Fillaut, Y. Gnanou, D. Astruc *J. Chem. Soc., Chem. Commun.* **1994**, 2433; (b) E. Cloutet, J.-L. Fillaut, D. Astruc, Y. Gnanou *Macromolecules* **1999**, *32*,1043.

129 M.K. Mishra, B. Wang, J.P. Kennedy *Polym. Bull.* **1987**, *17*, 307.

130 S. Jacob, I. Majoros, J.P. Kennedy *Macromolecules* **1996**, *39*, 8631.

131 E. Cloutet, J.-L. Fillaut, Y. Gnanou, D. Astruc *Chem. Commun.* **1997**, *17*, 2047.

132 J.E. Puskas, C.J. Wilds *J. Polym. Sci. Part A: Polym. Chem.* **1998**, *36*, 85.

133 J.E. Puskas, A. Michel *Macromol. Symp.* **2000**, *161*, 141.

134 K. Aoi, O. Masahiko *Prog. Polym. Sci.* **1996**, *21*, 151.

135 G. Cai, M.H. Litt *J. Polym. Sci. Part A: Polym. Chem.* **1989**, *27*, 3603.

136 Y. Chujo, K. Sada, T. Kawasaki, T. Saegusa *Polym. J.* **1992**, *24*, 1301.

137 A. Dworak, R.C. Schulz *Macromol. Chem. Phys.* **1991**, *192*, 437.

138 T.K. Bera, S. Sivaram *Macromol. Chem. Phys.* **1995**, *196*, 1515.

139 S. Kobayashi, H. Uyama, Y. Narita *Macromolecules* **1992**, *25*, 3232.

140 J.Y. Chang, HJ. Ji, M. H. Han, S.B. Rhee, S. Cheong, M. Yoon *Macromolecules* **1994**, *27*, 1376.

141 S. Angot, K.S. Murthy, D. Taton, Y. Gnanou *Macromolecules* **1998**, *31*, 7218.

142 C.J. Hawker *Angew. Chem. Int. Ed. Engl.* **1995**, *34*, 1456.

143 S. Robin, O. Guerret, J.-L. Couturier, Y. Gnanou *Macromolecules* **2002**, *35*, 2481.

144 Y. Miura, Y.Yoshida *Macromol. Chem. Phys.* **2002**, *203*, 879.

145 P. Gopalan, Y. Zhang, X. Li, U. Wiesner, C.K. Ober *Macromolecules* **2003**, *36*, 3357.

146 G. Chessa, A. Scrivanti, U. Matteoli, V. Castelvetro *Polymer* **2001**, *42*, 9347.

147 T. Kakuchi, A. Narumi, T. Matsuda, Y. Miura, N. Sugimoto, T. Satoh, H. Koga *Macromolecules* **2003**, *36*, 3914.

148 Y. Miura, H. Dote *J. Polym. Sci. Part A: Polym. Chem.* **2005**, *43*, 3689.

149 D. Taton, Y. Gnanou *Macromol. Symp.* **2001**, *174*, 333.

150 K. Matyjaszewski *Polym. Int.* **2003**, *52*, 1559

151 J. Ueda, M. Matsuyama, M. Kamigaito, M. Sawamoto *Macromolecules* **1998**, *31*, 557.

152 A.M. Kasko, A.M. Heintz, C. Pugh *Macromolecules* **1998**, *31*, 256.

153 J. Ueda, M. Kamigaito, M. Sawamoto *Macromolecules* **1998**, *31*, 6762.

154 S. Angot, K.S. Murthy, D. Taton, Y. Gnanou *Macromolecules* **2000**, *33*, 7261.

155 K. Matyjaszewski, P.J. Miller, J. Pyun, G. Kickelbick, S. Diamanti *Macromolecules* **1999**, *32*, 6526.

156 (a) A. Heise, J.L. Hedrick, M. Tröllsas, R.D. Miller, C.W. Franck *Macromolecules*

1999, *32*, 231; (b) A. Heise, J. L. Hedrick, C. Franck, R. D. Miller *J. Am. Chem. Soc.* **1999**, *121*, 8647; (c) A. Heise, C. Nguyen, R. Malek, J. L. Hedrick, C. W. Franck, R. D. Miller *Macromolecules* **2000**, *33*, 2346.
157 D. M. Haddleton R. Edmonds, A. M. Heming, E. J. Kelly, D. Kukulj *New J. Chem.* **1999**, *23*, 477.
158 F. A. Plamper, H. Becker, M. Lanzendörfer, M. Patel, A. Wittemann, M. Ballauff, A. H. E. Müller *Macromol. Chem. Phys.* **2005**, *206*, 1813.
159 D. M. Haddleton, A. M. Heming, A. P. Jarvis, A. Khan, A. Marsh, S. Perrier, S. A. Bon, S. G. Jackson, R. Edmonds, E. Kelly, D. Kukulj, C. Waterson, *Macromol. Symp.* **2000**, *157*, 201.
160 J. Li, H. Xiao, Y. S. Kim, T. L. Lowe *J. Polym. Sci. Part A: Polym. Chem.* **2005**, *43*, 6345.
161 K. Karaky, S. Reynaud, L. Billon, J. François, Y. Chreim *J. Polym. Sci. Part A: Polym. Chem.* **2005**, *43*, 5186.
162 X. Wang, H. Zhang, G. Zhong, X. Wang *Polymer* **2004**, *45*, 3637.
163 K. Jankova, M. Bednarek, S. Hvilsted *J. Polym. Sci. Part A: Polym. Chem.* **2005**, *43*, 3748.
164 V. Percec, B. Barboiu, T. K. Bera, Van der Sluis, R. B. Grubbs, J. M. J. Fréchet, *J. Polym. Sci.: Part A: Polym. Chem.* **2000**, *38*, 4776.
165 S. Strandman, M. Luostarinen, S. Niemelä, K. Rissanen, H. Tenhu *J. Polym. Sci. Part A: Polym. Chem.* **2004**, *42*, 4189.
166 P. Moschogianni, S Pispas, N. Hadjichristidis, *J. Polym. Sci.: Part A: Polym. Chem.* **2001**, *39*, 650.
167 C.-F. Huang, H.-F. Lee, S.-W. Kuo, H. Xua, F-C. Chang *Polymer* **2004**, *45*, 2261.
168 N. J. Hovestad, G. van Koten, S. A. F. Bon, D. M. Haddleton *Macromolecules* **2000**, *33*, 4048.
169 Y. Zhao, Y. Chen, C. Chen, F. Xi *Polymer* **2005**, *46*, 5808.
170 G. Deng, Y. Chen *Macromolecules* **2004**, *37*, 18.
171 (a) R. M. Johnson, P. S. Corbin, C. Ng, C. L. Fraser, *Macromolecules* **2000**, *33*, 7404; (b) X. Wu, C. L. Fraser, *Macromolecules* **2000**, *33*, 7776; (c) X. Wu, J. E. Collins, J. E. McAlvin, R. W. Cutts, C. L. Fraser, *Macromolecules*, **2001**, *34*, 2812.
172 (a) D. Moinard, D. Taton, Y. Gnanou, C. Rochas, R. Borsali *Macromol. Chem. Phys.* **2003**, *204*, 89–97; (b) D. Moinard, R. Borsali, D. Taton, Y. Gnanou *Macromolecules* **2005**, *38*, 7105.
173 Y. K. Chong, T. P. Le, G. Moad, E. Rizzardo, S. H. Thang *Macromolecules* **1999**, *32*, 2071.
174 (a) M. H. Stenzel-Rozenbaum, T. P. Davis, V. Chen, A. G. Fane *J. Polym. Sci.: Part A Polym. Chem.* **2001**, *39*, 2777; (b) M. H. Stenzel-Rozenbaum, T. P. Davis, A. G. Fane, V. Chen *Angew. Chem. Int. Ed.* **2001**, *40*, *18*, 3428.
175 M. H. Stenzel, T. P. Davis *J. Polym. Sci.: Part A Polym. Chem.* **2002**, *40*, 4498.
176 C.-Y. Hong, Y.-Z. You, J. Liu, C.-Y. Pan *J. Polym. Sci. Part A: Polym. Chem.* **2005**, *43*, 6379.
177 M. H. Stenzel, T. P. Davis, C. Barner-Kowollik *Chem. Commun.* **2004**, 1546.
178 J. F. Quinn, R. Chaplin, T. P. Davis, *J. Polym. Sci.: Part A Polym. Chem.* **2002**, *40*, 2956.
179 Y. Z. You, C. Y. Hong, C.-Y. Pan *Adv. Mater.* **2004**, *16*, 1953.
180 J. Xu, S. Luo, W. Shi, S. Liu *Langmuir* **2006**, *22*, 989.
181 M. Chen, K. P. Ghiggino, S. H. Thang, G. J. Wilson *Angew. Chem. Int. Ed.* **2005**, *44*, 4368.
182 W. Risse, D. R. Wheeler, L. F. Cannizzo, R. H. Grubbs *Macromolecules* **1989**, *22*, 3205.
183 H. Beerens, F. Verpoort, L. Verdonck *J. Mol. Cat A.: Chem.* **2000**, *151*, 279.
184 K. Onitsuka, T. Joh, S. Takahashi *Angew. Chem., Int. Ed. Engl.* **1992**, *31*, 851.
185 (a) K. Kanki, T. Masuda *Macromolecules* **2003**, *36*, 1500–1504; (b) N. Minaki, K. Kanki, T. Masuda *Polymer* **2003**, *44*, 2303.
186 (a) H. R. Kricheldorf, T. Adebahr *Makromol. Chem.* **1993**, *194*, 2103; (b) H. R. Kricheldorf, T. Stukenbrock *Polymer* **1997**, *38*, 3373; (c) H. R. Kricheldorf, T. Stukenbrock, C. Friedrich *J. Polym. Sci., Part A: Polym. Chem.* **1998**, *36*, 1387.

187 See for instance: (a) T. Yokozawa, T. Asai, R. Sugi, S. Ishigooka, S. Hiraoka *J. Am. Chem. Soc.* **2000**, *122*, 8313; (b) A.Yokoyama, R. Miyakoshi, T. Yokozawa *Macromolecules* **2004**, *37*,1169.

188 See for instance: T. Yokozawa, M. Ogawa, A. Sekino, A.R. Sugi, A. Yokoyama *J. Am. Chem. Soc.* **2002**, *124*, 15158.

189 R. Sugi, Y. Hitaka, A. Yokoyama, T. Yokozawa *Macromolecules,* **2005**, *38*, 5526.

190 R. Pelton *Adv. Colloid Interface Sci.* **2000**, *85*, 1.

191 S. Nayak, L.A. Lyon *Angew. Chem. Int. Ed.* **2005**, *44*, 7686.

192 N.A. Peppas *Hydrogels in Medicine and Pharmacy* CRC Press, Boca Raton **1986**.

193 See for instance: K.L. Wooley *J. Polym. Sci. Part A: Polym. Chem.* **2000**, *38*, 1397.

194 (a) H. Staudinger, W. Heuer, *Chem. Ber.* **1934**, *67*, 1164; (b) H. Staudinger, E. Husemann, *Chem. Ber.* **1935**, *68*, 1618.

195 W.H. Stockmayer *J. Chem. Phys.* **1943**, *12*, 125.

196 P.J. Flory *Principles of Polymer Chemistry* Cornell University Press, Ithaca, NY, **1953**.

197 R.C.S. Dias, M.R.P.F.N. Costa *Polymer* **2005**, *46*, 6163.

199 A. Baltsas, D.S. Achilias, C. Kiparissides *Macromol. Theory Simul.* **1996**, *5*, 477.

200 (a) M.R.P.F.N. Costa, R.C.S. Dias *Chem. Eng. Sci.* **1994**, 49, 491; (b) M.R.P.F. N. Costa, R.C.S. Dias *Macromol. Theory Simul.* **2003**, *12*, 560; (c) R.C.S. Dias, M.R.P.F.N. Costa *Macromolecules* **2003**, *36*, 8853–8863; (d) M.R.P.F.N. Costa, R.C.S. Dias *Chem. Eng. Sci.* **2005**, *60*, 423; (e) R.C.S. Dias, M.R.P.F.N. Costa. *Macromol. Theory Simul.* **2005**, *14*, 243.

201 D.L. Kurdikar, J. Somvarsky , K. Dusek, N.A. Peppas *Macromolecules* **1995**, *28*, 5910.

202 J.P. He, H.D. Zhang, J.M. Chen, Y.L. Yang *Macromolecules* **1997**, *30*, 8010.

203 H. Tobita *Macromolecules* **1993**, *26*, 5427.

204 J.E. Elliott, C.N. Bowman *Macromolecules* **1999**, *32*, 8621.

205 J. Prescott *Macromolecules* **2003**, *36*, 9608.

206 M. Wen, L.E. Scriven, A. McCormick *Macromolecules* **2003**, *36*, 4140.

207 H. Tobita, N. Tani, T. Dakai *J. Polym. Sci., Part B: Polym. Phys.* **2000**, *38*, 2342.

208 (a) F. Teymour, J.D. Campbell *Macromolecules* **1994**, *27*, 2460; (b) G. Papavasiliou, F. Teymour *Macromol. Theory Simul.* **2003**, *12*, 543.

209 P.D. Iedema, H.C.J. Hoefsloot *Macromolecules* **2004**, *37*, 10155.

210 K. Dusek, M. Duskova-Smrckova *Prog. Polym. Sci.* **2000**, *25*, 1215.

211 A. Matsumoto *Adv. Polym. Sci.* **1995**, *123*, 41.

212 F. Li, Z. Liu, X. Liu, X. Yang, S. Chen, Y. An, J. Zuo, B. He *Macromolecules* **2005**, *38*, 69.

213 J.B. Hutchison, K.S. Anseth *Macromol. Theory Simul.* **2001**, *10*, 600.

214 (a) M. Ghiass, A.D. Rey, B. Dabir *Macromol. Theory Simul.* **2001**, *10*, 657; (b) M. Ghiass, A.D. Rey, B. Dabir *Polymer* **2002**, *43*, 989.

215 S. Lu, K.S. Anseth *J. Controlled Release* **1999**, *57*, 291.

216 J. Jakubiak, J. Nie, L.A. Linden, J.F. Rabek *J. Polym. Sci. Part A: Polym. Chem.* **2000**, *38*, 876.

217 (a) L. Rey, J. Galy, H. Sautereau *Macromolecules* **2000**, *33*, 6780; (b) L. Rey, J. Duchet, J. Galy, H. Sautereau, D. Vouagner, L. Carrion *Polymer* **2002**, *43*, 4375.

218 J.D. Campbell, F. Teymour, M. Morbidelli *Macromolecules* **2005**, *38*, 752.

219 O. Okay *Polymer* **1999**, *40*, 4117.

220 See for instance: (a) O. Okay, M. Kurz, K. Lutz, W. Funke *Macromolecules* **1995**, *28*, 2728; (b) Y. Huang, U. Seitz, W. Funke *Makromol. Chem.* **1985**, *186*, 273.

221 M. Antonietti , C. Rosenauer *Macromolecules* **1991**, *24*, 3434.

222 (a) H. Chen, K. Ishizu, T. Fukutomi, T. Kakurai *J. Polym. Sci. Chem.* **1984**, *22*, 2123; (b) K. Ishizu, S. Kuwabara, H. Chen, H. Mizuno, T. Fukutomi *J. Polym. Sci., Polym. Chem. Ed.* **1986**, *24*, 1735; (c) K. Ishizu, M. Nunomura, T. Fukutomi *J. Polym. Sci., Polym. Lett. Ed.* **1986**, *24*, 607; *(d)* K. Ishizu, M. Nu-

nomura, T. Fukutomi *J. Polym. Sci., Polym. Chem. Ed.* **1987**, *25*, 1163.
223 N. B. Graham, A. Cameron *Pure Appl. Chem.* **1998**, *70*, 1271; (b) N. B. Graham, C. M. G. Hayes, *Macromol. Symp.* **1995**, *93*, 293; (c) N. B. Graham *Colloid Surf. A* **1996**, *118*, 211; (d) N. B. Graham, J. Mao, A. Urquhart *Angew. Makromol. Chem.* **1996**, *240*, 113.
224 J. S. Downey, G. McIsaac, R. S. Frank, H D. H. Stöver *Macromolecules* **2001**, *34*, 4534.
225 R. S. Frank, J. S. Downey, K. Yu, H. D. H. Stöver *Macromolecules* **2002**, *35*, 2728.
226 L. Y. Tsarik, O. N. Novikov, V. V. Magdinets *J. Polym. Sci. A* **1998**, *36*, 371.
227 A. Biffis *J. Mol. Catal. A* **2001**, *165*, 303.
228 C. Schunicht, A. Biffis, G. Wulff *Tetrahedron* **2000**, *56*, 1693.
229 P. L. Nayak, S. Alva, K. Yang, K. Dhal Pradeep, J. Kumar, S. K. Tripathy *Macromolecules* **1997**, *30*, 7351.
230 (a) N. O'Brien, A. McKee, D. C. Sherrington, A. T. Slark, A. Titterton *Polymer* **2000**, *41*, 6027; (b) P. A. Costello, I. K. Martin, A. T. Slark, D. C. Sherrington, A. Titterton *Polymer*, **2002**, *43*, 245; (c) F. Isaure, P. A. G. Cormack, D. C. Sherrington *Macromolecules* **2004**, *37*, 2096; (d) S. Graham, P. A. G. Cormack, D. C. Sherrington *Macromolecules* **2005**, *38*, 86.
231 The terms "hyperbranched polymers" generally refers to highly branched macromolecules obtained by step-growth polymerization of AB_n-type monomers, although more recent synthetic developments have been proposed; for an overview, see Ref. [8].
232 (a) T. Hirano, K. Tanaka, H. Wang, M. Seno, T. Sato *Polymer* **2005**, 46, 8964; (b) T. Sato, Y. Arima, M. Seno, T. Hirano *Macromolecules*, **2005**, 38, 1627; (c) T. Sato, A. Ono, T. Hirano, M. Seno *J. Polym. Sci. Part A: Polym. Chem.* **2006**, *44*, 2328; (d) T. Hirano, H. Ihara, T. Miyagi, H. Wang, M. Seno, T. Sato *Macromol. Rapid Commun.* **2005**, *206*, 860.
233 T. Sato, N. Higashida, T. Hirano, M. Seno *J. Polym. Sci. Part A: Polym. Chem.* **2004**, *42*, 1609.
234 S. Rimmer, S. Collins P. Sarker *Chem. Commun.* **2005**, 6029.
235 (a) N. Ide, T. Fukuda *Macromolecules* **1997**, *30*, 4268; (b) N. Ide, T. Fukuda Macromolecules 1999, 32, 95.
236 F. Isaure, P. A. G. Cormack, S. Graham, D. C. Sherrington, S. P. Armes, V. Bütün *Chem Commun.* 2004, 1138.
237 D. Taton, J.-F. Baussard, L. Dupayage, J. Poly, Y. Gnanou, V. Ponsinet, M. Destarac, C. Mignaud, C. Pitois *Chem. Commun.* **2006**, 1955.
238 B. Liu, A. Kazlauciunas, J. T. Guthrie, S. Perrier *Macromolecules* **2005**, *38*, 2131.
239 A. Narumi, H. Kaga, Y. Miura, T. Satoh, N. Kaneko, T. Kakuchi *Polymer* **2006**, *47*, 2269.
240 S. Kirsch, A. Doerk, E. Bartsch, H. Sillescu, K. Landfester, H. W. Spiess, W. Maechtle *Macromolecules* **1999**, *32*, 4508.
241 Y.-J. Park, M. J. Monteiro, S. van Es, A. L. German *Eur. Polym. J.* **2001**, *37*, 965.
242 M. Antonietti, R. A. Caruso, C. G. Goltner, M. P. C. Weissenberg *Macromolecules* **1999**, *32*, 1383.
243 X. Z. Zhang, R. X. Zhuo, Y. Y. Yang *Biomaterials* **2002**, *23*, 1313.
244 H. Kast, W. Funke *Makromol. Chem.* **1981**, *182*, 1567 and reference of the same group cited therein.
245 (a) H. Tobita K. Yamamoto *Macromolecules* **1994**, *27*, 3389; (b) H. Tobita, M. Kumagai, N. Aoyagi, *Polymer* **2000**, *41*, 481.
246 E. Jabbari *Polymer* **2001**, *42*, 4873.
247 A. Matsumoto, N. Murakami, H. Aota, J. Ikeda, I. Capek *Polymer* **1999**, *40*, 5687.
248 A. Mura-Kuentz, G. Riess *Macromol. Symp.* **2000**, *150*, 229.
249 J. S. Lowe, B. Z. Chowdhry, J. R. Parsonage, M. J. Snowden *Polymer* **1998**, *39*, 1207.
250 A. Laukkanen, S. Hietala, S. L. Maunu, H. Tenhu *Macromolecules* **2000**, *33*, 8703.
251 D. J. Evans, A. Williams, R. J. Pryce *J. Mol. Catal. A* **1995**, *99*, 41.
252 K. Ishii *Colloid Surf. A* **1999**, *153*, 591.
253 D. Dupin, S. Fujii, S.P Armes, S. M. Baxter *Langmuir* **2006**, *22*, 3381.
254 (a) A. Matsumoto, K. Kodama, Y. Mori, H. Aota, *J. Macromol. Sci., Pure Appl. Chem. A* **1998**, *35*, 1459; (b) A. Matsu-

moto, K. Kodama, H. Aota, I. Capek *Eur. Polym. J.* **1999**, *35*, 1509; (c) A. Matsumoto, T. Shimatani, H. Aota *Polym. J.* **2000**, 32, 871; (d) A. Matsumoto, M. Fujihashi, H. Aota *Polym. J.* **2001**, *33*, 636; (e) A. Matsumoto *Macromol. Symp.* **2002**, *179*, 141.
255 (a) M. Antonietti, W. Bremser, M. Schmidt *Macromolecules* **1990**, *23*, 3796; (b) E. Bartsch, M. Antonietti, W. Schupp, H. Sillescu *J. Chem. Phys.* **1992**, *97*, 3950.
256 (a) F. Groehn, M. Antonietti, *Macromolecules* **2000**, *33*, 5938; (b) M. Antonietti, A. Briel, F. Groehn *Macromolecules* **2000**, *33*, 5950.
257 L.-W. Chen, B.-Z. Yang, M.-L. Wu *Progr. Org. Coatings* **1997**, *31*, 393.
258 S.-Y. Lin, K.-S. Chen, L. Run-Chu *Polymer* **1999**, *40*, 6307.
259 S. Neyret, B. Vincent *Polymer* **1997**, *38*, 6129.
260 (a) D. Duracher, A. Elaissari, C. Pichot *Macromol. Symp.* **2000**, *150*, 305; (b) A. Guillermo, J. P. Cohen Addad, J. P. Bazile, D. Duracher, A. Elaissari, C. Pichot *J. Polym. Sci. B* **2000**, *38*, 889.
261 (a) C. D. Jones, L. A. Lyon *Macromolecules* **2000**, *33*, 8301; (b) J. D. Debord, S. Eustis, S. B. Debord, M. T. Lofye, L. A. Lyon *Adv. Mater.* **2002**, 14, 658.
262 Y. Gao, S. C. F. Au-Yeung, C. Wu *Macromolecules* **1999**, *32*, 3674.
263 (a) S. Peng, C. Wu *Macromol. Symp.* **2000**, *159*, 179; (b) S. Peng, C. Wu *Macromolecules* **2001**, *34*, 6795; (c) S. Peng C. Wu *Polymer* **2001**, *42*, 6871; (d) S. Peng, Ch. Wu *J. Phys. Chem. B* **2001**, *105*, 2331; (e) S. Peng, C. Wu *Macromolecules* **2001**, *34*, 568.
264 (a) H. Kawaguchi, M. Kawahara, N. Yaguchi, F. Hoshino, Y. Ohtsuka *Polym. J.* **1988**, *20*, 903; (b) M. Kashiwabara, K. Fujimoto, H. Kawaguchi *Colloid Polym. Sci.* **1995**, *273*, 339.
265 G. M. Eichenbaum, P. F. Kiser, D. Shah, S. A. Simon, D. Needham *Macromolecules* **1999**, *32*, 8996; (b) G. M. Eichenbaum, P. F. Kiser, A. V. Dobrynin, S. A. Simon, D. Needham *Macromolecules* **1999**, *32*, 4867; (c) G. M. Eichenbaum, P. F. Kiser, D. Shah, W. P. Meuer, D. Needham, S. A. Simon *Macromolecules* **2000**, *33*, 4087.
266 T. Fukutomi, K. Asakawa, N. Kihar *Chem. Lett.* **1997**, *1997*, 783.
267 N. B. Graham, J. Mao *Colloids Surf.* **1996**, *118*, 211.
268 (a) F. Baumann, M. Schmidt, B. Deubzer, M. Geck, J. Dauth *Macromolecules* **1994**, 27, 6102; (b) F. Baumann, B. Deubzer, M. Geck, J. Dauth, M. Schmidt *Macromolecules* **1997**, *30*, 7568.
269 G. Lindenblatt, W. Schaertl, T. Pakula, M. Schmidt *Macromolecules* **2000**, *33*, 9340.
270 T. Batzilla, W. Funke *Makromol. Chem. Rapid Commun.* **1987**, *8*, 261.
271 Y. Shindo, T. Sugimura, K. Horie, I. Mita *J. Photopolym. Sci. Technol.* **1988**, *1*, 155.
272 J. Jiang, S. Thayumanavan *Macromolecules* **2005**, *38*, 5886.
273 X. Lu, Z. Hu, J. Gao *Macromolecules* **2000**, *33*, 8698.
274 S. Al-Assaf, G. O. Phillips, D. J. Deeble, B. Parsons, H. Starnes, C. von Sonntag, *Radiat. Phys. Chem.* **1995**, *46*, 207.
275 U. Brasch, W. Burchard *Macromol. Chem. Phys.* **1996**, *197*, 223.
276 D. Zhao, G. Liao, G. Gao, F. Liu *Macromolecules* **2006**, *39*, 1160.
277 B. Gebben, H. W. A. van der Berg, D. Bargeman, C. A. Smolders *Polymer* **1985**, *26*, 1737.
278 W. Arbogast, A. Horvath, B. Vollmert *Makromol. Chem.* **1980**, *181*,1513.
279 M. Frank, W. Burchard *Makromol. Chem., Rapid Commun.* **1991**, *12*, 645.
280 P. Ulanski, J. M. Rosiak *Nucl. Instrum. Methods B* **1999**, *151*, 356.
281 S. Sabharval, H. Mohan, Y. K. Bhardwaj, A. B. Majali *Radiat. Phys. Chem.* **1999**, *54*, 643.
282 K.-F. Arndt, T. Schmidt, R. Reichelt *Polymer* **2001**, *42*, 6785.
283 B. R. Saunders, B. Vincent *Adv. Colloid Interf. Sci.* **1999**, *80*, 1.
284 R. H. Pelton, P. Chibante *Colloids Surf.* **1986**, *20*, 247.
285 K. Kratz, T. Hellweg, W. Eimer *Colloids Surf. A* **2000**, *170*, 137.
286 X. Wu, R. H. Pelton, A. E. Hamielec, D. R. Woods, W. McPhee *Colloid Polym. Sci.* **1994**, *272*, 467.

287 S. Ito, K. Ogawa, H. Suzuki, B. Wang, R. Yoshida, E. Kokufuta *Langmuir* **1999**, *15*, 4289.

288 T. Ngai, S. H. Behrens, H. Auweter *Chem. Commun.* **2005**, 331.

289 S. Meyer, W. Richtering *Macromolecules* **2005**, *38*, 1517–1519.

290 (a) I. Berndt, W. Richtering *Macromolecules* **2003**, *36*, 8780–8785; (b) I. Berndt, J. S. Pedersen, W. Richtering *J. Am. Chem. Soc.* **2005**, *127*, 9372; (c) I. Berndt, J. S. Pedersen, P. Lindner, W. Richtering *Langmuir* **2006**, *22*, 459.

291 M.-F. Leung, J. Zhu, F. W. Harris, P. Li *Macromol. Rapid Commun.* **2004**, *25*, 1819.

292 X. Li, J. Zuo, Y. Guo, X. Yuan *Macromolecules* **2004**, *37*, 1042.

293 (a) C. D. Jones, A. Lyon *Macromolecules* **2000**, *33*, 8301; (b) D. Gan, L. A. Lyon *J. Am. Chem. Soc.* **2001**, *123*, 7511; (c) D. Gan, L. A. Lyon *J. Am. Chem. Soc.* **2001**, *123*, 8203; (d) J. Wang, D. Gan, L. A. Lyon, M. A. El-Sayed *J. Am. Chem. Soc.* **2001**, *123*, 11284.

294 C. B. Agbugba, B. A. Hendriksen, B. Z. Chowdhry, M. J. Snowden *Colloid Surf. A* **1998**, *137*, 155.

295 (a) S. Kara, O. Pekcan *Polymer* **2000**, *41*, 6335; (b) S. Kara, O. Pekcan *Polymer* **2000**, *41*, 3093; (c) O. Pekcan, S. Kara *Polymer* **2001**, *42*, 7411.

296 J. F. Miller, K. Schaetzel, B. Vincent *J. Colloid Interface Sci.* **1991**, *143*, 532.

297 B. H. Zimm, F. P. Price, J. P. Bianchi *J. Phys. Chem.* **1958**, *62*, 979.

298 D. Held, A. H. E. Muller *Macromol. Symp.* **2000**, *157*, 225.

299 (a) B. Wang, S. Mukataka, M. Kodama, E. Kokufuta *Langmuir* **1997**, *13*, 6108; (b) B. Wang, S. Mukataka, E. Kokufuta, M. Ogiso, M. Kodama *J. Polym. Sci. B* **2000**, *38*, 214.

300 (a) H. Senff, W. Richtering, *J. Chem. Phys.* **1999**, *111*, 1705; (b) H. Senff, W. Richtering *Colloid Polym. Sci.* **2000**, *278*, 830.

301 M. Antonietti, A. Briel, S. Foerster *Chem. Phys.* **1996**, *105*, 7795.

302 M. Antonietti *Angew. Chem.* **1988**, *100*, 1813.

303 G. Van Assche, E. Verdonck, B. Van Mele *Polymer* **2001**, *42*, 2959.

304 (a) I. M. Krieger, T. J. Dougherty *Trans. Soc. Rheol.* **1959**, *3*, 137; (b) S. P. Meeker, W. C. K. Poon, P. N. Pusey *Phys. Rev. E.* **1997**, *55*, 5719.

305 M. Stieger, W. Richtering, J. S. Pedersen, P. Lindner *J. Chem. Phys.* **2004**, *120*, 6197.

306 D. Saatweber, B. Vogt-Birnbrich *Prog. Org. Coat.* **1996**, 28, 33.

9
Molecular Design and Self-assembly of Functional Dendrimers

Wei-Shi Li, Woo-Dong Jang, and Takuzo Aida

9.1
Introduction

Dendrimers are monodisperse macromolecules with a regularly branched three-dimensional architecture. In general, they are composed of three basic structural elements: a focal core, peripheral functional groups and repeating branched units, as illustrated in Fig. 9.1. The number of repeating layers of branching units from the focal core to the periphery is defined as the generation number of the dendrimer. Due to such a unique architecture, dendrimers are considered

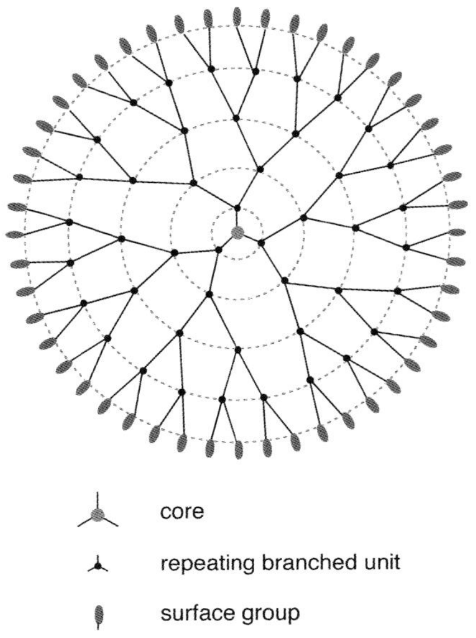

Fig. 9.1 Schematic representation of the structure of a dendrimer.

Macromolecular Engineering. Precise Synthesis, Materials Properties, Applications.
Edited by K. Matyjaszewski, Y. Gnanou, and L. Leibler

ISBN: 978-3-527-31446-1

as the fourth major class of macromolecules after traditional polymers with (1) linear, (2) cross-linked and (3) statistically branched architectures [1].

The discovery of dendrimers dates back to 1978, when Vögtle and coworkers reported the synthesis of an oligoamine having a second-generation branched structure by an iterative divergent growth with a branching strategy called "cascade synthesis" [2]. Subsequently, Denkewalter and coworkers patented L-lysine-based dendrimers in the early 1980s [3]. However, the first dendritic structures that have been thoroughly investigated and received widespread attention are poly(amidoamine) dendrimers (PAMAMs), reported by Tomalia et al. in 1985 [4]. In that paper, the term "dendrimer", originating from the Greek word "*dendron*", which means "tree", was used for the first time to represent this new class of hyperbranched macromolecules. In the same year, Newkome et al. reported "arborols" [5]. Another milestone was achieved by Fréchet and coworkers, who invented the convergent synthetic methodology in the early 1990s [6]. This method allows the synthesis of defect-free dendrimers and triggered a rapid progress in the research field of dendrimers, as reported in books [7], special issues of journals [8] and reviews [9]. To date, a variety of fascinating dendrimers have been designed for, e.g., biomimetics, light harvesting, sensors, catalyses, optoelectronics and medicinal/biomedical applications. Dendrimers have also attracted attention as promising candidates for controlled self-assembly to construct higher order structures. This chapter considers several appropriate research topics and focuses attention on recent progress in the development of functional dendrimers and their assembly.

9.2
Synthesis and Fundamental Properties of Dendrimers

As described above, dendrimers can be synthesized by two different methods [10], i.e. divergent [1–5] and convergent approaches (Scheme 9.1) [6]. Both approaches involve iterative reaction steps, with each repetition leading to the creation of an additional generation, i.e. a layer of branched units. In a divergent synthesis, a dendrimer is grown in a stepwise manner from a central core towards the periphery. In each repetition step, numerous reactions are performed simultaneously on a number of peripheral functional groups of a single

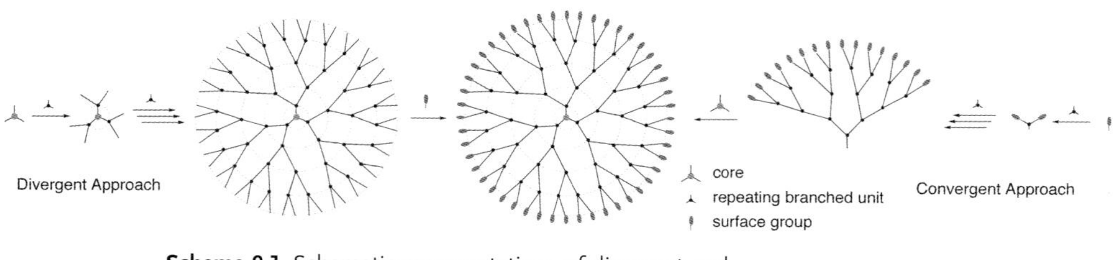

Scheme 9.1 Schematic representations of divergent and convergent syntheses of dendrimers.

molecule. For each successive generation, the number of reactions required increases exponentially. Due to a statistical problem and difficult purification, the divergent approach often suffers from the accumulation of impurities or unreacted functionalities, which results in structural defects in the final product [11]. Such potential problems can be overcome using the convergent approach, which starts the synthesis from the periphery and ends it at the core. In principle, during each step in a convergent synthesis, only the functionality at the focal core is involved in chemical transformation, resulting in a small number of side-products that are easily separable. As a consequence, each new generation can be purified and defects in the final product can in principle be avoided. Later, a number of improvements have been reported in an attempt to increase the throughput and efficacy of the convergent method. Examples include the double-stage convergent method [12], double-exponential growth strategy [13], orthogonal coupling strategy [14] and utilization of click chemistry [15].

In comparison with traditional polymers with linear, statistically branched and cross-linked architectures, dendritic macromolecules generally have the following unique features [16]:

1. By definition, dendrimers should have a uniform molecular weight and no molecular weight distribution.
2. Morphologies of dendrimers may be predicted: lower generation dendrimers are generally floppy and adopt an open architecture, whereas higher generation dendrimers are relatively robust and less deformable spheroids, ellipsoids or cylinders, with the form depending on the shape and directionality of the core, along with the rigidity of the branching units [17]. Thus, the focal core of a large dendrimer may be isolated from its outer environment, leading to specific site-isolation effects on their properties [18].
3. Dendrimers become larger as a function of the generation number, ranging from several to tens of nanometers in diameter. Unlike linear polymers, entanglement or interpenetration of dendritic branches is generally unfavorable. Thus, dendrimers have smaller hydrodynamic volumes and provide lower solution viscosities than linear polymers of the same molecular weight.
4. Large dendrimers possess a highly congested surface, while occasionally retaining a free space in their interior, thereby allowing them to serve as a specific pocket for accommodating a variety of guest molecules.
5. Iterative synthesis of dendrimers allows site-selective functionalization.
6. Peripheral functionalities of dendrimers predominantly determine their solubility.

9.3 Redox-active Dendrimers

Redox-active materials are useful for molecular batteries, memory media, sensors, catalyses and electronic devices. To date, a variety of redox-active groups, such as 4,4′-bipyridinium salts (viologens), tetrathiafulvalene derivatives, ferrocenes, porphyrins, fullerenes and transition metal complexes, have been successfully integrated with dendrimers either at their periphery, within their branches or at their focal core, resulting in interesting redox properties [19].

Dendrimers functionalized with a redox-active core can be used to investigate how the core is isolated by the dendritic architecture [18]. Spatial isolation of active species is sometimes very important in nature. For example, in metalloproteins, the active metal centers are often embedded in a protein matrix and given a specific microenvironment to exert desired functions. As discussed extensively in the literature [20–22], a change in the generation number of a dendrimer shell around the redox-active core generally results in changes of redox potential and electron transfer kinetics. Obviously, these trends do not mean that the dendrimer shell alters the inherent electronic properties of the core, since electrochemical methods just visualize the electron transfer process between the core and the electrode (heterogeneous electron transfer) through the dendrimer shell. Therefore, the results are not as simple as those one might expect. The heterogeneous electron transfer, evaluated based on cyclic voltammetry, can be affected by a variety of structural parameters such as chemical structure and flexibility of the dendrimer shell along with its dimensions (generation number) and morphology, i.e. either spherical or conical [23].

Dendrimer **1** having a zinc porphyrin core attached to four third-generation poly(ether–amide) dendritic substituents (Fig. 9.2), in comparison with its lower generation homologues, exhibits in cyclic voltammetry a negative shift for both the first reduction and oxidation potentials [24]. Furthermore, it shows a notable decrease in electrochemical reversibility, indicating that the heterogeneous electron transfer is slow. The attenuation of the heterogeneous electron transfer rate, thus observed, reflects how the electroactive core is spatially shielded from

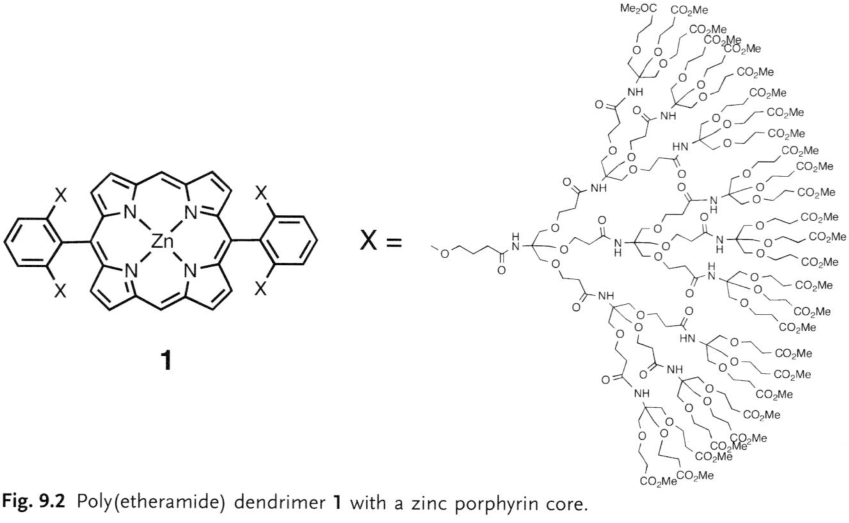

Fig. 9.2 Poly(etheramide) dendrimer **1** with a zinc porphyrin core.

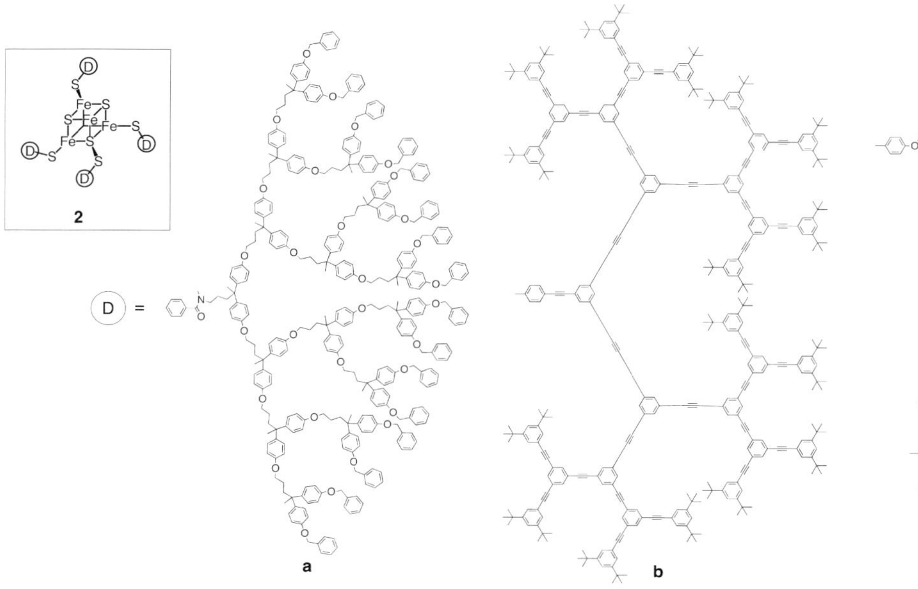

Fig. 9.3 Dendrimers **2** with a redox-active iron–sulfur cluster focal core.

its outer environment. When the zinc porphyrin unit in **1** is placed at the core of a third-generation poly(benzyl ether) dendrimer, the electrochemical irreversibility is more pronounced [25]. The electrochemistry of dendrimers **2a–d** with a redox-active iron–sulfur cluster focal core (Fig. 9.3) has been extensively studied. From the heterogeneous electron transfer rates of **2a** and **b**, the iron–sulfur cluster core of **2b** appears to be electrochemically more shielded than that in **2a** [26]. In conformity with this trend, the hydrodynamic radius of **2a** is smaller than that of **2b**, as evaluated by their diffusion coefficients [27]. On the other hand, when dendrimer **2c** is compared with its isomer **2d**, the former shows a higher electron transfer rate than the latter, while the dendritic unit in **2c** can be more extended than that in **2d**. From this contrasting trend, the flexibility of the dendritic shell, allowing the core to expose more to its outer environment, has been suggested to play a role in the electron transfer events of **2c** and **a** also. The heterogeneous electron transfer through the dendrimer shell is probably affected by the compatibility of the dendrimer with solvent molecules. In better solvents, dendrimers are able to swell, whereas in poor solvents, dendrimers tend to shrink. For example, in aprotic solvents such as $CHCl_3$, CH_2Cl_2 and THF, the redox potential of a poly(ether–amide) dendrimer with a ferrocene core shifts positively upon increment of the generation number. In contrast, in MeOH, which can form a hydrogen bond with the repeating units of the dendrimer, the redox potential is virtually immune to the generation number [21a].

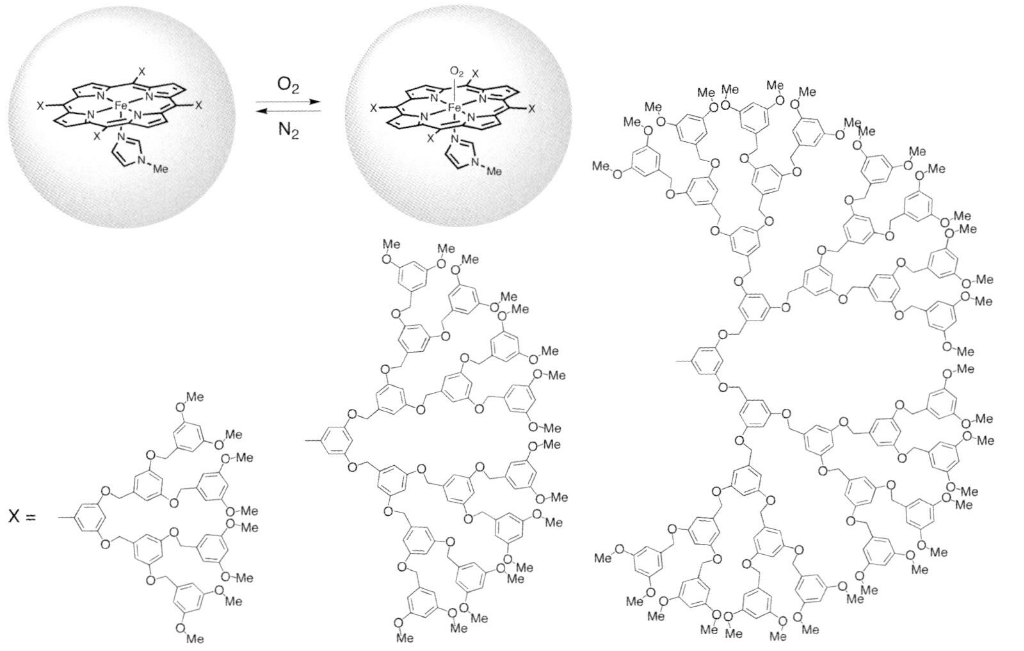

3a 3b 3c

Fig. 9.4 Reversible O_2 binding of dendritic iron porphyrins **3** for mimicking hemoproteins.

Considering the structural analogy with biological metalloproteins, biomimetic applications of dendrimers with a redox-active core are an interesting subject. Figure 9.4 shows iron(II) porphyrins caged in a large poly(benzyl ether) dendrimer shell. Analogous to hemoglobin and myoglobin, these dendrimer porphyrin complexes can bind O_2 reversibly in the presence of 1-methylimidazole in toluene [28]. The size of the dendritic shell greatly affects the stability of the dioxygenated species and the reversibility of the O_2-binding event. Non-dendritic iron(II) porphyrin, when exposed to O_2, is instantly and irreversibly oxidized to give a μ-oxo dimer – a reaction typical of iron(II) porphyrins without any steric protection. In sharp contrast, the largest dendrimer (**3c**), with a fourth-generation poly(benzyl ether) dendritic shell, forms a stable O_2 adduct with a half-life ($t_{1/2}$) of more than 2 months. Bubbling with N_2 allows regeneration of the starting deoxygenated form of **3c**. Such an oxygenation–deoxygenation cycle is repeatable many times. Likewise, compounds **3a** and **b**, which are the lower generation homologues of **3c**, display a reversible O_2-binding activity. However, the O_2 adducts of **3a** and **b** are substantially less stable than that of **3c**, with $t_{1/2}$ values of only 1.5 and 6 h, respectively. Hence the large and spherical dendrimer shell can efficiently protect the interior active site from bimolecular irreversible oxidation. It is worth pointing out that the large dendrimer shell

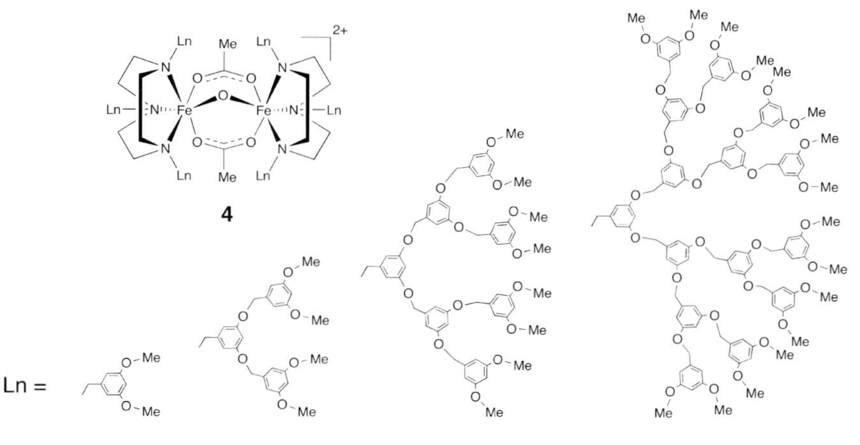

a b c d

Fig. 9.5 Dendritic (μ-O)(μ-OAc)$_2$ diiron(III) complexes **4** as non-heme metalloprotein mimics.

shows a low gas permeability, resulting in unusually high resistance of the O_2 adduct of **3c** towards carbonylation in a CO atmosphere: the $t_{1/2}$ value is as long as 50 h. A reversible O_2-binding activity has also been observed for Newkome-type PEGylated poly(ether–amide) dendritic iron(II) porphyrins in the presence of 1,2-dimethylimidazole in toluene [29].

As non-heme metalloprotein mimics, poly(benzyl ether) dendrimers with mononuclear copper(I) and iron(II) complexes at their focal core have been reported [30]. For example, dendrimers $[Fe^{II}(\eta^2\text{-OAc})(D_3TACN)]^+$ (D = dendritic wedge) are oxidized under aerobic conditions to form the corresponding dinuclear (μ-O)(μ-OAc)$_2$ diiron(III) species **4a–d** (Fig. 9.5). The formation of the dinuclear complexes obeys second-order kinetics with respect to the starting dendritic iron(II) complexes, where the observed rate constants are clearly dependent on the generation number of the dendritic substituents. The oxo-bridged diiron(III) complexes show such a good stability that they can be successfully isolated from the reaction mixture and purified in moderate to good yield. This is rarely possible when using non-dendritic bulky capping ligands. The size of the dendrimer shell has a great influence on the stability of the diiron(III) center against alkaline hydrolysis. Furthermore, the large dendrimer shell electrochemically insulates the redox-active core. Of further interest is the fact that diiron(II) species, formed by photochemical reduction of the diiron(III) complexes, can be reoxidized under aerobic conditions to revert to the diiron(III) complexes, in an analogous way to the oxygenation of non-heme metalloproteins.

Dendrimers functionalized with a large number of redox-active units at the periphery or the branching units [31] may serve as potential electron reservoirs [32], which are attractive as components for molecular batteries and nanoscale

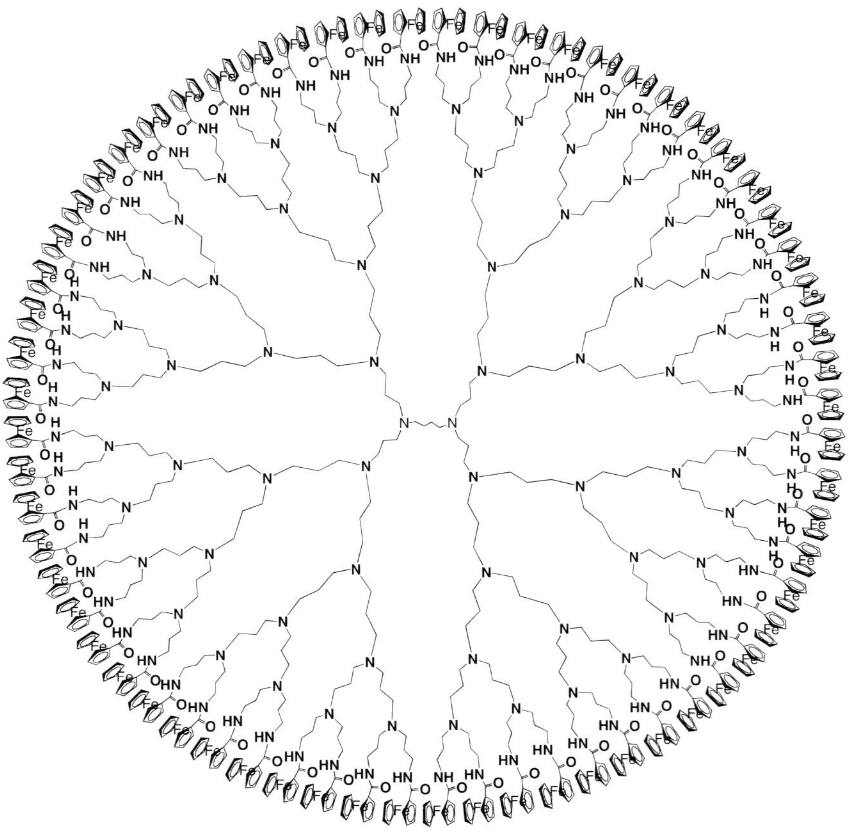

5

Fig. 9.6 Poly(propylenimine) dendrimer **5** with ferrocene units at its periphery.

electronics. For example, dendrimer **5** (Fig. 9.6) has been reported, which bears 64 ferrocene units at the periphery of a fifth-generation poly(propylenimine) dendrimer [33]. Cyclic voltammetry of **5** shows a reversible redox profile with a single oxidation wave, indicating the absence of any electronic communication among the surface ferrocene units. As an alternative to the covalent synthetic approach, a hydrogen-bonding interaction has been used for the synthesis of a dendrimer with multiple ferrocene units [34]. More recently, assembly of a thiol-core dendron bearing multiple ferrocene units on the surface of a gold nanocluster has been reported, which is one of the most efficient methods to form a ferrocene cluster [35]. In fact, the estimated number of the ferrocene units per nanoparticle is as large as 360. Ferrocene-decorated dendrimers have been shown to detect sensitively anions such as $H_2PO_4^-$ and ATP^{2-} [36]. Electrochemical properties of metallocenes, transition metal clusters and tetrathiafulvalene located at the periphery of dendrimers [37] have been reported, along with those of viologen [38] and transition metal complexes of polypyridines [39] positioned at their branching units [19 e]. Of interest, upon electrochemical reduction of a dendrimer having 21 viologen units, only 14 viologen units are reduced [38]. Similarly to ordinary viologens, this dendronized viologen interacts with erosin, a red dye, where one molecule of erosin is incorporated into each viologen unit.

9.4 Light-harvesting Dendrimers

Artificial light-harvesting antenna systems are fascinating synthetic targets for chemists. In the photosynthetic complexes of purple bacteria, numerous bacteriochlorophyll units are spatially organized into wheel-like assemblies, to form light-harvesting antenna complexes LH1 and LH2, which play a vital role in the capture of dilute sunlight and funnel the acquired energy into the photosynthetic reaction center with an efficiency of 100% [40]. For the design of artificial light-harvesting molecules, a large number of chromophore units have to be incorporated for a large absorption cross-section. Furthermore, they have to be spatially organized in a proper way for the energy transfer to occur directionally. Such requirements may be addressed by a dendritic architecture, in which a large number of chromophore units can be placed at their periphery and/or branch units, together with a single energy trap. Elaborate molecular designs will allow for the realization of the "antenna effect" essential for biological photosynthetic systems. The first paper on this subject was published in 1992 [41], when Balzani and coworkers investigated the emission properties of metallodendrimers composed of site-selectively positioned ruthenium and osmium polypyridine complexes. Later, multichromophoric dendrimers that are better suited for exploring intramolecular energy migration and transfer events, referred to as the "antenna effect", were designed [42]. Representative examples include poly(phenylacetylene) dendrimers with a focal perylene unit (**6**) [43], reported by Moore and coworkers, poly(benzyl ether) dendrimers with a focal porphyrin core (**7**) [44] and more recently designed multiporphyrin dendrimers with a focal free-base porphyrin core attached to branched zinc porphyrin arrays (**10**) [45], synthesized by Aida and coworkers.

9.4.1 Unique Features of Dendrimer Structures in the Process of Intramolecular Energy Transfer

Excitation energy transfer can occur non-radiatively by the Dexter-type through-bond mechanism [46] or the Förster-type through-space mechanism [47]. The Dexter-type mechanism requires a spatial overlap between the donor and acceptor orbitals. This interaction is necessarily short range and falls off exponentially. On the other hand, the efficiency of Förster-type energy transfer is directly

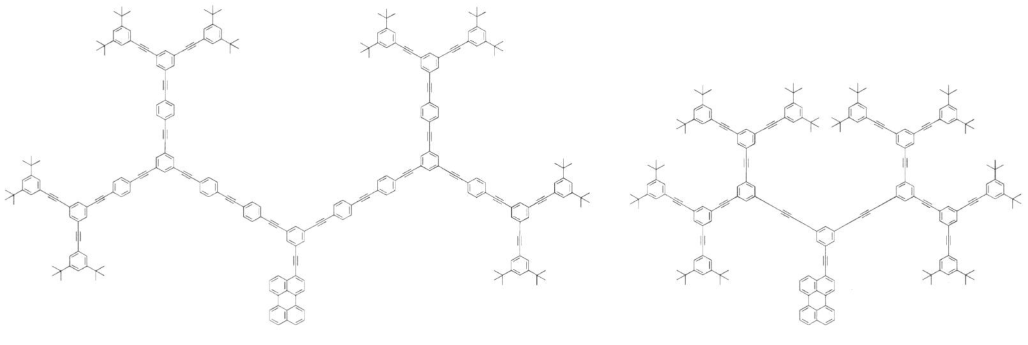

6a 6b

Fig. 9.7 Poly(phenylacetylene) dendrimers **6** with a focal perylene core.

proportional to the overlap of the emission spectrum of the donor and the absorption spectrum of the acceptor, but inversely proportional to the sixth power of the distance between the donor and acceptor units. Studies on intramolecular energy transfer events in dye-containing dendrimers have shown a variety of interesting aspects.

By taking advantage of the iterative synthesis, a well-designed energy gradient, essential for directional energy transfer, has been constructed. In monodendron **6a** (Fig. 9.7), the length of the linear segments between the tri-branched aromatic units decreases by one unit when proceeding from the focal point to the periphery [43]. This architecture provides an energy gradient that decreases as a function of position from the periphery to the core. Hence a directional energy transduction can be expected from the periphery to the core. In fact, the efficiency of the energy transfer (Φ_{ENT}) from the peripheral phenylacetylene units to the focal perylene terminus (98%) is higher than that of **6b** (85%) without any designed energy gradients. The energy transfer rate, observed for **6a** (1.9×10^{11} s^{-1}), is nearly two orders of magnitude greater than that for **6b** (2.27×10^{9} s^{-1}). Müllen and coworkers reported a rigid polyphenylene dendrimer with a designed energy gradient by site-selective incorporation of three different chromophores into the dendritic scaffold [48]. Namely, a terylene tetracarboxydiimide (TDI) unit is placed at the core, while perylene dicarboxymonoimide (PDI) and naphthalene dicarboxymonoimide (NDI) units are positioned in a mid-layer and at the periphery, respectively. Upon selective photoexcitation of NDI, a directional energy transfer from peripheral NDI to PDI and then to central TDI occurs. Similar synthetic strategies have also been reported by other groups in order to realize efficient vectorial energy transfers [49].

Energy transfer events in non-conjugated dendrimers are significantly affected by their morphology [44]. For example, dendrimer **7e** bearing four fourth-generation poly(benzyl ether) dendritic wedges, upon excitation at 280 nm, exhibits an energy transfer efficiency to the focal porphyrin core (Φ_{ENT}) of 80% in CH_2Cl_2 at 20 °C (Fig. 9.8). In sharp contrast, the Φ_{ENT} values of partially substi-

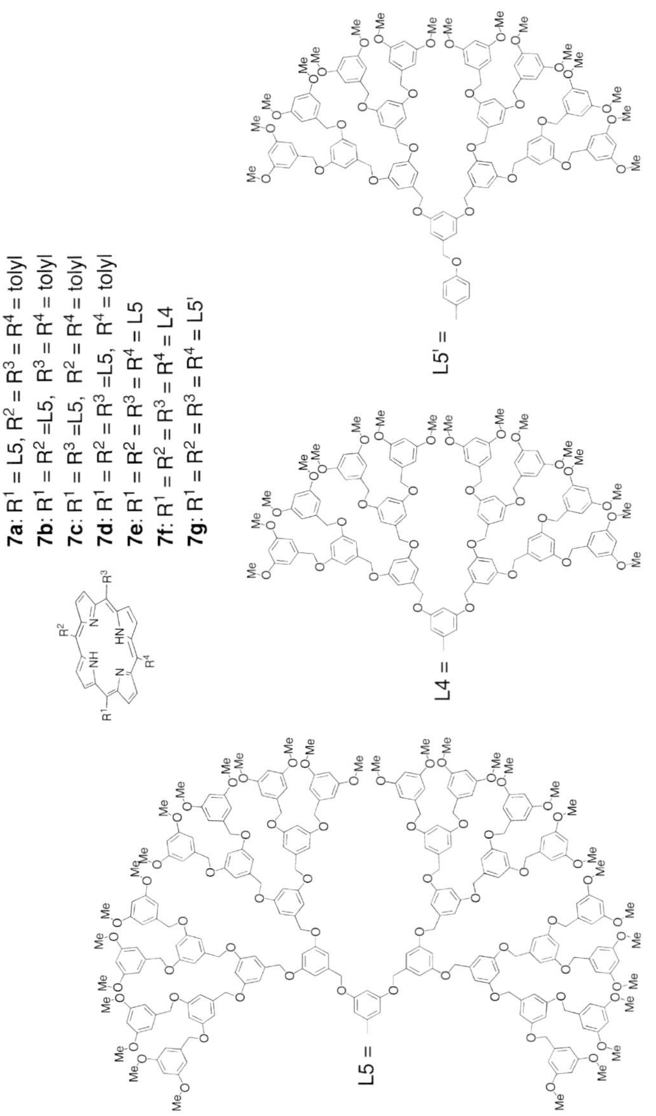

Fig. 9.8 Poly(benzyl ether) dendrimers **7** with a porphyrin core.

tuted analogues **7a**–d are only 10.1 (**7a**), 10.1 (**7b**), 19.7 (**7c**) and 31.6% (**7d**), respectively. Although the third-generation homologue, **7f**, shows a comparable Φ_{ENT} value to **7e** at 20 °C, this value decreases to only 35.6% when the temperature is elevated to 80 °C, while **7e** maintains a Φ_{ENT} value of around 80%. The introduction of a non-branching benzyl ether unit between the dendron wedges and the porphyrin core (**7g**) also results in a significant drop in Φ_{ENT} from 79.2 to 53.7% at 20 °C. All these observations suggest that the enhanced energy transfer of dendrimer **7e** originates from a dense packing of the dendron subunits, which permits a more efficient energy-migration process, as evidenced from the ^{1}H NMR pulse-relaxation time (T_1) and fluorescence anisotropy measurements. An essential difference from the case of conjugated dendrimers is that the energy transfer in **7** occurs in a through-space Förster mechanism, where the excitation energy hops from one chromophore unit to the other, suspended in a dendritic architecture. In this sense, the energy transfer event in **7** is more like those in biological light harvesting.

When a chromophore is isolated by a large dendrimer shell, its light-emitting activity is enhanced (site-isolation effect). For example, poly(phenyleneethynylene) has a strong aggregation tendency due to its rigid conformation, which results in quenching of the excited state and prevents its use as a light emitter. However, when encapsulated by the large shell of a third-generation poly(benzyl ether) dendrimer (**8b**) (Fig. 9.9), the fluorescence quantum yield of poly(phenyleneethynylene), upon direct excitation, is almost 100% [50]. Such a high efficiency of light emission is retained even when the solution is more concentrated. In sharp contrast, the fluorescence quantum yield of **8a** having smaller dendritic wedges is only 56%, even under dilute conditions, and drops further

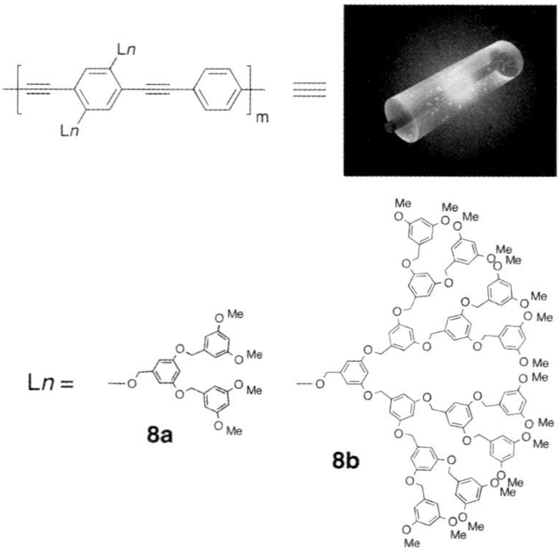

Fig. 9.9 Poly(phenylethynylenes) **8** with poly(benzyl ether) dendritic side-groups.

when the concentration is increased. On the other hand, upon excitation of the dendritic wedges, the conjugated backbone of dendrimer **8b** emits an 11 times more intense fluorescence than that of **8a**. Both **8a** and **b** show 100% efficiency for the energy transfer from the dendritic wedges to the conjugated backbone. The much stronger emission from **8b** is due not only to a larger absorption cross section of its dendritic antenna parts but also to their perfect site-isolation effect on the light-emitting backbone. Lanthanide-core poly(benzyl ether) dendrimers show a similar site-isolation effect on their luminescence activity [51].

9.4.2 Multiporphyrin Dendrimers

In relation to natural photosynthesis, the design of artificial light-harvesting molecules, containing a large number of porphyrin units for absorbing visible light, is an interesting subject. Early examples of branched multiporphyrin arrays include pentamers, nonamers and 21-mers of a zinc porphyrin [52, 53]. Later, fifth-generation poly(L-lysine) dendrimer **9a** was reported, which is functionalized with 16 free-base porphyrin (P_{Fb}) units and 16 zinc porphyrin (P_{Zn}) units at its periphery on separate hemispheres (Fig. 9.10) [54]. Excitation of the P_{Zn} units on one hemisphere results in an energy transfer to the P_{FB} units on the other hemisphere with an efficiency of 43%. The energy transfer efficiency increases to 85% in **9b**, where the P_{Zn} and P_{Fb} units are randomly mixed on the dendrimer surface [55]. These results indicate that the excitation of a certain P_{Zn} unit in **9a** is followed by an energy migration to the other P_{Zn} units in the same domain and then by an energy transfer to a P_{Fb} unit at the boundary of the two different domains. Energy-migration processes in multiporphyrin arrays, anchored on the periphery of dendrimers, have been studied with time-resolved fluorescence-anisotropy measurements [56].

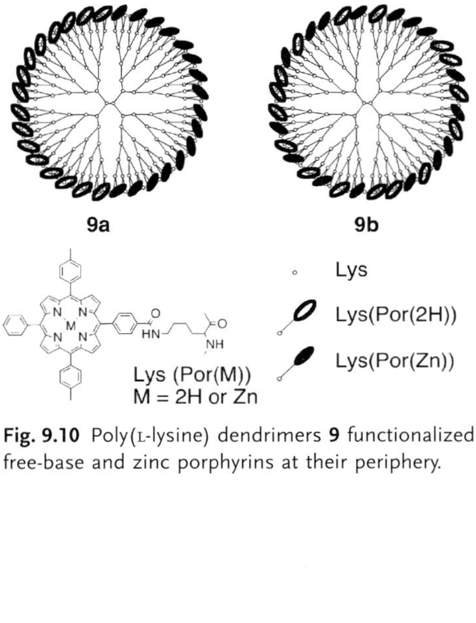

Fig. 9.10 Poly(L-lysine) dendrimers **9** functionalized with free-base and zinc porphyrins at their periphery.

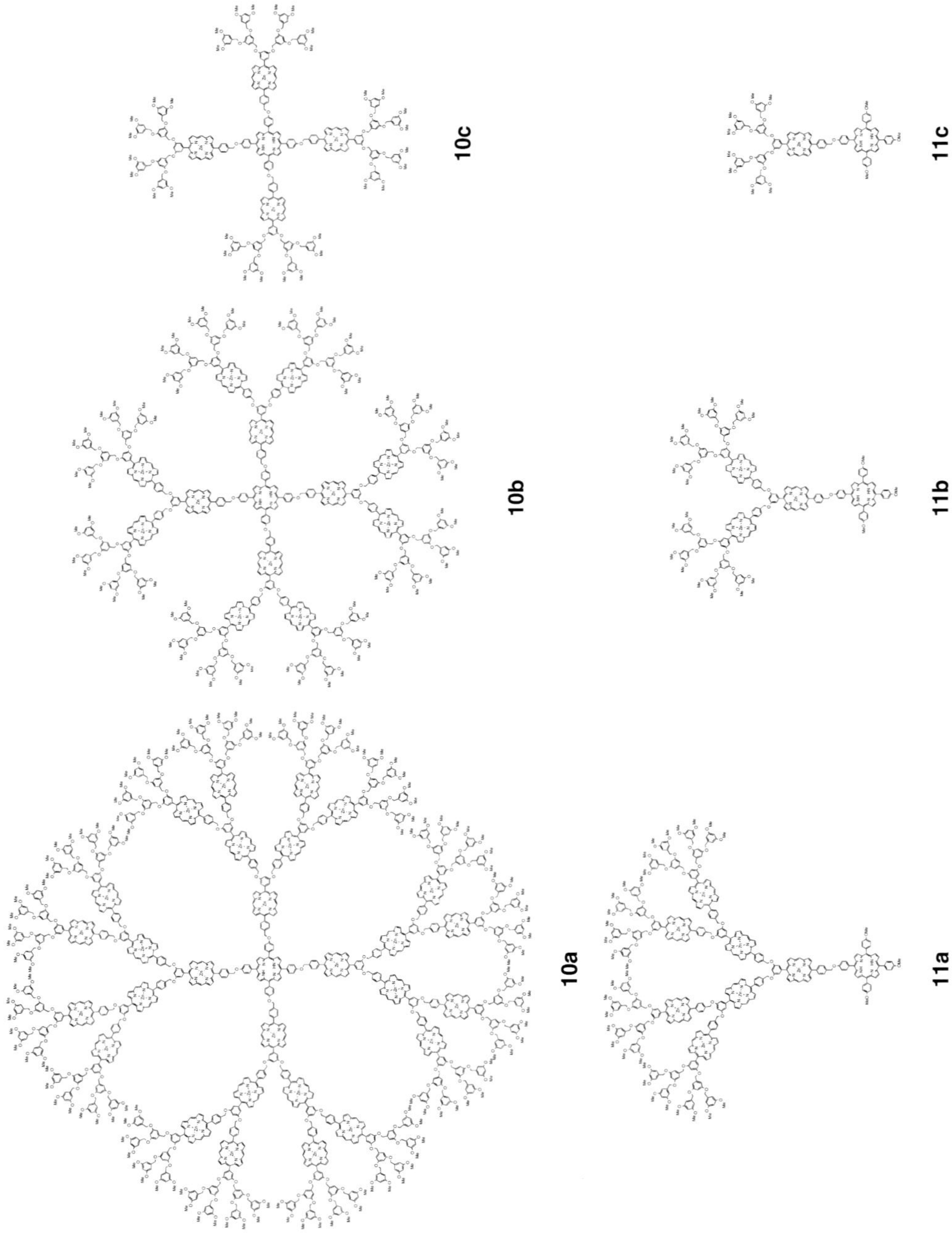

Fig. 9.11 Star and conically shaped multiporphyrin arrays **10** and **11**, respectively.

More recently, giant multiporphyrin dendrimer **10a** (Fig. 9.11) has been reported, which is composed of four dendritic wedges of the P_{Zn} heptamer as energy-donating units attached to a focal P_{Fb} unit as an energy trap [45]. Upon excitation of a P_{Zn} unit at its periphery, the focal P_{Fb} core emits a significantly enhanced fluorescence, indicating an efficient P_{Zn}-to-P_{Fb} energy transfer, with an estimated efficiency (Φ_{ENT}) of 71%. By contrast, **11a**, a conical version of **10a**, having a single dendritic wedge of the P_{Zn} heptamer, shows an extremely low P_{Zn}-to-P_{Fb} energy transfer efficiency (19%). Furthermore, the two series of dendritic arrays **10a–c** and **11a–c** show different generation-number dependences of energy transfer. In the conically shaped series (**11a–c**), the Φ_{ENT} value drops dramatically with an increase in the generation number (86% for **11c**, 66% for **11b** and 19% for **11a**). In contrast, the Φ_{ENT} values of the star-shaped series (**10a–c**) shows a much smaller generation-number dependence, ranging from 87% for **10c** to 80% for **10b** and 71% for **10a**. Hence the morphology of the chromophore array plays an important role in efficient intramolecular energy transfer. A fluorescence depolarization study has proved that the energy transfer in **10a** is facilitated by cooperative energy migration over the dendritic P_{Zn} array – a situation analogous to the photochemical events in bacterial light-harvesting complexes.

9.4.3 Harvesting of Low-energy Photons

Recently, there has been growing interest in the design of dendritic light-harvesting antennae for collecting low-energy photons. Such systems would have a great impact on many biomedical applications, including bio-imaging and photodynamic therapy, since longer wavelength light is hardly absorbed by tissues and can penetrate deeply into the body. Furthermore, the risk associated with laser hyperthermia is much smaller that those with ultraviolet and visible light. For this purpose, dendrimer **12** (Fig. 9.12) has been designed, which consists of a porphyrin core integrated with eight energy-donating chromophore units of AF-343, known as a two-photon absorption (TPA) dye (8100 GM, where $1\ GM = 10^{-50}\ cm^4\ s$) [57]. Under two-photon excitation conditions using a laser for excitation of the AF-343 units at 780 nm, the porphyrin core fluoresces, similar to the case with single-photon excitation at 385 nm. The porphyrin emission is obviously the consequence of a fluorescence-resonance energy transfer from a two-photon excited AF-343 dye in the dendrimer periphery to the porphyrin core.

As can be seen from the above example, utilization of low-energy photons for electronic transitions and reactions always requires coherent light sources, because vibrational excitation energy thermalizes instantaneously and does not localize. In relation to this issue, an azobenzene unit encapsulated in a fourth-generation poly(benzyl ether) dendrimer (**13**) has been reported to undergo cis-to-trans isomerization (Fig. 9.13) upon exposure to 1600-cm^{-1} non-coherent light absorbed by the dendrimer framework [58]. Since lower generation homologues,

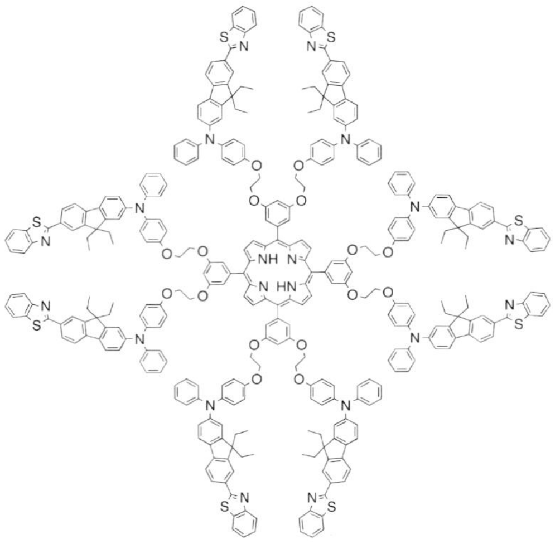

12

Fig. 9.12 Two-photon-absorption dye-based dendrimer **12** with a porphyrin core.

in addition to non-dendritic azobenzene, do not show any response to infrared irradiation, one has to assume an unusual process, i.e. retardation of the vibronic energy dissipation in such a large poly(benzyl ether) dendrimer. In order to clarify the mechanism behind this phenomenon, continuous-wave (CW) anti-Stokes Raman spectroscopy has been conducted on **7e**$_{FeCl}$, a chloroiron(III) complex of **7e** having fourth-generation poly(benzyl ether) dendritic wedges (Fig. 9.8) [59]. Upon exposure of **7e**$_{FeCl}$ to 1600-cm^{-1} non-coherent infrared radiation, certain bonds in the chloroiron(III) porphyrin core undergo a mode-selective vibronic excitation, while the dendrimer shell itself is totally silent. This intriguing observation again indicates an unusual mechanism in which the vibrational excitation energy, acquired by the dendrimer shell, is rapidly channeled to the central core and long-lived to heat up the metalloporphyrin unit.

9.4.4
Photoinduced Electron Transfer

Photoinduced electron transfer is of great interest in relation to the conversion of solar energy into chemical energy in natural photosynthesis. In order to stabilize the charge-separation state resulting from photoinduced electron transfer, electron donor and acceptor units have to be properly arranged. For the design of photoinduced electron transfer systems, the dendritic architecture again provides an ideal platform.

Early examples of photoinduced electron transfer through dendritic frameworks include a dendritic zinc porphyrin with a negatively charged surface [60].

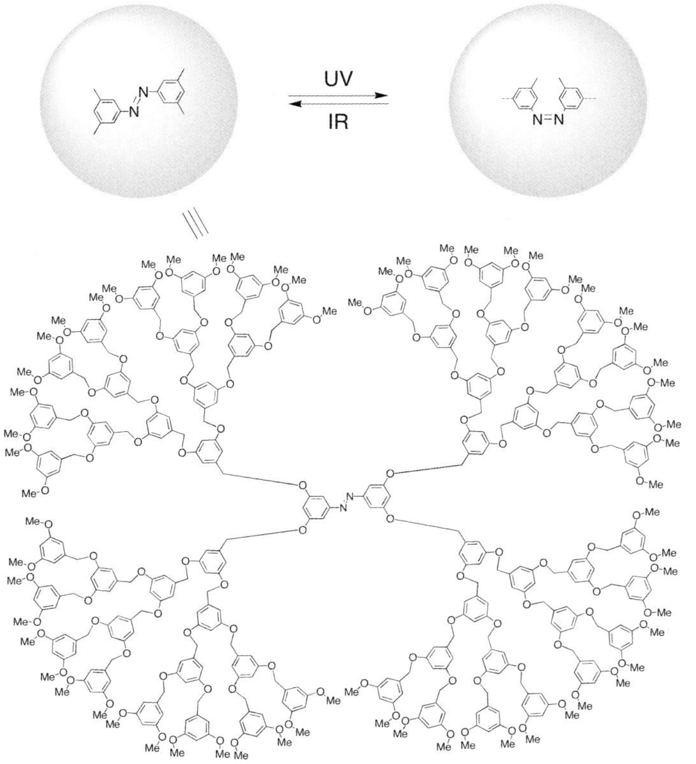

Fig. 9.13 Poly(benzyl ether) dendrimer **13** with an azobenzene core.

This dendrimer electrostatically traps a number of methyl viologen (MV^{2+}) molecules on its surface, spontaneously forming a spatially separated donor–acceptor system. Upon excitation of the zinc porphyrin core, a long-range electron transfer occurs from the singlet-excited state of the core to trapped MV^{2+}, resulting in efficient quenching of the zinc porphyrin fluorescence. Subsequently, several systems were reported using electron acceptors, such as quinines [61, 62], viologens [63] and fullerenes [64], or electron donors, such as triphenylamines [65] and tetrathiafulvalenes [66], covalently integrated into a dendritic framework.

Recently, to mimic more closely the natural photosynthetic system, studies on the combination of two fundamental processes – energy transfer and electron transfer – have been targeted [67]. For example, excitation of dendrimer **14** at its peripheral diarylaminopyrene units results in an energy transfer (ENT) to the focal benzothiadiazole moiety, which then receives an electron (ET) from an amino group of the dendrimer (Fig. 9.14) [68]. The energy transfer occurs on the picosecond timescale, whereas the electron transfer takes place in the nano-

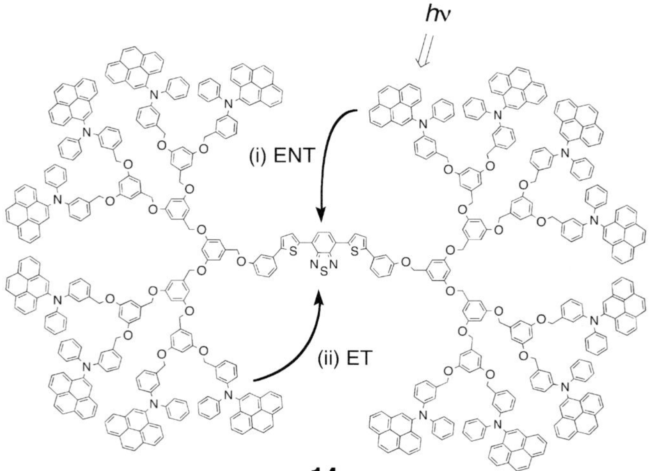

14

Fig. 9.14 Aromatic amine dendrimer **14** with a benzothiadiazole core.

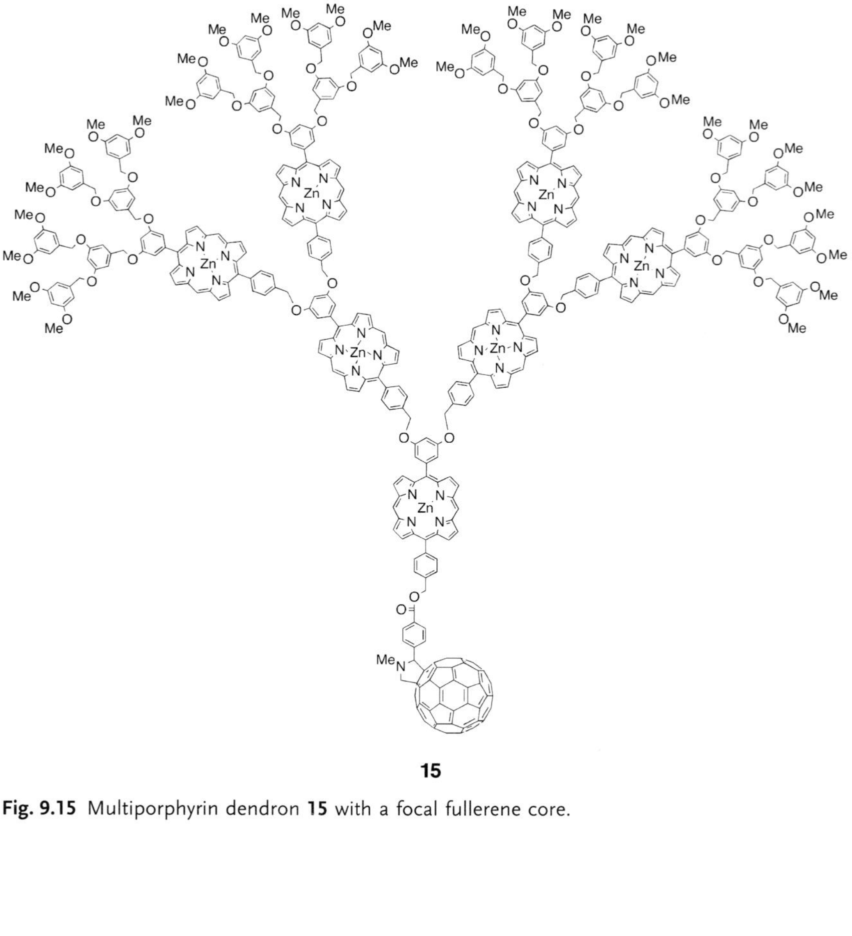

15

Fig. 9.15 Multiporphyrin dendron **15** with a focal fullerene core.

second range. Dendrimer **15** has been synthesized, which carries an electron-accepting C_{60} unit at the focal core of a light-harvesting multiporphyrin dendron consisting of seven zinc porphyrin units (Fig. 9.15) [69]. Upon excitation with visible light, the dendritic zinc porphyrin array in **15** captures the light energy efficiently and then channels it to the focal zinc porphyrin unit, giving rise to an electron transfer to C_{60}. More interestingly, the dendritic antenna in **15** also allows possible intramolecular hopping of a hole generated by the electron transfer, resulting in a longer charge-separation lifetime (0.66 μs in PhCN at 22 °C) than those of its lower generation homologues.

Very recently, a concentric photoactive structure consisting of segregated donor and acceptor arrays (**16**) has been synthesized by complexation of a rigid-backbone dendrimer carrying 24 zinc porphyrin units at its periphery (**16a**) with a bipyridine-attached fullerene cluster (**16b**) (Fig. 9.16) [70]. The complex is stable enough to allow isolation by chromatography. Their lower generation homologues also form similar supramolecular complexes with engineered fullerene clusters, some of which can be visualized by UHV-STM. Upon increment of the numbers of the donor and acceptor units, the zinc porphyrin-to-fullerene

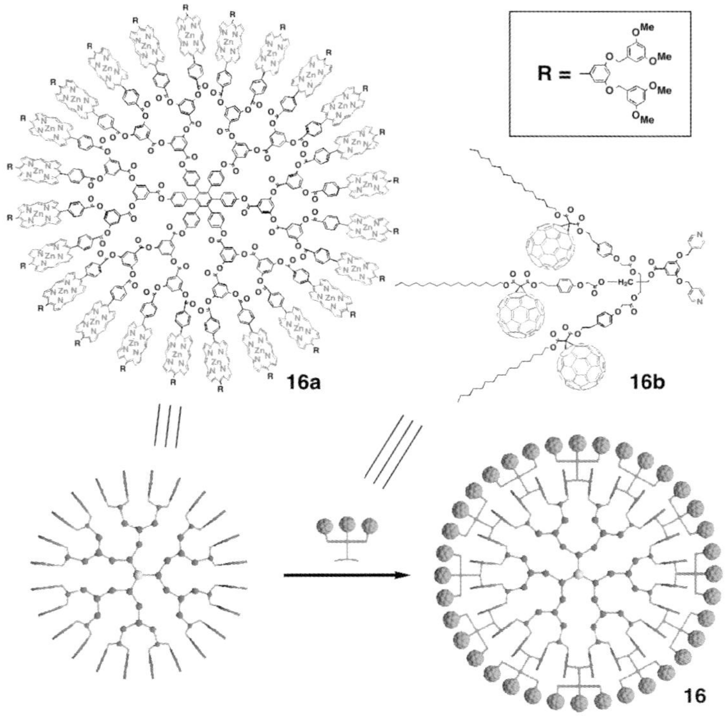

Fig. 9.16 Photoactive dendrimer **16** with concentric zinc porphyrin donor and fullerene acceptor arrays at its periphery.

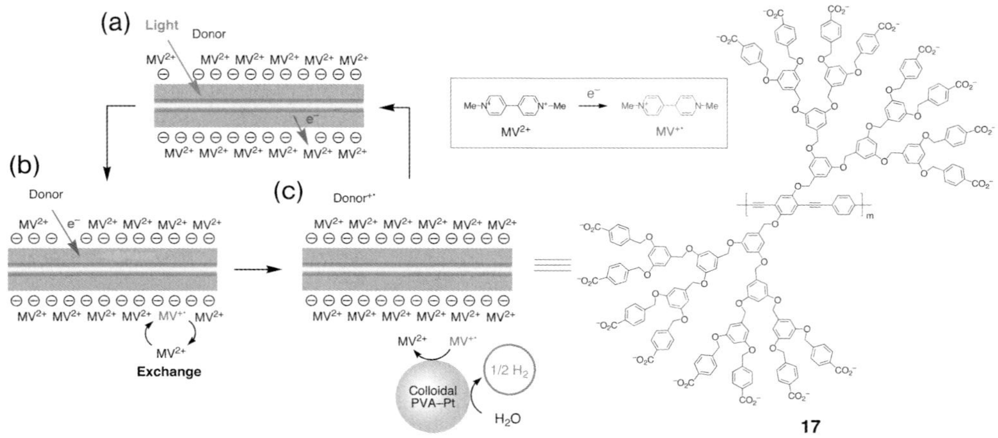

Fig. 9.17 Photoinduced evolution of H_2 from water sensitized by dendritic poly(phenyleneethylene) **17**.

electron transfer is remarkably accelerated, while the recombination rate of the resulting charge-separated state remains virtually unchanged. Consequently, the largest system, which contains 24 zinc porphyrin units and roughly 30 fullerene units on the dendrimer periphery, furnishes the best photochemical performance among the family.

In 2004, the first example of photochemical hydrogen evolution from water using a dendritic photosensitizer was reported [71]. The system is composed of a conjugated polymer wrapped in a water-soluble dendrimer (**17**), methyl viologen (MV^{2+}) and a sacrificial donor. Upon mixing with MV^{2+}, dendrimer **17** can trap this electron acceptor electrostatically on its negatively charged surface (Fig. 9.17), affording a spatially separated donor and acceptor supramolecular assembly. Upon photoexcitation at the conjugated backbone of **17**, an electron is transferred from its backbone to MV^{2+}, affording $MV^{+\bullet}$, a reduced species of MV^{2+}. Since $MV^{+\bullet}$ has a lower affinity than MV^{2+} to the negatively charged surface of **17**, it can rapidly exchange with MV^{2+} in the bulk solution, resulting in a lower probability of back-electron transfer. Subsequently, the $MV^{+\bullet}$ accumulates in solution and reduces water into H_2 under catalysis with colloidal PVA-Pt particles. Triethanolamine, used as a sacrificial electron donor, enables the conjugated backbone to revert to the original neutral state. Thus, repetition of the sequence of these events results in turnover of the H_2 evolution. The overall quantum efficiency of the H_2 evolution is 13%, which is one of the largest values reported to date.

9.5
Stimuli-responsive Dendrimers

Responsiveness is one of the most important and fundamental properties of biological systems. Bio-organisms show surprisingly high sensitivity to external stimuli, such as environmental variation, light, heat or certain chemicals and change their appearances, output certain signals or release certain molecules in response. The realization of such stimuli-responsive properties in an artificial system is a promising proposition in the development of "smart" materials.

Some amphiphilic dendrimers can change their morphology or shape in response to a change in the surrounding medium. For example, star-shaped copolymer **18**, with four dendritic poly(benzyl ether) wedges attached to the ends of a star polymer with long hydrophilic poly(ethylene glycol) (PEG) arms, changes its morphology depending on the solvent polarity (Fig. 9.18) [72]. In THF, **18** exists as a monomolecular micelle with a coagulated PEG core surrounded by a loose shell of the dendritic wedges adopting an extended conformation (**18a**). In contrast, when THF is replaced by a protic solvent such as MeOH, **18** inverts its morphology in such a way that a hydrophobic core composed of the dendritic wedges is surrounded by a hydrophilic shell of the PEG chains (**18b**). On the other hand, in halogenated solvents, **18** most likely adopt a fully extended conformation (**18c**), since the PEG and dendritic building

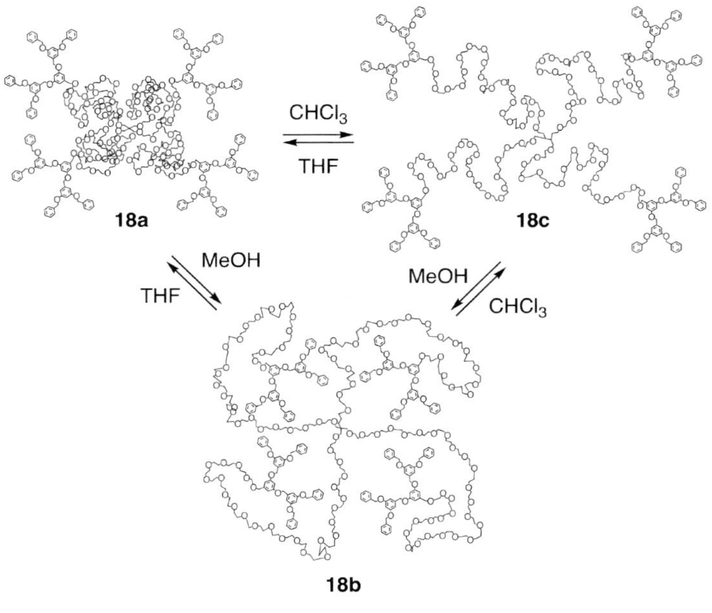

Fig. 9.18 Solvent-responsive star-shaped amphiphilic dendrimer **18**.

blocks are both highly soluble in those solvents. A phosphorus dendrimer with a hydrophobic inner domain and a positively charged hydrophilic surface has been claimed to behave like a dry sponge [73]. In water, this dendrimer is soluble due to its charged surface, whereas the hydrophobic inner domain is frozen. Upon addition of THF, the inner domain of the dendrimer is progressively solvated, as observed by ^1H NMR spectroscopy, allowing the dendrimer to swell entirely and accommodate hydrophobic guests.

Dendrimers with weakly acidic or basic building blocks are sensitive to the pH of the medium. Typical examples include PAMAM and poly(propylenimine) dendrimers, which change their conformations in response to media acidity, since the pH dependences of the protonation of primary amino groups at their termini and tertiary-amino groups at their branching points are different from one another [74, 75]. Such pH-responsive shape changes allow dendritic hosts to perform controlled inclusion and release of guest molecules. Poly(propylenimine) dendrimers bearing diaminobutane derivatives at their periphery have been reported to trap pyrene under basic to neutral conditions [76]. The fluorescence of entrapped pyrene is quenched due to a charge-transfer interaction with the tertiary amino groups at the branching points. On the other hand, when the pH drops from 10 to 6, the dendritic host releases pyrene, resulting in enhancement of the fluorescence. Poly(propylenimine) dendrimers modified with methyl orange at their periphery can trap and release pertechnetate anions in response to the media acidity [77].

Photoresponsive materials have attracted considerable attention due to their numerous potential applications in electronic and photonic devices, such as switches, memory media and sensors. Thus, many attempts have been made to introduce photochromic functionalities into designated positions in dendrimer architectures [78]. Since the first report dating back to 1993 [79], dendrimers possessing azobenzene groups either in the focal core [80], at the periphery [81] or throughout the dendritic architecture [82] have been synthesized for various purposes. Photoinduced trans-to-cis (*E*/*Z*) isomerization of the azobenzene units results in changing the entire conformation of the dendrimers. This type of photoinduced shape change can be used for trapping and releasing guest molecules. For instance, a poly(propylenimine) dendrimer (**19**), functionalized with azobenzene groups at its periphery, reversibly traps guest molecules such as eosin in response to the irradiation with UV and visible light (Fig. 9.19) [83]. The azobenzene units undergo reversible isomerization between the *E* (**19a**) and *Z* (**19b**) forms upon irradiation with UV and visible light, respectively. Interestingly, the *Z*-isomer exhibits a larger affinity than the *E*-isomer towards eosin. In aqueous media, azobenzene dendrimer **19** forms micrometer-scale, large vesicles, whose diameter changes upon photoisomerization of the azobenzene functionalities [84].

Heat is another important stimulus. Poly(benzyl ether) dendrimer **20** (Fig. 9.20), which encapsulates an iron triazole coordination polymer, reversibly changes its color from purple to off-white upon heating [85]. This color change is due to the thermally induced spin transition of the iron species between low-

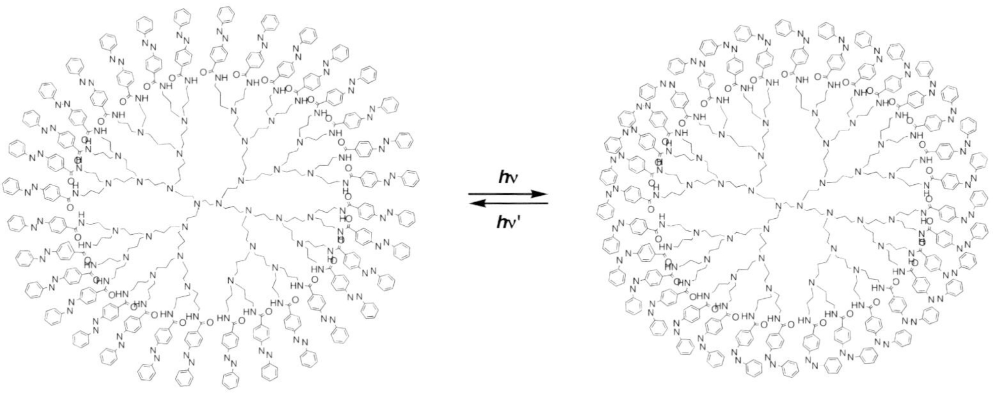

19a 19b

Fig. 9.19 Poly(propylenimine) dendrimer **19** with azobenzene groups on its surface.

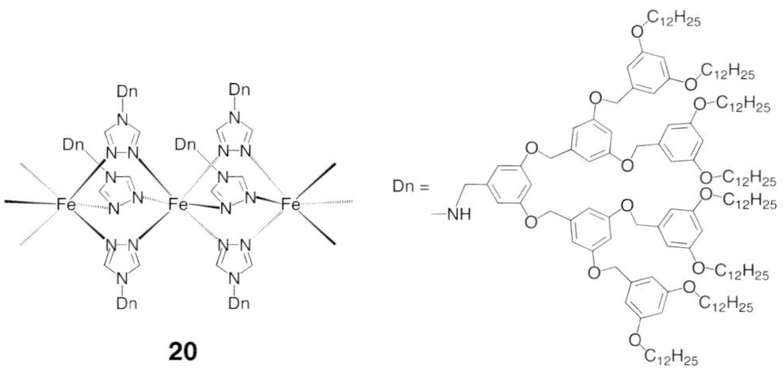

20

Fig. 9.20 Spin-crossover dendrimer **20** with an iron triazolate coordination polymer at its core.

and high-spin states. The dendrimer not only contributes to the reversibility of the spin crossover but also allows the spin transition to occur in a narrow temperature range.

Recognition of substrates and subsequent responses are important functions of bio-organisms. For mimicking these biological events, dendrimers functionalized with various receptor moieties have been synthesized, which exhibit molecular recognition capabilities toward certain substrates [86]. Molecular recognition arises from hydrophobic, hydrogen-bonding, metal-coordination and electrostatic interactions. Examples based on hydrophobic interactions include water-soluble dendrimers containing at their core β-cyclodextrin [87] and cyclophane [88], which can recognize adamantane and steroids, respectively. Using hydrogen-bonding interactions, "dendroclefts" with an enantiomerically pure 9,9′-spirobi[9*H*-fluorene] core is capable of recognizing glucopyranosides [89]. Phenyleneethynylene-based dendrons functionalized with an optically active binaphthyl group at their focal core can recognize the enantiomers of various chiral amino alcohols, as sensitively observed by an enhanced fluorescence response of the light-emitting dendritic wedge [90]. Multiporphyrin dendrimers containing 12, 24 (**16a**) (Fig. 9.16), 18 and 36 peripheral zinc porphyrin units are capable of chiroptically sensing asymmetric dipyridyl compounds [91], whereas a non-dendronized bisporphyrin zinc complex, equivalent to the branch unit of the above dendrimers, is totally inactive. The capability of chiroptical sensing is highly dependent on the generation number and branching pattern of the dendritic scaffold.

9.6
Catalytic Dendrimers

After the discovery of dendrimers, chemists realized that the introduction of catalytic functionalities into dendrimer architectures [92] would lead to a new class of catalysts that could combine the merits of homogeneous catalysts, such as high activity, good reproducibility and accessibility to all catalytic sites, and those of heterogeneous catalysts (easy recovery and reuse). To date, a variety of excellent dendritic catalysts have been reported, based on the covalent integration of active metal complexes [93], phosphines [94] and non-metal-based catalytic groups into the dendritic scaffold or based on the encapsulation of metal nanoparticles within dendritic cages [95]. These studies have demonstrated that dendrimer-based catalysts have their own unique features compared with traditional unsupported catalysts and those that are supported by organic polymers such as polystyrene or inorganic polymers such as silica. The unique architecture of dendrimers offers excellent control of the number, shape and structure of the catalytic sites and also of their microenvironment, which has a great influence on the catalytic activity and selectivity.

9.6.1
Dendrimer Effects on Catalytic Activity

Dendrimer catalysts bearing catalytic functionalities on their surface usually allow high accessibility to substrates, resulting in catalytic activities and reaction rates comparable to those of their parent monomeric catalytic compounds [96]. However, when they are too large, the multiple catalytic sites on the dendrimer surface are congested, leading to either negative effects or cooperative effects on their catalytic behaviors. For example, a negative effect on catalytic activity upon increment of the dendrimer size has been observed for the Kharasch addition catalyzed by carbosilane dendrimers with peripheral diamino arylnickel(II) species [97] and for the Suzuki reaction catalyzed by poly(propylenimine) dendrimers with (diphosphino)(diacetato)palladium(II) functionalities [98]. By contrast, a positive effect has been demonstrated with dendritic diphosphine–dimethylpalladium(II) complex **21** in its use for the Heck reaction and hydroformylation

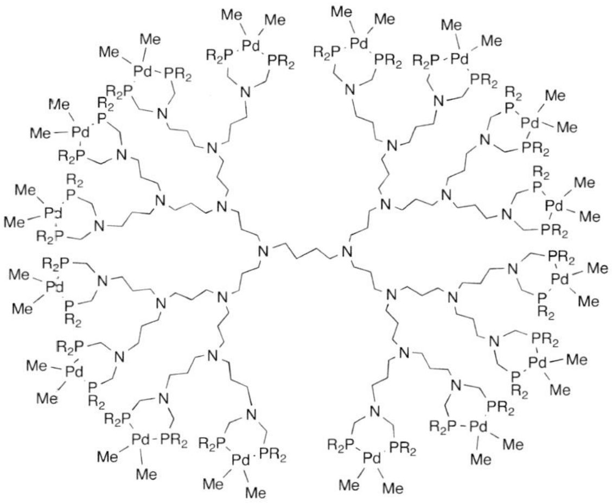

21

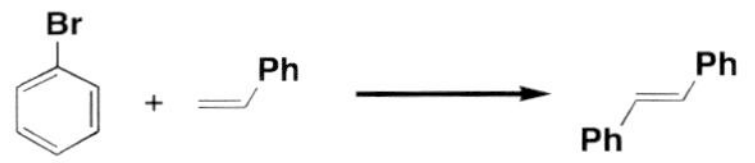

Fig. 9.21 Poly(propylenimine) dendritic catalyst **21** with diphosphino Pd^{II} species at its periphery for catalyzing the Heck reaction.

(Fig. 9.21) [99]. Dendritic catalyst **21** exhibits a catalytic turnover number of 50, whereas that of a non-dendritic reference is only 16. The positive dendrimer effect has been attributed to a high thermal stability of dendrimer complex **21**. A strong positive dendrimer effect has also been observed for the bromination of cyclohexene with NaBr–H_2O_2, catalyzed by poly(benzyl ether) dendrimers with phenylselenide units on their surface [100]. The autocatalysis, which becomes more efficient with a high local concentration of the organoselenide species on the dendrimer surface, is likely to play a role in this significant dendrimer size effect.

When using dendrimers bearing a catalytic functionality at the core, the surrounding dendritic shell more significantly affects the catalytic activity. A negative dendrimer effect has often been observed, due to the reduction in mass transport caused by the enhanced steric bulk around the active site. For example, carbosilane dendrimers having a ferrocenylphosphine–palladium(II) complex at their core show a decrease in catalytic activity for allylic alkylation upon increment of the size of the dendritic shell [101]. Conversely, a large dendritic shell can control the local microenvironment of the active site and can be designed to benefit catalysis. For example, reverse micelle-type dendrimer **22a** (Fig. 9.22) exhibits a positive effect on the elimination of HI from tertiary alkyl iodides in the presence of $NaHCO_3$ to form olefins [102]. The hydrophilic interi-

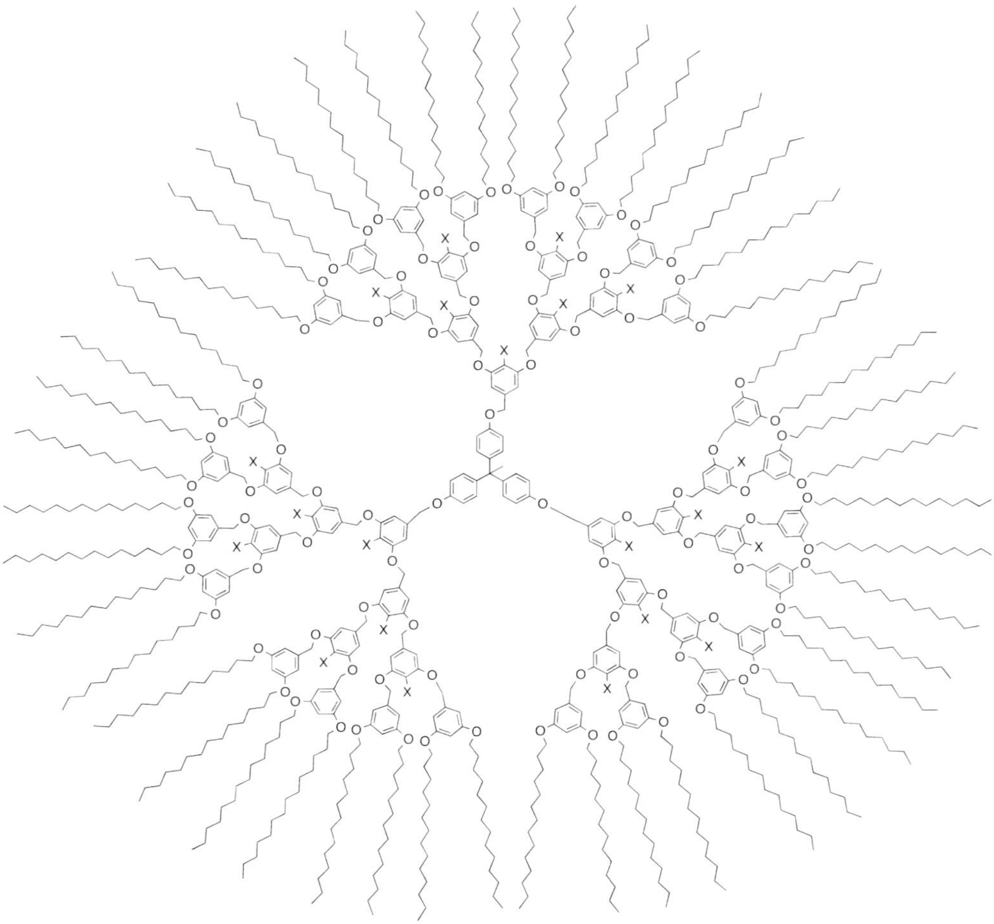

22a X = CH_2OH
22b X = CO_2Me

Fig. 9.22 Amphiphilic dendritic catalyst **22a** with a pump effect and its non-amphiphilic version **22b**.

or of **22a** can trap polar substrates, stabilize the transition states and then drive the non-polar products to the external hydrophobic media in a so-called "pump effect". Consequently, **22a** exhibits a catalytic turnover number of 17 400, which is much higher than that of 149 realized by **22b**, an esterified version of **22a**. Decreasing the size of the dendritic shell from fourth to third generation results in a 15–20% reduction in catalytic activity. More recently, this strategy has successfully been extended to the photoinduced [4+2] cycloaddition of singlet oxygen to dienes, with subsequent reduction to allylic diols [103].

9.6.2
Dendrimer Effects on Selectivity

The unique morphology of dendrimers can lead to peculiar effects on the selectivity of reactions. For example, a poly(aromatic ester) dendrimer with a manganese porphyrin core has been reported to catalyze shape selectively the epoxidation of olefins using iodoisobenzene as an oxidant [104]. Compared with a non-dendritic reference, the dendritic catalyst preferentially epoxidizes less sterically hindered olefins. More recently, the AIBN-mediated alkenylation of a cobalt(II) porphyrin with propargyl alcohol (Fig. 9.23) has been reported to proceed very selectively in a large poly(aromatic ester) dendrimer (**23**) [105]. The reaction is initiated by the transient formation of a cobalt(III) hydride species from **23** and AIBN with elimination of acrylonitrile. Then, propargyl alcohol is inserted into the Co^{III}–H bond to give an alkenylcobalt(III) species, Co^{III}–C(=CH_2)CH_2OH. When lower generation homologues of **23** are subjected to this reaction, other organocobalt(III) species such as Co^{III}–C(CH_3)=CHOH and Co^{III}–CH(CH_3)-CHO result. These organocobalt(III) species are isomerized products of Co^{III}–C(=CH_2)CH_2OH by the reaction with the cobalt(III) hydride species. In con-

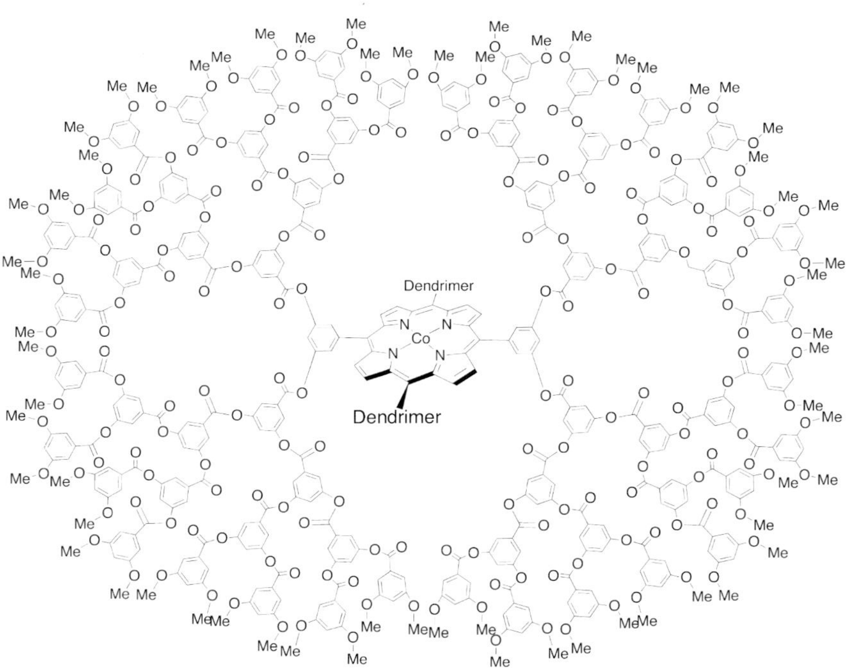

23

Fig. 9.23 Poly(aromatic ester) dendrimer **23** with a cobalt(II) porphyrin core for alkenylation with propargyl alcohol.

trast, when **23** is subjected to the same reaction, only Co^{III}–C(=CH_2)CH_2OH forms in 91% yield (Fig. 9.23). This selectivity is due to the suppression of the side-reaction between Co^{III}–C(=CH_2)CH_2OH and the cobalt hydride species by their steric protection with the large dendritic shell.

Chiral dendrimer catalysts can bring about enantioselective reactions. For example, in allylic amination, a fourth-generation PAMAM dendrimer bearing at its periphery 64 pyriphos–Pd^{II} species furnishes an enantiomeric excess of 69%, whereas it is only 9% when a non-dendritic reference is used [106]. A phenylacetylene dendrimer with an (*S*)-1,1′-bi-2-naphthol core mediates asymmetric addition of diethylzinc to benzaldehyde to give a chiral alcohol, where the major enantiomer of the product has an opposite configuration to that obtained with a non-dendritic reference [107]. The stereochemical inversion with the dendrimer catalyst has been explained by its ability to prevent aggregation of the organozinc species.

9.6.3 Recyclable Dendrimer Catalysts

One of the major advantages of dendrimer catalysts is that they are easily recoverable and reusable [108]. Similarly to their monomeric analogues, dendritic catalysts can be recycled by precipitation [109], phase separation [110] or immobilization on insoluble supports [111]. In addition, due to their large molecular size compared with substrates and products, dendritic catalysts can be recovered by ultrafiltration or nanofiltration. For example, PAMAM dendrimers with peripheral D-gluconolactone functionalities are recovered by nanofiltration through a Millipore microporous membrane after their use as chiral ligands for the $NaBH_4$ reduction of prochiral aromatic ketones [112]. Through regeneration with HCl–MeOH, the dendritic catalysts can be reused up to 10 times without a substantial decrease in their catalytic activity. An even more interesting application of dendrimer catalysts is their use for reactions in a continuous-flow membrane reactor, since they are retained in the reactor when combined with nanoporous membranes (Fig. 9.24). Kragl and coworkers have demonstrated allylic substitution in a continuous-flow reactor using a fourth-generation poly(propylenimine) dendrimer functionalized on its surface with palladium phosphine complexes [113]. In this case, >99.9% of the catalyst is retained in the reactor, resulting in a six-fold increase in the total turnover number of the catalyst. Later, successful examples with other dendritic catalysts were reported [114].

9.7 Self-assembled Dendrimers

Self-assembly is fundamental in nature for constructing the hierarchical levels of biological systems. From a limited number of biomolecules, self-assembly can result in entities ranging from molecular-level organized structures, such as

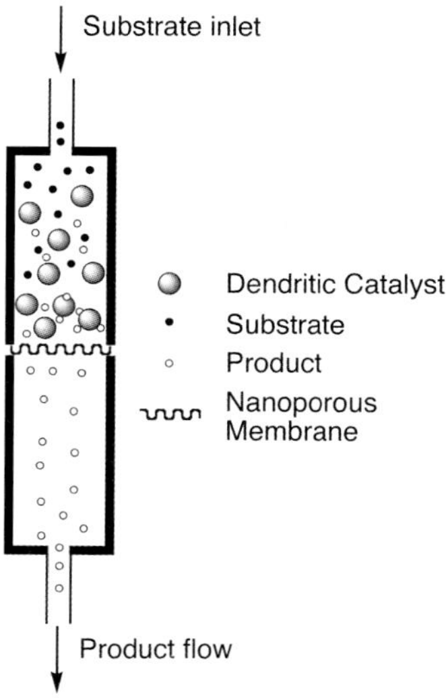

Fig. 9.24 Schematic representation of a continuous-flow membrane reactor.

DNA, proteins and enzymes, to assembled functional structures, such as membranes and ribosomes, to cells and finally to organisms. Learning from examples in nature, scientists have recognized it as a simple but powerful "bottom-up" approach to the development of functional materials and used it in the fabrication of interesting nanostructures. As unique nanometer-scale molecular objects with predictable and controllable three-dimensional structures and precise size regimes, dendrimers and dendrons are promising building blocks for the construction of nano- and microscale functional structures. Furthermore, due to their morphological similarity to biomolecules, the self-assembly of dendrimers allows the understanding of biological assembly events.

9.7.1
Supramolecular Chemistry of Dendrimers

A hexameric supramolecular dendritic entity (**24b**) can be assembled from a dendritic isophthalic acid derivative (**24a**) containing two isophthalic acid units covalently attached to a rigid aromatic spacer, by hydrogen bonding-mediated dimerization of the carboxylic acid functionalities (Fig. 9.25) [115]. In sharp contrast, a non-dendritic isophthalic acid derivative self-assembles to form both an infinite linear aggregate and a cyclic hexamer. Clearly, the large dendritic wedges of **24a** play an important role in the stabilization of the hexameric aggregate of **24b**. In a similar vein, first- and third-generation poly(benzyl ether) dendrons with a focal assembly motif consisting of fixed 60° angle donor–donor–acceptor and acceptor–acceptor–donor hydrogen-bonding arrays have been

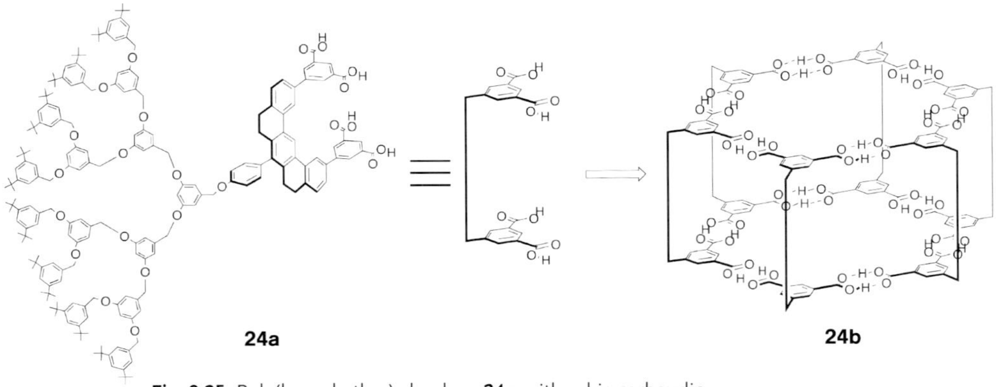

24a

24b

Fig. 9.25 Poly(benzyl ether) dendron **24a** with a bis-carboxylic acid focal core for hydrogen bond-mediated self-assembly.

reported to self-assemble into the corresponding cyclohexameric structures. Interestingly, mixing of these two dendrons results in co-assembly to form a hexameric ring with an alternating arrangement of the dendrons [116]. Analogous to the folding of proteins, secondary structures of large dendrimers can be controlled when they are designed in such a way that intramolecular hydrogen-bonding interactions are properly operative [117].

Supramolecular functionalization of dendrimers is an effective way to modify their properties and functions and leads to various new aspects. From a fundamental point of view, the supramolecular chemistry here involves an important issue of "multivalency". Meijer and coworkers reported supramolecular functionalization of the periphery of dendrimers (Fig. 9.26) [118]. For example, a urea adamantyl poly(propylenimine) dendrimer can be assembled with a phosphine derivative equipped with a urea acetic acid functionality, to form a dendrimer with a supramolecularly engineered surface, which serves as a multidentate ligand for transition metal ions. They have also reported that this phosphine-modified supramolecular dendrimer can mediate palladium-catalyzed allylic amination [118b]. Such urea adamantyl poly(propylenimine) dendrimers can

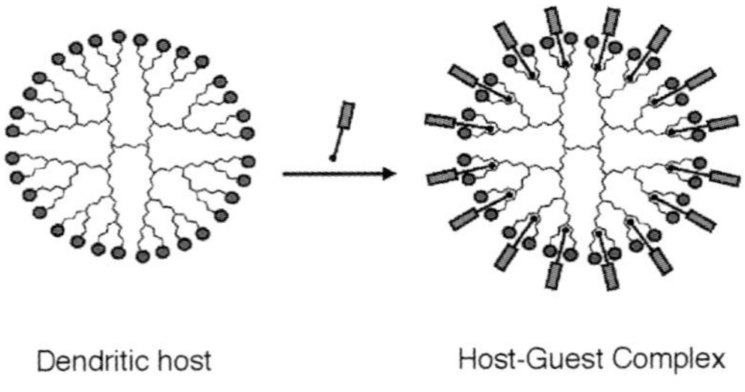

Fig. 9.26 Multivalent supramolecular functionalization of dendrimers.

also trap at their periphery oligo(*p*-phenylenevinylene) (OPV) derivatives bearing an aryl urea glycine unit, affording highly emissive supramolecular dendrimers. The OPV molecules on the dendrimer surface are 10-fold more emissive that those in the solid state [118c].

Dendrimers can accommodate guest molecules in their interior. For example, PAMAM dendrimers can host small organic acids due to electrostatic (acid–base) interactions [119]. A poly(benzyl ether) dendrimer with carboxylate surface groups can solubilize pyrene in water [120]. Meijer and coworkers proposed the concept of a "dendritic box" with a flexible poly(propylenimine) dendrimer modified with 64 *t*-BOC-protected L-phenylalanine end groups on its surface [121].

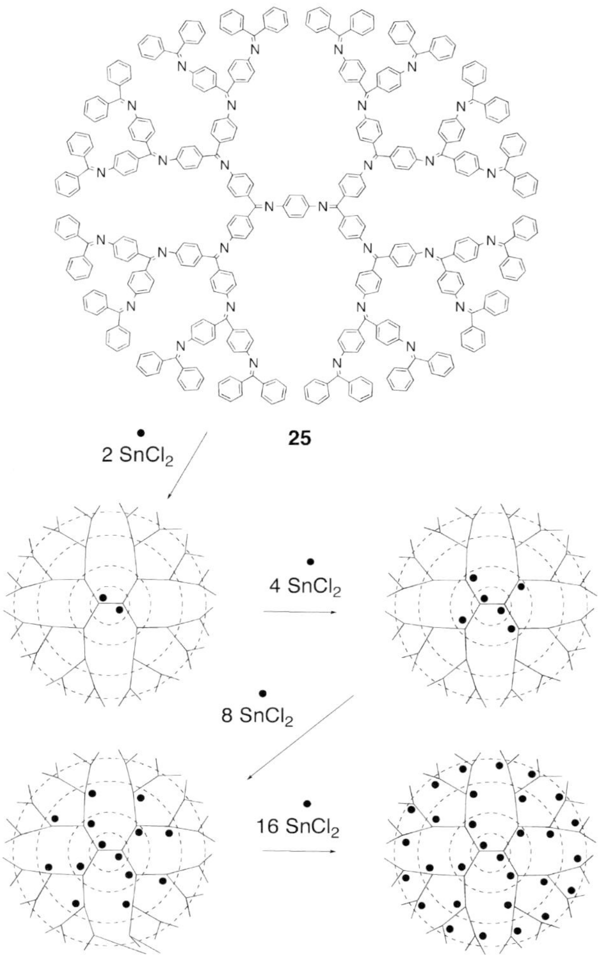

Fig. 9.27 Polyphenylazomethine dendrimer **25** for stepwise radial complexation with $SnCl_2$.

A precursor amine-terminated dendrimer can accommodate a variety of guest molecules such as Bengal Rose and *p*-nitrobenzoic acid in its interior. On the other hand, when such a guest-included dendrimer is functionalized with *t*-BOC-protected L-phenylalanine, the guest release occurs extremely slowly due to a steric effect of the bulky *t*-BOC groups on the dendrimer surface. Namely, the dendrimer behaves like a box. Aida and coworkers utilized a metal–ligand interaction for investigating the interpenetrating interaction between dendritic macromolecules [122]. Of interest, a zinc porphyrin caged in a large fourth-generation dendrimer can even bind, at its axial position, a large dendritic imidazole with a third-generation dendritic wedge. Yamamoto and coworkers synthesized a series of polyphenylazomethine dendrimers such as **25** and reported an interesting observation that these dendrimers coordinate transition metal ions such as Sn^{2+} and Fe^{3+} at azomethine units in a stepwise radial fashion from the core to the periphery (Fig. 9.27) [123]. They have also reported controlled release of bound metal ions via electrochemical reduction [123c]. The groups of Crooks [95] and Tomalia [124] reported concurrently that dendrimer-encapsulated metal nanoparticles can be prepared by sequestering metal ions within poly(propylenimine) and PAMAM dendrimers, followed by chemical reduction. The dimensions and solubilities of the resulting nanoparticles are dependent on those of the dendrimer templates. Some of those dendrimer-encapsulated metal nanoparticles can be used as catalysts for various reactions [95].

Interlocked molecular assemblies, such as rotaxanes and catenanes, are members of another class of interesting discrete objects formed in solution [125]. One notable example is the pseudorotaxane formation between a diaminobutane-terminated poly(propylenimine) dendrimer and cucurbituril. This takes great advantage of the high affinity of cucurbituril toward aminoalkanes due to the hydrophobic inner cavity for accommodating the guest tail and the aligned carbonyl functionalities at the edge parts of its cavity for hydrogen-bonding interactions with the guest amino group [125b].

9.7.2
Hierarchical Self-assembly of Dendrimers

Gels are interesting soft materials. Dendritic macromolecules are of particular interest in the formation of physical gels, as their three-dimensional branched structure can endow multiple cross-linking points [126]. Poly(benzyl ether) dendrons with a dipeptide core exhibit significant gelation ability in several organic solvents through a hydrogen-bonding interaction of the dipeptide part (Fig. 9.28) [127]. FE-SEM observation of dried gels revealed a hierarchical fibrous assembly formation. Notably, the gelation of the dendrimers is observable only with dendrons **26c** and **d** bearing large dendritic substituents, indicating the participation of van der Waals interactions among the dendritic wedges into the fiber formation. Majoral and coworkers reported that an organophosphorus dendrimer serves as an excellent gelator for water [128]. Polylysine dendrimers have been reported to form a physical gel upon mixing with diaminoalkanes in aprot-

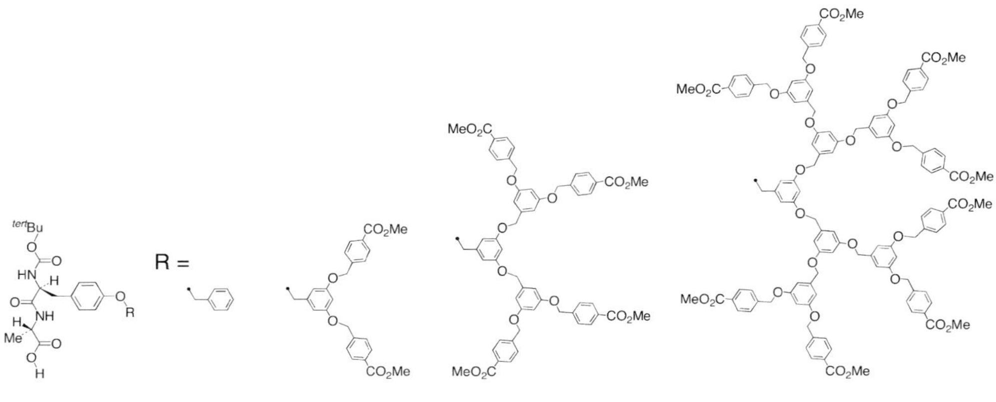

26a **26b** **26c** **26d**

Fig. 9.28 Dendrons **26** with a hydrogen-bonding dipeptide focal core for physical gelation.

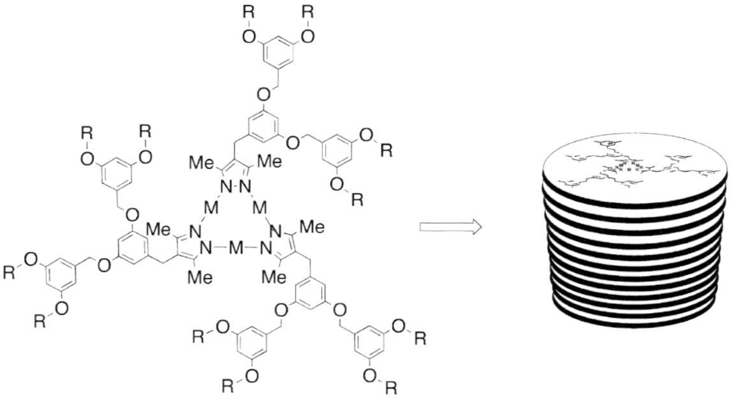

27

R = CH_3, $C_{18}H_{37}$ M = Cu^+, Ag^+, Au^+

Fig. 9.29 Self-assembly of dendritic pyrazolate complexes **27** mediated by metallophilic interactions.

ic solvents such as toluene, CH_2CH_2 and acetonitrile [129]. Furthermore, a certain aliphatic polyamide dendrimer, decorated with dodecyl groups on its surface, forms a physical gel in organic solvents such as $CHCl_3$ and THF but vesicles in aqueous media [130].

Metallophilic interactions are known to operate among group XI metal ions and have attracted attention, as they are usable for the fabrication of phosphorescent solid materials. Trimeric metallocycles formed by coordination of pyrazole-anchored dendrimer **27** (R=Me) with Cu^I, Ag^I and Au^I (Fig. 9.29) show an interesting self-assembly behavior [131]. Through metal–metal metallophilic interactions, these dendritic metallocycles are further assembled to form phosphorescent superhelical fibers when paraffin suspensions of these complexes are heated to 200 °C and then allowed to cool to room temperature. By replacing the peripheral methyl groups with long alkyl chains such as octadecyl groups, lower generation homologues of **27** form light-emitting organogels in hexane [132], where the photoluminescence color can be changed synchronously with the sol–gel transition as a consequence of the reversible formation and disruption of the metallophilic interactions. In the solid state, the trinuclear Cu^I complex of **27** ($R=C_{18}H_{37}$) displays dichroic luminescence [133]. When cooled from its hot melt, the complex emits at two different wavelengths, depending on the cooling rate. The dichroic luminescence is due to the difference in phosphorescence color between the kinetically and thermodynamically favored self-assembled structures. Either of these two structures can be selected by tuning the cooling rate of the hot melt. Papers coated with this material have been demonstrated to be usable for rewritable media, where written information can be visible only under illumination. Such luminescent rewritable papers are certainly important for the next-generation security technology for information handling.

Stupp and coworkers designed a so-called dendron rod–coil molecule consisting of coil-like, rod-like and dendritic segments, which self-assembles into nano-ribbons in dilute organic solvents [134]. Of interest, the nano-ribbons give highly stable dispersions of ZnO nanocrystals in organic solvents. The nano-ribbons themselves and the ZnO nanocrystals, hybridized with the nano-ribbons, are aligned in an electric field. Aida and coworkers found that an acyclic zinc porphyrin dimer (**28**), carrying a large dendritic wedge along with six carboxylic acid functionalities, co-assembles with C_{60} or C_{70}, to form "supramolecular peapods" that consist of aligned fullerenes in a hydrogen-bonded zinc porphyrin nanotube (Fig. 9.30) [135]. Without fullerenes, porphyrin dimer **28** gives a heavily entangled, irregular assembly. Percec and coworkers reported a library of poly(benzyl ether) dendrimers bearing a hydrogen-bonding dipeptide focal core and peripheral long alkyl chains. These dendrimers self-assemble in hydrocarbons and in bulk into a columnar structure with one-dimensional helical channels [136]. The channels are hydrophobic and capable of transporting protons when they are integrated into phospholipids membranes.

Vesicles and capsules are of interest because of their potential applications to drug delivery, in gene therapy and as models of biomembranes. Several amphiphilic dendrimers have been reported to form vesicles or capsules [137]. For example, an amphiphilic zinc porphyrin–fullerene dyad bearing a poly(benzyl ether) dendritic wedge, modified at its periphery with hydrophilic triethylene glycol chains, forms multi-lamellar vesicles with a uniform diameter of 100 nm, which are thermally stable and robust against membrane lysis with a surfactant [138].

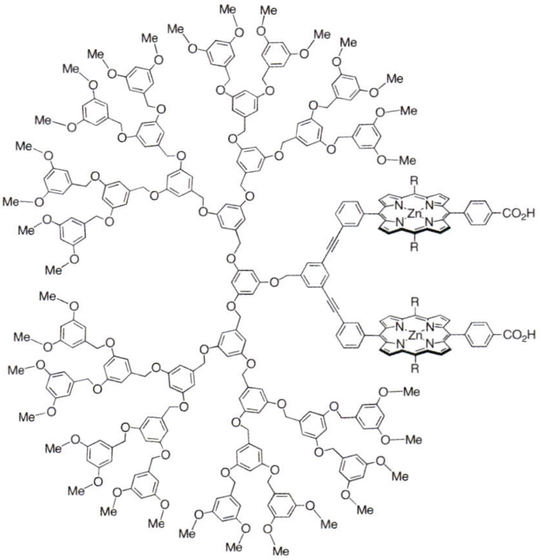

28

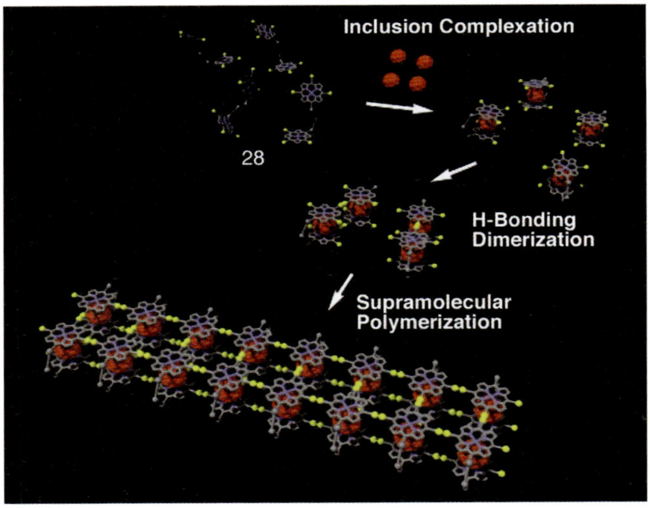

Fig. 9.30 Tubular co-assembly of poly(benzyl ether) dendritic zinc porphyrin dimer **28** with fullerenes such as C_{60} and C_{70} to form "supramolecular peapods". Yellow parts represent carboxylic acid functionalities.

9.7.3
Liquid Crystalline Dendrimers

The combination of mesogenic properties and dendritic structures is an interesting synthetic approach to a new class of liquid crystalline (LC) materials. Owing to their well-defined branched structures, LC dendrimers possess properties half way between those of polymers and those of single monomeric species: low viscosity, higher mesophase stabilization and good processing ability. Numerous LC dendrimers have been exploited by the integration of various mesogenic groups at the periphery of different dendrimers [139]. In this way, poly(propylenimine) [140], PAMAM [141], carbosilanes [142] and other dendrimers [143] have been used as soft cores. Later, an alternative approach to LC dendrimers was developed, in which mesogens linked by long flexible chains are placed in the branching units [144]. More recently, supramolecular LC dendrimers based on the self-assembly of T-shaped mesogens with poly(propylenimine) dendrimers have been reported [145].

Percec and coworkers reported that poly(benzyl ether) monodendrons, bearing long alkyl chains or semifluorinated alkyl chains in their periphery, along with a carboxylic acid, ester or crown ether unit at their focal core, show LC properties [146]. It is interesting that the assembly pattern is significantly dependent on the molecular shape of the dendrons. Flat, tapered monodendrons such as **29a** form a cylindrical aggregate and further assemble to afford a $p6mm$ Φ_h LC lattice, whereas conical monodendrons such as **29b** form a spherical aggregate and further assembles to give a $Pm\bar{3}n$ Cub LC lattice (Fig. 9.31). Percec and coworkers also synthesized dendrons **29b** bearing at their focal core polymerizable functionalities such as styrene ($X{=}CO_2CH_2PhCH{=}CH_2$) and methacrylate [$X{=}CH_2OCOC(CH_3){=}CH_2$] [147]. Interestingly, the polymerization of these conically shaped dendritic monomers is closely controlled by their self-assembly. At the initial stage, the polymerization gives a polymer adopting a spherical shape with a random-coil backbone conformation. On the other hand, when the polymerization proceeds, the sphere undergoes a transition to a cylindrical shape with an extended backbone. They also developed a new class of supramolecular electronic materials by the co-assembly of a monodendron having an electron-donating focal core with that carrying an electron-accepting focal core. Linear polymers carrying compatible donor or acceptor side chains can co-assemble with these dendrons. Some of the resulting materials consisting of donor–acceptor co-assembled cylinders show a high charge-carrier mobility [148].

9.7.4
Dendritic Monolayers and Multilayers

Dendrimers are also attractive building blocks for the controlled formation of monolayer or multilayer thin films, which have been demonstrated to be useful in a variety of potential applications, such as electron-transfer mediators, catalysts, sensors and electronic devices [149]. For example, highly oriented electro-

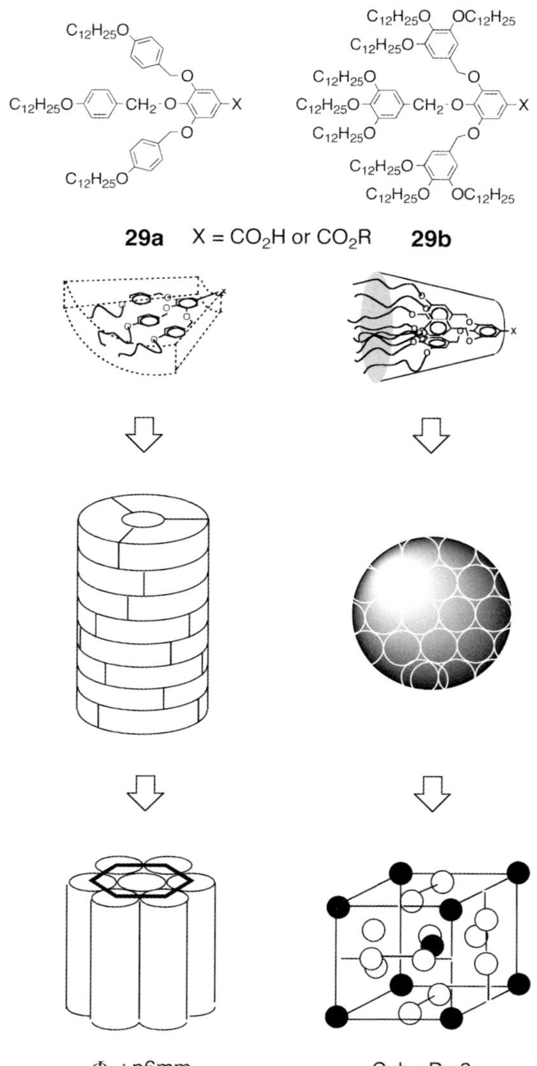

Fig. 9.31 Self-assembly behaviors of tapered and conical dendrons **29a** and **b**, respectively.

chemically active monolayers on pyrolytic graphite surfaces have been fabricated using terpyridine-terminated dendrimers together with bridging ligands and transition metal ions such as Fe^{2+} or Co^{2+} [150]. The formation of a monolayer film of PAMAM on a gold surface has been reported using multiple amine–Au interactions [151]. Successive exposure of the monolayer film to hexadecanethiol results in compression of the dendrimer structure, due to the additional attachment of the alkanethiol molecules to the gold surface. Interestingly, the dendrimer molecules in this compressed monolayer act like ion gates, which control the penetration of ion species in response to H^+. Thus, when the dendrimer is fully deprotonated at pH 11, the monolayer film allows both $[Fe(CN)_6]^{3-}$ and $[Ru(NH_3)_6]^{3+}$ to pass through. In contrast, at pH 6.3, only negatively charged $[Fe(CN)_6]^{3-}$ can permeate through the film.

By means of high-resolution lithography using a scanning probe, a 60-nm scale resist has been developed on the surface of a silicon wafer, covalently modified with a poly(benzyl ether) dendrimer [152]. A zinc porphyrin bearing two poly(benzyl ether) dendritic wedges and two carboxylic acid groups at its opposing meso positions has been reported to self-assemble in aprotic solvents such as benzene and $CHCl_3$, forming a porphyrin *J*-aggregate via hydrogen-bonding and van der Waals interactions along with a π-electronic interaction [153]. Of interest, when the solution is spin-coated on to a glass surface, the resulting thin film displays a chiroptical activity, whose sense is selected by the spinning direction. In sharp contrast, a cast film of the same solution does not show any dominance in chirality.

9.8 Conclusions

The unique structural features of dendritic macromolecules with precise three-dimensional size regimes endow these exotic molecules with unexpectedly versatile characteristics. These features make dendrimers attractive building blocks for nanotechnology and materials sciences. As exemplified by the selected topics in this chapter, much of the molecular design and application of dendrimers are based on the lessons and concepts learned from nature, as a consequence of the structural similarity of dendrimers to natural biomacromolecules. Naturally occurring macromolecular systems, which have had millions of years to evolve, provide chemists with powerful and sophisticated systems from which to draw inspiration and will continue to stimulate the studies on dendritic macromolecules in the future.

References

1 D.A. Tomalia, J.M.J. Fréchet, *J. Polym. Sci., Part A: Polym. Chem.* **2002**, *40*, 2719–2728.

2 E. Buhleier, W. Wehner, F. Vögtle, *Synthesis* **1978**, 155–158.

3 (a) R.G. Denkewalter, J. Kolc, W.J. Lukasavage, *US Patent 4289872*, **1981**; (b) R.G. Denkewalter, J. Kolc, W.J. Lukasavage, *US Patent 4360646*, **1982**; (c) R.G. Denkewalter, J. Kolc, W.J. Lukasavage, *US Patent 4410688*, **1983**.

4 D.A. Tomalia, H. Baker, J. Dewald, M. Hall, G. Kallos, S. Martin, J. Roeck, J. Ryder, P. Smith, *Polym. J.* **1985**, *17*, 117–132.

5 G.R. Newkome, Z.-Q. Yao, G.R. Baker, V.K. Gupta, *J. Org. Chem.* **1985**, *50*, 2003–2004.

6 (a) C. Hawker, J. M. J. Fréchet, *J. Chem. Soc., Chem. Commun.* **1990**, 1010–1013; (b) C. J. Hawker, J. M. J. Fréchet, *J. Am. Chem. Soc.* **1990**, *112*, 7638–7647; (c) C. J. Hawker, J. M. J. Fréchet, *Macromolecules* **1990**, *23*, 4726–4729.

7 For selected books, see: (a) J. M. J. Fréchet, D. A. Tomalia (Eds.), *Dendrimers and Other Dendritic Polymers*, Wiley, New York, **2001**; (b) G. R. Newkome, C. N. Moorefield, F. Vögtle, *Dendrimers and Dendrons: Concept, Syntheses, Applications*, Wiley-VCH, Weinheim, **2001**.

8 For special issues, see: (a) *Top. Curr. Chem.* **1998**, *197*, 1–228; **2000**, *210*, 1–308; **2001**, *212*, 1–194; **2001**, *217*, 1–238; **2003**, *228*, 1–236; (b) *Adv. Dendritic Macromol.* **1994**, *1*, 1–168; **1995**, *2*, 1–190; **1996**, *3*, 1–195; **1999**, *4*, 1–201; **2002**, *5*, 1–247; (c) *C. R. Chem.* **2003**, *6*, 715–1127; (d) *Prog. Polym. Sci.* **2005**, *30*, 217–473.

9 For selected reviews, see: (a) A. W. Bosman, H. M. Janssen, E. W. Meijer, *Chem. Rev.* **1999**, *99*, 1665–1688; (b) M. Fischer, F. Vögtle, *Angew. Chem. Int. Ed.* **1999**, *38*, 884–905; (c) A. D. Schlüter, J. P. Rabe, *Angew. Chem. Int. Ed.* **2000**, *39*, 864–883; (d) K. Inoue, *Prog. Polym. Sci.* **2000**, *25*, 453–571; (e) F. Vögtle, S. Gestermann, R. Hesse, H. Schwierz, B. Windisch, *Prog. Polym. Sci.* **2000**, *25*, 987–1041; (f) J. M. J. Fréchet, *Proc. Natl. Acad. Sci. USA* **2002**, *99*, 4782–4787; (g) J. M. J. Fréchet, *J. Polym. Sci., Part A: Polym. Chem.* **2003**, *41*, 3713–3725; (h) D. K. Smith, *Chem. Commun.* **2006**, 34–44.

10 (a) C. J. Hawker, K. L. Wooley, *Adv. Dendritic Macromol.* **1995**, *2*, 1–39; (b) C. J. Hawker, *Adv. Polym. Sci.* **1999**, *147*, 113–160; (c) S. M. Grayson, J. M. J. Fréchet, *Chem. Rev.* **2001**, *101*, 3819–3867.

11 J. C. Hummelen, J. L. J. van Dongen, E. W. Meijer, *Chem. Eur. J.* **1997**, *3*, 1489–1493.

12 (a) K. L. Wooley, C. J. Hawker, J. M. J. Fréchet, *J. Am. Chem. Soc.* **1991**, *113*, 4252–4272; (b) G. L'Abbé, B. Forier, W. Dehaen, *Chem. Commun.* **1996**, 2143–2144; (c) D. M. Junge, D. V. McGrath, *Tetrahydron Lett.* **1998**, *39*, 1701–1704.

13 T. Kawaguchi, K. L. Walker, C. L. Wilkins, J. S. Moore, *J. Am. Chem. Soc.* **1995**, *117*, 2159–2165.

14 (a) R. Spindler, J. M. J. Fréchet, *J. Chem. Soc., Perkin Trans. 1* **1993**, 913–918; (b) F. Zeng, S. C. Zimmerman, *J. Am. Chem. Soc.* **1996**, *118*, 5326–5327.

15 (a) P. Wu, A. K. Feldman, A. K. Nugent, C. J. Hawker, A. Scheel, B. Viot, J. Pyun, J. M. J. Fréchet, K. B. Sharpless, V. V. Fokin, *Angew. Chem. Int. Ed.* **2004**, *43*, 3928–3932; (b) B. Helms, J. L. Mynar, C. J. Hawker, J. M. J. Fréchet, *J. Am. Chem. Soc.* **2004**, *126*, 15020–15021; (c) P. Wu, M. Malkoch, J. N. Hunt, R. Vestberg, E. Kaltgrad, M. G. Finn, V. V. Fokin, K. B. Sharpless, C. J. Hawker, *Chem. Commun.* **2005**, 5775–5777; (d) M. Malkoch, K. Schleicher, E. Drockenmuller, C. J. Hawker, T. P. Russell, P. Wu, V. V. Fokin, *Macromolecules* **2005**, *38*, 3663–3678; (e) M. J. Joralemon, R. K. O'Reilly, J. B. Matson, A. K. Nugent, C. J. Hawker, K. L. Wooley, *Macromolecules* **2005**, *38*, 5436–5443.

16 D. A. Tomalia, J. M. J. Fréchet, Introduction to the dendritic state, in *Dendrimers and Other Dendritic Polymers*, J. M. J. Fréchet, D. A. Tomalia (Eds.), Wiley, New York, **2001**, pp. 3–44.

17 D. A. Tomalia, A. M. Naylor, W. A. Goddard III, *Angew. Chem. Int. Ed. Engl.* **1990**, *29*, 138–175.

18 (a) S. Hecht, J. M. J. Fréchet, *Angew. Chem. Int. Ed.* **2001**, *40*, 74–91; (b) C. B. Gorman, J. S. Smith, *Acc. Chem. Res.* **2001**, *34*, 60–71; (c) J.-F. Nierengarten, *C. R. Chim.* **2003**, *6*, 725–733.

19 For selected reviews, see: (a) M. R. Bryce, W. Devonport, *Adv. Dendritic Macromol.* **1996**, *3*, 115–149; (b) C. M. Casado, I. Cuadrado, M. Morán, B. Alonso, B. García, B. González, J. Losada, *Coord. Chem. Rev.* **1999**, *185/186*, 53–80; (c) C. M. Cardona, S. Mendoza, A. E. Kaifer, *Chem. Soc. Rev.* **2000**, 37–42; (d) C. B. Gorman, *C. R. Chim.* **2003**, *6*, 911–918; (e) M. Venturi, P. Ceroni, *C. R. Chim.* **2003**, *6*, 935–945; (f) A. Juris, *Annu. Rep. Prog. Chem., Sect. C* **2003**, *99*, 177–241.

20 (a) G. R. Newkome, R. Günther, C. N. Moorefield, F. Cardullo, L. Echegoyen, E. Pérez-Cordero, H. Luftmann, *Angew.*

Chem. Int. Ed. Engl. **1995**, *34*, 2023–2026; (b) Y. Rio, G. Accorsi, N. Armaroli, D. Felder, E. Levillain, J.-F. Nierengarten, *Chem. Commun.* **2002**, 2830–2831.
21 (a) D.L. Stone, D.K. Smith, P.T. McGrail, *J. Am. Chem. Soc.* **2002**, *124*, 856–864; (b) P.R. Ashton, V. Balzani, M. Clemente-León, B. Colonna, A. Credi, N. Jayaraman, F.M. Raymo, J.F. Stoddart, M. Venturi, *Chem. Eur. J.* **2002**, *8*, 673–684.
22 (a) P. Ceroni, V. Vicinelli, M. Maestri, V. Balzani, W.M. Müller, U. Müller, U. Hahn, F. Osswald, F. Vögtle, *New J. Chem.* **2001**, *25*, 989–993; (b) R. Toba, J.M. Quintela, C. Peinador, E. Román, A.E. Kaifer, *Chem. Commun.* **2001**, 857–858.
23 (a) C.M. Cardona, A.E. Kaifer, *J. Am. Chem. Soc.* **1998**, *120*, 4023–4024; (b) Y. Wang, C.M. Cardona, A.E. Kaifer, *J. Am. Chem. Soc.* **1999**, *121*, 9756–9757; (c) C.M. Cardona, T.D. McCarley, A.E. Kaifer, *J. Org. Chem.* **2000**, *65*, 1857–1864.
24 P.J. Dandliker, F. Diederich, M. Gross, C.B. Knobler, A. Louati, E.M. Sanford, *Angew. Chem. Int. Ed. Engl.* **1994**, *33*, 1739–1742.
25 K.W. Pollak, J.W. Leon, J.M.J. Fréchet, M. Maskus, H.D. Abrua, *Chem. Mater.* **1998**, *10*, 30–38.
26 C.B. Gorman, J.C. Smith, M.W. Hager, B.L. Parkhurst, H. Sierzputowska-Gracz, C.A. Haney, *J. Am. Chem. Soc.* **1999**, *121*, 9958–9966.
27 T.L. Chasse, R. Sachdava, Q. Li, Z. Li, R.J. Petrie, C.B. Gorman, *J. Am. Chem. Soc.* **2003**, *125*, 8250–8254.
28 D.-L. Jiang, T. Aida, *Chem. Commun.* **1996**, 1523–1524.
29 (a) J.P. Collman, L. Fu, A. Zingg, F. Diederich, *Chem. Commun.* **1997**, 193–194; (b) P. Weyermann, F. Diederich, *J. Chem. Soc., Perkin Trans. 1* **2000**, 4231–4233; (c) A. Zingg, B. Felber, V. Gramlich, L. Fu, J.P. Collman, F. Diederich, *Helv. Chim. Acta* **2002**, *85*, 333–351; (d) S.V. Doorslaer, A. Zingg, A. Schweiger, F. Diederich, *ChemPhysChem* **2002**, *3*, 659–667; (e) B. Felber, C. Calle, P. Seiler, A. Schweiger, F. Diederich, *Org. Biomol. Chem.* **2003**, *1*, 1090–1093; (f) B. Felber, F. Diederich, *Helv. Chim. Acta* **2005**, *88*, 120–153.
30 (a) M. Enomoto, T. Aida, *J. Am. Chem. Soc.* **1999**, *121*, 874–875; (b) M. Enomoto, T. Aida, *J. Am. Chem. Soc.* **2002**, *124*, 6099–6108.
31 (a) B. Alonso, I. Cuadrado, M. Morán, J. Losada, *J. Chem. Soc., Chem. Commun.* **1994**, 2575–2576; (b) C. Valério, J.-L. Fillaut, J. Ruiz, J. Guittard, J.-C. Blais, D. Astruc, *J. Am. Chem. Soc.* **1997**, *119*, 2588–2589; (c) I. Cuadrado, C.M. Casado, B. Alonso, M. Morán, J. Losada, V. Belsky, *J. Am. Chem. Soc.* **1997**, *119*, 7613–7614; (d) J. Alvarez, T. Ren, A.E. Kaifer, *Organometallics* **2001**, *20*, 3543–3549; (e) C.-O. Turrin, J. Chiffre, D. de Montauzon, G. Balavoine, E. Manoury, A.-M. Caminade, J.-P. Majoral, *Organometallics* **2002**, *21*, 1891–1897; (f) M.-C. Daniel, J. Ruiz, J.-C. Blais, N. Daro, D.A. Astruc, *Chem. Eur. J.* **2003**, *9*, 4371–4379.
32 D. Astruc, *Acc. Chem. Res.* **2000**, *33*, 287–298.
33 I. Cuadrado, M. Morán, C.M. Casado, B. Alonso, F. Lobete, B. García, M. Ibisate, J. Losada, *Organometallics* **1996**, *15*, 5278–5280.
34 M.-C. Daniel, J. Ruiz, D. Astruc, *J. Am. Chem. Soc.* **2003**, *125*, 1150–1151.
35 M.-C. Daniel, J. Ruiz, S. Nlate, J.-C. Blais, D. Astruc, *J. Am. Chem. Soc.* **2003**, *125*, 2617–2628.
36 (a) D. Astruc, J.-C. Daniel, S. Gatard, S. Nlate, J. Ruiz, *C.R. Chim.* **2003**, *6*, 1117–1127; (b) W. Ong, M. Gómez-Kaifer, A.E. Kaifer, *Chem. Commun.* **2004**, 1677–1683; (c) D. Astruc, M.-C. Daniel, J. Ruiz, *Chem. Commun.* **2004**, 2637–2649.
37 (a) H. Zheng, G.R. Newkome, C.L. Hill, *Angew. Chem. Int. Ed.* **2000**, *39*, 1771–1774; (b) F. Le Derf, E. Levillain, G. Trippé, A. Gorgues, M. Sallé, R.-M. Sebastían, A.-M. Caminade, J.-P. Majoral, *Angew. Chem. Int. Ed.* **2001**, *40*, 224–227; (c) N. Godbert, M.R. Bryce, *J. Mater. Chem.* **2002**, *12*, 27–36; (d) J.R. Aranzaes, C. Belin, D. Astruc, *Angew. Chem. Int. Ed.* **2006**, *45*, 132–136.
38 (a) S. Heinen, L. Walder, *Angew. Chem. Int. Ed.* **2000**, *39*, 806–809; (b) F. March-

ioni, M. Venturi, A. Credi, V. Balzani, M. Belohradsky, A. M. Elizarov, H.-R. Tseng, J. F. Stoddart, *J. Am. Chem. Soc.* **2004**, *126*, 568–573; (c) F. Marchioni, M. Venturi, P. Ceroni, V. Balzani, M. Belohradsky, A. M. Elizarov, H.-R. Tseng, J. F. Stoddart, *Chem. Eur. J.* **2004**, *10*, 6361–6368.

39 S. Serroni, S. Campagna, F. Puntoriero, C. D. Pietro, N. D. McClenaghan, F. Loiseau, *Chem. Soc. Rev.* **2001**, *30*, 367–375.

40 (a) G. McDermott, S. M. Prince, A. A. Freer, A. M. Hawthornthwaite-Lawless, M. Z. Papiz, R. J. Cogdell, N. W. Isaacs, *Nature* **1995**, *374*, 517–521; (b) S. Karrasch, P. A. Bullough, R. Ghosh, *EMBO J.* **1995**, *14*, 631–638; (c) T. Pullerits, V. Sundström, *Acc. Chem. Res.* **1996**, *29*, 381–389.

41 G. Denti, S. Campagna, S. Serroni, M. Ciano, V. Balzani, *J. Am. Chem. Soc.* **1992**, *114*, 2944–2950.

42 For selected reviews, see: (a) A. Adronov, J. M. J. Fréchet, *Chem. Commun.* **2000**, 1701–1710; (b) V. Balzani, P. Ceroni, M. Maestri, V. Vicinelli, *Curr. Opin. Chem. Biol.* **2003**, *7*, 657–665.

43 C. Devadoss, P. Bharathi, J. S. Moore, *J. Am. Chem. Soc.* **1996**, *118*, 9635–9644.

44 D.-L. Jiang, T. Aida, *J. Am. Chem. Soc.* **1998**, *120*, 10895–10901.

45 (a) M.-S. Choi, T. Aida, T. Yamazaki, I. Yamazaki, *Angew. Chem. Int. Ed.* **2001**, *40*, 3194–3198; (b) M.-S. Choi, T. Aida, T. Yamazaki, I. Yamazaki, *Chem. Eur. J.* **2002**, *8*, 2667–2678.

46 D. L. Dexter, *J. Chem. Phys.* **1953**, *21*, 836–850.

47 (a) T. Förster, *Ann. Phys.* **1948**, *2*, 55–75; (b) T. Förster, *Discuss. Faraday Soc.* **1959**, *27*, 7–17.

48 (a) T. Weil, E. Reuther, K. Müllen, *Angew. Chem. Int. Ed.* **2002**, *41*, 1900–1904; (b) R. Gronheid, J. Hofkens, F. Köhn, T. Weil, E. Reuther, K. Müllen, F. C. De Schryver, *J. Am. Chem. Soc.* **2002**, *124*, 2418–2419; (c) R. E. Bauer, A. C. Grimsdale, K. Müllen, *Top. Curr. Chem.* **2005**, *245*, 253–286.

49 (a) U. Hahn, M. Gorka, F. Vögtle, V. Vicinelli, P. Ceroni, M. Maestri, V. Balzani, *Angew. Chem. Int. Ed.* **2002**, *41*, 3595–3598; (b) W. R. Dichtel, S. Hecht, J. M. J. Fréchet, *Org. Lett.* **2005**, *7*, 4451–4454.

50 T. Sato, D.-L. Jiang, T. Aida, *J. Am. Chem. Soc.* **1999**, *121*, 10658–10659.

51 M. Kawa, J. M. J. Fréchet, *Chem. Mater.* **1998**, *10*, 286–296.

52 (a) S. Prathapan, T. E. Johnson, J. S. Lindsey, *J. Am. Chem. Soc.* **1993**, *115*, 7519–7520; (b) D. L. Officer, A. K. Burrell, D. C. W. Reid, *Chem. Commun.* **1996**, 1657–1658; (c) T. Norsten, N. Branda, *Chem. Commun.* **1998**, 1257–1258; (d) M. del R. Benites, T. E. Johnson, S. Weghorn, L. Yu, P. D. Rao, J. R. Diers, S. I. Yang, C. Kirmaier, D. F. Bocian, D. Holten, J. S. Lindsey, *J. Mater. Chem.* **2002**, *12*, 65–80.

53 (a) C. C. Mak, N. Bampos, J. K. M. Sanders, *Angew. Chem. Int. Ed.* **1998**, *37*, 3020–3023; (b) C. C. Mak, D. Pomeranc, M. Montalti, L. Prodi, J. K. M. Sanders, *Chem. Commun.* **1999**, 1083–1084.

54 N. Maruo, M. Uchiyama, T. Kato, T. Arai, H. Akisada, N. Nishino, *Chem. Commun.* **1999**, 2057–2058.

55 T. Kato, M. Uchiyama, N. Maruo, T. Arai, N. Nishino, *Chem. Lett.* **2000**, 144–145.

56 (a) E. K. L. Yeow, K. P. Ghiggino, J. N. H. Reek, M. J. Crossley, A. W. Bosman, A. P. H. J. Schenning, E. W. Meijer, *J. Phys. Chem. B* **2000**, *104*, 2596–2606; (b) S. Cho, W.-S. Li, M.-C. Yoon, T. K. Ahn, D.-L. Jiang, J. Kim, T. Aida, K. Kim, *Chem. Eur. J.* **2006**, *12*, 7576–7584.

57 (a) W. R. Dichtel, J. M. Serin, C. Edder, J. M. J. Fréchet, M. Matuszewski, L.-S. Tan, T. Y. Ohulchanskyy, P. N. Prasad, *J. Am. Chem. Soc.* **2004**, *126*, 5380–5381; (b) M. A. Oar, J. M. Serin, W. R. Dichtel, J. M. J. Fréchet, *Chem. Mater.* **2005**, *17*, 2267–2275.

58 D.-L. Jiang, T. Aida, *Nature* **1997**, *388*, 454–456.

59 Y.-J. Mo, D.-L. Jiang, M. Uyemura, T. Aida, T. Kitagawa, *J. Am. Chem. Soc.* **2005**, *127*, 10020–10027.

60 R. Sadamoto, N. Tomioka, T. Aida, *J. Am. Chem. Soc.* **1996**, *118*, 3978–3979.

61 (a) G. J. Capitosti, S. J. Cramer, C. S. Rajesh, D. A. Modarelli, *Org. Lett.* **2001**, *3*, 1645–1648; (b) C. S. Rajesh, G. J. Capitosti, S. J. Cramer, D. A. Modarelli, *J. Phys. Chem. B* **2001**, *105*, 10175–10188; (c) G. J. Capitosti, C. D. Guerrero, D. E.

Binkley Jr., C. S. Rajesh, D. A. Modarelli, *J. Org. Chem.* **2003**, *68*, 247–261.
62 X. Camps, E. Dietel, A. Hirsch, S. Pyo, L. Echegoyen, S. Hackbarth, B. Röder, *Chem. Eur. J.* **1999**, *5*, 2362–2373.
63 T. H. Ghaddar, J. F. Wishart, D. W. Thompson, J. K. Whitesell, M. A. Fox, *J. Am. Chem. Soc.* **2002**, *124*, 8285–8289.
64 (a) K. Feldrapp, W. Brütting, M. Schwörer, M. Brettreich, A. Hirsch, *Synth. Met.* **1999**, *101*, 156–157; (b) J. L. Segura, R. Gómez, N. Martín, C. Luo, A. Swartz, D. M. Guldi, *Chem. Commun.* **2001**, 707–708; (c) R. Kunieda, M. Fujitsuka, O. Ito, M. Ito, Y. Murata, K. Komatsu, *J. Phys. Chem. B* **2002**, *106*, 7193–7199; (d) S. Campidelli, E. Vázquez, D. Milic, M. Prato, J. Barberá, D. M. Guldi, M. Marcaccio, D. Paolucci, F. Paolucci, R. Deschenaux, *J. Mater. Chem.* **2004**, *14*, 1266–1272; (e) M. Gutierrez-Nava, G. Accorsi, P. Masson, N. Armaroli, J.-F. Nierengarten, *Chem. Eur. J.* **2004**, *10*, 5076–5086.
65 (a) G. M. Stewart, M. A. Fox, *J. Am. Chem. Soc.* **1996**, *118*, 4354–4360; (b) T. D. Selby, S. C. Blackstock, *J. Am. Chem. Soc.* **1998**, *120*, 12155–12156; (c) M. Lor, J. Thielemans, L. Viaene, M. Cotlet, J. Hofkens, T. Weil, C. Hampel, K. Müllen, J. W. Verhoeven, M. V. der Auweraer, F. C. De Schryver, *J. Am. Chem. Soc.* **2002**, *124*, 9918–9925; (d) M. Lor, L. Viaene, R. Pilot, E. Fron, S. Jordens, G. Schweitzer, T. Weil, K. Müllen, J. W. Verhoeven, M. V. der Auweraer, F. C. De Schryver, *J. Phys. Chem. B* **2004**, *108*, 10721–10731.
66 (a) M. R. Bryce, W. Devonport, A. J. Moore, *Angew. Chem. Int. Ed. Engl.* **1994**, *33*, 1761–1763; (b) A. Kanibolotsky, S. Roquet, M. Cariou, P. Leriche, C.-O. Turrin, R. de Bettignies, A.-M. Caminade, J.-P. Majoral, V. Khodorkovsky, A. Gorgues, *Org. Lett.* **2004**, *6*, 2109–2112.
67 (a) D. M. Guldi, A. Swartz, C. Luo, R. Gómez, J. L. Segura, N. Martín, *J. Am. Chem. Soc.* **2002**, *124*, 10875–10886; (b) J. Qu, N. G. Pschirer, D. Liu, A. Stefan, F. C. De Schryver, K. Müllen, *Chem. Eur. J.* **2004**, *10*, 528–537.
68 K. R. J. Thomas, A. L. Thompson, A. V. Sivakumar, C. J. Bardeen, S. Thayumanavan, *J. Am. Chem. Soc.* **2005**, *127*, 373–383.
69 M.-S. Choi, T. Aida, H. Luo, Y. Araki, O. Ito, *Angew. Chem. Int. Ed.* **2003**, *42*, 4060–4063.
70 W.-S. Li, K. S. Kim, D.-L. Jiang, H. Tanaka, T. Kawai, J. H. Kwon, D. Kim, T. Aida, *J. Am. Chem. Soc.* **2006**, *128*, 10527–10532.
71 D.-L. Jiang, C.-K. Choi, K. Honda, W.-S. Li, T. Yuzawa, T. Aida, *J. Am. Chem. Soc.* **2004**, *126*, 12084–12089.
72 I. Gitsov, J. M. J. Fréchet, *J. Am. Chem. Soc.* **1996**, *118*, 3785–3786.
73 J. Leclaire, Y. Coppel, A.-M. Caminade, J.-P. Majoral, *J. Am. Chem. Soc.* **2004**, *126*, 2304–2305.
74 (a) W. Chen, D. A. Tomalia, J. L. Thomas, *Macromoleucles* **2000**, *33*, 9169–9172; (b) I. Lee, B. D. Athey, A. W. Wetzel, W. Meixner, J. R. Baker Jr., *Macromolecules* **2002**, *35*, 4510–4520; (c) M. H. Kleinman, J. H. Flory, D. A. Tomalia, N. J. Turro, *J. Phys. Chem. B* **2000**, *104*, 11472–11479.
75 G. Nisato, R. Ivkov, E. J. Amis, *Macromolecules* **1999**, *32*, 5895–5900.
76 (a) G. Pistolis, A. Malliaris, D. Tsiourvas, C. M. Paleos, *Chem. Eur. J.* **1999**, *5*, 1440–1444; (b) Z. Sideratou, D. Tsiourvas, C. M. Paleos, *Langmiur* **2000**, *16*, 1766–1769.
77 H. Stephan, H. Spies, B. Johannsen, C. Kauffmann, F. Vögtle, *Org. Lett.*, **2000**, *2*, 2343–2346.
78 (a) O. Villavicencio, D. V. McGrath, *Adv. Dendritic Macromol.* **2002**, *5*, 144; (b) V. Shibaev, A. Bobrovsky, N. Boiko, *Prog. Polym. Sci.* **2003**, *28*, 729–836; (c) A. Momotake, T. Arai, *Polymer*, **2004**, *45*, 5369–5390.
79 H.-B. Mekelburger, F. Vögtle, K. Rissanen, *Chem. Ber.* **1993**, *126*, 1161–1169.
80 (a) D. M. Junge, D. V. McGrath, *Chem. Commun.* **1997**, 857–858; (b) D. Grebel-Köhler, D. Liu, S. De Feyter, V. Enkelmann, T. Weil, C. Engels, C. Samyn, K. Müllen, F. C. De Schryver, *Macromolecules* **2003**, *36*, 578–590; (c) A. Momotake, T. Arai, *Tetrahedron Lett.* **2004**, *45*, 4131–4134.
81 (a) A. Cattani-Scholz, C. Renner, D. Oesterhelt, L. Moroder, *ChemBioChem* **2001**, *2*, 542–549; (b) K.-Y. Kay, K.-J. Han, Y.-J.

Yu, Y. D. Park, *Tetrahedron Lett.* **2002**, *43*, 5053–5056; (c) A. Dirksen, E. Zuidema, R. M. Williams, L. De Cola, C. Kauffmann, F. Vögtle, A. Roque, F. Pina, *Macromolecules* **2002**, *35*, 2743–2747.

82 (a) D. M. Junge, D. V. McGrath, *J. Am. Chem. Soc.* **1999**, *121*, 4912–4913; (b) S. Wang, X. Wang, L. Li, R. C. Advincula, *J. Org. Chem.* **2004**, *69*, 9073–9084.

83 A. Archut, G. C. Azzellini, V. Balzani, L. D. Cola, F. Vögtle, *J. Am. Chem. Soc.* **1998**, *120*, 12187–12191.

84 K. Tsuda, G. C. Dol, T. Gensch, J. Hofkens, L. Latterini, J. W. Weener, E. W. Meijer, F. C. De Schryver, *J. Am. Chem. Soc.* **2000**, *122*, 3445–3452.

85 T. Fujigaya, D.-L. Jiang, T. Aida, *J. Am. Chem. Soc.* **2005**, *127*, 54845489.

86 For selected reviews, see: (a) G. R. Newkome, *Pure Appl. Chem.* **1998**, *70*, 2337–2343; (b) S. C. Zimmerman, L. J. Lawless, *Top. Curr. Chem.* **2001**, *217*, 95–120.

87 G. R. Newkome, L. A. Godínez, C. N. Moorefield, *Chem. Commun.* **1998**, 1821–1822.

88 B. Kenda, F. Diederich, *Angew. Chem. Int. Ed.* **1998**, *37*, 3154–3158.

89 D. K. Smith, F. Diederich, *Chem. Commun.* **1998**, 2501–2502.

90 (a) V. J. Pugh, Q. S. Hu, L. Pu, *Angew. Chem. Int. Ed.* **2000**, *39*, 3638–3641; (b) L. Z. Gong, Q.-S. Hu, L. Pu, *J. Org. Chem.* **2001**, *66*, 2358–2367.

91 W.-S. Li, D.-L. Jiang, Y. Suna, T. Aida, *J. Am. Chem. Soc.* **2005**, *127*, 7700–7702.

92 (a) G. E. Oosterom, J. N. H. Reek, P. C. J. Kamer, P. W. N. M. van Leeuwen, *Angew. Chem. Int. Ed.* **2001**, *40*, 1828–1849; (b) D. Astruc, F. Chardac, *Chem. Rev.* **2001**, *101*, 2991–3023; (c) L. J. Twyman, A. S. H. King, I. K. Martin, *Chem. Soc. Rev.*, **2002**, *31*, 69–82.

93 P. A. Chase, R. J. M. K. Gebbink, G. van Koten, *J. Organomet. Chem.* **2004**, *689*, 4016–4054.

94 (a) J. N. H. Reek, D. de Groot, G. E. Oosterom, P. C. J. Kamer, P. W. N. M. van Leeuwen, *C. R. Chim.* **2003**, *6*, 1061–1077; (b) A.-M. Caminade, J.-P. Majoral, *Coord. Chem. Rev.* **2005**, *249*, 1917–1926.

95 (a) R. W. Crooks, M. Zhao, L. Sun, V. Chechik, L. K. Yeung, *Acc. Chem. Res.* **2001**, *3*, 181–190; (b) R. W. J. Scott, O. M. Wilson, R. M. Crooks, *J. Phys. Chem., B* **2005**, *109*, 692–704.

96 R. Laurent, A.-M. Caminade, J.-P. Majoral, *Tetrahydron Lett.* **2005**, *46*, 6503–6506.

97 (a) J. W. J. Knapen, A. W. van der Made, J. C. de Wilde, P. W. N. M. van Leeuwen, P. Wijkens, D. M. Grove, G. van Koten, *Nature* **1994**, *372*, 659–663; (b) A. W. Kleij, R. A. Gossage, R. J. M. K. Gebbink, N. Brinkmann, E. J. Reijerse, U. Kragl, M. Lutz, A. L. Spek, G. van Koten, *J. Am. Chem. Soc.* **2000**, *122*, 12112–12124.

98 J. Lemo, K. Heuzé, D. Astruc, *Org. Lett.* **2005**, *7*, 2253–2256.

99 M. T. Reetz, G. Lohmer, R. Schwickardi, *Angew. Chem. Int. Ed. Engl.* **1997**, *36*, 1526–1529.

100 (a) C. Francavilla, M. D. Drake, F. V. Bright, M. R. Detty, *J. Am. Chem. Soc.* **2001**, *123*, 57–67; (b) M. D. Drake, F. V. Bright, M. R. Detty, *J. Am. Chem. Soc.* **2003**, *125*, 12558–12566.

101 G. E. Oosterom, R. J. van Haaren, J. N. H. Reek, P. C. J. Kamer, P. W. N. M. van Leeuwen, *Chem. Commun.* **1999**, 1119–1120.

102 M. E. Piotti, F. Rivera Jr., R. Bond, C. J. Hawker, J. M. J. Fréchet, *J. Am. Chem. Soc.* **1999**, *121*, 9471–9472.

103 S. Hecht, J. M. J. Fréchet, *J. Am. Chem. Soc.* **2001**, *123*, 6959–6960.

104 (a) P. Bhyrappa, J. K. Young, J. S, Moore, K. S. Suslick, *J. Mol. Catal. A* **1996**, *113*, 109–116; (b) P. Bhyrappa, J. K. Young, J. S, Moore, K. S. Suslick, *J. Am. Chem. Soc.* **1996**, *118*, 5708–5711; (c) P. Bhyrappa, G. Vaijayanthimala, K. S. Suslick, *J. Am. Chem. Soc.* **1999**, *121*, 262–263.

105 (a) M. Uyemura, T. Aida, *J. Am. Chem. Soc.* **2002**, *124*, 11392–11403; (b) M. Uyemura, T. Aida, *Chem. Eur. J.* **2003**, *9*, 3492–3500.

106 Y. Ribourdouille, G. D. Engel, M. Richard-Plouet, L. H. Gade, *Chem. Commun.* **2003**, 1228–1229.

107 Q.-S. Hu, V. Pugh, M. Sabat, L. Pu, *J. Org. Chem.* **1999**, *64*, 7528–7536.

108 R. van Heerbeek, P. C. J. Kamer, P. W. N. M. van Leeuwen, J. N. H. Reek, *Chem. Rev.*, **2002**, *102*, 3717–3756.

109 (a) M.T. Reetz, G. Lohnmer, R. Schwickardi, *Angew. Chem. Int. Ed. Engl.* **1997**, *36*, 1526–1529; (b) K. Heuzé, D. Méry, D. Gauss, J.-C. Blais, D. Astruc, *Chem. Eur.J.* **2004**, *10*, 3936–3944; (c) X.Y. Liu, X.Y. Wu, Z. Chai, Y.Y. Wu, G. Zhao, S.Z. Zhu, *J. Org. Chem.* **2005**, *70*, 7432–7435.

110 (a) G.-J. Deng, Q.-H. Fan, X.-M. Chen, D.-S. Liu, A.S.C. Chan, *Chem. Commun.* **2002**, 1570–1571; (b) W.-J. Tang, N.-F. Yang, B. Yi, G.-J. Deng, Y.-Y. Huang, Q.-H. Fan, *Chem. Commun.* **2004**, 1378–1379.

111 (a) S.C. Bourque, F. Maltais, W.-J. Xiao, O. Tardif, H. Alper, P. Arya, L.E. Manzer, *J. Am. Chem. Soc.* **1999**, *121*, 3035–3038; (b) Y.-M. Chung, H.-K. Rhee, *Chem. Commun.* **2002**, 238–239; (c) S.-M. Lu, H. Alper, *J. Am. Chem. Soc.* **2005**, *127*, 14776–14784.

112 A.R. Schmitzer, S. Franceschi, E. Perez, I. Rico-Lattes, A. Lattes, L. Thion, M. Erard, C. Vidal, *J. Am. Chem. Soc.* **2001**, *123*, 5956–5961.

113 (a) U. Kragl, D. Gygax, O. Ghisalba, C. Wandrey, *Angew. Chem. Int. Ed. Engl.* **1991**, *30*, 827–828; (b) N. Brinkmann, D. Giebel, G. Lohmer, M.T. Reetz, U. Kragl, *J. Catal.*, **1999**, *183*, 163–168.

114 (a) E.B. Eggeling, N.J. Hovestad, J.T.B.H. Jastrzebski, D. Vogt, G. van Koten, *J. Org. Chem.* **2000**, *65*, 8857–8865; (b) D. de Groot, J.N.H. Reek, P.C.J. Kamer, P.W.N.M. van Leeuwen, *Eur.J. Org. Chem.* **2002**, *6*, 1085–1095.

115 S.C. Zimmerman, F. Zeng, D.E.C. Reichert, S.V. Kolotuchin, *Science* **1996**, *271*, 1095–1098.

116 Y. Ma, S.V. Kolotuchin, S.C. Zimmerman, *J. Am. Chem. Soc.* **2002**, *124*, 13757–13769.

117 (a) B. Huang, J.R. Parquette, *Org. Lett.* **2000**, *2*, 239–242; (b) J. Recker, D.J. Tomcik, J.R. Parquette, *J. Am. Chem. Soc.* **2000**, *122*, 10298–10307; (c) B. Huang, J.R. Parquette, *J. Am. Chem. Soc.* **2001**, *123*, 2689–2690.

118 (a) M.W.P.L. Baars, A.J. Karlsson, V. Sorokin, B.F.W. de Waal, E.W. Meijer, *Angew. Chem. Int. Ed.* **2000**, *39*, 4262–4265; (b) D. de Groot, B.F.M. de Waal, J.N.H. Reek, A.P.H.J. Schenning, P.C.J. Kamer, E.W. Meijer, P.W.N.M. van Leeuwen, *J. Am. Chem. Soc.* **2001**, *123*, 8453–8458; (c) F.S. Precup-Blaga, J.C. Garcia-Martinez, A.P.H.J. Schenning, E.W. Meijer, *J. Am. Chem. Soc.* **2003**, *125*, 12953–12960; (d) A. Dirksen, E.W. Meijer, W. Adriaens, T.M. Hackeng, *Chem. Commun.* **2006**, 1667–1669.

119 (a) A.M. Naylor, W.A. Goddard III, G.E. Kiefer, D.A. Tomalia, *J. Am. Chem. Soc.* **1989**, *111*, 2339–2341; (b) D.A. Tomalia, A.M. Naylor, W.A. Goddard III, *Angew. Chem. Int. Ed. Engl.* **1990**, *29*, 138–175.

120 (a) J.M.J. Fréchet, *Science* **1994**, *263*, 1710–1715; (b) A.I. Cooper, J.D. Londono, G. Wignall, J.B. McClain, E.T. Samulski, J.S. Lin, A. Dobrynin, M. Rubinstein, A.L.C. Burke, J.M.J. Fréchet, J.M. DeSimone, *Nature* **1997**, *389*, 368–371.

121 (a) J.F.G.A. Jansen, E.M.M. de Brabander-van den Berg, E.W. Meijer, *Science* **1994**, *266*, 1226–1229; (b) J.F.G.A. Jansen, E.W. Meijer, *J. Am. Chem. Soc.* **1995**, *117*, 4417–4418; (c) M.W.P.L. Baars, R. Kleppinger, M.H.J. Koch, S.-L. Yeu, E.W. Meijer, *Angew. Chem. Int. Ed.* **2000**, *39*, 1285–1288; (d) M.A.C. Broeren, B.F.M. de Waal, M.H.P. van Genderen, H.M.H.F. Sanders, G. Fytas, E.W. Meijer, *J. Am. Chem. Soc.* **2005**, *127*, 10334–10343.

122 Y. Tomoyose, D.-L. Jiang, R.-H. Jin, T. Aida, T. Yamashita, K. Horie, E. Yashima, Y. Okamoto, *Macromolecules* **1996**, *29*, 5236–5238.

123 (a) K. Yamamoto, M. Higuchi, S. Shiki, M. Tsuruta, H. Chiba, *Nature* **2002**, *415*, 509–511; (b) M. Higuchi, M. Tsuruta, H. Chiba, S. Shiki, K. Yamamoto, *J. Am. Chem. Soc.* **2003**, *125*, 9988–9997; (c) R. Nakajima, M. Tsuruta, M. Higuchi, K. Yamamoto, *J. Am. Chem. Soc.* **2004**, *126*, 1630–1631.

124 L. Balogh, D.A. Tomalia, *J. Am. Chem. Soc.*, **1998**, *120*, 7355–7356.

125 (a) D.B. Amabilino, P.R. Ashton, V. Balzani, C.L. Brown, A. Credi, J.M.J. Fréchet, J.W. Leon, F.M. Raymo, N. Spencer, J.F. Stoddart, M. Venturi, *J. Am. Chem. Soc.* **1996**, *118*, 12012–12020; (b) J.W. Lee, Y.H. Ko, S.-H.

Park, K. Yamaguchi, K. Kim, *Angew. Chem. Int. Ed.* **2001**, *40*, 746–749; (c) A. M. Elizarov, T. Chang, S.-H. Chiu, J. F. Stoddart, *Org. Lett.* **2002**, *4*, 3565–3568; (d) H. W. Gibson, N. Yamaguchi, L. Hamilton, J. W. Jones, *J. Am. Chem. Soc.* **2002**, *124*, 4653–4665.

126 A. R. Hirst, D. K. Smith, *Top. Curr. Chem.* **2005**, *256*, 237–273.

127 (a) W.-D. Jang, D.-L. Jiang, T. Aida, *J. Am. Chem. Soc.* **2000**, *122*, 3232–3233; (b) W.-D. Jang, T. Aida, *Macromolecules* **2003**, *36*, 8461–8469; (c) W.-D. Jang, T. Aida, *Macromolecules* **2004**, *37*, 7325–7330.

128 C. Marmillon, F. Gauffre, T. Gulik-Krzywicki, C. Loup, A.-M. Caminade, J.-P. Majoral, J.-P. Vors, E. Rump, *Angew. Chem. Int. Ed.* **2001**, *40*, 2626–2629.

129 K. S. Partridge, D. K. Smith, G. M. Dykes, P. T. McGrail, *Chem. Commun.* **2001**, 319–320.

130 C. Kim, K. T. Kim, Y. Chang, H. H. Song, T.-Y. Cho, H.-J. Jeon, *J. Am. Chem. Soc.* **2001**, *123*, 5586–5587.

131 M. Enomoto, A. Kishimura, T. Aida, *J. Am. Chem. Soc.* **2001**, *123*, 5608–5069.

132 A. Kishimura, T. Yamashita, T. Aida, *J. Am. Chem. Soc.* **2005**, *127*, 179–183.

133 A. Kishimura, T. Yamashita, K. Yamaguchi, T. Aida, *Nat. Mater.* **2005**, *4*, 546–549.

134 (a) E. R. Zubarev, M. U. Pralle, E. D. Sone, S. I. Stupp, *J. Am. Chem. Soc.* **2001**, *123*, 4105–4106; (b) E. R. Zubarev, S. I. Stupp, *J. Am. Chem. Soc.* **2002**, *124*, 5762–5773; (c) L. Li, E. Beniash, E. R. Zubarev, W. Xiang, B. M. Rabatic, G. Zhang, S. I. Stupp, *Nature Mater.* **2003**, *2*, 689–694; (d) B. W. Messmore, J. F. Hulvat, E. D. Sone, S. I. Stupp, *J. Am. Chem. Soc.* **2004**, *126*, 14452–14458.

135 T. Yamaguchi, N. Ishii, K. Tashiro, T. Aida, *J. Am. Chem. Soc.* **2003**, *125*, 13934–13935.

136 (a) V. Percec, A. E. Dulcey, V. S. K. Balagurusamy, Y. Miura, J. Smidrkal, M. Peterca, S. Nummelin, U. Edlund, S. D. Hudson, P. A. Heiney, H. Duan, S. N. Magonov, S. A. Vinogradov, *Nature* **2004**, *430*, 764–768; (b) V. Percec, A. E. Dulcey, M. Peterca, M. Ilies, J. Ladislaw, B. M. Rosen, U. Edlund, P. A. Heiney, *Angew. Chem. Int. Ed.* **2005**, *44*, 6516–6521; (c) V. Percec, A. E. Dulcey, M. Peterca, M. Ilies, M. J. Sienkowska, P. A. Heiney, *J. Am. Chem. Soc.* **2005**, *127*, 17902–17909; (d) V. Percec, A. E. Dulcey, M. Peterca, M. Ilies, S. Nummelin, M. J. Sienkowska, P. A. Heiney, *Proc. Natl. Acad. Sci. USA* **2006**, *103*, 2518–2523.

137 (a) K. Tusda, G. C. Dol, T. Gensch, J. Hofkens, L. Latterini, J. W. Weener, E. W. Meijer, F. C. De Schryver, *J. Am. Chem. Soc.* **2000**, *122*, 3445–3452; (b) C. Kim, S. J. Lee, I. H. Lee, K. T. Kim, *Chem. Mater.* **2003**, *15*, 3638–3642; (c) Y. Zhou, D. Yan, *Angew. Chem. Int. Ed.*, **2004**, *43*, 4896–4899; (d) B.-S. Kim, O. V. Lebedeva, D. H. Kim, A.-M. Caminade, J.-P. Majoral, W. Knoll, O. I. Vinogradova, *Langmuir* **2005**, *21*, 7200–7206.

138 R. Charvet, D.-L. Jiang, T. Aida, *Chem. Commun.* 2004, 2664–2665.

139 J. Barberá, B. Donnio, L. Gehringer, D. Guillon, M. Marcos, A. Omenat, J. L. Serrano, *J. Mater. Chem.* **2005**, *15*, 4093–4105.

140 (a) K. Yonetake, T. Masuko, T. Morishita, K. Suzuki, M. Ueda, R. Nagahata, *Macromolecules* **1999**, *32*, 6578–6586; (b) F. S. Precup-Blaga, A. P. H. J. Schenning, E. W. Meijer, *Macromolecules* **2003**, *36*, 565–572.

141 M. Marcos, R. Giménez, J. L. Serrano, B. Donnio, B. Heinrich, D. Guillon, *Chem. Eur. J.* **2001**, *7*, 1006–1013.

142 S. A. Ponomarenko, N. I. Boiko, V. P. Shibaev, R. M. Richardson, I. J. Whitehouse, E. A. Rebrov, A. M. Muzafarov, *Macromolecules* **2000**, *33*, 5549–5558.

143 B. Dardel, R. Deschenaux, M. Even, E. Serrano, *Macromolecules* **1999**, *32*, 5193–5198.

144 (a) D. J. Pesak, J. S. Moore, *Angew. Chem. Int. Ed. Engl.* **1997**, *36*, 1636–1639; (b) H. Meier, M. Lehnmann, *Angew. Chem. Int. Ed.* **1998**, *37*, 643–645; (c) L. Gehringer, D. Guillon, B. Donnio, *Macromolecules* **2003**, *36*, 5593–5601; (d) L. Gehringer, C. Bourgogne, D. Guillon, B. Donnio, *J. Am. Chem. Soc.* **2004**, *126*, 3856–3867.

145 A. G. Cook, U. Baumeister, C. Tshierske, *J. Mater. Chem.* **2005**, *15*, 1708–1721.

146 (a) V. Percec, G. Johansson, G. Ungar, J. Zhou, *J. Am. Chem. Soc.* **1996**, *118*, 9855–9866; (b) V. S. K. Balagurusamy, G. Ungar, V. Percec, G. Johansson, *J. Am. Chem. Soc.* **1997**, *119*, 1539–1555; (c) S. D. Hudson, H. T. Jung, V. Percec, W.-D. Cho, G. Johansson, G. Ungar, V. S. K. Balagurusamy, *Science* **1997**, *278*, 449–452; (d) V. Percec, W.-D. Cho, P. E. Mosier, G. Ungar, D. J. P. Yeardley, *J. Am. Chem. Soc.* **1998**, *120*, 11061–11070; (e) V. Percec, W.-D. Cho, M. Möller, S. A. Prokhorova, G. Ungar, D. J. P. Yeardley, *J. Am. Chem. Soc.* **2000**, *122*, 4249–4250; (f) G. Ungar, Y. Liu, X. Zeng, V. Percec, W.-D. Cho, *Science* **2003**, *299*, 1208–1211; (g) X. Zeng, G. Ungar, Y. Liu, V. Percec, A. E. Dulcey, J. K. Hobbs, *Nature* **2004**, *428*, 157–160.

147 (a) V. Percec, C.-H. Ahn, B. Barboiu, *J. Am. Chem. Soc.* **1997**, *119*, 12978–12979; (b) V. Percec, C.-H. Ahn, G. Ungar, D. J. P. Yeardley, M. Möller, S. S. Sheiko, *Nature* **1998**, *391*, 161–164.

148 V. Percec, M. Glodde, T. K. Bera, Y. Miura, I. Shiyanovskaya, K. D. Singer, V. S. K. Balagurusamy, P. A. Heiney, I. Schnell, A. Rapp, H.-W. Spiess, S. D. Hudson, H. Duan, *Nature* **2002**, *419*, 384–387.

149 D. C. Tully, J. M. J. Fréchet, *Chem. Commun.* **2001**, 1229–1239.

150 D. J. Díaz, G. D. Storrier, S. Bernhard, K. Takada, H. D. Abruña, *Langmuir* **1999**, *15*, 7351–7354.

151 (a) M. Zhao, H. Tokuhisa, R. M. Crooks, *Angew. Chem. Int. Ed. Engl.* **1997**, *36*, 2595–2598; (b) H. Tokuhisa, R. M. Crooks, *Langmuir* **1997**, *13*, 5608–5612.

152 D. C. Tully, K. Wilder, J. M. J. Fréchet, A. R. Trimble, C. F. Quate, *Adv. Mater.* **1999**, *11*, 314–318.

153 T. Yamaguchi, T. Kimura, H. Matsuda, T. Aida, *Angew. Chem. Int. Ed.* **2004**, *43*, 6350–6355.

10
Molecular Brushes – Densely Grafted Copolymers

Brent S. Sumerlin and Krzysztof Matyjaszewski

10.1
Introduction

A graft copolymer consists of several chains of one monomer being pendantly grafted from various points of a chain composed of another monomer [1–3]. The main chain is commonly referred to as the *backbone* and the branches as the *side-chains*. The fact that chains with at least two different monomer units are covalently linked makes graft copolymers, as block copolymers, of particular interest when attempting to prepare nanostructured materials with hybrid properties. The additional consideration of the topological arrangement of the branched structure is of further importance in many applications. An example of this is given by the enhanced potential of graft copolymers to compatibilize homopolymer blends [4, 5], control nucleation and growth of crystals from aqueous media [6] and stabilize colloidal dispersions [7].

Polymer brushes are characterized by dense layers of chains tethered to an interface such that the distance between the attached chains result in adoption of entropically unfavorable chain extension [8–10]. The interface, which serves as a common point of attachment, can be planar, curved or linear in nature. While the majority of research has focused on the preparation of brushes attached to planar or curved surfaces, the same concept can be extended to the aforementioned class of graft copolymers, such that the backbone chain acts as a one-dimensional anchoring interface. Thus, *molecular brushes* [11, 12] are a specific class of graft copolymers containing a high density of polymer/oligomer side-chains that are attached to a common polymer backbone at an approximate density of one chain per backbone monomer unit. This review focuses primarily on the synthesis and architecture of molecular brushes prepared predominantly from vinyl (macro)monomers. However, a brief discussion of the special properties originating from these unusual structures is also included.

Synthetically, a variety of approaches have been adopted to allow the preparation of such advanced macromolecular topologies. Within each approach, several considerations must be taken in order to approach the diverse available

Macromolecular Engineering. Precise Synthesis, Materials Properties, Applications.
Edited by K. Matyjaszewski, Y. Gnanou, and L. Leibler

ISBN: 978-3-527-31446-1

structures. Prior to discussing the preparation of molecular brushes, an overview of structure is offered in order to focus the synthetic approaches towards the specific type of brush being targeted.

10.2
Molecular Brush Topology

In contrast to common graft copolymers, the high grafting density in molecular brushes results in steric repulsion of the densely packed side-chains that forces the backbone polymer to extend from a statistical Gaussian coil into a conformation with an increased persistence length approaching the dimensions of the fully extended backbone [13–22]. If the length of the backbone is significantly longer than that of the side-chains, intramolecular excluded volume effects cause the polymer to adopt a cylindrical shape with the backbone polymer in the core from which the side-chains emanate radially. Conversely, molecular brushes with backbones on the order of the length of the side-chains generally adopt compact, spherical dimensions that resemble star polymers [23].

With sufficient backbone and side-chain length, molecular brushes can be visualized as individual molecules by atomic force microscopy (Fig. 10.1) [24–31]. In addition to visualizing individual side-chains [24], backbone architectures and conformations [26, 30, 32, 33], specific side-chain–surface interactions [25, 32–35] and encapsulated contents [36–38], precise microscopic analysis allows the calculation of the molecular weight and molecular weight distribution that are in accord with the values obtained by other absolute methods such as multi-angle laser light scattering (MALLS) [39].

Classifications of brush copolymer compositions must consider variations along both the backbone and the side-chains. Depending on the direction of

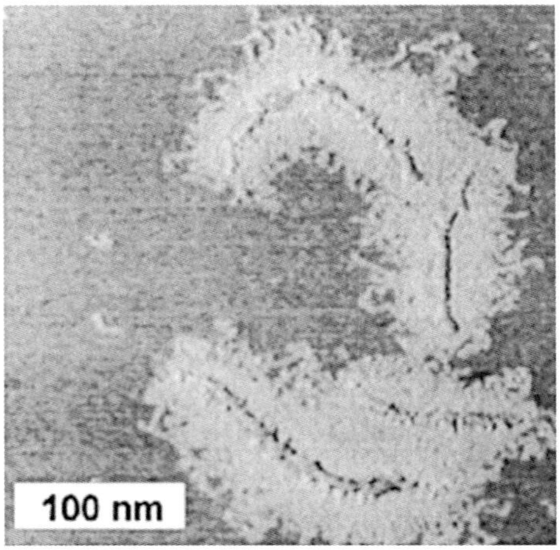

Fig. 10.1 AFM image of molecular brushes with poly(*n*-butyl acrylate) (PBA) side-chains of DP_n=52. Adapted with permission from [24]. Copyright 2001 American Chemical Society.

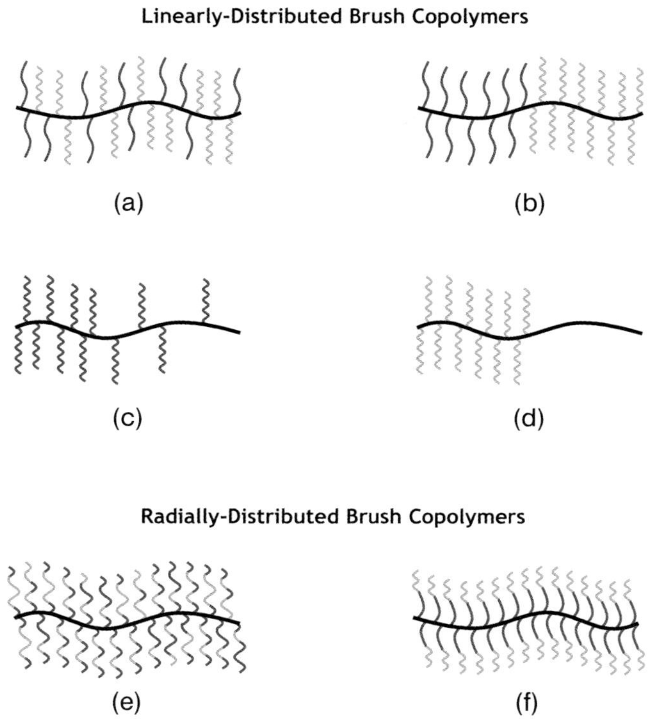

Fig. 10.2 Brush copolymer structures with the monomer composition varied linearly or radially. (a) Different types of side-chains arranged in a statistical format (heterograft brush copolymer); (b) block copolymer structure along the backbone with more than one type of side-chain (brush-*block*-brush copolymer); (c) gradient distribution of side-chains along backbone (gradient brush copolymer); (d) brush and linear block along backbone (brush-*block*-linear or brush–coil copolymer); (e) block copolymer side-chains (core–shell brush copolymer); (f) statistical copolymer side-chains.

chemical heterogeneity, it is possible to divide molecular brushes copolymers into two broad categories: brushes with block or statistical structures along (i) the length of the backbone (*linearly distributed copolymers*) or (ii) the side-chains (*radially distributed brushes*) (Fig. 10.2).

10.2.1 Linear Distribution

By constructing a brush with side-chains composed of different polymers, the functional structure of the brush is combined with the chemical diversity inherent to copolymers. If the identity of the side-chains is varied along the length of the backbone statistically, heterograft brush copolymers are obtained (Fig. 10.2 a) [40, 41]. Depending on the interaction parameter of the various side-chains with their surroundings, intermolecular phase separation may be expected due to preferential side-chain segregation. This could lead to conformations with randomly mixed side-chains or *Janus*-type structures in bulk, in solution or on surfaces. By distributing the homopolymer side-chains along the backbone in a completely asymmetric fashion, brush-*block*-brush copolymers are obtained (Fig. 10.2 b). These types of structures are similar to low molecular weight linear block copolymers, except that the aspect ratio of the fully extended chain is much lower for the molecular brushes.

Other types of linearly distributed molecular brushes can result from incorporation of low molecular weight monomers within the backbone. If this is done in an indiscriminate order with a low concentration of side-chains, loosely grafted copolymers result. These arguably do not fall within the realm of molecular brushes since the side-chains and backbone may not demonstrate significant extension. A different architecture can be envisioned in which the side-chains are present in a non-random fashion along the backbone such that one end of the chain is grafted with sufficient density to facilitate extension of a portion of the side-chains and backbone. Because the grafting density is varied gradually along the backbone, such structures can be referred to as *gradient copolymer brushes* (Fig. 10.2 c). With a linear distribution of homopolymer side-chains in a discontinuous manner such that at east one block of the backbone has side-chains and another portion does not, brush-*block*-linear or *brush–coil* copolymers result (Fig. 10.2 d).

10.2.2
Radial Distribution

The radially distributed copolymers involve molecular brush topologies with copolymer side-chains. Thus the chains projecting from the backbone core are composed of more than one type of repeat unit. The side-chains can be either statistical (Fig. 10.2 e) or block copolymers (Fig. 10.2 f). Statistical copolymer side-chains can lead to brush properties that are a hybrid of those expected for the corresponding homopolymers. Thus, the solubility, T_g, lower critical solution temperature (LCST), etc., can be tuned by appropriate selection of comonomers and their respective incorporation.

Brushes with block copolymer side-chains present particularly interesting structures since they can form covalently linked cylindrical micelles in solution [36–38, 42–45]. As opposed to the multimolecular micelles that are susceptible to a concentration-dependent equilibrium between molecularly dissolved unimers and micelles, these brushes maintain their core–shell structure regardless of concentration. Core–shell brushes have demonstrated utility for the preparation of a range of hybrid nano-objects [36–38]. Polyacrylic acid (PAA) cores have led to superparamagnetic or semiconducting nanocylinders. Metal cations were coordinated with the internal carboxylate functionalities and subsequently reduced to yield the nanohybrids. In each case, the side-chain outer blocks served as protective shells for the encapsulated contents. Thus, concepts such as nanowires can be envisioned in which the shell serves as insulating layers for the (semi)conducting cores. Similar approaches were employed for the preparation of gold [43], silver [46] and polypyrrole [47] nanowires.

10.2.3 Other Molecular Brush Topologies

This list of molecular brush copolymer structures is not exhaustive. Brushes with more than two types of homopolymer side-chains or repeat units within side-chains can also be envisioned. Another example is brushes containing side-chains that are *themselves* molecular brushes (*double-grafted*). Regardless of the specific brush structure, the characteristic feature of interest is the high density of side-chains and compact nature, which lead to materials that behave differently from linear polymers [11, 25–27, 48–54]. Molecular brushes have been described as biomimetics of proteoglycans present in articular cartilage that are similarly highly dense, supramolecular polymer brushes responsible for absorbing compression forces and lubricating joints [55–57]. Specific examples of the various copolymers will be discussed within the context of molecular brush synthesis.

10.3 Synthesis of Molecular Brushes

As with preparing polymers attached to surfaces, there are three main strategies for preparing molecular brushes: (i) "grafting to", (ii) "grafting through" and (iii) "grafting from" (Scheme 10.1). The following sections will discuss each of these methods, detailing their capabilities and limitations from a synthetic perspective. While each method has particular utility individually, it is often advantageous to employ a combination of methods to prepare brushes with structures unobtainable by a single technique.

10.3.1 Grafting Through

The grafting through route involves the polymerization of macromonomers – polymers with polymerizable end-groups – "through" their terminal functionality [29, 49, 58–73]. Perhaps the most attractive feature of this method is that each backbone repeat unit will contain a covalently bound side-chain, given a sufficient level of purity of the original macromonomer. Also, because the macromonomers are prepared separately, the side-chains can be characterized prior to incorporation into the higher molecular weight polymacromonomer structure. However, this method suffers from the degree of polymerization of the backbone being dependent on the macromonomer length and type [74]. Additionally, viscosity issues are inherent even at low conversion and, due to the necessarily low concentration of polymerizable end-groups and high steric hin-

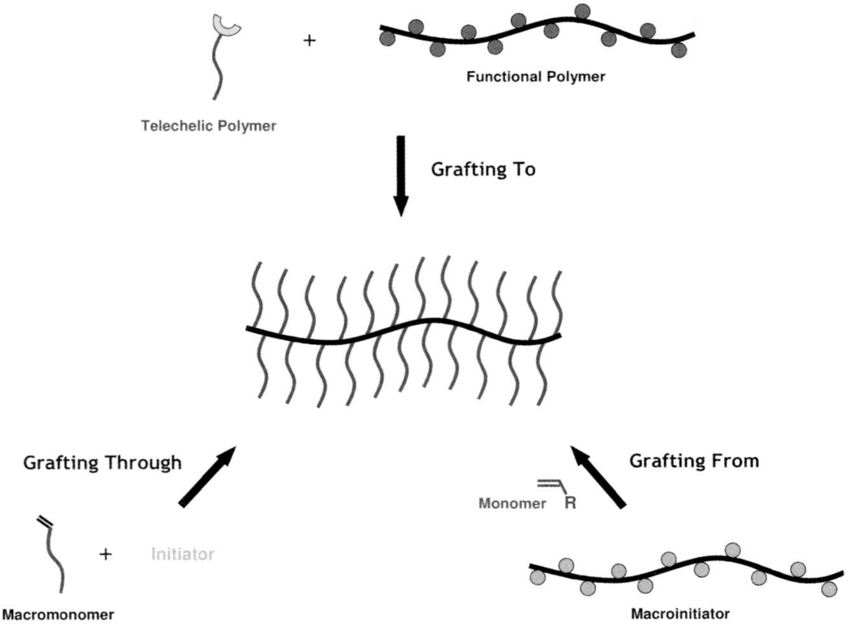

Scheme 10.1 Synthetic methods for preparing molecular brushes by grafting to, through and from. *Grafting to* requires the coupling of individual side-chains to a common backbone polymer. *Grafting through* consists of the polymerization of macromonomers. *Grafting from* involves preparing a backbone polymeric precursor with monomer units that contain functionalities capable of initiating polymerization of a second monomer.

drance of the propagating chain end, polymerizations can be slow and not proceed to high conversion. Low conversion can lead to tedious purification since fractionation or dialysis is required to remove unreacted macromonomer. Nonetheless, there have been several successful reports of the preparation of densely grafted copolymers by grafting through polymerizations.

10.3.1.1 Anionic Homopolymerization of Macromonomers

Due to the high level of control typically demonstrated in anionic polymerizations, there have been several reports of homopolymerizations of macromonomers by anionic mechanisms [75–77]. For instance, styryl- or methacryloyl-terminal polystyrene (PS) was polymerized to yield polymacromonomers with a DP_n of the backbone equal to ~ 40 when a side-chain of $DP_n = 20$ was used [76]. The styryl-based macromonomers demonstrated control over a large range of temperatures and, in several solvents, provided that LiCl was present and the appropriate initiator was employed. Due to significant steric hindrance, susceptibility to adventitious impurities and difficulty in obtaining macromonomers with high purity, most polymer brushes so prepared have been limited to low molecular weight and with moderate control over the molecular weight distribution [75]. In some cases, experimental precautions can overcome these limitations. For instance, Hadjichristidis and coworkers reported the anionic polymerization of macromonomers of polyisoprene, polybutadiene or PS with terminal styrenic functionality [78]. To prevent the introduction of impurities that could unfavorably affect the subsequent anionic polymerization to form the molecular brush, the macromonomers were prepared and immediately polymerized, without isolation, within the same reaction vessel.

10.3.1.2 Ring-opening Metathesis Homopolymerization of Macromonomers

The polymerizability of most vinyl macromonomers decreases with increasing molecular weight due to steric crowding in the vicinity of the propagating chain end in addition to that surrounding the unreacted vinyl moiety. Thus, an attractive alternative for molecular brush formation involves a mechanism by which the spacing between side-chains can be increased while not necessarily being diluted by the presence of a low molecular weight monomer. As an example of a polymerization mechanism that fits such criteria, ring-opening metathesis polymerization (ROMP) [79, 80] has proved to be an efficient means to prepare molecular brushes (Scheme 10.2). Ring strain in norbornenyl macromonomers enhances the thermodynamic driving force for polymerization and decreased side-chain density provides a kinetically more favorable environment for propagation reactions [81].

Gnanou and coworkers prepared norbornenyl-terminal PS [82] and poly(ethylene oxide) (PEO) [83] and subsequently employed ROMP as a means to prepare the molecular brushes. For the case of the PEO macromonomers, nearly full conversion was obtained, but the efficiency of initiation was less than ideal, leading to polymacromonomers of molecular weight higher than expected.

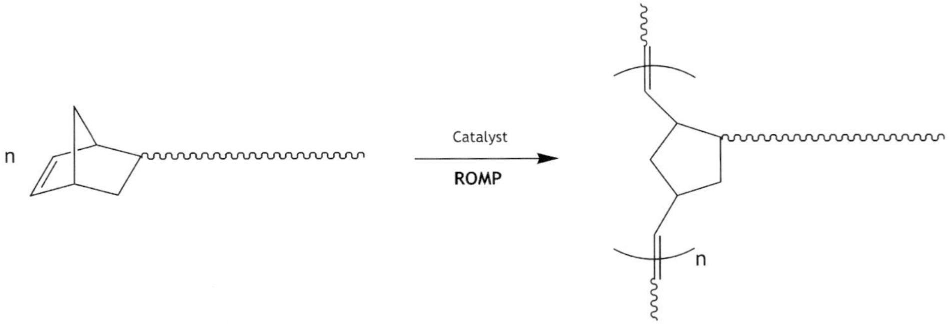

Scheme 10.2 Ring strain and the decreased ratio of side-chain to backbone carbon make ring-opening metathesis polymerization (ROMP) of a norbornenyl macromonomer attractive for the preparation of molecular brushes.

Other norbornenyl-containing macromonomers have been homopolymerized by ROMP, including those with chains of polyphosphazene [84], poly(ε-caprolactone) (PCL) [85] and polylactide [86], each of these resulting in high molecular weight and conversion. In the latter case, a molecular weight of $>5\times10^6$ g mol^{-1} ($M_w/M_n=1.19$) was obtained from a macromonomer of 4300 g mol^{-1}, corresponding to $DP_n=1160$ for the molecular brush.

10.3.1.3 Radical Homopolymerization of Macromonomers

The increased steric hindrance in the vicinity of the active chain end and the polymerizable moiety of the macromonomer results in a smaller rate constant of propagation compared with that of a low molecular weight monomer with an analogous polymerizable group. However, for radical polymerizations, the effect of sterics on the rate constant of bimolecular termination is even more significant [58, 59]. This results in radical polymerizations of macromonomers not being as exceedingly slow as might first be expected for a reaction in which the propagation step involves a polymer–polymer reaction. In fact, the majority of grafting through polymerizations has involved free radical polymerization of vinyl macromonomers. Polymacromonomers of high *DP* can be obtained, although this usually necessitates careful fractionation and appropriate selection of reaction conditions (high macromonomer and low radical concentrations) [58, 61, 87, 88].

Conventional radical polymerization has been employed to prepare a range of molecular brushes. In one example, block macromonomers were synthesized by anionic polymerization of oxiranes [89]. The two monomers were 1-ethoxyethyl glycidyl ether, which after hydrolysis formed hydrophilic glycidol blocks, and glycidyl phenyl ether, which led to hydrophobic blocks. By reacting these living polymers with *p*-(chloromethyl)styrene, polymerizable moieties were introduced on to the chain ends. These diblock macromonomers were then initiated with 2,2′-azoisobutyronitrile (AIBN) to yield core–shell unimolecular micelles. Because the polymerizations were conducted in water, the amphiphilic macromonomers self-assembled prior to polymerization and thus were able to reach $DP_n \approx 60$ at a limiting conversion near 50%. Djalali et al. [43] and Ishizu et al. [44] also reported the polymerization of diblock macromonomers to yield core–shell structures.

Although the chemistry associated with conventional radical polymerization allows the use of a wide range of monomers and reaction conditions, the relatively poor control over molecular weight and chain end functionality precludes its application to prepare well-defined structures. However, the evolution of the various methods of controlled/living radical polymerization (CRP) has facilitated better control over molecular weight and allowed the preparation of more advanced macromolecular architectures [90–95]. Despite the opportunity for enhanced control, reports of the successful homopolymerization of macromonomers by CRP have been relatively few [52, 74, 96–100].

After extensive purification by solvent extraction, well-defined homopolymers of poly(ethylene glycol) monomethacrylate (PEGMA) [101–104] ($M_w/M_n<1.1$)

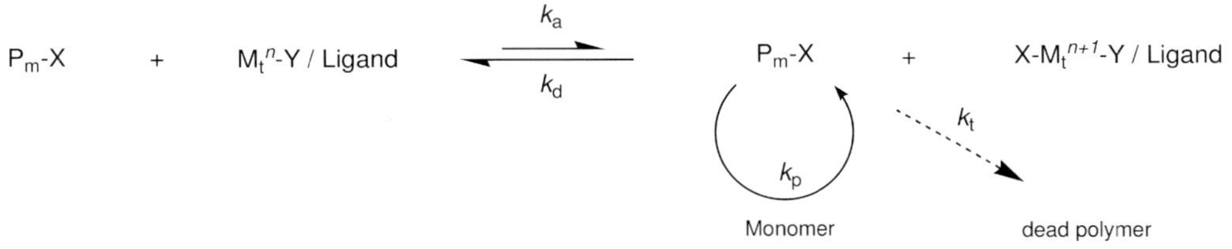

Scheme 10.3 General scheme of ATRP. P_m represents a polymer chain of $DP=m$; X is a (pseudo)halogen; M_t^n–Y is a low oxidation state transition metal salt with counter ion Y; and k_a, k_d, k_p, k_t are the rate constants for activation, deactivation, propagation and termination, respectively.

were prepared [97] by atom transfer radical polymerization (ATRP) [91, 92, 105, 106] (Scheme 10.3), but no polymerization was observed after conversions of 50–60%. Matyjaszewski and coworkers reported the grafting through polymerization of PEGMA (300 g mol^{-1}) by ATRP to reach $DP=425$ with $M_w/M_n=1.46$ at 90% monomer conversion [52]. When the polymerization was allowed to progress beyond this point, cross-linking occurred, which led to materials with super-soft elastomeric properties.

ATRP was used to prepare dendronized aliphatic polymers by grafting through of macromonomers with dendritic side-chains with a generation number of two or less [98]. Also employing dendritic macromonomers, Percec and coworkers demonstrated the dependence of final macromolecular conformation on the backbone DP of molecular brushes with short backbones yielding spheres and longer brushes leading to cylindrical topologies [107, 108]. Although not as bulky as dendritic side-chains, macromonomers with two chains of either poly(*tert*-butyl acrylate) (P*t*BA) and PS macromonomers were re-

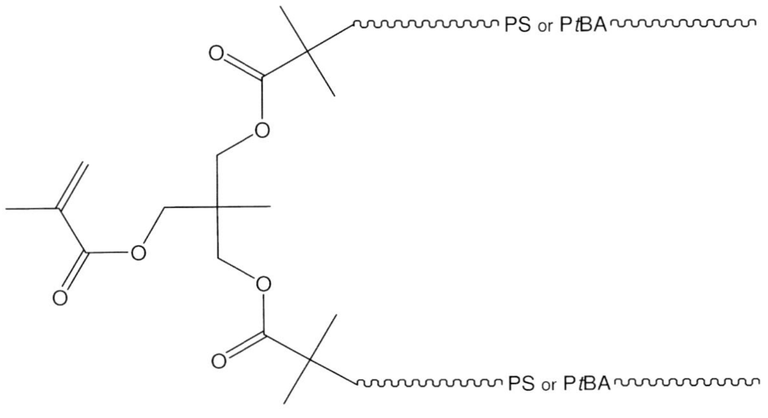

Fig. 10.3 Methacrylate macromonomer with two polystyrene (PS) or poly(*tert*-butyl acrylate) (P*t*BA) side-chains [109].

ported to polymerize by conventional radical polymerization (Fig. 10.3) [109]. While providing increased backbone rigidity due to increased excluded volume effects, the bulky nature of the doubly substituted macromonomer resulted in low monomer conversions.

10.3.1.4 **Copolymerization by Grafting Through**

The relative reactivity of the two macromonomers determines the distribution of the side-chains along the backbone. Such heterograft brush copolymers are expected to demonstrate interesting solution [110–114], bulk [115, 116] and surface [30] properties due to the interspersed nature of the side-chains. For example, brushes with PEG and PS side-chains demonstrated phase separation in solution to form large, cylindrical aggregates [112, 113]. Similarly, others reported the preparation of statistical copolymer brushes with poly(methyl methacrylate) (PMMA) and poly(2-vinylpyridine) (P2VP) side-chains. These heterograft copolymers could be manipulated to change shape from worm-like to curved structures when spin-cast on to mica [30]. In bulk, these brushes formed intermolecular filament-type structures as determined by scattering experiments [116]. Brushes with both statistical and gradient distributions of PEG and octadecyl side-chains were prepared by ATRP [117]. X-ray results indicated that inefficient packing of the helical PEG and hexagonal octadecyl crystallizable segments resulted in an amorphous fraction of octadecyl side-chains instead of the typical semicrystalline morphology.

In addition to the polymerizable moiety, the identity of the side-chain and the presence of additives can influence reactivity, as shown by Ishizu and coworkers [118–121]. Copolymerization of with maleate-terminated PEG (M_2) resulted in brushes with comonomer distributions that approached alternation ($r_1 = 0.765$, $r_2 = 0.064$) [118]. In addition to polar effects resulting from the electron-donating

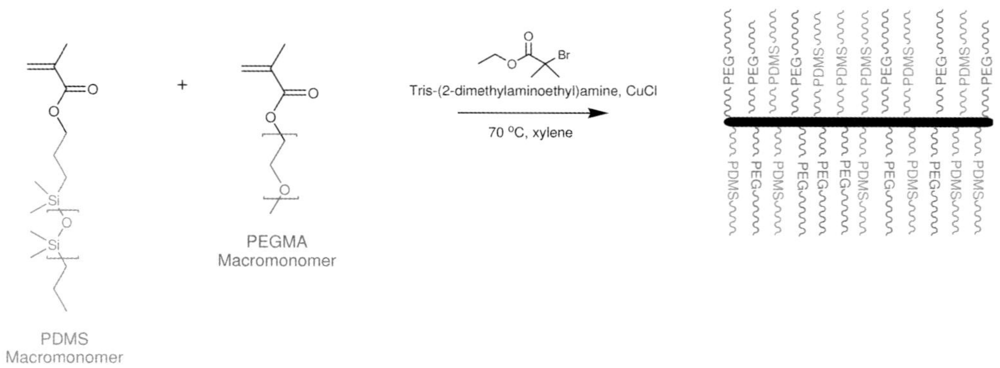

Scheme 10.4 Statistical ATRP of poly(ethylene glycol) methyl ether methacrylate and monomethacryloxypropyl poly (dimethylsilyloxane) to yield heterograft molecular brushes [122].

PS and the electron-deficient PEG macromonomers, monomer reactivities were influenced by phase separation during the copolymerization. It was concluded that the PS macromonomers were preferentially incorporated early in the polymerization and subsequent diffusion of the PEG macromonomers to the propagating radical chain end was hindered by the phase heterogeneity. In order to obtain molecular brushes with a higher fraction of alternating side-chains, vinylbenzyl-terminated PS (M_1) and PEGMA (M_2) were copolymerized in a variety of solvents [119]. While methacryloyl monomers are generally considered insufficiently electron-deficient to cause alternation with styrenic monomers, complexation with a Lewis acid ($SnCl_4$) led to reactivity ratios of $r_1=0.25$ and $r_2=0.02$ and the polymerizations were homogeneous to high conversion.

Heterograft copolymers were prepared by copolymerization of a poly(dimethylsiloxane) (PDMS) macromonomer ($1000\,g\,mol^{-1}$) with PEGMA ($1100\,g\,mol^{-1}$) (Scheme 10.4) [122]. Phase separation of the amorphous PDMS grafts and the crystallizable PEGMA grafts resulted in these materials also behaving as super-soft elastomers [123].

10.3.2 Grafting To

The grafting to strategy of molecular brush synthesis involves the reaction of end-functional polymers with a polymer backbone precursor containing complementary functionality on each monomer unit (Scheme 10.1). Because the backbone and side-chains are prepared independently, both can be synthesized by mechanisms appropriate to the respective monomer structures and subsequently characterized prior to coupling. However, owing to the relatively low yield associated with most polymer–polymer reactions due to steric and diffusion considerations, limited grafting density is often observed. Although similar to the case of grafting through, in which viscosity issues necessitate low functional group concentrations, typically an excess of side-chains is employed in order to drive the grafting reaction to high conversion. Although this approach can successfully increase the density of grafting, purification becomes problematic when trying to remove the unreacted side-chains. Additionally, as the extent of grafting increases, the backbone and side-chains must gradually adopt an increasingly extended conformation, for which the entropic penalty is significant. The requirement of reactive linking groups imposes limitations on the selection of functional groups that can be incorporated into the side-chain and backbone polymer structures. These factors have limited the grafting to approach to only the most efficient, high-yielding types of reactions.

10.3.2.1 Side-chain Attachment by Nucleophilic Substitution

Most grafting to reactions have involved the preparation of well-defined side-chains by living anionic polymerization and their subsequent reaction with a backbone of monomer units that are susceptible to nucleophilic attack. Examples of such functional groups have included esters, anhydrides, benzylic halides, nitriles, chlorosilanes and epoxides [124, 125].

The reaction of polystyryllithium with poly(chloroethyl vinyl ether) (synthesized by cationic polymerization) resulted in the successful preparation of linear and macrocyclic molecular brushes [124]. Because both the backbone and side-chains were prepared by living polymerizations, the resulting brush structures were well defined. In other reports, the reaction fidelity of such a mechanism was often too low, with significant contributions from side-reactions being observed because of the highly reactive nature of polyanion end-groups. For instance, reacting polystyryllithium with poly(chloromethylstyrene) is a fairly efficient means to prepare graft copolymers (Scheme 10.5), but unless specific precautions are taken, the grafting density rarely exceeds 60% [126, 127]. Approaches have been suggested to improve the yield of the coupling reactions by reducing the contribution of side-reactions. By using potassium rather than lithium counterions, near-quantitative conversion was observed during the preparation of four-arm star polymers from 1,2,4,5-tetrachloromethylbenzene [128]; however, extension of this approach to the synthesis of brushes from poly(chloromethylstyrene) resulted in significantly lower yields (~80%) [129]. Capping the living polymer with a single 1,1-diphenylethylene unit prior to reaction with the backbone proved to be a viable means of significantly increasing the reaction yield (>95%) between the polystyrene anions and the chloromethyl groups, but the reaction was conducted with polystyrene-*stat*-poly(chloromethylstyrene), so true molecular brushes were not prepared [130].

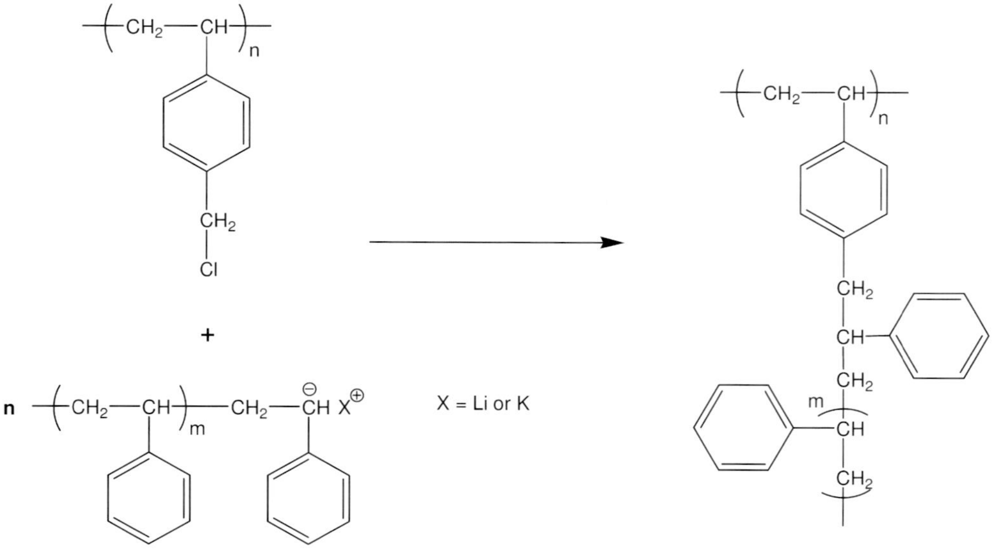

Scheme 10.5 Molecular brushes prepared by grafting living polystyrene anions to polychloromethylstyrene.

10.3.2.2 Side-chain Attachment by Click Chemistry

Dendronized polymers have also been prepared by a grafting to route [98, 131–133]. Due to the bulky nature inherent to functional dendrons, highly efficient reactions can potentially lead to increased grafting density. By employing highly efficient copper(I)-catalyzed azide–alkyne coupling reactions [134, 135], commonly classified as a form of *"click chemistry"* [136], post-polymerization modification reactions have been observed to proceed to high conversion [131, 133, 137–139].

Fréchet and coworkers employed this route as a means to couple azide-functionalized dendrons to linear backbones of poly(vinylacetylene) (Scheme 10.6) [133] and poly(*p*-hydroxystyrene) [131] (after being functionalized to contain acetylene groups). For the case of the reaction of poly(vinylacetylene) with benzyl ether dendritic azides, the coupling reaction was quantitative for generations 1 and 2 and greater than 98% conversion was observed for generation 3 [133]. Matyjaszewski and coworkers explored the modification of a poly(hydroxyethyl methacrylate) with 4-pentynoic acid in order to incorporate alkyne functionality along the backbone [140]. Subsequent reaction with polymers prepared by ATRP and substituted to contain terminal azido functionality enabled well-defined side-chains to be attached in an efficient manner.

10.3.3 Grafting From

Grafting from involves the preparation of a backbone polymer (macroinitiator) with a predetermined number of initiation sites that is subsequently used to initiate polymerization (Scheme 10.1). The backbone can be prepared directly (assuming that the initiating sites are retained) or by first preparing a precursor that is subsequently functionalized to include initiating moieties. In grafting from polymerizations, the polydispersity index of a brush polymer is governed by that of the macroinitiator. Thus, a brush with a narrow molecular weight distribution can be facilitated by a controlled polymerization of the backbone

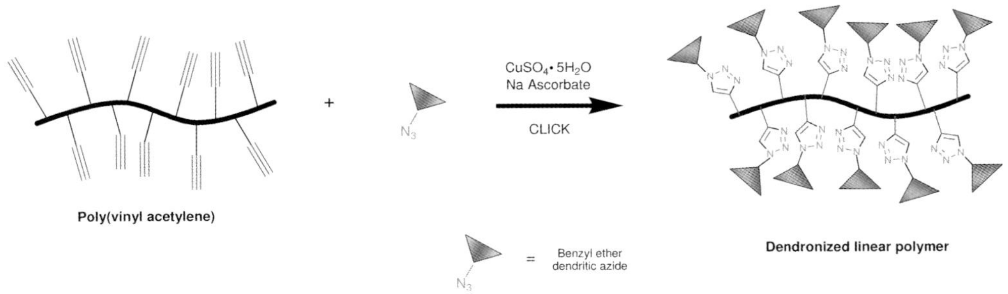

Scheme 10.6 Synthetic approach employed to prepare dendronized linear polymers by grafting to via click chemistry [133].

monomer. Long molecular brushes can be obtained because the molecular weight of the backbone is governed only by the limitations of the particular polymerization mechanism as applied to low molecular weight monomers. After preparing a suitable backbone that contains multiple initiating sites, the polymerization of the side-chains can be initiated and allowed to proceed under conditions that allow molecular weight control. Thus, the length of the side-chains increases gradually, alleviating concerns over steric issues that are limiting conditions in the grafting through and grafting to strategies.

Most reports of molecular brush synthesis by grafting from have involved the use of ATRP (Scheme 10.2), although other methods also appear promising [88, 141]. Like the other techniques of controlled/living polymerization, polymers prepared by ATRP demonstrate an increase in molecular weight with monomer conversion. Additionally, due to the low instantaneous concentration of radical species, the contribution of termination events is suppressed. These two factors are particularly important in the case of polymerization from multifunctional initiators, in which it is advantageous for all side-chains to grow simultaneously in a progressive fashion and intermolecular termination should be avoided since it can lead to macroscopic gelation. Due to the relative ease by which ATRP initiators can be prepared, a range of macroinitiators derived from synthetic and natural polymer scaffolds can be envisioned.

10.3.3.1 Variation of the Backbone Macroinitiator Structure (Linear Distribution)

The general mechanism for molecular brush synthesis by grafting from involves ATRP-active halogen atoms on the backbone being used as initiating sites to grow side-chains by the ATRP of a second monomer. These macroinitiators have been obtained by polymerizing a monomer containing an ATRP-initiator group using a different polymerization mechanism including conventional free radical polymerization [27], stable free radical polymerization [142], reversible addition–fragmentation chain transfer (RAFT) polymerization [141] or anionic polymerization [143, 144]. The majority of reports, however, employed ATRP of a monomer carrying a precursor that was subsequently transformed to an ATRP-initiating group either directly [49] or after deprotection (Scheme 10.7) [27].

The backbone of a molecular brush is essentially the template from which the remainder of the structure will be derived; therefore, it is beneficial to have the ability to tailor its composition in order to prepare increasingly advanced topologies.

Brush–coil block copolymers

Schmidt and coworkers have reported several examples of molecular brushes with block copolymer backbones [144–146]. Most of the examples are brush–coil block copolymers in which one block is a cylindrical brush while the other is composed of a linear polymer. Due to their asymmetric nature, these amphiphilic block copolymers exhibited interesting solution behavior in aqueous media. For example, a copolymer with a linear PS block and a brush block with

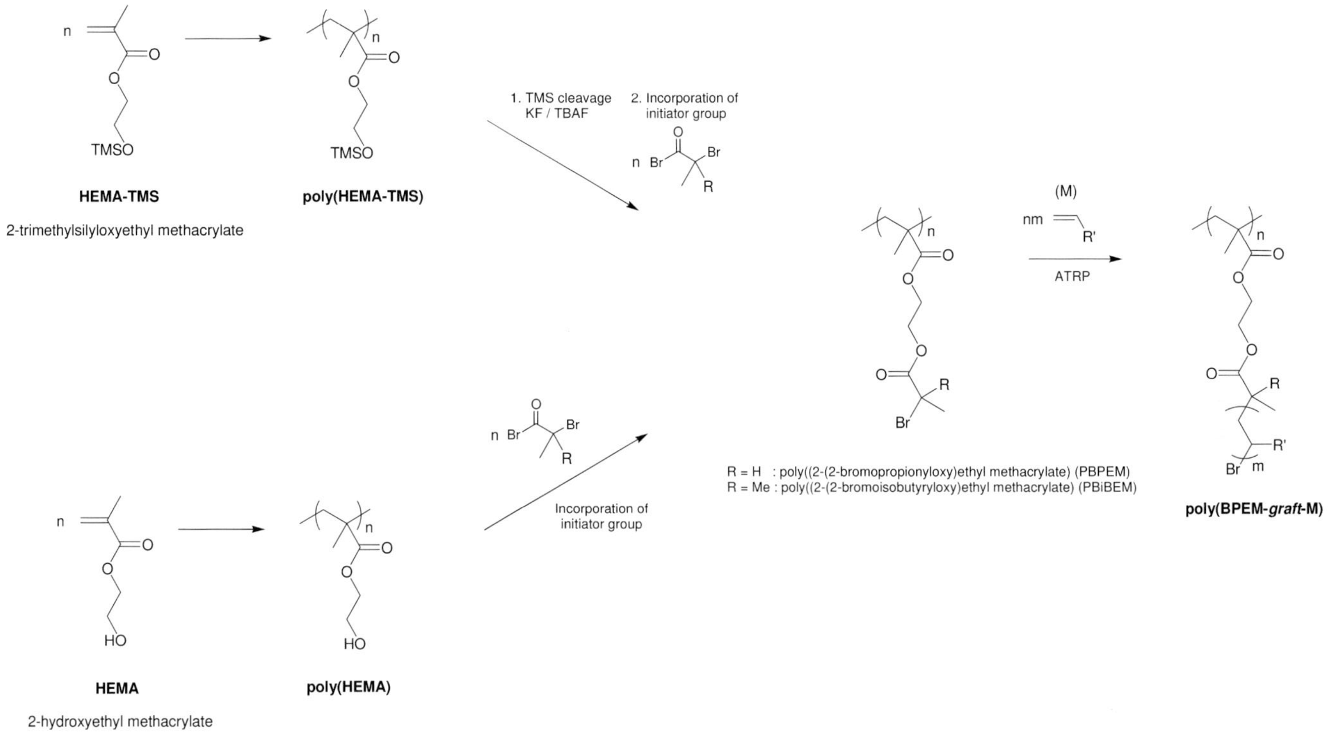

Scheme 10.7 Outline of backbone precursor preparation and derivatization to a macroinitiator (two routes) and subsequent side-chain growth by ATRP [27, 49].

PtBA side-chains was prepared by anionic polymerization of the backbone and subsequent ATRP for the side-chains [144]. After hydrolysis of the *tert*-butyl groups, the resulting PAA side-chains rendered the brush block hydrophilic. In aqueous solution, the amphiphilic brush–coil block copolymers aggregated to form micelles with sizes of $R_g = 54$ nm and aggregation numbers of ~4–5, as determined by static light scattering. AFM of the aggregates, in both the dry state after spin-casting and under water, both confirmed the size and aggregation numbers observed by AFM, although a large fraction of brush–coil unimers were also present. The authors attributed this inefficient micellization to the high degree of steric hindrance surrounding the PS linear block. Interestingly, when the opposite block copolymer structure was formed, i.e. PAA linear block and brush block with PS side-chains, micelles with a broad distribution of sizes, some of which were >300 nm, were observed by transmission electron microscopy (TEM) [145]. The latter brush–coil block copolymers were prepared by grafting through of methacryloyl-PS macromonomers via coordination polymerization with a metallocene catalyst followed by polymerization of tBA and subsequent hydrolysis to PAA. The living nature of the first polymerization to prepare the polymacromonomer block could not be confirmed, which might explain the broad distribution of micellar sizes observed in solution.

Brush–brush block copolymers

In order to prepare block copolymers in which both blocks contain side-chains (*brush–brush block copolymers*), the combination of grafting through and grafting from is employed. Block copolymerization of PEGMA macromonomer with 2-hydroxyethyl methacrylate (HEMA) resulted in a polymer that could be functionalized with halogenated initiating sites [111]. PHEMA side-chains were then grafted from via ATRP to yield poly(PEGMA-*block*-poly{[2-(2-bromoisobutyryloxy)ethyl methacrylate (PBIBEM)-*graft*-PHEMA]}). Similarly, the preparation of block-type brushes by a combination of grafting through and grafting from via ATRP was reported [53]. Polymerization of octadecyl methacrylate (ODMA) followed by blocking with HEMA–TMS resulted in AB and ABA block copolymers that were subsequently functionalized and employed as macroinitiators for the polymerization of BA. The resulting structures resembled block brushes with segments of saturated aliphatic side-chains derived from PODMA and PBA segments obtained by grafting from the macroinitiator (Scheme 10.8). Owing to the propensity of the PODMA segments to aggregate due to crystallization, self-assembly of the AB and ABA block copolymer brushes was observed by AFM. This type of material could potentially give rise to a new class of thermoplastic elastomers.

Heterograft copolymers

Brushes with intermixed side-chains of more than one identity can be obtained by copolymerization of a macromonomer, as discussed previously, or by copolymerization of a macromonomer with a low molecular weight monomer containing an initiator precursor functionality. This approach is a combination of graft-

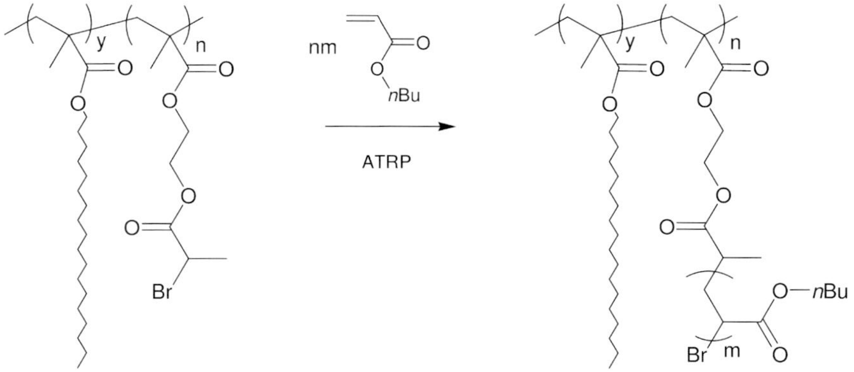

Scheme 10.8 Synthesis of PODMA-*block*-(PBPEM-*graft*-PBA) by a combination of grafting through and from. The macroinitiator was prepared by ATRP of ODMA followed by HEMA–TMS. The PHEMA–TMS block was transformed to contain macroinitiator groups and the PBA side-chains were grown by ATRP [53].

ing through and grafting from and has led to heterograft copolymers demonstrating interesting solution and bulk phase separation such that Janus-type structures can be envisioned [11, 51, 111].

Heterograft brush copolymers with PEG and poly(*n*-butyl acrylate) (PBA) chains were prepared in this manner [51]. PEGMA was copolymerized with HEMA–TMS and, after deprotection and transformation to 2-bromopropionate initiating groups, the BA chains were grown. Due to the near random distribution of side-chains, the crystallization of the PEG segments was suppressed and the brushes were homogeneous in the bulk. A similar method was used to prepare heterograft brushes with PEG and PHEMA side-chains [111].

Gradient brush/graft copolymers

In order to vary systematically the spacing between the grafting sites, Matyjaszewski and coworkers incorporated non-initiating side groups into a backbone by copolymerization of a protected monomer, 2-(trimethylsilyloxy)ethyl methacrylate (HEMA–TMS), with methyl methacrylate (MMA) [48]. The reactivity ratios of MMA and HEMA–TMS are close to unity, which generally leads to random copolymers being formed; however, the authors forced a gradient in backbone composition by continuous feeding of HEMA–TMS during the course of the polymerization. After transforming the trimethylsilyl groups into 2-bromopropionate groups, macroinitiators were obtained in which the backbone contained a gradient of initiating sites along its length. After polymerization of BA to form the side-chains, brushes with gradient grafting densities were observed by AFM. A backbone gradient can also be obtained spontaneously by copolymerization of monomers with significantly different reactivity ratios, e.g. acrylates and methacrylates [147]. When one monomer had a nonfunctional pendant group and the second carried the TMS-protected functionality from which the initiating centers would be derived, a gradient of active halogen sites along the macroinitiator was prepared. AFM analysis of the subsequently grafted copolymers demonstrated that the densely grafted portion of the chains preferentially undergo an asymmetric coil-to-globule transition upon monolayer compression on water, which led to tadpole-shaped molecules [148–150].

Other architectures

Higher order brush structures are accessible by tailoring the architecture of the backbone polymer. Star-shaped polymers with molecular brush arms have been prepared by grafting to [151] and grafting from [26, 152]. By first preparing three- and four-arm star macroinitiators, ATRP was used to prepare brushes with BA side-chains. Because the stars were visible as individual molecules, AFM allowed the evaluation of the molecular weight and molecular weight distribution [152] and also the initiation efficiency involved obtained during the preparation of the original backbone (Fig. 10.4) [26].

A combination of grafting through and grafting from was used to prepare brushes that contained side-chains that were also molecular brushes [52]. These double-grafted brushes were obtained by polymerizing PEGMA macromonomers from a well-defined macroinitiator backbone by ATRP. Due to severe steric crowding, polymerizations that achieved side-chain $DP > 50$ or macromonomer $DP > 23$ resulted in cross-linked systems with properties of the aforementioned super-soft elastomers.

Polymerization of a precursor monomer by grafting from a surface is a means to prepare immobilized molecular brushes [153]. A silicon surface modified with an ATRP initiator was used to prepare immobilized PHEMA chains that were subsequently derivatized to yield tethered macroinitiators (Scheme 10.9). The immobilized macroinitiator was employed to prepare ter-

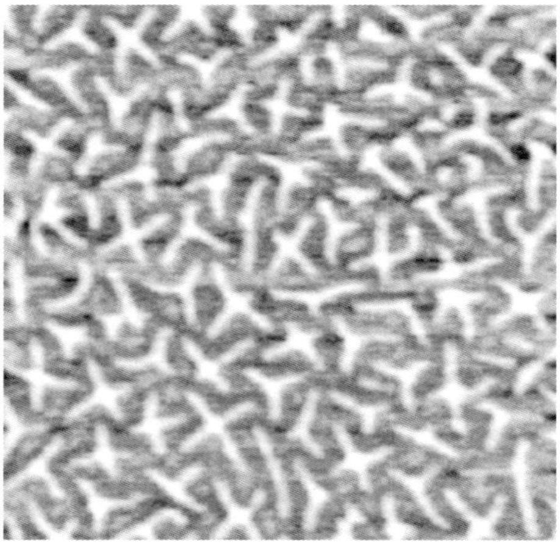

Fig. 10.4 AFM image of four-arm star poly(2-{2-bromopropionyloxy[ethyl methacrylate]}-*graft*-BA) prepared by ATRP. Adapted with permission from [26]. Copyright 2003 American Chemical Society.

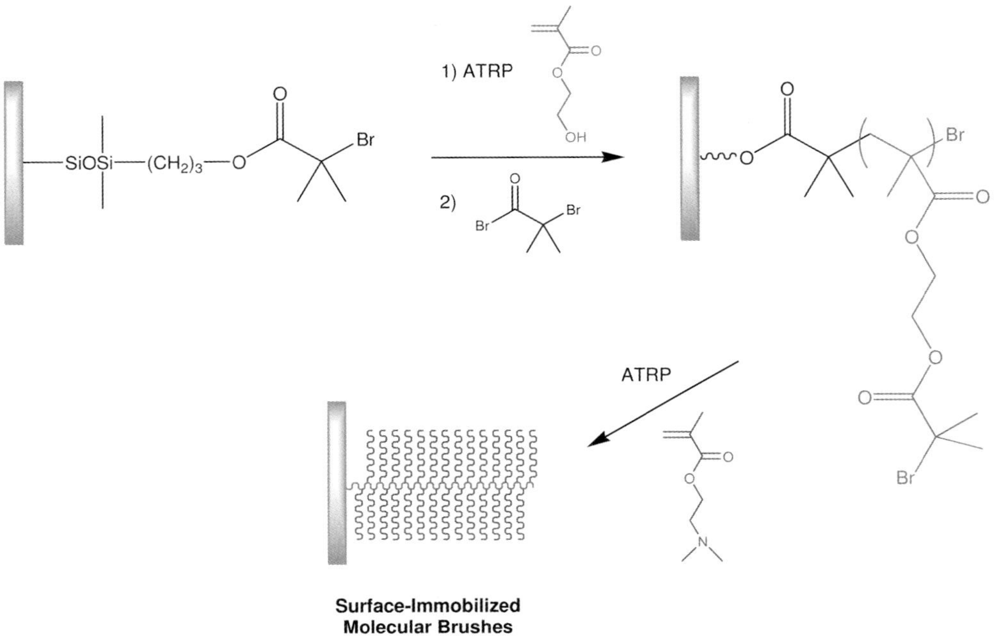

Scheme 10.9 Synthesis of PBIBEM-*graft*-PDMAEMA molecular brushes attached to silicon wafers [153].

minally attached poly(*N*,*N*-dimethylaminoethyl methacrylate) (PDMAEMA) molecular brushes from the silicon surface.

10.3.3.2 Variation of the Side-chain Composition (Radial Distribution)

As opposed to polymerizations started by low molecular weight initiators, polymerizations from interfaces may require precautions to be taken in order to achieve well-defined polymers. Due to the high rate constant for radical–radical coupling, low concentrations of active radical species are required and this necessity is of utmost importance when polymerizing from a multifunctional initiator. Intermolecular termination events occurring during grafting from by a radical mechanism can rapidly lead to macroscopic gelation. In order to prevent this, highly dilute macroinitiator concentrations are employed and low conversion is targeted [25, 27, 154, 155].

Homopolymer side-chains

Another variable available for molecular brush design is the tailoring of the side-chain identity and structure. A wide variety of monomers have been successfully homopolymerized from macroinitiator backbones by ATRP, including styrenics [25, 27, 49], acrylates [25, 26, 48, 49, 143, 147, 155], methacrylates [28,

52, 154, 156] and acrylamides [157]. Other methods of CRP have also been employed. For instance, poly(vinyl alcohol) was functionalized with xanthate moieties that facilitated macromolecular design via interchange of xanthate (MADIX)/RAFT polymerization of vinyl acetate [158]. The resulting poly(vinyl acetate) side-chains were hydrolyzed to yield poly(vinyl alcohol) brushes.

Copolymer side-chains

Core–hell molecular brushes have been prepared by sequential monomer addition to form block copolymer side-chains. Brushes with soft PBA cores and hard PS shells were prepared by ATRP [25]. The inverted structures were also prepared and the resulting surface interactions, as observed by AFM, were shown to depend on the blocking sequence.

Amphiphilic brushes were prepared by a similar methodology. ATRP was used to prepare PS-*block*-P*t*BA, P*t*BA-*block*-PS and P*t*BA-*block*-PBA [49] side-chains [143]; after hydrolysis of the *tert*-butyl groups, acrylic acid units were obtained. Due to the responsive nature of the PAA blocks, stimuli-responsive behavior was observed as a function of solvent quality [49]. Compared with con-

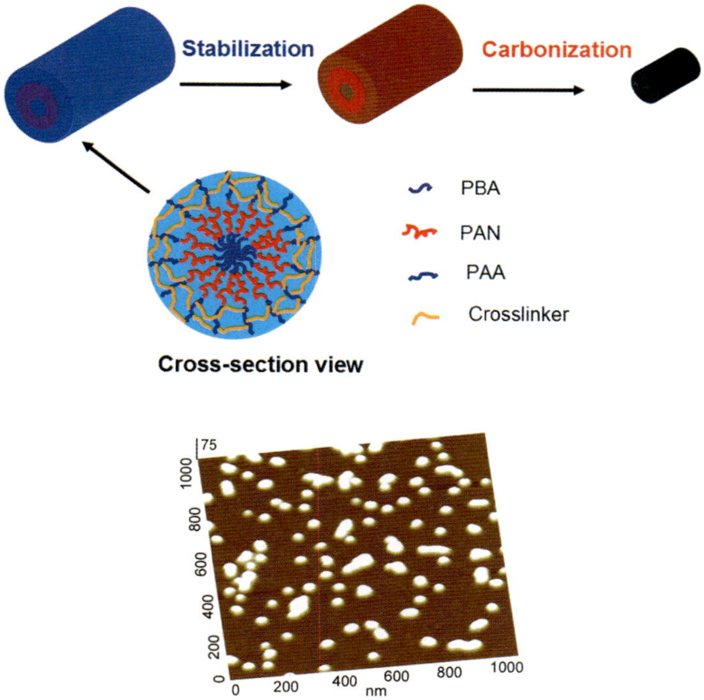

Fig. 10.5 Method of shell cross-linked molecular brushes and their conversion into nanostructured carbons through thermal treatment and AFM image the carbon nanorods by stabilization of precursors at 250 °C [45].

ventional multimolecular micelles, these core–shell structures formed unimolecular cylindrical micelles that were not susceptible to dissociation on dilution. As mentioned previously, brushes with block copolymer side-chains that coordinate metal ions can act as templates for the preparation of novel nanocomposites. Similarly, the cylindrical shape of brushes with triblock copolymer side-chains composed of PBA-*b*-polyacrylonitrile-*b*-PAA facilitated the preparation of carbon nanorods [45]. After cross-linking the outer PAA blocks to ensure shape integrity while heating, these single shell cross-linked molecular brushes showed good stability during transfer from solution to flat substrates and have been successfully converted into nanostructured carbons through pyrolysis. The resulting pyrolyzed materials exhibited characteristics of glassy carbon.

10.3.4
Special Precautions during Synthesis of molecular brushes

Regardless of the mechanism employed, brush synthesis is often complicated by several factors. The preparation of such high molecular weight polymer can result in significantly high viscosity even at low conversion. During the polymerization of macromonomers, this can be especially problematic and special precautions (high dilution, increased temperature, etc.) must be taken in order to facilitate a more manageable polymerization. Both the grafting through and grafting to methods can result in difficult purification steps being necessary since unreacted side-chains must be removed from the polymeric product. Selective precipitation, dialysis or ultrafiltration are often required. Purification is less of an issue during the grafting from approach since low molar mass monomer is more readily removed from the brush product. However, special precautions are often necessary, especially during radical polymerizations, in order to limit the intra- and intermolecular reactivity of side-chains that can result in cross-linked systems. In homogeneous solution, conversion and high initiator dilution are typically utilized in order to limit these coupling events, so as to reduce the chance of macroscopic gelation. While these precautions are typically successful in preventing detrimental network formation, the low concentrations of active species lead to decreased rates of polymerization. Further, the necessity for limited monomer conversion complicates purification and may prove not to be effective for commercialization.

Exploiting the compartmentalization afforded by miniemulsion polymerization has provided an alternative synthetic approach to increase the polymerization rate and level of conversion obtainable while avoiding macroscopic gelation from cross-linking [159]. While preparing a brush by grafting from via ATRP, the number of macroinitiator backbones in each droplet is highly limited. This allows higher initiator concentrations to be employed and increased conversion to be obtained, while any increased rate of termination results in the coupling of only a few brush macromolecules, as opposed to macroscopic gelation observed during solution polymerization.

When considering brush synthesis by grafting from in homogeneous media, it is vital to understand the fundamentals affecting grafting density, as this is largely responsible for the resulting unique material properties. Due to the high local concentration of initiation sites present on the backbone, steric interactions may adversely affect the efficiency of the grafting process. This would also affect the uniformity of the resulting side-chains and possibly limit the ability to prepare well-defined core–shell molecules with block copolymer side-chains. The initiation efficiency for the ATRP of BA and MMA from a poly{2-[2-bromopropionyloxy(ethyl methacrylate)} (PBPEM) macroinitiator backbone was investigated by Sumerlin and coworkers [154, 155]. In order to isolate the effects of the congested environment on the initiation process, the results were compared with those obtained during an analogous linear polymerization initiated with a similar low molar mass ATRP initiator. The polymerizations from the macroinitiator backbones demonstrated significantly lower initiation efficiency, which is in accordance with the findings of others [28]. The differences were attributed to the congested environment (high local concentration of initiation sites), which led to slower deactivation of growing radicals at low conversion. Two methods were demonstrated for improving the initiation efficiency of the brush polymerization at low conversion: (i) reduced concentration of initiator and (ii) increased concentration of the deactivating Cu(II) catalyst [155]. These results may not be limited to the synthesis of molecular brushes, but could possibly also provide valuable insight into the initiation phenomena of polymerizations conducted in other sterically congested environments (e.g. from surfaces, within pores, etc.).

10.4 Structure–Property Correlation

The specific architecture and characteristic shape of molecular brushes that define their intramolecular density distribution give rise to a variety of interesting physical phenomena. Due to the ability to prepare brushes with a range of diverse molecular characteristics, these materials demonstrate potential utility in macromolecular engineering for preparing new advanced materials [114]. Therefore, it is important to consider the fundamental structure–property correlation inherent to molecular brushes.

10.4.1 Solution Properties

In addition to the previously discussed capability of molecular brushes with block copolymer side-chains to yield unimolecular core–shell micelles, it is important to consider the structural factors that facilitate the cylindrical morphology typically observed. Regular branching with high local densities of side-chains leads to rigid backbone solution conformations. In a good solvent, molecular brushes are extended with their densely grafted side-chains radiating relatively perpendicular to the polymer backbone, such that the intramolecular excluded volume effects yield shape-persistent cylindrical structures. Several studies have been conducted to elucidate the effect of brush parameters on the resulting solution conformations. The findings suggest that the specific conformations observed are largely dependent on the identity and characteristics of the backbone and side-chains.

Nakamura and coworkers employed static light scattering and viscosity measurements to study the effect of side-chains on the underlying backbone conformation of brushes with backbone and side-chains composed of polystyrene [160–163]. The contour length per backbone monomer unit (l_b) approached 0.25 nm, essentially the limiting value for a fully extended backbone, and was independent of the side-chain length and number. Brushes with poly(alkyl methacrylate) backbones and flexible PBA side-chains were investigated by static light scattering and small angle neutron scattering and were also shown to have an l_b=0.25 nm that was independent of the side-chain length [114]. Also, solutions of brushes in certain concentrations ranges form lyotropic hexagonal phases. These results indicate that the backbone stretches to allow the side-chains to adopt entropically more favorable, coiled conformations. Because the large majority of the molecular brush mass is derived from the side-chains, the loss in entropy by stretching of the backbone is easily compensated by allowing the side-chains to adopt a coiled conformation. Others have reported that l_b is often significantly lower than that corresponding to a fully extended backbone for brushes with poly(methyl methacrylate) backbone and side-chains [29] and for brushes with poly(alkyl methacrylate) backbones and polystyrene side-chains [164]. These results indicate that, whereas sterics lead to backbone extension, the degree of stretching is dependent on a number of factors including the flexibility of the side-chains and backbone, in addition to the side-chain–backbone interactions.

The extended nature of molecular brushes in solution can also be tuned by incorporation of stimuli-responsive polymer segments. Reversible collapse and expansion of the densely grafted side-chains in response to an appropriate stimulus can result in the entire brush transforming its shape and/or solubility. In aqueous solution, brushes with poly(*N*-isopropylacrylamide) side-chains undergo rod-to-globule transitions in response to changes in temperature due to the LCST susceptibility of the side-chains [157]. PDMAEMA side-chains confer similar temperature-responsive behavior to molecular brushes and the nature of the LCST transition was found to be photo-tunable by both statistical and block copolymerization with 4-methacryloyloxyazobenzene (MOAB) (Fig. 10.6) [165]. Upon irradiation with UV light, the azobenzene groups of the MOAB units undergo a trans to cis isomerization that is accompanied by an increase in hydrophilicity. The increased solubility of the MOAB units in response to photo-stimulation resulted in brushes with temperature- *and* light-responsive behavior.

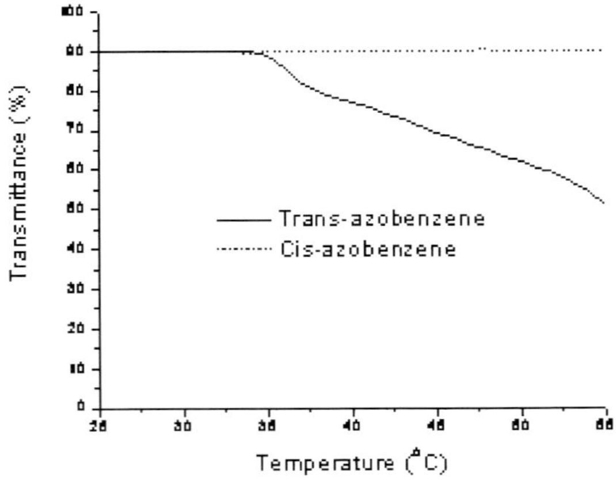

Fig. 10.6 Transmission at 600 nm of an aqueous solutions of poly[BPEM-*graft*-(DMAEMA-*stat*-MOAB)] as a function of temperature. The LCST (indicated by decrease in transmission) is dependent on the isomerization of the azobenzene groups in the MOAB units [165].

10.4.2 Bulk Properties

The high density and proportion of relatively short side-chains present in molecular brushes also have an important effect on their resulting bulk properties. Due to the radial distribution and extended nature of the backbone, chain packing can be significantly hindered, leading to morphologies different than that expected for simple linear polymers with the same identity as the side-chains.

Whereas PEG is a highly crystalline polymer, polymacromonomers of PEGMA and double-graft brushes (PEGMA side-chains grafted from a well-defined backbone) were shown to be amorphous. Analysis of the bulk structure and mechanical properties indicated that the specific architecture of the copolymers had completely suppressed crystallization of the PEG, leading to amorphous, homogeneous materials. When the brushes were transformed to a cross-linked network structure by heat treatment (or by spontaneous cross-linking during polymerization), high local mobility and sufficient macroscopic mechanical stability were achieved simultaneously, giving rise to a class of materials, termed *super-soft elastomers*, with a rubbery plateau of the shear modulus (G') of $>10^4$ Pa (Fig. 10.7). Other materials with moduli in this range (e.g. hydrogels) depend on small molecule solvation of network structures in order to retain a morphology that facilitates the soft elastomeric properties. For the polymeric brushes, these unusual mechanical properties were observed for the first time in the bulk phase and were the result of the molecular network of the backbone being diluted by short side-chains that were below the entanglement limit. Because the side-chains do not entangle and are covalently attached to the matrix, stability against evaporation or deforma-

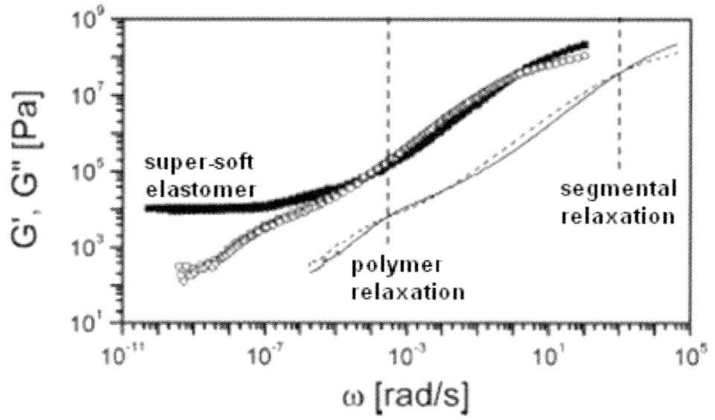

Fig. 10.7 Frequency dependences of the real (G') and imaginary (G'') shear modulus components for double-graft brushes prepared by grafting PEGMA ($DP_n=30$) from a PBPEM backbone ($DP_n=428$) and annealing at 120 °C under vacuum for 4 h. Adapted with permission from [52]. Copyright 2003 American Chemical Society.

tion is possible and the networks resisted collapsing. Due to high molecular mobility of the PEG side-chains, these brushes have been considered as solvent-free cation conducting materials for use in lithium battery applications [123].

10.4.3
Interfacial Properties

The relatively high molecular weight and molecular density of polymer brushes facilitate their investigation by surface microscopy techniques. Specifically, with sufficient backbone and side-chain length, molecular brushes can be visualized as individual molecules by AFM, providing information about the nature of the interactions with the underlying substrate and allowing the study of adsorption, ordering and diffusion of individual macromolecules [24–39]. For instance, brushes with soft PBA cores and hard PS shells were prepared by ATRP and it was demonstrated that the morphology of the adsorbed brushes was dependent on the interaction of the blocks with the mica substrate and with one another [25]. The interplay of the interfacial interactions and entropic elasticity of the side-chains resulted in a necklace-like morphology with PS globuli resting on top of the PBA chain segments that were tightly adsorbed to the substrate.

10.4.3.1 Controlling Conformation on Surfaces

The molecular brush surface conformation depends on the fraction of adsorbed side-chains [31, 149, 150, 166]. Steric repulsion between adsorbed side-chains causes extension of the backbone, while desorption and attraction between desorbed side-chains induces conformational changes from worm-like to globular. Adsorbed side-chains reduce the interfacial energy of the system but causes an entropic penalty due to extension of the side-chains and backbone [150].

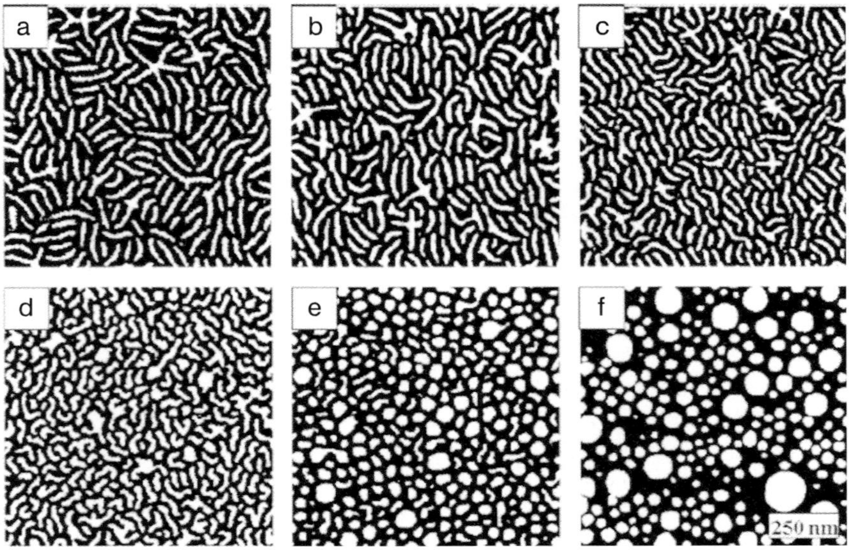

Fig. 10.8 AFM images of PBA brush monolayers transferred on mica at different percentages of methanol in water: (a) 0% (no methanol), (b) 5%, (c) 10%, (d) 20%, (e) 21% and (f) 22%. As the surface energy dropped from 89.7 mJ m^{-2} (**a**) to 69.1 mJ m^{-2} (**e**) a rod-to-globule transition was observed. Adapted with permission from [150]. Copyright 2003 American Chemical Society.

Sheiko and coworkers transferred Langmuir–Blodgett monolayers to characterizable surfaces to show that this enthalpy/entropy competition causes rod-to-globule transitions upon desorption of PBA side-chains. These transitions can be induced by pressure changes due to lateral compression of a monolayer on water [149] or by controlling the surface energy of the monolayers by varying the composition of the underlying non-solvent surface (Fig. 10.8) [150].

10.4.3.2 **Probing Surface Dynamics**

A major benefit of AFM is the combination of high spatial resolution and the ability to observe the dynamics of polymer systems in real time. For example, individual surface-adsorbed molecular brushes with PBA side-chains were shown to undergo conformational transitions when exposed to vapors of varying polarity and this process was visualized in a step-by-step manner [32]. Single molecular brushes assumed a compact globular conformation when exposed to vapor of amphiphilic compounds that adsorbed on mica and lowered the surface energy of the substrate (e.g. alcohols) [33]. By contrast, the brushes extended into two-dimensional worm-like conformations when exposed to vapors of compounds with higher surface tension (e.g. water). It was determined that the observed tendency was due to competition in spreading on the substrate surface between the brushes and the co-adsorbed vapor molecules [32] and that the characteristics of the side-chains and backbone helped dictate conformation [31]. If brush–substrate interactions are favorable for spreading, extended conformations result; when interactions are unfavorable, collapse to globuli reduces the surface area per macromolecule. Therefore, conformation transitions of individual molecules can be induced and observed via appropriate variation in environment and real-time monitoring by AFM.

The dynamics of diffusion for individual brushes, in addition to the spreading phenomenon of monolayer drops, can also be visualized microscopically [34, 35]. By observing the motion of a group of individual brushes, the front of a spreading drop of brushes with PBA side-chains was visualized as a function of time, thus providing information regarding the nature of thin-film spreading and the extent to which molecular conformations can affect flow uniformity. The rate of diffusion of individual molecules within a drop was shown to be significantly less than the spreading rate of the drop itself, thus indicating that plug flow (collective sliding of unentangled chains) was the predominant mass transport mechanism for this particular example of monolayer spreading [34]. Other spreading phenomena can also be observed, as evidenced by the determination that fingering instability can be triggered by conformational transitions of brush macromolecules induced by the pressure gradient driving the flow [35].

10.4.3.3 Controlling Long-range/Epitaxial Ordering

Self-organization/orientation of molecular brushes can be facilitated by specific interactions with the underlying surface. Sheiko and coworkers have shown that the crystalline nature of highly oriented pyrolytic graphite (HOPG) can induce long-range order within a monolayer of PBA brushes through surface epitaxy mechanisms. Molecular motion provided during spreading can significantly enhance this degree of orientation [167]. Although diffusion during spreading facilitates enhanced ordering, the direction of orientation was independent of the flow direction, being determined solely by the underlying crystallographic lattice of the HOPG surface (Fig. 10.9).

10.5 Conclusion

As with polymer brushes attached to surfaces, molecular brushes give rise to interesting properties because of the highly extended nature of the grafted side-chains. A variety of synthetic methods are available to prepare these advanced macromolecular architectures and each is associated with advantages and precautions that should be considered when targeting specific topologies. Not surprisingly, the living polymerization techniques appear to be the most promising methods for precise structural control. Specifically, recent advances in CRP have facilitated a rapid increase in the available complexity of molecular brushes by allowing enhanced molecular weight control and the incorporation of increased functional-

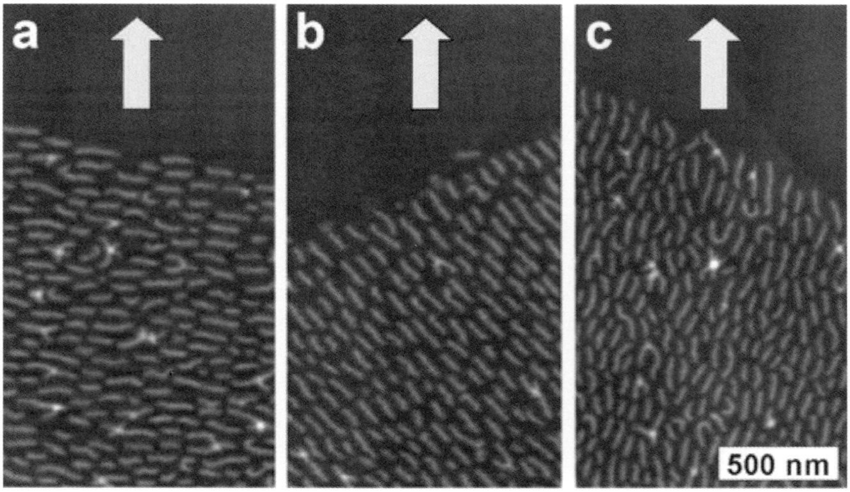

Fig. 10.9 AFM images reveal a lack of correlation between flow direction and orientation of the flowing PBA brushes on HOPG. Adapted with permission from [167]. Copyright 2003 American Chemical Society.

ity. The high molecular density, well-defined shape, high aspect ratio and encapsulation capabilities of molecular brushes confer promise for their utility as molecular shock-absorbers, high-performance elastomers and nanotemplates for inorganic materials. Their esthetically appealing characteristic of being visible as individual molecules facilitates their use as tools for studying fundamentals that include polymer–surface interactions, the nature of thin-film spreading and long-range/epitaxial ordering, among others.

References

1 Bhattacharya, A., Misra, B. N. *Prog. Polym. Sci.* **2004**, *29*, 767–814.
2 Gnanou, Y. *J. Macromol. Sci., Rev. Macromol. Chem. Phys.* **1996**, *C36*, 77.
3 Jenkins, D. W., Hudson, S. M. *Chem. Rev.* **2001**, *101*, 3245–3273.
4 Leger, L., Raphael, E., Hervet, H. *Adv. Polym. Sci.* **1999**, *138*, 185.
5 Peiffer, D. G., Rabeony, M. *J. Appl. Polym. Sci.* **1994**, *51*, 1283.
6 Wegner, G., Baum, P., Mullert, M., Norwig, J., Landfester, K. *Macromol. Symp.* **2001**, *175*, 349.
7 Napper, D. H. *Polymeric Stabilization of Colloidal Dispersions.* Academic Press, London, **1983**.
8 Advincula, R. C., Brittain, W. J., Baster, K. C., Ruhe, J. *Polymer Brushes: Synthesis, Characterization, Applications.* Wiley-VCH, Weinheim, **2004**.
9 Pyun, J., Kowalewski, T., Matyjaszewski, K. *Macromol. Rapid Commun.* **2003**, *24*, 1043–1059.
10 Pakula, T., Minkin, P., Matyjaszewski, K. *ACS Symp. Ser.* **2003**, *854*, 366–382.
11 Zhang, M., Müller, A. H. E. *J. Polym. Sci., Part A: Polym. Chem.* **2005**, *43*, 3461–3481.

12 Ishizu, K. *Polymer J.* **2004**, *36*, 775–792.

13 Lecommandoux, S., Checot, F., Borsali, R., Schappacher, M., Deffieux, A., Brulet, A., Cotton, J. P. *Macromolecules* **2002**, *35*, 8878–8881.

14 Feuz, L., Leermakers, F. A. M., Textor, M., Borisov, O. *Macromolecules* **2005**, *38*, 8891–8901.

15 Elli, S., Ganazzoli, F., Timoshenko, E. G., Kuznetsov, Y. A., Connolly, R. *J. Chem. Phys.* **2004**, *120*, 6257–6267.

16 Potemkin, I. I., Khokhlov, A. R., Reineker, P. *Eur. Phys. J. E: Soft Matter* **2001**, *4*, 93–101.

17 Subbotin, A., Saariaho, M., Stepanyan, R., Ikkala, O., ten Brinke, G. *Macromolecules* **2000**, *33*, 6168–6173.

18 Subbotin, A., Saariaho, M., Ikkala, O., Ten Brinke, G. *Macromolecules* **2000**, *33*, 3447–3452.

19 Saariaho, M., Subbotin, A., Ikkala, O., Ten Brinke, G. *Macromol. Rapid Commun.* **2000**, *21*, 110–115.

20 Saariaho, M., Subbotin, A., Szleifer, I., Ikkala, O., Ten Brinke, G. *Macromolecules* **1999**, *32*, 4439–4443.

21 Saariaho, M., Szleifer, I., Ikkala, O., Ten Brinke, G. *Macromol. Theory Simul.* **1998**, *7*, 211–216.

22 Saariaho, M., Ikkala, O., Szleifer, I., Erukhimovich, I., ten Brinke, G. *J. Chem. Phys.* **1997**, *107*, 3267–3276.

23 Vlassopoulos, D., Fytas, G., Loppinet, B., Isel, F., Lutz, P., Benoit, H. *Macromolecules* **2000**, *33*, 5960–5969.

24 Sheiko, S. S., Möller, M. *Chem. Rev.* **2001**, *101*, 4099–4123.

25 Boerner, H. G., Beers, K., Matyjaszewski, K., Sheiko, S. S., Moeller, M. *Macromolecules* **2001**, *34*, 4375–4383.

26 Matyjaszewski, K., Qin, S., Boyce, J. R., Shirvanyants, D., Sheiko, S. S. *Macromolecules* **2003**, *36*, 1843–1849.

27 Beers, K. L., Gaynor, S. G., Matyjaszewski, K., Sheiko, S. S., Moeller, M. *Macromolecules* **1998**, *31*, 9413–9415.

28 Muthukrishnan, S., Zhang, M., Burkhardt, M., Drechsler, M., Mori, H., Mueller, A. H. E. *Macromolecules* **2005**, *38*, 7926–7934.

29 Gerle, M., Fischer, K., Roos, S., Mueller, A. H. E., Schmidt, M., Sheiko, S. S., Prokhorova, S., Moeller, M. *Macromolecules* **1999**, *32*, 2629–2637.

30 Stephan, T., Muth, S., Schmidt, M. *Macromolecules* **2002**, *35*, 9857–9860.

31 Potemkin, I. I., Khokhlov, A. R., Prokhorova, S., Sheiko, S. S., Moeller, M., Beers, K. L., Matyjaszewski, K. *Macromolecules* **2004**, *37*, 3918–3923.

32 Gallyamov, M. O., Tartsch, B., Khokhlov, A. R., Sheiko, S. S., Boerner, H. G., Matyjaszewski, K., Moeller, M. *J. Microsc.* **2004**, *215*, 245–256.

33 Gallyamov, M. O., Tartsch, B., Khokhlov, A. R., Sheiko, S. S., Boerner, H. G., Matyjaszewski, K., Moeller, M. *Chem. Eur. J.* **2004**, *10*, 4599–4605.

34 Xu, H., Shirvanyants, D., Beers, K., Matyjaszewski, K., Rubinstein, M., Sheiko, S. S. *Phys. Rev. Lett.* **2004**, *93*, 206103/1–206103/4.

35 Xu, H., Shirvanyants, D., Beers, K. L., Matyjaszewski, K., Dobrynin, A. V., Rubinstein, M., Sheiko, S. S. *Phys. Rev. Lett.* **2005**, *94*, 237801/1–237801/4.

36 Zhang, M., Teissier, P., Krekhova, M., Cabuil, V., Mueller, A. H. E. *Prog. Colloid Polym. Sci.* **2004**, *126*, 35–39.

37 Zhang, M., Estournes, C., Bietsch, W., Mueller, A. H. E. *Adv. Funct. Mater.* **2004**, *14*, 871–882.

38 Zhang, M., Drechsler, M., Mueller, A. H. E. *Chem. Mater.* **2004**, *16*, 537–543.

39 Sheiko, S. S., da Silva, M., Shirvaniants, D., LaRue, I., Prokhorova, S., Moeller, M., Beers, K., Matyjaszewski, K. *J. Am. Chem. Soc.* **2003**, *125*, 6725–6728.

40 Yagci, Y., Ito, K. *Macromol. Symp.* **2005**, *226*, 87–96.

41 Ishizu, K., Sawada, N., Satoh, J., Sogabe, A. *J. Mater. Sci. Lett.* **2003**, *22*, 1219–1222.

42 Tsubaki, K., Ishizu, K. *Polymer* **2001**, *42*, 8387–8393.

43 Djalali, R., Li, S. Y., Schmidt, M. *Macromolecules* **2002**, *35*, 4282–4288.

44 Ishizu, K., Tsubaki, K. I., Uchida, S. *Macromolecules* **2002**, *35*, 4282–4288.

45 Tang, C., Dufour, B., Kowalewski, T., Matyjaszewski, K. *Macromolecules* submitted, **2006**.

46 Ishizu, K., Kakinuma, H., Ochi, K., Uchida, S., Hayashi, M. *Polym. Adv. Tech.* **2006**, *16*, 834–839.

47 Ishizu, K., Tsubaki, K., Uchida, S. *Macromolecules* **2002**, *35*, 10193–10197.
48 Boerner, H.G., Duran, D., Matyjaszewski, K., da Silva, M., Sheiko, S.S. *Macromolecules* **2002**, *35*, 3387–3394.
49 Cheng, G., Boeker, A., Zhang, M., Krausch, G., Mueller, A.H.E. *Macromolecules* **2001**, *34*, 6883–6888.
50 Mori, H., Mueller, A.H.E. *Prog. Polym. Sci.* **2003**, *28*, 1403–1439.
51 Neugebauer, D., Zhang, Y., Pakula, T., Matyjaszewski, K. *Polymer* **2003**, *44*, 6863–6871.
52 Neugebauer, D., Zhang, Y., Pakula, T., Sheiko, S.S., Matyjaszewski, K. *Macromolecules* **2003**, *36*, 6746–6755.
53 Qin, S., Matyjaszewski, K., Xu, H., Sheiko, S.S. *Macromolecules* **2003**, *36*, 605–612.
54 Zhang, W., Zhang, X. *Prog. Polym. Sci.* **2003**, *28*, 1271–1295.
55 Bathe, M., Rutledge, G.C., Grodzinsky, A.J., Tidor, B. *Biophys. J.* **2005**, *89*, 2357–2371.
56 Seog, J., Dean, D.M., Frank, E.H., Ortiz, C., Grodzinsky, A.J. *Macromolecules* **2004**, *37*, 1156–1158.
57 Dean, D., Seog, J., Ortiz, C., Grodzinsky, A.J. *Langmuir* **2003**, *19*, 5526–5539.
58 Tsukahara, Y., Mizuno, K., Segawa, A., Yamashita, Y. *Macromolecules* **1989**, *22*, 1546–1522.
59 Tsukahara, Y., Tsutsumi, K., Yamashita, Y., Shimada, S. *Macromolecules* **1990**, *23*, 5201–5208.
60 Wintermantel, M., Schmidt, M., Tsukahara, Y., Kajiwara, K., Kohjiya, S. *Macromol. Rapid Commun.* **1994**, *15*, 279.
61 Wintermantel, M., Gerle, M., Fisher, K., Schmidt, M., Wataoka, I., Urakawa, H., Kajiwara, K., Tsukahara, Y. *Macromolecules* **1996**, *29*, 978.
62 Hatada, K., Kitayama, T., Masuda, E., Kamachi, M. *Makromol. Chem., Rapid Commun.* **1990**, *11*, 101.
63 Masuda, E., Kishiro, S., Kitayama, T., Hatada, K. *Polym. J.* **1991**, *23*, 847.
64 Sheiko, S.S., Gerle, M., Fischer, K., Schmidt, M., Möller, M. *Langmuir* **1997**, *13*, 5368.
65 Dziezok, P., Sheiko, S.S., Fischer, K., Schmidt, M., Möller, M. *Angew. Chem. Int. Ed.* **1997**, *36*, 2812.
66 Wintermantel, M., Fischer, K., Gerle, M., Ries, R., Schmidt, M., Kajiwara, K., Urakawa, H., Wataoka, I. *Angew. Chem. Int. Ed.* **1995**, *34*, 1472.
67 Tsukahara, Y., Ohta, Y., Senoo, K. *Polymer* **1995**, *36*, 3413.
68 Ito, K., Tanaka, K., Tanaka, H., Imai, G., Kawaguchi, S., Itsuno, S. *Macromolecules* **1991**, *24*, 2348.
69 Nomura, E., Ito, K., Kajiwara, A., Kamachi, M. *Macromolecules* **1997**, *30*, 2811.
70 Ishizu, K., Yukishima, S., Saito, R. *J. Polym. Sci., Part A: Polym. Chem.* **1993**, *31*, 3073.
71 Ishizu, K., Tsubaki, K., Ono, T. *Polymer* **1997**, *39*, 2935.
72 Grassl, B., Rempp, S., Galin, J.C. *Macromol. Chem. Phys.* **1998**, *199*, 239.
73 Ishizu, K., Furukawa, T. *Polymer* **2001**, *42*, 7233–7236.
74 Yamada, K., Miyazaki, M., Ohno, K., Fukuda, T., Minoda, M. *Macromolecules* **1999**, *32*, 290–293.
75 Ederle, Y., Isel, F., Grutke, S., Lutz, P. *J. Macromol. Symp.* **1998**, *132*, 197.
76 Peruch, F., Lahitte, J.-F., Isel, F., Lutz, P.J. *Macromol. Symp.* **2002**, *183*, 159.
77 Ishizu, K., Satoh, J. *J. Appl. Polym. Sci.* **2003**, *87*, 1790–1793.
78 Pantazis, D., Chalari, I., Hadjichristidis, N. *Macromolecules* **2003**, *36*, 3783–3785.
79 Grubbs, R.H., Tumas, W. *Science* **1989**, *243*, 907.
80 Schrock, R.R. *Acc. Chem. Res.* **1990**, *23*, 165–172.
81 Cheng, C., Khoshdel, E., Wooley, K.L. *Macromolecules* **2005**, *38*, 9455–9465.
82 Heroguez, V., Breunig, S., Gnanou, Y., Fontanille, M. *Macromolecules* **1996**, *29*, 4459–4464.
83 Breunig, S., Heroguez, V., Gnanou, Y., Fontanille, M. *Macromol. Symp.* **1995**, *95*, 151.
84 Allcock, H.R., Denus, C.R. D., Prange, R., Laredo, W.R. *Macromolecules* **2001**, *34*, 2757–2765.
85 Mecerreyes, D., Dahan, D., Lecomte, P., Dubois, P., Demonceau, A., Noels, A.F., Jérôme, R. *J. Polym. Sci., Part A: Polym. Chem.* **1999**, *37*, 2447–2455.
86 Jha, S., Dutta, S., Bowden, N.B. *Macromolecules* **2004**, *37*, 4365–4374.

87 Muehlebach, A., Rime, F. *J. Polym. Sci., Part A: Polym. Chem.* **2003**, *41*, 3425–3439.

88 Zhang, B., Fischer, K., Schmidt, M. *Macromol. Chem. Phys.* **2005**, *206*, 157–162.

89 Mendrek, A., Mendrek, S., Trzebicka, B., Kuckling, D., Walach, W., Adler, H.-J., Dworak, A. *Macromol. Chem. Phys.* **2005**, *206*, 2018–2026.

90 Matyjaszewski, K., Davis, T. P. *Handbook of Radical Polymerization.* Wiley, New York, **2002**.

91 Matyjaszewski, K., Xia, J. *Chem. Rev.* **2001**, *101*, 2921–2990.

92 Kamigaito, M., Ando, T., Sawamoto, M. *Chem. Rev.* **2001**, *101*, 3689–3745.

93 Perrier, S., Takolpuckdee, P. *J. Polym. Sci., Part A: Polym. Chem.* **2005**, *43*, 5347–5393.

94 Hawker, C. J., Bosman, A. W., Harth, E. *Chem. Rev.* **2001**, *101*, 3661–3688.

95 Moad, G., Rizzardo, E., Thang, S. H. *Aust. J. Chem.* **2005**, *58*, 379–410.

96 Buathong, S., Peruch, F., Isel, F., Lutz, P. J. *Designed Monom. Polym.* **2004**, *7*, 583–601.

97 Ali, M. M., Stöver, H. D. H. *Macromolecules* **2004**, *37*, 5219–5227.

98 Malkoch, M., Carlmark, A., Woldegiorgis, A., Hult, A., Malmstrom, E. E. *Macromolecules* **2004**, *37*, 322–329.

99 Boerner, H. G., Matyjaszewski, K. *Macromol. Symp.* **2002**, *177*, 1–15.

100 Beers, K. L., Matyjaszewski, K. *J. Macromol. Sci., Pure Appl. Chem.* **2001**, *A38*, 731–739.

101 Ydens, I., Degee, P., Haddleton, D. M., Dubois, P. *Eur. Polym. J.* **2005**, *41*, 2255–2263.

102 Batt-Coutrot, D., Haddleton, D. M., Jarvis, A. P., Kelly, R. L. *Eur. Polym. J.* **2003**, *39*, 2243–2252.

103 Wang, X. S., Malet, F. L. G., Armes, S. P., Haddleton, D. M., Perrier, S. *Macromolecules* **2001**, *34*, 162–164.

104 Haddleton, D. M., Perrier, S., Bon, S. A. F. *Macromolecules* **2000**, *33*, 8246–8251.

105 Kato, M., Kamigaito, M., Sawamoto, M., Higashimura, T. *Macromolecules* **1995**, *28*, 1721–1723.

106 Wang, J.-S., Matyjaszewski, K. *J. Am. Chem. Soc.* **1995**, *117*, 5614–5615.

107 Hudson, S. D., Jung, H.-T., Percec, V., Cho, W.-D., Johansson, G., Ungar, G., Balagurusamy, V. S. K. *Science* **1997**, *278*, 449–452.

108 Percec, V., Ahn, C.-H., Yeardley, P., Möller, M., Sheiko, S. S. *Nature* **1998**, *39*, 161–164.

109 Deng, G., Chen, Y. *J. Polym. Sci., Part A: Polym. Chem.* **2004**, *42*, 3887–3896.

110 Bo, Z., Rabe, J. P., Schlüter, A. D. *Angew. Chem. Int. Ed.* **1999**, *38*, 2370–2372.

111 Ishizu, K., Satoh, J., Sogabe, A. *J. Colloid Interface Sci.* **2004**, *274*, 472–479.

112 Ishizu, K., Satoh, J., Tsubaki, K. I. *J. Mater. Sci. Lett.* **2001**, *20*, 2253–2256.

113 Tsubaki, K. I., Kobayashi, H., Sato, J., Ishizu, K. *J. Colloid Interface Sci.* **2001**, *241*, 275–279.

114 Rathgeber, S., Pakula, T., Wilk, A., Matyjaszewski, K., Beers, K. L. *J. Chem. Phys.* **2005**, *122*, 124904/1–124904/13.

115 Liu, Y., Abetz, V. Muller, A. H. E. *Macromolecules* **2003**, *36*, 7894–7898.

116 Zhang, B., Zhang, S., Okrasa, L., Pakula, T., Stephan, T., Schmidt, M. *Polymer* **2004**, *45*, 4009–4015.

117 Neugebauer, D., Theis, M., Pakula, T., Wegner, G., Matyjaszewski, K. *Macromolecules* **2006**, *39*, 584–593.

118 Ishizu, K., Shen, X. X. *Polymer* **1999**, *40*, 3251–3254.

119 Ishizu, K., Shen, X. X., Tsubaki, K. I. *Polymer* **2000**, *41*, 2053–2057.

120 Ishizu, K., Sawada, N., Satoh, J., Sogabe, A. *J. Mater. Sci. Lett.* **2003**, *22*, 1219–1222.

121 Ishizu, K., Toyoda, K., Furukawa, T., Sogabe, A. *Macromolecules* **2004**, *37*, 3954–3957.

122 Neugebauer, D., Zhang, Y., Pakula, T., Matyjaszewski, K. *Macromolecules* **2005**, *38*, 8687–8693.

123 Zhang, Y., Costantini, N., Mierzwa, M., Pakula, T., Neugebauer, D., Matyjaszewski, K. *Polymer* **2004**, *45*, 6333–6339.

124 Schappacher, M., Billaud, C., Christophe, P., Deffieux, A. *Macromol. Chem. Phys.* **1999**, *200*, 2377.

125 Ryu, S., Hirao, A. *Macromolecules* **2000**, *33*, 4765–4771.

126 Gauthier, M., Moller, M. *Macromolecules* **1991**, *24*, 4548–4553.

127 Takaki, M., Asami, R., Kuwata, Y. *Macromolecules* **1979**, *12*, 378–382.

128 Yen, S. S. *Makromol. Chem., Rapid Commun.* **1965**, *81*, 152.

129 Takaki, M., Asami, R., Ichikawa, M. *Macromolecules* **1977**, *10*, 850.

130 Gauthier, M., Tichagwa, L., Downey, J. S., Gao, S. *Macromolecules* **1996**, *29*, 519–527.

131 Mynar, J. L., Choi, T.-L., Yoshida, M., Kim, V., Hawker, C. J., Fréchet, J. M. J. *Chem. Commun.* **2005**, *41*, 5169–5171.

132 Desai, A., Atkinson, N. F., Rivera, J., Devonport, W., Rees, I., Branz, S. E., Hawker, C. J. *J. Polym. Sci., Part A: Polym. Chem.* **2000**, *38*, 1033.

133 Helms, B., Mynar, J. L., Hawker, C. J., Fréchet, J. M. J. *J. Am. Chem. Soc.* **2004**, *126*, 15020.

134 Tornoe, C. W., Christensen, C., Meldal, M. *J. Org. Chem.* **2002**, *67*, 3057.

135 Rostovtsev, V. V., Green, L. G., Fokin, V. V., Sharpless, K. B. *Angew. Chem. Int. Ed.* **2002**, *41*, 2596.

136 Kolb, H. C., Finn, M. G., Sharpless, K. B. *Angew. Chem. Int. Ed.* **2001**, *40*, 2004–2021.

137 Gao, H., Louche, G., Sumerlin, B. S., Jahed, N., Golas, P., Matyjaszewski, K. *Macromolecules* **2005**, *38*, 8979–8982.

138 Sumerlin, B. S., Tsarevsky, N. V., Louche, G., Lee, R. Y., Matyjaszewski, K. *Macromolecules* **2005**, *38*, 7540–7545.

139 Tsarevsky, N. V., Sumerlin, B. S., Matyjaszewski, K. *Macromolecules* **2005**, *38*, 3558–3561.

140 Golas, P., Gao, H., Tsarevsky, N. V., Matyjaszewski, K. Unpublished results.

141 Venkatesh, R., Yajjou, L., Koning, C., Klumperman, B. *Macromol. Chem. Phys.* **2004**, *205*, 2161–2168.

142 Grubbs, R. B., Hawker, C. J., Dao, J., Fréchet, J. M. J. *Angew. Chem., Int. Ed. Engl.* **1997**, *36*, 270.

143 Zhang, M., Breinerb, T., Moria, H., Müller, A. H. E. *Polymer* **2003**, *44*, 1449–1458.

144 Khelfallah, N., Gunari, N., Fischer, K., Gkogkas, G., Hadjichristidis, N., Schmidt, M. *Macromol. Rapid Commun.* **2005**, *26*, 1693–1697.

145 Neiser, M. W., Muth, S., Kolb, U., Harris, J. R., Okuda, J., Schmidt, M. *Angew. Chem. Int. Ed.* **2004**, *43*, 3192–3195.

146 Kawaguchi, S., Maniruzzaman, M., Katsuragi, K., Matsumoto, H., Iriany, Ito, K., Hugenberg, N., Schmidt, M. *Polym. J.* **2002**, *34*, 253–260.

147 Lee, H.-i., Matyjaszewski, K., Yu, S., Sheiko, S. S. *Macromolecules* **2005**, *38*, 8264–8271.

148 Lord, S. J., Sheiko, S. S., LaRue, I., Lee, H.-i., Matyjaszewski, K. *Macromolecules* **2004**, *37*, 4235–4240.

149 Sheiko, S. S., Prokhorova, S. A., Beers, K. L., Matyjaszewski, K., Potemkin, I. I., Khokhlov, A. R., Moeller, M. *Macromolecules* **2001**, *34*, 8354–8360.

150 Sun, F., Sheiko, S. S., Moeller, M., Beers, K., Matyjaszewski, K. *J. Phys. Chem. A* **2004**, *108*, 9682–9686.

151 Schappacher, M., Deffieux, A. *Macromolecules* **2000**, *33*, 7371–7377.

152 Boyce, J. R., Shirvanyants, D., Sheiko, S. S., Ivanov, D. A., Qin, S., Boerner, H., Matyjaszewski, K. *Langmuir* **2004**, *20*, 6005–6011.

153 Pietrasik, J., Bombalski, L., Cusick, B., Huang, J., Pyun, J., Kowalewski, T., Matyjaszewski, K. *ACS Symp. Ser.* **2005**, *912*, 28–42.

154 Neugebauer, D., Sumerlin, B. S., Matyjaszewski, K. *Polymer* **2004**, *45*, 8173–8179.

155 Sumerlin, B. S., Neugebauer, D., Matyjaszewski, K. *Macromolecules* **2005**, *38*, 702–708.

156 Ishizu, K., Kakinuma, H. *J. Polym. Sci., Part A: Polym. Chem.* **2005**, *43*, 63–70.

157 Li, C., Gunari, N., Fischer, K., Janshoff, A., Schmidt, M. *Angew. Chem. Int. Ed.* **2004**, *43*, 1101–1104.

158 Bernard, J., Favier, A., Davis, T. P., Barner-Kowollik, C., Stenzel, M. H. *Polymer* **2006**, *47*, 1073–1080.

159 Min, K., Yu, S., Lee, H.-i., Mueller, L., Matyjaszewski, K. *Macromolecules* submitted, **2006**.

160 Terao, K., Nakamura, Y., Norisuye, T. *Macromolecules* **1999**, *32*, 711.

161 Terao, K., Hokajo, T., Nakamura, Y., Norisuye, T. *Macromolecules* **1999**, *32*, 3690.

162 Terao, K., Takeo, Y., Tazaki, M., Nakamura, Y., Norisuye, T. *Polym. J.* **1999**, *31*, 193.

163 Hokajo, T., Terao, K., Nakamura, Y., Norisuye, T. *Polym. J.* **2001**, *33*, 481.

164 Fischer, K., Schmidt, M. *Macromol. Rapid Commun.* **2001**, *22*, 787.

165 Lee, H.-i., Pietrasik, J., Matyjaszewski, K. *Macromolecules* submitted, **2006**.

166 Stepanyan, R., Subbotin, A., ten Brinke, G. *Phys. Rev. E* **2001**, *63*, 061805.

167 Xu, H., Sheiko, S. S., Shirvanyants, D., Rubinstein, M., Beers, K. L., Matyjaszewski, K. *Langmuir* **2006**, *22*, 1254–1259.

11
Grafting and Polymer Brushes on Solid Surfaces

Takeshi Fukuda, Yoshinobu Tsujii, and Kohji Ohno

11.1
Introduction

Polymers end-grafted on a solid surface can dramatically change various surface properties, including optical, electrical, chemical, thermodynamic, mechanical, tribological and rheological properties, hence they play highly important roles in many areas of science and technology [1–5]. There are essentially two methods to end-graft polymers on a solid surface. One is the grafting-to technique, in which a preformed polymer with a functional group at the chain end or a block copolymer with an adsorbing segment is chemically or physically attached on the surface. The other is the grafting-from technique, also referred to as surface-initiated polymerization, in which polymerization is conducted from the initiating groups chemically or physically bound on the surface. In many cases, chemically grafted polymers are more stable mechanically and environmentally (e.g. towards solvents and temperature) than physically grafted polymers. The grafting-from method has the clear advantage over the grafting-to method that it potentially provides much higher graft densities. This difference in graft density is because, in a grafting-to process, free polymers to be grafted on the surface have to diffuse against the concentration barrier formed by the already grafted polymers, which becomes more and more difficult as the density and chain length of the already grafted polymers increase, thus giving a limited graft density that decreases sharply with increasing chain length. In a grafting-from process, it is usually low-mass compounds such as monomer and catalyst that have to diffuse across the barrier of grafted polymers and, for this reason, chain growth can proceed even on a highly congested surface, to a very high degree of polymerization. The chain growth will especially be so effective when the surface-initiated polymerization proceeds in a living/controlled fashion so that all the graft chains grow more or less simultaneously with their active chain ends concentrated near the outermost surface of the graft layer (see below).

The conformation of end-grafted polymers in a good solvent can change dramatically with graft density [6–8]. In the "dilute" regime of graft density, in

Macromolecular Engineering. Precise Synthesis, Materials Properties, Applications.
Edited by K. Matyjaszewski, Y. Gnanou, and L. Leibler

ISBN: 978-3-527-31446-1

which no overlap of graft chains occurs, each chain will assume a conformation similar in dimensions to that of a free chain but somewhat skewed due to the presence of the surface (the "mushroom" conformation; Fig. 11.1). When the graft density exceeds a critical value, graft chains will overlap each other and, to avoid the thermodynamically unfavorable local increase in segment concentration, graft chains will make themselves stretch away from the surface, forming the so-called "polymer brush". Analogous to solutions of free polymers, polymer brushes may be divided into two categories that differ in graft density. One category is related to the "semi-dilute" regime of graft density, in which graft chains overlap but their volume fraction is still low enough that the free energy of interaction may be approximated by a binary interaction and the elastic free energy of chains by that of a Gaussian chain. Polymer brushes in this regime are termed "semi-dilute brushes". Scaling theoretical analysis [9–14] predicted that the equilibrium thickness (L_e) of a semi-dilute brush in a good solvent varies as

$$L_e \sim \nu N \sigma^{\frac{1}{3}} \tag{1}$$

where ν is a parameter characterizing the polymer–solvent interaction and N and σ are the degree of polymerization and the surface density of graft chains, respectively. The most interesting feature of this expression is that L_e depends on N in a linear way, in contrast to the dimensions of free chains being scaled as $N^{3/5}$. This means that the graft chain will assume a conformation of unusually large dimensions with its dimensional difference from the free chain becoming larger and larger with increasing N. The 1/3 dependence of L_e on graft density is another characteristic feature of semi-dilute brushes predicted by the theory. Extensive effort has been made to characterize the well-defined semi-dilute brushes prepared by grafting-to methods. For example, their equilibrium thickness and segmental

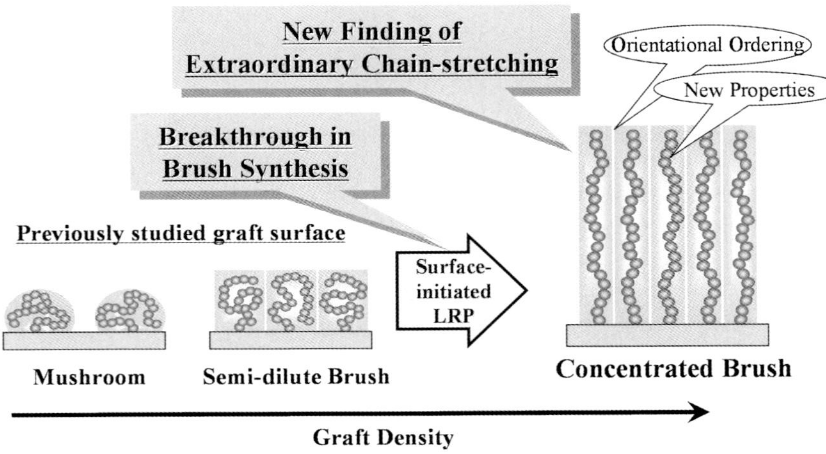

Fig. 11.1 Development of concentrated polymer brush.

density profiles in good solvents were studied by neutron reflectometry [15–19] and the interaction forces between brush surfaces were directly measured with a surface force apparatus [13, 20–22] and by atomic force microscopy (AFM) [14, 23, 24]. These studies supported the theoretical predictions.

The other category is related to the "concentrated" regime of graft density. According to scaling theoretical arguments, the mean distance between the nearest-neighbor graft points, which is measured by $\sigma^{-1/2}$ and defines the "blob" size, is so small that the excluded-volume effect is totally screened out, as in the "concentrated" solution of free chains. This consideration leads to the following prediction [11, 12]:

$$L_e \sim N\sigma^{\frac{1}{2}} \tag{2}$$

According to this relation, L_e should show a 1/2 dependence on graft density, independent of solvent. Polymer brushes in this regime are termed "concentrated brushes". Theoretical analyses, in which interactions higher than binary ones are explicitly taken into account, also predict that the repulsive forces of concentrated brushes will increase much more steeply with increasing graft density than those of semi-dilute brushes [25, 26]. Until recently, however, few experimental studies on the structure and properties of concentrated brush systems had been conducted because of the unavailability of well-defined concentrated brush samples. The recent development of surface-initiated living polymerization techniques, including anionic, cationic, ring-opening, ring-opening metathesis and radical methods, has opened up synthetic routes to well-defined concentrated and semi-dilute brushes. Above all, controlled/living radical polymerization (LRP), with its versatility with a wide range of monomers, simplicity and robustness in the experimental environment and conditions and fine controllability of chain length, chain length distribution and copolymer architecture has proved to be a particularly powerful method when applied to surface-initiated polymerizations on various organic, inorganic and metallic solids. Such surface-initiated living polymerizations, in particular LRP, have brought about a dramatic increase in graft density, allowing us to explore systematically deep into the concentrated brush regime and find that the structure and properties of concentrated brushes are very different and, in some cases, even unpredictable compared with those of the previously known semi-dilute brushes.

This chapter describes the preparation, structure and properties of polymer brushes with particular emphasis on those of concentrated brushes prepared by surface-initiated LRP: Section 11.2 deals with the controlled synthesis of polymer brushes by surface-initiated LRP and other living polymerization methods. Section 11.3 is concerned with experimental studies characterizing the structure and properties and functions of polymer brushes in solvent-swollen and dry states. Section 11.4 is specifically devoted to concentrated polymer brushes afforded on nanoparticles, which will be partly complementary to the descriptions in Chapter 12 in this volume. The reader is also referred to recent review articles on surface-initiated polymerizations and polymer brushes [27–31].

11.2
Controlled Synthesis of Polymer Brushes by Surface-initiated Polymerizations

11.2.1
Living Ionic and Ring-opening Polymerizations

Although classical methods of graft polymerization such as grafting-to techniques and surface-initiated conventional polymerizations are of versatile use, they do not meet both or either of the requirements of high graft density and well-defined chain structure. In recent years, a number of attempts have been made to apply living polymerization techniques to surface-initiated polymerization to meet both of those requirements simultaneously. Most successful results are from living radical systems, which will be described in Section 11.2.2. Notable results are also reported with other living polymerizations, some examples of which are given below.

In a surface-initiated anionic polymerization study by Jordan et al. [32], a self-assembled monolayer (SAM) containing bromobiphenyl groups was fixed on a gold substrate (see also Section 11.2.2.1) and converted, by reaction with *sec*-butyllithium, to the one with biphenyllithium groups to initiate polymerization of styrene (S). A uniform polymer film with a thickness of 18 nm was obtained after reaction for 3 days. Advincula et al. [33] noted that the addition of tetramethylethylenediamine to a similar system effectively increased the film thickness, and hence the rate of polymerization. By the sequential addition of monomers, they succeeded in preparing diblock copolymer brushes of polystyrene (PS)-*block*-polyisoprene and polybutadiene-*block*-PS segments, demonstrating the living nature of the systems. The need for rigorous purification of the system and reagents, long reaction times and rather small thicknesses of the polymer films produced limit the use of surface-initiated anionic polymerization.

A surface-initiated cationic polymerization of S was conducted by Zhao and Brittain [34] using a SAM with cumyl methyl ether groups, which was fixed on a silicon wafer (see also Section 11.2.2.1). Polymerization at –78 °C in the presence of $TiCl_4$ (activator) and di-*tert*-butylpyridine (proton scavenger) gave a 30-nm PS film within 1 h. However, the graft polymerization unavoidably produced free (unbound) PS chains, presumably initiated by protons (even in the presence of the proton scavenger). A merit of this technique may be its applicability to monomers such as isobutene [35], to which other methods are difficult to apply.

In an early study of surface-initiated ring-opening polymerization by Jordan and Ulman [36], the 11-hydroxyundecanethiol SAM on a gold substrate was reacted with $O(SO_2CF_3)_2$ in the vapor phase to introduce OSO_2CF_3 groups, which initiated the cationic ring-opening polymerization of 2-ethyl-2-oxazoline. The polymerization proceeded so slowly that it gave a 7-nm thick poly(*N*-propionylethylenimine) film after 7 days of reaction in refluxing chloroform. On the other hand, Husseman et al. [37] conducted the ring-opening polymerization of ε-caprolactone initiated from the primary hydroxyl moieties of the SAM formed on a gold substrate. Reaction for a few hours at room temperature with diethylaluminum alkoxide catalysis yielded a brush film as thick as 70 nm. However, to achieve good control of film thickness, it was necessary to add a suitable amount of free initiator (benzyl alcohol) to the reaction system.

A poly(lactic acid) (PLA)-grafted surface is interesting as it shows both optical activity and biodegradability. Choi and Langer [38] conducted polymerization of L-lactide initiated from the amino groups of the SAM on a silicon wafer. Reaction for 3 days at 80 °C catalyzed by tin(II) octoate produced a 70-nm thick PLA brush. The polymerization required no free initiator.

N-Carboxy anhydrides (NCAs) of amino acid esters also undergo ring-opening polymerization, whose early application to surface-initiated polymerization produced graft films with a thickness of only 10 nm or less [39]. In a more recent study, a 40-nm thick poly(L-glutamate) (PLG) brush was successfully produced by several hours of polymerization on an amine SAM fixed on a silicon wafer [40]. It was shown that the chains in this brush film assume an α-helix conformation with a tilt angle ranging from 32 to 48° depending on glutamate ester groups (methyl, benzyl, etc.). This success in synthesizing fairly thick PLG brushes may be attributed to the use of *N*,*N*-dimethylformamide as solvent, which is believed to suppress coagulation of chains.

Mechanistically different from the above-mentioned systems, the ring-opening metathesis polymerization of functionalized norbornenes has been applied to surface-initiated polymerization [41–43]. Targets of such studies include electronics applications.

11.2.2
Controlled/Living Radical Polymerization (LRP)

As detailed in Chapter 5 of Volume 1, the basic concept of LRP is the reversible activation process (Scheme 11.1). The dormant (end-capped) chain P–X is supposed to be activated to the polymer radical P$^\bullet$ by thermal, photochemical and/or chemical stimuli. In the presence of monomer M, P$^\bullet$ will undergo propagation until it is deactivated back to P–X. If a living chain experiences activation–deactivation cycles frequently enough over a period of polymerization time, all living chains will have a nearly equal chance to grow, yielding a low-polydispersity product. (The "living" chain denotes the sum of the dormant and active chains.) Capping agents X used for LRP include stable nitroxides [nitroxide-mediated polymerization (NMP)] [44–47], iodine (iodide-mediated polymerization) [48, 49], halogens with transition metal catalysts [often termed atom transfer radical polymerization (ATRP)] [50, 51], dithio ester and other unsaturated compounds

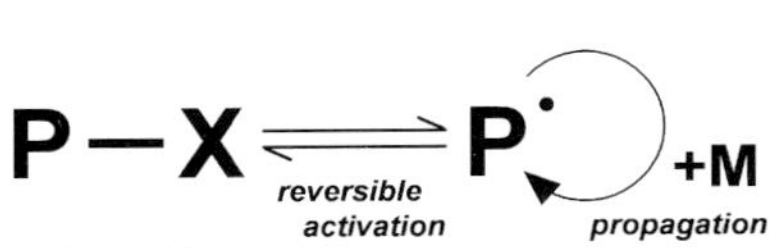

Scheme 11.1

[iniferter polymerization [52] or reversible addition-fragmentation chain transfer (RAFT) polymerization] [53, 54] and tellanyls [organotellurium-mediated radical polymerization (TERP)] [55]. The reversible activation reactions in most successful LRPs are classified into three types: (a) the dissociation–combination (DC) (for NMP and TERP), (b) the atom transfer (AT) (for ATRP) and (c) the degenerative chain transfer (DT) (for iodide, RAFT and TERP) mechanisms.

Of these LRPs, NMP, ATRP and RAFT polymerizations have already been applied to surface-initiated graft polymerization by immobilizing either a dormant species or a conventional radical initiator on the surface. In the latter case, a capping agent is added in the solution phase (reverse LRP). Complementary review articles concerning surface-initiated LRP are now available [28, 29, 31, 56, 57]. The use of a surface-bound dormant species is more promising for obtaining well-defined high-density polymer brushes. Figure 11.2 shows examples of surface-immobilizable dormant species including alkoxyamines for NMP and chlorosulfonyl, chlorobenzyl- and halo ester compounds for ATRP. The RAFT-mediated graft polymerization was achieved in most cases by the combined use of a surface-bound azo initiator and a free RAFT agent.

Surface-initiated (or surface-confined) polymerization brings about different situations from those in solution polymerization, which come from the immobilization of initiating (dormant) species and tethering and crowding of polymer chains on the surface. The following question arises: does the initiation occur efficiently on the surface? The initiation efficiency will be directly reflected on the graft density, which is one of the most important parameters of brush surfaces. Among other questions is whether the polymerization is controllable on the surface (within a graft layer) as in solution. The limited surface area of, for example, a flat substrate will lead to a low overall concentration of initiating (dormant) and capping species, resulting in poor control of polymerization. This problem must be circumvented to obtain well-defined polymer brushes. Tethering and crowding of polymers will affect the local concentration of reactants and hence their reaction rates. This surface confinement will have the most important effect on intermolecular reactions among graft chains, for example, the termination between graft radicals and the degenerative chain transfer between a graft radical and a graft dormant. For discussing this effect, it will be helpful to compare the chain length and chain-length distribution of graft polymers with those of free polymers polymerized under similar conditions; an interesting system is the polymerization with an added free initiator, since the polymerization simultaneously proceeds in both the graft layer and solution phases. These issues, specific to surface-initiated polymerization, will be discussed below.

11.2.2.1 Atom Transfer Radical Polymerization (ATRP)

ATRP has been widely applied to surface-initiated graft polymerization on a variety of materials, including flat substrates [58–71], fine particles [72–95] and porous materials [96–100]. Ejaz et al. [58] first succeeded in synthesizing a dense brush

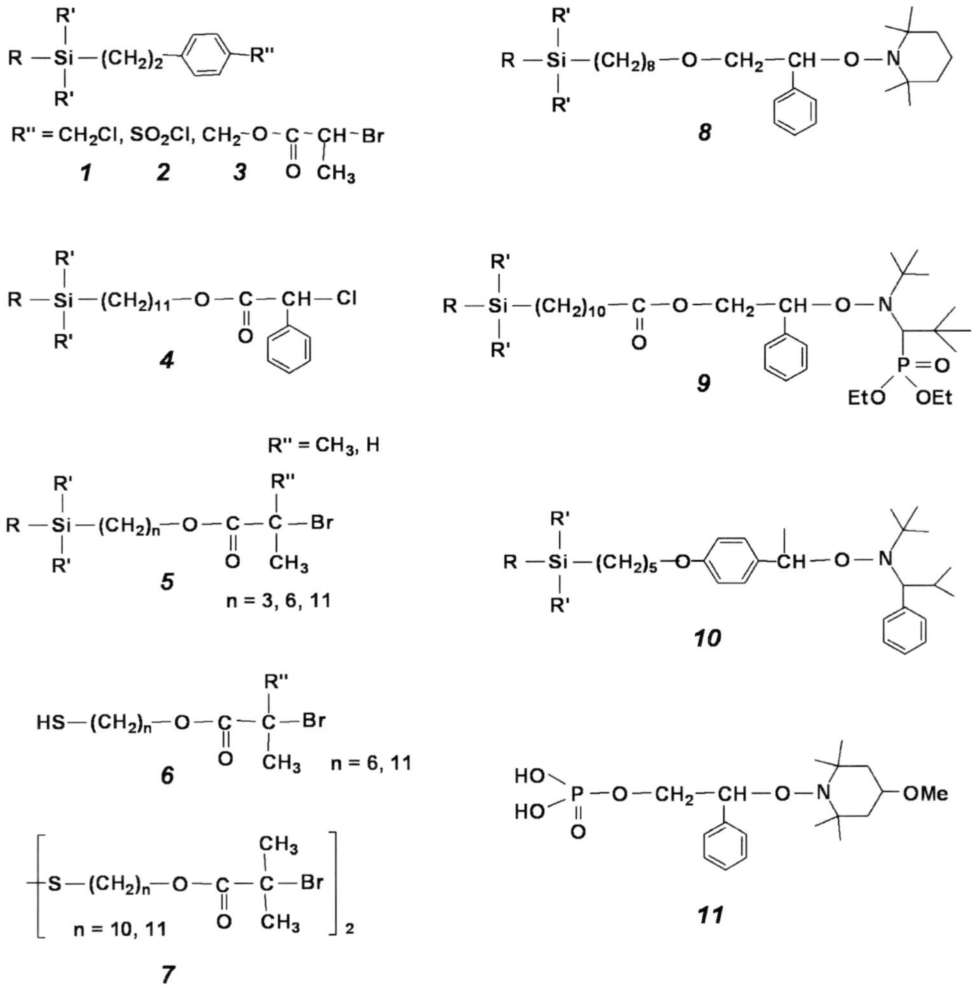

Fig. 11.2 Examples of surface-immobilizable dormant species.

of low-polydispersity poly(methyl methacrylate) (PMMA) by copper-mediated surface-initiated ATRP. They deposited a commercially available silane coupling agent, 2-(4-chlorosulfonylphenyl)ethyltriethoxysilane (CTS, **2** in Fig. 11.2), on a silicon wafer by the Langmuir–Blodgett technique to form a covalent bond by the coupling reaction with the silanol group on the silicon surface. Figure 11.3 schematically illustrates the graft polymerization process: the activator (A) such as a Cu(I) complex abstracts the halogen atom of the immobilized initiating dormant species (e.g. CTS) or the growing dormant chain, giving a propagating radical, to which some monomer units are added until it is recapped to be a dormant chain. This cycle occurs repeatedly and randomly on the halogenated sites on the surface,

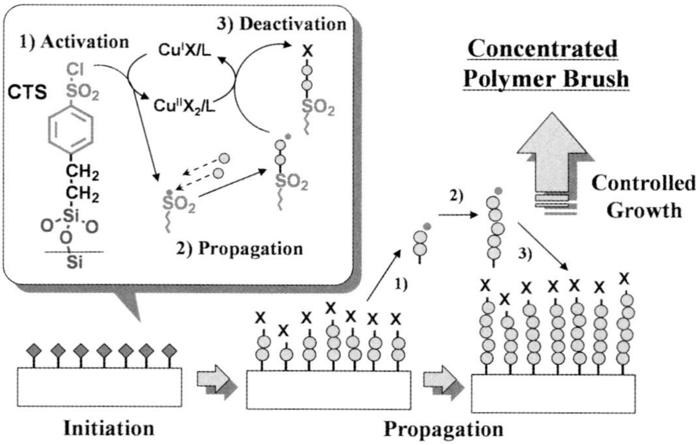

Fig. 11.3 Schematic illustration of surface-initiated atom transfer radical polymerization.

thus allowing all graft chains to grow slowly and nearly simultaneously, when viewed in a longer time-scale. This contrasts with the conventional radical graft polymerization, in which a radical formed on the surface instantly (≤ 1 s) grows to a high molecular weight polymer and the graft polymerization proceeds with an increase of the number of graft chains.

Hawker and coworkers [61] synthesized a halo ester type of silane coupling agent and successfully grafted PMMA using an Ni complex on the initiator-immobilized silicon wafer. The graft polymerization will hardly be controlled when the overall concentration of dormant species immobilized on the surface is too low to provide a sufficient amount of the capping agent in solution. This is almost always the case with a flat substrate. In these cases, therefore, a free initiator can be added to produce free polymers, thereby adjusting the concentration of the capping agent automatically by the so-called persistent radical effect [101–103]. Another advantage of adding a free initiator is that the free polymer produced can be a useful measure of the chain length and the chain-length distribution of the graft polymer. Good agreement of the number-average molecular weight (M_n) and the polydispersity index (M_w/M_n) between the graft and free polymers has been observed by several groups by directly characterizing the graft polymers cleaved off the silica particles (SiPs) [72, 73, 75, 104]. In some cases, slightly larger M_w/M_n values of the graft polymers were reported and discussed from the viewpoint of the confinement effect near the surface. Figure 11.4 confirms good agreement of M_n for SiPs with a wider range of diameters (ranging from 12 nm to 1.5 μm), suggesting that the surface curvature has little effect on the growth of graft chains. In the surface-initiated ATRP with an added free initiator, the amount (mass per unit area) of graft polymer increased proportionally to the M_n of the free polymer produced from the free initiator (see Fig. 11.5). This proportionality means that the graft density was kept constant during the course of polymerization.

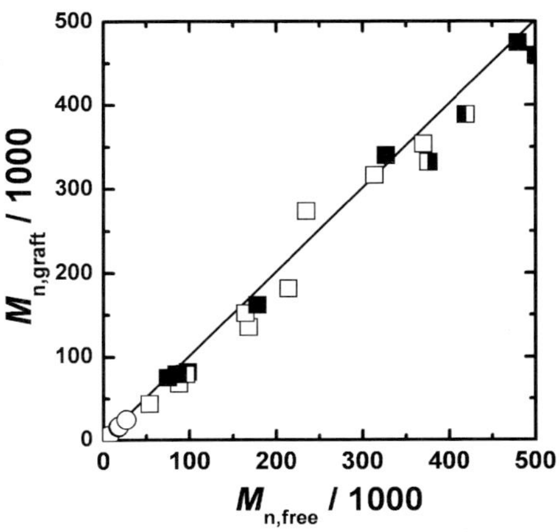

Fig. 11.4 Relationship of M_n of the cleaved and free polymers for the silica particles with diameters of 12 (○), 130 (□), 290 (◨), 790 (◧) and 1550 nm (■).

An alternative method to bring a system with a low overall concentration of dormant species under control is to add an appropriate amount of a capping agent X such as $CuBr_2$ prior to polymerization. In this case, no free polymer is produced and hence no additional process is required to remove the free polymer possibly contaminating the graft layer. This method was first demonstrated by ATRP on a silicon wafer by Matyjaszewski et al. [60] and subsequently by other researchers [62, 63, 68]. They observed a linear increase in the amount of

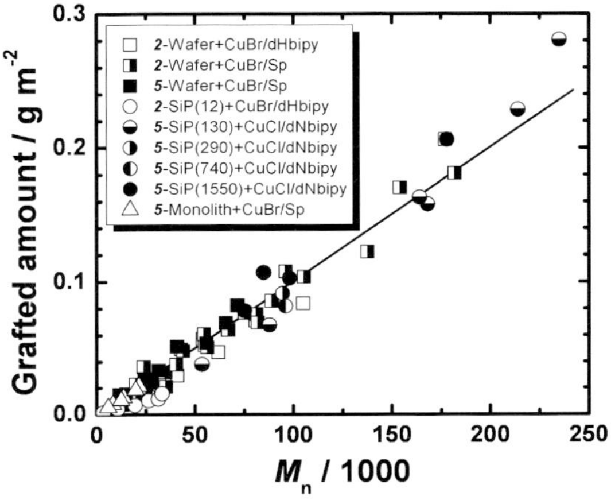

Fig. 11.5 Relationship between grafted amount and M_n of free polymers. The graft polymerization was carried out under various conditions on silicon wafer (squares), silica particles with varying diameters ($d=12$, 130, 290, 740, 1550 nm) (circles) and silica monolith with 50-nm mesopores (triangles). Two types of immobilized initiators, **2** and **5** ($n=6$ and $R''=CH_3$) in Fig. 11.2, two types of copper halides, CuBr and CuCl, and two types of ligands, sparteine (Sp) and dipyridyl derivatives [4,4′-diheptyl–2,2′-dipyridyl (dHbipy) and 4,4′-dinonyl–2,2′-dipyridyl (dNbipy)] were used.

graft polymer with polymerization time, which suggested that the graft polymerization proceeded in a living fashion with a constant graft density. This constant rate of chain growth was reasonable, since a flat substrate had so small a specific area that the overall concentrations of the monomer and catalytic species hardly changed with time.

An Au-coated substrate is another model surface to which many surface characterization methods can be applied. To achieve surface-initiated ATRP on Au-coated substrates, some halo ester compounds with thiol or disulfide group were developed [67–71]. Their SAMs were successfully prepared on an Au-coated substrate and used for surface-initiated ATRP. Because of the limited thermal stability of the sulfur–gold bond, the ATRP was carried out at a relatively low temperature, mostly at room temperature, by using a highly active catalyst system and water as a (co)solvent (water-accelerated ATRP).

Surface-initiated ATRP was applied not only on planar substrates but also on various kinds of fine particles. The latter systems will be described separately in Section 11.4. Porous materials are also fascinating targets for chromatographic application, making use of the unique structure and properties of high-density polymer brushes. Wirth and coworkers [96, 97] were the first to report the grafting of polyacrylamide (PAAm) on a porous silica gel.

Here, we focus on the surface-initiated ATRP of methyl methacrylate (MMA), which has been the most extensively studied. Figure 11.5 shows the relationship between the amount of graft polymer and the M_n of the free polymer (approximately equal to the M_n of the graft polymer; see above). These data were obtained by the authors' group under various polymerization conditions on a variety of silicate materials including silicon wafer, SiPs of differing diameters and monolithic silica gel with ca. 50-nm mesopores [58, 76, 105, 106]. These have flat, convex and concave/convex surfaces, respectively. Detailed experimental conditions are given in the captions. In all cases, the free dormant initiator was added to control the polymerization. This figure suggests that all the data fall on the same straight line passing through the origin, which means that the graft density was constant throughout the course of polymerization and that the graft polymerization proceeded in a living fashion. From the slope of the line, the graft density σ was estimated to be about 0.6 chains nm^{-2}. Table 11.1 summarizes the M_n, M_w/M_n and σ values available in the published papers. The graft density is reasonably close to the value obtained in Fig. 11.5, ranging from about 0.5 to about 0.8 chains nm^{-2}. These values are much higher than those obtained by the conventional techniques. It can be concluded that surface-initiated ATRP afforded not only control of the chain length and chain-length distribution of graft chains but also dramatically high graft densities. The question arises as to why such a high graft density was obtained. The most important difference between LRP and conventional RP might be the initiation efficiency. It should be essentially 100% in LRP. This must be so, since an LRP run usually experiences tens to hundreds of activation–deactivation cycles to give a low-polydispersity polymer and the initiation reaction in LRP is just one of those cycles in essence [102, 103].

Table 11.1 Surface-initiated ATRP and RAFT polymerization of MMA.

Method	Immobilized dormant	Activator (or RAFT agent)	Free initiator	Substrate	M_n	M_w/M_n	σ/chains nm^{-2}	Ref.
ATRP	**5** (n=6, R″=CH_3)	$NiBr_2$–PPh_3	EBiB	Si wafer	10 000–50 000 [a]		0.7	61
ATRP	2	CuBr–dHbipy	TsCl	Si wafer	45 400–83 300 [a]	1.1–1.3 [a]	0.8–0.6	142
ATRP	**5** (n=3, R″=CH_3)	CuBr–PMDETA	EBiB	Si wafer	30 000–90 000	1.15–1.27 [a]	0.3	64
ATRP	**5** (n=11, R″=CH_3)	CuBr–HMTETA	EBiB	H-Si wafer	10 000–20 000 [a]	~1.2 [a]	0.3	66
ATRP	**6**	Cur–Me_6TREN	–	Au-coated wafer	33 100–68 900	1.29–1.45	0.7–0.4	67
ATRP	**7** (n=11)	CuBr–bipy	–	Au substrate	35 000	1.59	0.6	163
ATRP	**8** (n=10)	$FeBr_2$–PPh_3	EBiB	Au-coated glass slide	6 000–60 000 [a]	1.1–1.4 [a]	0.5	167
RAFT	Azo initiator	2-Phenylprop-2-yl dithiobenzoate	AIBN	Si wafer	21 300 [a]	1.1 [a]	0.8	107
					25 400 [a]	1.72 [a]	0.5	

a) M_n and M_w/M_n values measured for free polymers.

Nevertheless, these graft densities are, in many cases, smaller than the density of the initiating dormant species (σ_i) on the surface, suggesting lower initiation efficiency of surface-initiated ATRP than that of the solution system. As a possible reason for this lower initiation efficiency, one may argue that at an early stage of polymerization, graft radicals will be strictly localized near the substrate surface and effectively terminated with each other and only the surviving chains will grow into longer chains. However, this is unlikely, because the graft density of PMMA brushes seems to be independent of the polymerization conditions and hence the concentration of active radicals.

Figure 11.5 and Table 11.1 contain the data obtained for different types of dormant species, copper halides and ligands and different types of surfaces (flat, convex and concave/convex) and different kinds of materials (silicate and gold). Interestingly, Baum and Brittain [107] achieved nearly the same graft density by the RAFT system. Another possible reason for the relatively low initiation efficiency of surface-initiated ATRP is concerned with the excluded volume effect of the monomer on the surface. To discuss this effect, let us consider the dimensionless graft density σ^*, defined as $\sigma^*=a^2\sigma$, where a^2 is the cross-sectional area per monomer unit given by $a^2=\nu_0/l_0$, with ν_0 being the molecular volume per monomer unit (estimated from the bulk density of monomer in this case) and l_0 the chain contour length per monomer unit ($l_0=0.25$ nm for vinyl polymers). The maximum value of σ^* is unity, corresponding to the close packing of graft chains with an all-trans conformation. A graft densities of 0.5–0.8 chains nm^{-2} correspond to $\sigma^*=0.4$–0.6. These values are close to the average surface coverage ($\sigma^*=0.6$) of spheres that are simulated to adsorb randomly and irreversibly on a flat surface [108].

In relation to the initiation efficiency on the surface, the dependence of σ on σ_i was explored by some researchers. Yamamoto and coworkers [106, 109] observed an almost constant graft density of PMMA brushes at initiator densities higher than a critical value, as shown in Fig. 11.6, where the initiator density was changed by photodecomposition and its concentration was estimated by FT-

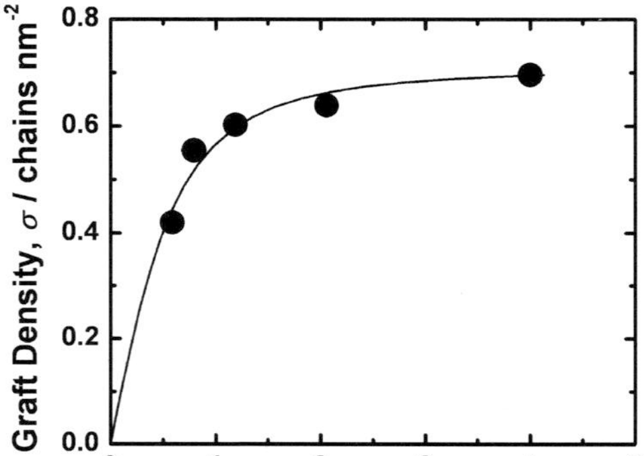

Fig. 11.6 Plot of graft density (σ) vs. initiator density (σ_i). The solid curve is drawn to guide the eye. Data reprocessed from [106].

IR spectroscopy. This figure suggests that the initiation efficiency is high and close to 100%, when σ_i is lower than the critical value equal to the maximum or cutoff graft density. On the other hand, Jones et al. [70] used mixed SAMs, formed on an Au-coated mica, composed of an initiation-active halo ester thiol and an inert thiol. They observed that the thickness of PMMA brushes obtained after a fixed polymerization time increased approximately linearly with increasing σ_i for fractions of active thiol ranging from 10 to 100%. At 100% of the active (halo ester) thiol fraction, σ and σ_i were ca. 0.6 chains nm^{-2} and ca. 5 molecules nm^{-2}, respectively. This result suggests that σ is proportional to σ_i, that is, the initiation efficiency is approximately constant and stays at a low value (about 10%), independent of the initiator density for $0.5\ \text{nm}^{-2} \leq \sigma_i \leq 5\ \text{nm}^{-2}$. These authors assumed the mixed SAMs to be compositionally uniform. However, a possibility has been suggested that two kinds of thiols can undergo a micro-scaled two-phase separation [110]. More recently, Ma et al. [111] reported a similar result to that given in Fig. 11.6 for oligo(ethylene glycol) methyl methacrylate (OEGMA) polymerized on an Au-coated substrate with a mixed SAM.

To discuss the kinetic aspects of surface-initiated ATRP, the system should be modeled as a confined polymerization, confined in the graft-layer phase. In the system producing free polymers, the polymerization will proceed simultaneously in the graft polymer phase and solution phase. The unbound reactants such as free polymer radicals, monomer, catalytic species and other additives will be partitioned between these two phases. The polymerization is usually carried out under good solvent conditions, under which the graft chains with sufficiently high σ are highly stretched, comparably to their full length, as will be discussed later. Therefore, the graft-layer phase can be assumed to have the thickness approximated by the full length of the graft chain. Then, the average concentration of polymer segments in the graft-layer phase will be about σ^* and the average concentration of graft-chain ends or graft dormants is inversely proportional to the chain length. The segmental density profile of the brush should also be taken into account. Matyjaszewski et al. [60] and Milchev et al. [112] estimated the density distributions of polymer segments and chain ends and also the chain-length distribution by computer simulation assuming an ideal living (termination free) polymerization without activation/deactivation reactions. They suggested that the chain ends are more populated in the outer region of the brush layer and that longer chains propagate favorably because of the gradient of polymer segment density, and hence of monomer concentration within the brush, giving a broader chain-length distribution of the graft chains than the free chains produced in the solution phase. Regarding termination, one may expect an enhanced termination because of graft chain-ends being highly localized in the outer region of the brush. Experimentally, however, no definite differences in average chain length and chain-length distribution have been observed between the graft polymer and the free polymer (see above). This suggests that in both the graft-layer and solution phases, termination has a negligible or at least a similar effect on the average chain length and chain-length distribution. Moreover, the termination reactions in the graft layer between graft chains and also between graft and free chains are believed to be much reduced owing to the tethering and high polymer concentration. Kinetic simulations including the activation/deactivation equilibrium and termination were presented by Xiao and Wirth [62] and Kim et al. [68] to predict the time evolution of grafted amount for surface-initiated ATRP without an added free initiator. However, these simulations did not take account of the segmental density profile causing the variation in local concentrations of reactants within the brush layer. More experimental and theoretical studies will be needed for a better quantitative understanding of the kinetics of surface-initiated ATRP.

11.2.2.2 **Nitroxide-mediated Polymerization (NMP)**

The first application of NMP to surface-initiated graft polymerization was reported by Husseman et al. in 1999 [61]. They succeeded in densely grafting S using a surface-bound dormant species with a 2,2,6,6-tetramethyl-1-piperidinyloxy (TEMPO) moiety. The free dormant chain with TEMPO moiety was added to control the polymerization and the living nature of the grafting was suggested by the formation of a block copolymer brush. They observed a proportional relationship between the thickness of the PS brush on an Si wafer and the M_n of the free polymer produced from the free alkoxyamine, calculating a graft density to be ca. 0.5 chains nm^{-2}. Using silica gel particles, they analyzed the graft polymer after cleavage, indicating that it had nearly the same M_n and a slight lower M_w/M_n than the corresponding free polymer. Subsequently, Devaux et al. [113] reported a higher graft density by a similar system except that the alkoxyamine moiety was immobilized on an Si wafer by the Langmuir–Blodgett technique. The controlled deposition of the initiators was claimed to play a key role to increase the graft density. They made an effort to measure directly the M_n and M_w/M_n of the polymers grafted on a flat surface and found an M_w/M_n value slightly lower and a gel permeation chromatographic (GPCy (GPC)“,4,1> peak molecular weight ca. 25% higher than those of the free polymers. Beyou and coworkers [114, 115] synthesized a silane coupling agent with an *N-tert*-butyl-*N*-(1-diethylphosphono-2,2-dimethylpropyl)-*N*-oxy (DEPN) moiety and applied it to grafting of PS on silica nanoparticles. Parvole et al. [116] succeeded in preparing a polyacrylate brush on SiPs using a surface-bound azo initiator along with free DEPN. In an attempt to improve the dispersibility of magnetite nanoparticles by grafting PS, Matsumoto [117] synthesized a new alkoxyamine and successfully immobilized it on the nanoparticle surface by reaction of a phosphonic acid group with the Fe–OH groups.

11.2.2.3 **Reversible Addition–Fragmentation Chain Transfer (RAFT) Polymerization**

Baum et al. [107] applied RAFT polymerization to synthesize brushes of PS, PMMA, poly(*N*,*N*-dimethylacrylamide) (PDMA) and their copolymers on azo initiator-bound silicate surfaces. 2-Phenylprop-2-yl dithiobenzoate was added as a free (unbound) RAFT agent to control the graft polymerization. Because of a very low concentration of surface-bound initiator, a free radical initiator, 2,2′-azobisisobutyronitrile, was needed to increase the polymerization rate to a practical level, as in solution RAFT polymerization. The PMMA and PS homopolymers grafted on an SiP were cleaved and analyzed to have M_n and M_w/M_n values comparable to those of the corresponding free polymers. The graft layer grew linearly and stepwise by sequential addition of monomers to give block copolymers. Subsequently, Zhai et al. [118] reported the synthesis of a polybetaine brush on a hydrogen-terminated silicon wafer by a similar approach using a surface-bound azo initiator and a free RAFT agent.

The graft polymerization based on the RAFT process is mechanistically different from those based on other types of LRP. Since the graft chains in a high-density polymer brush are highly stretched in a good solvent with their chain ends concentrated near the free surface of the graft layer (see Section 11.2.2.1), RAFT reactions would occur extraordinarily effectively among the graft polymers and a sequence of RAFT processes may be viewed as a migration or reaction–diffusion process of the otherwise strictly localized graft radicals (Fig. 11.7). This surface migration of the graft radicals would certainly increase the termination reactions among them. Tsujii et al. [119] studied the graft polymerization of S on SiP grafted with probe oligomeric PS chains with a terminal dithiobenzoyl (X) group introduced by surface-initiated ATRP followed by an exchange reaction of the terminal halogen atom with the X group. After the chain-extension

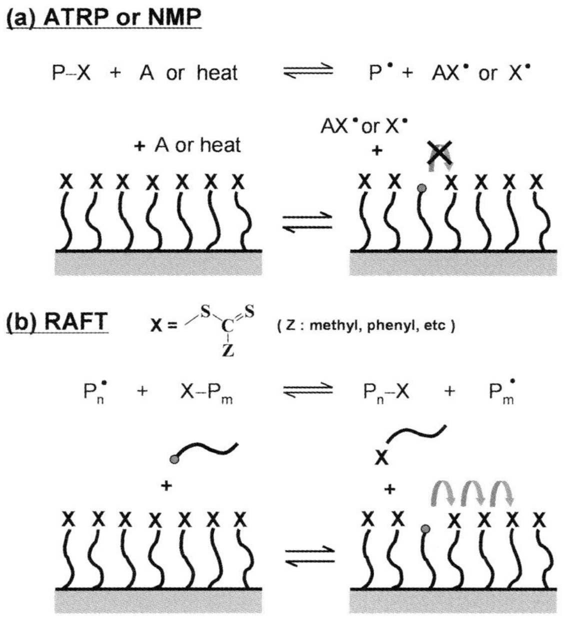

Fig. 11.7 Comparison of the key processes in (a) the ATRP- or NMP- and (b) RAFT-mediated graft polymerizations [119].

polymerization of S, the graft chain was cleaved from the SiP by treatment with HF and characterized. Polystyryl radicals are predominantly terminated by recombination to give dead chains of doubled molecular weight. The GPC analysis has shown two facts. First, the chain-length distribution of the main-peak component, which could be assigned to living or unterminated chains, was narrower than that of the free polymer. This suggested more frequent occurrence of the RAFT reaction on the surface than in the solution phase. Second, the minor-peak component, which could be assigned to the terminated chains of doubled molecular weight, was unusually large in quantity. This means a fast migration of the graft radicals on the surface. There was observed a critical value of the surface graft density below which the surface migration of the graft radicals hardly occurred. Below this limit, a RAFT system would behave similarly to, for example, an ATRP system.

11.3
Structure and Properties of Polymer Brushes

11.3.1
Swollen Brushes

11.3.1.1 Compressibility and Conformation of Graft Chains

A high-density PMMA brush prepared on a silicon wafer by surface-initiated ATRP and swollen in toluene was studied by AFM [105, 106]. The interaction force (F) between the graft layer and a silica probe attached on the AFM cantilever was measured as a function of separation (D) between the silicon substrate and silica probe surfaces (see Fig. 11.8). The measured force F can be reduced to the free energy of interactions (G_f) between two parallel plates according to the Derjaguin approximation [120], $F/R=2\pi G_f$, where R is the radius of the probe sphere (10 μm). Figure 11.8 shows a typical F/R versus D curve. Note that the true distance D between the substrate surface and the silica probe, which is usually difficult to define in AFM experiments, was successfully determined by AFM imaging the sample surface across the boundary of grafted and ungrafted regions of it. A notable feature of this F/R versus D curve is a rapid increase in the repulsive force with decreasing D. The observed repulsive forces originate from the steric interaction between the solvent-swollen brush and the probe sphere. Using the scaling approach, de Gennes [121] derived the equation concerning the interaction force between two parallel plates with the semi-dilute polymer brush layer, predicting that the force-distance profiles should be scaled by plotting $(F/R)\sigma^{-3/2}$ against D/L_e (for the definition of L_e, see below). Results for block-copolymer semi-dilute brushes were consistent with this scaling theory. The present high-density PMMA brushes ($\sigma>0.4$ chains nm^{-2}) prepared by surface-initiated ATRP, however, were poorly represented by this scaling theory [105, 106]. With increasing L_c and σ, the scaled force curve became steeper, meaning that the brush layer was more resistant to compression. The strong re-

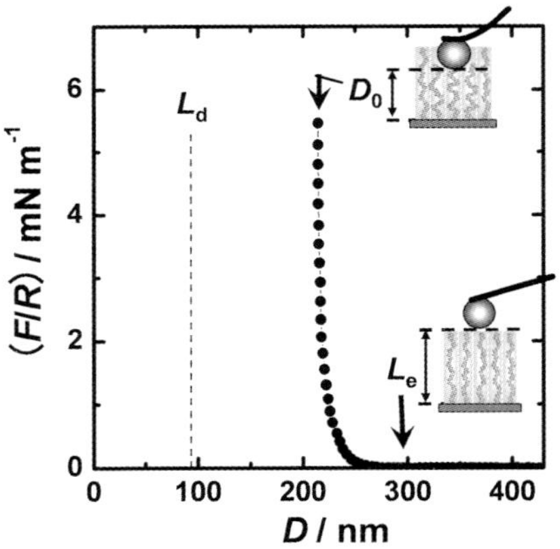

Fig. 11.8 Typical F/R vs D curve between the PMMA brush (L_d=87 nm, M_n=121 700, M_w/M_n=1.39) and the silica probe (attached on an AFM cantilever). The arrowheads indicate critical distances: L_e is the equilibrium thickness at which a repulsive force is detectable and D_0 is the offset distance beyond which the brush was no longer compressible.

sistance against compression is characteristic of concentrated polymer brushes, being ascribed to the extremely large osmotic pressure in this brush regime.

The breakdown of the semi-dilute brush theory was also revealed in the brush structure. The equilibrium thickness (L_e) of the solvent-swollen brushes was determined as the critical distance from the substrate surface beyond which no repulsive force was detectable (cf. Fig. 11.8). As mentioned in Section 11.1, the scaling and self-consistent mean field approaches predict, for both semi-dilute and concentrated brushes, that L_e is proportional to L_c for a constant σ, where L_c is the contour length (full length) of the graft chain. This relationship was confirmed by experimental data for semi-dilute brushes. High-density PMMA brushes with nearly equal density and different chain lengths (prepared by surface-initiated ATRP) also followed this proportional relationship [105]. To study the dependence of L_e on σ, a series of PMMA brushes with the same chain length and different graft densities ($0.07 < \sigma$ (chains nm^{-2})<0.7) were prepared by the photodecomposition of the surface initiator followed by the ATRP grafting [106]. Figure 11.9 shows the plot of $L_e/L_{c,w}$ versus σ^* on a logarithmic scale, where $L_{c,w}$ is the weight-average full length of the graft chain (in an all-trans conformation). The weight average, rather than the number average, was adopted by referring to the studies of Milner et al. [122, 123]. The slope of the curve in Fig. 11.9 is much larger than the 1/3 expected for the "semi-dilute" brush (Eq. 1: shown by the dashed line in the figure) and close to the 1/2 expected for the concentrated brush (Eq. 2). Figure 11.9 also shows that for the brush with $\sigma^*=0.4$ ($\sigma=0.7$ chains nm^{-2}), the value of $L_e/L_{c,w}$ even approaches 0.8–0.9. This large value certainly indicates that the graft chains in this brush regime are extended to an extraordinarily high extent, even comparable to their full lengths. The 1/2 dependence of L_e on σ as well as the above-mentioned characteristic compressibility of the brushes will confirm that we are in the concentrated brush regime. Thus, the surface-initiated LRP has enabled us to prepare well-defined high-density brushes with which to explore experimentally deep into the concentrated brush regime for the first time.

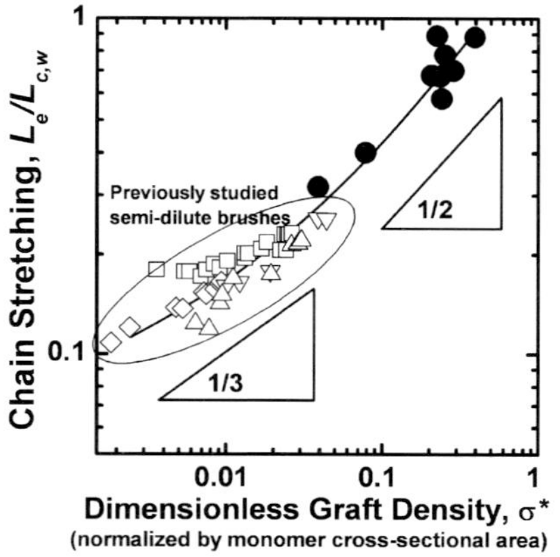

Fig. 11.9 Plots of $L_e/L_{c,w}$ vs. dimensionless graft density σ^*. (1) PS-*block*-polydimethyl siloxane: □, $M_{w,PS}=60\,000$; ◇, $M_{w,PS}=169\,000$ [25, 26]. (2) PEO-*block*-PS: △, $M_{w,PEO}=30\,800$; ▽, $M_{w,PEO}=19\,600$ [198]. (3) PMMA brushes: ●, $M_w=31\,300$–$267\,400$ [105, 106].

The high extension of graft chains will make the graft-chain ends localize near the outermost (free) surface of a swollen layer. Some theories and simulations predict that this trend becomes more and more prominent with increasing σ [112, 124]. To verify this localization of graft-chain ends, a short poly(4-vinylpyridine) (P4VP) hydrophilic segment with $M_{n,\,P4VP} \approx 2700$ was introduced, as a probe segment, at the chain ends of a concentrated PMMA brush with $M_{n,\,PMMA}=22\,000$ and $\sigma=0.62$ chains nm^{-2} by ATRP block copolymerization and the AFM force measurements were carried out in toluene [125]. The brushes with and without terminal P4VP segments gave nearly the same advancing-mode force profiles on compression by the silica probe sphere. This suggests that the P4VP segment introduced is short enough to have little effect on the equilibrium thickness and repulsion forces against compression. More interesting were the force profiles measured in the retracting mode, in which a strong attractive force was observed at large separations for the brush with P4VP segments but not for the precursory PMMA brush (Fig. 11.10). Such a long-range attractive force can be attributed to the bridging of graft chains between the probe sphere and the substrate; the P4VP segment would be adsorbed on the hydrophilic surface of the silica sphere, resulting in an attractive force due to an extension of graft chain in the retracting cycle, whereas the PMMA chains would not. This confirms that the P4VP segments are localized near the free surface, forming an outermost surface. In other words, an effective modification in chemical properties without a significant change in their physical properties can be achieved by the introduction of a short terminal block for the swollen, concentrated brush surface. Such a concept specific to concentrated brushes can be useful for their biointerface application, as will be discussed in Section 11.3.3.3.

A combinatorial approach was proposed by Wu and coworkers to study the structure of wet polymer brushes [126, 127]. The initiating dormant species with a gradient in surface density was immobilized on a silicon wafer by the vapor-diffusion method. The ATRP of acrylamide was carried out using a copper–li-

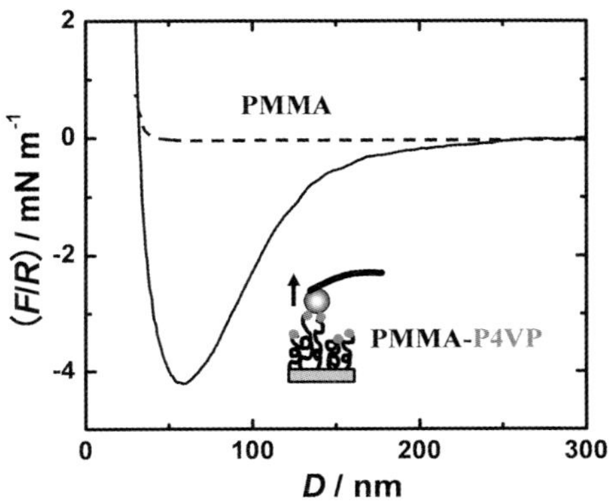

Fig. 11.10 Retracting-mode force profiles of concentrated PMMA brushes with and without P4VP terminal segment measured in toluene using a hydrophilic silica probe sphere.

gand complex on the functionalized wafer, successfully giving a PAAm-grafted surface with a gradient in graft density ranging from 0.001 to 0.2 chains nm^{-2}. The M_w and M_w/M_n values of the graft polymer were estimated to be ca. 17 000 and 1.7, respectively, by cleaving and analyzing the graft polymer prepared on SiPs under the same conditions. The characteristics of the graft polymer were reasonably assumed to be independent of the initiator density. The ellipsometric mapping of the brush height in dry and wet states clearly revealed the gradient brush structure on the wafer. Figure 11.11 shows the brush height in water, a good solvent, as a function of graft density, which is proportional to the dry

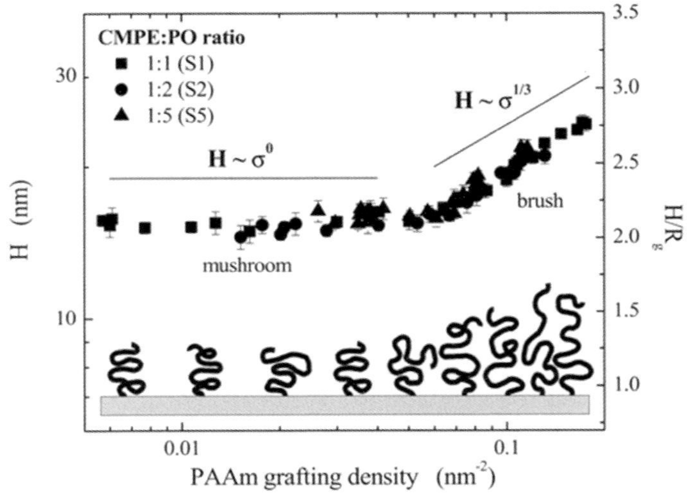

Fig. 11.11 Wet thickness (H) of PAAm as a function of the PAAm graft density for samples prepared on substrates containing the initiator gradients made of CMPE–OTS mixtures of (w/w) 1 : 1 (squares), 1 : 2 (circles) and 1 : 5 (triangles). The inset illustrates the polymer behavior. Reproduced with permission from [127] (Copyright 2003 American Chemical Society).

thickness in this case. This figure clearly indicates the mushroom-to-brush crossover at around σ=0.065 chains nm^{-2}, beyond which the height of the wet brush was proportional to σ^n with $n \approx 1/3$.

Other structural studies have been carried out for brushes with thermally responsive properties. Balamurugan et al. [128] investigated the thermally induced hydration of poly(*N*-isopropylacrylamide) (PNIPAM) brush synthesized by surface-initiated ATRP on an Au substrate. Surface plasmon resonance spectroscopy revealed the hydration transition occurring over a broad range of temperatures, while contact angle measurements suggested a sharp transition. These results were interpreted by taking account of the density profile that the polymer segment is highly solvated in the outermost region but less solvated within the brush. Kizhakkedathu and coworkers [94, 129] synthesized PS latex particles with polymer brushes of PNIPAM, PDMA and polymethoxyethylacrylamide by surface-initiated aqueous ATRP and characterized their M_n, M_w/M_n and σ after cleaving the graft polymers from the latex surface by hydrolysis. The hydrodynamic thickness of the PNIPAM brush was measured by a particle-size analyzer to scale as $L_c^{0.66}$ at a constant graft density. A broader range of transition temperatures was also observed. These particles were physisorbed on a glass substrate and the interaction forces between the brush and a silicon nitride tip were measured by AFM as a function of σ and M_n. The PDMA brushes exhibited a long-range attractive force due to bridging of the graft chains at σ=0.012 chains nm^{-2}. However, they showed only a repulsive force at higher values of σ with the force increasing with increasing M_n and σ, when the brush is compressed by the AFM tip. The critical density above which no bridging force was detected was larger for the PNIPAM brush. These critical σ values may be in the semi-dilute brush regime.

11.3.1.2 **Tribological Properties**

The strong resistance against compression and dense anchoring of graft chains in concentrated brushes are expected to improve drastically tribological properties such as friction, wear and lubrication as compared with those of semi-dilute brushes. The interaction forces between a brush-modified silicon wafer and a brush-modified silica probe sphere (R=5 μm) glued on the cantilever were measured by AFM [130]. The concentrated and semi-dilute PMMA brushes were fabricated on both surfaces by the grafting-from (surface-initiated ATRP) and grafting-to technique, respectively. The lateral (frictional) force F_L was evaluated according to the equation $F_L = k_L\theta_L/2R$, where k_L is the torsional spring constant and θ_L is the angular (torsional) displacement of the cantilever. The value of θ_L was monitored as a function of F_N, while the sample substrate was loaded with the silica probe particle and slid back and forth at a velocity of 20 μm s^{-1} along the direction normal to the long axis of the cantilever. Figure 11.12 plots the frictional coefficients μ (=F_L/F_N) as a function of normal force F_N (load). The semi-dilute brush (σ=0.04 chains nm^{-2}, M_n=79 000, M_w/M_n=1.2) had two different regimes of friction: at low applied loads, F_L and hence the μ value are very low (<0.001), and in

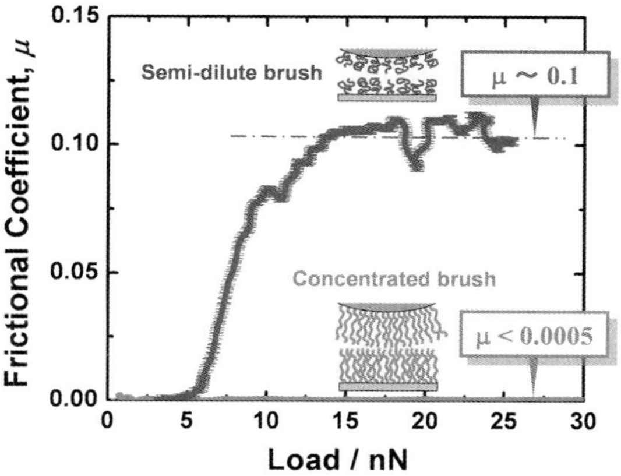

Fig. 11.12 Plot of frictional coefficient vs normal load between polymer brushes in solvent.

the threshold region, F_L increases steeply with increasing applied load, approaching the limiting constant value of about 0.1 (for higher loads). Such a transition from low to high frictional regions had been reported by Klein and Kumacheva [4], who measured the frictional properties between semi-dilute PS brushes in toluene using a surface force apparatus. This transition was ascribed to the interpenetration of the brushes at high loads. Most interestingly, the μ between concentrated brushes (σ=0.7 chains nm^{-2}, M_n=88 000, M_w/M_n=1.2) showed no such transition, staying at low values, lower than 5×10^{-4} in the whole range of loads studied. This μ value is one of the lowest of all materials and comparable to that achieved for polyelectrolyte semi-dilute brushes with the help of a charge effect [131]. This extremely low frictional property was reasonably ascribed to the fact that swollen concentrated brushes would hardly interpenetrate each other due to the large osmotic pressure and highly stretched chain conformation (entropic interaction). This mechanism of "super lubrication" of concentrated brushes should be effective not only on a microscopic scale as verified by AFM but also on a macroscopic scale. Kobayashi and coworkers [132, 133] studied the frictional properties of the concentrated brushes by sliding a stainless-steel or glass probe with R=5 mm on the brush surface in air and in solvents. Although μ did not reach the low value obtained on a microscopic scale, the concentrated PMMA brushes were found to have better frictional properties and better wear resistance than the corresponding spin-coated PMMA film [132]. These properties may be improved by increasing the brush thickness and decreasing the surface roughness of substrates.

11.3.1.3 **Size-exclusion Properties**

Because of the above-mentioned highly stretched conformation and strong resistance against compression, concentrated polymer brushes are expected to give unique interactions with solute molecules in solutions. For example, large molecules, sufficiently large compared with the distance between graft chains, would be difficult to get into the brush layer for entropic reasons. A concentrated brush would have this "size-exclusion" effect most effectively, since it is characterized by a high orientational/positional order of the graft chains. To demonstrate this size-exclusion effect chromatographically, a concentrated PMMA brush with $M_n = 15\,000$, $M_w/M_n = 1.2$ and $\sigma = 0.6$ chains nm^{-2} was prepared on the inner surface of a silica monolith column, which had a through-pore of a few μm in diameter and a continuous silica skeleton with ca. 50-nm mesopores. Because of the large porosity and high surface area, such a monolith with a bimodal pore structure has been extensively investigated as a new column system for high-performance liquid chromatography [134]. Figure 11.13 shows the size-exclusion chromatograms for standard PSs with molecular weights ranging from 10^2 to 10^6 as a function of elution volume normalized by the total column volume. In the unmodified monolith, the size exclusion was observed by mesopores, giving an exclusion-limit molecular weight of $>10^6$. The PMMA-grafted monolith gave a size exclusion by mesopores in a lower range of elution vol-

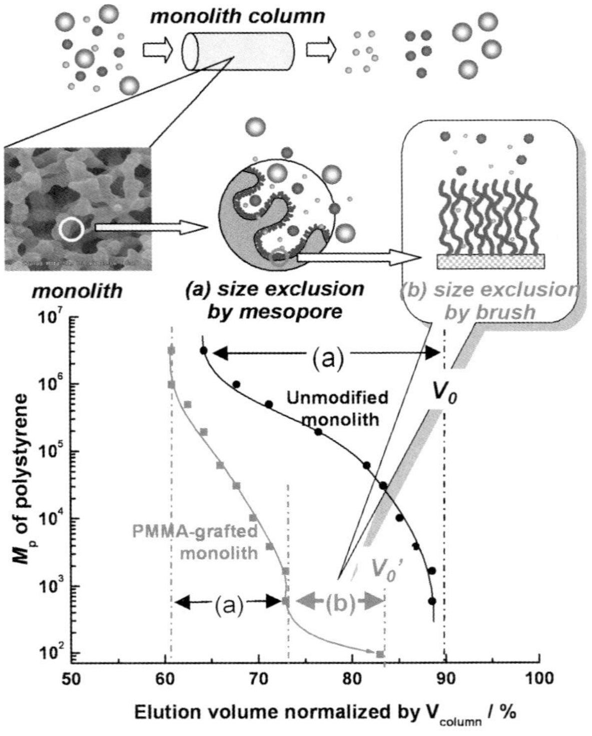

Fig. 11.13 Size-exclusion chromatogram of standard PSs for monolith columns with and without concentrated PMMA brush on an inner surface. The inset illustrates the size-exclusion mode by polymer brushes.

ume. Notably, a sharp separation was newly observed in a molecular weight range between 10^2 and 10^3, which was ascribed to the mentioned size exclusion by the polymer brushes. The dense structure of the concentrated brush presumably explains the sharp resolution and the low molecular-weight exclusion limit. Such a size-exclusion effect of (conformational) entropic origin presents a new notion of biocompatibility, as will discussed in Section 11.3.3.3.

11.3.2 Dry Brushes

Ultra-thin polymer films on a solid substrate (supported films) are extremely interesting objects, both scientifically and practically. A detailed knowledge of their structure and properties is essential for the design of advanced materials. Dry polymer brushes are also interesting as a kind of supported film. However, effects of end-grafting on their structure and properties have been studied only for low- to moderate-density (or semi-dilute) brushes. For example, the above-described concentrated PMMA brushes with $\sigma=0.7$ chains nm^{-2} should be highly anisotropic because their thickness in the dry state (L_d) reaches about 40% of the fully extended chain length ($L_d/L_{c,w}=0.35$), as illustrated in Fig. 11.14. Since the size of the unperturbed chain end-grafted on a repulsive surface is proportional to the square-root of chain length, this large value of L_d means that the chains are already

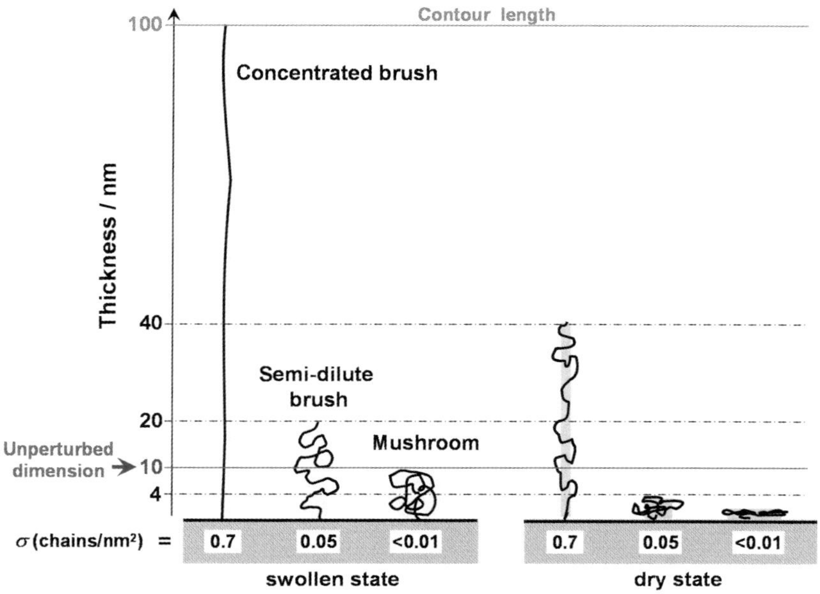

Fig. 11.14 Schematic illustration of conformations of end-grafted polymer chains in wet and dry states as a function of graft density.

highly extended in the dry state as compared with their unperturbed dimensions. This extended conformation of chains may result in structure and properties very different from those of previously studied polymer brushes.

11.3.2.1 Glass Transition

The glass transition temperature (T_g) of a thin polymer film has been extensively studied [135–140] and was shown to depend strongly on the thickness of the film and the nature of the substrate. The first observation of the T_g of concentrated PMMA brushes was made on a silicon wafer using temperature-variable spectroscopic ellipsometry [141]. Figure 11.15 shows the plot of the T_gs of the PMMA brushes and the corresponding cast films. The series of PMMA brushes studied have a nearly constant graft density (ca. 0.7 chains nm^{-2}) and different chain lengths (and hence different L_d). Remarkable is the difference in the T_g behavior between these two types of ultra-thin films. The molecular characteristics of the polymer forming each cast film is closely similar to those of the polymer forming the brush of (nearly) the same thickness and therefore the T_g difference cannot be ascribed to differences in molecular characteristics such as chain length, chain length distribution and stereoregularities. It may be completely ascribed to the effects of grafting, that is, chemically binding one of the chain ends on the substrate surface.

In the range of L_d smaller than 50 nm, cast films suffer a significant T_g depression, which was ascribed not only to the molecular weight effect but also to the interfacial effect. In contrast, the T_g of the brushes increases steeply with decreasing L_d. Obviously, end-grafting restricts the mobility of the chains. One may expect, however, that the effect of end-grafting on chain mobility would become less and less significant as the chain length increases and, in the limit of long chains, the T_g of the graft film would become equal to that of the cast film and hence that of the bulk polymer, since all surface effects should be unimportant in the long-chain limit. Figure 11.15 shows, however, that this is not the case. As L_d increases over about 50 nm, the T_g of the brushes reaches an almost

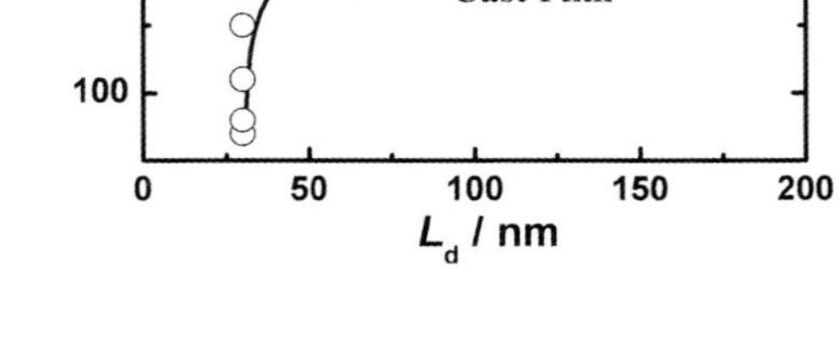

Fig. 11.15 L_d dependence of T_g measured by temperature-variable ellipsometry [141]. The solid and open circles represent the data for the brushes and the cast films, respectively.

constant value of about 119 °C, which is about 8 °C higher than that of the corresponding cast films. The figure strongly suggests that this difference in T_g between the brushes and cast films would be retained in the long-chain limit. This marked increase in the T_g of the long enough brushes was ascribed to an anisotropic structure/chain conformation in concentrated brushes. In fact, for low- to moderate-density (semi-dilute) PMMA brushes with $\sigma \leq 0.2$ chains nm^{-2}, no T_g increment was observed [140].

Tanaka et al. studied the surface molecular motion of PS film coated on a solid substrate by lateral force microscopy and found that the T_g at the surface was much lower than the corresponding bulk value [142]. Possible reasons for this included an excess free volume induced by localized chain ends, reduced cooperativity for the α-relaxation process, reduced entanglement and a unique chain conformation at the surface. From these viewpoints, they examined the surface relaxation behavior of high-density PMMA brushes with a highly distorted chain conformation. Figure 11.16 shows the temperature dependence of lateral forces measured at a scanning rate of 10^3 nm s^{-1} for a high-density PMMA brush and a spin-coated PMMA film. The α-relaxation process was clearly observed accompanying a small peak, which was assigned to a surface β-process. They commented that the surface molecular motion of the brush layer possibly differs from that of the spin-coated film but that it was rather difficult to conclude this because of slightly scattered data.

The effect of tethering and constraint on the glass transition of polymer brushes is expected to depend on the geometry of surfaces. Savin et al., who prepared PS brushes on an SiP with an average diameter of ca. 20 nm by surface-initiated ATRP, reported that by differential scanning calorimetry, the T_g of the brush sample with $M_n = 5230$ was 13 K higher than the ungrafted polymer with nearly the same molecular weight, while the T_g difference reduces to ca. 2 K for the sample with $M_n = 32\,670$ [143]. These results suggest that the effect of conformational constraint was mitigated for segments residing further away from the immobilized surface, in contrast to the case with a flat surface.

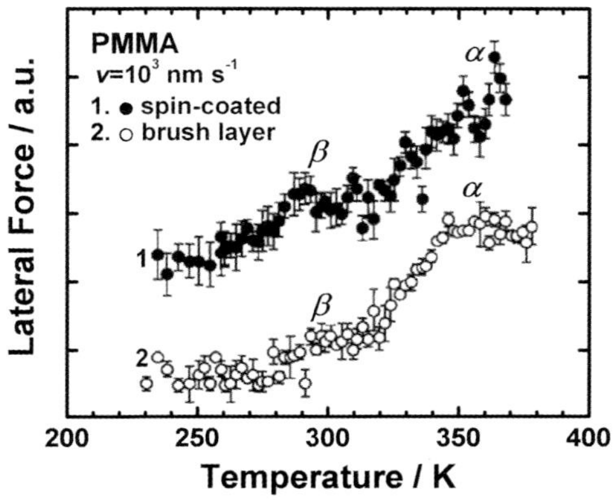

Fig. 11.16 Typical lateral force–temperature curves for the spin-coated film and the brush layer of PMMA at a scanning rate of 10^{-3} nm s^{-1}. Reproduced with permission from [142] (Copyright 2003 The Society of Polymer Science, Japan).

11.3.2.2 **Mechanical Properties**

The anisotropic structure of concentrated brushes is also reflected on their elastic properties [144]. Changes in the thickness of a graft film induced by an applied electric field (electrostriction) were measured with a Nomarski optical interferometer as a function of temperature [145–147]. The analysis of the electromechanical and dielectric data yields the plate compressibility (κ_p) of the brushes in the glassy and molten states (Fig. 11.17). Comparison of the results for concentrated PMMA brushes and a reference (spin-coated) PMMA layer will reveal the differences in elastic and dielectric properties between the brush composed of highly stretched graft chains and the layer formed by equivalent chains of random-coil conformation. In the glassy state, there was no appreciable difference in κ_p between the brush and the spin-coated layer, whereas in the molten state, κ_p of the brush was markedly (ca. 30–40%) lower than that of the spin-coated layer. This proves that the molten concentrated PMMA brush is more resistant against compression than the equivalent PMMA melt. An attempt was made to interpret the low compressibility of the molten concentrated brushes in terms of a rubber elasticity theory of a stretched polymer network with entanglements.

11.3.2.3 **Miscibility with Polymer Matrix**

Another example showing the unique properties of concentrated brushes concerns the miscibility of a polymer brush with a chemically identical polymer matrix [148]. Neutron reflectometry was applied to a series of deuterated PMMA ($PMMA_d$) brushes with a constant chain length (M_n=46 000, M_w/M_n=1.08) and different graft densities ($\sigma \approx 0.7$ and 0.06 chains nm^{-2}), on which hydrogenated PMMA ($PMMA_h$) with various molecular weights (2400<$M_{n,cast}$<780 000) was spin-coated. After annealing at 150 °C for 5 days in vacuum, the neutron reflectivity (NR) data were collected at room temperature and analyzed to elucidate

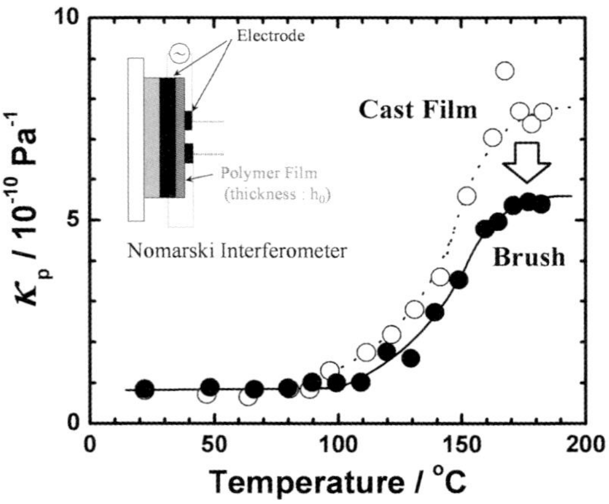

Fig. 11.17 Plate compressibility (κ_p) as a function of temperature (T) for the PMMA brushes G_1–G_4 and the spin-coated PMMA layer (M_n=106 000, M_w/M_n=1.16, L_d=110 nm) [144]. The modulation frequency is 9930 Hz.

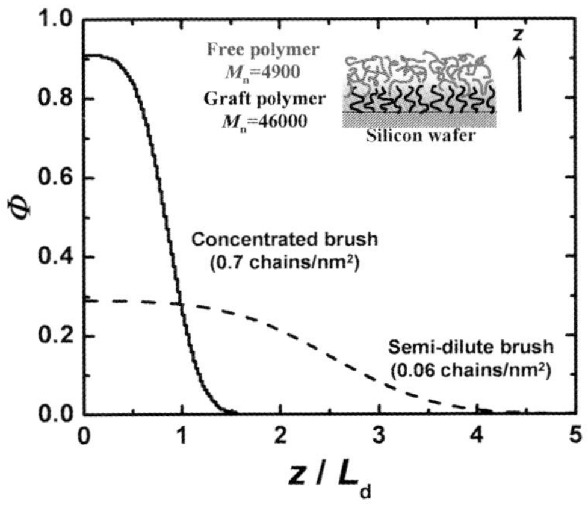

Fig. 11.18 Density profiles of polymer brushes faced with chemically identical polymer matrix. The polymer brushes have a constant chain length (M_n=46 000, M_w/M_n=1.08) and different graft densities ($\sigma \approx 0.7$ and 0.06 chains nm^{-2}) and the free polymer has M_n=4910 and M_w/M_n=1.1.

the brush concentration profile as a function of the distance (z) from the substrate surface. A representative result is shown in Fig. 11.18, where the brush fraction Φ is plotted against z/L_d. The figure clearly suggests that the miscibility between brush and free polymer depends strongly on graft density. The low-density (semi-dilute) polymer brush (M_n=46 000, σ=0.06 chains nm^{-2}) was swollen by an oligomeric PMMA (M_n=4910) to a thickness about four times that of the dry brush. On the other hand, the concentrated brush (M_n=46 000, σ=0.7 chains nm^{-2}) was hardly swollen by the same oligomeric PMMA, maintaining its "dry brush" structure. Similarly to the size exclusion of swollen brushes, this phenomenon was reasonably ascribed to a (conformational) entropic origin, which had been theoretically predicted [149] but never experimentally verified before.

11.3.3
Applications of Polymer Brushes

11.3.3.1 Brushes with Functional Polymers

The simplicity and versatility of LRP enable us to graft densely a variety of well-defined functional polymers, which include glycopolymers, polyelectrolytes, polymacromonomers, hyperbranched polymers and cross-linked polymers [111, 150–155]. Here, we focus on polyelectrolyte brushes, which have attracted considerable interest because the electrostatic interactions bring about new physical properties different from non-charged polymer brushes. Rühe et al. reviewed recent progress on the theory, synthesis and properties of a variety of polyelectrolyte brush systems including cylindrical (bottle) brushes and surface brushes at solid substrates [156]. Especially, they discussed the swelling behavior of "strong" and "weak" polyelectrolyte brushes as a function of external conditions such as the pH value and the ionic strength of the surrounding medium and the addition of multivalent ions. These brushes were mostly prepared on solid surfaces by the grafting-from method via conventional radical polymerization and the graft density might be classified in the semi-dilute regime. A weak polyelectrolyte brush gave a markedly different behavior from a strong polyelectrolyte brush, because the former charge density was not constant but changed as a function of the local concentration of protons inside the brush. One of the most notable features is a salt-induced swelling of poly(methacrylic acid) (PMAA) brush at relatively low concentrations of salt. The local pH value in the brush is governed by the requirement of charge neutrality. An added salt produces a countercation, which promotes the dissociation with keeping charge neutrality by exchange with protons in the brush. Theoretical considerations predicted that the charge neutralization would result in a difference in the pH values inside and outside the brush. The difference will be so large, especially for a brush with a high local concentration of ionizable groups, as to give a large shift in the critical pH value, beyond which brushes are charged and swollen. This effect is also called the "charge regulation" [157]. Rühe et al. [156] reported that the PMAA brushes synthesized by surface-initiated conventional radical polymerization showed a critical pH at around 4–5, slightly higher than the pK_a value of carboxylic acid groups.

Recently, surface-initiated LRP was applied to synthesize well-defined, concentrated polyelectrolyte brushes. Here, we discuss the swelling behavior of a concentrated weak polyelectrolyte brush to demonstrate that dense grafting can produce a unique property. A concentrated PMAA brush was prepared on a silicon wafer by the surface-initiated ATRP of 1-propoxyethyl methacrylate (PEMA) followed by deprotection (Fig. 11.19) [158]. The analysis of the simultaneously formed free polymers suggested that the graft polymerization proceeded in a controlled fashion to give a well-defined concentrated polymer brush with a graft density about 0.4 chains nm^{-2}. FT-IR analysis revealed that the hemiacetal ester group of poly(PEMA) grafts was quantitatively deprotected by heating at 120 °C in *p*-xylene containing zinc 2-ethylhexanoate as a catalyst, giving a concentrated polyelectrolyte brush. When glycidyl methacrylate (GMA) was copolymerized with PEMA, the deprotection of PEMA induced cross-linking between

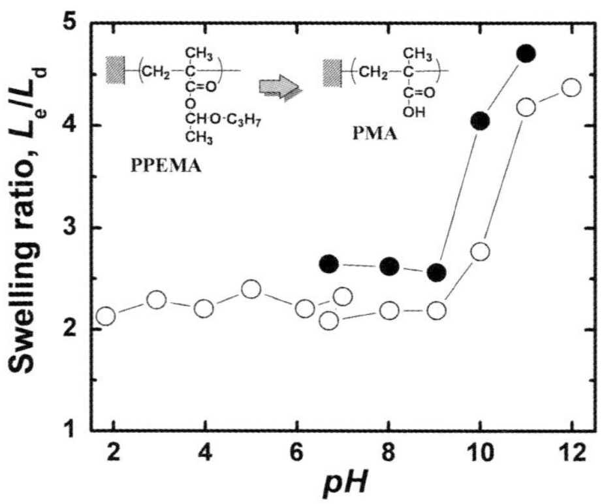

Fig. 11.19 Plots of swelling ratio of polyelectrolyte brushes with (closed circles) and without cross-links (open circles) vs. pH of aqueous solution.

graft chains by the reaction of the liberated carboxylic acid with the epoxide group of GMA. The swelling behavior of the polyelectrolyte brushes with and without cross-links was studied in an aqueous solution with different acidity. Figure 11.19 shows the swelling ratio of these brushes as a function of pH. The ratio increased steeply at around pH 10, nearly up to the maximum possible value corresponding to the fully-stretched chain length. This means that the apparent pK_a was shifted to about 10, a value much higher than the original value ($pK_a = 4–5$) of the carboxylic acid. The successful preparation of the concentrated polyelectrolyte brush has thus enabled us to make the first observation of such a large pK_a shift. The high density also brought about the almost fully stretched conformation of the graft chain. The cross-linking introduced by the copolymerization of 3 mol% GMA enhanced the chemical stability of the graft layer even in strong acidic/basic conditions.

11.3.3.2 **Morphological Control**

Artificially designed fine patterning of polymer films is often demanded in various fields of science and technology such as those related to microelectronics and functional sensor devices. For this purpose, a polymer resist layer is usually spin-coated on a substrate and patterned by lithographic techniques. However, such a polymer film has limited applicability as a functional surface because of its insufficient stability against temperature, solvents and mechanical forces. Thus, a number of different approaches have been made to fabricate a patterned polymer layer that is stable even in a wet system. For example, Rühe et al. prepared a patterned graft layer by selectively photoinitiating polymerization from an azo compound chemically immobilized on a surface via a conventional free radical polymerization process. The surface-initiated LRP can make a concentrated polymer brush grow on a patterned surface of initiator by the combined use of a variety of lithographic techniques [159–170]. The morphology of the grafted surface is also an important factor determining such surface properties as chemical reactivity, wettability, permeability, lubricity, biocompatibility and electrical properties.

Surface-initiated LRP technique makes it possible to control precisely and widely the structural parameters of polymer brushes including not only the chain length and chain length distribution but also the monomer sequence along the graft chain. The grafting of block or gradient copolymers will give a layered or gradient structure even in a solvent. This will give a new route to precise and effective tuning of surface properties, since such a structure is expected to be stabilized not only by the incompatibility of different polymer segments but also by the highly stretched conformation. Block copolymer brush surfaces have attracted much attention for their stimulus-responsive properties [69, 171–180], forming a variety of characteristic surface morphologies caused by the phase separation on a nanometer scale.

Another strategy to control the surface properties and morphologies by self-assembly of polymer chains is to graft randomly different kinds of homopolymers on a surface. Such a "mixed homopolymer brush" has been extensively investigated both theoretically and experimentally. It is predicted that different morphologies can be induced by lateral and perpendicular segregation in wet and dry states, depending on the structural parameters of brushes, incompatibility of the two polymers, solvent quality and so on. Experimentally, mixed homopolymer brushes were realized by the grafting-to and grafting-from techniques. Sidorenko et al. first synthesized mixed homopolymer brushes of PS and P2VP by the grafting-from technique via conventional radical polymerization on an azo initiator-immobilized silicon wafer and studied their morphologies and switching properties by treatment with selective and non-selective solvents. Recently, surface-initiated LRP was also applied to control precisely the structural parameters of mixed polymer brushes. PS and PMMA were randomly grafted by the combined use of different LRPs, ATRP of MMA followed by NMP of S [181–183]. These two LRP techniques are based on different activation mechanisms. Therefore, it is possible to carry out these two LRPs independently and selectively by controlling the temperature. Zhao and coworkers immobilized a mixed SAM of two kinds of silane coupling agents with ATRP and NMP initiators and prepared mixed brushes consisting of PMMA with $M_n = 26\,200$ and PS with various molecular weights. They measured the water contact angles on these brushes after treatment with dichloromethane and observed a transition with increasing PS molecular weight. Ejaz et al. synthesized a series of PS–PMMA mixed brushes with different compositions and chain lengths by a similar system except that a three-component SAM containing an inactive silane coupling agent was immobilized. The addition of the inactive species enabled them to control more precisely the graft densities of each component in a wider range. As shown in Fig. 11.20, the topographic AFM studies revealed that PMMA–PS mixed homopolymer brushes with a nearly equal total graft density gave, after treatment with acetone, which is a selectively good solvent for PMMA, characteristic morphologies depending on the fraction of each component. With increasing PS fraction, the surface morphology changed from a circular-domain structure to a honeycomb one via a lamellar-like one for a symmetrical brush. Depending on solvent quality, the morphology was dramatically changed: for example, the treatment of the mixed homopolymer brush ($\sigma_{PS} = 0.08$ chains nm^{-2} and $\sigma_{PMMA} = 0.17$ chains nm^{-2}) with acetone, a selectively good solvent for PMMA, gave the circular-domain structure, whereas treatment with cyclohexane, a selectively good solvent for PS, converted the surface morphology to a honeycomb structure. The height contrast was reversed depending on the selectivity of solvents. On the other hand, no characteristic morphology was observed after treatment with dichloromethane, a good solvent for both PMMA and PS. These morphological changes were reproducible and quite similar to those reported previously by Minko et al. Another approach was made by Zhao et al., who newly synthesized an asymmetric bifunctional silane-coupling agent with both ATRP and NMP initiators to avoid possible preferential adsorption and cluster formation in the SAM of a multicomponent system. They prepared a series of PS–PMMA brushes of PMMA with a fixed M_n and PS with systematically changed M_n and observed relatively ordered nanoscale domains after treatment with acetic acid.

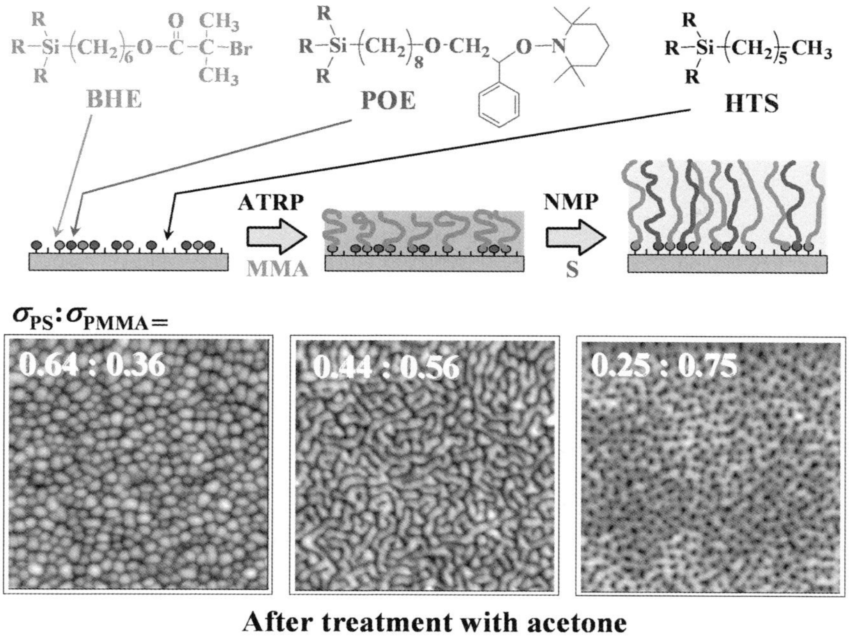

Fig. 11.20 Topographic AFM images of PMMA–PS mixed homopolymer brushes with different ratios of σ_{PMMA} and σ_{PS} but nearly equal $\sigma_{PMMA}+\sigma_{PS}$ ($\sim$0.5 chains nm^{-2}) after treatment with acetone, where σ_{PMMA} and σ_{PS} are the graft densities of PMMA and PS, respectively. The inset schematically illustrates the sequential growth of PMMA and PS by surface-initiated ATRP and NMP on a mixed SAM of BHE, POE and HTS.

11.3.3.3 **Novel Biointerfaces**

Biointerfaces to tune interactions of solid surfaces with biologically important materials are among the most interesting applications of polymer brushes. For example, proteins will adsorb on surfaces through non-specific interactions, often triggering a bio-fouling, for example, the deposition of biological cells, bacteria and so on. To prevent protein adsorption, attempts have been made to modify surfaces with polymer brushes. The interactions between proteins and brush-coated surfaces can be modeled by the three generic modes illustrated in Fig. 11.21 a. One is the primary adsorption, in which a protein diffuses into the brush and adsorbs on the substrate surface. The secondary adsorption is the one occurring at the outermost surface of the swollen brush film. The tertiary adsorption is caused by the interaction of protein with the polymer segments within the brush layer. For relatively small proteins, the primary and tertiary adsorptions would be particularly important, but they should become less important with increasing protein size and increasing σ, since a larger protein would be more difficult to diffuse against the concentration gradient formed by the polymer brush and this gradient, clearly, is a function of graft density. The size and density dependence of protein adsorption would manifest themselves much

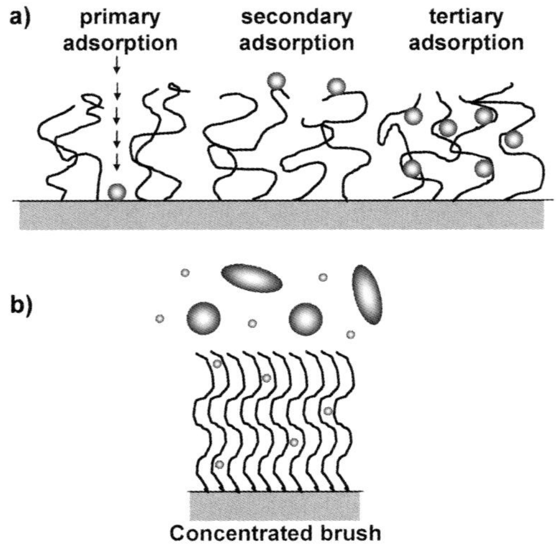

Fig. 11.21 Schematic illustration of (a) possible interactions of probe molecules with a polymer brush and (b) size-exclusion effect of concentrated brush. After Curie et al. [184], with some modifications.

more clearly for concentrated brushes due to the size-exclusion effect discussed in Section 11.3.1.3.

In the past, poly(ethylene oxide) (PEO)-grafted surfaces were extensively studied to discuss their protein adsorption behavior [184]. Since those surfaces were prepared by physisorption of block copolymers or comb-like polymers using self-assembly or the Langmuir–Blodgett technique, the graft density should have been in a semi-dilute regime. To increase graft density further, surface-initiated LRP has been applied. So far, well-defined brushes of PAAm [185], poly[2-(dimethylamino)ethyl methacrylate)] [186], poly[oligo(ethylene glycol)methyl methacrylate] [111], poly(2-methacryloxyethylphosphorylcholine) (PMPC) [187–189] and poly(2-hydroxyethyl methacrylate) (PHEMA) [190, 191] have been synthesized and observed to show good adsorption resistance for proteins and cells. From a mechanistic point of view, Yoshikawa and coworkers [190, 191] prepared a series of PHEMA brushes with different graft densities, demonstrating that the size-exclusion effect of concentrated brushes (see Section 11.3.1.3) plays an essential role in the biocompatibility of brush-modified surfaces.

11.4 Polymer Brushes on Fine Particles

11.4.1 Preparation

Silica particle (SiP) is among the most extensively studied particles for the application of surface-initiated ATRP [72–80]. Various ATRP initiators that can be fixed on silicon oxide surfaces have been synthesized by various groups [60, 61,

63, 65, 72–75, 77, 78, 96, 97, 143, 153, 167, 192–194]. All of them were mono- or trichlorosilane derivatives, except the monoethoxysilane derivative synthesized by Patten et al. [72, 73]. In a search for a better route to the modification of SiP by surface-initiated ATRP, Ohno et al. synthesized a new triethoxysilane derivative to introduce ATRP initiation sites on to SiP surfaces without causing any aggregation of the particles [76]. Even though the chlorosilane group has the advantage of high reactivity, the reaction with it must be carried out in a dried aprotic solvent to prevent unfavorable side-reactions. On the other hand, the reaction with the triethoxysilane group can be carried out in a protic solvent, *even in the presence of water.* This is an essential factor for achieving a homogeneous modification of SiP surface because SiP exhibits higher dispersibility in protic than in aprotic solvents due to the hydrophilic character of the SiP surface with silanol groups. In addition, the reaction with triethoxysilane group forms a stable Si–O–Si network via the trivalent reaction. The triethoxysilane derivative with an initiating site for ATRP was synthesized via a two-step reaction: in short, 5-hexen-1-ol was acylated with 2-bromoisobutyryl bromide to obtain 1-(2-bromo-2-methyl)propionyloxy-5-hexene (BPH), the aryl group of which was subsequently hydrosilylated with triethoxysilane in the presence of Karstedt's catalyst to give the final product (2-bromo-2-methyl)propionyloxyhexyltriethoxysilane (BHE). The ATRP initiator BHE was fixed on the SiP surface in ethanol solution with ammonia (NH_3) added as an alkaline catalyst [76]. The NH_3 concentration is a key parameter in this reaction: if it is too high, it will cause particle aggregation due to the high ionic strength, whereas if it is too low, it will make the reaction too slow. The final NH_3 concentration was optimized as 1 M.

The initiator-coated SiP was subsequently used for the copper-mediated ATRP of methyl methacrylate (MMA) in bulk [76]. To obtain a satisfactory result, the following points were important. First, the initiator-coated SiP should never be dried before the polymerization. Once dried, the particles are difficult to be homogeneously redispersed in the polymerization medium even with the aid of ultrasonication. Second, the polymerization should be carried out in the presence of the "sacrificial" free initiator ethyl 2-bromoisobutylate (EBiB). Free polymers produced by EBiB will produce an entangled network structure, through which the particles can hardly diffuse. This would prevent interparticle coupling causing gelation or particle aggregation. {Another role of the free initiator is, of course, to accumulate an appropriate amount of Cu(II) species via the termination of polymer radicals, and thus to control the polymerization by the so-called persistent radical effect [58, 101, 155].} The polymerization proceeded in a living manner, producing SiPs grafted with well-defined PMMA of a target molecular weight up to about 500 000 with a graft density as high as 0.65 chains nm^{-2}. DLS measurements showed that the particles retained their high dispersibility throughout the experimental processes, that is, before and after the initiator fixation and also before and after the graft polymerization.

11.4.2
Two-dimensional Ordered Arrays

Monodisperse SiPs grafted with a concentrated PMMA brush (PMMA–SiP) synthesized by the above-mentioned method showed exceptionally good dispersibility in organic solvents. When a defined amount of suspension of PMMA–SiP in toluene was deposited and spread on the surface of purified water, a surface film of the PMMA–SiP was formed at the air–water interface as the toluene was evaporated off [195]. The film suffered some cracks during its formation for some kind of stress and the mean size of the cracked pieces of film was about 5 cm^2. Fig. 11.22 a shows the TEM image of the transferred film of the PMMA–SiP. The SiP cores visible as dark circles are uniformly dispersed throughout the film with a constant interparticle distance, whereas the PMMA chains, which should be forming fringes surrounding the SiP cores, are hardly visible because of their much lower electron density. This technique for the fabrication of PMMA–SiP monolayers with a two-dimensional lattice structure allows us to control the mean nearest-neighbor center-to-center distance D_p by simply changing the length of PMMA graft chain. For example, Fig. 11.22 b shows that by increasing the M_n of the PMMA grafts by 4.4 times for the same SiP core as in Fig. 11.22 a, the D_p value increases about 1.5 times.

To investigate the correlation between the M_n value of the PMMA graft chain and the D_p value in more detail, the same experiments were carried out using a series of PMMA–SiP samples with a fixed diameter (290 nm) of the SiP core, nearly the same graft density and varying lengths of the PMMA graft chain. The D_p value increased with increasing M_n value of the graft, but the increment became smaller as M_n increased. These data were compared with the D_p values

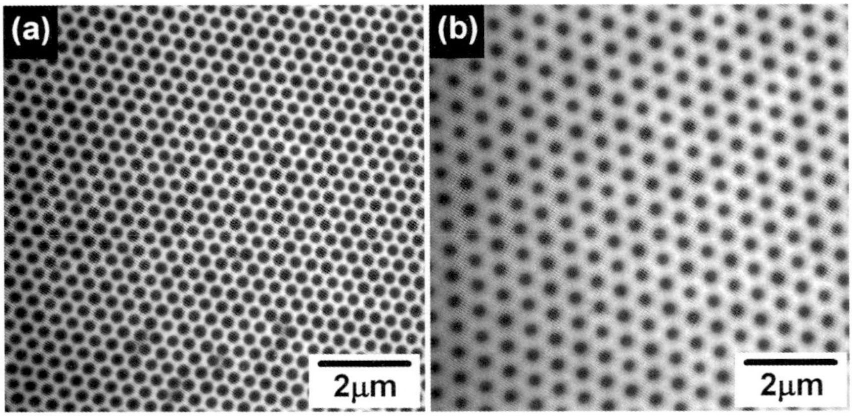

Fig. 11.22 Transmission electron microscopic images of the transferred films of PMMA–SiPs originally formed on the air–water interface. The diameter of silica particle core is 290 nm. The M_n of the PMMA grafts are (a) 69 000 and (b) 304 000.

calculated for the hybrid particle model consisting of an SiP core and a PMMA shell of the bulk density, showing that the PMMA grafts of the hybrid particle form a compact shell of its bulk density in the surface film.

The optimization of the concentration and amount of the suspension of PMMA–SiP to be deposited on water surface was necessary to obtain homogeneous surface monolayers such as shown above. When the concentration was too high or the amount of particles was too large, an inhomogeneous, partially multilayered surface film was obtained. When the concentration was too low or the amount of particles was too small, no good film formation was observed.

The surface morphologies of the PMMA–SiP monolayer were then used as a master to produce a patterned surface on a PDMS elastomer [195]. According to a standard procedure of replica molding, a liquid prepolymer of PDMS elastomer (Sylgard 184, Dow Corning) was cast on the PMMA–SiP monolayer on the silicon wafer and, after thermal treatment, the PDMS film was peeled off from the master. An AFM observation of the PDMS surface showed the successful formation of a negative pattern reflecting the shape and morphology of the master surface.

Hybrid particles with a core of gold nanoparticles (AuNP) and a shell of PMMA brush, deposited on the water surface from a dilute benzene suspension and compressed to a defined surface pressure in a Langmuir trough, also formed a two-dimensional ordered array. In this system, the PMMA grafts were highly extended, unlike the PMMA–SiP system [84]. The mechanism by which the PMMA grafts swollen in a good solvent become contracted to their bulk density (as in the PMMA–SiP system in toluene) or stay more or less in the same, highly extended conformation (as in the PMMA–AuNP on the water surface) remains to be explored.

11.4.3 Colloidal Crystals

As already described, a concentrated PMMA brush formed on a silicon wafer is swollen in a good solvent to give a film thickness as large as 80–90% of the full contour length of the PMMA grafts, suggesting that the chains are extended nearly as highly and the frictional coefficient between swollen concentrated brushes is extremely small even at high pressures, suggesting that concentrated brushes would hardly interpenetrate each other. These surprising phenomena led Ohno et al. [95] to postulate that concentrated brushes formed on spherical particles would also be highly swollen in a good solvent, exerting a (non-penetrating) quasi-hard-sphere potential of long range between particles. If so, such hybrid particles dispersed in a good solvent for the polymer brush can form a colloidal crystal in a concentration regime dependent on the chain length and density of the grafts. Of course, the effective graft density and hence the mentioned concentrated brush effects would decrease with increasing radial distance or increasing graft chain length relative to the particle radius. In other words, the interparticle potential in this system can be tuned, from a quasi-hard type of concentrated brush through a "soft" type of randomly coiled graft chains, by changing graft density and chain length. Such a system should be new.

In fact, a colloidal crystal was newly identified for a liquid suspension of the hybrid particles (PMMA–SiPs) having a core of monodisperse SiP and a shell of PMMA concentrated brush [95]. With increasing particle concentration, the suspension progressed from a (disordered) fluid to a fully crystallized system, going through a narrow crystal–fluid coexisting regime. This spontaneous phase transition could be interpreted by the idea of a Kirkwood–Alder transition [196, 197]. The PMMA–SiP suspension was subjected to confocal laser scanning microscopic (CLSM) measurements to observe the crystal structure *in situ*. Observations were made on an inverted-type CLSM with a 458-nm wavelength Ar laser and 63× objective in reflection mode. Figure 11.23 a shows a CLSM image of a two-dimensional slice in the sample bulk. The SiP cores of the hybrid particles are clearly visible as white circles forming a two-dimensional hexagonal array, while the PMMA brushes that should be surrounding the SiP cores are hardly visible because of their much lower reflectivity. The mean nearest-neighbor center-to-center distance measured 560 nm. This value, while close to the hydrodynamic diameter of 520 nm, was much larger than the diameter of 260 nm of the "compact core-shell model" [84, 95], which consisted of an SiP core and a PMMA shell of the bulk density, and was as large as about 52% of the diameter of 1070 nm of the "fully stretched core-shell model" [84, 95], which consisted of an SiP core and a PMMA shell whose size was equal to that of the PMMA chains radially stretched in an all-trans conformation. This meant that the PMMA shell was not compact but had a surprisingly large extension in the radial direction, even though the degree of extension of the graft chain would be less significant in the periphery of the shell than in the vicinity of the core, since the net graft density should decrease radially. A large advantage of this system was that the interparticle distance could be controlled by changing the length of graft chain. Figure 11.23 b shows that by grafting

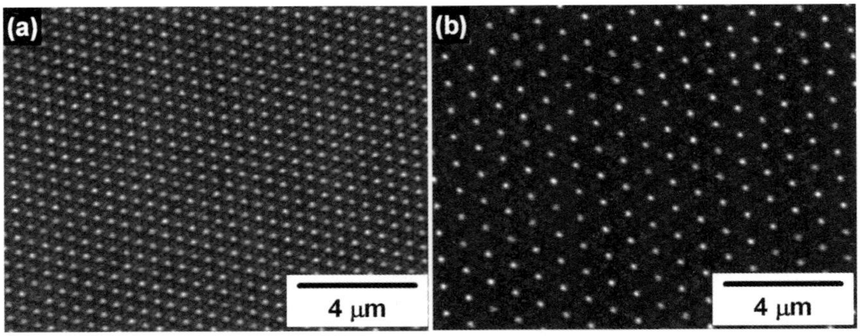

Fig. 11.23 Confocal laser scanning microscopic images of PMMA–SiP crystals. Observations were performed using an Ar laser of wavelength 488 nm and 63× objective in reflection mode. The distance of the focal plane from the inside of the cover slip was 100 μm. The diameter of SiP core is 130 nm. The M_n pf PMMA grafts were (a) 158000 and (b) 432000. The mean nearest-neighboring center-to-center distances were (a) 560 and (b) 950 nm.

PMMA of M_n=432 000 (about 2.7 times the M_n for the Fig. 11.23 a sample), the D_p value doubled, reaching nearly 1 μm. A three-dimensional CLSM image of the crystal showed that the crystal is characterized by a face-centered cubic structure.

Unlike the previously known *soft* and *hard* colloidal crystals, which can be achieved by electrostatic long-range interaction and by steric short-range interaction, respectively, the driving force of crystallization in the above-mentioned system was a long-range repulsive (non-interpenetrating) interaction between the highly swollen concentrated brush layers. The crystallization concentration was situated between those of typical *soft* and *hard* systems. Hence it was a new type of colloidal crystal and termed the "*semi-soft* colloidal crystal" by the authors. The main advantages of this new system over the conventional ones include: (1) controllability of the interparticle distance and interaction by controlling the graft chain length and density; (2) applicability to various monomers to have various graft polymers (e.g. hydrophilic, hydrophobic and electrolytic polymers); (3) applicability to various particles (organic, inorganic and metallic particles); and (4) usability of various solvents and solvent mixtures to adjust refractive index, light and X-ray absorbance and density. Advantage (2) also suggests the possibility of the fixation of the colloidal crystal by, for example, cross-linking all or part of the graft layers via the cross-linkable units introduced in the graft chain. These advantages will offer unique possibilities to fundamental and applied research on colloidal crystals.

11.5 Conclusion

The recent successful application of living polymerizations has made it possible to modify solid surfaces with various end-grafted polymers of well-defined structure, including simple homopolymers, end-functionalized polymers, block/random/gradient copolymers, blend polymers and functional polymers. In particular, surface-initiated controlled/living radical polymerization (SILRP) has brought about a dramatic increase in graft density, allowing us to investigate deep into the concentrated brush regime of graft density. Graft chains in this regime were found to be highly extended in good solvents, even to the order of their full lengths. Moreover, these concentrated brushes have highly characteristic properties, in both swollen and dry states, very different from those of low-density or semi-dilute polymer brushes and even unpredictable. In this regard, SILRP and the concentrated polymer brushes thereby obtained are opening up a versatile route to the creation of new surfaces that have never been realized by the conventional techniques of surface modifications. Hopefully, this progress will lead to a wealth of applications.

The highest graft density that has been reached by SILRP is no higher than about 40% in reduced graft density σ^*. There is no clear theoretical reason why this value must be the limiting density. A further increase in graft density will certainly bring about dramatic changes in brush properties, for example those of dry brushes. We anticipate much progress in the synthesis of polymer brushes also.

References

1 D.H. Napper, *Polymeric Stabilization of Colloidal Dispersions*. Academic Press, London, **1983**.

2 E. Raphael, P.G. Degennes, *J. Phys. Chem.* **1992**, *96*, 4002–4007.

3 J. Klein, *Annu. Rev. Mater. Sci.* **1996**, *26*, 581–612.

4 J. Klein, E. Kumacheva, *Science* **1995**, *269*, 816–819.

5 R.S. Parnas, Y. Cohen, *Rheol. Acta* **1994**, *33*, 485–505.

6 J.N. Israelachvili, *Intermolecular and Surface Forces*, 2nd edn. Academic Press, London, **1992.**

7 A. Halperin, M. Tirrell, T.P. Lodge, *Adv. Polym. Sci.* **1992**, *100*, 31–71.

8 M. Kawaguchi, A. Takahashi, *Adv. Colloid Interface Sci.* **1992**, *37*, 219–317.

9 S. Alexander, *J. Phys. (Paris)* **1977**, *38*, 983–987.

10 P.G. de Gennes, *Macromolecules* **1980**, *13*, 1069–1075.

11 M. Daoud, J.P. Cotton, *J. Phys. (Paris)* **1982**, *43*, 531–538.

12 T.M. Birstein, E.B. Zhulina, *Polymer* **1984**, *25*, 1453.

13 H.J. Taunton, C. Toprakcioglu, L.J. Fetters, J. Klein, *Macromolecules* **1990**, *23*, 571–580.

14 A. Courvoisier, F. Isel, J. Francois, M. Maaloum, *Langmuir* **1998**, *14*, 3727–3729.

15 S.K. Satija, C.F. Majkrzak, T.P. Russell, S.K. Sinha, E.B. Sirota, G.J. Hughes, *Macromolecules* **1990**, *23*, 3860–3864.

16 R. Levicky, N. Koneripalli, M. Tirrell, S.K. Satija, *Macromolecules* **1998**, *31*, 3731–3734.

17 T. Cosgrove, T.G. Heath, J.S. Phipps, R.M. Richardson, *Macromolecules* **1991**, *24*, 94–98.

18 J.B. Field, C. Toprakcioglu, R.C. Ball, H.B. Stanley, L. Dai, W. Barford, J. Penfold, G. Smith, W. Hamilton, *Macromolecules* **1992**, *25*, 434–439.

19 D.L. Anastassopoulos, A.A. Vradis, C. Toprakcioglu, G.S. Smith, L. Dai, *Macromolecules* **1998**, *31*, 9369–9371.

20 G. Hadziioannou, S. Patel, S. Granick, M. Tirrell, *J. Am. Chem. Soc.* **1986**, *108*, 2869–2876.

21 H. Watanabe, M. Tirrell, *Macromolecules* **1993**, *26*, 6455–6466.

22 M.A. Ansarifar, P.F. Luckham, *Polymer* **1988**, *29*, 329–335.

23 T.W. Kelley, P.A. Schorr, K.D. Johnson, M. Tirrell, C.D. Frisbie, *Macromolecules* **1998**, *31*, 4297–4300.

24 S.J. Oshea, M.E. Welland, T. Rayment, *Langmuir* **1993**, *9*, 1826–1835.

25 D.F.K. Shim, M.E. Cates, *J. Phys. (Paris)* **1989**, *50*, 3535–3551.

26 P.Y. Lai, A. Halperin, *Macromolecules* **1991**, *24*, 4981–4982.

27 B. Zhao, W.J. Brittain, *Prog. Polym. Sci.* **2000**, *25*, 677–710.

28 S. Edmondson, V.L. Osborne, W.T.S. Huck, *Chem. Soc. Rev.* **2004**, *33*, 14–22.

29 R.C. Advincula, W.J. Brittain, K.C. Caster, J. Ruehe (Eds) *Polymer Brushes*. Wiley-VCH, Weinheim, **2004**.

30 R. Jordan, *Adv. Polym. Sci.* **2006**, *197*, 1–171 and *198*, 1–183.

31 Y. Tsujii, K. Ohno, S. Yamamoto, A. Goto, T. Fukuda, *Adv. Polym. Sci.* **2006**, *197*, 1–45.

32 R. Jordan, A. Ulman, J.F. Kang, M.H. Rafailovich, J. Sokolov, *J. Am. Chem. Soc.* **1999**, *121*, 1016–1022.

33 R. Advincula, Q.G. Zhou, M. Park, S.G. Wang, J. Mays, G. Sakellariou, S. Pispas, N. Hadjichristidis, *Langmuir* **2002**, *18*, 8672–8684.

34 B. Zhao, W.J. Brittain, *Macromolecules* **2000**, *33*, 342–348.

35 I.J. Kim, R. Faust, *J. Macromol. Sci. Pure Appl. Chem.* **2003**, *A40*, 991–1008.

36 R. Jordan, A. Ulman, *J. Am. Chem. Soc.* **1998**, *120*, 243–247.

37 M. Husseman, D. Mecerreyes, C.J. Hawker, J.L. Hedrick, R. Shah, N.L. Abbott, *Angew. Chem. Int. Ed.* **1999**, *38*, 647–649.

38 I.S. Choi, R. Langer, *Macromolecules* **2001**, *34*, 5361–5363.

39 H. Menzel, P. Witte, in *Polymer Brushes*, Chapter 4, R.C. Advincula, W.J. Brittain, K.C. Caster, J. Ruehe (Eds.). Wiley-VCH, Weinheim, **2004**.

40 R.H. Wieringa, E.A. Siesling, P.F.M. Geurts, P.J. Werkman, E.J. Vorenkamp, V. Erb, M. Stamm, A.J. Schouten, *Langmuir* **2001**, *17*, 6477–6484.

41 N. L. Jeon, I. S. Choi, G. M. Whitesides, N. Y. Kim, P. E. Laibinis, Y. Harada, K. R. Finnie, G. S. Girolami, R. G. Nuzzo, *Appl. Phys. Lett.* **1999**, *75*, 4201–4203.

42 A. Juang, O. A. Scherman, R. H. Grubbs, N. S. Lewis, *Langmuir* **2001**, *17*, 1321–1323.

43 J. H. Moon, T. M. Swager, *Macromolecules* **2002**, *35*, 6086–6089.

44 E. H. Solomon, E. Rizzardo, P. Cacioli, *Eur. Pat. Appl. EP135280*, **1985**.

45 M. K. Georges, R. P. N. Veregin, P. M. Kazmaier, G. K. Hamer, *Macromolecules* **1993**, *26*, 2987–2988.

46 D. Benoit, V. Chaplinski, R. Braslau, C. J. Hawker, *J. Am. Chem. Soc.* **1999**, *121*, 3904–3920.

47 D. Benoit, S. Grimaldi, S. Robin, J. P. Finet, P. Tordo, Y. Gnanou, *J. Am. Chem. Soc.* **2000**, *122*, 5929–5939.

48 Y. Yutani, M. Tatemoto, *Eur. Pat. Appl. EP489370A1*, **1992**.

49 K. Matyjaszewski, S. Gaynor, J. S. Wang, *Macromolecules* **1995**, *28*, 2093–2095.

50 M. Kato, M. Kamigaito, M. Sawamoto, T. Higashimura, *Macromolecules* **1995**, *28*, 1721–1723.

51 J. S. Wang, K. Matyjaszewski, *J. Am. Chem. Soc.* **1995**, *117*, 5614–5615.

52 T. Otsu, M. Yoshida, *Makromol. Chem. Rapid Commun.* **1982**, *3*, 127–132.

53 J. Krstina, G. Moad, E. Rizzardo, C. L. Winzor, C. T. Berge, M. Fryd, *Macromolecules* **1995**, *28*, 5381–5385.

54 J. Chiefari, Y. K. Chong, F. Ercole, J. Krstina, J. Jeffery, T. P. T. Le, R. T. A. Mayadunne, G. F. Meijs, C. L. Moad, G. Moad, E. Rizzardo, S. H. Thang, *Macromolecules* **1998**, *31*, 5559–5562.

55 S. Yamago, K. Iida, J. Yoshida, *J. Am. Chem. Soc.* **2002**, *124*, 2874–2875.

56 J. Pyun, T. Kowalewski, K. Matyjaszewski, *Macromol. Rapid Commun.* **2003**, *24*, 1043–1059.

57 C. J. Hawker, A. W. Bosman, E. Harth, *Chem. Rev.* **2001**, *101*, 3661–3688.

58 M. Ejaz, S. Yamamoto, K. Ohno, Y. Tsujii, T. Fukuda, *Macromolecules* **1998**, *31*, 5934–5936.

59 W. X. Huang, M. J. Wirth, *Macromolecules* **1999**, *32*, 1694–1696.

60 K. Matyjaszewski, P. J. Miller, N. Shukla, B. Immaraporn, A. Gelman, B. B. Luokala, T. M. Siclovan, G. Kickelbick, T. Vallant, H. Hoffmann, T. Pakula, *Macromolecules* **1999**, *32*, 8716–8724.

61 M. Husseman, E. E. Malmstrom, M. McNamara, M. Mate, D. Mecerreyes, D. G. Benoit, J. L. Hedrick, P. Mansky, E. Huang, T. P. Russell, C. J. Hawker, *Macromolecules* **1999**, *32*, 1424–1431.

62 D. Q. Xiao, M. J. Wirth, *Macromolecules* **2002**, *35*, 2919–2925.

63 J. D. Jeyaprakash, S. Samuel, R. Dhamodharan, J. Ruhe, *Macromol. Rapid Commun.* **2002**, *23*, 277–281.

64 A. Ramakrishnan, R. Dhamodharan, J. Ruhe, *Macromol. Rapid Commun.* **2002**, *23*, 612–616.

65 J. Parvole, G. Laruelle, C. Guimon, J. Francois, L. Billon, *Macromol. Rapid Commun.* **2003**, *24*, 1074–1078.

66 W. H. Yu, E. T. Kang, K. G. Neoh, S. P. Zhu, *J. Phys. Chem. B* **2003**, *107*, 10198–10205.

67 J. B. Kim, M. L. Bruening, G. L. Baker, *J. Am. Chem. Soc.* **2000**, *122*, 7616–7617.

68 J. B. Kim, W. X. Huang, M. D. Miller, G. L. Baker, M. L. Bruening, *J. Polym. Sci. Part A: Polym. Chem.* **2003**, *41*, 386–394.

69 W. X. Huang, J. B. Kim, M. L. Bruening, G. L. Baker, *Macromolecules* **2002**, *35*, 1175–1179.

70 D. M. Jones, A. A. Brown, W. T. S. Huck, *Langmuir* **2002**, *18*, 1265–1269.

71 S. M. Desai, S. S. Solanky, A. B. Mandale, K. Rathore, R. P. Singh, *Polymer* **2003**, *44*, 7645–7649.

72 T. von Werne, T. E. Patten, *J. Am. Chem. Soc.* **1999**, *121*, 7409–7410.

73 T. von Werne, T. E. Patten, *J. Am. Chem. Soc.* **2001**, *123*, 7497–7505.

74 J. Pyun, K. Matyjaszewski, T. Kowalewski, D. Savin, G. Patterson, G. Kickelbick, N. Huesing, *J. Am. Chem. Soc.* **2001**, *123*, 9445–9446.

75 J. Pyun, S. J. Jia, T. Kowalewski, G. D. Patterson, K. Matyjaszewski, *Macromolecules* **2003**, *36*, 5094–5104.

76 K. Ohno, T. Morinaga, K. Koh, Y. Tsujii, T. Fukuda, *Macromolecules* **2005**, *38*, 2137–2142.

77 G. Carrot, S. Diamanti, M. Manuszak, B. Charleux, I. P. Vairon, *J. Polym. Sci. Part A: Polym. Chem.* **2001**, *39*, 4294–4301.

78 H. Mori, D.C. Seng, M.F. Zhang, A.H.E. Muller, *Langmuir* **2002**, *18*, 3682–3693.
79 X.Y. Chen, S.R. Armes, *Adv. Mater.* **2003**, *15*, 1558–1562.
80 X.Y. Chen, S.P. Armes, S.J. Greaves, J.F. Watts, *Langmuir* **2004**, *20*, 587–595.
81 S. Nuss, H. Bottcher, H. Wurm, M.L. Hallensleben, *Angew. Chem. Int. Ed.* **2001**, *40*, 4016–4018.
82 T.K. Mandal, M.S. Fleming, D.R. Walt, *Nano Lett.* **2002**, *2*, 3–7.
83 K. Ohno, K. Koh, Y. Tsujii, T. Fukuda, *Macromolecules* **2002**, *35*, 8989–8993.
84 K. Ohno, K. Koh, Y. Tsujii, T. Fukuda, *Angew. Chem. Int. Ed.* **2003**, *42*, 2751–2754.
85 Y. Wang, X.W. Teng, J.S. Wang, H. Yang, *Nano Lett.* **2003**, *3*, 789–793.
86 E. Marutani, S. Yamamoto, T. Ninjbadgar, Y. Tsujii, T. Fukuda, M. Takano, *Polymer* **2004**, *45*, 2231–2235.
87 S.C. Farmer, T.E. Patten, *Chem. Mater.* **2001**, *13*, 3920–3926.
88 B. Gu, A. Sen, *Macromolecules* **2002**, *35*, 8913–8916.
89 T.Q. Liu, S. Jia, T. Kowalewski, K. Matyjaszewski, R. Casado-Portilla, J. Belmont, *Langmuir* **2003**, *19*, 6342–6345.
90 D. Holzinger, G. Kickelbick, *Chem. Mater.* **2003**, *15*, 4944–4948.
91 M.M. Guerrini, B. Charleux, J.P. Vairon, *Macromol. Rapid Commun.* **2000**, *21*, 669–674.
92 G.D. Zheng, H.D.H. Stover, *Macromolecules* **2002**, *35*, 7612–7619.
93 K.N. Jayachandran, A. Takacs-Cox, D.E. Brooks, *Macromolecules* **2002**, *35*, 4247–4257.
94 J.N. Kizhakkedathu, R. Norris-Jones, D.E. Brooks, *Macromolecules* **2004**, *37*, 734–743.
95 K. Ohno, T. Morinaga, S. Takeno, Y. Tsujii, T. Fukuda, *Macromolecules* **2006**, *39*, 1245–1249.
96 X.Y. Huang, M.J. Wirth, *Anal. Chem.* **1997**, *69*, 4577–4580.
97 X.Y. Huang, L.J. Doneski, M.J. Wirth, *Anal. Chem.* **1998**, *70*, 4023–4029.
98 S. Habaue, O. Ikeshima, H. Ajiro, Y. Okamoto, *Polym. J.* **2001**, *33*, 902–905.
99 A. Feldmann, U. Claussnitzer, M. Otto, *J. Chromatogr. B* **2004**, *803*, 149–157.
100 M. Ejaz, Y. Tsujii, T. Fukuda, *Polymer* **2001**, *42*, 6811–6815.
101 H. Fischer, *Chem. Rev.* **2001**, *101*, 3581–3610.
102 A. Goto, T. Fukuda, *Prog. Polym. Sci.* **2004**, *29*, 329–385.
103 T. Fukuda, *J. Polym. Sci. Part A: Polym. Chem.* **2004**, *42*, 4743–4755.
104 S. Blomberg, S. Ostberg, E. Harth, A.W. Bosman, B. Van Horn, C.J. Hawker, *J. Polym. Sci. Part A: Polym. Chem.* **2002**, *40*, 1309–1320.
105 S. Yamamoto, M. Ejaz, Y. Tsujii, M. Matsumoto, T. Fukuda, *Macromolecules* **2000**, *33*, 5602–5607.
106 S. Yamamoto, M. Ejaz, Y. Tsujii, T. Fukuda, *Macromolecules* **2000**, *33*, 5608–5612.
107 M. Baum, W.J. Brittain, *Macromolecules* **2002**, *35*, 610–615.
108 Y. Tsujii, unpublished work, **2005**.
109 S. Yamamoto, Y. Tsujii, T. Fukuda, *Macromolecules* **2000**, *33*, 5995–5998.
110 T. Kakiuchi, M. Iida, N. Gon, D. Hobara, S. Imabayashi, K. Niki, *Langmuir* **2001**, *17*, 1599–1603.
111 H.W. Ma, J.H. Hyun, P. Stiller, A. Chilkoti, *Adv. Mater.* **2004**, *16*, 338–341.
112 A. Milchev, J.P. Wittmer, D.P. Landau, *J. Chem. Phys.* **2000**, *112*, 1606.
113 C. Devaux, J.P. Chapel, E. Beyou, P. Chaumont, *Eur. Phys. J. E* **2002**, *7*, 345–352.
114 E. Beyou, J. Humbert, P. Chaumont, *E-polymers* **2003**, http://www.e-polymers.org/papers/beyou_170403.pdf.
115 C. Bartholome, E. Beyou, E. Bourgeat-Lami, P. Chaumont, N. Zydowicz, *Macromolecules* **2003**, *36*, 7946–7952.
116 J. Parvole, L. Billon, J.P. Montfort, *Polym. Int.* **2002**, *51*, 1111–1116.
117 A. Matsumoto, *Polym. J.* **2003**, *35*, 93–121.
118 G.Q. Zhai, W.H. Yu, E.T. Kang, K.G. Neoh, C.C. Huang, D.J. Liaw, *Ind. Eng. Chem. Res.* **2004**, *43*, 1673–1680.
119 Y. Tsujii, M. Ejaz, K. Sato, A. Goto, T. Fukuda, *Macromolecules* **2001**, *34*, 8872–8878.
120 B.V. Derjaguin, *Kolloid Z.s* **1934**, *69*, 155–164.
121 P.G. de Gennes, *Adv. Colloid Interface Sci.* **1987**, *27*, 189–209.

122 S.T. Milner, T.A. Witten, M.E. Cates, *Macromolecules* **1988**, *21*, 2610–2619.

123 S.T. Milner, T.A. Witten, M.E. Cates, *Macromolecules* **1989**, *22*, 853–861.

124 T. Pakula, *Makromol. Chem.* **1999**, *139*, 49–56.

125 S. Yamamoto, Y. Tsujii, T. Fukuda, to be published.

126 T. Wu, K. Efimenko, J. Genzer, *J. Am. Chem. Soc.* **2002**, *124*, 9394–9395.

127 T. Wu, K. Efimenko, P. Vlcek, V. Subr, J. Genzer, *Macromolecules* **2003**, *36*, 2448–2453.

128 S. Balamurugan, S. Mendez, S.S. Balamurugan, M.J. O'Brien, G.P. Lopez, *Langmuir* **2003**, *19*, 2545–2549.

129 D. Goodman, J.N. Kizhakkedathu, D.E. Brooks, *Langmuir* **2004**, *20*, 2333–2340.

130 Y. Tsujii, K. Okayasu, K. Ohno, T. Fukuda, to be published.

131 U. Raviv, S. Giasson, N. Kampf, J.F. Gohy, R. Jerome, J. Klein, *Nature* **2003**, *425*, 163–165.

132 H. Sakata, M. Kobayashi, H. Otsuka, A. Takahara, *Polym. J.* **2005**, *37*, 767–775.

133 M. Kobayashi, A. Takahara, *Chem. Lett. (Jpn.)* **2005**, *34*, 1582–1583.

134 N. Ishizuka, H. Minakuchi, K. Nkanishi, K. Hirao, N. Tanaka, *Colloids Surf. A* **2001**, *187/188*, 273–279.

135 J.A. Forrest, K. Dalnoki-Veress, *Adv. Colloid Interface Sci.* **2001**, *94*, 167–196.

136 D.S. Fryer, P.F. Nealey, J.J. de Pablo, *Macromolecules* **2000**, *33*, 6439–6447.

137 D.S. Fryer, R.D. Peters, E.J. Kim, J.E. Tomaszewski, J.J. de Pablo, P.F. Nealey, C.C. White, W.L. Wu, *Macromolecules* **2001**, *34*, 5627–5634.

138 R.A.L. Jones, *Cur. Opin. Colloid Interface Sci.* **1999**, *4*, 153–158.

139 J.L. Keddie, R.A.L. Jones, *Isr. J. Chem.* **1995**, *35*, 21–26.

140 O. Prucker, S. Christian, H. Bock, J. Ruhe, C.W. Frank, W. Knoll, *Macromol. Chem. Phys.* **1998**, *199*, 1435–1444.

141 S. Yamamoto, Y. Tsujii, T. Fukuda, *Macromolecules* **2002**, *35*, 6077–6079.

142 K. Tanaka, K. Kojio, R. Kimura, A. Takahara, T. Kajiyama, *Polym. J.* **2003**, *35*, 44–49.

143 D.A. Savin, J. Pyun, G.D. Patterson, T. Kowalewski, K. Matyjaszewski, *J. Polym. Sci. Part B Polym. Phys.* **2002**, *40*, 2667–2676.

144 K. Urayama, S. Yamamoto, Y. Tsujii, T. Fukuda, D. Neher, *Macromolecules* **2002**, *35*, 9459–9465.

145 K. Urayama, O. Kircher, R. Bohmer, D. Neher, *J. Appl. Phys.* **1999**, *86*, 6367–6375.

146 K. Urayama, M. Tsuji, D. Neher, *Macromolecules* **2000**, *33*, 8269–8279.

147 H.J. Winkelhahn, T. Pakula, D. Neher, *Macromolecules* **1996**, *29*, 6865–6871.

148 S. Yamamoto, Y. Tsujii, T. Fukuda, N. Torikai, M. Takeda, *KENS Rep.* **2001/2002**, *14*, 204.

149 M. Aubouy, G.H. Fredrickson, P. Pincus, E. Raphael, *Macromolecules* **1995**, *28*, 2979–2981.

150 D.E. Bergbreiter, C.L. Tao, *J. Polym. Sci. Part A: Polym. Chem.* **2000**, *38*, 3944–3953.

151 S. Edmondson, W.T.S. Huck, *J. Mater. Chem.* **2004**, *14*, 730–734.

152 W.X. Huang, G.L. Baker, M.L. Bruening, *Angew. Chem. Int. Ed.* **2001**, *40*, 1510–1512.

153 H. Mori, A. Boker, G. Krausch, A.H.E. Muller, *Macromolecules* **2001**, *34*, 6871–6882.

154 J.Y. Wang, W. Chen, A.H. Liu, G. Lu, G. Zhang, J.H. Zhang, B. Yang, *J. Am. Chem. Soc.* **2002**, *124*, 13358–13359.

155 M. Ejaz, K. Ohno, Y. Tsujii, T. Fukuda, *Macromolecules* **2000**, *33*, 2870–2874.

156 J. Ruhe, M. Ballauff, M. Biesalski, P. Dziezok, F. Grohn, D. Johannsmann, N. Houbenov, N. Hugenberg, R. Konradi, S. Minko, M. Motornov, R.R. Netz, M. Schmidt, C. Seidel, M. Stamm, T. Stephan, D. Usov, H.N. Zhang, *Adv. Polym. Sci.* **2004**, *165*, 79–150.

157 T. Abe, S. Hayashi, N. Higashi, M. Niwa, K. Kurihara, *Colloids Surf. A* **2000**, *169*, 351–356.

158 Y. Tsujii, Y. Hirose, M. Ejaz, T. Fukuda, M. Ishidoya, *Polym. Prepr. Am. Chem. Soc.* **2002**, *43*, 317–318.

159 B. de Boer, H.K. Simon, M.P.L. Werts, E.W. van der Vegte, G. Hadziioannou, *Macromolecules* **2000**, *33*, 349–356.

160 S.F. Hou, Z.C. Li, Q.G. Li, Z.F. Liu, *Appl. Surf. Sci.* **2004**, *222*, 338–345.
161 F. Hua, J. Shi, Y. Lvov, T. Cui, *Nano Lett.* **2002**, *2*, 1219–1222.
162 M. Husemann, M. Morrison, D. Benoit, K.J. Frommer, C.M. Mate, W.D. Hinsberg, J.L. Hedrick, C.J. Hawker, *J. Am. Chem. Soc.* **2000**, *122*, 1844–1845.
163 D.M. Jones, W.T.S. Huck, *Adv. Mater.* **2001**, *13*, 1256–1259.
164 D.M. Jones, J.R. Smith, W.T.S. Huck, C. Alexander, *Adv. Mater.* **2002**, *14*, 1130–1134.
165 I.S. Maeng, J.W. Park, *Langmuir* **2003**, *19*, 4519–4522.
166 U. Schmelmer, R. Jordan, W. Geyer, W. Eck, A. Golzhauser, M. Grunze, A. Ulman, *Angew. Chem. Int. Ed.* **2003**, *42*, 559–563.
167 R.R. Shah, D. Merreceyes, M. Husemann, I. Rees, N.L. Abbott, C.J. Hawker, J.L. Hedrick, *Macromolecules* **2000**, *33*, 597–605.
168 F. Zhou, W.M. Liu, J.C. Hao, T. Xu, M. Chen, Q.J. Xue, *Adv. Funct. Mater.* **2003**, *13*, 938–942.
169 M. Ejaz, S. Yamamoto, Y. Tsujii, T. Fukuda, *Macromolecules* **2002**, *35*, 1412–1418.
170 Y. Tsujii, M. Ejaz, S. Yamamoto, T. Fukuda, K. Shigeto, K. Mibu, T. Shinjo, *Polymer* **2002**, *43*, 3837–3841.
171 S.G. Boyes, B. Akgun, W.J. Brittain, M.D. Foster, *Macromolecules* **2003**, *36*, 9539–9548.
172 S.G. Boyes, W.J. Brittain, X. Weng, S.Z.D. Cheng, *Macromolecules* **2002**, *35*, 4960–4967.
173 A. Carlmark, E.E. Malmstrom, *Biomacromolecules* **2003**, *4*, 1740–1745.
174 J.B. Kim, W.X. Huang, M.L. Bruening, G.L. Baker, *Macromolecules* **2002**, *35*, 5410–5416.
175 X.X. Kong, T. Kawai, J. Abe, T. Iyoda, *Macromolecules* **2001**, *34*, 1837–1844.
176 V.L. Osborne, D.M. Jones, W.T.S. Huck, *Chem. Commun.* **2002**, 1838–1839.
177 R.A. Sedjo, B.K. Mirous, W.J. Brittain, *Macromolecules* **2000**, *33*, 1492–1493.
178 B. Zhao, W.J. Brittain, *Macromolecules* **2000**, *33*, 8813–8820.
179 B. Zhao, W.J. Brittain, W.S. Zhou, S.Z.D. Cheng, *J. Am. Chem. Soc.* **2000**, *122*, 2407–2408.
180 B. Zhao, W.J. Brittain, W.S. Zhou, S.Z.D. Cheng, *Macromolecules* **2000**, *33*, 8821–8827.
181 M. Ejaz, K. Ohno, Y. Tsujii, T. Fukuda, *Polym. Prepr. Am. Chem. Soc.* **2003**, *44*, 532–533.
182 B. Zhao, *Polymer* **2003**, *44*, 4079–4083.
183 B. Zhao, T. He, *Macromolecules* **2003**, *36*, 8599–8602.
184 E.P.K. Currie, W. Norde, M.A.C. Stuart, *Adv. Colloid Interface Sci.* **2003**, *100*, 205–265.
185 D.Q. Xiao, H. Zhang, M. Wirth, *Langmuir* **2002**, *18*, 9971–9976.
186 S.B. Lee, R.R. Koepsel, S.W. Morley, K. Matyjaszewski, Y.J. Sun, A.J. Russell, *Biomacromolecules* **2004**, *5*, 877–882.
187 W. Feng, J. Brash, S.P. Zhu, *J. Polym. Sci. Part A: Polym. Chem.* **2004**, *42*, 2931–2942.
188 R. Iwata, P. Suk-In, V.P. Hoven, A. Takahara, K. Akiyoshi, Y. Iwasaki, *Biomacromolecules* **2004**, *5*, 2308–2314.
189 W. Feng, S.P. Zhu, K. Ishihara, J.L. Brash, *Langmuir* **2005**, *21*, 5980–5987.
190 C. Yoshikawa, A. Goto, Y. Tsujii, T. Fukuda, T. Kimura, K. Yamamoto, A. Kishida, *Macromolecules* **2006**, *39*, 2284–2290.
191 C. Yoshikawa, A. Goto, Y. Tsujii, N. Ishizuka, K. Nakanishi, T. Fukuda, to be published.
192 Y.P. Wang, X.W. Pei, K. Yuan, *Mater. Lett.* **2005**, *59*, 520–523.
193 A. El Harrak, G. Carrot, J. Oberdisse, C. Eychenne-Baron, F. Boue, *Macromolecules* **2004**, *37*, 6376–6384.
194 P. Liu, J. Tian, W.M. Liu, Q.J. Xue, *Polym. Int.* **2004**, *53*, 127–130.
195 T. Morinaga, K. Ohno, Y. Tsujii, T. Fukuda, *Polym. J.* **2006**, in press.
196 W.G. Hoover, F.H. Ree, *J. Chem. Phys.* **1968**, *49*, 3609–3617.
197 B.J. Alder, W.G. Hoover, D.A. Young, *J. Chem. Phys.* **1968**, *49*, 3688–3696.
198 H.D. Bijsterbosch, V.O. Dehaan, A.W. Degraaf, M. Mellema, F.A.M. Leermakers, M.A.C. Stuart, A.A. Vanwell, *Langmuir* **1995**, *11*, 4467–4473.

12
Hybrid Organic Inorganic Objects

Stefanie M. Gravano and Timothy E. Patten

12.1
Introduction

Well-defined polymer-grafted inorganic nano-objects typically consist of an inorganic core with an end-grafted, tailored polymer shell. These hybrids combine the magnetic, electronic or optical properties of the inorganic core with the versatile mechanical and chemical properties of the polymer outer layer. Many inorganic materials in the nano-regime possess properties not realized in their bulk counterparts, so additional characteristics can be engineered into the hybrid. The core–shell particles can also be used as nano building blocks to form complex higher order structures that can contain multiple functions depending on the component particles. The assembly of well-defined core–shell particles not only provides an opportunity to understand interactions at the nanometer level, but also the means to build devices with new structures and functions or devices that mimic those already present in nature. Applications of these core–shell particles include the encapsulation of drugs or dyes, bearing sealants, imaging agents, compatibilizers, biological markers, stimulus-sensitive surfaces and even nano-sized environments for molecule selective reactions.

The synthesis, characterization and application of nanomaterials require interdisciplinary work across the fields of chemistry, physics, biology and engineering. For example, in the area of semiconductor nanoparticles, synthesis is addressed by chemists, the physical properties are measured by physicists and engineers and their interactions in living systems are studied by biologists. Collectively, chemists, biologists, engineers and physicists take part in the study, development and application of these nanoparticles.

There are three structural components to the core–shell particles: the inorganic core, the interface and the polymer shell (Fig. 12.1). Each section requires special attention when determining the design and synthesis of the material. The inorganic core can impart properties to the overall structure, such as photoluminescence, magnetism and mechanical reinforcement, which cannot easily be obtained using just organic materials. The particular type of inorganic mate-

Macromolecular Engineering. Precise Synthesis, Materials Properties, Applications.
Edited by K. Matyjaszewski, Y. Gnanou, and L. Leibler

ISBN: 978-3-527-31446-1

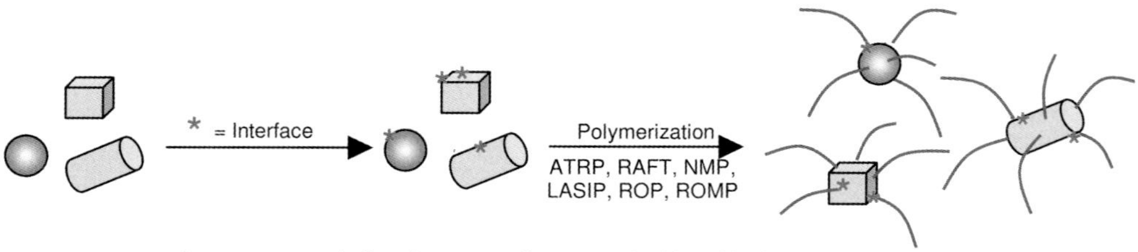

Fig. 12.1 Core–shell architectures for use as building blocks in nanoscale systems formed by core functionalization and controlled polymerization.

rial will dictate the chemistry required to tether initiators to the surface and potentially can interfere with the chemistry of the polymerization. The interface where the core and shell meet is another key component in the design of the core–shell nanoparticle. The polymeric shell must be tethered to the core for optimum stability of the structure and to overcome potential incompatibilities between the two phases. The distribution of initiator tethers is also a variable that can be used to alter the density of chains on the surface. There are several alternatives to chain end attachment of the polymer shell to the core that can be employed, including emulsion polymerization, *in situ* encapsulation and self-assembly techniques [1]. However, this chapter will focus primarily on "grafting from" methods utilizing living polymerizations [2–5]. The final key component in the design of core–shell nanoparticles is the grafted polymer shell, which can vary considerably in thickness (molar mass) and density. The grafted polymer can also add function to the overall hybrid, as chemical functionality in the side-chains can assist in particle self-assembly or serve as a scaffold for the attachment of biological molecules. The polymer itself can serve as a protective barrier, a matrix for the composite or a solubility/dispersibility enhancer. Variations in the length of the polymer chains also can affect the mechanical and morphological properties of the composite materials.

12.2
Synthetic Methods

There are three general methods that can be used to graft a polymer shell to a core particle: "grafting from", "grafting onto" and "grafting through". The "grafting from" technique involves conducting a chain growth polymerization from an initiator monolayer on the particle surface. This method is also referred to as "surface initiated polymerization" (SIP). "Grafting from" methods yield the highest grafting densities of chains, because they avoid the entropically costly process of elongating and packing polymer chains into the brush layer [6]. Consequently, "grafting from" methods are the most widely applied to the formation of core–shell structures. The "grafting onto" technique entails bonding a chain end functionalized polymer or a block copolymer segment to the particle surface. This method has the advantage that the polymer can be characterized before attachment to the surface, but suffers from a limit on the density of grafted chains achievable. The "grafting through" technique involves attaching monomer to the particle surface and then conducting a polymerization in the presence of the surface-modified particles. Actively growing chains can react with the surface-bound monomer to graft the polymer covalently to the surface. The grafted layer of chains contains a range of attachment sites along the polymer backbones. Encapsulating the inorganic cores in a polymer layer not only allows for further modification, that is, chemical linkage or another shell formation, but also prevents the particles from aggregating [7].

Gold (Au) and silica (SiO_2) nanoparticles are the most commonly used inorganic cores in the synthesis of core–shell nanoparticles, because surface modification chemistries are well established for both materials. Consequently, this chapter will focus on these materials in particular. Alkanethiol monolayers form readily on Au surfaces and SiO_2 surfaces can be readily derivatized using mono-, di- or trisiloxanes. Thus, initiators or polymers containing these functional groups can be attached to the particle surfaces. In addition to chemical modification, polymers and initiators can be physically adsorbed on particle surfaces. Obviously, this type of linkage can be less robust than chemical modification. Both types of surface functionalization methods allow one to use either the "grafting from" or "grafting on to" approach to forming core–shell structures.

Many of the techniques used in core–shell structure synthesis are analogous to those used to form polymer brushes on flat surfaces. For example, controlled radical polymerization methods have been applied to grafting polymers from flat SiO_2 and Au surfaces [8, 9] and concave and convex SiO_2 surfaces [10]. There are a few differences in methodology between the two systems. For example, it is known that triethoxysilanes can oligomerize, or even gel, before deposition on a surface. With a flat substrate, an oligomeric siloxane in the deposition solution can be easily removed by washing the surface. With particles, however, separation of a high molecular weight oligomeric siloxane from particles can be difficult, especially when very small nanoparticles are used that do not sediment easily from solution. Also, in radical polymerization, termination via coupling or disproportionation will not greatly affect the final material when the polymerization is conducted from a flat surface; however, when the polymerization is conducted from particle surfaces, even a small amount of termination can lead to gelation of the reaction mixture [11, 12]. Finally, for a brush on a curved surface the polymer chains will have more configurational volume the further the chain segment is from the surface, because the volume available for each chain is conical rather than cylindrical, as is the case for brushes formed on a flat surface. Some examples of polymerizations conducted on both particles and flat surfaces are those using atom transfer radical polymerization (ATRP) [13, 14], living anionic surface-initiated polymerization (LASIP) [15, 16] and nitroxyl radical-mediated polymerization (NMP) [17, 18].

12.2.1
Methods for Depositing Initiators on Particle Surfaces

The types of molecules needed to prepare an initiator-modified surface will depend on the inorganic material, the polymerization technique used and the robustness of attachment required (Table 12.1).

12.2.1.1 SiO_2

Siloxanes are often used to attach covalently organic functional groups to silica particle surfaces. Thus, a simple procedure for preparing initiator-modified surfaces is to treat the silica particles with tandem siloxane–initiator compounds. Indeed, for most controlled polymerization methods, the first application to core–shell particle synthesis utilized SiO_2 as the substrate for this reason. Another reported method for surface functionalization pertained to the "grafting onto" method. First the silica particles were capped with oleic acid and then an emulsion polymerization of styrene was performed. The intermediate styrene radicals reacted with the C=C bond in the oleic acid chain effectively to anchor the polystyrene chains to the core surface [2]. In another method, peroxides were bound to the surface of silica for use in reverse ATRP [19].

12.2.1.2 Au

Thiols are commonly used to attach organic functional groups to Au surfaces. Tandem thiol–polymerization initiators have been prepared and deposited on Au nanoparticles. Examples of controlled polymerizations used to graft polymers from initiator-modified Au nanoparticles include ATRP and reversible addition–fragmentation chain-transfer polymerization (RAFT) [20–22]. Another reported method for surface functionalization involved first covering the Au particles with 3-mercaptopropyltrimethoxysilane. Subsequently the silyl-terminated Au particles were treated with one of the tandem siloxane–polymerization initiators to yield an initiator shell about the nanoparticle [23].

12.2.1.3 Other Methods

In addition to chemical modification, polymers and initiators can be physically adsorbed to particle surfaces. Under basic conditions the surface of silica has a net negative charge. The counterions can be exchanged with a polycation as seen in layer-by-layer deposition methods. If the polycation is part of a block copolymer that also contains a segment with polymerization initiators as the side-chains, then this method becomes viable for depositing initiators on the silica particle surface (Fig. 12.2) [24]. This procedure has also been used to deposit small molecule initiators for NMP on montmorillonite clays [25].

Functionalization of surfaces with tandem initiator–thiol or initiator–siloxane compounds represents a direct, one-step method for creating initiator-modified

Table 12.1 Examples of tandem initiator/surface modifier molecules used to graft polymers from surfaces using controlled/living polymerization methods.

Method[a)]	SiO_2	Au
ATRP	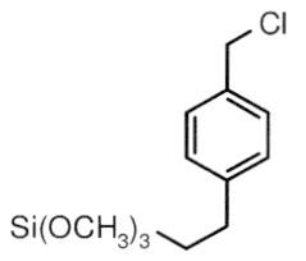 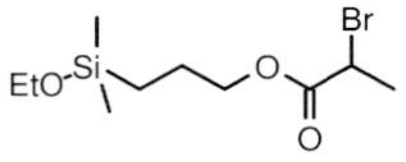	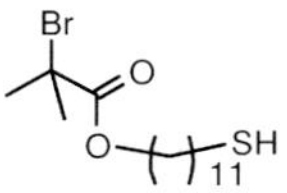
NMP	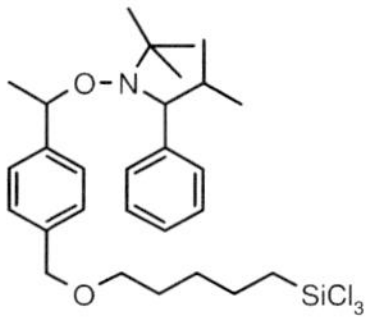	
RAFT	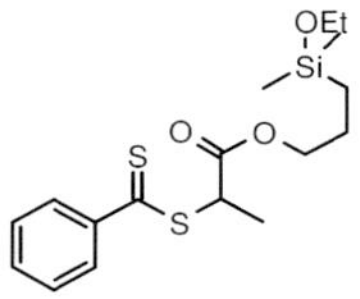	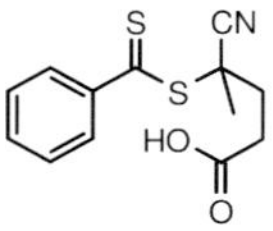
ROMP	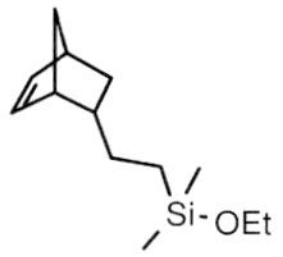	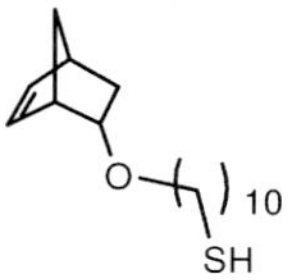
ROP	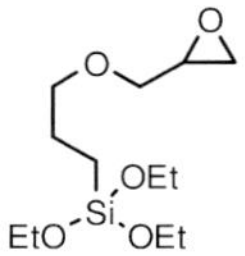	
LASIP	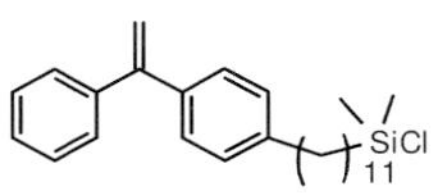	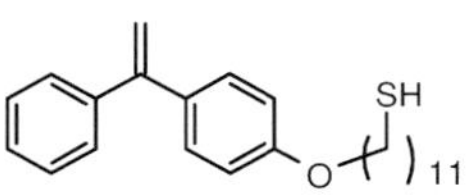

a) ATRP = atom transfer radical polymerization; NMP = nitroxyl radical-mediated polymerization; RAFT = reversible addition–fragmentation chain transfer polymerization; ROMP = ring-opening metathesis polymerization; ROP = ring-opening polymerization; LASIP = living anionic surface-initiated polymerization.

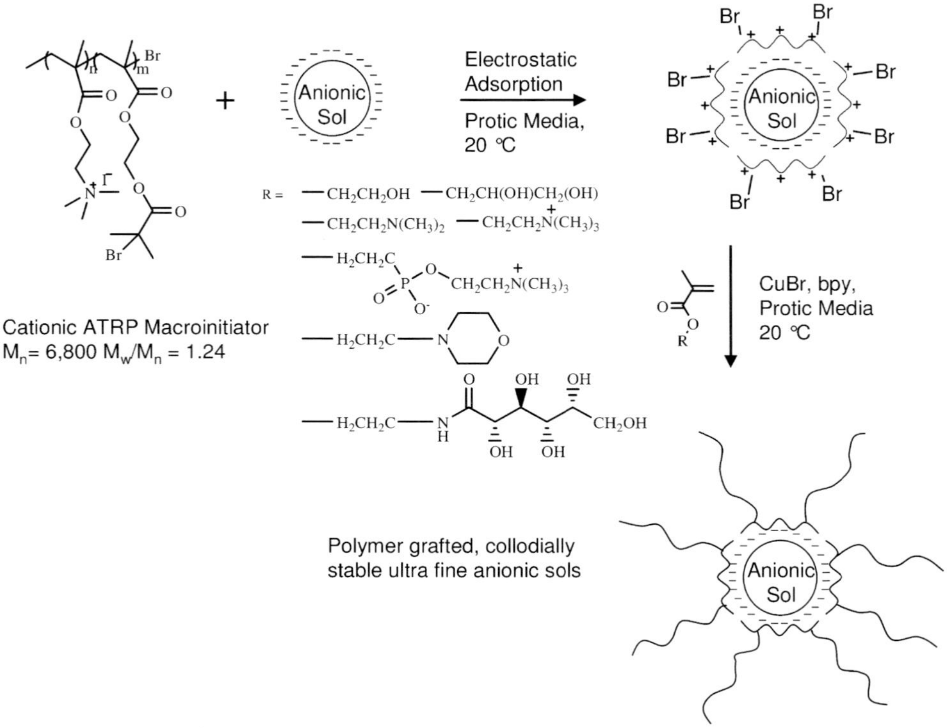

Fig. 12.2 Block copolymers containing polycationic segments can be used as an alternative strategy to modifying surface with small molecules [118].

surfaces. These surfaces can also be prepared in a two-step process. A siloxane or thiol containing a non-initiator functional group can be deposited on a surface. In a subsequent reaction, this functional group can be transformed into a polymerization initiator. For example, a vinylbenzylsiloxane was deposited on a silica surface followed by a free radical chlorination reaction to yield a benzylic chloride-derivatized surface capable of serving as an initiator for ATRP or reverse ATRP [26–29].

12.2.1.4 **Characterization of Initiators on Particles**

In each of these surface modification processes, it is important to determine the number of initiator sites per unit area of the surface. This value has been measured using a number of characterization methods including elemental analysis [11, 30], titration analysis [31], XPS [32, 33] and solid-state NMR spectroscopy [34, 35]. The core–shell structures were characterized using DLS, AFM, SPM and TEM and the polymer molecular weights were determined by degrafting the polymer using HF etching or basic hydrolysis of an ester-containing linker, for SiO_2 particles and via I_2 or CN^- etching for Au particles followed by GPC, FTIR, TGA and DSC analysis. With some manipulation of the core–shell system, common characterization techniques for polymers, inorganic structures and monomer can all be used to analyze these syntheses.

12.2.2
Methods for Polymerizing from the Initiator Monolayers

With initiator-modified particles in hand, one can then use these particles as macroinitiators for controlled/living polymerizations to graft a layer of well-defined polymer chains from the particle surface. The particular polymerization methods that have been used to make core–shell structures include controlled radical polymerizations (CRPs): (a) ATRP (atom transfer radical polymerization), (b) NMP (nitroxyl radical-mediated polymerization), (c) RAFT (reversible addition–fragmentation chain transfer) and other living polymerizations, (d) LASIP (living anionic surface-initiated polymerization), (e) ROP (ring-opening polymerization) and (f) ROMP (ring-opening metathesis polymerization). Each polymerization method requires its own conditions and exhibits different chemical and macromolecular behavior.

12.2.2.1 Atom Transfer Radical Polymerization

ATRP has been the most widely used CRP in the formation of core–shell structures. As with other controlled/living radical polymerization methods, the mild polymerization conditions disfavor degradation of the inorganic core's properties and favor retention of side-chain and end-group functionality. With a reasonable rate of polymerization, ATRP provides polymers with predictable molecular weights and narrow molecular weight distributions for a range of monomers. In 1999, polystyrene-coated SiO_2 core–shell nanoparticle2 core–shell nanoparticles“,4,1> systems prepared using ATRP were first reported [13] and were followed later with a more detailed study including methyl methacrylate (MMA) and styrene [36]. Common ATRP surface initiators include secondary and tertiary α-halo esters [13, 37, 38] and halomethyl- and (1-haloethyl)phenylsiloxane and thiol derivatives [39]. Styrene and MMA polymerized from the surface-bound initiator layers yielded polymers with molar masses ranging from 10^3 to 10^4 g mol^{-1} and molecular weight distributions as narrow as 1.2. Since then, other types of grafted polymers with various molecular weights and narrow molecular weight distributions have been reported.

In comparison with traditional ATRP done in bulk or in a relatively small amount of solvent, reaction conditions for the formation of core–shell structures are more dilute in order to suspend the particles and to minimize the effect of particle–particle coupling resulting from the small amount of termination during polymerization (Fig. 12.3). Despite the dilute reaction conditions, the polymerizations still progressed within a reasonable amount of time with the same

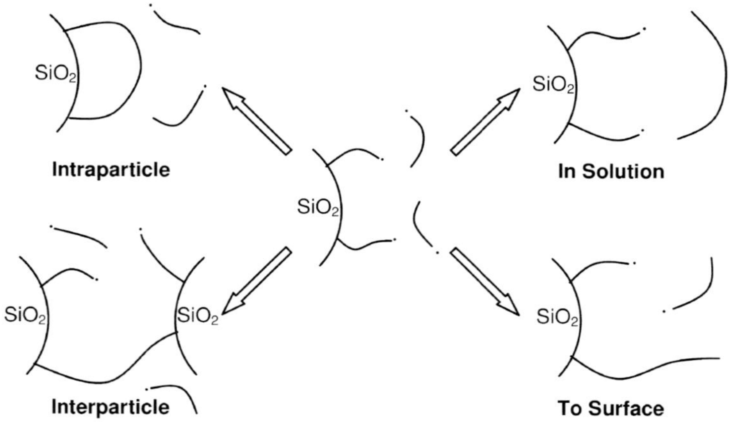

Fig. 12.3 Proposed termination modes in the formation of core–shell structures using radical-based polymerizations. The rates of particle-to-particle interactions are increased without a reasonable amount of solvent [36].

predictable molecular weights and narrow molecular weight distributions compared with traditional conditions [37].

Due to the relatively small amount of polymer when working with nanomaterials containing few initiator sites, a secondary method of polymer characterization has been used instead of directly characterizing the bound polymer. A secondary approach is the addition of free, non-surface-bound initiator to the polymerization solution [40]. The chains grown in solution are then separated from the inorganic substrate and analyzed. The underlying assumptions are that exchange between active centers on the inorganic surface and the chains in solution is fast and that the molecular weight characteristics of the free chains will match the surface grafted chains. Because small particles (10–100 nm in diameter) have a large surface area-to-volume ratio, the amount of grafted polymer is significant and these assumptions can be checked directly. Several of these studies have been performed and each has confirmed that solution polymerization and surface polymerization from small particles yield virtually identical polymers (Fig. 12.4) [36, 38, 41, 42].

This presence of growing free chains in the polymerization solution was also shown to have the beneficial effect of limiting particle–particle coupling by skewing the small amount of chain termination that occurs to faster solution processes over slower surface-based processes [36]. Due to the large number of chains growing from a particle in a typical brush synthesis, even a small percentage of chain–chain coupling can lead to complete gelation of the polymerization. The polymerization temperature for ATRP varies from room temperature to 110 °C depending on the monomer. Styrene is known to undergo thermally self-initiated polymerization at higher temperatures. This process has the effect of generating free chains in solution and, if it is not extensive over the

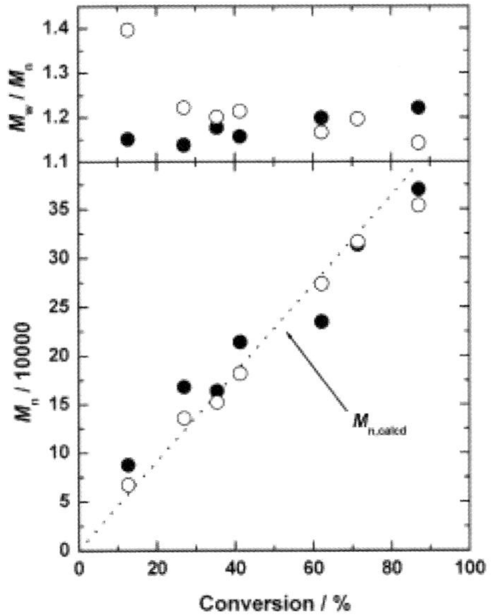

Fig. 12.4 Number-average molecular weight (M_n) and polydispersity (M_w/M_n) of graft (λ) and free (μ) polymers as a function of monomer conversion of methyl methacrylate (MMA) in bulk at 70 °C with initiator-coated silica particles of diameter 130 nm (2wt.%): $[43]_0$:[ethyl 2-bromoisobutylate]$_0$:[Cu(I)Cl]$_0$:[4,4′-dinonyl-2,2′-bipyridine]$_0$ = 6000 : 1 : 10 : 20. This graph shows that free initiator has no effect on the polymerization of MMA from the surface of a silica particle [38].

course of polymerization, it can contribute to the minimization of particle–particle coupling. Methacrylate monomers do not undergo thermal autopolymerization and chain terminate predominately by disproportionation (>90%). The unsaturated chain ends can add to chains growing on other particles, leading to particle–particle coupling. The addition of a small amount of free initiator to mimic chains generated via autopolymerization effectively mitigated chain termination-based problems [36].

Another procedure to limit particle–particle coupling and to increase molecular weight control in polymerization from nanoparticle surfaces is to add excess deactivating species [Cu(II) complexes] [34]. Studies of the kinetics and mechanism of ATRP have shown that increasing the concentration of deactivator reduces the steady-state concentration of radicals by increasing the rate of radical deactivation relative to other reactions that radicals can undergo, including radical termination. In one particle study, when styrene polymerizations were conducted from 75-nm diameter silica particles, the grafted polymer exhibited controllable molecular weights and narrow molecular weight distributions. In contrast, when styrene polymerization was conducted from 300-nm diameter particles, the molecular weight distributions were significantly higher (>1.5), indicating significantly less molecular weight control.

The difference in control between polymerizations using the two types of particles was attributed to the lower net amount of initiator for the larger particles. When a low initiator concentration is present [43] ($[M]_0 < 10^{-3}$ M), a small concentration of catalyst is required and the deactivator concentration in the polymerization is low. Consequently, if no free initiator is added, then deactivating species must be added to increase the rate of radical deactivation [36]. In another study on the preparation of core–shell structures from latex spheres, *N,N*-dimethylacrylamide (DMA) was grafted using aqueous ATRP and a similar decrease in reaction rate and molecular weight distribution with the addition of a deactivating species was observed. Both the addition of a deactivating species and free initiator are methods that impart control over the rate of polymerization and the molecular weight behavior of the polymerization [44].

A range of monomer types, including hydrophobic, multifunctional and water-soluble monomers, have been grafted from the surface of a nanoparticles. Depending on the type of polymer, interesting properties were observed. Polymerizations of common organic monomers have been performed to study the fundamental chemistry of grafting reactions and to compare solution versus surface polymerization characteristics. For example, the ATRP kinetics of styrene, *n*-butyl acrylate and methyl methacrylate under identical reaction conditions and with 10 mol% of deactivator added were investigated and whereas methyl methacrylate polymerized at the same rate from free initiator as from the surface, styrene and *n*-butyl acrylate polymerized from the surface at a rate four times slower than from free initiator [34].

Inimers are a class of monomers that possess both a polymerizable group and an initiating group. Their self-condensing vinyl polymerization (SCVP) results in the formation of macromolecules with branched architectures and with a large quantity of initiator groups within the same polymer chain. The ATRP of inimers has been used to amplify the initiator content on the particle surface. An inimer containing an acrylate and an α-bromo ester functionality was used in an SCVP with α-bromo ester-functionalized silica nanoparticles to achieve a thick layer of initiating sites from which *tert*-butyl acrylate was polymerized also using ATRP. The final particle had a dual shell structure of a hyperbranched layer followed by a polyacrylate brush layer [45]. The structure of these high density core–shell structures was confirmed using scanning force microscopy (SFM) (Fig. 12.5).

12.2.2.2 **Nitroxyl Radical-mediated Polymerization**

Nitroxyl radical-mediated polymerization (NMP) has also been applied to the synthesis of polymer brushes on nanoparticle surfaces. Initiators have been grafted to particle surface through a number of means. Tandem nitroxide–siloxane compounds have been synthesized and directly attached to silica surfaces in a similar procedure as used for ATRP grafting methods [32]. Additionally, a two-step functionalization process has been shown to be effective. For example, methacryloxypropyltriethoxysilane was deposited on a particle surface and then the particle was

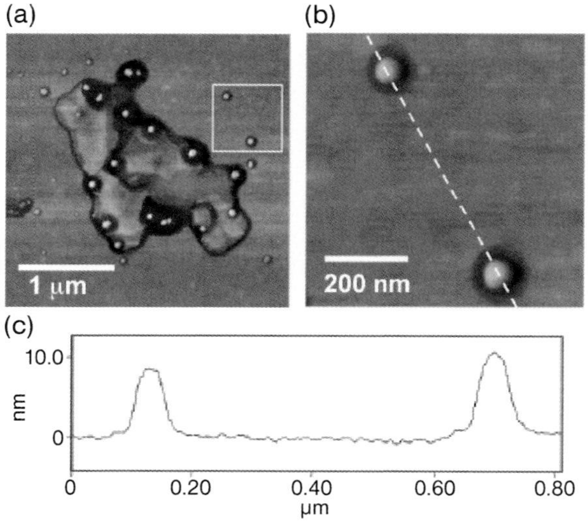

Fig. 12.5 Representative SFM images of the branched poly(*tert*-butyl acrylate)–silica hybrid particles obtained from SCVP-modified initiator particles: (a) phase image (50°) (b) higher magnification phase image taken from the area inside the box indicated in (a); (c) cross-section taken at the position indicated by the dashed line in (b) [45].

heated with a nitroxide. At high temperatures, the nitroxide dissociated into a stable nitroxyl radical and a reactive organic radical. The organic radical added to the methacrylate group on the particle surface and the free nitroxide trapped the new radical on the surface (Fig. 12.6) [46]. Subsequent thermal treatment of the particle with monomer led to controlled grafting of chains from the particle surface to form the core–shell structure. In the polymerization of bulk styrene without added free initiator, difficulties in purification were noted. The core–shell structures were embedded in a cross-linked network of polystyrene and could not be separated even with extensive washing. Most likely, particle–particle coupling was the cause of this cross-linking. When the polymerizations were conducted in solvent and with free initiator present, a controlled polymerization resulted with predictable molecular weights and narrow molecular weight distributions of both the surface-bound and free polymers [47].

12.2.2.3 **Reversible Addition–Fragmentation Chain Transfer**

Reversible addition–fragmentation chain transfer (RAFT) polymerization has been used in conjunction with both gold and silica particles to form core–shell structures [48]. The most common means of attaching the initiator to a silica particle surface is via deposition of a difunctional molecule containing a dithioester and a reactive siloxane [49, 50]. In the case of Au particles, the surface

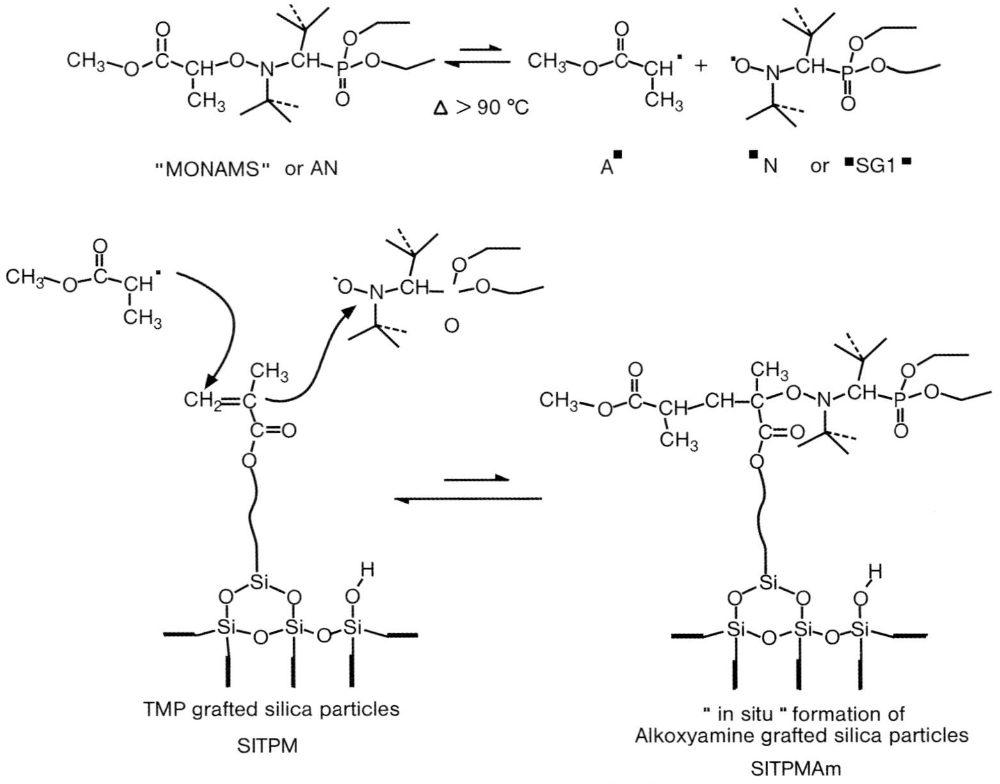

Fig. 12.6 Surface functionalization by nitroxide-mediated initiator formation [46].

must be passivated before exposure to the dithioester RAFT initiator, because the thioester could interact with the thiophilic Au surface rendering the initiator useless. A two-step surface functionalization method has been shown to be effective. First, the Au nanoparticles were covered with an ω-hydroxyalkanethiol and then the hydroxyl groups were coupled with a difunctional carboxylic acid–RAFT compound. These particles were used successfully in the polymerization of *N*-isopropylacrylamide (NIPAm), yielding poly(NIPAm)s with predictable molecular weights [21].

There are significant differences in the kinetics of RAFT initiated from a surface and RAFT from free dithioester transfer agents. Under identical reagent stoichiometries and concentrations, monomer was consumed at a constant rate in both cases; however, polymerizations from the surface were slower than those from free transfer agents. This decrease in rate of polymerization for surface-initiated RAFT was found to be a result of the dithioester transfer agents on the surface rapidly transferring with neighboring chains with free-radical centers. The rate of this process is faster than the rate of dithioester exchange in solution, thus slowing the observed rate of polymerization. After a significant incubation period, enough monomer added to each chain to reduce the local concentration of dithioester transfer agents around the free radicals and subsequently the polymerization proceeded at a rate more consistent with the polymerizations using free transfer agent. Ultimately this incubation period did not affect the final molecular weights or broaden the molecular weight distributions [49].

RAFT has also been used to prepare polymers for a "grafting on to" modification of metal nanoparticles. The dithioester-terminated polymers were reduced to thiol-terminated polymers using standard reduction techniques. When the reduction was performed in the presence of gold, silver, platinum or ruthenium precursor complexes, the metal nanoparticles formed and the polymer assembled around the exterior [51]. This "grafting on to" method to form the core–shell structures alleviates all the difficulties associated with the "grafting from" method; however, as with all "grafting on to" methods, steric hindrance could limit the maximum number of polymers that can be grafted to the core.

12.2.2.4 **Other Living Polymerization Methods**

In addition to controlled radical polymerization, other methods such as ionic, lactone ring-opening and ring-opening metathesis polymerizations (ROMP) have been applied to the preparation core polymer grafted nanoparticles, which has added to the variety of polymer backbone structures available for the synthesis of core–shell nanoparticles.

Living anionic surface-initiated polymerization (LASIP) has been used in the surface polymerization of vinyl monomers from silica nanoparticles. The initiating species was formed using a "two-step" method by adding *n*-butyllithium to diphenylethylene (DPE) groups anchored to the surface via a siloxane linkage. A color change associated with initiator formation from pale yellow to red was observed [15, 16]. Styrene or butadiene was grafted from the particle surface to give predetermined molar masses on the order of 10^3–10^5 g mol^{-1} and molecular weight distributions $M_w/M_n < 1.2$ [52]. The kinetics of polymerization from free initiators in solution showed faster rates of polymerization and a slightly lower M_w/M_n compared with chains grown from the surface; however, the decrease in reaction rate was not accompanied by a decrease in molecular weight control. Also, the molecular weight characteristics of the free polymer could not be used to determine the molecular weight characteristics of the grafted polymer. A degrafting step therefore needed to be employed. The lower rate of polymerization was attributed to limited diffusion of the monomer to activated sites on the particle. The polymerizations were conducted in a specially prepared glass apparatus composed of an air-free reaction flask, break-seal containers for the reagents and a filter frit, and the monomer and free polymer could be washed from the final particles (Fig. 12.7).

The ring-opening polymerization of ε-caprolactone was conducted from CdS [53] and silica [54] nanoparticles. Either hydroxyl groups or amine groups [31] served as the surface-bound initiators and the polymerization was catalyzed by

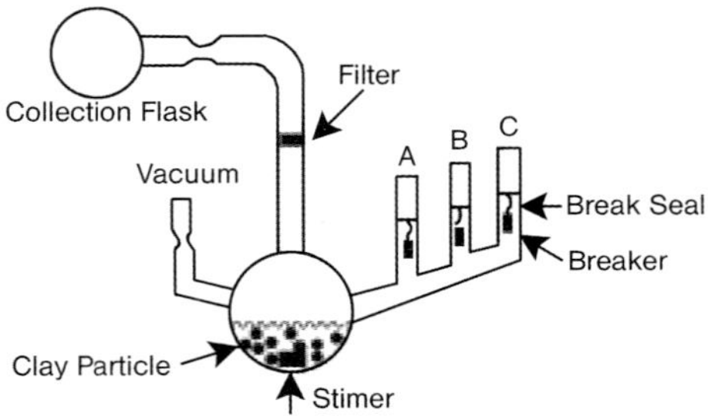

Fig. 12.7 Schematic diagram of the vacuum polymerization reactor having several compartments: (A) ampule containing styrene; (B) ampule containing *n*-BuLi; (C) ampule containing MeOH [16].

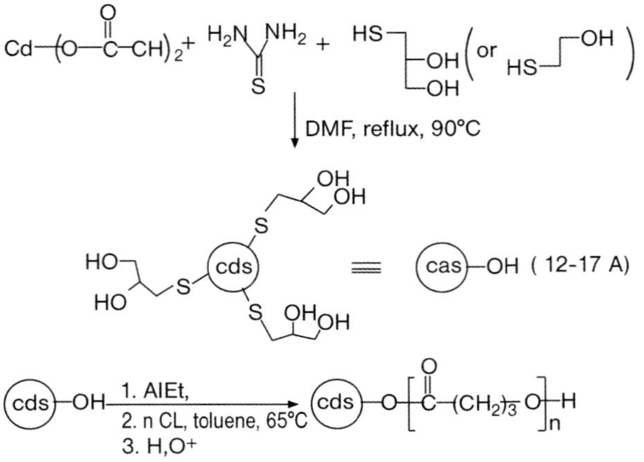

Fig. 12.8 Hydroxyl functionalization of CdS nanoparticles used as the initiating species for ROP of ε-caprolactone [31].

the addition of $AlEt_3$. After long periods of time (days), the polymerization was terminated with HCl (Fig. 12.8). To calculate the surface functionality of the hydroxyl functionalized nanoparticles, the authors treated the particles with a known amount of *n*-butyllithium, filtered off any unreacted *n*-butyllithium and then titrated the filtrate to back-calculate the number of hydroxyl groups on the surface. To validate that the polymerization reaction was indeed living, the authors grafted a block copolymer by adding a segment of valerolactone after the initial caprolactone graft. The molar masses of the block copolymer were characterized using ^{1}H NMR spectroscopy, since etching the polymer from the particle surface using acid or base would also potentially degrade the polymer. The ring-opening polymerization of caprolactone and valerolactone proved an interesting way to insert functionality in the polymer backbone of the core–shell structures.

Table 12.2 Chain density for 20-nm particles reacted for different lengths of time with catalyst [55].

Reaction time (min)	Sample	Chains/ particle	MW ($\times 10^3$ g mol^{-1})	PDI	nm^2/chain
5	1	31	126	1.24	38
10	1	43	118	1.29	27
15	1	62	96.6	1.47	19
3	2	43	266	1.11	30
60	2	75	217	1.35	17
420	2	42	274	1.13	31

Cyclic and bicyclic olefins, specifically norbornene, have been grafted from nanoparticle surfaces using ROMP. Norbornenyl groups have been anchored to silica [55, 56], Au [57], Co_3O_4 [58] and ZnO [59] particles using standard siloxane or thiol ligation strategies. Treatment with Grubb's catalyst and then monomer led to the grafting of polynorbornene from the surface. Another reported initiator modification strategy for CdSe nanoparticles involved the synthesis of the particles using a phosphine oxide that contained a vinylbenzyl group [60]. Treatment of the nanoparticle with Grubb's catalyst yielded catalyst-modified nanoparticles ready for use in subsequent ROMP reactions. A very detailed study of the ROMP grafting process was reported by Jordi and Seery [55], who characterized the density of surface-bound initiator, catalyst molecules and polymer chains, all of which are directly related to materials properties such as hydrolytic stability and structural dimensions. The polynorbornene resulting from these surface-bound polymerizations displayed high molar masses and relatively narrow molecular weight distributions (Table 12.2). The density of initiator groups on the surface ranged from one chain/17 nm^2 to one chain/38 nm^2, characterized by calibrated GC/MS, and the authors measured the amount of unbound initiator molecules left in solution after particle functionalization and washing. Since ROMP surface polymerization required a two-step initiation process (particle coating with a siloxy-containing ROMP initiator and reaction with a stoichiometric amount of Grubb's catalyst) where the catalyst is bound to the initiator as compared with a one-step initiation process for ATRP, the surface coverage was significantly reduced compared with ATRP where there was nearly one chain/nm^2. The density of catalyst bound to surface initiator sites in ROMP was found to be $<10\%$ of the total initiator sites, explaining the reduced polymer density.

12.2.2.5 Thermal Behavior of the Grafted Polymer

As seen in polymer brushes grafted from flat surfaces, the dense grafting of chains on a particle surface increases the glass transition temperature (T_g) of the polymer phase due to constricted mobility of the chains. However, because the polymer chain can adopt a more random coil-like conformation the further it is away from a spherical particle surface, this effect decreases with increasing molar mass. For example, a composite comprised of 20-nm SiO_2 with surface-grafted polystyrene (M_n 5000 g mol^{-1}) exhibited a T_g 13 °C higher than native polystyrene, whereas composite particles prepared with polystyrenes of M_n 15 000 and 31 000 g mol^{-1} exhibited increased T_gs of only 6 and 3 °C. Due to the tendency towards random coil conformation (just like the random coil conformation in bulk systems), the T_g is less impacted with higher molar mass core–shell structures [61].

12.3
Inorganic Cores other than Au and SiO_2 used in Core–Shell Nanoparticle Synthesis

Varying the inorganic core provides a means to alter the range of optical and magnetic properties that can be imparted to these composite particles. Additionally, the shapes of the inorganic nano-objects can be varied from spherical, tubular, rod-like, lamellar to fibrous [62].

12.3.1
Magnetic Cores

Polymer-grafted magnetic nanoparticles have potential applications as ferrofluids, imaging agents, bearing seals and dampers, in environmental applications [63] and as coolants. Paramagnetic materials such as gadolinium and iron oxide have been used as *in vivo* contrast agents for imaging tumors in tissue [64]. Their small size, paramagnetic character and selectivity for cancerous cells are ideal for their use as contrast agents. The same attributes are also important in the area of magnetic storage media.

Iron oxide–polymer core–shell structures were formed by first modifying the surface with a difunctional molecule containing a carboxylate group capable of binding to the surface and a secondary or tertiary halide that can act as an ATRP initiator [30, 65]. An alternative to this method involved forming the nanoparticle in a microemulsion and capping them with difunctional siloxane–ATRP coupling agent [66]. A host of monomers ranging from styrene to 2-hydroxyethyl acrylate were used in subsequent graft polymerization reactions. [67] In another report, ring-opening polymerization was used to graft poly(caprolactone) from an Fe_3O_4 core [68, 69]. In another variation, a small amount of divinylbenzene was used as the monomer to form a thin layer of cross-linked polymer around the nanoparticle. The formation of an initiator "cage" in this manner rendered the possibility of surface initiator ligand exchange improbable [70]. In another variant, block copolymers of poly(glyceryl monoacrylate) and poly(glyceryl monomethacrylate) prepared using living radical polymerizations were chemisorbed on Fe_3O_4 nanoparticle surfaces to form water-soluble particles ready for biological application. The polymer magnetic fluids were stable in 10% NaCl, 10% $CaCl_2$ or in the pH range 2–14 [71]. It is thought the shell can impact the magnetic properties of the core. A reduction in coercivity after the formation of a polystyrene shell has been observed, which has been ascribed to the reduction of magnetic surface anisotropy upon coating [72].

The "grafting on to" method has also been used in conjunction with a triblock copolymer containing an acid functionality to physisorb to the surface of an iron oxide (Fe_3O_4) core (100–200 nm in diameter) [63]. After attachment to iron nanoparticles, these core–shell structures act as targeting species for the destruction of polyhalogenated solvents [73–75]. These polymer/magnetic particles were easy to form, transportable in water, but also have an affinity for the organic/water interface and can remediate chlorinated groundwater using the reactive iron oxide.

12.3.2 Semiconductor Nanoparticles

Hybrid materials composed of rare earth metals or semiconductor nanoparticles have been used in display applications such as mercury free discharge lamps, plasma displays, organic and polymeric light-emitting diodes (LEDs) and novel laser materials [76]. Semiconductor nanoparticles are optimal for use in display materials, because they have small sizes which translate to increased resolution and because they display size-dependent photoemission, rendering the display color tunable [77]. As an example of the latter property, as the particle size of CdSe decreases from 6 to 2 nm, the photoemission wavelength also decreases (that is, from red to purple). CdS has been encased in SiO_2 and then the method applied to the formation of core–shell SiO_2 nanoparticles were applied to yield CdS–SiO_2–poly(methyl methacrylate) core–shell–shell nanoparticles [78]. These nanoparticles were applied to the creation of an aggregation-based biological sensor [79]. CdS nanoparticles were surface modified with a thioglycerol compound and then a ring-opening polymerization was performed to yield polycaprolactone-grafted nanoparticles [53].

12.3.3 Miscellaneous Cores

Titania (TiO_2) has shown promising characteristics as a high-efficiency conductive material and as a catalyst. Of the three crystalline forms of titania, anatase powders have been highlighted due to their high efficiency in the photocatalytic decomposition of harmful organic components. Nanosized TiO_2 encased in an electron-conducting polymer has promising applications as a conductive material. To from these composites, TiO_2 was surface modified with a difunctional pyrrole–acetylacetonate (acac) molecule. The acac group coordinated to the TiO_2 surface and the pyrrole group served as a polymerization initiator to yield polypyrrole [80].

Interest in zeolites complexed with poly(ethylene oxide) (PEO) stems from their application in lithium ion batteries. In its semicrystalline form, PEO has low ion conductivity due to the high degree of crystallinity; however, incorporation of regularly dispersed nanoparticles into the PEO matrix prevents crystalline domains from forming. To form this structure, amine-terminated zeolites were reacted with a difunctional molecule to form ATRP initiator sites that were successively used to graft PEO methacrylates [81]. The resulting composites were shown to be completely amorphous.

12.3.4 Inorganic Cores with Non-spherical Shapes

Most inorganic particles studied to date in core–shell particle synthesis have been spherically shaped. A few examples have been reported, however, of polymer grafting from non-spherical objects, such as nano-rods, tubes, lamellae, cubes and needles. POSS cubes offer interesting structural features because the eight corners of the cube can be modified with organic groups such as monomers or ATRP initiators. When used in polymerization reactions in the octamonomer form they create cross-linking points and when used in the octainitiator form they generate multi-arm star polymers [82].

The exfoliation of sodium montmorillonite lamellae and formation of an orientationally ordered nanocomposite was prepared by a "grafting from" method. Polystyrene was grafted to the exfoliated montmorillonite surface via NMP, ATRP and RAFT [83–85]. These polymer-bound sheets were then incorporated with a styrene–butadiene–styrene triblock copolymer (SBS). The polystyrene-covered sheets ordered preferentially parallel to the polystyrene blocks in the SBS phase [86]. Needle-shaped alumina–polystyrene core–shell composites have been prepared by the "grafting from" method with the use of 3-methacryloxypropyltrimethoxysilane and emulsion polymerization techniques [87].

Most polymerizations have been conducted from the surface of flat or convex objects; however, acrylonitrile was polymerized by ATRP from concave silica templates, filling a nanosized cavity and then carbonized to form cylindrical carbon rods (11 nm in diameter) with mesopores 2–50 nm in diameter [10]. This procedure shows an alternative way to form mesoporous carbon with primary mesopore diameter ranging from 8 to 12 nm, which is much larger than most ordered mesoporous carbons. Mesoporous carbon formed by polymerization from a silica template and then carbonized shows how the process of core–shell structure formation is applicable in the areas of adsorbents [88, 89], catalyst supports [90] and fuel cells [91, 92].

Carbon nanotubes were also used in the formation of a core–shell structures. The tube surface was modified by carbanion addition and then introduction of styrene resulted in polymer grafts. It is thought that the negative charge induced by initiator formation with *sec*-butyllithium helps to separate the tube aggregates, yielding individual structures [93]. Rod-like carbon nanotubes are of interest due to their electrical conductivity, high mechanical strength and distinctive geometric attributes, such as differentiation of the inner and outer surfaces [94]. Carbon nanotubes have been rendered water soluble by the installation of phenol groups during tube formation. Further functionalization of the hydroxyl groups and polymerization yielded poly(acrylic acid)-functionalized nanotubes which potentially have biological applications [95]. Biochemically functionalized silica nanotubes have shown promise for use in biocatalysis in the oxidation of glucose [96].

12.4 Different Types of Grafted Polymer Layers

The novel properties of nanoparticles have slowly begun to find their way into biological, electronic, non-linear optic and magnetic applications [97]. The small size of these particles makes them easier to apply to *in vivo* applications, and also more easily incorporated into other applications where miniaturization is important. The grafted polymer layer in core–shell nanoparticles is a handle through which the compatibility of the inorganic core can be tailored to suit a range of applications and, therefore, many studies have been reported on methodologies for grafting different classes of polymers from nanoparticle surfaces.

12.4.1 Polyelectrolytes and Water-soluble Polymers

Much attention is being given to grafting hydrophilic polymer and polyelectrolytes to render the particles water soluble, because as water-soluble shell structures are key for application of core–shell nanoparticles in biological applications. Armes and coworkers have recently developed a procedure that uses a polyelectrolytic initiator in conjunction with ATRP to form water-soluble SiO_2 grafted with hydrophilic methacrylates [98]. In an attempt to adapt composites for biological applications, hydroxyapatite (HAp) particles were adhered to a silk fibroin (SF) by surface grafting the fibroin with methacryloxypropyltrimethoxysilane via a free-radical polymerization. Upon treatment of the polymer-grafted surface with hydroxyapatite, the free siloxanes in the polymer layer were bound to the hydroxyl groups of the inorganic material. Hydroxyapatite shows a strong affinity for cell adhesion, so the overall material could be used to prevent infections in percutaneous applications (Fig. 12.9) [99].

Core–shell structures have allowed researchers to compatibilize traditionally synthetic materials by encasing them in a biocompatible polymer. Polyethylene glycol (PEG)-coated silica particles have been evaluated as a model for encapsulation of other reagents, possibly small molecules or specific proteins for *in vivo* diagnosis, analysis and measurement [100]. Core–shell nanostructures have been used in conjunction with surface-enhanced Raman scattering (SERS)-like spectra as tags for ultrasensitive detection of biomolecules [101].

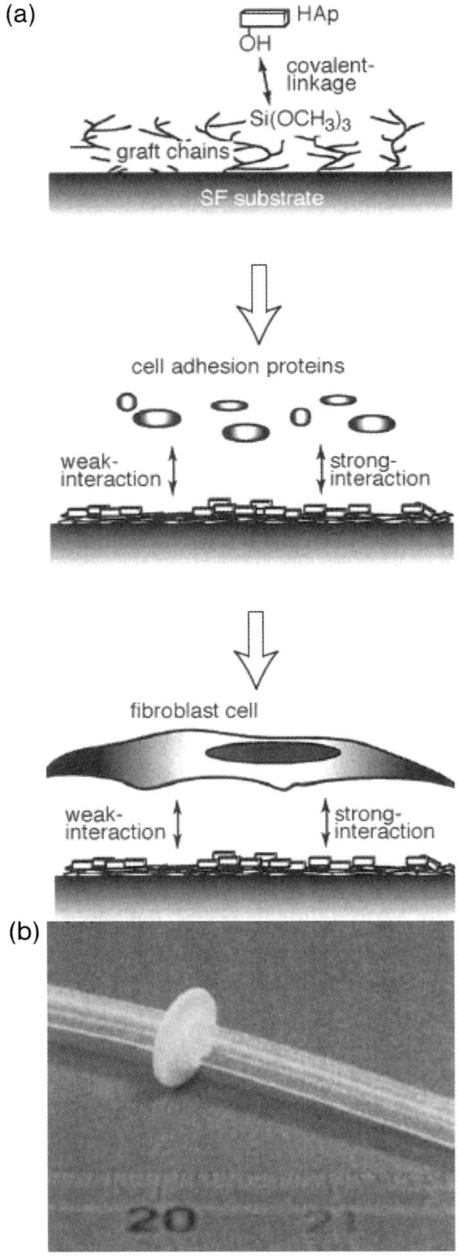

Fig. 12.9 (a) Preparation of composite used as skin adhesive for prevention of infection and expected mechanism of cell adhesion on HAp-coated SF and (b) prototype of percutaneous device [99].

12.4.2 Block Copolymers and Polymer Brushes

Organic-inorganic hybrids with block copolymers have been formed using both the "grafting from" and "grafting on to" approach to form. In using the "grafting from" approach ATRP has been used for block copolymer formation [34, 102]. The adhesion of the copolymer to the particle was dependent on the grafting density of the initiator and particle size [103]. AFM micrographs of poly(styrene)-*b*-poly(benzyl acrylate) formed using ATRP from ~20-nm POSS cores have shown phase separation between the polymer shells on the nanometer scale [104, 105].

The "grafting on to" methodology allows block copolymers to function as selective adsorption agents for nanoparticles [106]. In the case of patterning nanoparticles to surfaces through the adhesive properties in the polymer, certain parameters have been found to be paramount: particle size, polymer–particle interactions [107, 108], polymer molecular weight (MW) [109–111] and the polymer grafting density. A study using poly(*N*-isopropylacrylamide) (PNIPAAm) brushes showed that absorption of ~16-nm Au particles remained on the surface of the brushes, because they were larger than the dimensions of the polymer, whereas 3.5-nm Au particles intercalated into the polymer brush until a saturation level was reached. Both the "grafting from" and "grafting on to" methodologies allow for interesting properties of block copolymers and their incorporation with nanoparticles.

12.5 Unusual Structures

Because the preparation of core–shell inorganic–polymer nanoparticles is a rapidly maturing field, these particles are now being used as building blocks and templates for the construction of more elaborate structures [112].

12.5.1 Hollow Spheres

Walt et al. [119] reported the synthesis of hollow spheres using core–shell nanoparticles as a template. First, poly(benzyl methacrylate) was grafted from silica nanoparticles and then the silica core was etched using HF. Because poly(benzyl methacrylate) has a high T_g, the polymer shell was left intact after etching [113], leaving a hollow sphere for the insertion of other chemicals such as metals to generate magnetic particles [114]. In another case, Au–air–poly(benzyl methacrylate) "sphere in a shell" structures were formed by first coating an Au nanoparticle with silica (via sol–gel chemistry with tetraethoxysilane) and then surface modifying the particle with an ATRP initiator. A poly(benzyl methacrylate) shell was formed around the silica layer, which after etching left an Au particle "floating" within the interior of the polymer shell [115]. Another interesting architecture that has been formed was an "inverse" core–shell structure with the polymer layer as the core and the inorganic phase as the shell. Micelles were formed using a silanol-functionalized amphiphilic block copolymer. The micelle was then mixed with sodium silicate to form a silica layer around the polymer core [116]. These types of hollow-sphere structures potentially could be used as nano-reactors or nano-containers.

12.5.2 Magnetic Rings

Core–shell polystyrene-coated Fe_3O_4 nanoparticles prepared using ATRP "grafting from" method were cast on to a self-assembled monolayer of hexadecanethiol disks within a 16-mercaptohexadecanoic acid (MHA) matrix [117]. After dip casting and drying the self-assembled monolayer surface from a chloroform solution of polystyrene-coated Fe_3O_4 particles, magnetic rings formed at the edges of the hexadecanethiol disks. Magnetic force microscopy indicated that the particles within the rings maintained their magnetic behavior.

12.6 Conclusions

Using mainly SiO_2 and Au nanoparticles both the "grafting on to" and "grafting from" polymerization methods have been used to make functional materials. Both controlled radical polymerizations and living polymerizations have been developed to coat the nano-objects. With the combination of different types of polymerizations allowing for many monomers with differing physical properties and the many types of nano-objects offering optical and magnetic properties, core–shell structures are ripe for incorporation into multifunctional systems from bio-devices to sensor applications. This marriage of organic, inorganic and physical sciences is the foundation for growth into the miniaturization of devices and the discovery and utilization of new materials properties.

References

1 E. Bourgeat-Lami, Organic–inorganic nanostructured colloids, *Journal of Nanoscience and Nanotechnology 2* (**2002**) 1–24.

2 X. Ding, J. Zhao, Y. Liu, H. Zhang, Z. Wang, Silica nanoparticles encapsulated by polystyrene via surface grafting and *in situ* emulsion polymerization, *Materials Letters 58* (**2004**) 3126–3130.

3 J. Pyun, K. Matyjaszewski, Synthesis of nanocomposite organic/inorganic hybrid materials using controlled/"living" radical polymerization, *Chemistry of Materials 13* (**2001**) 3436–3448.

4 J. Pyun, T. Kowalewski, K. Matyjaszewski, Synthesis of polymer brushes using atom transfer radical polymerization, *Macromolecular Rapid Communications 24* (**2003**) 1043–1059.

5 F. Caruso, Nanoengineering of particle surfaces, *Advanced Materials (Weinheim, Germany) 13* (**2001**) 11–22.

6 R.C. Advincula, Surface initiated polymerization from nanoparticle surfaces, *Journal of Dispersion Science and Technology 24* (**2003**) 343–361.

7 C. Bartholome, E. Beyou, E. Bourgeat-Lami, P. Chaumont, F. Lefebvre, N. Zydowicz, Nitroxide-mediated polymerization of styrene initiated from the surface of silica nanoparticles. *In situ* generation and grafting of alkoxyamine initiators, *Macromolecules 38* (**2005**) 1099–1106.

8 J. Pyun, S. Jia, T. Kowalewski, G.D. Patterson, K. Matyjaszewski, Synthesis and characterization of organic/inorganic hybrid nanoparticles: kinetics of surface-initiated atom transfer radical polymerization and morphology of hybrid nanoparticle ultrathin films, *Macromolecules 36* (**2003**) 6952.

9 A. Ramakrishnan, R. Dhamodharan, J. Ruhe, Controlled growth of PMMA brushes on silicon surfaces at room temperature, *Macromolecular Rapid Communications 23* (**2002**) 612–616.

10 M. Kruk, B. Dufour, E.B. Celer, T. Kowalewski, M. Jaroniec, K. Matyjaszewski, Synthesis of mesoporous carbons using ordered and disordered mesoporous silica templates and polyacrylonitrile as carbon precursor, *Journal of Physical Chemistry B 109* (**2005**) 9216–9225.

11 T.A. von Werne, D.S. Germack, E.C. Hagberg, V.V. Sheares, C.J. Hawker, K.R. Carter, A versatile method for tuning the chemistry and size of nanoscopic features by living free radical polymerization, *Journal of the American Chemical Society 125* (**2003**) 3831–3838.

12 H. Mori, D.C. Seng, M. Zhang, A.H.E. Mueller, Hybrid silica nanoparticles with hyperbranched polymer and polyelectrolyte shells, *Progress in Colloid and Polymer Science 126* (**2004**) 40–43.

13 T. von Werne, T.E. Patten, Preparation of structurally well-defined polymer-nanoparticle hybrids with controlled/living radical polymerizations, *Journal of the American Chemical Society 121* (**1999**) 7409–7410.

14 D.O.H. Teare, D.C. Barwick, W.C.E. Schofield, R.P. Garrod, L.J. Ward, J.P.S. Badyal, Substrate-independent approach for polymer brush growth by surface atom transfer radical polymerization, *Langmuir 21* (**2005**) 11425–11430.

15 R. Advincula, Q. Zhou, M. Park, S. Wang, J. Mays, G. Sakellariou, S. Pispas, N. Hadjichristidis, Polymer brushes by living anionic surface initiated polymerization on flat silicon (SiO_x) and gold surfaces: homopolymers and block copolymers, *Langmuir 18* (**2002**) 8672–8684.

16 Q. Zhou, S. Wang, X. Fan, R. Advincula, J. Mays, Living anionic surface-initiated polymerization (LASIP) of a polymer on silica nanoparticles, *Langmuir 18* (**2002**) 3324–3331.

17 C. Bartholome, E. Beyou, E. Bourgeat-Lami, P. Chaumont, N. Zydowicz, Nitroxide-mediated polymerizations from silica nanoparticle surfaces: "graft from" polymerization of styrene using a triethoxysilyl-terminated alkoxyamine initiator, *Macromolecules 36* (**2003**) 7946–7952.

18 L. Andruzzi, A. Hexemer, X. Li, C.K. Ober, E.J. Kramer, G. Galli, E. Chiellini, D.A. Fischer, Control of surface properties using fluorinated polymer brushes produced by surface-initiated controlled radical polymerization, *Langmuir 20* (**2004**) 10498–10506.

19 Y.-P. Wang, X.-W. Pei, X.-Y. He, K. Yuan, Synthesis of well-defined, polymer-grafted silica nanoparticles via reverse ATRP, *European Polymer Journal 41* (**2005**) 1326–1332.

20 K. Ohno, K.-m. Koh, Y. Tsujii, T. Fukuda, Synthesis of gold nanoparticles coated with well-defined, high-density polymer brushes by surface-initiated living radical polymerization, *Macromolecules 35* (**2002**) 8989–8993.

21 J. Raula, J. Shan, M. Nuopponen, A. Niskanen, H. Jiang, E.I. Kauppinen, H. Tenhu, Synthesis of gold nanoparticles grafted with a thermoresponsive polymer by surface-induced reversible-addition–fragmentation chain-transfer polymerization, *Langmuir 19* (**2003**) 3499–3504.

22 S. Nuss, H. Bottcher, H. Wurm, M.L. Hallensleben, Gold nanoparticles with

covalently attached polymer chains, *Angewandte Chemie, International Edition 40* (**2001**) 4016–4018.
23 A. Kotal, T.K. Mandal, D.R. Walt, Synthesis of gold–poly(methyl methacrylate) core–shell nanoparticles by surface-confined atom transfer radical polymerization at elevated temperature, *Journal of Polymer Science, Part A: Polymer Chemistry 43* (**2005**) 3631–3642.
24 X. Chen, D.P. Randall, C. Perruchot, J.F. Watts, T.E. Patten, T. von Werne, S.P. Armes, Synthesis and aqueous solution properties of polyelectrolyte-grafted silica particles prepared by surface-initiated atom transfer radical polymerization, *Journal of Colloid and Interface Science 257* (**2003**) 56–64.
25 D. Shah, G. Fytas, D. Vlassopoulos, J. Di, D. Sogah, E.P. Giannelis, Structure and dynamics of polymer-grafted clay suspensions, *Langmuir 21* (**2005**) 19–25.
26 A. El Harrak, G. Carrot, J. Oberdisse, C. Eychenne-Baron, F. Boue, Surface-atom transfer radical polymerization from silica nanoparticles with controlled colloidal stability, *Macromolecules 37* (**2004**) 6376–6384.
27 G. Zheng, H.D.H. Stoever, Grafting of polystyrene from narrow disperse polymer particles by surface-initiated atom transfer radical polymerization, *Macromolecules 35* (**2002**) 6828–6834.
28 P. Liu, W. Liu, Q. Xue, Preparation of comb-like styrene grafted silica nanoparticles, *Journal of Macromolecular Science, Pure and Applied Chemistry A41* (**2004**) 1001–1010.
29 Y.-P. Wang, X.-W. Pei, K. Yuan, Reverse ATRP grafting from silica surface to prepare well-defined organic/inorganic hybrid nanocomposite, *Materials Letters 59* (**2004**) 520–523.
30 S.M. Gravano, R. Dumas, K. Liu, T.E. Patten, Methods for the surface functionalization of γ-Fe_2O_3 nanoparticles with initiators for atom transfer radical polymerization and the formation of core–shell inorganic–polymer structures, *Journal of Polymer Science, Part A: Polymer Chemistry 43* (**2005**) 3675–3688.
31 G. Carrot, D. Rutot-Houze, A. Pottier, P. Degee, J. Hilborn, P. Dubois, Surface-initiated ring-opening polymerization: a versatile method for nanoparticle ordering, *Macromolecules 35* (**2002**) 8400–8404.
32 J. Parvole, G. Laruelle, A. Khoukh, L. Billon, Surface initiated polymerization of poly(butyl acrylate) by nitroxide mediated polymerization: first comparative polymerization of a bimolecular and a unimolecular initiator-grafted silica particles, *Macromolecular Chemistry and Physics 206* (**2005**) 372–382.
33 K. Koh, S. Sugiyama, T. Morinaga, K. Ohno, Y. Tsujii, T. Fukuda, M. Yamahiro, T. Iijima, H. Oikawa, K. Watanabe, T. Miyashita, Precision synthesis of a fluorinated polyhedral oligomeric silsesquioxane-terminated polymer and surface characterization of its blend film with poly(methyl methacrylate), *Macromolecules 38* (**2005**) 1264–1270.
34 J. Pyun, S. Jia, T. Kowalewski, G.D. Patterson, K. Matyjaszewski, Synthesis and characterization of organic/inorganic hybrid nanoparticles: kinetics of surface-initiated atom transfer radical polymerization and morphology of hybrid nanoparticle ultrathin films, *Macromolecules 36* (**2003**) 5094–5104.
35 Y. Kwak, A. Goto, K. Komatsu, Y. Sugiura, T. Fukuda, Characterization of low-mass model 3-arm stars produced in reversible addition–fragmentation chain transfer (RAFT) process, *Macromolecules 37* (**2004**) 4434–4440.
36 T. von Werne, T.E. Patten, Atom transfer radical polymerization from nanoparticles: a tool for the preparation of well-defined hybrid nanostructures and for understanding the chemistry of controlled/"living" radical polymerizations from surfaces, *Journal of the American Chemical Society 123* (**2001**) 7497–7505.
37 G. Carrot, S. Diamanti, M. Manuszak, B. Charleux, J.P. Vairon, Atom transfer radical polymerization of *n*-butyl acrylate from silica nanoparticles, *Journal of Polymer Science, Part A: Polymer Chemistry 39* (**2001**) 4294–4301.
38 K. Ohno, T. Morinaga, K. Koh, Y. Tsujii, T. Fukuda, Synthesis of monodisperse silica particles coated with well-defined, high-density polymer brushes by surface-in-

itiated atom transfer radical polymerization, *Macromolecules 38* (**2005**) 2137–2142.

39 J. Bai, J.-B. Pang, K.-Y. Qiu, Y. Wei, Synthesis and characterization of structurally well-defined polymer–inorganic hybrid nanoparticles via atom transfer radical polymerization, *Chinese Journal of Polymer Science 20* (**2002**) 261–267.

40 K. Matyjaszewski, P. J. Miller, N. Shukla, B. Immaraporn, A. Gelman, B. B. Luokala, T. M. Siclovan, G. Kickelbick, T. Vallant, H. Hoffmann, T. Pakula, Polymers at interfaces: using atom transfer radical polymerization in the controlled growth of homopolymers and block copolymers from silicon surfaces in the absence of untethered sacrificial initiator, *Macromolecules 32* (**1999**) 8716–8724.

41 D. Li, X. Sheng, B. Zhao, Environmentally responsive "hairy" nanoparticles: mixed homopolymer brushes on silica nanoparticles synthesized by living radical polymerization techniques, *Journal of the American Chemical Society 127* (**2005**) 6248–6256.

42 J.-B. Kim, W. Huang, M. D. Miller, G. L. Baker, M. L. Bruening, Kinetics of surface-initiated atom transfer radical polymerization, *Journal of Polymer Science, Part A: Polymer Chemistry 41* (**2003**) 386–394.

43 J. Adebahr, N. Byrne, M. Forsyth, D. R. MacFarlane, P. Jacobsson, Enhancement of ion dynamics in PMMA-based gels with addition of TiO_2 nano-particles, *Electrochimica Acta 48* (**2003**) 2099–2103.

44 K. N. Jayachandran, A. Takacs-Cox, D. E. Brooks, Synthesis and characterization of polymer brushes of poly(*N,N*-dimethylacrylamide) from polystyrene latex by aqueous atom transfer radical polymerization, *Macromolecules 35* (**2002**) 4247–4257.

45 H. Mori, D. C. Seng, M. Zhang, A. H. E. Mueller, Hybrid nanoparticles with hyperbranched polymer shells via self-condensing atom transfer radical polymerization from silica surfaces, *Langmuir 18* (**2002**) 3682–3693.

46 R. Inoubli, S. Dagreou, A. Khoukh, F. Roby, J. Peyrelasse, L. Billon, 'Graft from' polymerization on colloidal silica particles: elaboration of alkoxyamine grafted surface by in situ trapping of carbon radicals, *Polymer 46* (**2005**) 2486–2496.

47 S. Blomberg, S. Ostberg, E. Harth, A. W. Bosman, B. van Horn, C. J. Hawker, Production of crosslinked, hollow nanoparticles by surface-initiated living free-radical polymerization, *Journal of Polymer Science, Part A: Polymer Chemistry 40* (**2002**) 1309–1320.

48 P. Liu, J. Tian, W. Liu, Q. Xue, Surface-initiated atom transfer radical polymerization (ATRP) of styrene from silica nanoparticles under UV irradiation, *Polymer International 53* (**2004**) 127–130.

49 C. Li, B. C. Benicewicz, Synthesis of well-defined polymer brushes grafted onto silica nanoparticles via surface reversible addition–fragmentation chain transfer polymerization, *Macromolecules 38* (**2005**) 5929–5936.

50 C.-Y. Hong, Y.-Z. You, C.-Y. Pan, Synthesis of water-soluble multiwalled carbon nanotubes with grafted temperature-responsive shells by surface RAFT polymerization, *Chemistry of Materials 17* (**2005**) 2247–2254.

51 A. B. Lowe, B. S. Sumerlin, M. S. Donovan, C. L. McCormick, Facile preparation of transition metal nanoparticles stabilized by well-defined (co)polymers synthesized via aqueous reversible addition–fragmentation chain transfer polymerization, *Journal of the American Chemical Society 124* (**2002**) 11562–11563.

52 L. Zheng, A. F. Xie, J. T. Lean, Polystyrene nanoparticles with anionically polymerized polybutadiene brushes, *Macromolecules 37* (**2004**) 9954–9962.

53 D. Rutot-Houze, W. Fris, P. Degee, P. Dubois, Controlled ring-opening (co)polymerization of lactones initiated from cadmium sulfide nanoparticles, *Journal of Macromolecular Science, Pure and Applied Chemistry A41* (**2004**) 697–711.

54 M. Joubert, C. Delaite, E. B. Lami, P. Dumas, Synthesis of poly(epsilon-caprolactone)–silica nanocomposites: from hairy colloids to core–shell nanoparticles, *New Journal of Chemistry 29* (**2005**) 1601–1609.

55 M. A. Jordi, T. A. P. Seery, Quantitative determination of the chemical composi-

tion of silica–poly(norbornene) nanocomposites, *Journal of the American Chemical Society 127* (**2005**) 4416–4422.

56 A.-F. Mingotaud, S. Reculusa, C. Mingotaud, P. Keller, C. Sykes, E. Duguet, S. Ravaine, Ring-opening metathesis polymerization on well defined silica nanoparticles leading to hybrid core–shell particles, *Journal of Materials Chemistry 13* (**2003**) 1920–1925.

57 K. J. Watson, J. Zhu, S. T. Nguyen, C. A. Mirkin, Redox-active polymer–nanoparticle hybrid materials, *Pure and Applied Chemistry 72* (**2000**) 67–72.

58 S. R. Ahmed, P. Kofinas, Magnetic properties and morphology of block copolymer–cobalt oxide nanocomposites, *Journal of Magnetism and Magnetic Materials 288* (**2005**) 219–223.

59 R. F. Mulligan, A. A. Iliadis, P. Kofinas, Synthesis and characterization of ZnO nanoparticles within diblock copolymers, *Polymeric Materials Science and Engineering 83* (**2000**) 447.

60 H. Skaff, M. F. Ilker, E. B. Coughlin, T. Emrick, Preparation of cadmium selenide–polyolefin composites from functional phosphine oxides and ruthenium-based metathesis, *Journal of the American Chemical Society 124* (**2002**) 5729–5733.

61 D. A. Savin, J. Pyun, G. D. Patterson, T. Kowalewski, K. Matyjaszewski, Synthesis and characterization of silica-graft-polystyrene hybrid nanoparticles: effect of constraint on the glass-transition temperature of spherical polymer brushes, *Journal of Polymer Science, Part B: Polymer Physics 40* (**2002**) 2667–2676.

62 G. Schmidt, M. M. Malwitz, Properties of polymer–nanoparticle composites, *Current Opinion in Colloid and Interface Science 8* (**2003**) 103–108.

63 N. Saleh, T. Phenrat, K. Sirk, B. Dufour, J. Ok, T. Sarbu, K. Matyjaszewski, R. D. Tilton, G. V. Lowry, Adsorbed triblock copolymers deliver reactive iron nanoparticles to the oil/water interface, *Nano Letters 5* (**2005**) 2489–2494.

64 V. P. Torchilin, PEG-based micelles as carriers of contrast agents for different imaging modalities, *Advanced Drug Delivery Reviews 54* (**2002**) 235–252.

65 Y. Wang, X. Teng, J.-S. Wang, H. Yang, Solvent-free atom transfer radical polymerization in the synthesis of Fe_2O_3–polystyrene core–shell nanoparticles, *Nano Letters 3* (**2003**) 789–793.

66 D. Holzinger, G. Kickelbick, Hybrid inorganic–organic core–shell metal oxide nanoparticles from metal salts, *Journal of Materials Chemistry 14* (**2004**) 2017–2023.

67 P. Liu, L. Zhang, Z. Su, Surface-initiated ATRP of HEA from nanocrystal α-Fe_2O_3 under ultrasonic irradiation, *Journal of Nanoscience and Nanotechnology 5* (**2005**) 1713–1717.

68 A. M. Schmidt, The synthesis of magnetic core–shell nanoparticles by surface-initiated ring-opening polymerization of ε-caprolactone, *Macromolecular Rapid Communications 26* (**2005**) 93–97.

69 C. Flesch, E. Bourgeat-Lami, S. Mornet, E. Duguet, C. Delaite, P. Dumas, Synthesis of colloidal superparamagnetic nanocomposites by grafting poly(ε-caprolactone) from the surface of organosilane-modified maghemite nanoparticles, *Journal of Polymer Science, Part A: Polymer Chemistry 43* (**2005**) 3221–3231.

70 G. Li, J. Fan, R. Jiang, Y. Gao, Cross-linking the linear polymeric chains in the atrp synthesis of iron oxide/polystyrene core/shell nanoparticles, *Chemistry of Materials 16* (**2004**) 1835–1837.

71 S. Wan, Y. Zheng, Y. Liu, H. Yan, K. Liu, Fe_3O_4 nanoparticles coated with homopolymers of glycerol mono(meth)acrylate and their block copolymers, *Journal of Materials Chemistry 15* (**2005**) 3424–3430.

72 C. R. Vestal, Z. J. Zhang, Atom transfer radical polymerization synthesis and magnetic characterization of $MnFe_2O_4$/polystyrene core/shell nanoparticles, *Journal of the American Chemical Society 124* (**2002**) 14312–14313.

73 Y. Liu, H. Choi, D. Dionysiou, G. V. Lowry, Trichloroethene hydrodechlorination in water by highly disordered monometallic nanoiron, *Chemistry of Materials 17* (**2005**) 5315–5322.

74 Y. Liu, S. A. Majetich, R. D. Tilton, D. S. Sholl, G. V. Lowry, TCE dechlorination rates, pathways and efficiency of nanoscale iron particles with different proper-

ties, *Environmental Science and Technology 39* (**2005**) 1338–1345.

75 D.W. Elliott, W.-x. Zhang, Field assessment of nanoscale bimetallic particles for groundwater treatment, *Environmental Science and Technology 35* (**2001**) 4922–4926.

76 M.M. Lezhnina, H. Katker, U.H. Kynast, Rare-earth ions in porous matrices, *Physics of the Solid State 47* (**2005**) 1479–1484.

77 G. Bacher, H. Schomig, M.K. Welsch, M. Scheibner, J. Seufert, M. Obert, A. Forchel, A.A. Maksimov, S. Zaitsev, V.D. Kulakovskii, Nano-optics on individual quantum objects – from single to coupled semiconductor quantum dots, *Acta Physica Polonica, A 102* (**2002**) 475–494.

78 S.C. Farmer, T.E. Patten, Synthesis of electroluminescent organic/inorganic polymer nanocomposites, *Polymer Preprints 42* (**2001**) 578–579.

79 P.J. Costanzo, T.E. Patten, T.A.P. Seery, Protein–ligand mediated aggregation of nanoparticles: a study of synthesis and assembly mechanism, *Chemistry of Materials 16* (**2004**) 1775–1785.

80 S. Roux, G.J.A.A. Soler-Illia, S. Demoustier-Champagne, P. Audebert, C. Sanchez, Titania/polypyrrole hybrid nanocomposites built from *in-situ* generated organically functionalized nanoanatase building blocks, *Advanced Materials (Weinheim, Germany) 15* (**2003**) 217–221.

81 B.-Z. Zhan, M.A. White, P. Fancy, C.A. Kennedy, M. Lumsden, Functionalization of a nano-faujasite zeolite with PEG-grafted PMA tethers using atom transfer radical polymerization, *Macromolecules 37* (**2004**) 2748–2753.

82 D. Holzinger, G. Kickelbick, Modified cubic spherosilicates as macroinitiators for the synthesis of inorganic–organic starlike polymers, *Journal of Polymer Science, Part A: Polymer Chemistry 40* (**2002**) 3858–3872.

83 M.W. Weimer, H. Chen, E.P. Giannelis, D.Y. Sogah, Direct synthesis of dispersed nanocomposites by *in situ* living free radical polymerization using a silicate-anchored initiator, *Journal of the American Chemical Society 121* (**1999**) 1615–1616.

84 H. Boettcher, M.L. Hallensleben, S. Nuss, H. Wurm, J. Bauer, P. Behrens, Organic/inorganic hybrids by 'living'/controlled ATRP grafting from layered silicates, *Journal of Materials Chemistry 12* (**2002**) 1351–1354.

85 N. Salem, D.A. Shipp, Polymer-layered silicate nanocomposites prepared through in situ reversible addition–fragmentation chain transfer (RAFT) polymerization, *Polymer 46* (**2005**) 8573–8581.

86 Y.-H. Ha, Y. Kwon, T. Breiner, E.P. Chan, T. Tzianetopoulou, R.E. Cohen, M.C. Boyce, E.L. Thomas, An orientationally ordered hierarchical exfoliated clay–block copolymer nanocomposite, *Macromolecules 38* (**2005**) 5170–5179.

87 Z. Zeng, J. Yu, Z.-X. Guo, Preparation of functionalized core–shell alumina/polystyrene composite nanoparticles, 1. Encapsulation of alumina via emulsion polymerization, *Macromolecular Chemistry and Physics 206* (**2005**) 1558–1567.

88 S. Han, K. Sohn, T. Hyeon, Fabrication of new nanoporous carbons through silica templates and their application to the adsorption of bulky dyes, *Chemistry of Materials 12* (**2000**) 3337–3341.

89 M. Choi, R. Ryoo, Ordered nanoporous polymer–carbon composites, *Nature Materials 2* (**2003**) 473–476.

90 W.S. Ahn, K.I. Min, Y.M. Chung, H.K. Rhee, S.H. Joo, R. Ryoo, Novel mesoporous carbon as a catalyst support for Pt and Pd for liquid phase hydrogenation reactions, *Studies in Surface Science and Catalysis 135* (**2001**) 4710–4717.

91 S.H. Joo, S.J. Choi, I. Oh, J. Kwak, Z. Liu, O. Terasaki, R. Ryoo, Ordered nanoporous arrays of carbon supporting high dispersions of platinum nanoparticles, *Nature 412* (**2001**) 169–172.

92 K.-Y. Chan, J. Ding, J. Ren, S. Cheng, K.Y. Tsang, Supported mixed metal nanoparticles as electrocatalysts in low temperature fuel cells, *Journal of Materials Chemistry 14* (**2004**) 505–516.

93 G. Viswanathan, N. Chakrapani, H. Yang, B. Wei, H. Chung, K. Cho, C.Y. Ryu, P.M. Ajayan, Single-step in situ synthesis of polymer-grafted single-wall nanotube composites, *Journal of the*

American Chemical Society 125 (**2003**) 9258–9259.

94 S.J. Son, J. Reichel, B. He, M. Schuchman, S.B. Lee, Magnetic nanotubes for magnetic-field-assisted bioseparation, biointeraction and drug delivery, *Journal of the American Chemical Society 127* (**2005**) 7316–7317.

95 Z. Yao, N. Braidy, G.A. Botton, A. Adronov, Polymerization from the surface of single-walled carbon nanotubes – preparation and characterization of nanocomposites, *Journal of the American Chemical Society 125* (**2003**) 16015–16024.

96 D.T. Mitchell, S.B. Lee, L. Trofin, N. Li, T.K. Nevanen, H. Soederlund, C.R. Martin, Smart nanotubes for bioseparations and biocatalysis, *Journal of the American Chemical Society 124* (**2002**) 11864–11865.

97 B.C. Sih, M.O. Wolf, Metal nanoparticle-conjugated polymer nanocomposites, *Chemical Communications* (**2005**) 3375–3384.

98 X.Y. Chen, S.P. Armes, S.J. Greaves, J.F. Watts, Synthesis of hydrophilic polymer-grafted ultrafine inorganic oxide particles in protic media at ambient temperature via atom transfer radical polymerization: use of an electrostatically adsorbed polyelectrolytic macroinitiator, *Langmuir 20* (**2004**) 587–595.

99 T. Furuzono, S. Yasuda, T. Kimura, S. Kyotani, J. Tanaka, A. Kishida, Nanoscaled hydroxyapatite/polymer composite IV. Fabrication and cell adhesion properties of a three-dimensional scaffold made of composite material with a silk fibroin substrate to develop a percutaneous device, *Journal of Artificial Organs 7* (**2004**) 137–144.

100 H. Xu, F. Yan, E.E. Monson, R. Kopelman, Room-temperature preparation and characterization of poly(ethylene glycol)-coated silica nanoparticles for biomedical applications, *Journal of Biomedical Materials Research, Part A 66A* (**2003**) 870–879.

101 X. Su, J. Zhang, L. Sun, T.-W. Koo, S. Chan, N. Sundararajan, M. Yamakawa, A.A. Berlin, Composite organic–inorganic nanoparticles (COINs) with chemically encoded optical signatures, *Nano Letters 5* (**2005**) 49–54.

102 G. Laruelle, J. Parvole, J. Francois, L. Billon, Block copolymer grafted-silica particles: a core/double shell hybrid inorganic/organic material, *Polymer 45* (**2004**) 5013–5020.

103 H. Li, H. Zhang, Y. Xu, K. Zhang, P. Ai, X. Jin, J. Wang, Study of thick-coated spherical polymer brushes grown from a silicon gel surface by atom transfer radical polymerization, *Materials Chemistry and Physics 90* (**2005**) 90–94.

104 J. Pyun, K. Matyjaszewski, T. Kowalewski, D. Savin, G. Patterson, G. Kickelbick, N. Huesing, Synthesis of well-defined block copolymers tethered to polysilsesquioxane nanoparticles and their nanoscale morphology on surfaces, *Journal of the American Chemical Society 123* (**2001**) 9445–9446.

105 J. Pyun, K. Matyjaszewski, Synthesis of hybrid polymers using atom transfer radical polymerization: homopolymers and block copolymers from polyhedral oligomeric silsesquioxane monomers, *Macromolecules 33* (**2000**) 217–220.

106 R.R. Bhat, J. Genzer, Combinatorial study of nanoparticle dispersion in surface-grafted macromolecular gradients, *Applied Surface Science 252* (**2006**) 2549–2554.

107 R.A. Gage, E.P.K. Currie, M.A.C. Stuart, Adsorption of nanocolloidal SiO_2 particles on PEO brushes, *Macromolecules 34* (**2001**) 5078–5080.

108 V. Pardo-Yissar, R. Gabai, A.N. Shipway, T. Bourenko, I. Willner, Gold nanoparticle/hydrogel composites with solvent-switchable electronic properties, *Advanced Materials (Weinheim, Germany) 13* (**2001**) 1320–1323.

109 Z. Liu, K. Pappacena, J. Cerise, J. Kim, C.J. Durning, B. O'Shaughnessy, R. Levicky, Organization of nanoparticles on soft polymer surfaces, *Nano Letters 2* (**2002**) 219–224.

110 R.R. Bhat, J. Genzer, B.N. Chaney, H.W. Sugg, A. Liebmann-Vinson, Controlling the assembly of nanoparticles using surface grafted molecular and

macromolecular gradients, *Nanotechnology 14* (**2003**) 1145–1152.
111 R.R. Bhat, M.R. Tomlinson, J. Genzer, Assembly of nanoparticles using surface-grafted orthogonal polymer gradients, *Macromolecular Rapid Communications 25* (**2004**) 270–274.
112 R.A. Caruso, A. Susha, F. Caruso, Multilayered titania, silica and laponite nanoparticle coatings on polystyrene colloidal templates and resulting inorganic hollow spheres, *Chemistry of Materials 13* (**2001**) 400–409.
113 G.-D. Fu, Z. Shang, L. Hong, E.-T. Kang, K.-G. Neoh, Nanoporous, ultra-low-dielectric-constant fluoropolymer films from agglomerated and cross-linked hollow nanospheres of poly(pentafluorostyrene)-block-poly(divinylbenzene), *Advanced Materials (Weinheim, Germany) 17* (**2005**) 2622–2626.
114 X. Xu, S.A. Asher, Synthesis and utilization of monodisperse hollow polymeric particles in photonic crystals, *Journal of the American Chemical Society 126* (**2004**) 7940–7945.
115 K. Kamata, Y. Lu, Y. Xia, Synthesis and characterization of monodispersed core–shell spherical colloids with movable cores, *Journal of the American Chemical Society 125* (**2003**) 2384–2385.
116 K. Koh, K. Ohno, Y. Tsujii, T. Fukuda, Precision synthesis of organic/inorganic hybrid nanocapsules with a silanol-functionalized micelle template, *Angewandte Chemie, International Edition 42* (**2003**) 4194–4197.
117 L. An, W. Li, Y. Nie, B. Xie, Z. Li, J. Zhang, B. Yang, Patterned magnetic rings fabricated by dewetting of polymer-coated magnetite nanoparticles solution, *Journal of Colloid and Interface Science 288* (**2005**) 503–507.
118 X. Chen, S.P. Armes, Surface polymerization of hydrophilic methacrylates from ultrafine silica sols in protic media at ambient temperature: a novel approach to surface functionalization using a polyelectrolytic macroinitiator, *Advanced Materials (Weinheim, Germany) 15* (**2003**) 1558–1562.
119 T.K. Maudal, M.S. Fleming, D.R. Walt, Production of hollow micro spheres by surface-confined living radical polymerization on silica templates, *Chem. Mater. 12* (**2000**) 3481–3487.

13
Core–Shell Particles [1)]

Anna Musyanovych and Katharina Landfester

13.1
Introduction

For more than 50 years, latex particles in the size range between 50 and 500 nm have been extensively used for both fundamental and practical purposes. One of the routes to designing colloidal particles with novel physical and chemical properties involves taking the core particles and modifying their surface with various materials. Core–shell particles typically refer to the structured composite particles consisting of at least two materials with different chemical composition, i.e. one material forms the core and the other the shell of the particles. Core–shell particles have attracted increasing interest in a wide variety of applications such as paints, coatings, adhesives, cosmetics, magnetic storage materials, catalysts, impact modifiers and biomedical-related areas. Mechanical, optical, thermal, electrical, magnetic and catalytic properties of the materials can vary in a broad range utilizing core–shell particles. This is due to the fact that, compared with materials made by blending or copolymerization, particles of such well-defined morphology have unique physical properties as different material attributes can be associated with well-defined compartments – the core and the shell. For example, core–shell particles differing in their glass transition temperatures (T_g) may be used to modify the properties of latex-based paints [1, 2]. Here, a high-T_g core imparts improved mechanical stability, whereas the low-T_g shell allows for good film-forming ability. Core–shell latexes with polymer phases differing in pH sensitivity have been made to manufacture void-containing particles that can be used as opaquifiers in coatings [3]. Particles with a soft core and hard shell can be applied as thermoplastic elastomers or as additives in high-impact plastics. Furthermore, the core–shell particles can be applied as a model system for studying the material properties, because they can be homogeneously distributed in the matrix and in the ideal case possess a regular and spherical shaped structure.

1) A List of Abbreviations can be found at the end of this chapter.

Macromolecular Engineering. Precise Synthesis, Materials Properties, Applications.
Edited by K. Matyjaszewski, Y. Gnanou, and L. Leibler

ISBN: 978-3-527-31446-1

Different kinds of organic and inorganic materials can be involved in the formation of core–shell particles. The core of the particle can be in the form of a solid or liquid and the shell can be designed according to the needs and application.

The aim of the present chapter is to give an overview of sub-micron size core–shell particles. First, we describe the theoretical prediction of the (thermodynamic) control of two-phase particle morphology. The influence of various parameters will be discussed in order to point out their role in the evolution of the (kinetically controlled) particle morphology. Then the synthetic methods and potential applications of core–shell particles are reviewed. Special attention is focused on the latest progress in the preparation of well-defined core–shell nanoparticles. Examples of core–shell particles with different compositions are presented. Finally, the main characterization methods to evidence the core–shell morphology are described.

13.2 Prediction of Core–Shell Particle Morphology

Composite particles can have the great variety of heterogeneous morphologies such as core–shell, inverted core–shell, hemispheres, sandwich structures, "confetti-like", "raspberry-like", etc. (Fig. 13.1).

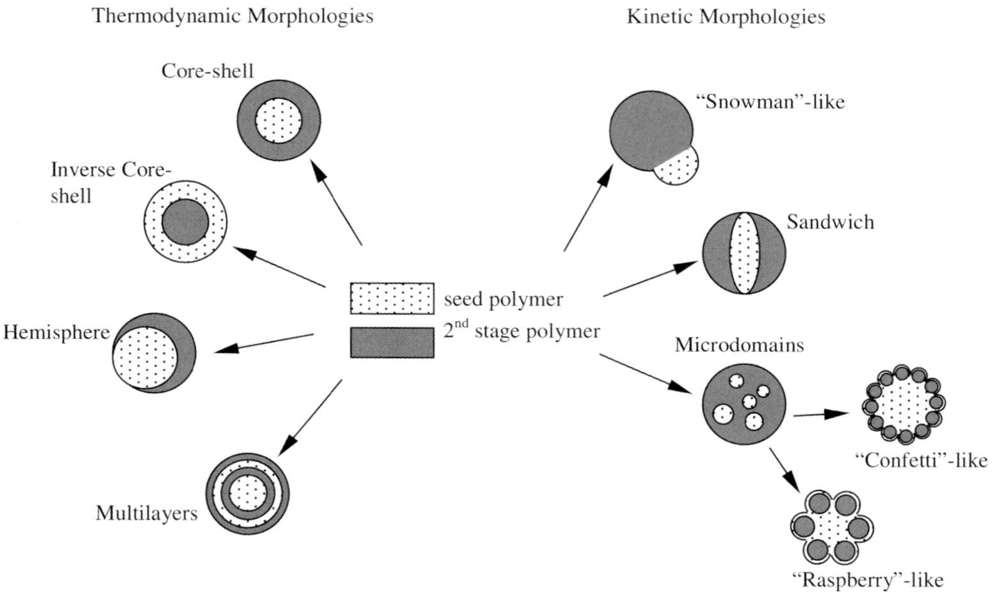

Fig. 13.1 Overview of two-phase particle morphologies divided into two classes: thermodynamically governed and kinetically governed morphologies.

In order to produce latex particles with a homogeneous and well-defined structure, it is essential to control the phase morphology during the particle synthesis. The physical and chemical properties of the particles largely depend on the morphological features of the particles, which are governed by various polymerization parameters and conditions. Excellent reviews on latex particle morphology have been published in the last decade [4, 5]. Therefore, here we shall focus mainly on the most important research work related directly to the subject.

The two main factors which control latex particle morphology are thermodynamic and kinetic. The thermodynamic factors describe the equilibrium morphology of the composite particle, whereas the kinetic factors represent the ease with which a thermodynamically stable arrangement can be achieved.

13.2.1 Thermodynamic Considerations

More than 30 years ago, Torza and Mason [6] made the first contribution to the problem of describing the theoretical prediction of three-phase interactions in shear and electrical fields. They studied the phase behavior of two immiscible organic liquids dispersed in water and proposed that the resulting equilibrium morphology can be readily predicted from the various interfacial tensions γ_{ij} and the spreading coefficient S_i, which can be defined as

$$S_i = \gamma_{ik} - (\gamma_{ij} + \gamma_{ik}) \quad (1)$$

Three different equilibrium states are possible by assuming that $\gamma_{12} > \gamma_{23}$, where subscripts 1, 2 and 3 are assigned for the first organic liquid, water and the second organic liquid, respectively:

$S_1, S_2 < 0, S_3 > 0$	Phase 1 is completely engulfed by phase 3	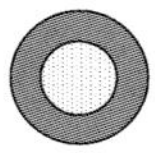	Core–shell morphology
$S_1, S_2, S_3 < 0$	Partial engulfing and formation of two-phase droplets with three interfaces	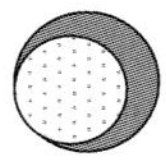	Hemispherical morphology
$S_2 > 0, S_1, S_3 < 0$	No engulfing	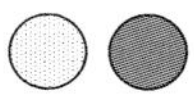	Individual particles

The mechanism of engulfing was established with high-speed cinematography and was shown to involve two competitive processes, penetration and spreading. By measuring the interfacial tensions and by using the calculations of S_i, Torza and Mason demonstrated the application of their approach to various three-phase liquid systems. The theoretical predictions were in good agreement with the experimental results obtained.

The thermodynamic analysis made by Torza and Mason was based on the liquids with a low viscosity, which can diffuse rapidly and reach the equilibrium state, i.e. the morphology with the lowest interfacial energy, within the experimental time. However, in the case of high molecular weight polymers, the diffusion of the polymer chains is limited and therefore the final equilibrium morphology may not always be achieved. In such a system, the driving force, which is equivalent to the Gibbs free energy change of the process, has the main influence on the final particle morphology. Berg et al. [7, 8] proposed the thermodynamic analysis of two-stage particle formation, demonstrating that the thermodynamically favored arrangement of the three-phase system (i.e. polymer 1, polymer 2 and water) would be the one which has the lowest value of free energy:

$$G = \sum \gamma_{ij} A_{ij} \qquad (2)$$

where G is the Gibbs free energy of the system, γ_{ij} is the interfacial tension between phases i and j and A_{ij} is their interfacial area.

A detailed thermodynamic analysis of the polymer particle morphology was presented by Sundberg et al. [9]. They discussed the high influence of the surfactant and the nature of the incompatible polymers on the interfacial tension. It has been found that, by changing the type of surfactant used in the emulsion, one could change the particle morphology from core–shell to hemispherical; this was in agreement with thermodynamic predictions. Several apparently different morphologies (hemispherical, sandwich and multiple lobes) have been found to coexist in the same location in the emulsion, suggesting that they may simply be different states of phase separation and not thermodynamically stable, unique morphologies [10]. The number of states of phase separation depends on the speed of the formation process. The authors designated these morphological structures "rate-limited" or "frozen" morphologies and considered them as non-equilibrium states.

Chen et al. [11, 12] developed a thermodynamically based model to describe the free-energy differences among various possible particle structures. In this analysis, the contact angles between interfaces were used to describe the degree of phase separation. The simulation study on the particle morphology development as a function of conversion was carried out by Winzor and Sundberg both for synthetic latexes which were obtained by seeded emulsion polymerization [13] and for artificial latexes which were obtained by phase separation from polymer solution without any further reaction [14]. The particle morphology for a system consisting more than three phases was studied by Sundberg and Sundberg [15].

González-Ortiz and Asua [16–18] proposed cluster dynamics to simulate and estimate the whole process in the evolution of composite particle morphology. By applying Sundberg et al.'s theory [9], Huo and Liu [19] proposed a quantitative simulation method for the prediction of the equilibrium morphology of poly(butyl acrylate)–poly[styrene-*co*-(methyl methacrylate)] [PBA–P(S-*co*-MMA)] core–shell particles prepared by a semicontinuous seeded emulsion polymerization under monomer-starved conditions. The polarity of the copolymer phase in the second-stage was controlled by varying the styrene (S) to methyl methacrylate (MMA) ratio. Sun et al. [20] investigated poly(vinyl acetate)–poly(butyl acrylate) (PVAc–PBA) inverted core–shell particle formation by emulsion polymerization. The time required to obtain equilibrium morphology was predicted with the cluster dynamics proposed by González-Ortiz and Asua [16–18]. Zhao et al. [21] analyzed the accuracy of the prediction of equilibrium morphologies for several types of core–shell latexes obtained via two-stage seeded emulsion polymerization and calculated the related sensitivity range for the prediction. The related sensitivity is defined as the relationship between the possible equilibrium morphology of a core–shell particle and interfacial tensions as an effect of emulsifier and initiator used for the synthesis of the particles. It was found that for systems such as PS–PMMA, PBA–PMMA and PBA–PS, the equilibrium morphology is sensitive to the experimental conditions and the theoretical prediction does not agree well with the experiment. On the other hand, systems such as PVAc–PBA and PVAc–PS are not sensitive and the morphology of core–shell particles can be predicted before performing the experiment.

The first computer algorithm for the prediction of the equilibrium morphology in latex particles was presented by Durant and Sundberg [22]. The calculations were performed to simulate the conversion-dependent morphologies for several latex systems composed of PS–PMMA and two different surfactants, i.e. sodium dodecyl sulfate (SDS) and natural pectin. Later, a simulation model based on the software UNHLATEX^{TM} KMORPH was developed by Stubbs et al. [23] which allows the prediction of the morphology in two-phase latex particles obtained in a seeded emulsion polymerization process. The resulting morphology is predicted taking into account the diffuse penetration of radicals into the seed particles. The system studied consisted of poly(methyl acrylate-*co*-methyl methacrylate) [P(MA-*co*-MMA)] as polymer for the seed particles and PS as the second-stage core polymer. The experimentally observed changes in the particle morphology as a function of the T_g of the seed polymer and the mode by which the second-stage monomer was added to the reactor, i.e. as batch, semi-batch and different feed rates, were in good agreement with the model predictions.

13.2.2
Thermodynamically and Kinetically Controlled Structures

However, in some cases even when the thermodynamics allow the formation of stable particle morphology, thermodynamically unfavorable structures can be produced. This is possible, for example, when either the surface properties of polymers differ from the properties of the bulk or when kinetic factors during the synthesis become significant. During the reaction the following parameters may have a great influence on the interfacial energies and may result in different structure formation:

- differences in the hydrophilicity of the monomers and the polymers and the different solubilities of the monomers and polymers in the continuous phase [24–26];
- degree of compatibility of the polymers formed [27];
- mobility of the polymer chains formed and the inter-diffusion [28];
- phase volume ratio [9, 11];
- molecular weight and degree of cross-linking of both polymers [26];
- types and amounts of the emulsifiers and initiators used [9, 29, 30];
- mode and feed rate of monomer addition [25, 29, 31–33];
- reaction temperature [33, 34].

Regarding the monomer hydrophilicity, three groups can be specified if subsequent monomer addition is performed: (1) monomer I has the same hydrophilicity as monomer II; (2) monomer II is more hydrophilic than monomer I; and (3) monomer I is more hydrophilic than monomer II.

13.2.2.1 **Monomer I has the Same Hydrophilicity as Monomer II**

If both monomers possess the same hydrophilicity and polymer I is immiscible with polymer II, the morphology of the resulting structured particle will depend on the order of monomer addition in an emulsion polymerization process. If monomer II is added after monomer I, the particles consist of a core made with polymer I and a shell predominantly formed from polymer II. If monomer I is added after monomer II, an inverted structure with polymer II as core and polymer I as shell is formed. As an example, latex particles from PMMA and poly(ethyl acrylate) (PEA) can be considered. The solubility of both monomers in water is very similar (MMA = 1.2 g L^{-1}; EA = 1.5 g L^{-1}) and therefore both PMMA and PEA can act as a core particle.

13.2.2.2 **Monomer II is More Hydrophilic than Monomer I**

Fabrication of the core–shell morphology is also possible when monomer II is more hydrophilic than monomer I. In this case, the core will consist of polymer I and the shell will be formed by growth of polymer II or through precipitation of newly generated particles from monomer II (in the form of microdomains) and also through the capture of those oligomers. The polymerization will take place basically on the surface of the particle. The amount of polymer II microdomains increases the less compatible they are with polymer I. Above a certain number of microdomains they can coalescence, resulting in a closed shell from polymer II. A core–shell structure was evidenced for the system PS–P(S-*co*-AN), as reported by Dimonie et al. [35]. In order to increase the compatibility between two phases, graft polymerization between core and shell was proposed by Merkel et al. [36].

13.2.2.3 **Monomer I is More Hydrophilic than Monomer II**

When monomer I is more hydrophilic than monomer II, the most thermodynamically stable morphology is the inverse core–shell structure. However, different degrees of hemispherical morphology or "confetti"-like structures can also often appear. To the already studied systems belong, among others, PEA–PS [31], PBA–PS [37], P(EA-*co*-MMA)–P(S-*co*-B) (B = butadiene) [26] and PMMA–PS [38]. Radicals which are generated in the aqueous phase can diffuse inside the core particles swollen from monomer II. Inside the core, due to the incompatibility of the two polymers and mainly due to the high viscosity of the core, polymer II forms microdomains, which can appear near the surface or in the center of the core particle. Generally, monomer II is more soluble in polymer II than in polymer I, so that the domains become large throughout the reaction time. In order to avoid the formation of microdomains, the polymerization can be carried out with a small part of grafted copolymers, as already studied with the systems PMA–PS [PMA = poly(methyl acrylate)] [39] and PBA–PS [40]. Such copolymers are compatible with both phases and thus play the role of stabilizers. Furthermore, Min et al. [37] found that the PBA–PS core–shell particles with a high degree of grafting are more stable. Phase separation and the formation of two separate particles were observed with non- or low-grafted particles.

In the following, a few parameters, which have a great influence on the morphology formation, will be discussed in more details.

13.2.3 Effect of Weight Ratio

The effect of the weight ratio of seed polymer to monomer on the morphology of the PMMA–PS composite particles obtained in a methanol–water medium was studied by Okubo et al. [41]. It was found that, when the PMMA to PS weight ratios was 3 : 1 and 2 : 1, the composite particles had a clear morphology consisting of a PMMA core and a PS shell. When the ratio was 1/1, a lot of small PS domains were observed in the PMMA core although the PS shell was still formed (Fig. 13.2).

By stepwise addition of the styrene monomer, the formation of the small PS domain was depressed and a complete core–shell morphology was formed. It was also found that the core–shell structure obtained by seeded dispersion polymerization in a polar medium is thermodynamically unstable. The morphology of the PMMA–PS (1 : 1, w/w) composite particles can be changed from a core–shell to a multilayered structure by the absorption–release treatment of toluene (Fig. 13.3).

13.2.4 Influence of Viscosity

Lee and Ishikawa [26] studied the formation of the inverted core–shell morphology using two polymer pairs: a soft polymer pair, P(EA-*co*-MMA)–P(S-*co*-B), and a hard polymer pair, P(EA-*co*-S-*co*-MAA)–PS. It was found that in the case of the

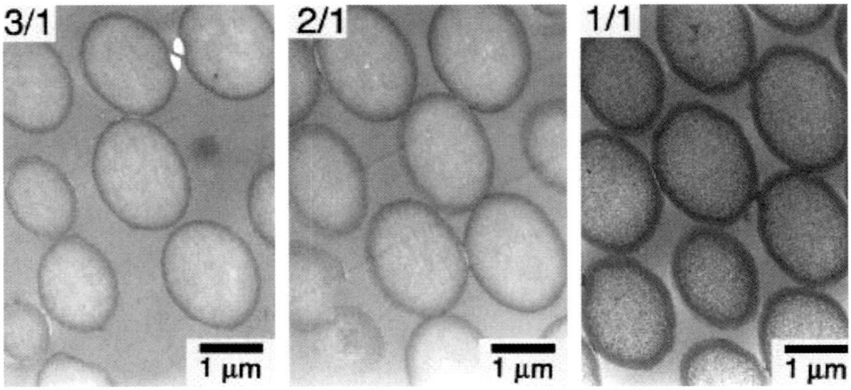

Fig. 13.2 TEM images of ultrathin cross-sections of PMMA–PS composite particles produced by batch seeded dispersion polymerizations stained with RuO_4 vapor. The ratios on the images indicate the weight ratio of PMMA to PS. Reproduced from [41], with permission.

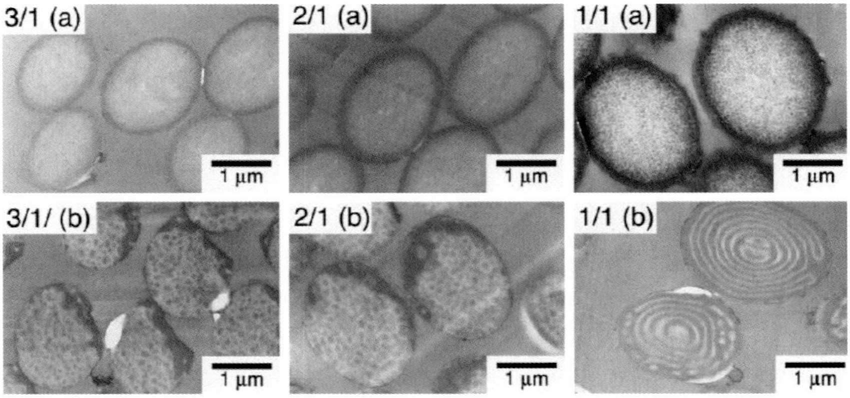

Fig. 13.3 TEM images of ultrathin cross-sections of PMMA–PS composite particles stained with RuO_4 vapor: (a) before and (b) after absorption–release treatment with toluene (particle : toluene = 1 : 10, w/w). The ratios on the images indicate the weight ratio of PMMA to PS. Reproduced from [41], with permission.

soft polymer pair, the formation of an inverted core–shell morphology was equally complete regardless of the molecular weight of the hydrophilic polymer molecules. Here, the viscosity is low enough that interdiffusion of the chains is always possible. However, in the case of the hard polymer pair, the molecular weight of both hydrophilic and hydrophobic polymers influences the inversion efficiency. Here, the increase in the viscosity on increasing the molecular weight is more effective in hindering polymer diffusion. Therefore, the thermodynamically favored structure is not preferably formed.

13.2.5
Effect of Cross-linking Agent

In the presence of a cross-linking agent, the free mobility of the polymer chains will be significantly limited and the diffusion of a second polymer inside an already formed core will be prevented. Sperling and coworkers [42, 43] studied the systems PS–PB, PB–PS, PVC–P(B-*co*-AN) (AN = acrylonitrile) and PA–PMA (PA = polyacrylate). The existence of a shell richer in polymer II and a core richer in polymer I was confirmed by electron microscopy. It was found that the core exhibits a cellular structure or a structure with small inclusions (diameter ~10 nm), confirming the interpenetration of both networks. The cellular structure preferentially appears when (1) monomer II is better soluble in its own polymer as in polymer I, (2) both polymers are partially compatible and (3) the size of the particles is small. In contrast, small inclusions will be formed when monomer II is more soluble in polymer I and both polymers are not compatible. Sometimes cross-linking was also employed to increase the colloidal stability [44].

13.2.6
Effect of Type of Initiator

The morphology of the final composite particles can also be controlled by the initiator type used in the synthesis. Several groups [29, 34] have described the synthesis of PMMA–PS particles composed of a PMMA core which was synthesized with an ionic initiator and the shell polymerization of PS was carried out with a non-ionic initiator. Due to the difference in the polarity of the polymers and the presence of ionic groups on the PMMA surface, which make the core surface even more hydrophilic, the formation of inverse core–shell particles was preferable. The dependence of the phase morphology on the applied initiator was also observed in the system PB–P(S-*co*-MMA) by Aerdts et al. [45]. Particles with a "mushroom"-like structure were obtained by using a water-soluble initiator, i.e. $K_2S_2O_8$. When, instead of $K_2S_2O_8$, a redox initiating system such as $C_9H_{12}O_2$–$CH_3NaO_3S \cdot H_2O$ (cumulene hydroperoxide–sodium formaldehyde sulfoxylate), $FeSO_4$ and EDTA (ethylenediaminetetraacetic acid) was used, the formation of a PB core and small microdomains consisting of the PS–PMMA copolymer was observed.

Jönsson et al. [29] studied the effect of initiators with different polarity on the morphology of PMMA–PS particles. It was observed that with batchwise addition of the second monomer, a more occluded structure was obtained in the presence of the hydrophobic initiator *tert*-butyl hydroperoxide (*t*-BHP) as compared with water-soluble $K_2S_2O_8$ (Fig. 13.4).

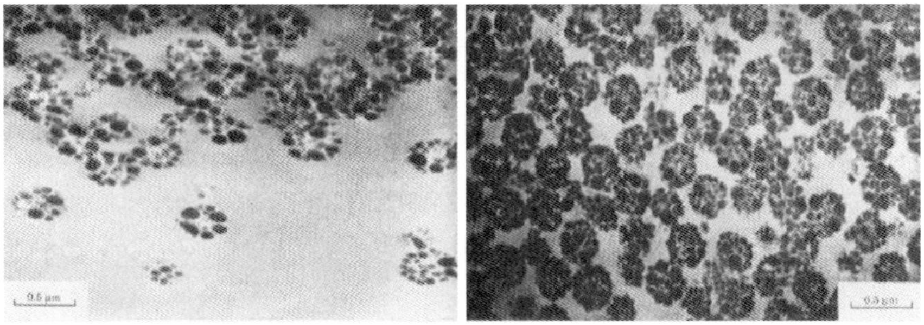

(a) (b)

Fig. 13.4 TEM micrographs of thin-sectioned particles prepared by batchwise addition of styrene with different initiators: (a) $K_2S_2O_8$ and (b) *t*-BHP. Reproduced from [29], with permission.

13.2.7
Effect of Addition Mode

In a two-stage polymerization process, the second monomer, an initiator and emulsifier solution will be added in the second stage. Two different modes have usually been considered for the second-stage polymerization: a batch process (seeded swelling method) and a continuous addition process. The addition of the second monomer can be done either all at one time during the later stages of polymerization (shot addition) or over a longer period (continuous addition). The continuous addition mode is preferred if the corresponding polymer is insoluble in the continuous phase. In this case, the accumulation of the monomer in the continuous phase and secondary nucleation can be sufficiently overcome.

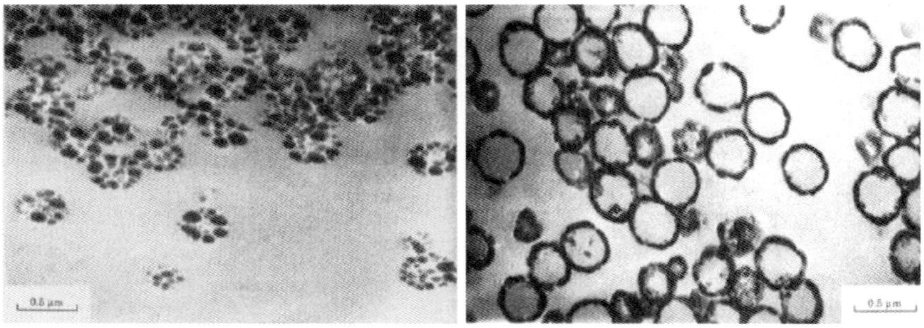

(a) (b)

Fig. 13.5 TEM micrographs of thin-sectioned particles prepared by (a) batchwise and (b) continuous addition of styrene. $K_2S_2O_8$ was used to initiate the polymerization. Reproduced from [29], with permission.

Jönsson et al. [29] showed that in the system PMMA–PS, microdomains from PS in PMMA-core are formed with batchwise monomer addition, whereas the core–shell particle morphology can be obtained under continuous addition of the second monomer [29] (Fig. 13.5).

Another factor affecting the particle morphology, especially in the second stage of the polymerization, is the feed rate of monomer addition. The slower the addition speed under "starved-feed" conditions, the higher is the kinetic barrier. The thermodynamically preferred phase inversion can be effectively avoided in this way. It is very important that the feed rate of the monomers is in agreement with the rate of polymerization. Schellenberg et al. [46] performed the synthesis of structured core–shell latex particles consisting of a low-viscosity core poly(2-ethylhexyl methacrylate) (PEHMA) covered with a thin shell of a cross-linked rubber PBA using a two-stage emulsion polymerization process. The authors applied reaction calorimetry in order to determine online the rate of polymerization and monomer conversion at both stages.

13.3 Synthesis of Core–Shell Particles

Different processes have been developed that permit the synthesis of composite particles with core–shell morphology. Both organic and inorganic cores and shells have been formed by various techniques. The choice of method mostly depends on the combination of materials used for core–shell particle formation and on the final needs and application of these particles. This section provides examples of synthetic methods to obtain core–shell particles from different materials and their potential applications.

13.3.1 Polymer Core–Polymer Shell

Core–shell particles with a polymeric core can be produced by a single- or multi-stage route (Fig. 13.6). Each route can involve different types of liquid-phase polymerization, e.g. emulsion, dispersion, miniemulsion, microemulsion polymerization.

13.3.1.1 Two-step Emulsion Polymerization Process

Two-stage seeded emulsion polymerization performed under different conditions is the most commonly used technique to synthesize particles with structured morphologies and have been reviewed earlier [4, 5]. In the first stage, the core particles are prepared by any kind of polymerization either separately or *in situ*. In the second stage, a different monomer or monomer mixture is added and a cross-linked shell is polymerized around the core particle. Monodisperse core–shell latex particles bearing different functional groups on the surface were

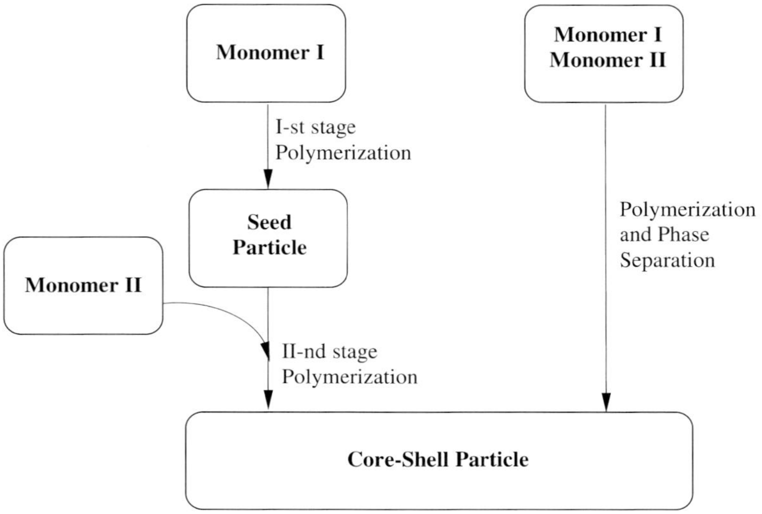

Fig. 13.6 Commonly used routes for the fabrication of core–shell particles.

synthesized by a two-step emulsion polymerization process in a batch reactor. The core was prepared via emulsion polymerization of styrene and the shell was formed in the second stage via emulsion copolymerization of styrene and different monomers. For instance, a shell of acetal-functionalized particles was composed of styrene, methacrylic acid and methacrylamidoacetaldehyde di(*N*-methyl) acetal [47] and in the case of chloromethyl functionality styrene and chloromethylstyrene form the shell layer [48].

13.3.1.2 **Dispersion Polymerization Process**

Several groups have reported the synthesis of core–shell particles by dispersion polymerization. In a dispersion polymerization, an initially homogeneous mixture of monomer, initiator and solvent becomes heterogeneous during the reaction as growing oligomeric radicals after reaching a critical length become insoluble and aggregate to form a separate polymer phase. This method gives the advantage of preparing core–shell particles in the 0.5–10-μm size range. To keep the particle stable in the dispersed medium, a suitable steric stabilizer should be used. Therefore, by carrying out the cross-linking reaction between the core and molecules of the stabilizer, particles with core–shell morphology can be obtained. Casagrande and coworkers described dispersion polymerization carried out in alcohol medium. Polystyrene microspheres in the size range 2–10 μm covered with a shell of the steric stabilizer polyepichlorohydrin [49] or a double shell formed by the statistical copolymer of methacrylic acid and ethyl acrylate (MAA–EA) in a ratio of 1:1 (Eudragit) [50] were successfully prepared and used for the selective adsorption of enzymes [51].

Watanabe et al. [52] described the preparation of PS–PIm (PIm = polyimide) core–shell particles using aromatic poly(amic acid) (PAmA) as a stabilizer. First, the PAmA was incorporated into the particle by using the cationic comonomer vinylbenzyltrimethylammonium chloride. Then, imidization of PAmA on the particle was carried out by acetic anhydride and *N,N*-dimethylaminopyridine in order to obtain PS–PI particles.

Okubo and coworkers prepared monodisperse composites having chloromethyl groups [53], vinyl groups [54] or a PANI (PANI = polyaniline) shell [55] on the surface. The particles obtained were synthesized by either copolymerization of styrene and chloromethylstyrene or of styrene and divinylbenzene in an ethanol–water medium. In the case of PS–PANI particles, the approach involves the chemical oxidative seeded dispersion polymerization of aniline at 0 °C in the presence of seed particles in aqueous HCl solution, keeping the pH of the reaction mixture acidic.

Another route, which is more environmentally friendly and attractive, to obtain particles with high purity is polymerization in supercritical CO_2. Highly reactive epoxy- or isocyanate-containing groups, both of which are sensitive to nucleophilic solvents (i.e. water and protic organic solvents), are ideally suited for synthesis in CO_2 due to its inertness.

Syntheses of PS–PGMA [PGMA = poly(glycidyl methacrylate)], PS–PIEM [PIEM = poly(2-isocyanatoethyl methacrylate)] and PVP–PS [PVP = poly(1-vinyl-2-pyrrolidone)] by a two-stage dispersion polymerization in supercritical CO_2 were reported by Young and DeSimone [56]. CO_2 provides advantageous solubility properties for composite particle synthesis by dispersion polymerization since most hydrophilic and hydrophobic vinyl monomers are soluble in CO_2 whereas their respective polymers are not. It was found that (89 : 11 and 58 : 42 mol%) formed kinetically trapped nanodomains of PS in PVP and PVP–PS (26 : 74 mol%) formed the more thermodynamically favorable inverted core–shell morphology with a PVP shell and PS core. The main disadvantage of the process is the high cost of the special emulsifiers that are suitable for a CO_2 medium.

13.3.1.3 **Microemulsion Polymerization Process**

Microemulsion polymerization is an alternative approach to emulsion and miniemulsion polymerization for synthesizing polymer particles of small size (< 50 nm) and high molecular mass (> 10^6 g mol^{-1}) [57]. As already mentioned, the main disadvantages of the microemulsion method are the large amounts of surfactants required to perform the polymerization and the low solid content of the final latexes. To overcome these drawbacks, synthesis by a two-stage microemulsion polymerization process resulting in core–shell particles was proposed. Aguiar et al. [58] prepared PS–PBA core–shell particles. In the first stage, PS seed particles were obtained. In the second stage, butyl acrylate (BA) and in some cases itaconic acid were added semicontinuously to form the shell. The process was optimized by adding more styrene in a semicontinuous mode to the latex resulting from the polymerization of the parent microemulsion to produce high-solid-content polystyrene latex (ca. 40% solids). Later, the mechanical properties of PS–PBA and PBA–PS core–shell latexes were examined as a function of the location of soft and hard polymer and it was concluded that the hardness increases and impact energy decreases with increasing PS content, regardless of its location [59].

Water-borne PA–PUR (PUR = polyurethane) particles with different core : shell ratios and cross-linking structures were synthesized by microemulsion polymerization [60]. The core part was a copolymer of styrene, BA and diacetoneacrylamide and the shell layer was composed of PUR.

13.3.1.4 **Emulsifier-free Polymerization Process**

The preparation of core–shell latex particles can also be achieved without the use of emulsifiers. Greci et al. [61] described the preparation of monodisperse core–shell poly(*p*-hydroxystyrene) particles through emulsifier-free emulsion polymerization. The formation of the shell layer was carried out by grafting of the *p*-acetoxystyrene monomer on to the polystyrene core particles during the last 30 min of the reaction, followed by hydrolysis of the acetoxy group in the presence of base.

13.3.1.5 **Graft Polymerization Approach**

Another interesting example is the preparation of monodisperse core–shell particles by graft polymerization. The grafting of the shell layer on to the core particle can be achieved by free-radical polymerization between the second-stage monomer and functional surface groups present on the core particle. For example, cross-linked microspheres containing double bonds on the particle surface can be used directly to graft polymers from the surface. Styrene has been grafted from cross-linked poly(divinylbenzene) (PDVB) core microspheres by both reversible addition fragmentation chain-transfer (RAFT) polymerization and conventional free-radical polymerization [62].

For a grafting process, amphiphilic initiators can also be used. Tokarev et al. [63] used specially designed reactive peroxide surfactants (i.e. inisurf) (Fig. 13.7)

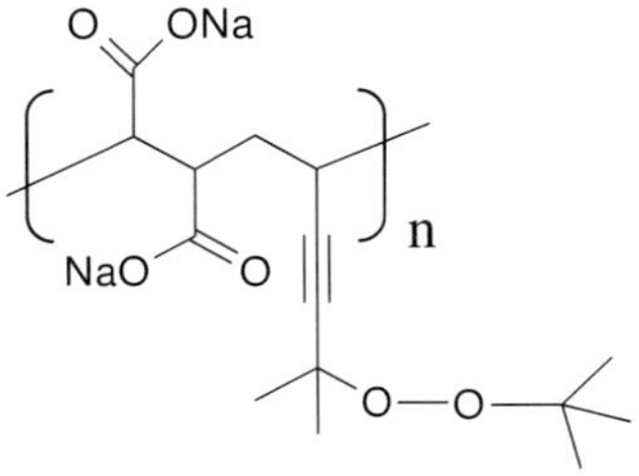

Fig. 13.7 Structure of peroxide inisurf.

in the synthesis of core–shell PS–PBA particles. In the first stage, seed particles were prepared via emulsion polymerization in the presence of an inisurf. The polymer particles obtained contained surface peroxide groups which allow the growth of the shell polymer chains from the surface. The initiation was started thermally or by using a redox mechanism, resulting in a shell layer covalently grafted to the core.

13.3.1.6 Amphiphilic Core–Shell Particles

The production of amphiphilic core–shell particles is a special challenge owing to solubility problems. Via the miniemulsion process, it was possible to start from very different spatial monomer distributions, resulting in very different amphiphilic copolymers in dispersion [64]. The monomer, which is insoluble in the continuous phase, was miniemulsified to form stable and small droplets with a low amount of surfactant. The monomer with the opposite hydrophilicity dissolved in the continuous phase (and not in the droplets). As examples, the formation of AAm–MMA (AAm = acrylamide) and AAm–S copolymers through the use of the miniemulsion process was chosen. In all cases, the syntheses were performed both in water and in cyclohexane as the continuous phase. When the synthesis is performed in water, a hydrophobic monomer with low water solubility (S or MMA) mainly forms monomer droplets, whereas the hydrophilic monomer AAm, with a high water solubility, will be mainly dissolved in the water phase. In the case of an inverse miniemulsion, the hydrophilic monomer is expected to form droplets, whereas the hydrophobic monomer is dissolved in the continuous phase. Starting from those two dispersion situations, the locus of initiation is expected to have a great influence on the reaction products, the quality of the copolymers obtained and the morphology of the nanoparticles.

In the system AAm–MMA, copolymerization in a direct miniemulsion with water as continuous phase via the interfacial initiator PEGA200 {azodi-[poly(ethylene glycol)isobutyrate] with an ethylene oxide molecular weight of 200 g mol^{-1}} results in a higher homo-AAm content. This can be attributed to the fact that the initiator, due to its hydrophilicity, has a slightly higher tendency to be in the water phase, in which AAm units can be captured. As can be seen from Fig. 13.8, the core–shell structure of the resulting aggregates or latexes is nicely depicted. Note that the occurrence of core–shell structures is not in contradiction with larger amounts of homopolymer.

Microspheres with an amphiphilic core–shell structure were prepared by a graft copolymerization reaction in aqueous medium. In the presence of hydrophobic vinyl monomer, the water-soluble synthetic or biopolymers (e.g. casein, gelatin) which contain amino groups were treated with *tert*-butyl hydroperoxide in order to generate free radicals and initiate the graft copolymerization of vinyl monomer [65]. Using this approach, PMMA–PEI (PEI = polyethylenimine), PMMA–PAlA (PAlA = polyallylamine) [66], PMMA–casein [67] and PBA–chitosan particles with well-defined core–shell morphologies were synthesized. PBA–

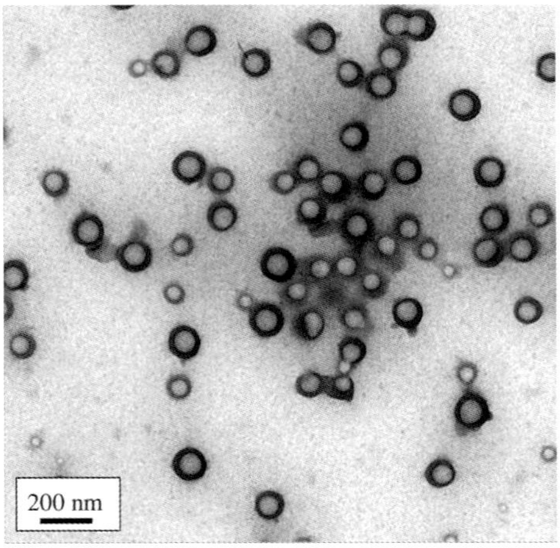

Fig. 13.8 TEM micrograph of PAAm–PMMA copolymer obtained in a direct miniemulsion with interfacial initiation stained with RuO_4 [64].

chitosan particles were utilized as an antibacterial coating for cotton fabrics, showing bacterial reductions of more than 99% [68].

13.3.1.7 **Core–Shell Particles with Fluorinated Monomers**

Coatings with excellent water and oil repellency can be obtained by using acrylate copolymers with fluorinated side-chains. Typically these copolymers are fairly expensive, hence the synthesis of core–shell particles having a fluorinated monomer in the shell is a very promising way to overcome this drawback. The groups of Lee [69] and Lang [70] used a semicontinuous emulsion polymerization process to synthesize core–shell particles containing a fluoropolymer in the shell. This technique, however, relies on the compatibility between the first-stage polymer and the fluoromonomer. Another possibility is to carry out the polymerization from a fluorinated microemulsion [71]. Here, a large amount of fluorinated surfactants was required to perform the synthesis. Latexes consisting of fluorinated polymers in the size range 100–250 nm stabilized by a low amount of protonated surfactants were prepared through the miniemulsion polymerization process [72]. The miniemulsification of mixed monomer systems allows efficient copolymerization reactions to be performed using a fluorinated monomer with standard hydrophobic or hydrophilic protonated monomers. Depending on the reaction conditions, such as the type and amount of initiator and amounts of surfactant and comonomer, homogeneous particles, but also different particle morphologies, such as core–shell, multiblobs and cups, can be obtained.

13.3.1.8 Core–Shell Particles Containing Conducting Polymers

In recent years there has been increased interest in the fabrication of transparent conductive thin films, which could find application in liquid-crystal displays, photovoltaic cells, electromagnetic-interference shielding and so on. Poor mechanical properties of the conducting polymers can be significantly improved by the creation of core–shell particles which consist of conducting polymer and a polymer possessing better film formation properties. Wiersma et al. [73] synthesized water-borne colloidal particles based on PPy–PUR (Ppy = polypyrrole) and PPy–alkyd dispersions. Monodisperse PS–PANI core–shell latexes employing cationic spheres as the polymer core were synthesized.

In order to produce electrically conductive core–shell particles, poly(*N*-vinylpyrrolidone)-stabilized PS latexes have been coated with a thin layers of PANI [74] or PPy [75]. It was observed that PANI-coated particles exhibit a non-uniform morphology in comparison with the relatively smooth PPy layers. A more uniform deposition of PANI and a corresponding improvement in colloid stability were obtained using aniline hydrochloride monomer in the absence of added acid.

Another interesting example is the preparation of monodisperse core–shell particles containing a conductive core and a dielectric shell. PANI–PMMA microparticles were prepared by an *in situ* suspension polymerization method and then adopted as an electrorheological material [76]. The conductivity of PANI–PMMA particles was much lower than that of the rod-like PANI. Jang and Oh [77] reported the synthesis of PPy–PMMA core–shell particles with diameters of several tens of nanometers by a two-step microemulsion polymerization. The highly transparent conductive film was fabricated by using the composite nanoparticles as fillers in the PMMA matrix.

Li and Kumacheva [78] employed controlled electrostatically driven heterocoagulation between the large PPy cores (200 nm) and small PMMA or PBMA-PMMA (in the size range 25–60 nm), which form a polymeric shell upon annealing.

13.3.2
Inorganic Core–Polymer Shell or Polymer Core–Inorganic Shell

Inorganic particles possess specific properties, such as magnetism, conductivity and catalysis. Inorganic nanoparticles covered with a layer of polymer are of great interest in terms of their chemical resistance and potential applications in electronic, optical and biomedical fields and in coatings. The polymeric shell is responsible for the chemical properties and interaction with the surrounding medium, whereas the physical properties are mainly governed by the inorganic core. On the other hand, colloid particles consisting of a polymer (e.g. dielectric) core coated with a metal shell are promising candidates for the design of materials with new optical properties [79]. Depending on the size and shape and on the chemical composition of core and shell, various new materials with novel properties can be fabricated. The preparation of both types of core–shell particles can be mainly achieved either by reactions directly performed on the surface or by controlled precipitation of the shell material on the surface of the core particle.

13.3.2.1 **Encapsulation of Inorganic Materials**

Titanium dioxide inorganic core and polymer shell composite P(MMA-*co*-BA-*co*-MAA) particles with an encapsulation efficiency up to 80% were prepared via emulsion copolymerization [80]. Bourgeat-Lami and coworkers described the formulation of SiO_2–PMMA latex particles by emulsion polymerization of MMA using a non-ionic surfactant, nonylphenol poly(oxyethylene), and three differently charged initiators, 2,2′-azobis(2-amidinopropane) dihydrochloride (AIBA), potassium persulfate ($K_2S_2O_8$) and azobisisobutyronitrile (AIBN) [81, 82]. The high coating efficiency and polymer content were obtained with the cationic initiator (AIBA), which is due to the electrostatic attraction between a negatively charged seed particle and positively charged polymeric chains. The heterocoagulation is mainly involved in the mechanism of coating the silica particles with a polymer layer.

Quaroni and Chumanov [83] reported the encapsulation of silver nanoparticles into a polymer shell via emulsion polymerization in order to produce a protective polymer layer. The shell layer was produced via emulsion polymerization of styrene and/or methacrylic acid in the presence of oleic acid, which works as a stabilizer. In such a system, silver particles are coated with a uniform, well-defined polymer layer and are present in a non-aggregated suspension. After encapsulation, particles retain their optical properties, possess chemical resistance toward etching and can be easily purified and modified.

Vestal and Zhang [84] prepared magnetic nanoparticles (size <15 nm) consisting of an $MnFe_2O_4$ core and PS shell employing atom transfer radical polymerization (ATRP). The core was prepared by the reverse (water in toluene) micelle microemulsion procedure and after polymerization possess initiator groups on the surface. The modified core nanoparticles were then used as macroinitiators in the second-stage polymerization.

Jang and Lim [85] reported on the use of one-step vapor deposition polymerization in order to fabricate various inorganic–polymer core–shell particles. Either silica or titania nanoparticles were covered with a layer of PMMA or PDVB.

A very elegant approach for the efficient encapsulation of inorganic particles is the miniemulsion process. For the encapsulation of pigments by miniemulsification, two different approaches can be used. In both cases, the pigment/polymer interface and the polymer/water interface have to be chemically adjusted carefully to obtain encapsulation as a thermodynamically favored system. The design of the interfaces is mainly dictated by the use of two surfactant systems, which govern the interfacial tensions, and also by the employment of appropriate functional comonomers, initiators or termination agents. The sum of all the interface energies has to be minimized.

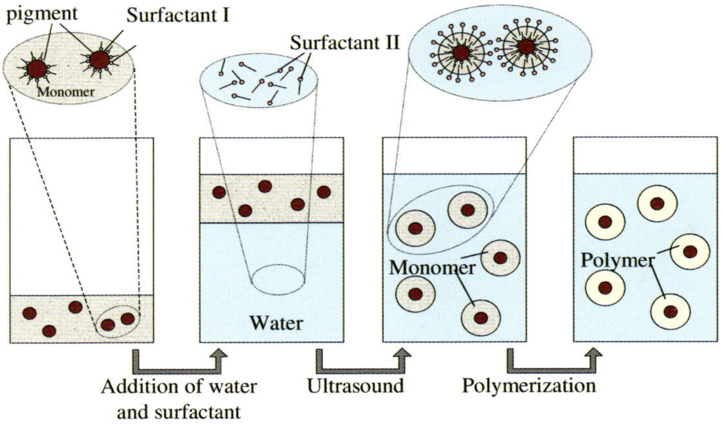

Fig. 13.9 Principle of encapsulation by miniemulsion polymerization.

For successful incorporation of a pigment into latex particles, both the type and amount of surfactant systems have to be adjusted to yield monomer particles, which have the appropriate size and chemistry to incorporate the pigment according to its lateral dimension and surface chemistry. For the preparation of miniemulsions, two steps have to be controlled (Fig. 13.9). First, the already hydrophobic or hydrophobized particulate pigment with a size up to 100 nm has to be dispersed in the monomer phase. Hydrophilic pigments require a hydrophobic surface to be dispersed into the hydrophobic monomer phase, which is usually promoted by a surfactant system I with low hydrophilic–lipophilic balance (HLB) value. Then, this common mixture is miniemulsified in the water phase, with the use of a surfactant system II with high HLB, which has a higher tendency to stabilize the monomer (polymer)/water interface.

Erdem et al. [86] described the encapsulation of TiO_2 particles via a miniemulsion in two steps. First, they dispersed TiO_2 in the monomer, using a polybutene–succinimide pentamine (OLOA 370) as stabilizer; then, this phase was dispersed in an aqueous solution to form stable submicron droplets. The presence of TiO_2 particles within the droplets limited the droplet size. Complete encapsulation of all the TiO_2 in the colloidal particles was not achieved. Also, the amount of encapsulated material, at about 3%, was very low.

Nanoparticulate hydrophilic $CaCO_3$ was effectively coated with a layer of stearic acid as surfactant system I prior to dispersing of the pigments into the oil phase [87]. The –COOH groups acted as good linker groups to the $CaCO_3$ and the tendency of the stearic acid to go to the second polymer/water interface was low. About 5 wt% of $CaCO_3$ based on monomer could be completely encapsulated into PS particles. Also, pigments such as the organic phthalocyanine blue pigment [88] or copper phthalocyanine dyes could be encapsulated [89]. Silica could be encapsulated through the use of silica with a double layer of CTMA-Cl

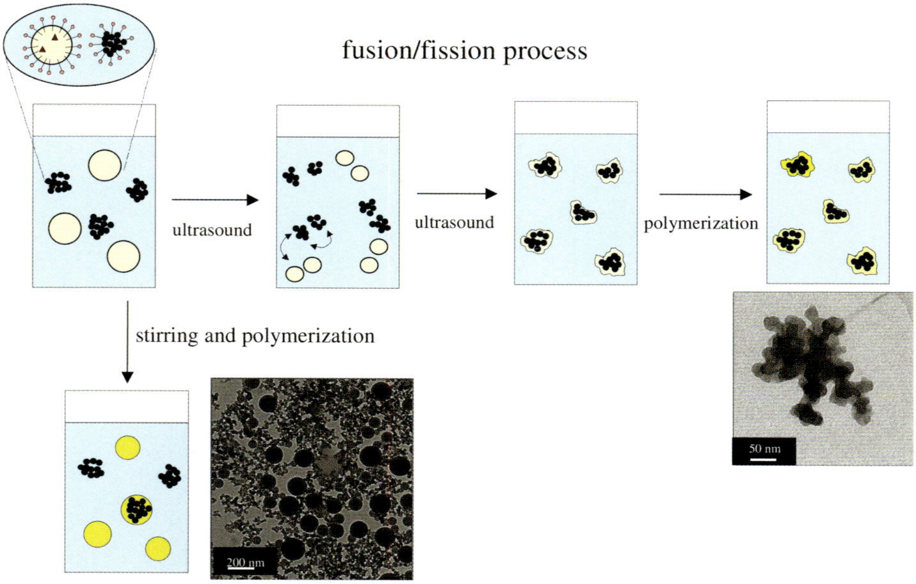

Fig. 13.10 Co-miniemulsification process for the encapsulation of carbon black.

[90]. Fleischhaker and Zentel incorporated CdS–ZnS-coated CdSe quantum dots into PS–PMMA particles by the miniemulsion process [91].

Because carbon black is a rather hydrophobic pigment (depending on the preparation conditions), the encapsulation of carbon black in the latexes by direct dispersion of the pigment powder in the monomer phase prior to emulsification is again a suitable way to achieve encapsulation [92]. However, this direct dispersion only allows for the incorporation of 8 wt% carbon black because the carbon is still highly agglomerated in the monomer. At higher amounts, the carbon cluster breaks the miniemulsion and less defined systems with encapsulation rates lower than 100%, which also contain pure polymer latexes, are obtained.

To increase the amount of encapsulated carbon to up to 80 wt%, Tiarks et al. developed another approach [92] in which both the monomer and carbon black are independently dispersed in water through the use of SDS as a surfactant and subsequently mixed in any ratio between the monomer and carbon. Then, this mixture is cosonicated, the controlled fission–fusion process characteristic for miniemulsification destroys all aggregates and liquid droplets and only hybrid particles composed of carbon black and monomer remain owing to their higher stability (Fig. 13.10). This controlled droplet fission and heteroaggregation process can be realized by high-energy ultrasound or high-pressure homogenization. TEM and ultracentrifuge results showed that this co-miniemulsion process results in effective encapsulation of the carbon with practically quantita-

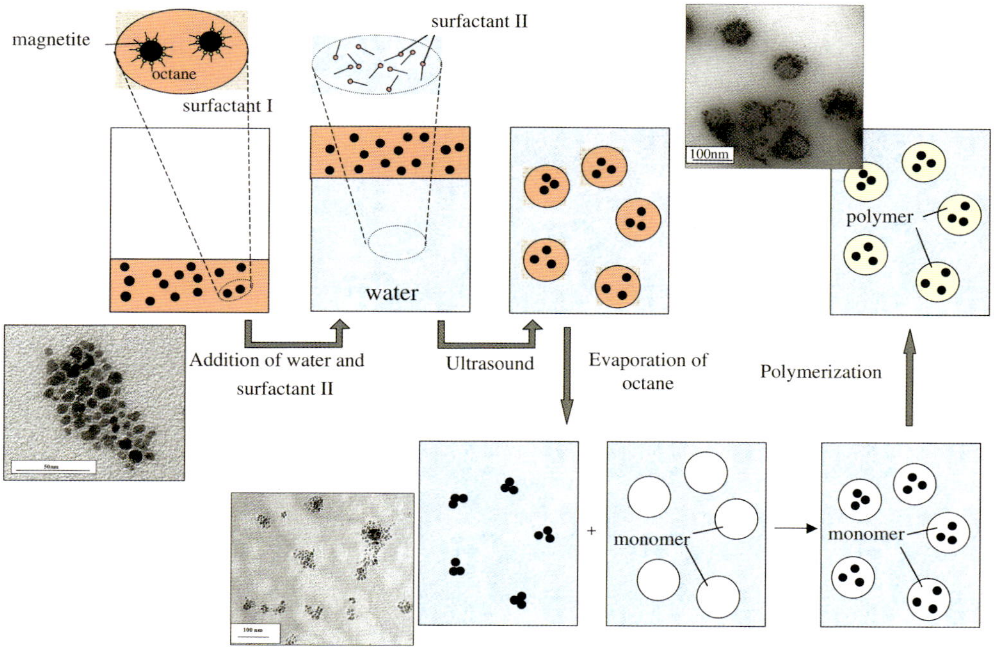

Fig. 13.11 Encapsulation of magnetic particles via a three-step process.

tive yield. Only rather small hybrid particles, and no free carbon or empty polymer particles, were found. Hybrid particles with high carbon contents do not possess a spherical core–shell structure but adopt the typical fractal structure of carbon clusters, coated with a thin, but homogeneous, polymer film.

On changing the hydrophilicity of the encapsulated material from hydrophobic to hydrophilic, the material has to be hydrophobized prior to the encapsulation process. The encapsulation of hydrophilic magnetite particles into PS particles is efficiently achieved by a new three-step preparation route including two miniemulsion processes [93] (Fig. 13.11). In the first step, a dispersion of oleic acid-coated magnetite particles in octane is obtained. In the second step, magnetite aggregates in water are produced in a miniemulsion process with SDS as the surfactant. In the third step, the dispersion with the magnetite aggregates that are covered by an oleic acid–SDS bilayer was mixed with a monomer miniemulsion and a second miniemulsion process, an ad-miniemulsification process, is performed to obtain full encapsulation. Such particles can be used for biomedical applications, e.g. detection by magnetic resonance tomography and destruction of tumor cells by hyperthermia.

13.3.2.2 Encapsulation by Inorganic Materials

The self-assembling layer-by-layer technique was applied for the formation of core–shell particles with PS as a core and different inorganic shells [94]. Titanium dioxide, silica and Laponite® nanoparticles were used as the inorganic building blocks for shell formation via controlled assembly of the preformed nanoparticles in alternation with oppositely charged polyelectrolytes on to the PS core.

Conductive core–shell particles were prepared by the self-assembling technique [95, 96]. First, negatively charged PS latex particles (size ~700 nm) were coated with a positively charged layer of PEI, followed by adsorption of the negatively charged gold nanoparticles, resulting in a conductive gold shell layer.

13.3.3 Liquid Core–Polymer Shell (Capsule)

Core–shell particles with a liquid core have attracted considerable attention owing to their application as submicron containers for microencapsulation.

13.3.3.1 Capsules by Heterophase Polymerization

One of the earliest processes for making hollow latex particles was developed at Rohm and Haas [3, 97]. Their concept involved making a structured particle with a carboxylated core polymer and one or more outer shells. The ionization of the carboxylated core with base under the appropriate temperature conditions expands the core by osmotic swelling to produce hollow particles with water and polyelectrolyte in the interior [98]. Itou et al. [99] prepared submicron crosslinked hollow polymer capsules composed of P(MMA-*co*-DVB) by means of a seeded emulsion polymerization. McDonald et al. [100] found that the modification of an emulsion polymerization with a water-miscible alcohol and a hydrocarbon non-solvent for the polymer can influence the morphology and permits the formation of monodisperse particles with a hollow structure or diffuse microvoids. Both kinetic and thermodynamic aspects of the polymerization dictate particle morphology. Complete encapsulation of the hydrocarbon occurs provided low molecular weight polymer is formed initially in the process. Hollow monodisperse particles with diameters from 0.2 to 1 μm can be obtained and void fractions as high as 50% are feasible.

Nanometer-sized hollow polymer particles were successfully prepared by miniemulsion techniques using different formation mechanisms. In the first case, the formation mechanism was based on phase separation during the polymerization time [101]. Non-polymerizable hydrophobic oil (hexadecane) and monomer were mixed together in miniemulsion droplets before polymerization, whereas the polymer is immiscible with the oil and demixes throughout polymerization, resulting in polymer particles with a morphology consisting of a hollow polymer structure surrounding the oil. The differences in the hydrophilicity of the oil and the polymer turned out to be the driving force for the

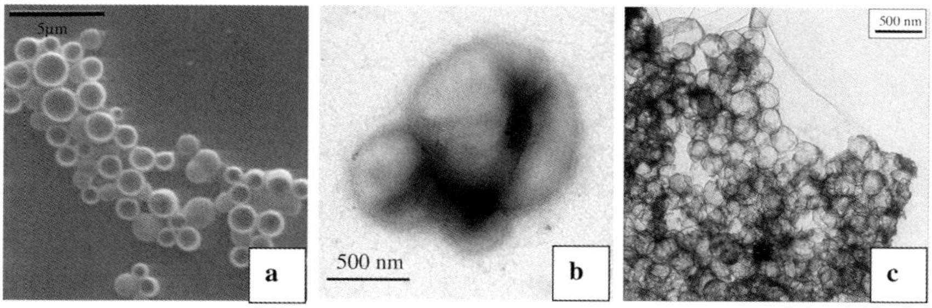

Fig. 13.12 Formation of nanocapsules with (a) polymer PS shell, (b) biodegradable chitosan shell and (c) inorganic SiO_2 shell.

formation of nanocapsules. Direct nanocapsule formation was favorable for a PMMA and hexadecane system. On the other hand, by employing styrene as monomer, the hydrophilicity of the polymer phase has to be adjusted in order to obtain the nanocapsule structure, which was done by addition of either an appropriate comonomer or an initiator (Fig. 13.12 a). The second mechanism involves the cross-linking reaction carried out at the water/droplet interface. Chitosan-containing nanocapsules consisting of hybrid polyaddition polymers as the shell and hydrophobic oil as the core were elaborated [102] (Fig. 13.12 b). These capsules are biocompatible and biodegradable and can find applications in drug delivery.

13.3.3.2 **Capsules by Precipitation**

Microcapsules with PMMA, P(*iso*-BMA) or poly(tetrahydrofuran) shells have been prepared via internal phase separation from water-in-oil emulsions, formulated with the shell polymer dissolved in the aqueous phase by adding a low boiling-point cosolvent. Removal of the cosolvent results in phase separation and formation of a polymer shell around the aqueous core [103]. The selective solvent evaporation approach was used to obtain microcapsules with a polymer shell and an oil interior. The shell polymer, i.e. PS, dichloromethane (a good solvent for PS) and hexadecane (a poor solvent for PS) were mixed and formed oil droplets. The subsequent evaporation of the good solvent led to the phase separation of the polymer within the oil droplets and formation of the polymer shell encapsulating the oil core [104].

A very recent approach involves the modified nanoprecipitation of polymers on to stable nanodroplets of miniemulsions to prepare well-defined nanocapsules whose core is composed of an antiseptic agent, i.e. chlorhexidine digluconate aqueous solution [105]. The stable nanodroplets were obtained by inverse miniemulsions with an aqueous antiseptic solution dispersed in an organic medium of solvent–non-solvent mixture containing an oil-soluble surfactant and the polymer for the shell formation. The change of gradient of the solvent–non-solvent mixture of dichloromethane–cyclohexane led to the precipitation of the polymer in the organic continuous phase and deposition on to the large interface of the miniemulsion aqueous droplets. The nanocapsules could easily be transferred into water as continuous phase, resulting in aqueous dispersions with nanocapsules containing an aqueous core with the antiseptic agent. The amount of the antiseptic agent encapsulated was evaluated to indicate the durability of the nanocapsule's wall. In addition, the use of different types of polymers having glass transition temperatures (T_g) ranging from 10 to 100 °C in this process has also been successful.

13.3.3.3 Capsules by Surface Adsorption

Nanocapsules can also be obtained by a miniemulsion process with an inorganic shell. Here, the employment of crystalline inorganic materials, such as clay sheets of thickness 1.5 nm, can be recommended. Since those clay sheets are fixed like scales on the soft, liquid miniemulsion droplet, the resulting objects are called "armored latexes" [106]. Since clays carry a negative surface charge, miniemulsions stabilized with cationic sulfonium surfactants, which are obtained by the ring opening of epoxy derivatives with thioethers [107], present a convenient way to generate those armored latexes or crystalline nanocapsules. Due to their high stability also against changes in the chemical environment, it is possible to use miniemulsion droplets themselves, but also already polymerized latex particles as templates for such a complexation process. As a result, the liquid droplets or the polymer particles are then completely covered with clay plates and therefore film formation or coalescence is prevented (Fig. 13.12 c).

13.3.3.4 Capsules by Interfacial Polymerization

Another well-established method is the preparation of polyalkylcyanoacrylate (PACA) nanocapsules. The size and shell thickness of the capsule mainly depends on the concentrations of the oil and the monomer components, and also on the type of monomer and surfactant [108, 109]. Several groups have used the interfacial polymerization approach to synthesize PACA capsules containing an aqueous [110–112] or an oil core [113–115]. PACA nanocapsules are biodegradable, possess low toxicity and enhance the intracellular penetration of drugs. Therefore, they are widely used as nanocontainers of biomolecules in a drug delivery system [116, 117]. Polymer capsules were also synthesized by employing other interfacial cross-linking reactions, such as polyaddition [118], polycondensation [119, 120] and radical polymerization [121, 122].

Nanocapsules made by the polymerization of rather complex entities were reported by Stewart and Liu [123]. They stabilized block copolymer vesicles made of polyisoprene-*b*-poly(2-cinnamoylethyl methacrylate) diblock copolymers by UV cross-linking. Jang and Ha [124] prepared nanosized hollow PS nanospheres via a ternary microemulsion polymerization. First, PMMA–cross-linked PS core–shell nanospheres were formed by using triblock copolymers of poly(oxyethylene)–poly(oxypropylene)–poly(oxyethylene) as surfactant. In the second stage, the PMMA core was etched by methylene chloride, resulting in PS hollow nanospheres of size 15–30 nm with a shell thickness of ca. 2–5 nm.

13.3.3.5 Capsules by the Template Method

Another approach to fabricate the polymer nanocapsules is to form a polymer shell around a preformed template particle that can subsequently be removed, thus resulting in an empty polymer shell. Feldheim and coworkers used gold nanoparticles as templates for the synthesis of hollow PPy and poly(*N*-methylpyrrole) shells [125]. After etching the gold, the hollow polymer capsules with a shell thickness governed by the polymerization time were obtained. Polymer micro- and nanocapsules were generated by alternate adsorption of oppositely charged polyelectrolytes on the surface of colloidal particles. Cores of different nature (organic or inorganic) with sizes varying from 0.1 to 10 μm can be used as a template [126, 127]. Hollow silica spheres of size ranging from 30 nm to 2 μm were obtained after treating the CdS–SiO_2 core–shell particles with concentrated nitric acid [128]. An overview of different techniques, based on a combination of colloidal templating and self-assembly processes, developed for synthesizing uniform hollow capsules of a broad range of materials, was published by Caruso [129].

13.3.3.6 Capsules by the Double Emulsion Technique

Microencapsulation of peptides and proteins is achieved by preparing microcapsules by using a double emulsion technique [130]. In this method, an aqueous drug solution is intensively mixed with an organic polymer solution and then an aqueous surfactant solution is added slowly to the water-in-oil (W/O) emulsion. The water-in-oil-in-water (W/O/W) emulsion obtained was stirred under reduced pressure until the organic solvent was removed and the microcapsule was built up. The solvent evaporation approach in the W/O/W system was used by Kawano and coworkers [131] in order to produce capsules composed of biodegradable polymers.

13.4 Methods of Characterization

A large variety of techniques can be used to characterize core–shell particles with respect to their morphology and composition. Some of these methods (for example, contact angle measurements, conductometric or potentiometric titration) give information about the changes in the particle surface properties if the shell layer was formed in a separate stage to the core. The most commonly used techniques which allow evidence of core–shell morphology are described in the following section.

13.4.1 Microscopy

Microscopy is the most frequently used technique to characterize the morphology of composite particles. There are different microscopy methods available which can resolve details on the nanometer scale, such as the light optical microscopy, scanning electron microscopy (SEM), transmission electron microscopy (TEM) and atomic force microscopy (AFM).

Optical techniques enhance the contrast between polymers that are transparent but that have different optical properties, such as the refractive index and the thickness. Optical bright field imaging of composite particles has the potential to resolve the details in a sample which is less than 1 μm thick; however, if the contrast between the phases is low, it is very difficult to see the differences. Optical microscopy produces images with a small depth of focus, low resolution and low degree of magnification and therefore is not usually used for the investigation of core–shell morphology.

SEM offers a high degree of magnification and a good depth of focus, but not good contrast between two phases. Therefore, in order to study the morphology of particles, either they can be cut into two parts or the inner part of the particle can be etched by using a solvent selective for the core. Samples for SEM should be electrically conductive in order to produce secondary electrons and to minimize charge build-up, as polymers generally are non-conductive. This can be achieved by covering the probe with a conductive metal (e.g. gold–palladium) before the measurements.

The size and structural morphology of individual particles can be directly visualized by TEM, which was one of the first methods to be used for this pur-

Table 13.1 Polymer functional groups and commonly used stains [134].

Polymers/functional group	Stain
Unsaturated hydrocarbons (e.g. polybutadiene, isoprene)	OsO_4
Saturated hydrocarbons (e.g. polyethylene, polypropylene)	Phosphotungstic acid or RuO_4
Chlorinated polyethylene	Bicyclic amine, then OsO_4
Amines, ethers, aldehydes, alcohols (e.g. poly(vinyl alcohol))	OsO_4 or RuO_4
Acids, esters (e.g. butyl acrylate, polyester)	Phosphotungstic acid, uranyl acetate, silver sulfide or hydrazine, then OsO_4
Amides (e.g. nylon)	Phosphotungstic acid
Aromatics (e.g. styrene)	RuO_4 or silver sulfide
High-impact polystyrene	OsO_4
Acrylonitrile–butadiene–styrene	OsO_4
Acrylonitrile–styrene–acrylate	OsO_4
Epoxy resin	RuO_4
Epoxy thermosets	Tetrahydrofuran, then OsO_4

pose more than 30 years ago [132, 133]. To minimize the damage caused by the electron beam to polymers with a low T_g, they are usually placed on the grid covered with a carbon film or the probe is frozen at liquid nitrogen temperature and the observations are performed in a cold stage. In addition, since most of the polymers have a low atomic number and display a little scattering, they show low contrast in TEM. By staining of the polymer with heavy-metal compounds, the contrast between phases can be improved. Two kinds of staining can be used: *negative staining* (e.g. uranyl acetate or phosphotungstic acid), when the metal layer is placed onto the particle surface and *positive staining* (e.g. RuO_4, OsO_4), when the metal reacts with a specific part inside the polymer. The staining agents exhibit a high selectivity for certain polymers and therefore are very suitable for morphological studies. Table 13.1 gives an overview of the most frequently used staining agents for selective staining.

In the past decade, AFM has been extensively developed, providing new possibilities to investigate particle morphology. Three-dimensional images, subnanometer resolution and a broad variation of the observation conditions (i.e. at atmospheric pressure, in air, in solvent vapors and in liquid medium) make this technique very attractive for characterization of nanocolloids.

Pioneering work in AFM studying PBA–PMMA latex particles with a core–shell morphology was done by Sommer et al. [135]. From the results obtained, it was concluded that the shell was formed by an aggregation mechanism of PMMA microdomains. Kim et al. [136] employed AFM for investigating the morphology of PS–PBA composite particles. The pure PS particles and PS–PBA with a low content of BA monomer were almost spherical and regular. With increase in the BA monomer content, particles with a "golf ball"-like morphology were produced (Fig. 13.13).

Gerharz et al. [137] examined the morphology, size and viscoelastic inhomogeneities of PBA–PMMA composite particles in order to deduce their behavior during film formation.

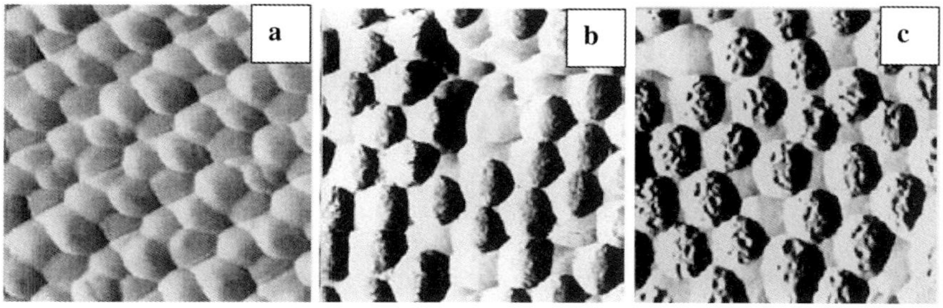

Fig. 13.13 AFM images of (a) PS particles, (b) PS–PBA composite particles (9:1) and (c) PS–PBA composite particles (7:3). The scanning dimension is 2 × 2 μm. The images were recorded in the deflection mode. Reproduced from [136], with permission.

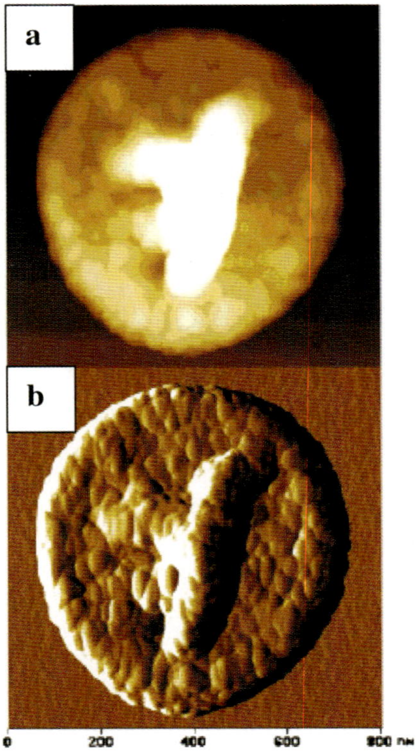

Fig. 13.14 AFM micrographs of polymerized 10-layer hollow DZR–PSS capsules prepared from coated 640-nm PS particles: (a) height image (height scale: 300 nm) and (b) amplitude image (after filtering). Reproduced from [138], with permission.

AFM is also a very suitable method to confirm the formation of hollow particles or with a liquid core. For example, Caruso and coworkers [138] made AFM observations on polyelectrolyte microcapsules obtained via the layer-by-layer self-assembly of a diazo resin (DZR) and poly(styrenesulfonate) (PSS) (Fig. 13.14). The mechanical deformation of polyelectrolyte multilayer capsules in an aqueous environment was studied using direct-deformation measurements by colloidal probe AFM [139, 140].

Several groups have examined the surface morphology of microcapsules in dry form and under ambient conditions, in an aqueous environment [141, 142] (Fig. 13.15). However, to perform the measurements in liquids, the particle should be relatively larger in order to have slow Brownian motion or be strongly adsorbed on the surface.

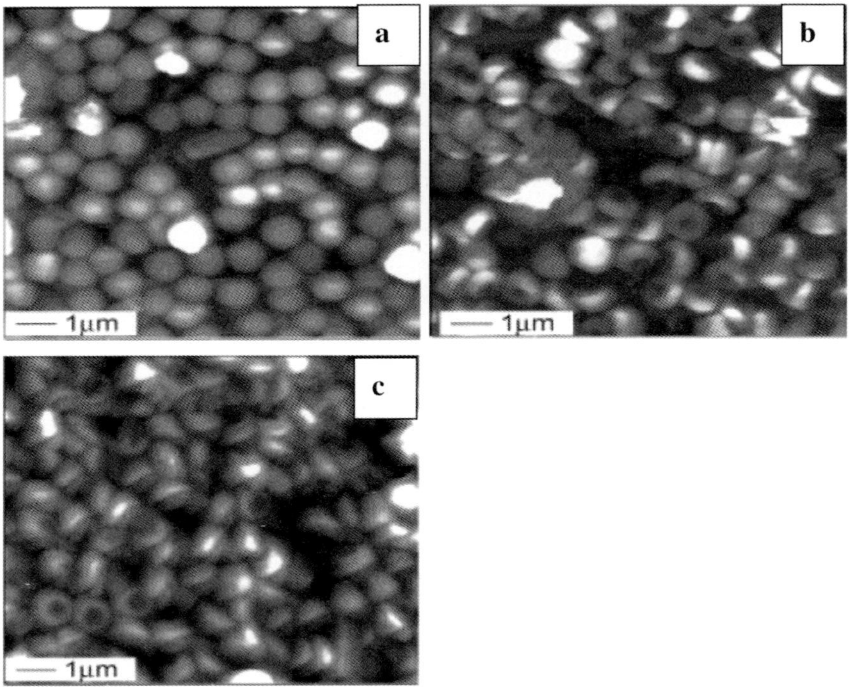

Fig. 13.15 AFM images of 740-nm hollow polymeric particles: (a) measured in water, after etching silica core with HF and filling with water, (b) measured under air, after drying at room temperature for 1 h, and (c) measured in water, after immersion in deionized water for 24 h. Reproduced from [142], with permission.

13.4.2
Nuclear Magnetic Resonance (NMR)

Nuclear magnetic resonance (NMR) spectroscopy is today one the most valuable tools for the elucidation of the molecular structure, order and dynamics of chemical systems. For the characterization of macromolecules, NMR can be divided into two major areas: high-resolution NMR provides detailed information about the chain microstructure in solution and solid-state NMR allows one to characterize the molecular structure and the organization of the macromolecules in the bulk. Various solid-state NMR techniques offer new possibilities of studying domain sizes and interphase structures in heterogeneous polymers and latex particles, e.g. the miscibility and the interfacial region can be characterized by the various relaxation times [143].

The interphase thickness and its average composition were studied using a combination of differential scanning calorimetry (DSC) and solid-state NMR relaxation methods [144–146]. Moreover, the influence of core–shell ratio and surfactant coverage of the seed latex on the interphase characteristics of PBA–PMMA latexes was examined. In contrast to DSC, which could only be successfully used at PMMA contents between 30 and 75 wt%, solid-state NMR is a suitable method to examine the interphase in the complete range of composition. The interfacial region was also determined by relaxation measurements in PDVB–PBA core–shell latexes synthesized with different process parameters, such as the mode of addition of the second-stage monomer, the rate of addition and the extent of conversion of the PDVB seed latex at the time of addition of the second-stage monomer [147].

Advanced solid-state NMR techniques allowed the identification of the components in the interphase of core–shell latex particles due to gradients of concentration reflected in gradients of mobility within the latex particles. Experiments were developed to detect the differences in mobility [148].

The *degree of phase separation* can be determined *qualitatively* by the *2D WISE experiment* [149]. It allows one to characterize whether there is an extended interphase between the two phases in the core–shell particle. In an ideal core–shell structure, the WISE experiment detects the mobile core component as sharp lines, whereas the rigid shell component shows broad lines, reflecting the differences in their dipole–dipole couplings. In reality, most of the core–shell systems show a superposition of both types of line shapes, indicating an interphase with immobilized core component and mobilized shell component.

Filter experiments with ^{13}C NMR detection [148] characterize the *amount of shell polymer diffused into the core.* The intensity of the most intense signal of the rigid component in the ^{13}C spectra is evaluated and plotted versus the filter strength. At a weak filter strength (filter strength = 1), the amount of mobilized shell component in the interphase between the two phases is obtained. With increasing filter strength, the decrease in magnetization gives information about the size of structures in the interphase. In a simplified description of the experiment, each filter strength corresponds to a certain T_2 relaxation filter. Because the T_2 values are related to the mobility, the data in the filter experiment graph reflect the amounts of the shell material at different mobilities. The more difficult it is to suppress the magnetization of the shell component, the higher are the mobilities in the interphase and the smaller the structures. This is shown schematically in Fig. 13.16 for different interphase structures. For an ideal core–shell system with effective phase separation, curve A is expected. Even with a weak filter, the magnetization of the shell component can be completely suppressed. If the shell component is partially mobilized by the core component, curves B–D are obtained. The different morphologies of interphases are sketched in Fig. 13.16. For cases B and C, the values of detected PMMA for a weak filter are similar due to similar amounts of immobilized component in the interphase. In C, it is more difficult to suppress the magnetization of the rigid component, indicating smaller structures in the interphase. This leads to a higher mobility of the rigid phase, as shown by the finer steps in the gradient of gray shades. In the case of D, much more of the rigid component is mobi-

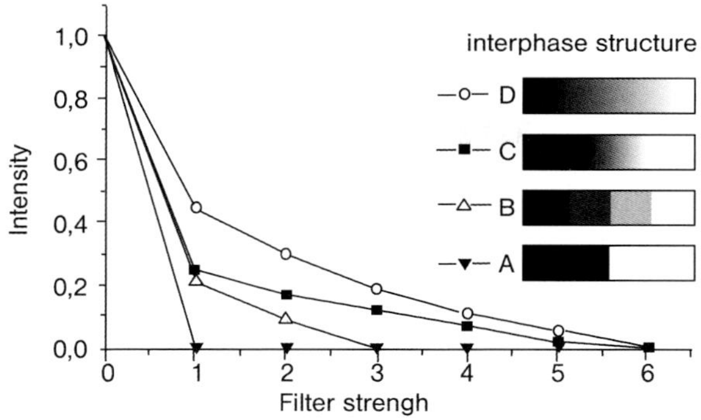

Fig. 13.16 Scheme of filter experiment: (A) ideal phase separation, (B) interphase with two different mobilities, (C) interphase with a gradient in mobility and (D) expanded interphase with the same gradient in mobility as in (C). Reproduced from [148], with permission.

lized because the interphase as a whole is larger. For C and D, the magnetization of the rigid component is suppressed at strong filters (filter strength = 6).

Spin-diffusion experiments allow the determination of the *interphase thickness* and the quantification of immobilized soft component by the soft component–hard component contact area [148]. To this end, quantitative analysis of spin-diffusion curves with ^{1}H detection is employed. Specifically, here the decay of the retained ^{1}H signal after application of the dipolar filter as a function of spin-diffusion time is considered. Since a diffusive redistribution of magnetization is probed, the signal intensity is plotted as a function of mixing time $t_m^{1/2}$. In most cases, the spin-diffusion curve will decay monotonically to a plateau (final) value. Both the initial slope of this decay and the plateau contain important information about the interphase. In Fig. 13.17, an interface structure without a concentration gradient is considered. For an effective phase separation in idealized core–shell particles with diameters between 100 and 400 nm, slow spin diffusion is expected (upper curve). Fast spin diffusion results for small structures located in the interphase of core–shell particles (intermediate curve).

For real core–shell particles with dispersed small structures of the core as indicated, a superposition of at least two spin-diffusion processes is expected due to the entire structure and the substructure in the particles (lowest curve). In actual measurements only the first decay can usually be evaluated quantitatively and the second decay allows in most cases only a qualitative interpretation.

NMR techniques were used to characterize an extended series of core–shell particles composed of PBA and PMMA [150] synthesized in semicontinuous two-step emulsion polymerization processes. The influence of process parameters such as synthesis temperature, cross-linking density in the core, shell content, particle size and annealing were investigated in detail [148].

NMR spin-diffusion techniques could also be used to investigate the degree of coverage and the structure and thickness of interphases in composite latex

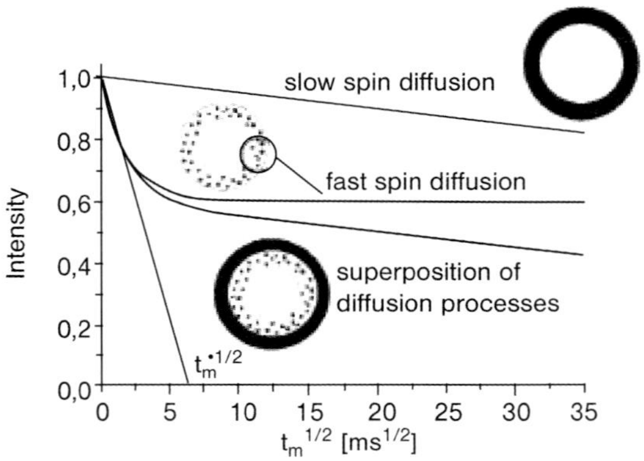

Fig. 13.17 Schematic spin-diffusion curves demonstrating the sensitivity to different structures in the latex particles. The entire core–shell structure leads to a slow spin diffusion; substructures in the interphase allow a fast spin-diffusion process. With characteristic core–shell particles, a superposition of diffusion processes is observed. Reproduced from [148], with permission.

particles consisting of a PBA core incorporated with PMMA macromonomer of different molecular weight and a PMMA shell [151]. By changing the temperature at a given filter strength or by adjusting the filter strength at a given temperature, the spin-diffusion experiment could be performed in such a way that the overall structure could be detected. At filter strengths lower than necessary for the detection of the overall structure, information regarding larger structures can be obtained, whereas for stronger filters detailed information about small structures in the interphase region can be extracted. The data obtained by NMR were related to previous TEM and dynamic mechanical studies on these particles.

13.4.3
Scattering Methods

Scattering methods such as small-angle X-ray scattering (SAXS) and small-angle neutron scattering (SANS) have also been extensively used to explore the structure of the particles down to a resolution of several nanometers. The main advantage of these methods is that it is possible to perform the analysis in a dispersed medium, i.e. without changing the natural environment of the particles.

SAXS measurements are based on the differences in contrast between the core and surrounded shell layer of the particle [152]. Therefore, the main requirement for SAXS experiments is a reasonable difference in electron density between two phases. For example, the analysis of thermosensitive PS–PNIPAM core–shell latex particles was successfully performed owing to the low SAXS contrast of PS towards water (6.4 $e^- nm^{-3}$) compared with the PNIPAM (45.8 e/nm^{-3}). Therefore, the SAXS intensity mainly originated from the shell layer of the particle [153].

The method was applied to evidence the structure of PS–PMMA core–shell particles [154]. The inverted core–shell morphology was shown when PMMA latex particles were swollen by styrene [155]; however, the PMMA–PMA system does not show this trend [156]. Hourston et al. [157] analyzed the morphological structure of PS–PMA core–shell latexes by combination of SAXS and modulated-temperature differential scanning calorimetry (M-TDSC).

SANS is another frequently used technique for the structural analysis of composite particles [158, 159]. The neutrons used in SANS experiments have zero charge and can well penetrate inside the sample. Neutrons are scattered by nuclei in samples or by the magnetic moments associated with unpaired electron spins (dipoles) in magnetic samples. As in SAXS, a high contrast between core and shell is the main requirement for the SANS technique. This can be achieved through the substitution of hydrogen atoms by deuterium atoms. Due to the absence of a neutron in the atom of hydrogen, the difference in the scattering length between hydrogen and deuterium is very large. In particular, variable contrast between the core and the shell in the composite latex particle can be obtained by employing one of the monomers in the deuterated form. The contrast variation technique was used to investigate the structure of PMMA–PS and PMMA–PMMA latex particles [160], thermosensitive core–shell particles consisting of a solid PS core and a cross-linked PNIPAM shell [161], the internal structure of PNIPAM–*N,N*″-methylenebisacrylamide core–shell microgel [162] and the inner structure of poly(D,L-lactide) nanocapsules [163].

13.4.4 Fluorescence Spectroscopy

The application of fluorescence techniques to the study of polymer colloids was reviewed by Winnik [164]. Information about the core–shell morphology, including information about the phase boundaries, can be obtained by fluorescence non-radiative energy transfer (NRET) processes through labeling the core polymer with a donor and the shell polymer with an acceptor. Several groups have reported the application of NRET to study the internal structure of core–shell particles. Winnik et al. [165] studied PMMA–PMMA core–shell particles synthesized by different emulsion polymerization methods. Lang and coworkers employed NRET to study the morphology of core–shell particles with different compositions such as PMMA–PBMA [166], PBMA–poly(BMA-*co*-BA-*co*-TFEMA) [70, 167] and PBMA–PPy [168].

13.5 Conclusion

This chapter provides a survey on the subject of core–shell particles. The main idea in fabricating particles with core–shell morphology is to combine materials with different physical and chemical properties. First, the theoretical prediction of the particle morphology and its practical realization have been outlined. Then, a large range of monomers, polymers, oxides, metals, etc., can be used as core or shell material in order to develop optimal performances for final needs. Different synthetic methods for preparing core–shell nanoparticles have been presented. In particular, the two-stage emulsion polymerization process is widely used in industry to fabricate core–shell particles on the tonnage level. On the other hand, for lower amounts, the miniemulsion technique is very suitable for generating well-defined structured nanoparticles with a solid or liquid core. The technique allows the efficient encapsulation of various inorganic and organic, solid and liquid materials. It is also possible to create copolymer structures of high perfection.

A wide variety of core–shell latex particles have been designed in the last 50 years. There is no doubt that a wealth of findings have been constantly contributing to the creation of core–shell particles and their use in a wide range of applications. However, there is always room for further optimization and new explorations.

List of Abbreviations

AAm	acrylamide
AFM	atomic force microscopy
AIBA	2,2′-azobis(2-amidinopropane) dihydrochloride
AIBN	azobisisobutyronitrile
AN	acrylonitrile
B	butadiene
BA	butyl acrylate
DVB	divinylbenzene
EA	ethyl acrylate
EDTA	ethylenediaminetetraacetic acid
HLB	hydrophilic–lipophilic balance
NRET	non-radiative energy transfer
P(*iso*-BMA)	poly(isobutyl methacrylate)
P(S-B)	poly(styrene–butadiene)
PA	polyacrylate
PACA	polyalkylcyanoacrylate
PAlA	polyallylamine
PAmA	poly(amic acid)
PANI	polyaniline
PB	polybutadiene
PBA	poly(butyl acrylate)
PBMA	poly(butyl methacrylate)
PDVB	polydivinylbenzene
PEA	poly(ethyl acrylate)
PEHMA	poly(2-ethylhexyl methacrylate)

PEI	polyethylenimine
PGMA	poly(glycidyl methacrylate)
PI	polyimide
PIEM	poly(2-isocyanatoethyl methacrylate)
PMA	poly(methyl acrylate)
PMMA	poly(methyl methacrylate)
PNIPAM	poly(*N*-isopropylacrylamide)
PPy	polypyrrole
PS	polystyrene
PSS	poly(styrene sulfonate)
PUR	polyurethane
PVAc	poly(vinyl acetate)
PVP	poly(1-vinyl-2-pyrrolidone)
SANS	small-angle neutron scattering
SAXS	small-angle x-ray scattering
SEM	scanning electron microscopy
t-BHP	*tert*-butyl hydroperoxide
TEM	transmission electron microscopy
TFEMA	trifluoroethyl methacrylate

References

1 Eliseeva, V. I. *Prog. Org. Coat.* **1985**, *13*, 195.

2 Devon, M. J., Gardon, J. L., Roberts, G., Rudin, A. *J. Appl. Polym. Sci.* **1990**, *39*, 2119.

3 Blankenship, R., Kowalski, A. *US Patent 4594363*, **1986**.

4 Sundberg, D. C., Durant, Y. G. *Polym. React. Eng.* **2003**, *11*, 379.

5 Dimonie, V. L., Daniels, E. S., Shaffer, O. L., El-Aasser, M. S., Control of particle morphology, in *Emulsion Polymerization and Emulsion Polymers*, Lovell, P. A., El-Aasser, M. S. (Eds.), Wiley, New York, **1997**, pp. 293–326.

6 Torza, S., Mason, S. G. *J. Colloid Interface Sci.* **1970**, *33*, 67.

7 Berg, J., Sundberg, D., Kronberg, B. *Proc. ACS Div. Polym. Mater.* **1986**, *54*, 367.

8 Berg, J., Sundberg, D., Kronberg, B. *J. Microencapsulation* **1989**, *6*, 327.

9 Sundberg, D. C., Casassa, A. P., Pantazopoulos, J., Muscato, M. R., Kronberg, B., Berg, J. *J. Appl. Polym. Sci.* **1990**, *41*, 1425.

10 Muscato, M. R., Sundberg, D. C. *J. Polym. Sci., Part B: Polym. Phys.* **1991**, *29*, 1021.

11 Chen, Y.-C., Dimonie, V., El-Aasser, M. S. *J. Appl. Polym. Sci.* **1991**, *42*, 1049.

12 Chen, Y.-C., Dimonie, V., El-Aasser, M. S. *Macromolecules* **1991**, *24*, 3779.

13 Winzor, C. L., Sundberg, D. C. *Polymer* **1992**, *33*, 3797.

14 Winzor, C. L., Sundberg, D. C. *Polymer* **1992**, *33*, 4269.

15 Sundberg, E. J., Sundberg, D. C. *J. Appl. Polym. Sci.* **1993**, *47*, 1277.

16 González-Ortiz, L. J., Asua, J. M. *Macromolecules* **1995**, *28*, 3135.

17 González-Ortiz, L. J., Asua, J. M. *Macromolecules* **1996**, *29*, 4520.

18 González-Ortiz, L. J., Asua, J. M. *Macromolecules* **1996**, *29*, 383.

19 Huo, D., Liu, D. *Polym. Int.* **2002**, *51*, 585.

20 Sun, P., Zhao, K., Liu, D., Huo, D. *J. Appl. Polym. Sci.* **2002**, *83*, 2930.

21 Zhao, K., Sun, P., Liu, D., Wang, L. *J. Appl. Polym. Sci.* **2004**, *92*, 3144.

22 Durant, Y.G., Sundberg, D.C. *J. Appl. Polym. Sci.* **1995**, *58*, 1607.
23 Stubbs, J.M., Carrier, R., Karlsson, O.J., Sundberg, D.C. *Prog. Colloid Polym. Sci.* **2004**, *124*, 131.
24 Muroi, S., Hashimoto, H., Hosoi, K. *J. Polym. Sci., Part A: Polym. Chem.* **1984**, *22*, 1365.
25 Okubo, M., Yamada, A., Matsumoto, T. *J. Polym. Sci., Part A: Polym. Chem.* **1980**, *18*, 3219.
26 Lee, D.I., Ishikawa, T. *J. Polym. Sci., Part A: Polym. Chem.* **1983**, *21*, 147.
27 Okubo, M., Kanaida, K., Matsumoto, T. *Colloid Polym. Sci.* **1987**, *265*, 876.
28 Okubo, M., Katsuta, Y., Matsumoto, T. *J. Polym. Sci., Polym. Lett. Ed.* **1982**, *20*, 45.
29 Jönsson, J.E., Hassander, H., Törnell, B. *Macromolecules* **1994**, *27*, 1932.
30 Chen, Y.C., Dimonie, V., El-Aasser, M.S. *J. Appl. Polym. Sci.* **1992**, *45*, 487.
31 Okubo, M., Seike, M., Matsumoto, T. *J. Appl. Polym. Sci.* **1983**, *28*, 383.
32 Okubo, M., Ando, M., Yamada, A., Katsuta, Y., Matsumoto, T. *J. Polym. Sci., Polym. Lett. Ed.* **1981**, *19*, 143.
33 Karlsson, L.E., Karlsson, O.J., Sundberg, D.C. *J. Appl. Polym. Sci.* **2003**, *90*, 905.
34 Lee, S., Rudin, A. *J. Polym. Sci., Part A: Polym. Chem.* **1992**, *30*, 2211.
35 Dimonie, V., El-Aasser, M.S., Klein, A., Vanderhoff, J.W. *J. Polym. Sci., Part A: Polym. Chem.* **1984**, *22*, 2197.
36 Merkel, M.P., Dimonie, V.L., El-Aasser, M.S., Vanderhoff, J.W. *Proc. ACS Div. Polym. Mater.* **1986**, *54*, 598.
37 Min, T.I., Klein, A., El-Aasser, M.S., Vanderhoff, J.W. *J. Polym. Sci., Part A: Polym. Chem.* **1983**, *21*, 2845.
38 Cho, I., Lee, K.-W. *J. Appl. Polym. Sci.* **1985**, *30*, 1903.
39 Hughes, L.J., Brown, G.L. *J. Appl. Polym. Sci.* **1961**, *5*, 580.
40 Gasperowicz, A., Kolendowicz, M., Skowronski, T. *Polymer* **1982**, *23*, 839.
41 Okubo, M., Izumi, J., Takekoh, R. *Colloid Polym. Sci.* **1999**, *277*, 875.
42 Sperling, L.H., Chiu, T.-W., Thomas, D.A. *J. Appl. Polym. Sci.* **1973**, *17*, 2443.
43 Sionakidis, J., Sperling, L.H., Thomas, D.A. *J. Appl. Polym. Sci.* **1979**, *24*, 1179.
44 Hattori, M., Sudol, E.D., El-Aasser, M.S. *J. Appl. Polym. Sci.* **1993**, *50*, 2027.
45 Aerdts, A.M., de Krey, J.E.D., Kurja, J., German, A.L. *Polymer* **1994**, *35*, 1636.
46 Schellenberg, C., Akari, S., Regenbrecht, M., Tauer, K., Petrat, F.M., Antonietti, M. *Langmuir* **1999**, *15*, 1283.
47 Santos, R.M., Forcada, J. *Prog. Colloid Polym. Sci.* **1996**, *100*, 87.
48 Sarobe, J., Molina-Bolívar, J.A., Forcada, J., Galisteo, F., Hidalgo-A lvarez, R. *Macromolecules* **1998**, *31*, 4282.
49 Laus, M., Lelli, M., Casagrande, A. *J. Polym. Sci., Part A: Polym. Chem.* **1997**, *35*, 681.
50 Laus, M., Dinnella, C., Lanzarini, G., Casagrande, A. *Polymer* **1996**, *37*, 343.
51 Dinnella, C., Laus, M., Lanzarini, G., Doria, M. *Polym. Adv. Technol.* **1996**, *7*, 548.
52 Watanabe, S., Ueno, K., Kudoh, K., Murata, M., Masuda, Y. *Macromol. Rapid Commun.* **2000**, *21*, 1323.
53 Okubo, M., Yamamoto, Y., Iwasaki, Y. *Colloid Polym. Sci.* **1991**, *267*, 1126.
54 Okubo, M., Nakagawa, T. *Colloid Polym. Sci.* **1994**, *272*, 530.
55 Okubo, M., Fujii, S., Minami, H. *Colloid Polym. Sci.* **2001**, *279*, 139.
56 Young, J.L., DeSimone, J.M. *Macromolecules* **2005**, *38*, 4542.
57 López-Quintela, M.A. *Curr. Opin. Colloid Interface Sci.* **2003**, *8*, 137.
58 Aguiar, A., González-Villegas, S., Rabelero, M., Mendizábal, E., Puig, J.E., Domínguez, J.M., Katime, I. *Macromolecules* **1999**, *32*, 6767.
59 Rabelero, M., López-Cuenca, S., Puca, M., Mendizábal, E., Esquena, J., Solans, C., López, R.G., Puig, J.E. *Polymer* **2005**, *46*, 6182.
60 Dong, A., Feng, S., Sun, D. *Macromol. Chem. Phys.* **1998**, *199*, 2635.
61 Greci, M.T., Pathak, S., Mercado, K., Surya Prakash, G.K., Thompson, M.E., Olah, G.A. *J. Nanosci. Nanotechnol.* **2001**, *1*, 3.
62 Barner, L., Li, C., Hao, X., Stenzel, M.H., Barner-Kowollik, C., Davis, T.P. *J. Polym. Sci., Part A: Polym. Chem.* **2004**, *42*, 5067.
63 Tokarev, V.S., Voronov, S.A., Adler, H.-J.P., Datzuk, V.V., Pich, A.Z., Shev-

chuk, O. M., Myahkostupov, M. V. *Macromol. Symp.* **2002**, *187*, 155.
64 Willert, M., Landfester, K. *Macromol. Chem. Phys.* **2002**, *203*, 825.
65 Li, P., Zhu, J., Sunintaboon, P., Harris, F. W. *Langmuir* **2002**, *18*, 8641.
66 Li, P., Zhu, J., Sunintaboon, P., Harris, F. W. *J. Dispers. Sci. Technol.* **2003**, *24*, 607.
67 Zhu, J., Li, P. *J. Polym. Sci., Part A: Polym. Chem.* **2003**, *41*, 3346.
68 Ye, W., Leung, M. F., Xin, J., Kwong, T. L., Lee, D. K. L., Li, P. *Polymer* **2005**, *46*, 10538.
69 Ha, J.-W., Park, I. J., Lee, S.-B., Kim, D.-K. *Macromolecules* **2002**, *35*, 6811.
70 Marion, P., Beinert, G., Juhué, D., Lang, J. *J. Appl. Polym. Sci.* **1997**, *64*, 2409.
71 Degorgio, V., Piazza, R., Bellini, T., Visca, M. *Adv. Colloid Interface Sci.* **1994**, *48*, 61.
72 Landfester, K., Rothe, R., Antonietti, M. *Macromolecules* **2002**, *35*, 1658.
73 Wiersma, A. E., vd. Steeg, L. M. A., Jongeling, T. J. M. *Synth. Met.* **1995**, *71*, 2269.
74 Barthet, C., Armes, S. P., Lascelles, S. F., Luk, S. Y., Stanley, H. M. E. *Langmuir* **1998**, *14*, 2032.
75 Lascelles, S. F., Armes, S. P., Zhdan, P. A., Greaves, S. J., Brown, A. M., Watts, J. F., Leadley, S. R., Luk, S. Y. *J. Mater. Chem.* **1997**, *7*, 1349.
76 Park, S. Y., Cho, M. S., Kim, C. A., Choi, H. J., Jhon, M. S. *Colloid Polym. Sci.* **2003**, *282*, 198.
77 Jang, J., Oh, J. H. *Adv. Funct. Mater.* **2005**, *15*, 494.
78 Li, H., Kumacheva, E. *Colloid Polym. Sci.* **2003**, *281*, 1.
79 Oldenburg, S. J., Averitt, R. D., Westcott, S. L., Halas, N. J. *J. Chem. Phys. Lett.* **1998**, *288*, 243.
80 Yu, D.-G., An, J. H., Bae, J. Y., Lee, Y. E., Ahn, S. D., Kang, S.-Y., Suh, K. S. *J. Appl. Polym. Sci* **2004**, *92*, 2970.
81 Luna-Xavier, J.-L., Bourgeat-Lami, E., Guyot, A. *Colloid Polym. Sci.* **2001**, *279*, 947.
82 Luna-Xavier, J.-L., Guyot, A., Bourgeat-Lami, E. *J. Colloid Interface Sci.* **2002**, *250*, 82.
83 Quaroni, L., Chumanov, G. *J. Am. Chem. Soc.* **1999**, *121*, 10642.
84 Vestal, C. R., Zhang, Z. J. *J. Am. Chem. Soc.* **2002**, *124*, 14312.
85 Jang, J., Lim, B. *Angew. Chem.* **2003**, *115*, 5758.
86 Erdem, B., Sudol, E. D., Dimonie, V. L., El-Aasser, M. S. *J. Polym. Sci., Part A: Polym. Chem.* **2000**, *38*, 4431.
87 Bechthold, N., Tiarks, F., Willert, M., Landfester, K., Antonietti, M. *Macromol. Symp.* **2000**, *151*, 549.
88 Lelu, S., Novat, C., Graillat, C., Guyot, A., Bourgeat-Lami, E. *Polym. Int.* **2003**, *52*, 542.
89 Takasu, M., Shiroya, T., Takeshita, K., Sakamoto, M., Kawaguchi, H. *Colloid Polym. Sci.* **2003**, *282*, 119.
90 Tiarks, F., Landfester, K., Antonietti, M. *Langmuir* **2001**, *17*, 5775.
91 Fleischhaker, F., Zentel, R. *Chem. Mater.* **2005**, *17*, 1346.
92 Tiarks, F., Landfester, K., Antonietti, M. *Macromol. Chem. Phys.* **2001**, *202*, 51.
93 Ramírez, L. P., Landfester, K. *Macromol. Chem. Phys.* **2003**, *204*, 22.
94 Caruso, R. A., Susha, A., Caruso, F. *Chem. Mater.* **2001**, *13*, 400.
95 Kaltenpoth, G., Himmelhaus, M., Slansky, L., Caruso, F., Grunze, M. *Adv. Mater.* **2003**, *15*, 1113.
96 Ji, T., Lirtsman, V. G., Avny, Y., Davidov, D. *Adv. Mater.* **2001**, *13*, 1253.
97 Kowalski, A., Blankenship, R. *US Patent 4468498*, **1984**.
98 Kaino, M., Takagishi, Y., Toda, H. *US Patent 5360827*, **1994**.
99 Itou, N., Masukawa, T., Ozaki, I., Hattori, M., Kasai, K. *Colloids Surf. A* **1999**, *153*, 311.
100 McDonald, C. J., Bouck, K. J., Chaput, A. B., Stevens, C. J. *Macromolecules* **2000**, *33*, 1593.
101 Tiarks, F., Landfester, K., Antonietti, M. *Langmuir* **2001**, *17*, 908.
102 Marie, E., Landfester, K., Antonietti, M. *Biomacromolecules* **2002**, *3*, 475.
103 Atkin, R., Davies, P., Hardy, J., Vincent, B. *Macromolecules* **2004**, *37*, 7979.
104 Dowding, P. J., Atkin, R., Vincent, B., Bouillot, P. *Langmuir* **2004**, *20*, 11374.
105 Paiphansiri, U., Tangboriboonrat, P., Landfester, K. *Macromol. Biosci.* **2006**, *6*, 33.

106 zu Putlitz, B., Landfester, K., Fischer, H., Antonietti, M. *Adv. Mater.* **2001**, *13*, 500.

107 zu Putlitz, B., Hentze, H.-P., Landfester, K., Antonietti, M. *Langmuir* **2000**, *16*, 3214.

108 Chouinard, F., Kan, F.W.K., Leroux, J.-C., Foucher, C., Lenaerts, V. *Int. J. Pharm.* **1991**, *72*, 211.

109 Puglisi, G., Fresta, M., Giammona, G., Ventura, C.A. *Int. J. Pharm.* **1995**, *125*, 283.

110 Lambert, G., Fattal, E., Pinto-Alphandary, H., Gulik, A., Couvreur, P. *Int. J. Pharm.* **2001**, *214*, 13.

111 Pitaksuteepong, T., Davies, N.M., Tucker, I.G., Rades, T. *Eur. J. Pharm. Biopharm.* **2002**, *53*, 335.

112 Watnasirichaikul, S., Davies, N.M., Rades, T., Tucker, I.G. *Pharm. Res.* **2000**, *17*, 684.

113 Al Khouri Fallouh, N., Roblot-Treupel, L., Fessi, H., Devissaguet, J.P., Puisieux, F. *Int. J. Pharm.* **1986**, *28*, 125.

114 Aboubakar, M., Puisieux, F., Couvreur, P., Vauthier, C. *Int. J. Pharm.* **1999**, *183*, 63.

115 Wohlgemuth, M., Mächtle, W., Mayer, C. *J. Microencapsulation* **2000**, *17*, 437.

116 Vauthier, C., Dubernet, C., Fattal, E., Pinto-Alphandary, H., Couvreur, P. *Adv. Drug Deliv. Rev.* **2003**, *55*, 519.

117 Allemann, E., Leroux, J.-C., Gurny, R. *Adv. Drug Deliv. Rev.* **1998**, *34*, 171.

118 Yamazaki, N., Du, Y.-Z., Nagai, M., Omi, S. *Colloids Surfaces B* **2003**, *29*, 159.

119 Arshady, R. *J. Microencapsulation* **1989**, *6*, 13.

120 Danicher, L., Frere, Y., Le Calve, A. *Macromol. Symp.* **2000**, *151*, 387.

121 Scott, C., Wu, D., Ho, C.-C., Co, C.C. *J. Am. Chem. Soc.* **2005**, *127*, 4160.

122 Sarkar, D., El-Khoury, J., Lopina, S.T., Hu, J. *Macromolecules* **2005**, *38*, 8603.

123 Stewart, S., Liu, G. *Chem. Mater.* **1999**, *11*, 1048.

124 Jang, J., Ha, H. *Langmuir* **2002**, *18*, 5613.

125 Marinakos, S.M., Novak, J.P., Brousseau, L.C., III, House, A.B., Edeki, E.M., Feldhaus, J.C., Feldheim, D.L. *J. Am. Chem. Soc.* **1999**, *121*, 8518.

126 Donath, E., Sukhorukov, G.B., Caruso, F., Davis, S.A., Möhwald, H. *Angew. Chem. Int. Ed.* **1998**, *37*, 2202.

127 Sukhorukov, G.B., Donath, E., Davis, S., Lichtenfeld, H., Caruso, F., Popov, V.I., Möhwald, H. *Polym. Adv. Technol.* **1998**, *9*, 759.

128 Teng, F., Tian, Z., Xiong, G., Xu, Z. *Catal. Today* **2004**, *93–95*, 651.

129 Caruso, F. *Chem. Eur. J.* **2000**, *6*, 413.

130 Hildebrand, G.E., Tack, J.W. *Int. J. Pharm.* **2000**, *196*, 173.

131 Kiyoyama, S., Ueno, H., Shiomori, K., Kawano, Y., Hatate, Y. *J. Chem. Eng. Jpn.* **2001**, *34*, 1182.

132 Grancio, M.R., Williams, D.J. *J. Polym. Sci., Part A: Polym. Chem.* **1970**, *8*, 2617.

133 Kanig, G., Neff, H. *Colloid Polym. Sci.* **1975**, *253*, 29.

134 Sawyer, L.C., Grubb, D.T., Specimen preparation methods, in *Polymer Microscopy*, Sawyer, L.C., Grubb, D.T. (Eds.), Chapman and Hall, London, **1987**, pp. 75–109.

135 Sommer, F., Minh Duc, T., Pirri, R., Meunier, G., Quet, C. *Langmuir* **1995**, *11*, 440.

136 Kim, S.H., Son, W.K., Kim, Y.J., Kang, E.-G., Kim, D.-W., Park, C.W., Kim, W.-G., Kim, H.-J. *J. Appl. Polym. Sci.* **2003**, *88*, 595.

137 Gerharz, B., Butt, H.J., Momper, B. *Prog. Colloid Polym. Sci.* **1996**, *100*, 91.

138 Pastoriza-Santos, I., Schöler, B., Caruso, F. *Adv. Funct. Mater.* **2001**, *11*, 122.

139 Lulevich, V.V., Radtchenko, I.L., Sukhorukov, G.B., Vinogradova, O.I. *J. Phys. Chem. B* **2003**, *107*, 2735.

140 Dubreuil, F., Shchukin, D.G., Sukhorukov, G.B., Fery, A. *Macromol. Rapid Commun.* **2004**, *25*, 1078.

141 Xu, K., Hercules, D.M., Lacik, I., Wang, T.G. *J. Biomed. Mater. Res.* **1998**, *41*, 461.

142 Xu, X., Asher, S.A. *J. Am. Chem. Soc.* **2004**, *126*, 7940.

143 Voelkel, R. *Angew. Chem. Int. Ed. Engl.* **1988**, *27*, 1468.

144 Tembou Nzudie, D., Delmotte, L., Riess, G. *Macromol. Chem. Phys.* **1994**, *195*, 2723.

145 Constaninescu, C., Tembou Nzudie, D., Riess, G. *Int. Polym. Colloids Newsl.* **1993**, *24*, 61.

146 Tembou Nzudie, D., Delmotte, L., Riess, G. *Makromol. Chem., Rapid Commun.* **1991**, *12*, 251.

147 Nelliappan, V., El-Aasser, M. S., Klein, A., Daniels, E. S., Roberts, J. E. *J. Appl. Polym. Sci.* **1995**, *58*, 323.

148 Landfester, K., Spiess, H. W. *Acta Polym.* **1998**, *49*, 451.

149 Schmidt-Rohr, K., Clauss, J., Spiess, H. W. *Macromolecules* **1992**, *25*, 3273.

150 Landfester, K., Boeffel, C., Lambla, M., Spiess, H. W. *Macromolecules* **1996**, *29*, 5972.

151 Landfester, K., Dimonie, V. L., El-Aasser, M. S. *Macromol. Chem. Phys.* **2002**, *203*, 1772.

152 Dingenouts, N., Bolze, J., Pötschke, D., Ballauff, M. *Adv. Polym. Sci.* **1999**, *144*, 1.

153 Dingenouts, N., Norhausen, C., Ballauff, M. *Macromolecules* **1998**, *31*, 8912.

154 Grunder, R., Urban, G., Ballauff, M. *Colloid Polym. Sci.* **1993**, *271*, 563.

155 Bolze, J., Hörner, K. D., Ballauff, M. *Langmuir* **1997**, *13*, 2960.

156 Dingenouts, N., Pulina, Ballauff, M. *Macromolecules* **1994**, *27*, 6133.

157 Hourston, D. J., Song, M., Pang, Y. *J. Braz. Chem. Soc.* **2001**, *12*, 87.

158 Hergeth, W.-D., Bittrich, H.-J., Eichhorn, F., Schlenker, S., Schmutzler, K., Steinau, U.-J. *Polymer* **1989**, *30*, 1913.

159 Saunders, B. R., Electrokinetic and small-angle neutron scattering studies of thermally sensitive polymer colloids, in *Colloidal Polymers: Synthesis and Characterization*, Vol. 115, Elaissari, A. (Ed.), Marcel Dekker, New York, **2003**, pp. 419–437.

160 Fisher, L. W., Melpolder, S. M., O'Reilly, J. M., Ramakrishnan, V., Wignall, G. D. *J. Colloid Interface Sci.* **1988**, *123*, 24.

161 Dingenouts, N., Seelenmeyer, S., Deike, I., Rosenfeldt, S., Ballauff, M., Lindner, P., Narayanan, T. *Phys. Chem. Chem. Phys.* **2001**, *3*, 1169.

162 Saunders, B. R. *Langmuir* **2004**, *20*, 3925.

163 Rübe, A., Hause, G., Mäder, K., Kohlbrecher, J. *J. Control. Release* **2005**, *107*, 244.

164 Winnik, M. A. *Polym. Eng. Sci.* **1983**, *24*, 87.

165 Winnik, M. A., Xu, H., Satguru, R. *Makromol. Chem., Macromol. Symp.* **1993**, *70/71*, 107.

166 Pérez, E., Lang, J. *Langmuir* **1996**, *12*, 3180.

167 Marion, P., Beinert, G., Juhué, D., Lang, J. *Macromolecules* **1997**, *30*, 123.

168 Huijs, F., Lang, J. *Colloid Polym. Sci.* **2000**, *278*, 746.

14
Polyelectrolyte Multilayer Films – A General Approach to (Bio)functional Coatings*

Nadia Benkirane-Jessel, Philippe Lavalle, Vincent Ball, Joëlle Ogier, Bernard Senger, Catherine Picart, Pierre Schaaf, Jean-Claude Voegel, and Gero Decher

14.1
Introduction

Layer-by-layer (LBL) assembly is an easy-to-use method for the fabrication of multicomposite films and has kindled widespread interest in such nanohybrids [1–23]. Electrostatic interactions between anionic and cationic compounds (e.g. synthetic or natural polyions such as polyelectrolytes, DNA, proteins or even colloids) offer four major advantages:

- layer-by-layer construction due to surface charge reversal in each layer
- restriction to single layers due to repulsion between the last layer and excess material in each deposition cycle
- deposition on almost any solvent accessible surface
- easy access to (bio)functional multicomposite films.

Here we review some of the guiding principles of LBL deposition (Fig. 14.1) and some existing and potential applications in the fields of materials science and applied biosciences for which the technological advantages of this method can be put to use. Further, we discuss the major structural features and the dynamics of multilayer films. Finally, we describe how LBL assembly can be used to design and functionalize surfaces as large as implants or as small as nanoparticles.

14.1.1
Multilayer Formation, Structure and Dynamics

The charge reversal with each deposition step is shown in Fig. 14.2, which depicts the zeta potential as resulting from a capillary measurement as a function of the number of layers [24]. Despite their important role for layer adsorption

* A List of Abbreviations can be found at the end of this chapter.

Macromolecular Engineering. Precise Synthesis, Materials Properties, Applications.
Edited by K. Matyjaszewski, Y. Gnanou, and L. Leibler

ISBN: 978-3-527-31446-1

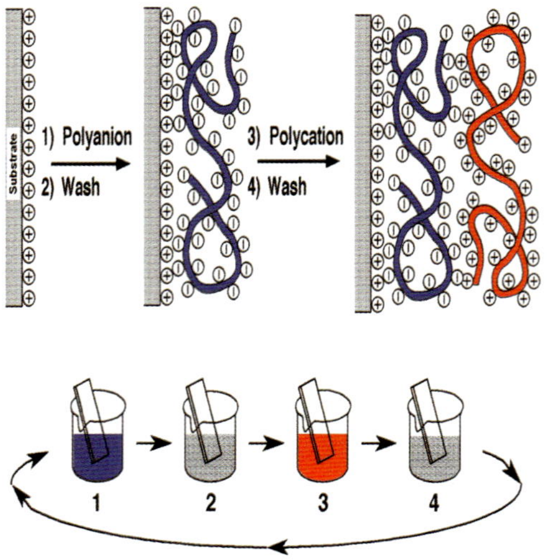

Fig. 14.1 Simplified "molecular" picture of the first two adsorption steps depicting film deposition as starting with a positively charged substrate, the counterions are omitted for clarity. The polyion conformation is highly idealized and layer interpenetration is not shown in order to better represent the surface charge reversal with each adsorption step.

and film cohesion in many cases, it should be noted that electrostatic interactions are not the sole type of interaction involved in this process. While in most cases the film deposition proceeds with equal increments for each polyanion/polycation pair, it was recently found that film growth can also be superlinear [25]. In linearly growing films, especially in the case of strong polyelectrolytes, individual layers are kinetically trapped in their positions [26] (Fig. 14.3). However, in exponentially growing films (see Section 14.2), it has been shown that some polyelectrolytes [25], especially weak polyions, can diffuse almost freely within the films.

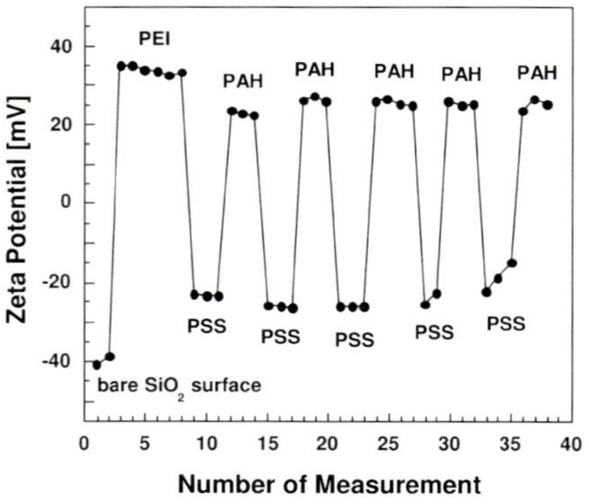

Fig. 14.2 Adsorption of polycations poly (ethylene imine) (PEI) and poly(allyl amine) (PAH) renders the surface positively charged. The deposition of poly(styrene sulfonate) (PSS) yields a negative surface charge (reproduced from [24]).

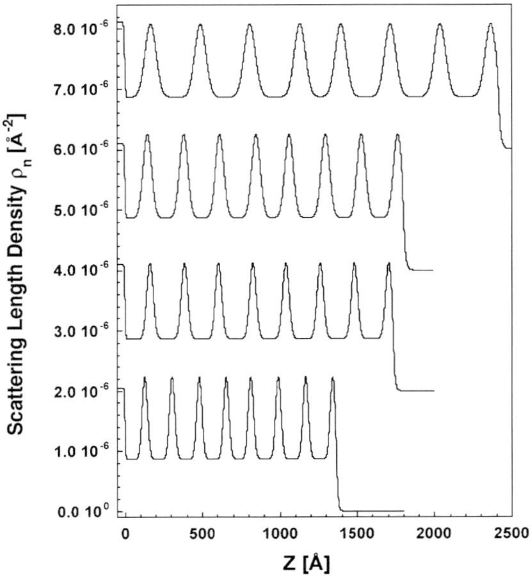

Fig. 14.3 Scattering length density profiles of four different multilayer samples obtained from neutron reflectivity data, each multilayer sample containing eight layers of deuterated PSS embedded in films composed of undeuterated PSS and PAH (reproduced from [26]). The layer structure of such films remains unchanged over years in the dry state.

14.1.2
Multilayers by Solution Dipping, Spraying or Spin Coating

In a variation to the deposition by adsorption from solution, the application of layers by spraying was introduced by Schlenoff [27] and the use of spin-coaters was demonstrated by Hong [28, 29] and also by Wang [30, 31]. Both spraying and spin coating have the advantage that only small amounts of liquids are needed to coat large surface areas. Figure 14.4 shows schematically the spray deposition process.

Figure 14.5 shows an earlier comparison of these new methods with solution dipping. We have recently been working on the classic polyelectrolyte pair PSS/PAH where we found [32] that the spraying process can speed-up the multilayer deposition by a factor of up to 250. Interestingly, there are cases where spray deposition yields excellent films while dipping in unstirred solutions leads to highly irregular adsorption. We have also seen that films deposited by spraying are thinner than films deposited by dipping, but this does not seem to originate from the shorter contact times. Even in experiments with repeated spray cycles the film thickness remains smaller than that found for classic dipping. Apparently, the local hydrodynamics of droplets arriving on the surface and fusing with the adhering layer may lead to a different adsorbed state of the polyions

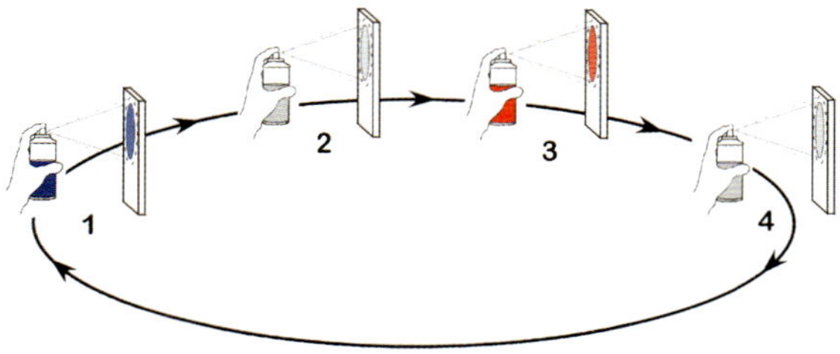

Fig. 14.4. Layer-by-layer deposition can also be carried out as a spray process. As in Fig. 14.1, the numbers 1 to 4 indicate the spray deposition of a polyanion, a rinsing step the spray deposition of a polycation and a rinsing step. This is the minimum sequence that is required for constructing multilayers. Additional materials are incorporated by using six or more spray cans per deposition cycle (or beakers in the case of dipping).

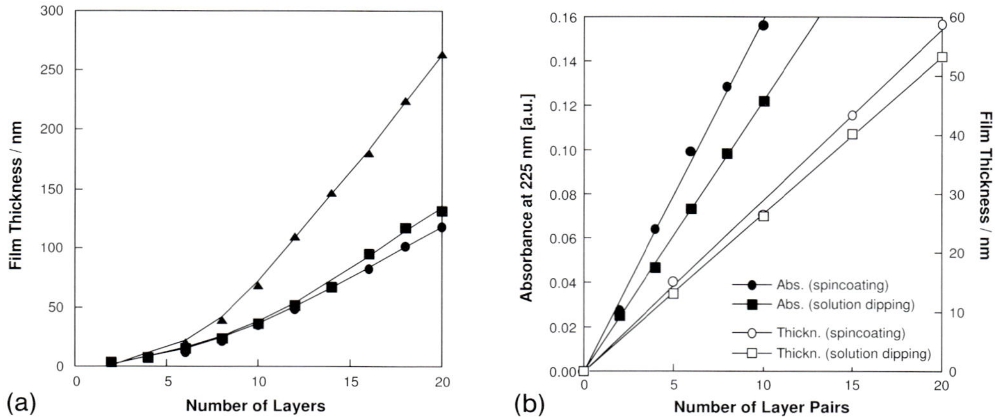

Fig. 14.5 (a) Ellipsometric monitoring of the growth of a multilayer film composed of poly(diallyldimethylammonium chloride) and poly(styrene sulfonate), both deposited from 1.0 M NaCl. Circles represent data on a film prepared by solution dipping, squares are from spraying. In both cases the contact time was about 10 s. Triangles were also obtained by solution dipping, but with a contact time of 5 min. The solid lines are a guide to the eye. All data are taken from the original work by Schlenoff [27]. (b) Comparison of spin-coated and solution-dipped films with respect to optical absorbance and film thickness. The films are composed of poly(((4,4′-bis(6-dimethylammonio)hexyl)-oxy) azobenzene bromide) and i-carrageenan, they were deposited on a precursor film composed of five layer pairs of PAH and i-carrageenan. All data taken from the original work of Hong [28].

for this spray deposition process. Repeated spray cycles of identical polyions followed by a rinsing step with pure water lead to thicker films than repeated spray cycles of identical polyions without the rinsing step. Such an effect has never been observed for deposition by classic dipping [32].

As also evident from Fig. 14.5, the deposition conditions play an important role with respect to the final film characteristics. Spraying and spin coating extend the parameter space of LBL deposition even further. However, it is to be expected that both methods will contribute to the general acceptance of the technology.

14.1.3
Multilayer Films on Nanoparticle Templates

It has been shown by Möhwald et al. that layer-by-layer deposition can also be used to fabricate micron-sized core/shell particles or even hollow microcapsules [7, 18, 33–37]. Caruso later extended this method to even smaller scales, using gold nanoparticles as templates [38, 39]. Although highly interesting, these results were also somewhat discouraging with respect to using LBL for the functionalization of nanoparticles. LBL was not performing with the usual ease and robustness, the assembly of a thiol layer seemed to be required prior to starting with LBL deposition. Without this thiol functionalization, polyelectrolyte layers were reported to desorb from the highly curved nanoparticle surface [38, 39]. This desorption of polyelectrolytes from nanoparticle surfaces may be related to the wrapping/unwrapping of the DNA–histone complex, which depends on polymer rigidity, the charge density on the polymer and on the nanoparticle, the nanoparticle size and the ionic strength (see for example [40]).

We have recently carried out a series of experiments in which we have tested the influence of a large number of parameters in order to optimize the conditions for LBL deposition on gold nanoparticles [41] (Fig. 14.6). One such parameter is the determination of suspension stability vs. flocculation as a function of the stoichiometry between nanoparticles and polyelectrolytes.

With respect to colloidal templates, we describe robust conditions for the coating of nanoparticles with multilayer shells. Using conditions as outlined in detail in [41], one obtains very stable suspensions of nanoparticles coated with up to 20 consecutively adsorbed layers of polyelectrolytes. In fact, the suspensions are so stable that they can be centrifuged and redispersed easily. The recovery yield of such core/shell particles is as high as 95% for the deposition of each layer if averaged over 20 deposition cycles. Thus the full toolbox of surface functionalization, available through the layer-by-layer deposition technique, can now be used to obtain new core/shell nanoparticle suspensions possessing extremely high surface areas. This opens a new path to the functionalization of nanoparticles in aqueous solutions, which is a strict prerequisite for using biomacromolecules such as proteins or DNA as components of the multilayer coating.

As a first proof for such a simple functionalization of nanoparticles, we have shown, in collaboration with Blanchard et al., that the fluorescence of a fluorescently labeled polymer deposited at different layer numbers above the metallic core is a function of the distance between the fluorophore layer and the gold surface [42]. Such a distance-dependent quenching of fluorescence is more or less a routine experiment for films on planar surfaces or for demonstrating interactions within or between biomolecules (in these cases fluorescence reso-

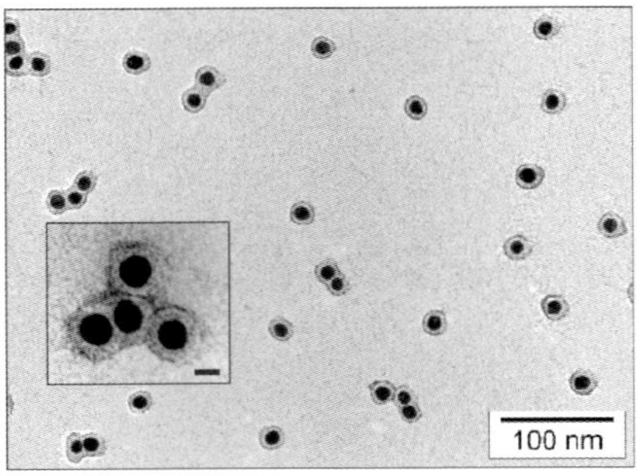

Fig. 14.6 Electron micrograph showing individual gold nanoparticles covered with a multilayer shell. The metallic cores are seen as dark spots, their diameter being 13.5 nm (reproduced from [41]). The inset is a further magnification (bar = 10 nm).

nance energy transfer, FRET), but for nanoparticles it can only be realized with a robust construction of a precise layer architecture around the core. Figure 14.7 shows electron micrographs and fluorescence spectroscopy data of fluorescently labeled gold nanoparticles with a diameter of 13.5 nm.

The following sections elaborate in more detail the dynamics within polyelectrolyte multilayers which are associated with superlinear growth, the effects of adsorbing mixtures of different polyelectrolytes, the fabrication of barrier layers and the embedding of liposomes into multilayer films which act as versatile reservoirs for incorporating water-soluble molecules, the fabrication of biofunctional films and finally, the interaction of living cells with layer-by-layer assembled coatings.

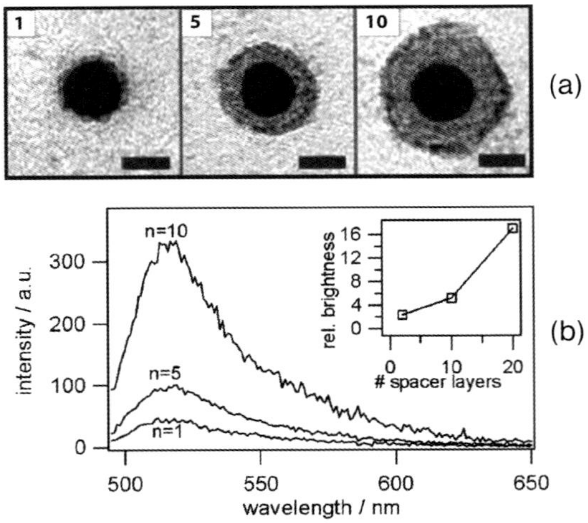

Fig. 14.7 (a) Transmission electron micrographs of individual LBL-ensheathed nanoparticles stained with uranyl acetate. The architecture of the objects is (AuNP-(PAH/PSS)$_n$(PAHFITC/PSS) with n = 1, 5, and 10. (b) Fluorescence spectra of the same species (reproduced from [42]). The dependence of the fluorescence intensity on the distance of the fluorescent layer from the metal core demonstrates that the LBL deposition around very small objects also leads to a stratified multilayer system.

14.2
Exponentially Growing Films

The first polyelectrolyte multilayers that were described exhibited a linear increase in the thickness or deposited mass with the number, *n*, of adsorption cycles. The motor of the growth process is the excess of charges that appears after each new polyanion or polycation deposition step (the excess is alternatively positive and negative). Beside these linearly growing films, another class of multilayers has been discovered more recently. These films are characterized by an exponential increase in their thickness and mass with *n*. Poly(L-lysine) (PL^L)/alginate [43], PL^L/hyaluronan (HA) [25, 44], PL^L/poly(L-glutamic acid) (PGA^L) [45] or chitosan/HA [46] constitute a few examples of exponentially growing films. Due to the exponential increase in the film thickness, these films can typically reach a thickness of the order of several micrometers after 20 polyanion/polycation deposition steps. Then, this allows one to use direct visualization techniques such as confocal laser scanning microscopy (CLSM) to investigate the film buildup process.

14.2.1
Buildup Process Leading to Exponentially Growing Films

The explanation of the buildup of exponentially growing films is based on the diffusion "in" and "out" of the whole film during each polyanion/polycation deposition step of at least one of the polyelectrolytes constituting the film. It may be mentioned that in the PGA^L/PL^L system both polyelectrolytes diffuse through the whole film in each cycle [45]. The existence of a diffusing polyelectrolyte introduces a marked contrast with linearly growing films where the polyelectrolytes deposited at step *n* interact mainly, if not solely, with the polyelectrolytes of opposite charge deposited at step (n–1). The (HA/PL^L) multilayer, where PL^L is the diffusing species, is certainly one of the examples of exponentially growing films whose growth mechanism was investigated in greatest detail. The proposed mechanism of its buildup process [47] is represented schematically in Fig. 14.8.

Consider a film after an HA deposition step. The HA solution is then rinsed and the film bears an outer negative excess charge. When this film is brought in contact with a PL^L solution, the PL^L chains first interact with the outer HA layer as would be the case for a linearly growing polyelectrolyte multilayer. In addition, PL^L chains diffuse into the film, down to the substrate. These chains were called "free PL^L chains" because it was expected that they would not interact strongly with the HA molecules that constitute the film. When the PL^L solution is then replaced by a pure buffer solution, some of the free PL^L chains that have diffused in the film, diffuse out of it. Nevertheless, part of these chains cannot overcome the electrostatic barrier due to the positive excess charge on top of the film, created during exposure to the PL^L solution. When the film is then further brought in contact with the HA solution, this electrostatic barrier

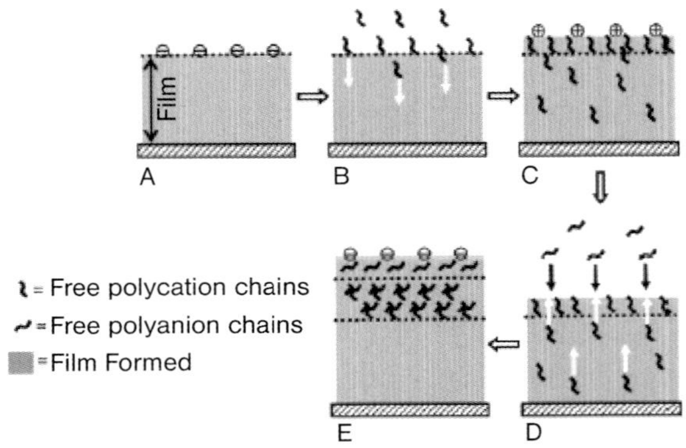

Fig. 14.8 Schematic representation of the buildup mechanism of a PL^L/HA film in which PL^L is the diffusing species. (A) The mechanism is assumed to start with a negatively charged HA-terminated film. (B) When PL^L is added, most of the molecules diffuse within the film, but some are adsorbed on the top, rendering the surface positive. (C) After rinsing some of the polycations remain in the film. (D) HA addition with diffusion out of the film of the free PL^L molecules. (E) End of step D resulting in a negative charge overcompensation and a film thicker than in step A.

disappears due to neutralization by the polyanions. Afterwards, a negative excess charge forms on top of the multilayer. All the remaining free PL^L chains then diffuse out of the film. As soon as they reach the film/solution interface, they are complexed by HA chains and remain bound to the interface. These HA/PL^L complexes form the new outer layer of the film. The HA deposition process stops when the film is empty of free PL^L chains, provided that it is not interrupted previously by a rinsing step.

This mechanism explains easily the exponential nature of the growth process. Indeed, during the nth deposition step, the mass increment, $\Delta Q(n)$, of the film corresponds to the new HA/PL^L complexes that have been formed at the boundary of the multilayer. This increment is thus proportional to the amount of free PL^L chains present in the film before contact with the HA solution. This amount is itself proportional to the film thickness and thus to the mass $Q(n-1)$ of the film at the end of deposition step $(n-1)$:

$$\Delta Q(n) = K\,Q(n-1)$$

where K is a constant. This equation corresponds to an exponential increase in $Q(n)$. However, because the mass is zero when $n=0$, this model can only be valid for n larger than some minimum value, n_0. The solution of the above equation is therefore written:

$$Q(n) = Q(n_0)\exp[K(n-n_0)]\,, \quad n \geq n_0$$

Now, a film cannot indefinitely grow following an exponential law. It enters necessarily into a linear growth regime after a given number of deposition steps.

Indeed, in a finite exposure time to the PL^L solution, the PL^L can only penetrate into the film down to a given depth. Conversely, in a finite exposure time to the HA solution, a limited quantity of PL^L can diffuse out of the film to form complexes at the film/solution interface. Moreover, even though still more PL^L could diffuse out of the film during this time lapse, the number of HA chains that are available at the interface is finite too and may play the role of limiting factor. The crossover from the exponential to the linear growth regime has actually been observed [48, 49]. However, its interpretation might be more complicated than suggested above. In particular, film restructuring within the earliest deposited layers, rendering the bottom architecture impermeable to the polyelectrolyte diffusion, could constitute another possible origin of the transition.

14.2.2 Experimental Facts Supporting the Exponential Buildup Mechanism

Several experimental facts support the buildup model based on the diffusion "in" and "out" of the multilayer during each deposition step of at least one of the polyelectrolytes constituting the film. The diffusion process was visualized directly by CLSM. Figure 14.9 shows a $(PL^L/HA)_{19}$-PL^L film. This film was constructed by using fluorescein isothiocyanate (FITC) labeled PLL and Texas Red (TR) labeled HA during the deposition step 19. One observes that the film is entirely green, which reveals the presence of $PL^{L\text{-}FITC}$ over the entire film, whereas only the upper part of the multilayer is red, indicating that the HA^{TR} is located only in the outer part of the film. These experiments prove that $PL^{L\text{-}FITC}$ has diffused, during the deposition step, down to the bottom of the film, whereas HA did not.

It was also well demonstrated that for linearly growing films, i.e. systems in which polyanions and polycations interact by means of electrostatic forces, the charge excess that appears in each new cycle constitutes the driving force of the growth. The streaming potential technique was also employed to follow the zeta potential variation of the HA/PL^L system after each new polyelectrolyte deposition (Fig. 14.10). After the deposition of the first PL^L layer the surface potential becomes positive (+70 mV). The following HA deposition renders the surface negative (–55 mV). A charge reversal is observed after each new deposited layer,

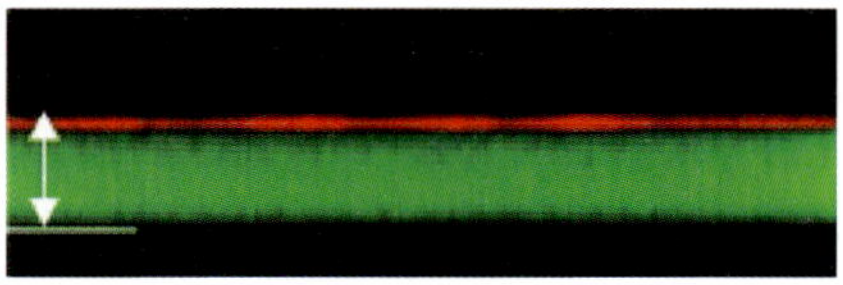

Fig. 14.9 Vertical section through a HA/PL^L film containing labeled PL^L and HA (HA^{TR}, red; $PL^{L\text{-}FITC}$, green) obtained with CLSM. The bottom of the chamber is visualized by a white line. The precise architecture is $(PL^L/HA)_{18}$-$(PL^{L\text{-}FITC}/HA^{TR})$-$PL^L$ which contains two labeled layers. The size of the image is 26.2×8.4 μm^2. The green fluorescence is seen over the whole film whereas the red line (HA) is localized on the film top.

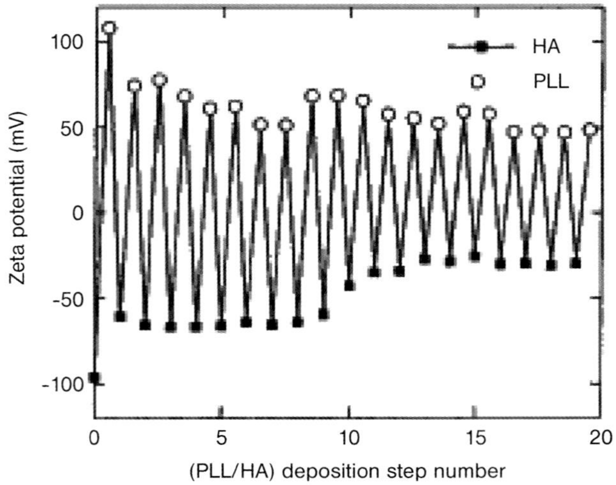

Fig. 14.10 Evolution of the ζ-potential for alternated HA and PL^L deposition during 20 deposition cycles. Positive (open symbols) values correspond to PL^L addition followed by rinsing, and negative values (closed symbols) correspond to HA addition after rinsing.

the top charge of the architecture being alternately negative and positive after addition of the polyanion or the polycation, respectively. This observation evidences the overcompensation of the charges, as was originally noted for the linearly growing polyanion/polycation couples.

Optical waveguide lightmode spectroscopy was also used to investigate the HA/PL^L film buildup process. In this technique, the film is sensed with an evanescent wave whose penetration depth is of the order of 300 nm. Thus, when a film reaches a thickness much larger than the penetration length of the evanescent wave, the detector becomes insensitive to what happens on top of the film (i.e. far from the waveguide) but remains sensitive to changes in the refractive index of the film near the substrate (waveguide). When this technique is used to follow the buildup of an exponentially growing film, an increase in the optical signal is observed during the first buildup steps. As the buildup process proceeds, the optical signal starts to saturate and enters into a cyclic evolution. Figure 14.11 represents a typical evolution of the effective refractive index measured by OWLS in the cyclic regime for a HA/PL^L film. When it enters in the cyclic regime, the film has reached a thickness that exceeds by far the penetration length of the evanescent wave. The cyclic variation of the optical signal results from a cyclic change in the refractive index at the bottom of the film during the successive buildup steps. These refractive index changes are directly related to the diffusion of the polyelectrolytes "in" and "out" of the film.

The vertical rapid diffusion of one of the polyelectrolytes also confers particular properties to the exponentially growing films. One of them is the ability to modify

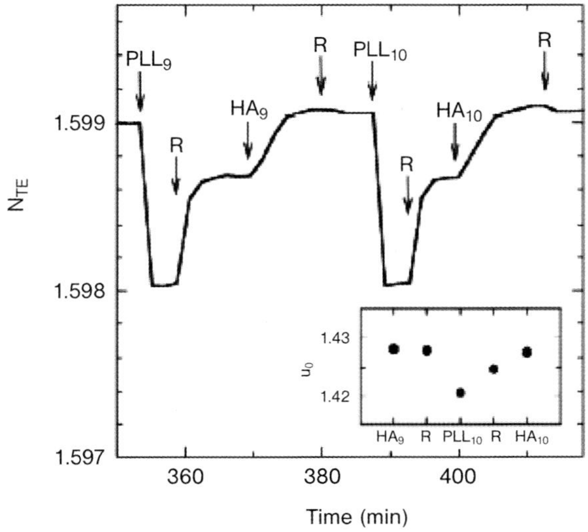

Fig. 14.11 Effective refractive indexes N_{TE} obtained by OWLS during HA/PL^L film buildup during two full cycles once the film is totally continuous. Each arrow corresponds to an injection step of HA, PL^L or rinsing (R). The numbers underneath PL^L or HA correspond to the cycle number of the construction. The inset shows the evolution of n_c, the refractive index close to the surface during a deposition cycle.

the secondary structure of these films after each polycation or polyanion addition. In the exponentially growing PGA^L/PAH films, PGA^L undergoes a random coil/α-helix transition once the film is brought in contact with PAH during the film buildup [50, 51]. This structural transition leads to a PGA^L/PAH film where the α-helix content switches regularly between 30% and 40% during the film construction when the multilayer is alternately brought into contact with the PAH and PGA^L solutions. Thus, the secondary structure of the film is entirely governed by the last deposited layer due to diffusion of the macromolecules.

14.2.3
Characteristic Properties of Exponentially Growing Films

Due to the diffusion of the polyelectrolytes "in" and "out" of the film, exponentially growing films are quite different from linearly growing ones. For example, in contrast to linearly growing ones, the exponentially growing films do not give rise to Bragg peaks in neutron reflection experiments when deuterated PE are used to label certain layers of the architecture.

Burke and Barrett performed swelling experiments on PL^L/HA multilayers [52]. They found that under certain buildup conditions these films can swell as much as 9 times when hydrated from the dry state. This shows that HA/PL^L films are very hydrated. This result is expected to reflect a general property of exponentially growing films. This hypothesis is supported by the high permeability of these films to small ions, in contrast to what is observed for linearly growing films. For example, recent cyclic voltammetry experiments have shown that whereas 4 pairs of poly(styrene sulfonate)/poly(allylamine) are sufficient to prevent almost totally the diffusion of ferrocyanide ions through the film, the deposition of as many as 18 pairs of poly(glutamic acid)/poly(allylamine) does not hinder this diffusion. The high degree of hydration of these films suggests that exponentially growing films in fact resemble gels more closely than structured films. The gel-like nature of these films was confirmed by two independent experimental results. First, Collin et al. determined the viscoelastic properties of HA/PL^L films made of 30 pairs of layers and slightly reticulated for experimental purposes [53]. They found that films exceeding a thickness of 30 μm present a G' value that is independent of frequency over a four orders of magnitude frequency range and values of G'' that are too small to be determined. These films behave as purely elastic hydrogels. Kulcsar et al. performed surface force apparatus experiments on both linearly and exponentially growing films [54, 55]. Whereas linearly growing films behaved as glassy materials, exponentially growing ones responded reversibly with respect to compression/retraction cycles, proving the elastic behavior of these films (Fig. 14.12). These were the

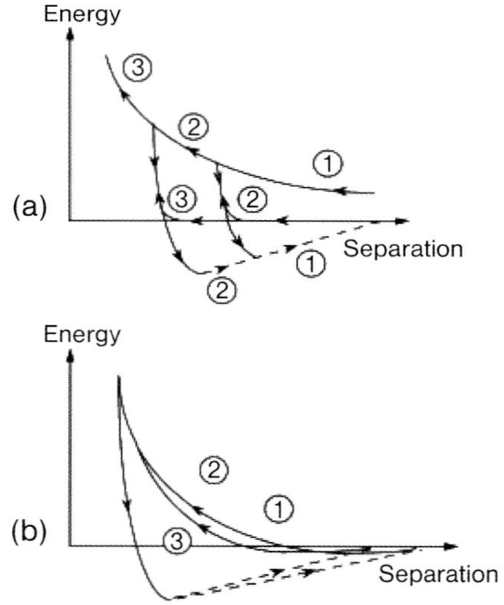

Fig. 14.12 Schematic representation of force–separation curves during consecutive load and unload cycles. (a) concerns the linearly growing PSS/PAH film; (b) corresponds to the exponentially growing PL^L/PGA^L film. The numbers 1, 2 or 3 close to the profiles correspond to the cycle number. The dashed lines indicate that the surfaces jump upon separation. For the PSS/PAH system, the force–separation curve follows the same path during the reload as during the preceding unload as long as the turning point where the direction of the surface is reversed is not passed. Conversely, for PL^L/PGA^L films, the reload is almost identical to the loading during the first path while subsequent load cycles would induce an interaction of shorter range.

first experimental data supporting the hypothesis of Cohen Stuart and co-workers that a film grows linearly when the polyanion/polycation complexes forming the multilayer are in a glassy state whereas the growth is exponential when the complexes are weak and thus the polyelectrolytes more mobile.

14.3 Mixtures of Polyelectrolytes

To extend the application possibilities of the polyelectrolyte multilayer (PEM) films, whether linearly or exponentially growing (see Section 14.2), it is appealing to examine their properties when they are built up using a polycation solution and a polyanion mixture solution or vice versa, or a mixed solution of both species. In this section, we shall summarize the striking features emerging from studies carried out on: (i) the growth of a PEM film, where a polycation solution and a mixed polyanion solution is used to build it [56]; (ii) the dependence of the β-sheet content on the relative amount of one of the two polyacids in the polyanion solution [57]; (iii) the use of mixtures of L- and D-enantiomers of both the polycation and the polyanion to tune the biological activity of a functionalized PEM film [58].

It is known that a film built up using PSS and PAH, dissolved in a 0.15 M NaCl aqueous solution, shows a thickness increasing linearly with the number of deposition cycles [59] (see Section 14.2). In contrast, the combination of PGA^L and PAH leads to an exponentially growing film [50] (see Section 14.2). How does a PEM film grow when the polyanion solution contains, at the same time, PSS and PGA^L?

To address this question, (PGA^L, PSS) mixtures, denoted $PGA^L{}_x$–PSS_{1-x}, where x represents the molar fraction of the monomer repeat unit of PGA^L, were prepared. Then, the evolution of the thickness of PAH–$(PGA^L{}_x\text{–}PSS_{1-x})_n$ films, for various values of x, was monitored with a quartz crystal microbalance, which allows measurement of both the frequency shifts of the resonance frequencies (5, 15, 25 and 35 MHz) of the bare crystal and the corresponding dissipation factors. On the basis of these experimental data, a viscoelastic model of the film [60], considered as a homogeneous layer, permits estimation of the thickness, d, of the film after the successive contacts with either the polycation or the polyanion solutions (Fig. 14.13).

For small values of x ($x \leq 0.1$), the thickness increases linearly with the number of exposures to the PE solutions, i.e. as though the polyanion solution contained only PSS. The interesting behavior of the film appears for $0.25 \leq x \leq 0.75$. For large enough PGA^L amount in the polyanion solution, the thickness increases exponentially up to a given value depending on x. Then, a transition point appears from where on the growth regime becomes linear. This transition is ineluctable because in a given exposure time the film cannot accumulate, and afterwards release an indefinitely increasing amount of diffusing molecules. In the present experiments, however, an additional feature deserves attention. The

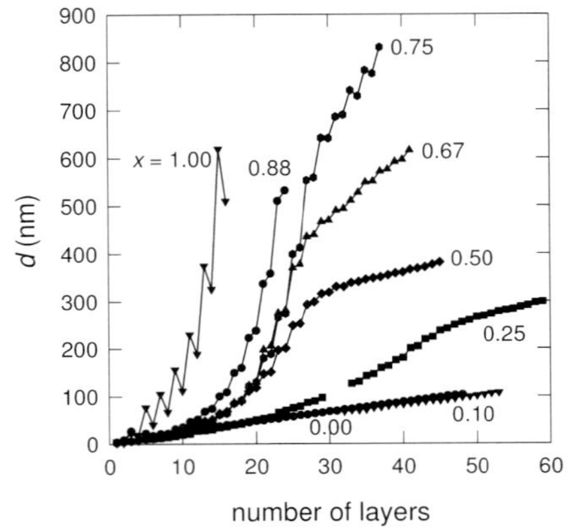

Fig. 14.13 Evolution of the thickness during the buildup of the PEM films using PAH and a mixture of PGA^L and PSS. Each curve is labeled by the value of the parameter *x* (see text). The thickness is derived from quartz crystal microbalance measurements processed by means of a viscoelastic model where the density of the film is arbitrarily fixed to 1 g cm^{-3}.

slope of the linear part is smaller than the final slope of the exponential branch. Moreover, the slope in the linear regime increases with the relative amount of PGA^L in the polyanion mixture. Therefore, it is anticipated that when x tends to unity, the exponential to linear crossover will not show a rupture of slope. Unfortunately, this could not be verified with the quartz crystal microbalance because the thickness corresponding to $x=0.88$ and 1 grows so rapidly that it exceeds the detection limit of the apparatus before the transition occurs. However, thickness measurements carried out with an ellipsometer on (PL^L/HA) films, built up by the spraying method (see Section 14.1), confirmed the expected behavior.

Assume that the two polyanions, PSS and PGA^L, have the same diffusion coefficient within the film. The immediate consequence of this hypothesis is that the growth law of the film thickness possesses a crossover from the exponential to the linear regime. However, under these circumstances, the slope in the linear regime is equal to the final slope of the exponential regime (for a detailed analysis, see [56]). In contrast, if the two polyanions have different diffusion coefficients (when they diffuse out of the film during the exposure to PAH), the film reaches a critical thickness beyond which the less mobile polyanion can no longer diffuse out of the entire film during the contact time with PAH. This induces a variation in the composition of the complexes forming at the film/PAH solution interface. Because the thickness measurements clearly evidence a reduction in the slope after the crossover from the exponential to the linear growth regime, the altered composition of the outermost part of the film must cause a significant reduction in the diffusion coefficient of one of the polyanions, with respect to its value during the exponential phase.

We proceed now to the examination of a second example where a single polycation, PL^L, was also used in association with a polyanion binary mixture composed of PGA^L and poly(L-aspartic acid) (PAA^L) [57]. It must be mentioned that the interaction of PL^L with PAA^L does lead to the formation of nearly no β-sheets, whereas the interaction of PL^L with PGA^L does. Therefore, the study was aimed at evidencing that the amount of β-sheets in a PEM film can be controlled through the composition of the polyanion mixture solution.

PEM films were built up using polyanion mixtures with different relative amounts of PGA^L and PAA^L. Two interesting features appear when the thickness of these films is measured and their β-sheet content evaluated. As to the thickness, quartz crystal microbalance measurements reveal that the growth of the film is insensitive to the proportion in mass, g_s, of PGA^L in the polyanion solution as long as g_s is smaller than about 0.5, the total polyanion concentration being fixed to 1 mg mL^{-1} in all experiments. For larger mass fractions of PGA^L, the thickness of the film grows with g_s for a fixed number of deposition steps. Moreover, infrared spectroscopy in the attenuated total reflection mode allows estimating the proportions of PGA^L and PAA^L in the film. It appears that the relative amount of PGA^L in the film, g_f, is smaller than g_s (Fig. 14.14 A). This observation evidences the preference toward the incorporation of PAA^L over PGA^L into the film.

As noted above, PL^L/PGA^L films contain intermolecular β-sheets, in contrast to PL^L/PAA^L films that are practically devoid of these structures. *A priori*, the amount of intermolecular β-sheets is expected to be proportional to the relative content of PGA^L in the film (g_f). Indeed, the infrared spectroscopy measurements reveal that the amount of β-sheets is always larger than expected if it

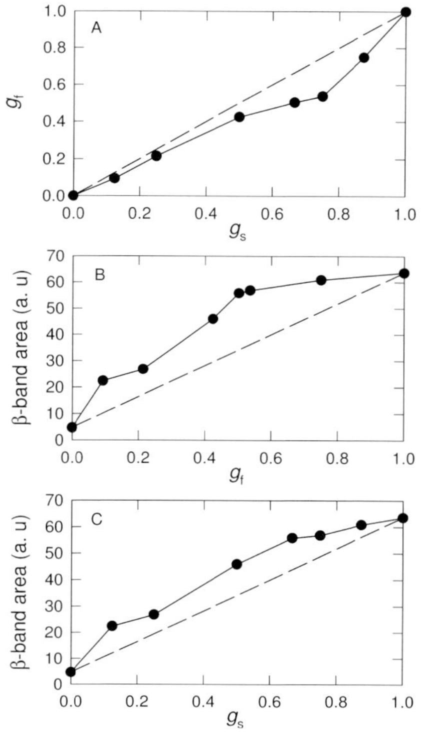

Fig. 14.14 (A) Relative contribution (in mass) of the PGA^L, g_f, to the total mass of polyanions incorporated in the film as a function of the relative PGA^L mass content, g_s, of the polyanion mixture solution. (B) Area of the β-band divided by the area of the amide-I band as a function of g_f. (C) Same as (B), except that g_s replaces g_f as the abscissa variable.

were proportional to the PGA^L content of the film and if PGA^L and PAA^L interacted independently with PL^L (Fig. 14.14 B). The reduction of g_f from 1 down to about 0.5 has a minor influence on the β-sheet content. Then, quite abruptly, the β-sheet content falls off and goes progressively to zero as g_f decreases further toward zero.

In summary, it appears that the mass of the film is independent of the PGA^L content of the polyanion mixture as long as it is smaller than about 0.5, although the film incorporates PGA^L. When the PGA^L content of the film increases from zero up to 0.5, the β-sheet content of the film increases roughly proportional to it; however, once the PGA^L content of the film exceeds approximately 0.5, the β-sheet content hardly increases further. Finally, because the only parameter, which can be controlled, is the PGA^L content of the buildup solution, it may be useful to correlate the β-sheet content of the film also to this parameter (Fig. 14.14 C). As can be seen, the β-sheet content follows quite closely the increase in the PGA^L concentration in the polyanion mixture solution.

As the last example, we shall now evoke the buildup of a PEM film aimed at tuning the response of cells deposited onto the film [58]. More precisely, macrophages were deposited on a film in which protein A was embedded at various depths and the production of TNF-α (TNF: tumor necrosis factor) induced by the contact of the cells with the protein A was monitored. The film itself was built up with polylysine and poly(glutamic acid). Both molecules were represented by their L- or D-form or by a mixture of the two enantiomers (L-enantiomers: PL^L and PGA^L; D-enantiomers: PL^D and PGA^D).

The PEM films were constructed using PE solutions of various compositions in L- and D-enantiomers. The polycation solution contained PL^D at x mg mL^{-1} and PL^L at $(1-x)$ mg mL^{-1} and the polyanion solution contained PGA^D at x mg mL^{-1} and PGA^L at $(1-x)$ mg mL^{-1}. The PEM films consisted of five (PL/PGA) pairs of layers followed by one PL layer, itself followed by the protein A layer. Then, $n=0$, 1, 5, 15 or 20 (PL/PGA) pairs of layers were added to the architecture, which was finally completed by the deposition of a PL layer, except in the case $n=0$ where the protein A remained on top of the construction.

The results are illustrated in Fig. 14.15 where the optical density (OD, measured at the wavelength of 450 nm), proportional to the amount of TNF-α produced, is represented as a function of n. For each value of n, the cellular activity was evaluated after various incubation times ranging from 1 to 6 h. Each frame in Fig. 14.15 corresponds to a given value of the composition parameter x (from top to bottom, $x=1$, 0.4 and 0.2 mg mL^{-1}).

The upper frame in Fig. 14.15 shows that as soon as even a unique (PL^D/PGA^D) pair of layers covers the protein A, its action on the cellular activity vanishes. This confirms that the enzymes produced by the cells cannot degrade D-molecules. At the intermediate D-content (middle frame in Fig. 14.15), the case $n=1$ reveals a regularly increasing activity with incubation time from 1 to 6 h. For $n=5$, the offset is more abrupt. During several hours, no significant activity is detectable. Then, after 6 h, it jumps to its maximum value. The PEM film acts in this case as a switch. Finally, at low D-content (bottom frame in

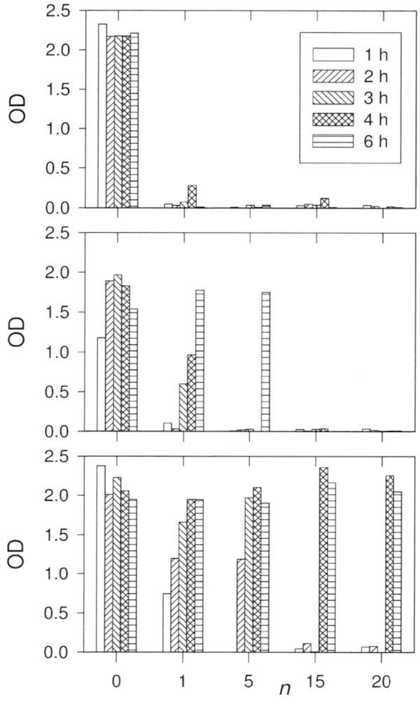

Fig. 14.15 Dependence of the TNF-α production (proportional to the optical density, OD) on the number, n, of pairs of PL/PGA layers deposited above the protein A and on the incubation time (1 to 6 h; see insert in the upper frame).

Fig. 14.15), the results are qualitatively the same as for the medium D-content, but the switch from no activity to full activity occurs for higher numbers of (PL/PGA) pairs of layers deposited onto the protein A layer. The increased number of layers covering the protein A produces a clear retardation effect. This study proves that bioactive films with time-scheduled activity can be built up using mixtures of degradable (L-enantiomers) and non-degradable (D-enantiomers) polyelectrolytes.

14.4 Compartmentation, Barrier Layers and Reservoirs

14.4.1 Design of Compartmentalized Multilayer Films

One strategy to produce coatings with time-controlled activity is to assemble multicompartment films [61, 62]. The films will be composed of strata of polyelectrolyte multilayers acting as reservoirs for various biomolecules separated by barriers avoiding vertical diffusion of these biomolecules from one compartment to another. Two strategies will be developed to design either degradable or non-degradable barriers. The main goal of the buildup of such multicompartment films is to apply a coating on a substrate that induces a cascade of timely

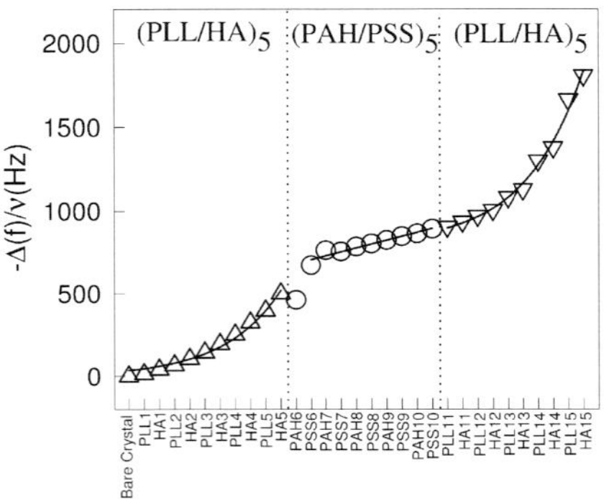

Fig. 14.16 Inverted frequency shift $-\Delta f/\nu$ measured with a quartz crystal microbalance during the buildup of a $(PL^L/HA)_5/(PAH/PSS)_5/(PL^L/HA)_5$ multilayer film on a quartz crystal previously coated with a SiO_2 layer. Measurements are performed at 15 MHz in solution (NaCl 0.15 M, pH 5.9).

scheduled events. The degradation of the barriers with time could allow a sequential contact of the cells anchored on the film with the active compounds and thus a sequential and time-scheduled biological activity.

Some of the differences between linearly and exponentially growing films are described above (see Section 14.2). Here we describe the combination of linearly and exponentially growing multilayers in the same film to investigate if this could be a way to control the diffusion of PL^L through the film section. If so, the presence of a stratum made of a linearly growing film would permit the fabrication of compartmentalized films.

We first investigated with a quartz crystal microbalance (QCM) the buildup of a multilayer film consisting of three parts: (i) a PL^L/HA multilayer in contact with the solid substrate, (ii) a PAH/PSS multilayer and (iii) again, a PL^L/HA multilayer. With QCM, it was possible to monitor the evolution of the frequency change of the crystal during the buildup process (Fig. 14.16). The variation of the inverted normalized frequency shift $-\Delta f/\nu$ (where ν is the overtone number) could be considered, in a first approximation, as proportional to the mass of material deposited on the crystal. We clearly observe in Fig. 14.16 that the buildup of the first $(PL^L/HA)_5$ part follows an exponential growth, whereas the second part consisting of $(PAH/PSS)_5$ follows a linear growth. However, the transition between the two parts is fuzzy for the first PAH and PSS layers. Deposition of the third part, $(PL^L/HA)_5$ on $(PAH/PSS)_5$, again results in an exponential growth in thickness. The two exponential growths separated by a linear one suggest that the PAH/PSS multilayer acts as an actual barrier against PL^L diffusion from one $(PL^L/HA)_5$ multilayer to the other. In the following, we will denote the PL^L/HA multilayers as "compartments" and the PAH/PSS multilayers as "barriers" that prevent PL^L diffusion from one compartment to the next.

14.4.1.1 Diffusion of PL^L Chains in PL^L/HA Compartments

To check the efficiency of the PAH/PSS barrier and of the diffusion of PL^L in each PL^L/HA compartment, we performed confocal laser scanning microscopy studies (CLSM). This technique allows imaging of a virtual section of the fluorescently labeled thick multilayer films. To this purpose, we built a film containing three PL^L/HA compartments separated by two PAH/PSS barriers. On the top of compartments 1 and 3 (the first and the last), PL^L labeled with fluorescein isothiocyanate (PL^L-FITC), emitting in the green, was deposited. The film was finally composed of $(PL^L/HA)_{30}/PL^L$-FITC/HA/$(PAH/PSS)_{30}/(PL^L/HA)_{30}/(PAH/PSS)_{30}/(PL^L/HA)_{30}/PL^L$-FITC (Fig. 14.17). We observe that the compartments 1 and 3 are fully green whereas compartment 2 where no PL^L-FITC has been deposited stays black. This proves that PL^L-FITC from the compartments 1 and 3 cannot cross the barriers composed of $(PAH/PSS)_{30}$. We also notice that the fluorescence in compartment 1 is more intense than in compartment 3. This results probably from the fact that compartment 3 was directly in contact with the solution and a slow release of PL^L-FITC from this compartment could occur.

Fluorescence recovery after photobleaching (FRAP) experiments was performed to make sure that the PL^L-FITC chains were still able to diffuse freely in each compartment as was observed previously for PL^L-FITC in a simple PL^L/HA multilayer film. The film section observed in Fig. 14.17 was bleached in its central part (Fig. 14.18a). After 35 min, the same section was observed and the green fluorescence was almost entirely recovered in compartments 1 and 3, demonstrating that the PL^L-FITC from both sides of the bleached area could diffuse (Fig. 14.18b–d). Compartment 2 remaining non-labeled, the recovery could not be due to vertical diffusion from one compartment to another.

The possibility to build up in a simple way a polyelectrolyte film with compartments and barriers using strata of exponentially and linearly growing multilayers has been demonstrated. We will now focus on the degradability of such barriers by specific cells to prove that such systems are powerful tools for future biological applications.

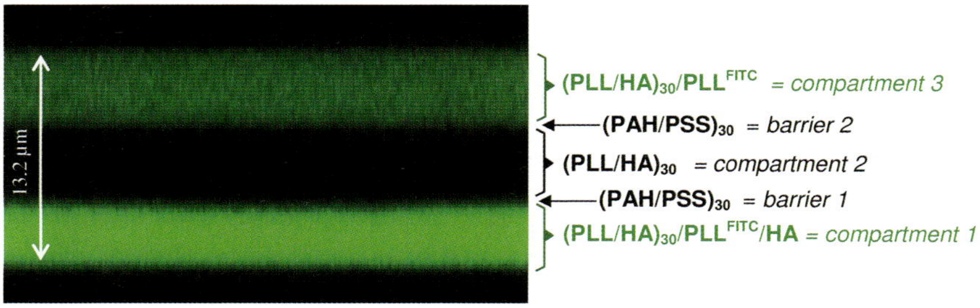

Fig. 14.17 Confocal microscopy vertical section image of a multilayer film composed of three (PL^L/HA) compartments and two (PAH/PSS) barriers. The glass slide is located at the bottom of the image and the compartments 1 and 3 are labeled in green with PL^L-FITC. The film maintained in solution (NaCl 0.15 M, pH 5.9) consists of $(PL^L/HA)_{30}/PL^L$-FITC/HA/$(PAH/PSS)_{30}/(PL^L/HA)_{30}/(PAH/PSS)_{30}/(PL^L/\ HA)_{30}/PL^L$-FITC. The total thickness is about 13.2 µm.

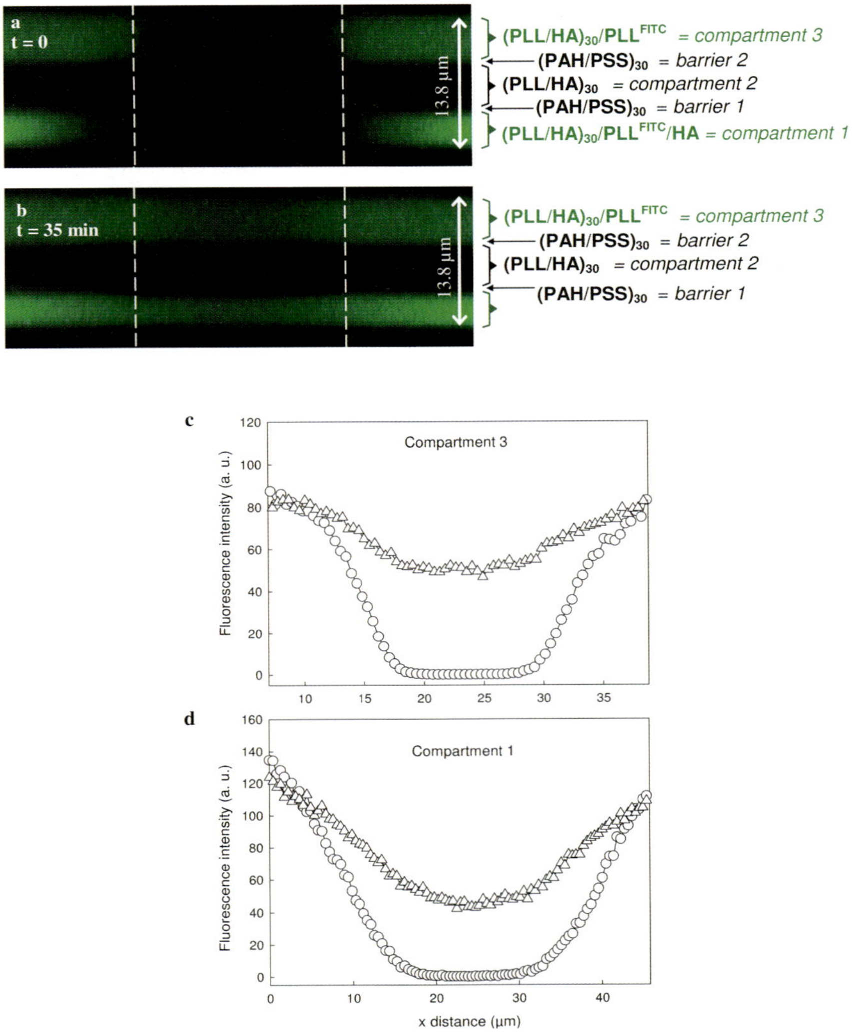

Fig. 14.18 Fluorescence recovery after photobleaching experiment performed with confocal microscope on a $(PL^L/HA)_{30}/PL^L$-FITC/HA/$(PAH/PSS)_{30}/(PL^L/HA)_{30}/(PAH/PSS)_{30}/(PL^L/HA)_{30}/PL^L$-FITC film in solution (NaCl 0.15 M, pH 5.9). The vertical dashed lines indicate the location of the bleached area in the central part of the image. Image (a) corresponds to observation immediately after performing the bleach and image (b) is taken 35 min later. The glass slide is located at the bottom of the image. The fluorescence intensity versus the horizontal position, *x*, in the film is plotted in (c) for compartment 3 and in (d) for compartment 1, immediately after the bleach (○) and 35 min later (△).

14.4.1.2 Cell Accessibility to Compartments

Murine bone marrow cells isolated from femurs were seeded on the films to check if these cells were able to cross PAH/PSS barriers and to interact with polyelectrolytes located in an underlying compartment. The bone marrow cell population is mainly composed of monocytic cells. These cells are characterized by their great phagocytosis capacity. Recently, degradation of PL^L and PGA^L multilayers by such cells has been described [58]. Here, we built up a film constituted of one PL^L/HA compartment labelled with PL^L-FITC and a PAH/PSS barrier deposited on the top, leading to a $(PL^L/HA)_{20}/PL^L$-FITC/HA/$(PAH/PSS)_{30}/PL^L$ multilayer film (Fig. 14.19a). The $(PL^L/HA)_{20}$ multilayer appeared entirely green due to PL^L-FITC diffusion as described before and the $(PAH/PSS)_{30}$ multilayer deposited on the top was non-labeled. After 17 h of seeding on the film, the cells internalize absolutely no PL^L-FITC and the film remains unaltered (Fig. 14.19a and a′). Control experiments with bone marrow cells seeded on a $(PL^L/HA)_{20}/PL^L$-FITC film without (PAH/PSS) barrier showed that after only a few minutes cells exhibit a green cytoplasm due to PL^L-FITC internalization. This proves that the (PAH/PSS) multilayer act as a non-degradable barrier for at least 17 h of contact with bone marrow cells.

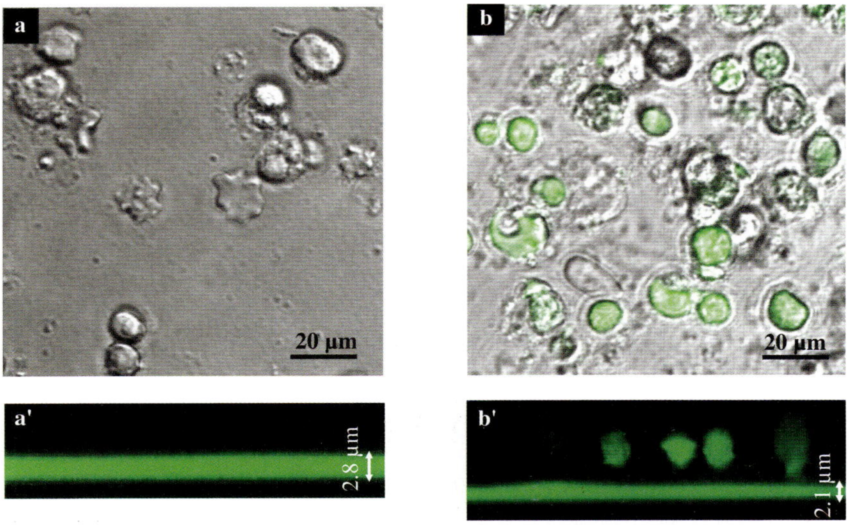

Fig. 14.19 CLSM images of bone marrow cells seeded 17 h on a $(PL^L/HA)_{20}/PL^L$-FITC/HA/$(PAH/PSS)_{30}/PL^L$ multilayer film (a and a′) or on a $(PL^L/HA)_{20}/PL^L$-FITC/PLGA (b and b′). (a) and (b): (*x*, *y*) image in the bright field and green channel. (a′) and (b′): (*x*, *z*) image of a film section in the green channel crossing area where cells were located. The glass slide is located at the bottom of the image.

14.4.1.3 Polylactide-co-glycolide Layers as Barriers

A (PAH/PSS) multilayer acted as a barrier for PL^L diffusion but monocytic cells could not degrade such a multilayer composed of synthetic polyelectrolytes. A new strategy was developed to design degradable barriers. Instead of polyelectrolyte multilayers as barriers, we tested barriers composed of a biocompatible and biodegradable polymer, poly(lactic-*co*-glycolic) acid (PLGA) [63]. An appropriate amount of this polymer dissolved in chloroform was sprayed on the (PL^L/HA) multilayer to obtain a thickness of the PLGA layer ranging between 3 and 5 μm. Figure 14.20 shows a CLSM vertical section of a $(PL^L/HA)_{30}/PL^{L\text{-}FITC}/$ $PLGA/(PL^L/HA)_{30}/PL^{L\text{-}Rho}$ where $PL^{L\text{-}Rho}$ corresponds to poly(L-lysine) chains labeled in red with rhodamine probes. The PLGA layer corresponds to the non-labeled black stripe between the red and the green PL^L/HA compartments and appears as a very homogeneous layer in terms of thickness. This image demonstrates that (i) the PLGA layer was well deposited on top of the first (green) compartment and the buildup of a second (red) compartment on top of a PLGA barrier was possible, (ii) the PLGA layer prevents PL^L diffusion from a PL^L/HA compartment to another one. We also checked with FRAP experiments that labeled PL^L chains diffuse freely in each compartment as observed before in the case of (PAH/PSS) barriers.

14.4.1.4 Cell Degradation of PLGA Barriers

First, degradation of PLGA barriers was tested on a $(PL^L/HA)_{30}/PL^{L\text{-}FITC}/PLGA$ film (one compartment and one barrier on the top). Most of the bone marrow cells seeded 17 h on such a film internalized $PL^{L\text{-}FITC}$ (Fig. 14.19 b and b′). In addition, degradation of the PLGA barriers was investigated with cells seeded on a $(PL^L/HA)_{30}/PL^{L\text{-}FITC}/PLGA/(PL^L/HA)_{30}/PL^{L\text{-}Rho}$ film (two compartments separated by one barrier) (Fig. 14.21). The compartment 1 was labeled in green with $PL^{L\text{-}FITC}$ and the compartment 2 in red with $PL^{L\text{-}Rho}$ (same film as imaged in Fig. 14.20). Most of the cells deposited on top of the film internalized $PL^{L\text{-}Rho}$ after 1 h (Fig. 14.20 a) but no green fluorescence is visualized in the cytoplasms. Cell fluorescence appeared qualitatively similar when observed after 24 h and

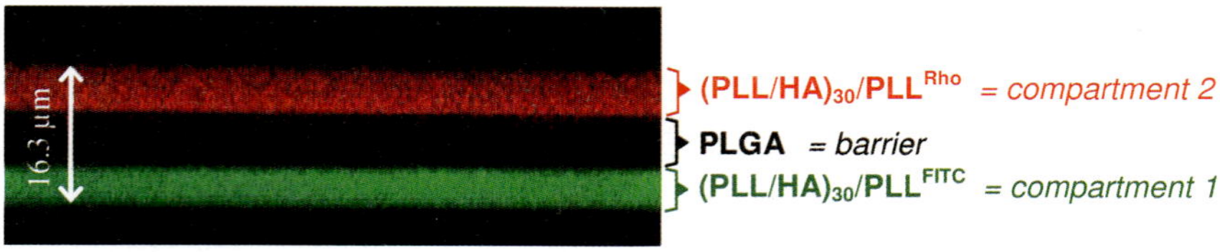

Fig. 14.20 CLSM vertical section of a multilayer film composed of two (PL^L/HA) compartments separated by a PLGA barrier. The glass slide is located at the bottom of the image. The compartments 1 and 2 were labeled respectively in green with $PL^{L\text{-}FITC}$ and in red with PL^L-Rho. The film maintained in solution (NaCl 0.15 M, pH 5.9) consists of $(PL^L/HA)_{30}/PL^{L\text{-}FITC}/PLGA/$ $(PL^L/HA)_{30}/PL^{L\text{-}Rho}$. The total thickness is about 16.3 μm.

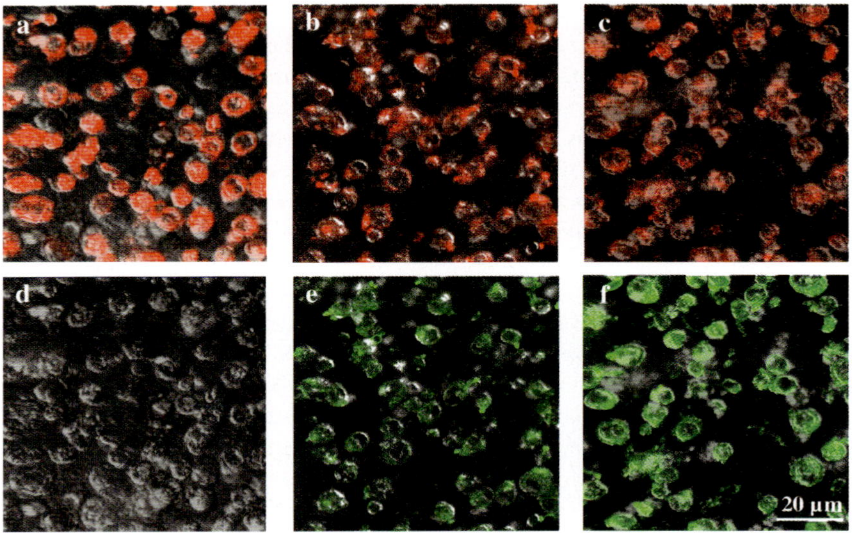

Fig. 14.21 CLSM (*x*, *y*)-images of bone marrow cells seeded 1 h (a, d), 5 days (b, e), and 10 days (c, f) on a $(PL^{L}/HA)_{30}/PL^{L\text{-}FITC}/PLGA/(PL^{L}/HA)_{30}/PL^{L\text{-}Rho}$ film. Images (a–c) correspond to sections in the bright field and red channels (image sizes are 76.8×76.8 μm^2). Images (d–f) correspond to (*x*, *y*)-sections in the bright field and green channels (image sizes are 76.8×76.8 μm^2).

2 days (data not shown). After 5 days, cells internalized $PL^{L\text{-}Rho}$ and also $PL^{L\text{-}FITC}$ from the lower compartment (compartment 1) (Fig. 14.20 b and e). Green intensity in the cells was stronger after 10 days of seeding (Fig. 14.20 c and f). These observations evidence that cells successively enter into contact with the upper compartment and then, after several days, with the lower one, leading to a cascade of successive internalization of each kind of labeled PL^{L} chains. It should be noticed that the high molecular weight of the PLGA used in the present study avoids a spontaneous full hydrolysis of the chains after a few days of contact with the medium [63]. We checked the absence of degradation of the similar films after 2 months of storage in an incubator without any cell seeding but in the presence of medium. Thus, the mechanism described above corresponds actually to the active degradation of the PLGA layer by the cells. By changing the thickness of the PLGA layer and/or the molecular weight of the PLGA, it will be possible to fine tune the kinetics of the cell accessibility to each compartment. Similar films could also be designed for other applications to release a molecule only under passive degradation due to spontaneous hydrolysis of the PLGA layers.

In conclusion, multicompartment films appear as a very promising tool to coat various substrates and to induce finely a cascade of cellular events in a controlled way. One of the main advantages of such designs are their simplicity and versatility. Exponentially growing polyelectrolyte multilayers act as compartments and are separated by barriers either composed of synthetic polyelectrolyte multilayers or of poly(lactic-*co*-glycolic) acid layers. These barriers prevent diffusion of polylysine chains from one compartment to another. Bone marrow cells seeded on such films can access sequentially to the upper and then to the lower compartment only if the barrier is composed of poly(lactic-*co*-glycolic) acid layers. Finally, synthetic polyelectrolyte multilayers acted as a kind of protective barrier to the underlying compartment against degradation by monocytic cells.

In the future, these model systems will allow one to induce time-scheduled cascades of biological activities for various types of cells. Compartments will be filled with proteins or peptides, either by direct diffusion into the hydrated exponentially growing multilayers or by a covalent linking of the biomolecules to a "carrier" polyelectrolyte like polylysine. Degradation of the poly(lactic-*co*-glycolic) acid layers will result from an active mechanism in the case of macrophages or from a spontaneous hydrolysis in the case of cells without any monocytic activity.

14.4.2
Immobilization and Embedding of Intact Phospholipid Vesicles in Polyelectrolyte Multilayers

One of the goals of surface modification of solid substrates with polyelectrolyte multilayer (PEM) films is to incorporate active compounds into these films. Thus, the PEM film plays the role of an active reservoir, as already described in Section 14.4.1. This functionalization can be achieved either by depositing polymers carrying covalently linked active groups or molecules [64, 65] or by depositing the desired molecules or particles directly during the layer by layer (LBL) deposition process [66–71]. In both these strategies, the amount of deposited active compounds can only be increased by increasing the number of adsorption steps of the active molecules. Indeed, it has to be noted that a polyelectrolyte can only be covalently modified to a certain extent with non-charged molecules because only a critical fraction of its charged monomers can be modified.

Therefore, it is of interest to elaborate a method in which a high load of active molecules, either hydrophilic or hydrophobic in nature, can be incorporated into PEM films *in a one-step process* without the need for covalent modification. To this aim, it seems straightforward to immobilize entire and intact capsules containing the molecules of interest in the LBL film. This seems possible owing to the possibility to both embed colloidal charged particles in such architectures and to produce different kinds of hollow particles into which the active substances may be encapsulated. At first glance it would have been tempting to use hollow polyelectrolyte capsules to reach this aim [72, 73] owing to their charged nature, their tunable mechanical properties [74] and to the possibility of encapsulating active substances in their interior [75, 76]. However, the high permeability of these capsules to small organic or inorganic solutes invalidates their use as embedded reservoirs impermeable to a broad spectrum of solutes. Therefore, we focused on immobilizing phospholipid vesicles. Such kinds of containers can be designed in a wide range of diameters, can incorporate charged lipids, and are impermeable to practically all neutral solutes, whatever their polarity, that have a molecular mass above 300 g mol^{-1}. The main obstacle to adsorption of this kind of capsules at a solid/liquid interface, which is mandatory to embed them inside a PEM film, is the possibility of spontaneous vesicle rupture, which happens on various surfaces [77–79]. This spontaneous rupture produces either a lipidic bilayer or a mixed bilayer–immobilized vesicles system. The occurrence of the rupture depends, among other parameters, on the physicochemical properties of the substrate, on the lipidic composition of the vesicles, on the temperature at which the experiment is performed and on the presence of divalent cations. To avoid such a rupture upon adsorption, which would result in an instantaneous release of all the encapsulated molecules, the rigidity of the lipidic membrane must be increased. Some trials have already successfully immobilized vesicles in a LBL architecture: the lipidic molecules, *N*,*N*-dihexadecyl,*N'*-3(triethoxysilyl)propylsuccinamide, were modified with ethoxysilyl head groups which allowed a polycondensation between these headgroups and the formation of siloxane bonds between the lipidic molecules [80, 81]. To simplify the preparation of the vesicles we used an interesting result, namely that the adsorption of just a monolayer of PAH increased remarkably the resistance of vesicles made of L-α-dimyristoylphosphatidic acid towards the addition of sodium dodecyl sulfate [82]. When the vesicles were modified with such a PAH monolayer, the addition of the detergent could not induce the disruption of the bilayer structure and so prevented the formation of detergent–lipid mixed micelles, whereas non-modified vesicles underwent a transition from a bilayer to a micellar structure.

We decided to use a similar strategy to modify negatively charged large unilamellar vesicles which were prepared from a mixture of 1 palmitoyl-2-oleoyl-sn-glycero-3-phosphatidylcholine (POPC) 91% w/w, 1 palmitoyl-2-oleoyl-sn-glycero-3-phosphatidyl-DL-glycerol (POPG), 4.5% w/w and cholesterol. The vesicles were then modified by the adsorption of relatively low molecular weight PL^D (MW = 27 200 g mol^{-1}). The PL^D adsorption on the surface of the vesicles was done at relatively low ionic strength (10 mM Tris buffer at pH 7.4 in the presence of 15 mM NaCl) in order to reduce the occurrence of aggregation. The lipid composition of the vesicles was chosen in such a way that vesicle formation was possible at room temperature. We first demonstrated the alternation of the zeta potential of the vesicles upon PL^D adsorption and the absence of significant aggregation as long as the concentration of added polycation remains below a critical value of about 0.5 mg mL^{-1} (Fig. 14.22). Additional details about the vesicle preparation can be found in [83]. In the following, the unmodified vesicles will be called uLUVs (unmodified large unilamellar vesicles) and the vesicles modified with a PL^D monolayer will be called mLUVs (modified large unilamellar vesicles).

The uLUVs and the mLUVS were then adsorbed on a PEI-$(PGA^L\text{-}PAH)_2$ and a PEI-$(PGA^L\text{-}PAH)_2$-PGA film, respectively. The deposition process was followed *in situ* with a quartz crystal microbalance with measurement of the dissipation

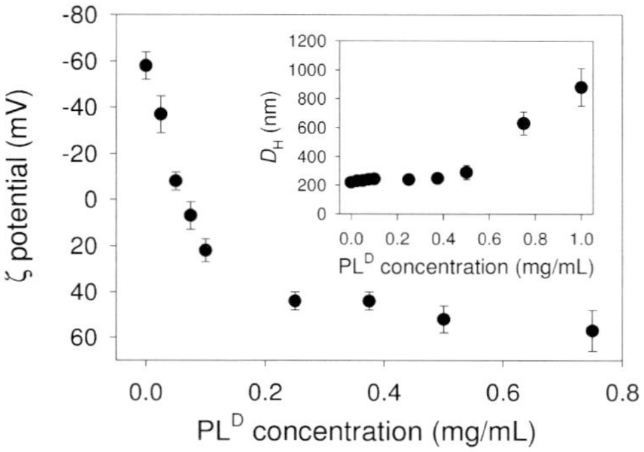

Fig. 14.22 Evolution of the zeta potential of the vesicle solution (in 10 mM Tris, 15 mM NaCl buffer at pH 7.4) with the concentration of added PL^D. Inset: Evolution of the average hydrodynamic diameter of the vesicles as a function of the concentration of added PL^D.

(QCM-D). The experimental details are given in [83]. The data are shown in Fig. 14.23 where panel A represents the normalized frequency change at the resonance of the third overtone of the quartz crystal (close to 15 MHz) in the case of the deposition experiment for uLUVs and mLUVs and panel B represents the dissipation changes for both kinds of vesicles with respect to the dissipation changes due to the buffer solution. The inset in Fig. 14.23 A shows that the deposition of a $(PAH\text{-}PGA^L)_2$ architecture on top of the lipidic film is possible, as expected on the basis of electrostatic interactions. When comparing the deposition kinetics of uLUVs and mLUVs on the surface of a PEI-$(PGA^L\text{-}PAH)_2$ and PEI-$(PGA^L\text{-}PAH)_2$-PGA^L film some differences appear: the adsorption kinetics of the mLUVs appears much faster than that of the uLUVS and is accompanied by a much smaller increase in dissipation (which represents the energy losses from the oscillating crystal in the film and in the aqueous solution). In addition, upon buffer rinse, the frequency increases slightly in the case where uLUVs have been deposited, whereas no frequency change is measured on the film with deposited mLUVs.

At first glance, the absence of a minimum in the frequency changes as well as the absence of a maximum in the dissipation curve points to an adsorption process of the vesicles without rupture [77–79]. However, the observation that the frequency increases upon buffer rinse in the case of uLUVs suggests either a buffer rinse-induced desorption of the vesicles or a partial vesicular disruption.

Hence, in order to have a "molecular" picture of the vesicular shape after their embedding under a PAH-$(PGA^L\text{-}PAH)_2$ and a $(PGA^L\text{-}PAH)_2$ film for the uLUVs and mLUVs, respectively, we performed some AFM experiments in the contact mode. In the case of uLUVs we observed a large fraction of fused vesicles (data not shown here), whereas practically no vesicle fusion was observed on the PEI-$(PGA^L\text{-}PAH)_2$-PGA^L-mLUVs-$(PGA^L\text{-}PAH)_2$ architecture (Fig. 14.24). In addition, the adsorbed vesicles have a diameter distribution extending from 200 to 300 nm, which is very close to the diameter distribution found for the modified vesicles in solution.

These experiments showed that it is possible to adsorb the mLUVs on a PEM film made from PGA^L and PAH and to embed them subsequently inside the

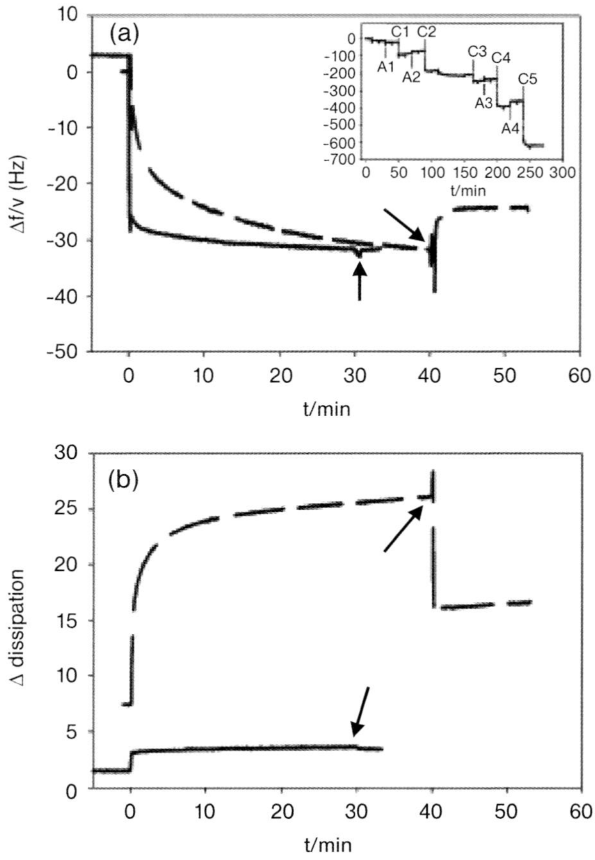

Fig. 14.23 Adsorption kinetics followed by QCM-D of uLUVs (dashed line) onto a PEI-$(PGA^L\text{-}PAH)_2$ multilayer and of mLUVs (solid line) onto a PEI-$(PGA^L\text{-}PAH)_2$-PGA^L polyelectrolyte multilayer. (A) normalized frequency change. (B) corresponding change in dissipation. The origin of the time axis corresponds to the injection of the large unilamellar vesicles. In both panels, the buffer rinse after the vesicle adsorption has reached a plateau is labeled with an arrow. For the sake of clarity, only the normalized frequency shifts corresponding to the third harmonic are represented. The inset in panel A represents the normalized frequency shift corresponding to the third harmonic during the whole buildup of the film. The arrows labeled with C correspond to the injection of the polycation, and the arrows labeled with A to the injection of the polyanion.

same architecture with the conservation of their spherical shape. The same experiments were carried out with PGA^L combined with PL^L as a polycation and it was found that even the mLUVs underwent important deformation and fusion upon adsorption and further embedding in the multilayer architecture. Hence, the nature of the PEM film is important in the successful incorporation of vesicles in their interior. Some systematic research is in progress in this field.

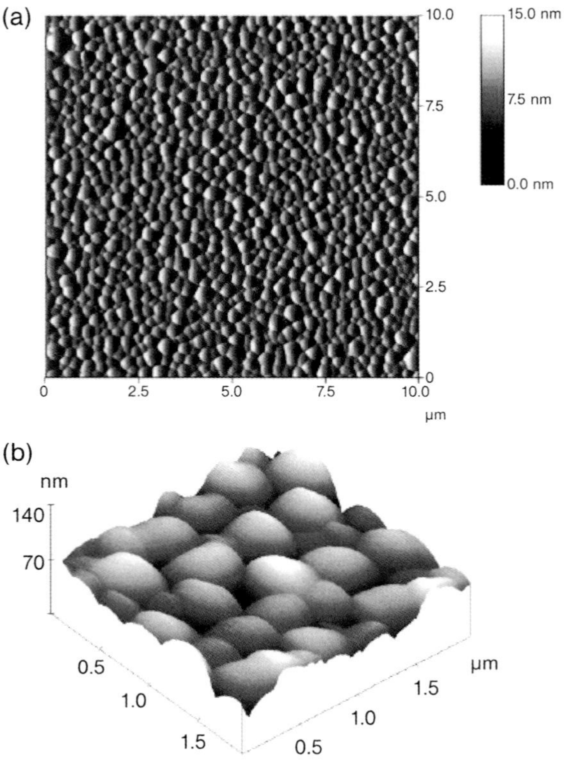

Fig. 14.24 Contact mode AFM images (taken in air) of a PEI-(PGA^L-PAH)$_2$-PGA^L-mLUVs-(PGA^L-PAH)$_2$ film. Panel A is a deflection image over a 10×10 μm^2 surface area. The root mean square roughness over this area was equal to 116 nm. Panel B is a 3D-view of the same surface but over a restricted area of 2×2 μm^2.

Nevertheless, the previously described experiments do not demonstrate the molecular integrity of the vesicles. They just show that the vesicles remain of spherical shape with a diameter distribution close to that of the vesicles in solution. To address the question of the molecular integrity, we encapsulated ferrocyanide ($Fe(CN)_6^{4-}$) anions in the aqueous internal compartment of the vesicles. This encapsulation was performed during the vesicle preparation and the non-encapsulated species were eliminated by dialysis [48]. The PEI-(PGA^L-PAH)$_2$-PGA^L-(mLUVs with ferrocyanide)-(PGA^L-PAH)$_2$ film was deposited on a gold electrode by the spray method in order to accelerate the deposition process. Indeed, in this way we were able to measure the release of ferrocyanide anions a few minutes after the mLUVs were adsorbed on the PEI-(PGA^L-PAH)$_2$-PGA^L film and further embedded under two (PGA^L-PAH) pairs of layers. The ferrocyanide release kinetics in the film and to the gold electrode was followed by means of cyclic voltammetry. The whole methodology relies on a previous study in which we demonstrated that $Fe(CN)_6^{4-}$ anions are able to diffuse in PEM films made from PGA^L and PAH whatever their surface charge [84]. Furthermore, these anions are not able to diffuse out of the PEM film when the ferrocyanide-containing solution is put in contact with a 10 mM Tris–150 mM NaCl buffer at pH 7.4. Hence the PGA^L and PAH-based films are a sink for ferrocyanide anions, probably owing to a positive Donnan potential. Based on these results, when ferrocyanide is leaving from the internal aqueous cavity of the vesi-

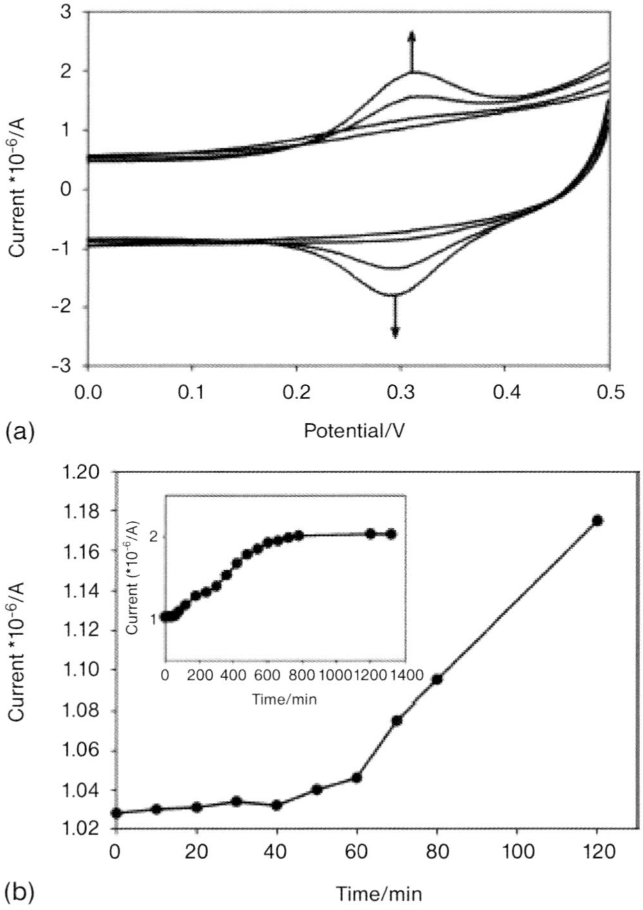

Fig. 14.25 (A) Time evolution of cyclic voltammograms on PEI-(PGA^L-PAH)$_2$-PGA^L-mLUVs-(PGA^L-PAH)$_2$ films sprayed onto the surface of a gold electrode. The mLUVs contained ferrocyanide anions. From bottom to top, the curves correspond to time $t=0$, 1, 6 and 12 h, where $t=0$ corresponds to a cyclic voltammogram taken just before the deposition of the vesicles. The arrows indicate the direction of time increase. The cyclic voltammograms were acquired at a scan rate of 200 mV s^{-1}. (B) Evolution of the maximum oxidation current obtained as a function of time for short time durations. Inset of B: the same over a time scale of 23 h.

cles, either by diffusion or after the opening of pores, it should be detected by means of its oxidation current when the potential between the gold and the reference electrode is scanned. This is indeed the case, and it was found that the release of the electroactive species out of the vesicles is a process taking place over a time scale of more than 12 h [48] with an initial lag phase (of about 1 h) during which no release at all is detected (Fig. 14.25). Control experiments were performed in the same manner but with uLUVs (hence in the absence of a protecting PL^D layer) and it was found that the maximum levels of oxidation and reduction currents were obtained in few minutes, indicating a fast release from the non-protected vesicles. This observation highlights, as suspected, the need to rigidify the vesicles before their embedding in a PEM film.

Based on these findings we also managed to embed two strata of intact vesicles in the same kind of architectures made from PGA^L and PAH (Fig. 14.26).

Finally, we have shown that enzymes can be encapsulated in the active state inside the embedded vesicles. The enzyme used, alkaline phosphatase (AP), coencapsulated with calcium and spermine inside the vesicles, was used to produce calcium phosphates in the internal lumen of the vesicles when paranitrophenol phosphate, a substrate of AP, was added on top of the multilayer architecture [85].

Hence our study opens the route to perform chemical reactions in these surface immobilized submicronic reactors. Further experiments will be performed in this direction by encapsulating two reacting species in two different strata of embedded vesicles: these species will then react upon their spontaneous diffusion out of the vesicles.

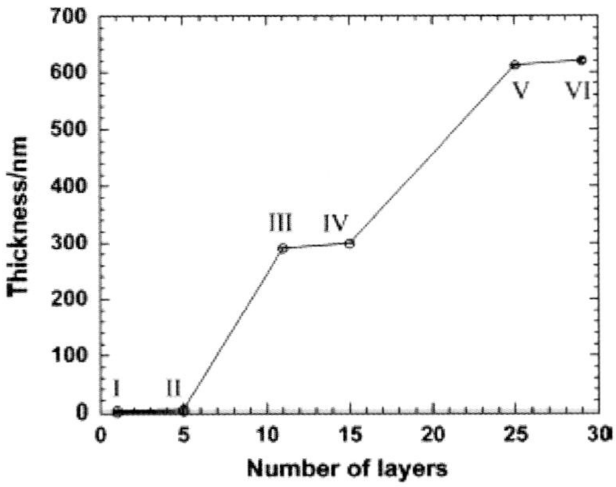

Fig. 14.26 Evolution of the thickness increase measured by ellipsometry (in the dry state) of PEI-$(PGA^L\text{-}PAH)_2$-PGA^L-mLUVs-$(PGA^L\text{-}PAH)_6$-PGA^L-mLUVs-$(PGA^L\text{-}PAH)_4$ multilayer film deposited on an oxidized silicon wafer by the spraying method. The different steps of the buildup are: step I: PEI, step II: step I + $(PGA^L\text{-}PAH)_2$, step III: step II + PGA^L-mLUVs + $(PGA^L\text{-}PAH)_2$, step IV: step III + $(PGA^L\text{-}PAH)_2$, step V: step IV + $(PGA^L\text{-}PAH)_2$-PGA^L-mLUVs-$(PGA^L\text{-}PAH)_2$, step VI: step V + $(PGA^L\text{-}PAH)_2$.

We will also try to embed vesicles made from saturated lipids, with the aim to increase their intrinsic stability. It seems also mandatory to extend the concept of vesicle embedding in PEM films made from natural polyelectrolytes, like hyaluronic acid and chitosan. Another possibility would be to embed vesicles made from triblock copolymers provided that the terminal hydrophilic parts carry a permanent charge [86, 87].

14.5 Biofunctional Films

The layer-by-layer adsorption of polyelectrolytes not only leads to the formation of films, but also allows protein adsorption or cell adhesion that can be controlled. It is especially interesting to notice how complex multilayer architectures can be used to control the access of cells to embed bioactive proteins [88], bioactive peptides [89], bioactive lipopolysaccharides [90] or drugs [91].

14.5.1 Multilayers Containing Active Proteins

Bioactive proteins can be directly integrated into the architecture without any covalent bonding with a polyelectrolyte [67, 68, 92–94] and keep a secondary structure close to their native form [95]. Degradable, layered structures would be advantageous for progressive delivery of associated active agents. Considering that the aim of the present study was to investigate whether a signal protein embedded in a polyelectrolyte multilayered film keeps its activity, we selected the PGA^L and PL^L system in which we embedded the Protein A (PA). This protein, issued from the cell wall of *Staphylococcus aureus*, possesses the ability to bind the Fc fragment of IgG and also has a large panel of biological activities such as antitumoral [96, 97], antitoxic [98], anticarcinogenic [99], antifungal [100] and antiparasitic [101] properties. Besides, PA stimulation of the human monocytic cell line THP-1 leads to the rapid expression of both the pro-inflammatory cytokine TNF-α and the anti-inflammatory cytokine (IL-10) [102]. To determine the ability for cells to respond to PGA^L/PL^L multilayers in which PA molecules were immobilized, we measured the production of the pro-inflammatory cytokine TNF-α by monocytic THP-1 cells. The main finding is that cells can interact with proteins incorporated in such polyelectrolyte multilayer films. This result opens the route for film functionalization with embedded proteins.

Buildup of polyelectrolyte multilayers including PA was checked by optical waveguide light mode spectroscopy (OWLS) [103]. PA molecules, negatively charged at pH 7.4, were adsorbed on an oppositely charged PL^L layer with an estimated coverage of about 35%, close to the expected coverage for a monolayer. PA remains biologically active when adsorbed at the surface or embedded in a multilayer, as shown by TNF-α secretion by monocytic THP-1 cells grown on the surface of a multilayered PEI-PGA^L-PL^L-PA-$(PGA^L\text{-}PL^L)_{10}$ architecture containing PA embedded under 10 (PGA^L-PL^L) bilayers. We examined the effect of the embedding depth on PA activity by comparing TNF-α amounts secreted by THP-1 cells grown on architectures where PA was incorporated under an increasing number of (PGA^L-PL^L) bilayers. Values obtained after 4 h and one night of cell interaction with films are similar, whatever the embedding depth of PA (up to $n=30$), and comparable to the value obtained when PA was adsorbed on the terminating layer (Fig. 14.27). This clearly indicates that a 4 h contact is sufficient for the cells to interact, even with deeply embedded PA molecules. Control experiments were performed in the absence of PA, indicating a basal secretion of TNF-α by THP-1 cells in contact with films without PA, which is in agreement with the fact that TNF-α was detected in supernatants from monocytes in the presence of PL^L [104].

The fact that a maximal amount of TNF-α was produced after 20 min of contact of THP-1 cells with free PA (data not shown), compared to 120 min needed to obtain a similar response in the presence of PA-containing films, suggests that a certain time is needed for a connection between cells and embedded PA to be established. Cells could come into contact with PA through local degradation of the film and cellular membrane extension up to the PA layer. We thus designed experiments reducing the possibility of polyelectrolyte degradation by replacing PL^L with PL^D in the polyelectrolyte architecture. We observed a very significant reduction of TNF-α production (Fig. 14.27 10-D), indicating that PL^D forms a barrier preventing the contact between the cells and embedded PA. A detailed analysis of the effect of the D-enantiomer of polylysine on the degradation of PGA/PL films can be found in Section 14.3.3.

To elucidate the process by which cells come into contact with the embedded protein A, we analysed by CLSM the evolution of the surface structure of a

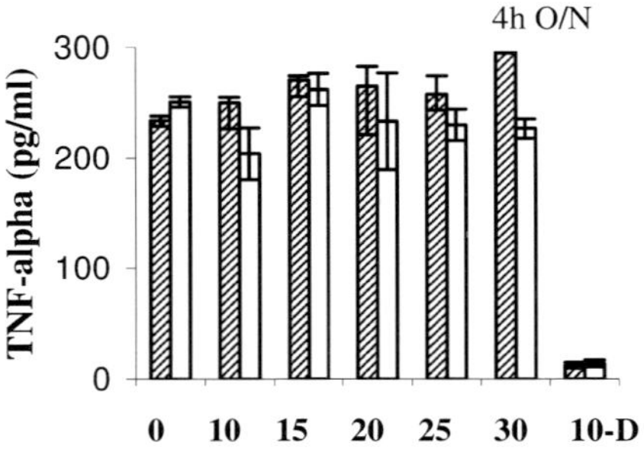

Fig. 14.27 TNF-α secretion by THP-1 cells grown on multilayer films containing PA. An increasing number n of (PGA^L-PL^L) bilayers ($n=0$, 10, 15, 20, 25, 30 or 10-D (10 bilayers where PL^D replaced PL^L) were assembled on PA adsorbed on the precursor film PEI-PGA^L-PL^L. Cells were deposited on multilayer films built in 24-well plates at a density of $5 \cdot 10^5$ cells/well. After 4 h or overnight (15 h) incubation, cells were centrifuged and total TNF-α levels were measured. The results are the means of three measurements and the vertical bars correspond to the standard deviation. Related values obtained with the same architectures in the absence of PA have been subtracted.

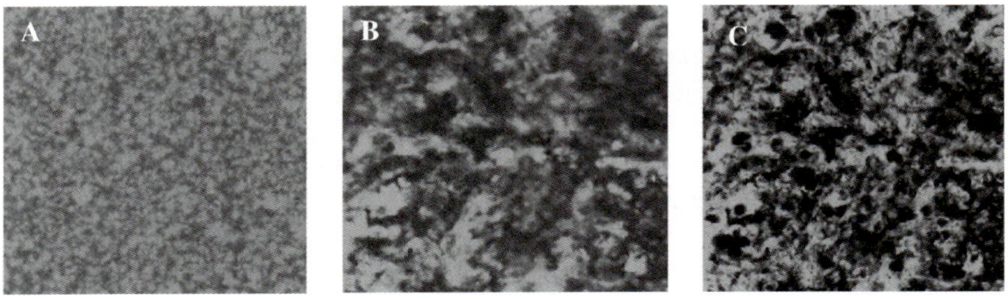

Fig. 14.28 Evolution of the surface structure, observed by confocal microscopy, of a $(PL^L\text{-}PGA)_5/PL^L\text{-}PA^{TR}\text{-}(PGA\text{-}PL^L)_{19}\text{-}PGA\text{-}PL^L\text{-}FITC$ film in contact with cells for different times (A: 0 min, B: 3 h, C: 12 h). Images sizes ≈ 50 μm×50 μm.

$(PEI\text{-}PGA^L\text{-}PL^L\text{-}PA^{TR}\text{-}(PGA^L\text{-}PL^L)_{19}\text{-}PGA^L\text{-}PL^L\text{-}FITC)$ film in contact with THP-1 cells for different times. We observed local surface degradation of the film after 180 min and a pronounced degradation after overnight contact with THP-1 cells (Fig. 14.28). This point was confirmed by internalization of Texas Red-conjugated PA by THP-1 cells after 180 min [58, 91], and by observation of the film, which shows that cells come into contact with the embedded active PA through membrane extensions (Fig. 14.29).

We have demonstrated that cells can interact with proteins incorporated in polyelectrolyte multilayer films and elucidated the mechanism by which cells come into contact with the active protein in the case of monocytic cells. Our results show that PL^D forms a barrier for cellular communication and that cells communicate with embedded proteins through local film degradation. Such functionalized coat-

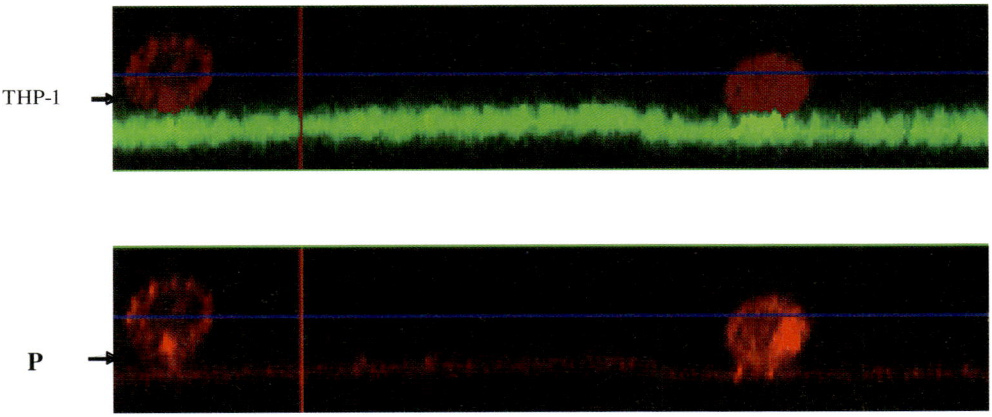

Fig. 14.29 Cell in contact with multilayer film containing the PA^{TR} embedded under $(PGA^L/PL^L)_{19}\text{-}PGA\text{-}PL^L\text{-}FITC$. Cells were deposited on multilayer film and observed by confocal microscopy (*z*-sections) after 4 h. P: Pseudopod. Images sizes = 146 μm×26 μm.

ings could, in the future, become potent tools for the modification of biomaterial surfaces as applied to implants, prostheses or in tissue engineering and could also, as such, have future clinical applications in cancer therapy.

14.5.2
Multilayers Containing Active Peptides

The biological response to tissue damage (acute inflammation) and especially to extracorporal material (chronic inflammation) is a major concern in human repair issues and limits the use of implants in many cases (for example, implanted electronic equipment, such as pacemakers, defibrillators, cochlear implants, pain modulators, and insulin pumps, bone substitutes, orthopedic prosthesis, etc.). One of the main undesirable effects of any foreign "object" like an implant concerns the chronic inflammatory response. Thus, an implant can be rejected by the body, accompanied by other side effects such as severe fever, and can lead to septic shock, which finally requires the removal of the implant. In consequence, implants are generally built using "biocompatible" materials such as TiO_2, stainless steel, and certain synthetic or semi-synthetic plastic materials that minimize chronic inflammation. However, such benign materials are typically found by trial and error and not by a systematic approach, that allows one to control inflammation in a predictable way.

In a simplistic view, inflammation may be caused by chemical products leaking from an implant to the surrounding tissue (e.g. residual monomers) or by stimulating inflammation as a direct response to the nature of the surface of the implant, which is the more general and interesting case and issue of the present report. The nature of a biomaterial surface governs the processes involved in biological response.

Thus, the control of inflammation and repair (fibrosis) in tissue has focused classically on the use of broad-spectrum drugs (e.g., steroid and non-steroid anti-inflammatory drugs) [105, 106]. However, long-term systemic use of these drugs is limited because of major side effects that develop with time. In consequence the development of more specific and local forms of treatment [63, 107] with fewer side effects is critical for the control of inflammation and fibrosis. Site-specific, controlled-release delivery has the advantage of reducing or eliminating systemic side effects and improving the therapeutic response through appropriate controlled dosing at the site of interest.

Considerable efforts have been undertaken over the past years to render materials biologically active. Different methods have been developed: active molecules have been incorporated directly into the material [108–110] or by chemical grafting [111–113]. Bioactive molecules such as insulin [114] or epidermal growth factor [115] have, for example, been chemically grafted and immobilized on surfaces. None of these methods is, however, without drawbacks: the incorporation of active molecules into the sub-surface regions of a given material is not always possible; adsorption of molecules often involves weak bonds so that the molecules rapidly desorb and chemical grafting can be very difficult to realize. Moreover, the irreversible attachment of molecules to a surface may also sometimes reduce their biological activity. The deposition of polyelectrolyte multilayers on charged surfaces [2, 10, 116–121] offers an interesting approach for the coating of surfaces in contact with a biological environment. These coatings are, for example, obtained by the subsequent immersion of the surface in polyanion and polycation solutions. This surface functionalization technique allows one to embed polyfunctional molecules in thin layers in which one functionality is used for surface attachment employing gentle methods and one or several others are used to implement the desired activity. Rather than optimizing the immobilization chemistry for each ligand and each substrate individually, the combination of solution coupling with a standardized deposition procedure presents a considerable advantage. In addition, layer-by-layer deposition allows easy fabrication of multi-material films in which different layers carry different functionalities or repeat the same functionality several times in order to control the quality or the quantity of active agents. Macromolecules coupled to active agents have already been shown to accomplish their purpose in multilayer architectures [45, 64, 122, 123]. Below we describe the biochemical and morphological response of monocytes in contact with surfaces functionalized with a melanocortin derivative (alpha-melanocyte stimulating hormone, α-MSH) incorporated in such multilayer assemblies. We will show that the presence of α-MSH in the multilayers confers anti-inflammatory properties to the coating.

In this study, a derivative of α-MSH was chosen as the active agent for inclusion in multilayer films. α-MSH is an endogenous linear tridecapeptide with potent anti-inflammatory properties [106]. It is already known that α-MSH can be covalently coupled to molecules using well-established procedures. From earlier experiments, it became clear that α-MSH coupled to poly(L-lysine) can be incorporated into multilayer films and that certain biological properties of such α-MSH derivatives remain active in this formulation [64]. In the present report, we investigate the anti-inflammatory properties of an α-MSH derivative using model surfaces for implants and monocytic cells *in vitro*. Here, we coupled the α-MSH derivative with the N-terminal sequence to the carrier polyion PGA^L, which leaves the anti-inflammatory C-terminal sequence Lys^{11}-Pro^{12}-Val^{13} of the peptide more accessible after attachment to the carrier polymer.

The use of active agents coupled to polyelectrolytes constitutes a major advantage in comparison with direct chemical immobilization methods. The conjugated polymers can be synthesized in classic fashion, purified and characterized. If the polymer does not meet a certain standard, it can be rejected prior to the surface treatment. In contrast, direct immobilization of active agents on a surface needs to be optimized for every individual agent/surface pair, the resulting surface structures are much more difficult to characterize and, if side reactions are detected, the whole batch of devices needs to be rejected. Melanocortin-derived peptides coupled to PGA^L (PGA^L-α-MSH) were used to fabricate multilayers with PL^L as counter-polyelectrolyte.

We verified through OWLS that PGA^L-α-MSH adsorbs readily to films terminating with a layer of PL^L and further that it can be embedded in PGA^L/PL^L

multilayers (data not shown), the growth of which has been described before [64]. Here we present data of the assembly of PL^L/PGA^L-α-MSH multilayers as determined by dissipation enhanced quartz crystal microbalance (QCM-D) (Fig. 14.30) where the normalized frequency shift $\Delta f/v$ is, to a first approximation, proportional to the mass of the film. The normalized frequency shift $\Delta f/v$ corresponds to the different resonance frequencies of the quartz crystal (15, 25 and 35 MHz) during the film buildup, Δf being the resonance frequency shift and v the overtone number. During the whole film buildup, $\Delta f/v$ is, to a first approximation, independent of v. This strongly indicates that the multilayer behaves as a rigid film and that the Sauerbrey relation [124] constitutes a valid approximation to describe the data. The evolution of $-\Delta f/v$ shows a regular film deposition starting with the first layer of PGA^L-α-MSH. Similarly, the polycation poly(ethylene imine) (PEI) can be used for the deposition of the first layer (data not shown). Films composed of 5, 10, 15 or 20 layer pairs were prepared and evaluated in cell activity tests. Such model surfaces for implants were brought into contact with human monocytic cells stimulated with lipopolysaccharide (LPS) bacterial endotoxin, in order to induce inflammation as an *in vitro* inflammatory model, and tested with respect to the production of tumor necrosis factor alpha (TNF-α and interleukin 10 (IL-10) using the respective antibody assays. The production of TNF-α is an indicator of inflammation and its inhibition is a positive response. The secretion of IL-10 is an indicator of an anti-inflammatory response and its induction is a positive response.

In order to first investigate whether the α-MSH peptide remains biologically active when covalently bound to PGA^L with its N-terminal sequence, we tested

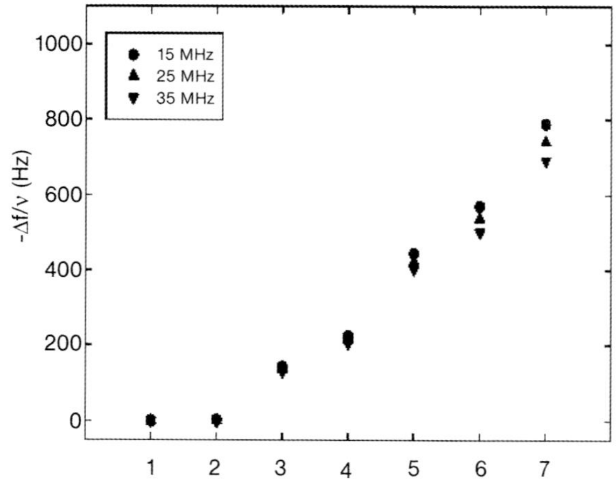

Fig. 14.30 Quartz crystal microbalance measurements. Normalized frequency shifts ($-\Delta f/v$ at 15, 25 and 35 MHz) are shown as a function of layer deposition. The abscissa labels denote the following deposition steps: (1) bare crystal, (2) first PL^L layer, (3) (PL^L/PGA^L-α-MSH), (4) (PL^L/PGA^L-α-MSH)/PL^L, (5) $(PL^L/PGA^L$-α-$MSH)_2$, (6) $(PL^L/PGA^L$-α-$MSH)_2/PL^L$, (7) $(PL^L/PGA^L$-α-$MSH)_3$.

TNF-α production by monocytic cells at different times in solution as a function of PGA^L-α-MSH concentration (data not shown). Moreover, we verified that the covalent coupling of α-MSH to PGA^L does not significantly change the time course of α-MSH activity. Indeed, the suppression of TNF-α by free α-MSH is rapid and continues for about 2 h while the induction of IL-10 is delayed by about 6 h and has duration of about 2 h, as for α-MSH coupled to PGA^L (data not shown).

In comparison to the solution, we also tested TNF-α and IL-10 secretion by cells in contact with multilayer films containing PGA^L-α-MSH. From QCM-D experiments and using the Sauerbrey relation we could estimate that a film corresponding to $n=5$ contains of the order of 25 μg of PGA^L-α-MSH which is comparable to the amount of PGA^L-α-MSH present in the solution (300 μL at 100 μg mL^{-1}) used for the IL-10 response in Fig. 14.28 B. After 1, 2, 3, 4, 6 or 8 h of deposition of stimulated cells on surfaces coated with PEI-(PGA^L-α-MSH/PL^L)$_n$ multilayers ($n=5$, 10, 15 or 20), the amounts of TNF-α and IL-10 produced by the cells were determined (Fig. 14.31).

One observes that, for all the investigated films containing PGA^L-α-MSH, the TNF-α activity suppression of PGA^L-α-MSH is preserved for longer times than in solution. Increasing the number of layer pairs to $n=15$ or 20 clearly reduces TNF-α production during the first 4 h when compared to films with $n=5$ and 10. This suggests that the multilayer film acts as a reservoir in the sense that the number of layers deposited, and thus the quantity of α-MSH, accessible to the cells, controls their response.

After demonstrating the anti-inflammatory response of our model surfaces for implants, the mechanism of action of multilayer films containing PGA^L-α-MSH was investigated. It is known that free α-MSH is only effective after inter-

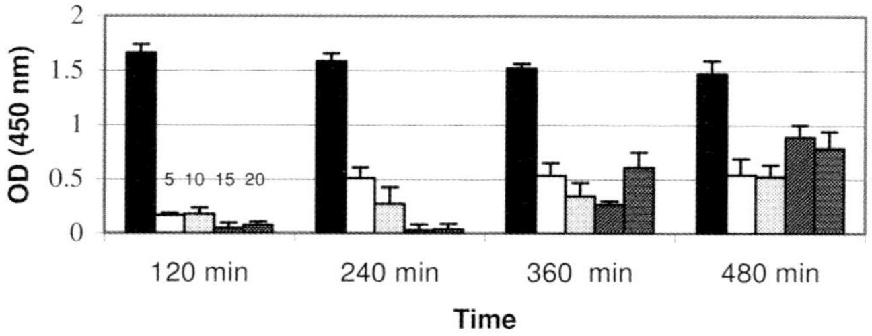

Fig. 14.31 Kinetics of TNF-α secretion by monocytic cells in contact with a multilayer film of the general architecture PEI/(PGA^L-α-MSH/PL^L)$_n$, where the index n corresponds to the number of deposited layer pairs. The black bars represent TNF-α levels measured for cells in contact with reference films not containing α-MSH. Each black bar represents the average over four experiments carried out on PEI/(PGA^L/PL^L)$_n$ films with $n=5$, 10, 15, 20, at each time. The other bars represent TNF-α levels measured for cells in contact with PEI/(PGA^L-α-MSH/PL^L)$_n$ films ($n=5$, 10, 15, 20 as indicated by the labels). Cells were stimulated with LPS (10 ng mL^{-1}) and deposited on the multilayer films. These results are the means of three measurements and the vertical bars represent the corresponding standard deviations.

nalization by monocytes following binding to surface receptors [105]. CLSM shows the internalization of FITC-labeled PGA^L-α-MSH (PGA^L-α-MSH^{FITC}) by THP-1 cells (data not shown). After 4 h of contact of cells with the film, internalization of PGA^L-α-MSH^{FITC} is observed. This is even more interesting as the fluorescently labeled layer was deposited at $n=7$ and covered by 20 further layer pairs without fluorescent label. If the time of contact is extended overnight, CLSM demonstrates a homogeneous fluorescence of the interior of the cells.

In addition to the anti-inflammatory effect of films containing α-MSH we observed, by *in situ* atomic force microscopy (AFM), morphological differences induced by the presence of α-MSH in the films [89]. More precisely, cells in contact with a film containing PGA^L-α-MSH undergo morphological changes that can be interpreted as the development of pseudopods. Number, diameter and fine structure of these rigid "fiber-like" extensions change with time. Whereas after 2 h of contact the extensions consist of mostly individual "fibers" protruding from the body of the cell, a longer contact time leads to the formation of extensions that resemble "bundled fibers" in addition to individual "fibers" (data shown in [89]). On the other hand, in the absence of PGA^L-α-MSH, cells do not develop such "fiber-like" protrusions. This indicates the absence of cellular stimulation in contact with non-functionalized films in contrast to functionalized ones. One must, however, point out that, from a general point of view, the presence of such protrusions does not constitute the exclusive signature of the activity of α-MSH but could be observed for other kinds of stimulation.

It is worth noting that the films are very stable with time and in the absence of polyelectrolytes in solution: there is no release of PGA^L or PGA^L-α-MSH from the film into the solution. Thus, such films do not act as drug delivery devices but the biological effect is rather based on the local degradation of the film by the cells. This was clearly shown for the biological activity of protein A embedded in similar $(PL^L/PGA^L)_n$ films [88] and is also strongly suspected in the present case.

We show that human monocytes exhibit an anti-inflammatory response to model surfaces for implants containing a melanocortin derivative. We show that the active agent PGA^L-α-MSH stimulates a response similar to that in solution with regard to a marker associated with inflammation (TNF-α) and to a marker associated with the reduction of inflammation (IL-10). The amount of PGA^L-α-MSH embedded in the multilayer film controls the intensity and the time course of the anti-inflammatory behavior. Whereas the internalization of fluorescently labeled PGA^L-α-MSH is clearly demonstrated, the exact mechanism for the access and uptake of the active agent by the cells remains speculative at the present time. However, it is evident that the presence of α-MSH in the film leads to an important morphological change of the THP-1 cells. As visualized by *in situ* atomic force microscopy, cells develop "fiber-like" extensions, probably so-called pseudopods, in contact with surfaces functionalized with PGA^L-α-MSH. This formulation, consisting of polyelectrolyte films functionalized with α-MSH, shows promise for the local prevention of inflammation, with the potential for progressive accessibility of the anti-inflammatory peptide by judicious choice of the polyelectrolytes. We presented one example where the architecture and composition of layer-by-layer assembled films control the cellular response *in vitro* with the prospect of potential applications *in vivo*. Such functionalized coatings could be used in the future as versatile systems for the modification of biomaterial surfaces applied to implants or prostheses and, as such, find real clinical applications. This potential is partially based on the fact that active agents can be coupled to a "carrier" polymer using conventional synthetic procedures and obtaining the surface functionalization *via* LBL-assembly in classic buffer solutions.

14.5.3
Multilayers Containing Active Drugs

We describe the buildup of biomaterial coatings based on polypeptide multilayers possessing anti-inflammatory properties. PL^L and PGA^L are used as polypeptides and piroxicam (Px) is used as the anti-inflammatory agent. In order to embed high enough amounts of Px, the drug is incorporated in the films in the form of complexes with a charged 6^A-carboxymethylthio-β-cyclodextrin (cCD). It is shown that this cyclodextrin can solubilize higher amounts of Px than the cyclodextrins used commercially. The anti-inflammatory properties are evaluated by determining the inhibition of TNF-α production by human monocytic THP-1 cells stimulated with lipopolysaccharide (LPS) bacterial endotoxin. Using FT-Raman spectroscopy, we show that Px is mainly in the neutral form in cCDPx complexes in solution and that it remains biologically active in this form, whereas up to now only the zwitterionic form was reported to possess anti-inflammatory properties. When incorporated in PL^L/PGA^L multilayers, Px in the cCDPx complexes changes from the neutral to the zwitterionic form. It is shown that these films present anti-inflammatory properties which can be delayed and whose duration can be tuned by changing the film architecture. This technique allows, in particular, preparation of supramolecular nanoarchitectures exhibiting specific properties in terms of cell activation control [64, 88, 125] and it could benefit the development of local drug delivery systems.

The objective of the present work was to make use of this concept to design coatings aimed at tuning the local inflammatory response to implants. Piroxicam (Px) (Fig. 14.32a) was chosen as an efficient non-steroidal anti-inflammatory drug. Its local use is, however, limited by its weak solubility in water. This drawback can be overcome in solution by including piroxicam into β-cyclodextrin to form piroxicam-β-cyclodextrin inclusion complexes.

β-cyclodextrin molecules constitute a group of cyclic oligosaccharides formed by seven D-glucopyranose units, which improve the physicochemical properties of many drugs through formation of inclusion complexes [126]. Cyclodextrins and their derivatives are widely used for their effects on solubility, dissolution rate, chemical stability and absorption of drugs. Piroxicam drug formulation is now available as a complex with native β-cyclodextrin or with hydroxypropyl-β-cyclodextrin [127, 128]. Both cyclodextrin molecules are however uncharged under weakly alkaline conditions.

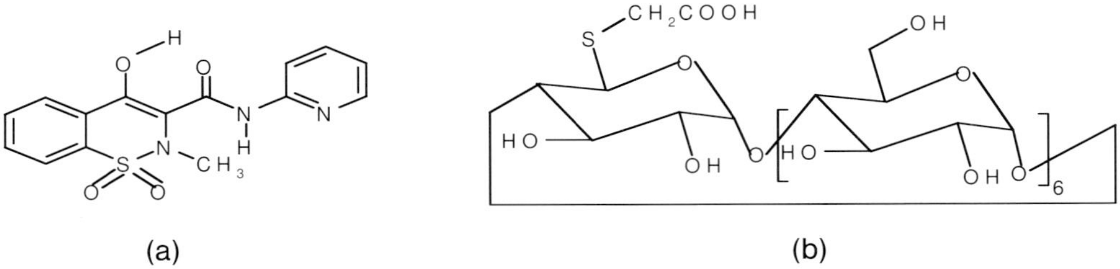

Fig. 14.32 Schematic representation of the chemical structure of (a) Piroxicam (Px) and (b) 6^A-carboxymethylthio-β-cyclodextrin (cCD).

The goal of this work was to develop multilayers containing piroxicam. In order to incorporate piroxicam into the multilayers we used a charged cyclodextrin molecule, 6^A-carboxymethylthio-β-cyclodextrin (cCD) (Fig. 14.32 b), to form piroxicam–cyclodextrin complexes (cCDPx). These complexes were first characterized in solution, by using UV-visible and fluorescence spectroscopy techniques, and their anti-inflammatory activity was assessed on monocytic (THP-1) cell cultures by the determination of the inhibition of the pro-inflammatory TNF-α production consecutive to lipopolysaccharide (LPS) endotoxin stimulation. The complexes were then embedded in PGA^L/PL^L multilayers. They were also used to construct $(PL^L/cCDPx)_n$ multilayers. The film buildup was followed by attenuated total reflection Fourier transform infrared spectroscopy (ATR-FTIR) and FT-Raman spectroscopy, as well as by QCM-D. The anti-inflammatory activity of these architectures was assessed by the decrease in the pro-inflammatory TNF-α production.

To study whether the cCDPx complexes remain biologically active in solution, we tested the inhibition of pro-inflammatory TNF-α production by THP-1 cells (data not shown). In the absence of cCD the concentration of the free solubilized Px is of the order of 10^{-5} M. When adding cCD at increasing concentrations in the 0–1.6×10^{-4} M range (0–0.2 mg mL^{-1}) the activity of Px increased to a plateau value of nearly 100% inhibition, clearly showing the anti-inflammatory activity of cCDPx complexes in solution [91].

We first constructed $(PGA^L\text{-}PL^L)_n$ polypeptide multilayers. The buildup of these multilayers was first followed by QCM-D [60, 129]. When such films, either ending with PL^L or PGA^L, were brought into contact with a Px solution no change in signal could be observed. This indicates that Px does not adsorb, at least not in amounts that are detectable by QCM-D. We then built two types of architectures with cCDPx complex: PEI-PGA^L-$(PL^L\text{-}PGA^L\text{-}PL^L\text{-}cCDPx)_n$ and PEI-PGA^L-$(PL^L\text{-}cCDPx)_n$, data not shown).

To study whether cCDPx complexes, adsorbed or embedded in multilayered films, remain biologically active, we tested the inhibition of pro-inflammatory TNF-α production by monocytic cells stimulated by LPS (Fig. 14.33).

In architecture (a) (Px adsorbed on the film surface; see legend of Fig. 14.33) the anti-inflammatory activity was less important than in solution, with a maximum of 16% inhibition after 1 h. This must result from the small amount of

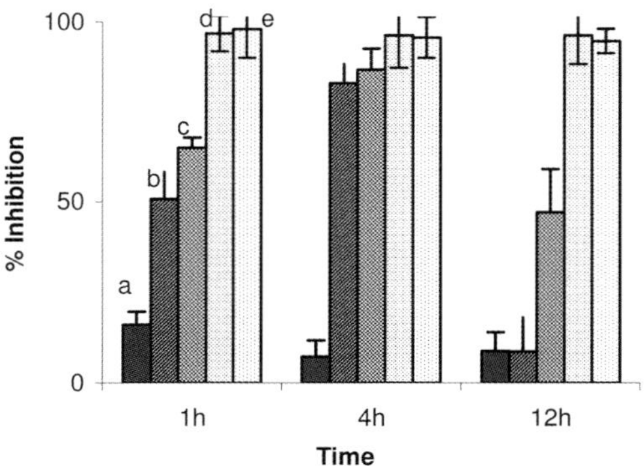

Fig. 14.33 Inhibition of TNF-α secretion by THP-1 cells, incubated with 10 ng mL^{-1} of LPS, grown on multilayered films containing cCDPx complexes. Cells were deposited on the multilayered films: (a) PEI-(PGA-PL^L)$_6$-Px, (b) PEI-(PGA^L-PL^L)$_6$-cCDPx, (c) PEI-(PGA^L-PL^L)$_6$-cCDPx-PL^L-(PGA-PL^L)$_3$, (d) PEI-(PGA^L-PL^L)$_3$-cCDPx-PL^L-(PGA^L-PL^L)$_3$-cCDPx-PL^L-(PGA^L-PL^L)$_3$, (e) PEI-(PGA^L-(PL^L-cCDPx)$_3$-PL^L. After 1, 4 or 12 h incubation, TNF-α levels were measured. The results are the means of three measurements and the vertical bars correspond to the standard deviation.

Px adsorbed on the film (not detectable by QCM-D). In architecture (b) (cCDPx complex adsorbed on the film surface; see legend of Fig. 14.33) the anti-inflammatory activity becomes quite important with a maximum of 83% inhibition after 4 h. The value is not entirely optimal, probably due to a dose effect. The activity decreases then significantly with time (9% after 12 h). This could be the consequence of the consumption, desorption or degradation of the exposed complex. When the complex is embedded under 3 (PGA^L-PL^L) bilayers (architecture (c); see legend of Fig. 14.33), a similar TNF-α inhibition level as that for adsorbed complexes is found at 4 h. The inhibition level also decreases over longer times but in a less significant way (47% after 12 h compared to 9% for adsorbed complexes) which can be explained by some protective effect of the embedding, delaying the degradation processes. Finally, no significant pro-inflammatory TNF-α production is found when the cells are deposited on PEI-(PGA^L-PL^L)$_3$-cCDPx-PL^L-(PGA^L-PL^L)$_3$-cCDPx -PL^L-(PGA^L-PL^L)$_3$ films (architecture (d); see legend of Fig. 14.33) for contact times lasting from 1 to 12 h. Here maximal efficiency is obtained already at 1 h, due to the increased dose of the complex. These results show the anti-inflammatory efficiency of such films and the long-term stability of their activity. Similar results were obtained on PEI-PGA^L-(PL^L-cCDPx)$_3$ (architecture (e); see legend of Fig. 14.33).

Recently, we demonstrated that cells can interact with proteins incorporated in (PGA^L-PL^L)$_n$ polyelectrolyte multilayer films and elucidated the mechanism by which they come into contact with the active protein in the case of monocytic cells. Our data showed that cells communicate with embedded proteins through local film degradation [88]. A similar mechanism could occur here. It is also possible that a slow cCDPx diffusion through the multilayer could continuously activate the adhering cells. In any case, the present results show that the cCDPx functionalized multilayers act as active coatings to control inflammation. Moreover, anti-inflammatory activity can be tuned in time by an appropriate architecture choice.

We investigated the solubility and stability of some charged cyclodextrin–piroxicam (cCD-Px) complexes together with the dynamics of the piroxicam complex formation in solution. We observed the presence of three different prototropic piroxicam forms in solution during the first 30 min after mixing Px with cCD. After 1 h, one obtains predominantly the neutral forms of the complexes in solution [91]. We demonstrated also that these neutral complexed Px forms preserve their anti-inflammatory activity.

Polyelectrolyte film architectures containing charged cyclodextrin–piroxicam complexes have been developed here as potential systems for biomaterial coatings with local anti-inflammatory drug delivery. The buildup of the films was followed by QCM-D and ATR-FTIR techniques and the presence of Px is well demonstrated by analyzing FT-Raman spectra of the buildups. The successive adsorption of (PL^L-cCDPx) bilayers or (PL^L-PGA^L-PL^L-cCDPx) layers seems to induce a gradual increase in the zwitterion marker band intensity, contrary to what was observed for piroxicam in solution.

We investigated the anti-inflammatory activity and long lasting efficiency of this formulation. These polypeptide multilayer films containing cyclodextrin–piroxicam complexes were shown to act as efficient reservoir devices for piroxicam. Moreover, by adjusting the architecture, one can control the time over which the film is active. Such modularly functionalized coatings could become, in the future, potent tools for the modification of biomaterial surfaces as applied to implants, prostheses or in tissue engineering.

14.6 Cross-linked Polysaccharide Multilayers: Control of Biodegradation and Cell Adhesion

Film bioactivity is an important aspect of the film properties. The PEM mechanical properties are another important aspect. In this section, emphasis will be put on the possibility of cross-linking films thereby considerably changing their physicochemical properties, biodegradability and cell adhesive properties.

In recent years, the use of natural polyelectrolytes such as dextran, alginate, heparin, hyaluronan and chitosan has emerged [43, 46, 130, 131]. On account of their biocompatibility and non-toxicity, these latter films constitute a rapidly expanding field with great potential for applications like preparation of bioactive and biomimetic coatings [46, 130, 132] and preparation of drug release vehicles [131, 133]. In terms of cell adhesion, linearly growing and stiff films (in a "glassy" state) are often good substrates [134], whereas exponentially growing and "gel-like" films were always found to be bad substrates for sustaining cell adhesion [43, 46]. It has to be noted that the polyamino acids and polysaccharide films are much more hydrated than synthetic PEM films, their hydration varying from 60 to 95% for hyaluronan-containing films [135, 136].

The aim of this section is not to survey the literature concerning cell adhesion (a brief review is given in [136]) but rather to examine, for a given type of film, how it is possible to tune its cell adhesive properties.

For thin films (nanometers to tens of nanometers in thickness), the film surface chemistry (nature of outermost layer, presence of adhesive peptides) does clearly influence cell adhesion [125], as does hydrophobicity/hydrophilicity [137], which can be controlled by carefully choosing the polymers used in the buildup. The grafting of adhesive peptides such as RGD is another way to change the film adhesive properties [138, 139]. Berg et al. grafted the RGD motif to PAH-ending micro-patterned films and showed an enhanced fibroblast adhesion. In our group, a 15-mer peptide containing the RGD motif was grafted to PL^{L}/PGA^{L} films. The PGA^{L}-15mer ending films were found to be highly adhesive for primary osteoblasts [139]. Thicker films (a few hundred nanometers to several micrometers) including films made of polysaccharides are highly non-adhesive for cells. Two hypotheses have been proposed to explain this lack of adhesion: (i) the high swellability of these films and their high hydration [125], (ii) the film softness: the films would be too soft for the cells to create stable and strong anchors [140]. This latter hypothesis is related to recent works on polyacrylamide, polydimethylsiloxane and alginate gels showing that the cell matrix should be sufficiently stiff to sustain cell anchoring [141].

In terms of biocompatibility, layer-by-layer films made of PEI, PSS, and PAH, although favorable to cell adhesion *in vitro*, may not be appropriate for surface modifications of medical devices for *in vivo* purposes [142]. For this reason, precisely biocompatible polymers (synthetic polypeptides or natural polysaccharrides) could be of greater interest. Chitosan is the second most abundant polysaccharide. Hyaluronan possesses lubricating functions in the cartilage, participates in the control of tissue hydration, water transport, and in the inflammatory response after a trauma [143]. Chitosan and hyaluronan can be easily chemically modified [144–146] and coupled to various molecules such as cell-targeted prodrugs [147], carbohydrates [148], which could be released during film hydrolysis.

In this section, we will more particularly focus on the properties of films made of polysaccharides with hyaluronan as polyanion and poly(L-lysine) or chitosan as polycations, i.e. (PL^{L}/HA and CHI/HA) films. We will show that film cross-linking via carbodiimide chemistry does completely change the film physicochemical properties, its stiffness, and also allows one to modulate its cell adhesive properties and its biodegradability.

14.6.1 Cross-linking of the Films and the Physicochemical Consequences

All the experimental details are given in the articles cited throughout this section. Recently, several studies suggested that mechanical properties of the films have a great influence on cell adhesion, both for cell lines and for primary cells

[149–151]. Strategies for increasing film stiffness include incorporation of nanocolloids in the films [149], building films at different pH for a given polycation/polyanion pair [151] or, as proposed here, chemical cross-linking.

For the chemical cross-linking, we used a protocol based on the carbodiimide chemistry, i.e. 1-ethyl-3-(3-dimethylamino-propyl)carbodiimide (EDC) in combination with *N*-hydroxysulfosuccinimide (s-NHS), where the ammonium groups are covalently linked to carboxylic groups via amide bonds. This reaction can be easily followed by Fourier transform infrared spectroscopy in total attenuated reflection mode (ATR-FTIR) [150]. The carboxylate, saccharides and amide bands can be clearly identified. Figure 14.34 shows a typical spectrum of a $(PL^L/HA)_8$ film prior to and after cross-linking. During the reaction, the carboxylic peaks (at 1606 and 1412 cm^{-1} and the band at 1675–1720 cm^{-1} decreased, whereas at the same time the intensity of other bands in the amide I (1600–1675 cm^{-1}) and amide II (1400–1500 cm^{-1}) regions increased. The decrease in the carboxylic peaks of HA and the concomitant increase in the amide bands suggest that reticulation did effectively occur due to the reaction between HA carboxyl groups and PL^L ammonium groups.

Such a cross-linking procedure can be applied to any other type of film, provided that it contains carboxylic and ammonium groups. Therefore, we also succeeded in cross-linking PL^L/PGA^L films [139] and CHI/HA films, which, beside amide linkage, exhibit ester linkages involving hydroxyl groups from polysaccharides and carboxylic or acidic anhydride formed between carboxylic groups [152].

As the PL^L/HA and CHI/HA films can be easily observed by CLSM using labeled polyelectrolytes as ending layers (respectively $PL^{L\text{-}FITC}$ and CHI^{FITC}), fluo-

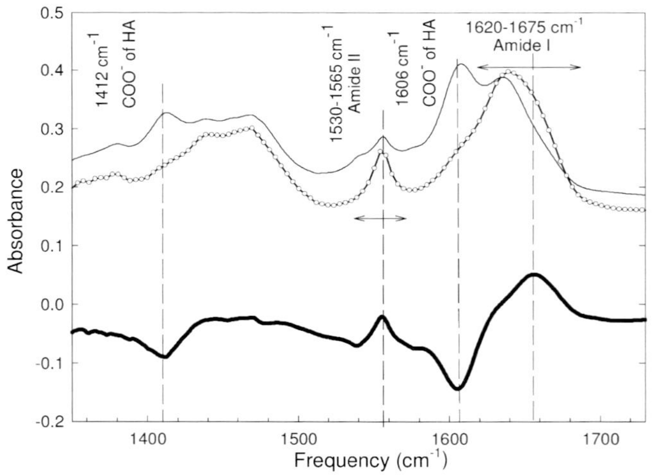

Fig. 14.34 ATR-FTIR spectra of a native and a cross-linked $(PL^L/HA)_8$ film before (—) and after the cross-linking procedure and the final rinsing step (–o–). The difference between the two spectra (before and after cross-linking) is also represented (thick black line).

rescent recovery after photobleaching (FRAP) experiments were performed by CLSM and a diffusion coefficient of $\approx 0.2\ \mu m^2\ s^{-1}$ was estimated for native films [153]. For CL films, diffusion is no longer visible on the time scale of the experiment (about 1 h).

An important change upon cross-linking is a change in film stiffness (represented by the film Young's modulus E_0) as measured by AFM nanoindentation experiments using a colloidal probe. Film stiffness is about eight times higher for a CL film as compared to a native film [140] (when EDC concentration is 35 mg mL^{-1}).

14.6.2 Control of Biodegradability

Films made of polysaccharides are intrinsically biodegradable, due to the presence *in vivo* in tissues and fluids of specific enzymes, which can cleave these natural polymers. This is the case for hyaluronan by hyaluronidase [154], for chitosan by chitosanase and other enzymes.

Native polysaccharide PEM films are degraded over a time scale ranging from a few hours to one day, depending on the enzyme concentration. In this section, we will focus on films CL at a precise EDC concentration (e.g. 35 mg mL^{-1}). These CL films are much more resistant to all enzymatic degradation (Fig. 14.35). No degradation was observed after contact with lysozyme. For α-amylase, a small superficial degradation was observable on the top view. For PL^L/HA films, THP-1 macrophages are able to degrade *in vitro* the native films but not the CL ones over a two-day period.

We performed *in vivo* experiments in the rat oral cavity, where $(CHI/HA)_{24}$-CHI^{FITC} coated PMMA disks were sutured to rats' cheek. After different time

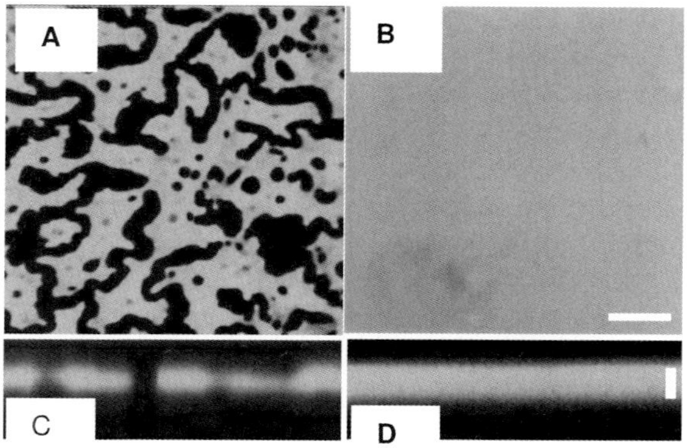

Fig. 14.35 Degradation of a $(CHI/HA)_{24}$-CHI^{FITC} film after contact with hyaluronidase (type IV, 500 U mL^{-1}) observed by CLSM after 4 h of contact for a native (A, C) and a cross-linked film (B, D). Top views ($56\times56\ \mu m^2$) of the films (A, B) (scale bar 10 μm) and vertical sections are shown (C, D). (Film thickness is about 6 μm, white line).

periods varying from 6 h to 3 days in the mouth, the disks were recovered and both sides were observed using a fluorescence stereomicroscope and CLSM. The native films are rapidly degraded in the mouth on both cheek and tongue side. On the cheek side, only 12% of the native film remains after 6 h. The cross-linked films are more resistant to degradation. Considering the cheek sides, after 6 h, more than 80% of the film remains. Observations of the disks implanted in different rats after 2 or 3 days show approximately 65% and 75% of film remaining, respectively. These data indicate that a large fraction of the cross-linked film is still on the polymer disks after 3 days of implantation *in vivo* [152]. It has to be noted that the film-coated polymer disks are not only subjected *in vivo* to saliva degradation but also to a mechanical degradation due to the rat's chewing.

In vivo experiments on cross-linked $(CHI/HA)_{24}$-CHI^{FITC} were also performed in the mouse peritoneal environment [155]. This location was chosen to study the degradability of the films in contact with macrophages *in vivo*, since macrophages are the main cell type in the mouse peritoneal cavity. After 6 days *in vivo*, the slides were recovered and the outer side was observed using microscopy techniques. Few macrophages adhered on the native films ($1.2 \pm 0.4 \times 10^5$ cells cm^{-2}) and about 40% more adhered on the film cross-linked at 5 mg mL^{-1} ($1.7 \pm 0.4 \times 10^5$ cells cm^{-2}). The films cross-linked at 35 mg mL^{-1} and at 150 mg mL^{-1} attracted the highest number of macrophages, with respectively $3.1 \pm 0.4 \times 10^5$ cells cm^{-2} and $4.5 \pm 0.2 \times 10^5$ cells cm^{-2}, which represents, for this latter condition, about fourfold the number of macrophages on the native films (Fig. 14.36). The film cross-linked at the highest EDC concentration (150 mg mL^{-1}) was the only one that led to fibrous tissue formation after 21 days *in vivo*. On the native film, only 95% of the film was degraded within 6 days, whereas about 85% of the CL films were still present. Film degradation also depended on EDC concentration, the degradation being enhanced at low EDC concentration (5 mg mL^{-1}). All together, these results suggest that macrophages adhesion and film resistance *in vivo* are increased when the EDC concentration is increased, with however an optimum at the 35 mg mL^{-1} EDC concentration.

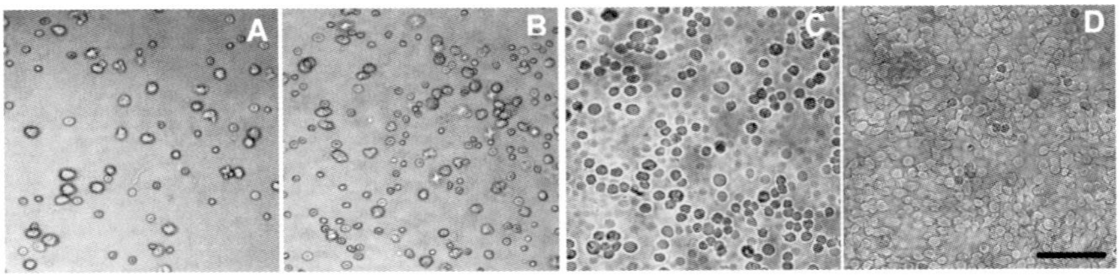

Fig. 14.36 Top-view observations in optical microscopy of macrophages that have adhered on the native and cross-linked $(CHI/HA)_{24}$-CHI^{FITC}-coated slides, which have been implanted in rat for 6 days. Images were taken after explantation. (A) a native film and cross-linked films with increased EDC concentration: 5 (B), 35 (C) and 150 mg mL^{-1} (D). Image sizes are 230×230 μm^2 (scale bar 50 μm).

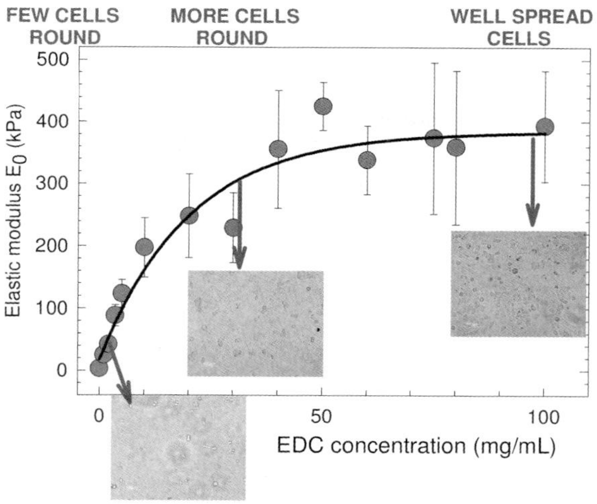

Fig. 14.37 Surface Young's modulus E_0 determined by the AFM nanoindentation technique for various EDC concentrations up to 100 mg mL^{-1}. An exponential asymptotic fit to the data is also represented (thick line). The error bars represent the standard deviation of 6 to 16 measurements of E_0 corresponding to various approach velocities. Chondrosarcoma cells images taken after three days in culture are given for some EDC concentrations.

The degradability properties of the CHI/HA films may also be related to their mechanical properties as was recently evidenced for cell adhesion on PL^L/HA films [140]. The film stiffness was estimated by AFM nano-indentation experiments using a colloidal probe. The Young's modulus E_0 increased as the EDC concentration and tended to a plateau (Fig. 14.37). Thus, it is possible to tune film stiffness over two orders of magnitude (from 3 to 400 kPa) simply by changing the cross-linker concentration [156].

14.6.3 Control of Cell Adhesion

With respect to cellular adhesion, we have evidenced that CL films are a much better substrate for cell adhesion than native ones. This was demonstrated for various cell types including chondrosarcoma cells, primary chondrocytes and motoneurons [157].

Whereas only a few primary chondrocytes remained on the native $(PL^L/HA)_{12}$ films after 5 days in culture, the CL films are nicely covered by cells (Fig. 14.38). In addition, the cells are able to deform the uncross-linked films in contrast to the CL ones. It has to be noted that adding a few uncross-linked layer pairs renders the film non-adhesive again. Noticeably, if one compares the effect of a cell adhesion peptide (such as a peptide containing the RGD motif) to that of pure cross-linked films, one can notice that cross-linking the film has

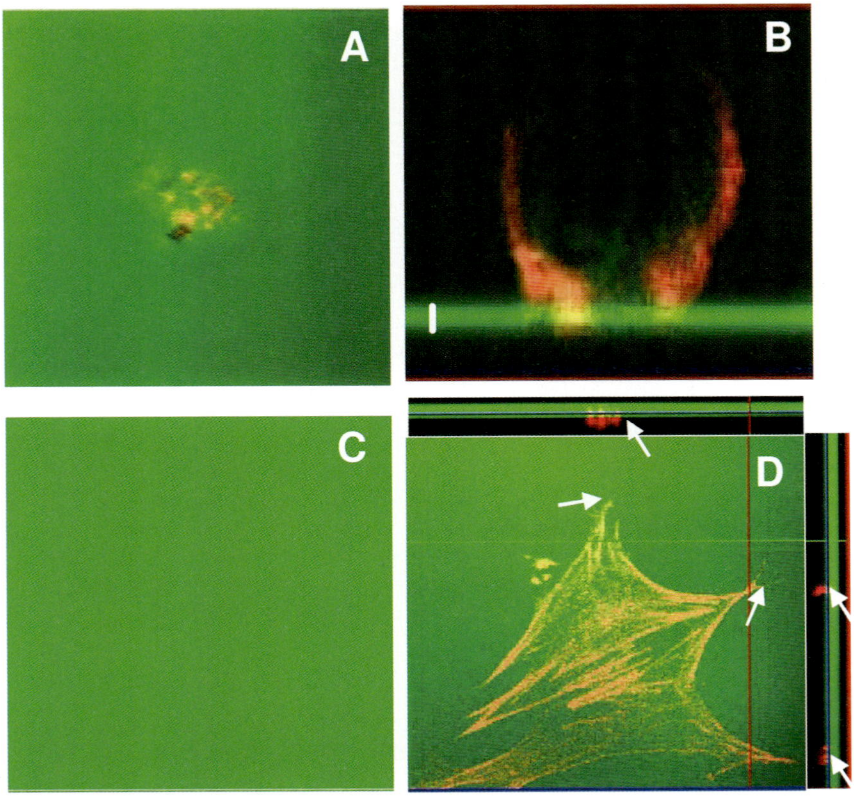

Fig. 14.38 CLSM observations of the adhesion of primary chondrocytes on top of native and cross-linked $(PL^{L}/HA)_{24}$-$PL^{L\text{-}FITC}$ films. Native films: (A) View in a *z*-section in the film, near the glass substrate, two channels being superposed. The green fluorescence comes from the film and the three red areas are due to cells. The cell was able to deform the film, as can be seen by the imprints of the cell in the film. (B) Side view of a cell reconstructed from a *z*-stack acquisition. Some green fluorescence is visible over the cell membrane as well as inside the cell suggesting the diffusion of the $PL^{L\text{-}FITC}$. Cross-linked films: (C) Similar view of the film near the glass substrate: the film appears uniformly green and no holes are visible. (D) View in a *z*-section inside the cell together with two (*x*, *y*)-sections over all the *z*-stack, performed along the two lines drawn on the image. On the upper image and on the side image the projected sections are shown, the (PL^{L}/HA) film appearing in green and the contact points of the cell in yellow (green and red co-localization). These images show the cell anchoring.

a similar or stronger effect than the chemical stimuli on cell proliferation. This was particularly striking for $(PL^{L}$/alginic acid$)_6$ and $(PL^{L}$/galacturonic acid$)_6$ films. Interestingly, it is possible to combine these two effects by adding a final layer of RGD grafted poly(glutamic) acid on top of a cross-linked film [139].

In our most recent work, we have focused on the possibility of varying the extent of cross-linking in the films. Toward this end, the EDC concentration has been varied over a large range from 1 to 200 mg mL^{-1}. For PL^{L}/HA films, ATR-FTIR spectra showed that the EDC consumption follows an exponential asymptotic law as a function of the initial EDC concentration. The plateau is reached at about 70 mg mL^{-1}, which indicates that all the EDC molecules are not consumed for the highest EDC concentrations. At high EDC concentration, new peaks show up (at 1410 cm^{-1} and 1620 cm^{-1}). This can explain why there is no monotonic increase in the peaks as a function of the EDC concentration. Similar results have been observed for CHI/HA films.

Noticeably, chondrosarcoma cell adhesion strongly depends on the EDC concentration: at low EDC concentration, only a few cells adhere to the films. At the highest EDC concentrations (above 100 mg mL^{-1}), all the cells are well spread. Not only did the initial cell number increase but also the cell spreading increased as a function of EDC concentration [158]. After 5 days, the number of living cells has increased for all the EDC concentrations. Here again, the trend for an enhanced viability at elevated EDC concentration is clearly visible (Fig. 14.37).

In summary, film cross-linking appears as an alternative strategy for modifying cell adhesion while at the same time increasing film stiffness. We have shown that it is possible to cross-link various types of PEM films provided that they possess carboxylic and amine groups. One of the major consequences of this modulated cross-linking is tunable film stiffness. For (PL^{L}/HA) and (CHI/HA) films, this stiffness can be varied over two orders of magnitude. Film degradation can be tuned by controlling film cross-linking. A moderate cross-linking seems to be of greater interest with respect to biodegradability and inflammatory response. We have shown that the film mechanical properties have a strong influence on cell behavior, stiff films being preferred in terms of cell adhesion and spreading. Such films are of interest for further studies including cytoskeletal organization, motility and cell differentiation.

Recently, we succeeded in combining mechanical, biodegradability and bioactivity properties. Cross-linked films were loaded with active molecules such as sodium diclofenac (an anti-inflammatory molecule) and paclitaxel (a cancer drug that has an action on the polymerization of the microtubules), the loading amount depending directly on the film thickness. Furthermore, the viability of human colonic adenocarcinoma cells (HT29) was found to be drastically decreased on cross-linked (PL^{L}/HA)$_{12}$ films loaded with paclitaxel [159].

List of Abbreviations

poly(ethylene imine)	PEI
poly(styrene sulfonate)	PSS
poly(allyl amine)	PAH
poly(L-lysine)	PL^{L}
poly(D-lysine)	PL^{D}
poly(L-glutamic acid)	PGA^{L}

poly(D-glutamic acid)	PGA^D
poly(lactic-*co*-glycolic) acid	PLGA
poly(L-aspartic acid)	PAA^L
hyaluronan	HA
chitosan	CHI

References

1 Decher, G. Layered Nanoarchitectures via Directed Assembly of Anionic and Cationic Molecules, in *Templating, Self-Assembly and Self-Organization*, Sauvage, J.-P., Hosseini, M. W., Eds., Pergamon Press Oxford, **1996**, Vol. 9, pp. 507–528.

2 Decher, G. Fuzzy nanoassemblies: Toward layered polymeric multicomposites, *Science* **1997**, *277*, 1232–1237.

3 Decher, G., Eckle, M., Schmitt, J., Struth, B. Layer-by-Layer assembled multicomposite films, *Curr. Opin. Colloid Interface Sci.* **1998**, *3*, 32–39.

4 Bertrand, P., Jonas, A., Laschewsky, A., Legras, R. Ultrathin polymer coatings by complexation of polyelectrolytes at interfaces: suitable materials, structure and properties, *Macromol. Rapid Commun.* **2000**, *21*, 319–348.

5 Hammond, P. T. Recent explorations in electrostatic multilayer thin film assembly, *Curr. Opin. Colloid Interface Sci.* **1999**, *4*, 430–442.

6 Paterno, L. G., Mattoso, H. L. C., de Oliveira, O. N. Ultrathin polymer films produced by the self-assembly technique: Preparation, properties and applications, *Quim. Nova* **2001**, *24*, 228–235.

7 Caruso, F. Nanoengineering of particle surfaces, *Adv. Mater.* **2001**, *13*, 11–22.

8 Freemantle, M. Science & Technology – Polyelectrolyte Multilayers, *Chem. Eng. News* **2002**, *80*, 44–48.

9 Gorman, J. Layered Approach: A simple technique for making thin coatings is poised to shift from curiosity to commodity, *Sci. News* **2003**, *164*, 91.

10 Decher, G., Schlenoff, J. B. *Multilayer Thin Films: Sequential Assembly of Nanocomposite Materials*, Wiley-VCH, Weinheim, **2003**.

11 Campas, M., O'Sullivan, C. Layer-by-layer biomolecular assemblies for enzyme sensors, immunosensing, and nano architectures, *Anal. Lett.* **2003**, *36*, 2551–2569.

12 Rusling, J. F., Forster, R. J. Electrochemical catalysis with redox polymer and polyion-protein films, *J. Colloid Interface Sci.* **2003**, *262*, 1–15.

13 von Klitzing, R., Wong, J. E., Jäger, W., Steitz, R. Short range interactions in polyelectrolyte multilayers, *Curr. Opin. Colloid Interface Sci.* **2004**, *9*, 158–162.

14 Burke, S. E., Barrett, C. J. Controlling the physicochemical properties of weak polyelectrolyte multilayer films through acid/base equilibria, *Pure Appl. Chem.* **2004**, *76*, 1387–1398.

15 Hammond, P. T. Form and function in multilayer assembly: New applications at the nanoscale, *Adv. Mater.* **2004**, *16*, 1271–1293.

16 Haynie, D. T., Zhang, L., Rudra, J. S., Zhao, W. H., Zhong, Y., Palath, N. Polypeptide multilayer films, *Biomacromolecules* **2005**, *6*, 2895–2913.

17 Dobrynin, A. V., Rubinstein, M. Theory of polyelectrolytes in solutions and at surfaces, *Prog. Polym. Sci.* **2005**, *30*, 1049–1118.

18 Sukhishvili, S. A. Responsive polymer films and capsules via layer-by-layer assembly, *Curr. Opin. Colloid Interface Sci.* **2005**, *10*, 37–44.

19 Kamande, M. W., Fletcher, K. A., Lowry, M., Warner, I. M. Capillary electrochromatography using polyelectrolyte multilayer coatings, *J. Sep. Sci.* **2005**, *28*, 710–718.

20 Ariga, K., Nakanishi, T., Michinobu, T. Immobilization of biomaterials to nanoassembled films self-assembled monolayers, Langmuir-Blodgett films, and layer-by-layer assemblies, and their related functions, *J. Nanosci. Nanotechnol.* **2006**, *6*, 2278–2301.

21 Haynie, D. T. Physics of polypeptide multilayer films, *J. Biomed. Mater. Res. B* **2006**, *78B*, 243–252.

22 Ariga, K., Vinu, A., Miyahara, M. Recent progresses in bio-inorganic nanohybrids, *Curr. Nanosci.* **2006**, *2*, 197–210.

23 Jiang, C. Y., Tsukruk, V. V. Freestanding nanostructures via layer-by-layer assembly, *Adv. Mater.* **2006**, *18*, 829–840.

24 Ladam, G., Schaad, P., Voegel, J.-C., Schaaf, P., Decher, G., Cuisinier, F. J. G. In situ determination of the structural properties of initially deposited polyelectrolyte multilayers, *Langmuir* **2000**, *16*, 1249–1255.

25 Picart, C., Mutterer, J., Richert, L., Luo, Y., Prestwich, G. D., Schaaf, P., Voegel, J.-C., Lavalle, P. Molecular basis for the explanation of the exponential growth of polyelectrolyte multilayers, *Proc. Natl. Acad. Sci. USA* **2002**, *99*, 12531–12535.

26 Lösche, M., Schmitt, J., Decher, G., Bouwman, W. G., Kjaer, K. Detailed structure of molecularly thin polyelectrolyte multilayer films on solid substrates as revealed by neutron reflectometry, *Macromolecules* **1998**, *31*, 8893–8906.

27 Schlenoff, J. B., Dubas, S. T., Farhat, T. Sprayed polyelectrolyte multilayers, *Langmuir* **2000**, *16*, 9968–9969.

28 Lee, S.-S., Hong, J.-D., Kim, C. H., Kim, K., Koo, J. P., Lee, K.-B. Layer-by-layer deposited multilayer assemblies of ionene-type polyelectrolytes based on the spin-coating method, *Macromolecules* **2001**, *34*, 5358–5360.

29 Cho, J., Char, K., Hong, J.-D., Lee, K.-B. Fabrication of highly ordered multilayer films using a spin self-assembly method, *Adv. Mater.* **2001**, *13*, 1076–1078.

30 Chiarelli, P. A., Johal, M. S., Casson, J. L., Roberts, J. B., Robinson, J. M., Wang, H. L. Controlled fabrication of polyelectrolyte multilayer thin films using spin-assembly, *Adv. Mater.* **2001**, *13*, 1167–1171.

31 Chiarelli, P. A., Johal, M. S., Holmes, D. J., Casson, J. L., Robinson, J. M., Wang, H. L. Polyelectrolyte spin-assembly, *Langmuir* **2002**, *18*, 168–173.

32 Izquierdo, A., Ono, S., Voegel, J.-C., Schaaf, P., Decher, G. Dipping versus spraying: Exploring the deposition conditions for speeding up layer-by-layer assembly, *Langmuir* **2005**, *21*, 7558–7567.

33 Donath, E., Sukhorukov, G. B., Caruso, F., Davis, S. A., Möhwald, H. Novel hollow polymer shells by colloid – Templated assembly of polyelectrolytes, *Angew. Chem. Int. Ed.* **1998**, *37*, 2202–2205.

34 Antipov, A. A., Sukhorukov, G. B. Polyelectrolyte multilayer capsules as vehicles with tunable permeability, *Adv. Colloid Interface Sci.* **2004**, *111*, 49–61.

35 Shi, X. Y., Shen, M. W., Möhwald, H. Polyelectrolyte multilayer nanoreactors toward the synthesis of diverse nanostructured materials, *Prog. Polym. Sci.* **2004**, *29*, 987–1019.

36 Vinogradova, O. I. Mechanical properties of polyelectrolyte multilayer microcapsules, *J. Phys.: Condens. Matter* **2004**, *16*, R1105–R1134.

37 Fery, A., Dubreuil, F., Möhwald, H. Mechanics of artificial microcapsules, *New J. Phys.* **2004**, *6*.

38 Gittins, D. I., Caruso, F. Multilayered polymer nanocapsules derived from gold nanoparticle templates, *Adv. Mater.* **2000**, *12*, 1947–1949.

39 Gittins, D. I., Caruso, F. Tailoring the polyelectrolyte coating of metal nanoparticles, *J. Phys. Chem. B* **2001**, *105*, 6846–6852.

40 Kunze, K. K., Netz, R. R. Salt-induced DNA-histone complexation, *Phys. Rev. Lett.* **2000**, *85*, 4389–4392.

41 Schneider, G., Decher, G. From functional core/shell nanoparticles prepared via layer-by-layer deposition to empty nanospheres, *Nano Lett.* **2004**, *4*, 1833–1839.

42 Schneider, G., Decher, G., Nerambourg, N., Praho, R., Werts, M. H. V., Blanchard-Desce, M. Distance-dependent fluorescence quenching on gold nanoparticles ensheathed with layer-by-layer assembled polyelectrolytes, *Nano Lett.* **2006**, *6*, 530–536.

43 Elbert, D. L., Herbert, C. B., Hubbell, J. A. Thin polymer layers formed by polyelectrolyte multilayer techniques on biological surfaces, *Langmuir* **1999**, *15*, 5355–5362.

44 Picart, C., Lavalle, P., Hubert, P., Cuisinier, F. J. G., Decher, G., Schaaf, P., Voegel, J.-C. Buildup mechanism for polyL-lysine/hyaluronic acid films onto a solid surface, *Langmuir* **2001**, *17*, 7414–7424.

45 Lavalle, P., Gergely, C., Cuisinier, F. J. G., Decher, G., Schaaf, P., Voegel, J.-C., Picart, C. Comparison of the structure of polyelectrolyte multilayer films exhibiting a linear and an exponential growth regime: An in situ atomic force microscopy study, *Macromolecules* **2002**, *35*, 4458–4465.

46 Richert, L., Lavalle, P., Payan, E., Shu, X. Z., Prestwich, G. D., Stoltz, J.-F., Schaaf, P., Voegel, J.-C., Picart, C. Layer-by-layer buildup of polysaccharide films: Physical chemistry and cellular adhesion aspects, *Langmuir* **2004**, *20*, 448–458.

47 Lavalle, P., Picart, C., Mutterer, J., Gergely, C., Reiss, H., Voegel, J.-C., Senger, B., Schaaf, P. Modeling the buildup of polyelectrolyte multilayer films having exponential growth, *J. Phys. Chem. B* **2004**, *108*, 635–648.

48 Michel, M., Izquierdo, A., Decher, G., Voegel, J.-C., Schaaf, P., Ball, V. Layer by layer self-assembled polyelectrolyte multilayers with embedded phospholipid vesicles obtained by spraying: Integrity of the vesicles, **2005**, *21*, 7854–7859.

49 Salomäki, M., Vinokurov, I. A., Kankare, J. Effect of temperature on the buildup of polyelectrolyte multilayers, *Langmuir* **2005**, *21*, 11232–11240.

50 Boulmedais, F., Ball, V., Schwinté, P., Frisch, B., Schaaf, P., Voegel, J.-C. Build-up of exponentially growing multilayer polypeptide films with internal secondary structure, *Langmuir* **2003**, *19*, 440–445.

51 Boulmedais, F., Bozonnet, M., Schwinté, P., Voegel, J.-C., Schaaf, P. Multilayered polypeptide films: secondary structures and effect of various stresses, *Langmuir* **2003**, *19*, 9873–9882.

52 Burke, S. E., Barrett, C. J. Swelling behavior of hyaluronic acid/polyallylamine hydrochloride multilayer films, *Biomacromolecules* **2005**, *6*, 1419–1428.

53 Collin, D., Lavalle, P., Méndez Garza, J., Voegel, J.-C., Schaaf, P., Martinoty, P. Mechanical properties of cross-linked hyaluronic acid/polyL-lysine multilayer films, *Macromolecules* **2004**, *37*, 10195–10198.

54 Kulcsár, A., Lavalle, P., Voegel, J.-C., Schaaf, P., Kékicheff, P. Interactions between two polyelectrolyte multilayers investigated by the surface force apparatus, *Langmuir* **2004**, *20*, 282–286.

55 Kulcsár, A., Voegel, J.-C., Schaaf, P., Kékicheff, P. Glassy state of polystyrene sulfonate/polyallylamine polyelectrolyte multilayers revealed by the surface force apparatus, *Langmuir* **2005**, *21*, 1166–1170.

56 Hübsch, E., Ball, V., Senger, B., Decher, G., Voegel, J.-C., Schaaf, P. Controlling the growth regime of polyelectrolyte multilayer films: Changing from exponential to linear growth by adjusting the composition of polyelectrolyte mixtures, *Langmuir* **2004**, *20*, 1980–1985.

57 Debreczeny, M., Ball, V., Boulmedais, F., Szalontai, B., Voegel, J.-C., Schaaf, P. Multilayers built from two component polyanions and single component polycation solutions: A way to engineer films with desired secondary structure, *J. Phys. Chem. B* **2003**, *107*, 12734–12739.

58 Benkirane-Jessel, N., Lavalle, P., Hübsch, E., Holl, V., Senger, B., Haïkel, Y., Voegel, J.-C., Ogier, J., Schaaf, P. Short-time tuning of the biological activity of functionalized polyelectrolyte multilayers, *Adv. Funct. Mater.* **2005**, *15*, 648–654.

59 Ramsden, J. J., Lvov, Y. M., Decher, G. Determination of optical constants of molecular films assembled via alternate polyion adsorption, *Thin Solid Films* **1995**, *254*, 246–251.

60 Voinova, M. V., Rodahl, M., Jonson, M., Kasemo, B. Viscoelastic acoustic response of layered polymer films at fluid-solid interfaces: Continuum mechanics approach, *Phys. Scr.* **1999**, *59*, 391–396.

61 Méndez Garza, J., Schaaf, P., Müller, S., Ball, V., Stoltz, J.-F., Voegel, J.-C., Lavalle, P. Multicompartment films made of alternate polyelectrolyte multilayers of exponential and linear growth, *Langmuir* **2004**, *20*, 7298–7302.

62 Méndez Garza, J., Jessel, N., Ladam, G., Dupray, V., Müller, S., Stoltz, J.-F., Schaaf, P., Voegel, J.-C., Lavalle, P. Polyelectrolyte multilayers and degradable polymer layers as multicompartment films, *Langmuir* **2005**, *21*, 12372–12377.

63 Grayson, A. C. R., Choi, I. S., Tyler, B. M., Wang, P. P., Brem, H., Cima, M. J., Langer, R. Multi-pulse drug delivery from a

resorbable polymeric microchip device, *Nat. Mater.* **2003**, *2*, 767–772.
64 Chluba, J., Voegel, J.-C., Decher, G., Erbacher, P., Schaaf, P., Ogier, J. Peptide hormone covalently bound to polyelectrolytes and embedded into multilayer architectures conserving full biological activity, *Biomacromolecules* **2001**, *2*, 800–805.
65 Thierry, B., Kujawa, P., Tkaczyk, C., Winnik, F. M., Bilodeau, L., Tabrizian, M. Delivery platform for hydrophobic drugs: Prodrug approach combined with self-assembled multilayers, *J. Am. Chem. Soc.* **2005**, *127*, 1626–1627.
66 Lvov, Y., Decher, G., Sukhorukov, G. B. Assembly of thin films by means of successive deposition of alternate layers of DNA and polyallylamine, *Macromolecules* **1993**, *26*, 5396–5399.
67 Lvov, Y., Ariga, K., Ichinose, I., Kunitake, T. Assembly of multicomponent protein films by means of electrostatic layer-by-layer adsorption, *J. Am. Chem. Soc.* **1995**, *117*, 6117–6123.
68 Lvov, Y., Haas, H., Decher, G., Möhwald, H., Mikhailov, A., Mtchedlishvily, B., Morgunova, E., Vainshtein, B. Successive deposition of alternate layers of polyelectrolytes and a charged virus, *Langmuir* **1994**, *10*, 4232–4236.
69 Kotov, N. A., Dékány, I., Fendler, J. H. Layer-by-layer self-assembly of polyelectrolyte-semiconductor nanoparticle composite films, *J. Phys. Chem.* **1995**, *99*, 13065–13069.
70 Caruso, F., Lichtenfeld, H., Giersig, M., Möhwald, H. Electrostatic self-assembly of silica nanoparticle-polyelectrolyte multilayers on polystyrene latex particles, *J. Am. Chem. Soc.* **1998**, *120*, 8523–8524.
71 Vazquez, E., Dewitt, D. M., Hammond, P. T., Lynn, D. M. Construction of hydrolytically-degradable thin films via layer-by-layer deposition of degradable polyelectrolytes, *J. Am. Chem. Soc.* **2002**, *124*, 13992–13993.
72 Sukhorukov, G. B., Donath, E., Davis, S., Lichtenfeld, H., Caruso, F., Popov, V. I., Möhwald, H. Stepwise polyelectrolyte assembly on particle surfaces: a novel approach to colloid design, *Polym. Adv. Technol.* **1998**, *9*, 759–767.
73 Sukhorukov, G. B., Donath, E., Lichtenfeld, H., Knippel, E., Knippel, M., Budde, A., Möhwald, H. Layer-by-layer self assembly of polyelectrolytes on colloidal particles, *Colloids Surf. A* **1998**, *137*, 253–266.
74 Heuvingh, J., Zappa, M., Fery, A. Salt softening of polyelectrolyte multilayer capsules, *Langmuir* **2005**, *21*, 3165–3171.
75 Radtchenko, I. L., Sukhorukov, G. B., Leporatti, S., Khomutov, G. B., Donath, E., Möhwald, H. Assembly of alternated multivalent ion/polyelectrolyte layers on colloidal particles. Stability of the multilayers and encapsulation of macromolecules into polyelectrolyte capsules, *J. Colloid Interface Sci.* **2000**, *230*, 272–280.
76 Volodkin, D. V., Larionova, N. I., Sukhorukov, G. B. Protein encapsulation via porous $CaCO_3$ microparticles templating, *Biomacromolecules* **2004**, *5*, 1962–1972.
77 Reimhult, E., Höök, F., Kasemo, B. Intact vesicle adsorption and supported biomembrane formation from vesicles in solution: Influence of surface chemistry, vesicle size, temperature, and osmotic pressure, *Langmuir* **2003**, *19*, 1681–1691.
78 Richter, R., Mukhopadhyay, A., Brisson, A. Pathways of lipid vesicle deposition on solid surfaces: A combined QCM-D and AFM study, *Biophys. J.* **2003**, *85*, 3035–3047.
79 Seantier, B., Breffa, C., Félix, O., Decher, G. In situ investigations of the formation of mixed supported lipid bilayers close to the phase transition temperature, *Nano Lett.* **2004**, *4*, 5–10.
80 Katagiri, K., Hamasaki, R., Ariga, K., Kikuchi, J.-i. Layer-by-layer self-assembling of liposomal nanohybrid "Cerasome" on substrates, *Langmuir* **2002**, *18*, 6709–6711.
81 Katagiri, K., Hamasaki, R., Ariga, K., Kikuchi, J.-i. Layered paving of vesicular nanoparticles formed with Cerasome as a bioinspired organic-inorganic hybrid, *J. Am. Chem. Soc.* **2002**, *124*, 7892–7893.
82 Ge, L., Möhwald, H., Li, J. Phospholipid liposomes stabilized by the coverage of polyelectrolyte, *Colloids Surf. A* **2003**, *221*, 49–53.
83 Michel, M., Vautier, D., Voegel, J.-C., Schaaf, P., Ball, V. Incorporation of lipo-

somes into layer by layer self assembled polyelectrolyte multilayers, *Langmuir* **2004**, *20*, 4835–4839.

84 Hübsch, E., Fleith, G., Fatisson, J., Labbé, P., Voegel, J.-C., Schaaf, P., Ball, V. Multivalent ion/polyelectrolyte exchange processes in exponentially growing multilayers, *Langmuir* **2005**, *21*, 3664–3669.

85 Michel, M., Arntz, Y., Fleith, G., Toquant, J., Haikel, Y., Voegel, J.-C., Schaaf, P., Ball, V. Layer-by-layer self-assembled polyelectrolyte multilayers with embedded liposomes: immobilized submicronic reactors for mineralization, *Langmuir* **2006**, *22*, 2358–2364.

86 Nardin, C., Winterhalter, M., Meier, W. Giant free-standing ABA triblock copolymer membranes, *Langmuir* **2000**, *16*, 7708–7712.

87 Nardin, C., Widmer, J., Winterhalter, M., Meier, W. Amphiphilic block copolymer nanocontainers as bioreactors, *Eur. Phys. J. E* **2001**, *4*, 403–410.

88 Jessel, N., Atalar, F., Lavalle, P., Mutterer, J., Decher, G., Schaaf, P., Voegel, J.-C., Ogier, J. Bioactive coatings based on polyelectrolyte multilayer architecture functionalised by embedded proteins, *Adv. Mater.* **2003**, *15*, 692–695.

89 Benkirane-Jessel, N., Lavalle, P., Meyer, F., Audouin, F., Frisch, B., Schaaf, P., Ogier, J., Decher, G., Voegel, J.-C. Control of monocyte morphology on and response to model surfaces for implants equipped with anti-inflammatory agents, *Adv. Mater.* **2004**, *16*, 1507–1511.

90 Benkirane-Jessel, N., Schwinté, P., Donohue, R., Lavalle, P., Boulmedais, F., Darcy, R., Szalontai, B., Ogier, J. Pyridylamino-beta-cyclodextrin as a molecular chaperone for lipopolysaccharide embedded in a multilayered polyelectrolyte architecture, *Adv. Funct. Mater.* **2004**, *14*, 963–969.

91 Benkirane-Jessel, N., Schwinté, P., Falvey, P., Darcy, R., Haïkel, Y., Schaaf, P., Voegel, J.-C., Ogier, J. Buildup of polyelectrolyte multilayer coatings with anti-inflammatory properties based on the embedding of piroxicam-cyclodextrin complexes, *Adv. Funct. Mater.* **2004**, *14*, 174–182.

92 Kong, W., Wang, L. P., Gao, M. L., Zhou, H., Zhang, X., Li, W., Shen, J. C. Immobilized bilayer glucose isomerase in porous trimethylamine polystyrene based on molecular deposition, *J. Chem. Soc., Chem. Commun.* **1994**, 1297–1298.

93 Ladam, G., Schaaf, P., Cuisinier, F. J. G., Decher, G., Voegel, J.-C. Protein adsorption onto auto-assembled polyelectrolyte films, *Langmuir* **2001**, *17*, 878–882.

94 Caruso, F., Niikura, K., Furlong, D. N., Okahata, Y. 2. Assembly of alternating polyelectrolyte and protein multilayer films for immunosensing, *Langmuir* **1997**, *13*, 3427–3433.

95 Schwinté, P., Voegel, J.-C., Picart, C., Haikel, Y., Schaaf, P., Szalontai, B. Stabilizing effects of various polyelectrolyte multilayer films on the structure of adsorbed/embedded fibrinogen molecules: an ATR-FTIR study, *J. Phys. Chem. B* **2001**, *105*, 11906–11916.

96 Kumar, S., Shukla, Y., Prasad, A. K., Verma, A. S., Dwivedi, P. D., Mehrotra, N. K., Ray, P. K. Protection against 7,12-dimethylbenzanthracene-induced tumour initiation by protein A in mouse skin, *Cancer Lett.* **1992**, *61*, 105–110.

97 Verma, A. S., Dwivedi, P. D., Mishra, A., Ray, P. K. Ehrlich's ascites fluid adsorbed over protein A containing Staphylococcus aureus Cowan I produces inhibition of tumor growth, *Immunopharmacol. Immunotoxicol.* **1999**, *21*, 89–108.

98 Subbulakshmi, V., Ghosh, A. K., Das, T., Ray, P. K. Mechanism of protein A-induced amelioration of toxicity of anti-AIDS drug, zidovudine, *Biochem. Biophys. Res. Commun.* **1998**, *250*, 15–21.

99 Ray, P. K., Raychaudhuri, S., Allen, P. Mechanism of regression of mammary adenocarcinomas in rats following plasma adsorption over protein A-containing Staphylococcus aureus, *Cancer Res.* **1982**, *42*, 4970–4974.

100 Srivastava, A. K., Singh, K. P., Ray, P. K. Protein A induced protection against experimental Candidiasis in mice, *Mycopathologia* **1997**, *138*, 21–28.

101 Ghose, A. C., Mookerjee, A., Sengupta, K., Ghosh, A. K., Dasgupta, S., Ray, P. K. Therapeutic and prophylactic uses

of Protein A in the control of Leishmania donovani infection in experimental animals, *Immunol. Lett.* **1999**, *65*, 175–181.

102 Frankenberger, M., Pechumer, H., Ziegler-Heitbrock, H.W. Interleukin-10 is upregulated in LPS tolerance, *J. Inflamm.* **1995**, *45*, 56–63.

103 Picart, C., Ladam, G., Senger, B., Voegel, J.-C., Schaaf, P., Cuisinier, F.J.G., Gergely, C. Determination of structural parameters characterizing thin films by optical methods: A comparison between scanning angle reflectometry and optical waveguide lightmode spectroscopy, *J. Chem. Phys.* **2001**, *115*, 1086–1094.

104 Strand, B.L., Ryan, L., Veld, P.I., Kulseng, B., Rokstad, A.M., Skjak-Braek, G., Espevik, T. Poly-L-lysine induces fibrosis on alginate microcapsules via the induction of cytokines, *Cell Transplant.* **2001**, *10*, 263–275.

105 Frölich, J.C. A classification of NSAIDs according to the relative inhibition of cyclooxygenase isoenzymes, *Trends Pharmacol. Sci.* **1997**, *18*, 30–34.

106 Getting, S.J. Melanocortin peptides and their receptors: New targets for anti-inflammatory therapy, *Trends Pharmacol. Sci.* **2002**, *23*, 447–449.

107 Miyata, T., Asami, N., Uragami, T. A reversibly antigen-responsive hydrogel, *Nature* **1999**, *399*, 766–769.

108 Murphy, W.L., Peters, M.C., Kohn, D.H., Mooney, D.J. Sustained release of vascular endothelial growth factor from mineralized polylactide-co-glycolide. Scaffolds for tissue engineering, *Biomaterials* **2000**, *21*, 2521–2527.

109 Arcos, D., Ragel, C.V., Vallet-Regí, M. Bioactivity in glass/PMMA composites used as drug delivery system, *Biomaterials* **2001**, *22*, 701–708.

110 Laurencin, C.T., Attawia, M.A., Lu, L.Q., Borden, M.D., Lu, H.H., Gorum, W.J., Lieberman, J.R. Polylactide-co-glycolide/hydroxyapatite delivery of BMP-2-producing cells: a regional gene therapy approach to bone regeneration, *Biomaterials* **2001**, *22*, 1271–1277.

111 Dee, K.C., Andersen, T.T., Bizios, R. Design and function of novel osteoblast-adhesive peptides for chemical modification of biomaterials, *J. Biomed. Mater. Res.* **1998**, *40*, 371–377.

112 Neff, J.A., Caldwell, K.D., Tresco, P.A. A novel method for surface modification to promote cell attachment to hydrophobic substrates, *J. Biomed. Mater. Res.* **1998**, *40*, 511–519.

113 Cannizzaro, S.M., Padera, R.F., Langer, R., Rogers, R.A., Black, F.E., Davies, M.C., Tendler, S.J.B., Shakesheff, K.M. A novel biotinylated degradable polymer for cell-interactive applications, *Biotechnol. Bioeng.* **1998**, *58*, 529–535.

114 Ito, Y., Zheng, J., Imanishi, Y., Yonezawa, K., Kasuga, M. Protein-free cell culture on an artificial substrate with covalently immobilized insulin, *Proc. Natl. Acad. Sci. USA* **1996**, *93*, 3598–3601.

115 Ito, Y., Chen, G., Imanishi, Y. Micropatterned immobilization of epidermal growth factor to regulate cell function, *Bioconjugate Chem.* **1998**, *9*, 277–282.

116 Decher, G., Hong, J.D., Schmitt, J. Buildup of ultrathin multilayer films by a self-assembly process: III. Consecutively alternating adsorption of anionic and cationic polyelectrolytes on charged surfaces, *Thin Solid Films* **1992**, *210–211*, 831–835.

117 Kleinfeld, E.R., Ferguson, G.S. Stepwise formation of multilayered nanostructural films from macromolecular precursors, *Science* **1994**, *265*, 370–373.

118 Caruso, F., Caruso, R.A., Möhwald, H. Nanoengineering of inorganic and hybrid hollow spheres by colloidal templating, *Science* **1998**, *282*, 1111–1114.

119 Ho, P.K.H., Kim, J.-S., Burroughes, J.H., Becker, H., Li, S.F.Y., Brown, T.M., Cacialli, F., Friend, R.H. Molecular-scale interface engineering for polymer light-emitting diodes, *Nature* **2000**, *404*, 481–484.

120 Hiller, J., Mendelsohn, J.D., Rubner, M.F. Reversibly erasable nanoporous anti-reflection coatings from polyelectrolyte multilayers, *Nat. Mater.* **2002**, *1*, 59–63.

121 Tang, Z., Kotov, N.A., Magonov, S., Ozturk, B. Nanostructured artificial nacre, *Nat. Mater.* **2003**, *2*, 413–418.

122 Braet, F., Seynaeve, C., De Zanger, R., Wisse, E. Imaging surface and submembranous structures with the atomic force microscope: A study on living cancer cells, fibroblasts and macrophages, *J. Microsc.* **1998**, *190*, 328–338.

123 Braet, F., Vermijlen, D., Bossuyt, V., De Zanger, R., Wisse, E. Early detection of cytotoxic events between hepatic natural killer cells and colon carcinoma cells as probed with the atomic force microscope, *Ultramicroscopy* **2001**, *89*, 265–273.

124 Sauerbrey, G. Verwendung von Schwingquartzen zur Wägung dünner Schichten und zur Mikrowägung, *Z. Phys.* **1959**, *155*, 206–222.

125 Mendelsohn, J. D., Yang, S. Y., Hiller, J., Hochbaum, A. I., Rubner, M. F. Rational design of cytophilic and cytophobic polyelectrolyte multilayer thin films, *Biomacromolecules* **2003**, *4*, 96–106.

126 Szente, L., Szejtli, J. Highly soluble cyclodextrin derivatives: chemistry, properties, and trends in development, *Adv. Drug. Deliv. Rev.* **1999**, *36*, 17–28.

127 Doliwa, A., Santoyo, S., Ygartua, P. Influence of piroxicam: hydroxypropyl-beta-cyclodextrin complexation on the in vitro permeation and skin retention of piroxicam, *Skin Pharmacol. Appl. Skin Physiol.* **2001**, *14*, 97–107.

128 McEwen, J. Clinical pharmacology of piroxicam-beta-cyclodextrin: Implications for innovative patient care, *Clin. Drug Investig.* **2000**, *19*, 27–31.

129 Höök, F., Kasemo, B., Nylander, T., Fant, C., Sott, K., Elwing, H. Variations in coupled water, viscoelastic properties, and film thickness of a Mefp-1 protein film during adsorption and cross-linking: A quartz crystal microbalance with dissipation monitoring, ellipsometry, and surface plasmon resonance study, *Anal. Chem.* **2001**, *73*, 5796–5804.

130 Serizawa, T., Yamaguchi, M., Akashi, M. Alternating bioactivity of polymeric layer-by-layer assemblies: Anticoagulation vs procoagulation of human blood, *Biomacromolecules* **2002**, *3*, 724–731.

131 Shenoy, D. B., Antipov, A., Sukhorukov, G. B., Möhwald, H. Layer-by-layer engineering of biocompatible, decomposable core-shell structures, *Biomacromolecules* **2003**, *4*, 265–272.

132 Thierry, B., Winnik, F. M., Merhi, Y., Silver, J., Tabrizian, M. Bioactive coatings of endovascular stents based on polyelectrolyte multilayers, *Biomacromolecules* **2003**, *4*, 1564–1571.

133 Balabushevich, N. G., Tiourina, O. P., Volodkin, D. V., Larionova, N. I., Sukhorukov, G. B. Loading the multilayer dextran sulfate/protamine microsized capsules with peroxidase, *Biomacromolecules* **2003**, *4*, 1191–1197.

134 Vautier, D., Karsten, V., Egles, C., Chluba, J., Schaaf, P., Voegel, J.-C., Ogier, J. Polyelectrolyte multilayer films modulate cytoskeletal organization in chondrosarcoma cells, *J. Biomater. Sci., Polym. Ed.* **2002**, *13*, 713–732.

135 Burke, S. E., Barrett, C. J. pH-responsive properties of multilayered polyL-lysine/hyaluronic acid surfaces, *Biomacromolecules* **2003**, *4*, 1773–1783.

136 Richert, L., Arntz, Y., Schaaf, P., Voegel, J.-C., Picart, C. pH dependent growth of polyL-lysine/polyL-glutamic acid multilayer films and their cell adhesion properties, *Surf. Sci.* **2004**, *570*, 13–29.

137 Salloum, D. S., Olenych, S. G., Keller, T. C., Schlenoff, J. B. Vascular smooth muscle cells on polyelectrolyte multilayers: hydrophobicity-directed adhesion and growth, *Biomacromolecules* **2005**, *6*, 161–167.

138 Berg, M. C., Yang, S. Y., Hammond, P. T., Rubner, M. F. Controlling mammalian cell interactions on patterned polyelectrolyte multilayer surfaces, *Langmuir* **2004**, *20*, 1362–1368.

139 Picart, C., Elkaim, R., Richert, L., Audoin, F., Da Silva Cardoso, M., Schaaf, P., Voegel, J.-C., Frisch, B. Primary cell adhesion on RGD functionalized and covalently cross-linked polyelectrolyte multilayer thin films, *Adv. Funct. Mater.* **2005**, *15*, 83–94.

140 Richert, L., Engler, A. J., Discher, D. E., Picart, C. Elasticity of native and cross-linked polyelectrolyte multilayer, *Biomacromolecules* **2004**, *5*, 1908–1916.

141 Engler, A., Bacakova, L., Newman, C., Hategan, A., Griffin, M., Discher, D. E. Substrate compliance versus ligand density in cell on gel responses, *Biophys. J.* **2004**, *86*, 617–628.

142 Ai, H., Meng, H., Ichinose, I., Jones, S. A., Mills, D. K., Lvov, Y. M., Qiao, X. Biocompatibility of layer-by-layer self-assembled nanofilm on silicone rubber for neurons, *J. Neurosci. Methods* **2003**, *128*, 1–8.

143 Laurent, T. C. *The Chemistry, Biology, and Medical Applications of Hyaluronan and Its Derivatives*, Cambridge University Press, Cambridge, U.K., **1998**, Vol. 72.

144 Desbrières, J., Martinez, C., Rinaudo, M. Hydrophobic derivatives of chitosan: characterization and rheological behaviour, *Int. J. Biol. Macromol.* **1996**, *19*, 21–28.

145 Prestwich, G. D., Marecak, D. M., Marecek, J. F., Vercruysse, K. P., Ziebell, M. R. Controlled chemical modification of hyaluronic acid: synthesis, applications, and biodegradation of hydrazide derivatives, *J. Controlled Release* **1998**, *53*, 93–103.

146 Sabnis, S., Block, L. H. Chitosan as an enabling excipient for drug delivery systems. I. Molecular modifications, *Int. J. Biol. Macromol.* **2000**, *27*, 181–186.

147 Luo, Y., Prestwich, G. D. Synthesis and selective cytotoxicity of a hyaluronic acid-antitumor bioconjugate, *Bioconjugate Chem.* **1999**, *10*, 755–763.

148 Morimoto, M., Saimoto, H., Usui, H., Okamoto, Y., Minami, S., Shigemasa, Y. Biological activities of carbohydrate-branched chitosan derivatives, *Biomacromolecules* **2001**, *2*, 1133–1136.

149 Koktysh, D. S., Liang, X., Yun, B. G., Pastoriza-Santos, I., Matts, R. L., Giersig, M., Serra-Rodríguez, C., Liz-Marzán, L. M., Kotov, N. A. Biomaterials by design: Layer-by-layer assembled ion-selective and biocompatible films of TiO_2 nanoshells for neurochemical monitoring, *Adv. Funct. Mater.* **2002**, *12*, 255–265.

150 Richert, L., Boulmedais, F., Lavalle, P., Mutterer, J., Ferreux, E., Decher, G., Schaaf, P., Voegel, J.-C., Picart, C. Improvement of stability and cell adhesion properties of polyelectrolyte multilayer films by chemical cross-linking, *Biomacromolecules* **2004**, *5*, 284–294.

151 Thompson, M. T., Berg, M. C., Tobias, I. S., Rubner, M. F., Van Vliet, K. J. Tuning compliance of nanoscale polyelectrolyte multilayers to modulate cell adhesion, *Biomaterials* **2005**, *26*, 6836–6845.

152 Etienne, O., Schneider, A., Taddei, C., Richert, L., Schaaf, P., Voegel, J.-C., Egles, C., Picart, C. Degradability of polysaccharides multilayer films in the oral environment: an in vitro and in vivo study, *Biomacromolecules* **2005**, *6*, 726–733.

153 Picart, C., Mutterer, J., Arntz, Y., Voegel, J.-C., Schaaf, P., Senger, B. Application of fluorescence recovery after photobleaching to diffusion of a polyelectrolyte in a multilayer film, *Microsc. Res. Tech.* **2005**, *66*, 43–57.

154 Alberts, B., Bray, D., Lewis, J., Raff, M., Roberts, K., Watson, J. D. *Molecular Biology of the Cell*, Garland Publishing Inc: New York, **1994**.

155 Picart, C., Schneider, A., Etienne, O., Mutterer, J., Egles, C., Jessel, N., Voegel, J.-C. Controlled degradability of polysaccharide multilayer films *in vitro* and *in vivo*, *Adv. Funct. Mater.* **2005**, *15*, 1771–1780.

156 Francius, G., Hemmerlé, J., Ohayon, J., Schaaf, P., Voegel, J.-C., Picart, C., Senger, B. Effect of cross-linking on the elasticity of polyelectrolyte multilayer films measured by colloidal probe AFM, *Microsc. Res. Tech.* **2006**, *69*, 84–92.

157 Richert, L., Schneider, A., Vautier, D., Vodouhê, C., Jessel, N., Payan, E., Schaaf, P., Voegel, J.-C., Picart, C. Imaging cell interactions with native and cross-linked polyelectrolyte multilayers, *Cell Biochem. Biophys.* **2006**, *44*, 273–286.

158 Schneider, A., Francius, G., Obeid, R., Schwinté, P., Hemmerlé, J., Frisch, B., Schaaf, P., Voegel, J.-C., Senger, B., Picart, C. Polyelectrolyte multilayers with a tunable Young's modulus: Influence of film stiffness on cell adhesion, *Langmuir* **2006**, *22*, 1193–1200.

159 Schneider, A., Vodouhê, C., Richert, L., Francius, G., Schaaf, P., Voegel, J.-C., Frisch, B., Picart, C. Cross-linked polysaccharide and polypeptide multilayer films: Combining mechanical resistance, biodegradability and bioactivity, *Biomacromolecules* **2006**, in press.

15
Bio-inspired Complex Block Copolymers/Polymer Conjugates and Their Assembly

Markus Antonietti, Hans G. Börner, and Helmut Schlaad

15.1
Introduction

Block copolymers are probably the best examined model system to study self-assembly or organization towards larger scale structures with controlled structural features on the nanometer scale [1–4]. However, compared with structural biology, e.g. the folding and assembly behavior of peptides, the control over tertiary structure by synthetic polymer chemistry is still rudimentary. One has to take this as a hint that polymer chemistry is still full of options and far from having reached limiting material properties. In contrast, using biomimetic approaches, one might expect a performance boost of present systems, just bypassing taking a statistical Gaussian shape as given.

A key element of natural self-interaction is reversible, secondary interaction. Some of those used include hydrophobic interactions, polarity gradients, coulombic interactions, dipolar interactions and H-bridges. The relative importance of such interactions for synthetic supramolecular structures has been discussed in detail elsewhere [5], but it is interesting to note that, usually, each interaction encodes one separate level of supramolecular organization. This is how a whole sequence of integration steps can be encoded already at the molecular level or in the well known, so-called alpha-structure, the primary sequence of monomer units. It is obvious that this level of control is beyond classical block copolymer chemistry, where control of alpha-structure is just a blocky arrangement of two or three different monomers and self-assembly is driven only by polarity differences or solvophobic interactions.

Some measures to fill this gap exist and all of them belong, in our opinion, to the most promising tasks in polymer science:

- The control of monomer sequence beyond the techniques of controlled polymerization to full sequential control of a library of monomers, analogous to peptides. These are (foreseeable) step-by-step techniques with high synthetic fidelity, usually performed in a robotic fashion in synthesizers.

Macromolecular Engineering. Precise Synthesis, Materials Properties, Applications.
Edited by K. Matyjaszewski, Y. Gnanou, and L. Leibler

ISBN: 978-3-527-31446-1

- The consideration of secondary interaction motifs beyond polarity differences in polymer synthesis, e.g. the repeated use of hydrogen-bridging systems and dipolar interactions, presumably even combined with block copolymer chemistry. For the choice of these interacting monomers biological or biomimetic structures are favorable, as they have already shown an extraordinary performance, but it is an advantage of polymer chemistry not to be restricted to natural availability and to extend systems to high temperatures and non-aqueous media.

This chapter describes current activities in one of these approaches to better controlled polymer superstructures, namely the synthesis and analysis of block copolymers comprising one or more bio-organic segments (polypeptide or polysaccharide) and another synthetic block made by classical means of controlled polymerization. To set some focus, we will not consider the related rich work on synthetic polymers with pendant amino acid or sugar residues and irregularly/hyperbranched structures. Apart from the central focus on synthesis, some emphasis is laid on the structural behavior of such systems.

15.2
Primary Structures: Polymer Synthesis

15.2.1
Polymers with Homopolymeric Bio-organic Segments

15.2.1.1 Polypeptides

Block copolymers containing homopolypeptide segments are traditionally prepared by a ring-opening polymerization of amino acid *N*-carboxyanhydrides (NCA, Leuch's anhydride [6]) using primary amine-functional (macro-) initiators [7]. The chain growth should proceed via the nucleophilic opening of the NCA ring through the primary amine chain ends ("amine mechanism"). However, the NCA can be deprotonated by the amine and thus transformed into a nucleophile that can alternatively initiate polymerization ("activated monomer mechanism"). Both pathways operate simultaneously: a propagation step of one mechanism is a side-reaction for the other and *vice versa*. The mechanism of the reaction is further complicated by the presence of impurities (water, acid chlorides or isocyanates in the NCA) and polypeptides therefore often exhibit a broad and/or multimodal molecular weight distribution [8]. Racemization of chiral amino acid units does not occur during the polymerization process, which is essential if the polypeptide chain is later to adopt a defined secondary structure (usually α-helix or β-sheet).

There exist a number of strategies to improve the control of NCA polymerization for the synthesis of well-defined polypeptide-based block copolymers, among which is the use of transition metal [9] and primary amine hydrochloride initiators [10]. Other important control parameters are temperature [11] and purity of reagents [12]. Polymerizations are considered to be "living" in nature, i.e. reactions other than nucleophilic ring-opening chain growth are absent and polymers can have a nearly monodisperse molecular weight distribution. NCAs of various protected or native L-amino acids, such as glutamate, aspartate and lysine or leucine, have been successfully polymerized. Diverse block copolymers comprising all-peptidic segments (block copolypeptides) or peptidic and synthetic segments (so-called peptide-based hybrid polymers or molecular chimeras [8]) have been prepared.

15.2.1.2 **Polysaccharides**

Polymers with linear amylose segments have been synthesized by phosphorylase-catalyzed enzymatic polymerization [13–15] and/or by coupling techniques [16–19]. The *in vitro* enzymatic polymerization approach, first demonstrated by Ziegast and Pfannemüller [13], involves the synthesis of a maltopentaose- or maltoheptaose-functional polymer [e.g. polystyrene, poly(ethylene oxide) or poly(L-glutamic acid)], the so-called primer, which is then reacted with α-D-glucose-1-phosphate in the presence of potato phosphorylase to grow the amylose block. The reaction proceeds analogously to "living" anionic polymerization and produces amyloses with a Poissonian molecular-weight distribution [20]. The final degree of polymerization, which can be in the range from a few tens to several hundreds or thousands, can be controlled simply by adjusting the molar ratio of glucose to primer.

Also worth mentioning is the synthesis of well-defined oligosaccharide-terminated polymers through controlled atom transfer radical polymerization [21]. The free-radical (co-)polymerization of styryl-terminal amylose macromonomers has been applied to produce brush-like structures with a synthetic backbone and pendent amylose side-chains [22].

15.2.2
Polymers with Sequenced Bio-organic Segments

15.2.2.1 **Peptides**

Homodisperse polypeptides, exhibiting a defined amino acid sequence in the absence of chemical and molecular weight distributions, are made by three major approaches: recombinant DNA methods, solid-phase supported peptide synthesis and ligation methods. These routes allow monomer sequence control in synthetic polypeptides and the integration of unnatural amino acids as selective loci to couple classical synthetic polymers.

Recombinant DNA methods

High-molecular weight polypeptides comprising native and progressively an increasing variety of non-natural amino acids [23] can be accessed via recombinant DNA methods using bacterial expression of corresponding artificial genes. Modern synthesis of DNA oligonucleotides can provide 100-mers or more and polymerase chain reaction (PCR)-based methods make it simple to assemble genes of essentially any sequence. In appropriate vectors, these genes can be used to express proteins in bacteria, yeast, plants, insect cells, mammalian cells or even living animals, making it possible to generate reasonable quantities of proteins. The effort of designing artificial genes, modifying host organisms, and in particular the isolation of the expressed protein from the host, limits the applicability of this technique to peptides with comparatively high molecular weights, much larger than 10 kDa. However, accuracy of peptide synthesis by the biosynthetic approach is unsurpassed and modern biotechnology makes large-scale production possible. Site-specific insertion of an expanding array of unnatural amino acids in small quantities of proteins can be achieved at any position using chemically acylated transfer-RNA (tRNA) [24]. Perhaps more excitingly, engineered tRNAs and synthetases in protein biosynthetic machineries of genetically modified bacteria or yeast now allow, without chemical acylation of tRNA, site-specific incorporation of unnatural amino acids in living cells with no chemical intervention. Methods of expanding the genetic code at the nucleic acid level, including four-base codons (sequence of adjacent nucleotides in the genetic code, determining the insertion of a specific amino acid in a polypeptide chain) and unnatural base pairs, are becoming useful for the addition of multiple amino acids to the genetic code [25–27]. Over 30 novel amino acids have been genetically encoded in response to unique triplet and quadruplet codons, including fluorescent, photoreactive and redox-active amino acids, glycosylated amino acids and amino acids with keto, azido, acetylenic and heavy atom-containing side-chains. By removing the constraints of the existing 20 amino acid code and expanding the repertoire of amino acids, it should be possible to generate proteins with new or enhanced properties [26, 28].

Solid-phase supported peptide synthesis

Modern synthetic methods based on stepwise solid-phase supported peptide synthesis (SPPS) were pioneered by Merrifield in 1963. These chemical synthesis approaches provide access to essentially any peptide of less than about 40 residues in length. Even if this is only the size of the smallest proteins known (about 6 kDa) [29], numerous oligopeptides have been described, showing distinct biological activity or defined functions. The developments in SPPS have overcome a number of difficulties related to the selectivity of chemical reactions, the insolubility of protected peptides, the optimization of the applied polymer supports concerning accessibility, diffusion properties and non-specific interactions with the peptide and also the difficulty of driving the stepwise reaction to quantitative conversion. Even if every coupling cycle could result in a 99% yield of desired product, the overall yield would only be about 60% after 50 cycles and 37% after 100 cycles. Hence the stepwise formation of extremely small amounts of side-products makes it essentially impossible to increase substantially the scope of SPPS. However, SPPS has greatly facilitated the synthesis of peptides and small proteins (< 100 amino acids) and has made peptides widely available through automated procedures.

The sequence-specific incorporation of unnatural amino acids via chemical synthesis is virtually unlimited. It allows versatile modifications of peptides ranging from single position mutants where one amino acid is substituted to the synthesis of pseudo-peptides comprising non-peptidic backbones and chemistry [30, 31]. SPPS methods are particularly useful for the incorporation of amino acid analogues that are toxic to cells or incompatible with the translational machinery (e.g. the synthesis of all D-amino acid proteins). Chemical synthesis also allows sequence-specific isotopic labeling (e.g. H-, C- or N-labeled amino acids) for spectroscopic studies. On the other hand, chemical synthesis can be problematic with peptides or proteins having poor solubility and becomes tedious, low yielding and expensive when applied to larger proteins. Finally, chemically synthesized proteins cannot be used for *in vivo* studies unless introduced mechanically or chemically into the cell.

Ligation methods

To overcome the size limitations of SPPS, efficient strategies have been developed to ligate synthetic peptides together. The combination of SPPS and enzymatic or chemoselective ligation now permits the construction of entirely synthetic proteins as large as 25 kDa. Moreover, recombinant fragments expressed by biosynthesis machinery can be ligated with chemically synthesized peptides utilizing selective fragment coupling reactions, e.g. thioester, oxime, thiazolidine/oxazolidine, thioether and disulfide bond formation of selectively addressable functional groups incorporated into peptides [32–34]. This expressed protein ligation allows unnatural modifications near protein termini in proteins of virtually any size.

An elegant strategy referred as "native chemical ligation" allows the selective coupling of peptide fragments under natural conditions to form a native peptide linkage [35]. In this approach, a peptide with a C-terminal thioester is temporarily ligated to an N-terminal cysteine residue of a second peptide through a trans-thioesterification reaction. The subsequent $S \rightarrow N$-acyl rearrangement of the thioester-linked intermediate leads to the native peptide bond at the ligation site. Native chemical ligation can be routinely used for the preparation of polypeptides of >100 amino acid residues in length. This method typically requires an N-terminal cysteine at the ligation site, but recently it was demonstrated that an auxiliary group such as *N*-α-(1-phenyl-2-mercaptoethyl) or sulfhydryl can be used that is removed after ligation, making the ligation possible at a non-Cys residue [36, 37]. Moreover, the approach was broadened by Muir, who described protein semisynthesis based on native chemical ligation, wherein the thioester is produced recombinantly from a partially-incapacitated protein splicing system [38].

The peptides made by the previous three methods can then be coupled to a synthetic polymer – this approach is called *conjugation*. Diverse strategies are available to introduce specifically polymers to polypeptides and even more challenging to proteins. Nature has solved the problem of specificity in post-translational modification using enzymatic recognition of extended motifs rather than unique reactive handles. However, the chemoselective conjugation of polymers to peptides is a more versatile method allowing the straightforward introduction of broad spectra of polymers applying often adverse conditions such as polar organic solvents, critical pH or ionic strength as well as increased temperatures.

In the progressively developing field of macromolecular conjugates – besides the peptide or protein conjugates made to enhance pharmaceutical indices of peptide-based drugs, which are outside the scope of this chapter – new functional materials are accessible following more than ever in material science the principles of rational design. These bio-inspired materials often show distinct hierarchical structures, responsiveness to external stimuli and potentially bioactivity. Hence they mimic bioorganic materials. In the following, the different synthetic strategies to construct polymer–peptide conjugates exhibiting oligopeptide segments will be summarized. Oligopeptides with up to 30 amino acids in length exhibit advantages, compared with high molecular weight proteins. They can be easily accessed by SPPS, if required up to the gram scale, and even more important the sequence–property relationship is still comparatively simple. Frequently this allows an accurate prediction by computer simulation methods and makes the rational design of functional segments more likely.

The different access routes to oligopeptide–polymer conjugates can be classified into two main approaches: (1) polymerization strategies, where the polymer segment is synthesized in the presence of the peptides using either a peptide macroinitiator or the polymerization of peptide macromonomers, and (2) coupling strategies, where peptide segments are coupled to polymers with one or multiple reactive sides (Fig. 15.1).

Controlled radical polymerization (CRP) techniques have proven to be suited for the synthesis of peptide–polymer conjugates. This is mainly due to the high

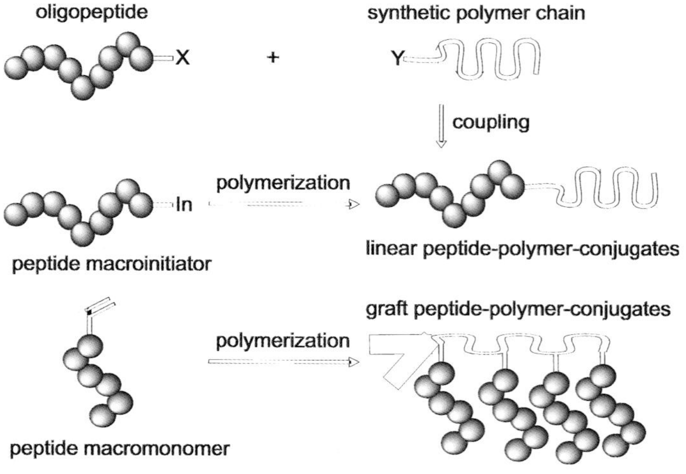

Fig. 15.1 Approaches to integrate sequence-controlled polypeptides into a synthetic polymer (e.g. X=NH_2, Y=COOH).

tolerance of the CRP systems to diverse functional groups and impurities. Moreover, CRP methods are well recognized to allow the controlled polymerization of diverse monomers, giving access to a variety of polymers with defined molecular weights with low polydispersity indexes. Since a broad spectrum of chain compositions, topologies and architectures can be realized, the incorporation of peptides extends the macromolecular toolbox by polymers with a flexible and (bio-)functional segment.

The three major CRP methods, atom transfer radical polymerization (ATRP), reversible addition-fragmentation chain transfer radical polymerization (RAFT) and nitroxide-mediated radical polymerization (NMP), were utilized to synthesize peptide–polymer conjugates via the oligopeptide macroinitiator approach. This comprises the solid-phase supported synthesis of an oligopeptide, followed by the sequence-specific introduction of the required CRP initiator or chain transfer moiety. Depending on the strategy, the oligopeptide CRP agent was either used directly to initiate solid-phase supported CRP in a regioselective manner or was liberated from the support followed by initiation of a homogeneous CRP in solution.

Wooley and coworkers demonstrated this concept for solid-phase supported block copolymer synthesis using NMP of different acrylates and styrene [39, 40]. Two triblock copolymers were synthesized combining on the one hand an HIV-1 Tat protein fragment (Tat_{46-57}) with poly(acrylic acid)-*block*-poly(methyl acrylate) [40] and on the other the antimicrobial tritrypticin [41] with poly(acrylic acid)-*block*-poly(styrene) [39]. However, no details were presented concerning the degree of control of the CRP reaction and only limited characterization data were accessible due to the complexity of the conjugates. ATRP was used by Washburn and coworkers [42] for the solid-phase supported polymerization of 2-hydroxyethyl methacrylate (HEMA). The resulting RGD–poly(HEMA) conjugate (RGD = arginine-glycine-aspartic acid) carried one of the functional binding domains of the fibronectin adhesion protein to modulate cell adhesion to polymer coatings. Analysis of the final product after liberation from the support revealed a polydispersity index of 1.47, but further details regarding polymerization kinetics were not presented.

Börner and coworkers investigated the interactions for the ATRP system with oligopeptide macroinitiators using homogeneous solution-phase ATRP instead of a solid-phase supported strategy [43]. Controlling a solid-phase supported polymerization remains comparably challenging. It has been shown for "grafting from" polymerizations from either silica supports or polystyrene microgel particles that diffusion limitation of the deactivation step in the CRP is difficult to overcome in both ATRP and NMP [44–47]. The investigation of the ATRP of *n*-butyl acrylate (*n*BA) in solution using a model peptide macroinitiator results in well-defined block copolymers with low polydispersity and predictable molecular weight. Although interactions between the catalyst and the oligopeptide were evident, it was demonstrated that they were not critical in terms of synthesis control. In all studies, the CRP initiator was attached to the N-terminal amine group of a supported oligopeptide. However, protecting group strategies are available to modify a ε-amino group of a lysine selectively. This enables one to make defined oligopeptides with a sequence-specific introduction of single or multiple initiator groups.

Furthermore, Börner and coworkers adapted the RAFT polymerization technique to the synthesis of well-defined conjugates [48]. Oligopeptide RAFT agents were synthesized by the utilization of novel solid-phase supported synthesis routes, making the usually required chromatographic purification step obsolete. The functionality switch of a resin-bound oligopeptide ATRP macroinitiator into an oligopeptide transfer agent was proven to be most suitable for the synthesis of a macro-RAFT agent with high purity. An additional route that involves the coupling of a carboxylic acid-functionalized RAFT agent to the N-terminal amine group of a supported oligopeptide yielded the macro-transfer agents in 76% purity, but suffered from side-product formation. Kinetic investigations, in which *n*BA was polymerized in solution using oligopeptide transfer agents, revealed efficient control of the polymerization processes. Peptide–polymer conjugates exhibiting a molecular weight distribution with a polydispersity index of about 1.1 and controllable molecular weight were obtained after retardation periods of about 4–8 h. The RAFT group remains at the end of the polymer chain after isolation of the conjugates, as was confirmed by ^{1}H NMR spectroscopy. This will allow the modification of the polymer chain end and further block extensions and thus enable polymers with advanced architectures to be made.

An interesting structure variation compared with the linear peptide-polymer conjugates that were described above are conjugates possessing multiple peptide segments in the side-chains. An elegant approach to these comb-like structures was published by van Hest and coworkers, who presented a "grafting through" route by polymerizing methacrylate-functionalized oligopeptide macromonomers via ATRP [49]. The resulting polymethacrylate-*graft*-oligopeptide exhibited a well-controlled number-average degree of polymerization and a low polydispersity index (1.25). The versatility of this approach was demonstrated by the incorporation of diverse oligopeptide segments including elastin repeats [49], short β-sheet builders [50] or gramicidin S analogues [51].

Similar comb-shaped structures were accessed by Godwin et al. via a coupling approach. The polymerization of *N*-methacryloxysuccinimide by ATRP leads to a polymer with active ester side-groups at each repeat unit [52]. The subsequent coupling of oligopeptides to these groups results in a comb conjugate, suitable for as a model therapeutic polymer for the poly[*N*-(2-hydroxypropyl)methacrylamide] (HPMA)–drug conjugate.

Alternatively, the coupling approach was used to introduce synthetic polymers region-selective to oligopeptides. The selectivity of the coupling reaction was achieved either by choosing chemoselective reactions between the polymer chain end-functionality and specifically addressable functionalities in the oligopeptide segment or by the utilization of protecting-group strategies. The latter are frequently combined with SPPS since fully side-chain protected oligopeptides were obtained. Klok and coworkers demonstrated the coupling of α-methoxy-ω-carboxy-functionalized poly(ethylene oxide) to the N-terminal amino functionality of a supported peptide [53, 54]. However, even if purification proceeds via simple washing, this approach is limited in the molecular weight of polymer which is coupled. This is due to a reduction of end-group reactivity, a decrease in diffusion rates of the polymer in the micro-gel supports and increased incompatibilities of some polymers with the support, leading to a strong decrease in the overall coupling rates. Klok and coworkers demonstrated that the coupling of a poly(ethylene glycol) (PEG) with a molecular weight of 0.7 kDa to a supported oligopeptide proceeds well, whereas that of a PEG of 1.8 kDa could be driven to only 40% conversion [53].

Diverse selective coupling reactions, adapted from peptide ligation [33, 55, 56], have been used, such as the formation of thioesters, oximes, thiazolidines/oxazolidines, thioethers and disulfides to attach polymers to peptides. More recently, non-proteinogenic amino acids with functionalities allowing the selective coupling of polymers to oligopeptides or even to proteins were introduced. Tirrell and coworkers used the copper-catalyzed Huisgen [3+2] dipolar cycloaddition of azides and alkynes, referred to as "click-coupling", to label selectively expressed recombinant proteins or membrane proteins in *Escherichia coli* bacteria in the native environment [57, 58]. This demonstrates impressively the bioorthogonality of this coupling reaction. Opsteen and Van Hest used this specific coupling reaction for the fusion of synthetic polymers [59], Lutz et al. introduced diverse functionalities to a model polymer [60] and Rutjes and coworkers fused alkyne-functionalized bovine serum albumin specifically with azide end-functionalized polystyrene, realizing a protein polymer conjugate [61].

Large polypeptides or proteins have also been conjugated to synthetic polymers, but this large field of research is outside the scope of this chapter. For some reviews and interesting research articles see [61–67].

15.2.2.2 **DNA–Polymer Conjugates**

The use of synthetic deoxyribonucleic acid (DNA) or oligonucleotide segments as programmable interconnects to guide structure formation processes in synthetic polymers might allow the preparation of materials with preconceived architectural parameters and new properties [68–70]. For this reason, efforts have been focused on the construction of DNA–polymer hybrid materials.

Linear conjugates of oligonucleotides and synthetic polymers have been successfully synthesized by Mirkin and coworkers on a 10-μmol scale [71]. The amphiphilic DNA–poly(styrene) conjugate was accessed by coupling a polystyrene having a phosphoramidite end-functionality to the 5′-OH group of a solid-phase supported single-strand DNA segment (20-mer). The oligonucleotide was prepared through conventional, fully automated SPPS on controlled-pore glass beads applying phosphoramidite coupling chemistry.

A similar strategy was applied to realize DNA–polymer conjugates with a multi-graft architecture [72]. Ring-opening metathesis polymerization (ROMP) of 5-*exo*-norbornen-2-ol yielded an alcohol-modified oligo(norbornene) with an average number of 17 repeat units. Subsequent transformation of the hydroxyl functionalities into phosphoramidites provided a polymer that could couple with multiple oligonucleotides, leading to a hybrid polymer with five DNA strands on average. Moreover, redox-active ferrocenyl or dibromoferrocenyl groups were integrated into such polynorbornene-*graft*-DNA conjugates. These conjugates exhibit predictable and tailorable electrochemical properties, a high DNA duplex stability and sharp melting transitions. Hence they meet desirable characteristics for DNA detection applications by electrochemical assays [71].

Unlike the enormous potential of oligonucleotides in bio-diagnostics [73], the applicability of DNA-mediated structure formation for material science appears to be rather limited. This is mainly due to the rapid degradation of single-strand oligonucleotides, occurring either enzymatically by DNases or chemically by hydrolysis. Another drawback is the accessibility of just too small quantities of synthetic oligonucleotides. Nowadays, oligonucleotides up to about 100 basepairs and of virtually any sequence can be easily synthesized in milligram quantities using commercially available, automated synthesizers that apply solid-phase supported phosphoramidite chemistry. However, the scale of synthesis is usually limited to μmol or nmol levels, hence accessibility of larger amounts of synthetic oligonucleotides in the 100-mg or even gram scale would be prohibitively expensive. Additional limitations result from the solubility of the polyelectrolytic backbone of DNA or oligonucleotides. This requires aqueous solutions in order to solubilize the oligonucleotide and, even more important, to obtain undisturbed hybridization events. The transfer of oligonucleotides into a non-aqueous environment has been realized, preserving the hybridization function [74]. However, these hybridization events usually lead to a duplex with decreased stability since the geometry of the nucleotide bases in the strands is altered and energy contributions from entropic effects (e.g. hydrophobic effect) are missing.

15.2.2.3 **PNA–Polymer Hybrids**

Recently, peptide–nucleic acids (PNA, Fig. 15.2) [75] have been developed, combining the stability and solubility of an achiral pseudo-peptide backbone with a defined sequence of nucleotide bases. Thus, PNAs with complementary base sequences are capable of hybridizing reversibly. The main driving force for the development of PNAs was their pharmacological potential [76–78], since cross-hybridization with DNA or RNA occurs [79]. This allows DNA double strand invasion to modulate gene activity or mRNA complexation to provide a useful tool for silencing protein expression [31]. However, PNA–PNA hybridization results in highly stable duplexes, requiring only 4–6 base pairs compared with 8–15 base pairs to form stable DNA–DNA duplexes [79]. This excellent stability, combined with good solubility in organic solvents and tolerance against changes of pH or ionic strength, makes the PNA an interesting segment to organize synthetic materials.

Like oligopeptides, PNAs are synthesized via SPPS following Fmoc or tBoc strategies. The main difficulty arises from the expensive synthesis that makes only small amounts feasible (~10 μmol).

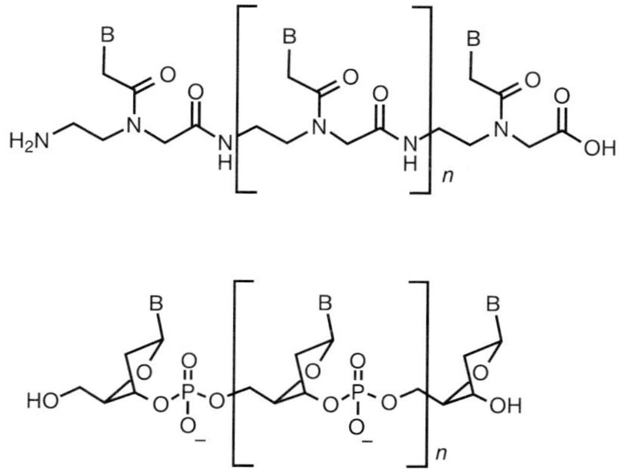

Fig. 15.2 Structures of "classical" PNA with an achiral poly-[*N*-(2-aminoethyl)glycine] backbone and polyelectrolytic DNA.

The conjugation methodologies described above to access peptide–polymer conjugates can be applied to introduce PNA segments to synthetic polymers. Armitage and coworkers [80] utilized the peptide macroinitiator approach [42, 43] and ATRP to synthesize a PNA-*block*-poly(2-hydroxyethyl acrylate) with a number-average molecular weight of 18.5 kDa and a polydispersity index of 1.06.

15.3
Secondary to Quaternary Structures: Polymer Self-assembly

15.3.1
Structure Formation Driven by Non-specific Interactions

As long as structure formation of block copolymers is driven solely by the counterbalance of thermodynamic demixing and chain elasticity, the size and the shape of superstructures can be controlled via the relative composition of polymer chains and the overall number of repeating units [81, 82]. The morphologies most commonly observed in solution are spherical and cylindrical micelles and vesicles. In the bulk, the corresponding phases are cubical, cylindrical and lamellar structures, plus some rarer structures. Deviations from the conventional phase behavior are expected when additional energy contributions come into play, such as from dipole–dipole interactions between helical chains of polypeptides or polysaccharides.

15.3.1.1 **Aggregates in Solution**

The morphology of aggregates produced by a hybrid block copolymer with a soluble polypeptide block is mainly determined by copolymer composition (or volume fraction of components) [83–85]. For instance, polybutadiene–poly(L-glutamate) diblock copolymers containing 70–75 mol% glutamate form small spherical micelles in aqueous solution, whereas samples containing 17–54 mol% glutamate form unilamellar vesicles [84, 85]. Whether or not the presence of α-helical polypeptide chains in the corona of aggregates leads to a shift in the phase diagram has not been elaborated. It is noteworthy that these aggregates should be in a thermodynamic equilibrium state, due to the low glass transition temperature of the core-forming polybutadiene. The occurrence of unprecedented structures of polypeptide copolymers with a glassy synthetic block (polylactide or polystyrene) [86, 87] might be explained in terms of non-equilibrium effects.

The occurrence of more complex phase behavior can be expected for hybrid copolymers with a core-forming polypeptide block because the establishment of hydrogen bridges and dipole–dipole interactions between α-helices is not disturbed or decoupled by solvent molecules. Indeed, several examples have been described where copolymers form complex coacervates or micellar clusters [88], rod-like assemblies [89], uni- or multilamellar vesicles [85, 90–93]; the structures found, however, might be non-equilibrium structures.

A secondary structure effect is clearly seen for an asymmetric polystyrene–poly[(*Z*)-L-lysine)] containing 30 mol% peptide, which forms in tetrachloromethane multilamellar cylindrical vesicles, where "ordinary" block copolymers would have formed spherical micelles instead [93]. Another example of secondary structure-induced assembly is that of comb-shaped copolymers made of a polyallylamine backbone and pendent poly(γ-methyl L-glutamate), which aggregates into amyloid-like fibrils in aqueous solution. The copolymer first assembles into small globules, which with time transform into long fibrils. Simultaneously and seemingly the driving force of the change in morphology, the polypeptide changes its secondary structure from α-helix to β-sheet (Fig. 15.3) [94]. Likewise, a γ-benzyl L-glutamate decamer with a phosphate head group was found to form a globular assembly in water, which slowly transforms into a ribbon and then a twisted ribbon-like aggregate. The formation of ribbons ought to be driven by the stacking of anti-parallel peptide β-sheets [95, 96].

Block copolypeptides, which are copolymers with two homopeptide blocks, also show a rich phase behavior due to secondary structure effects. Micelles and hexagonally shaped platelets have been observed for poly(L-leucine)–poly(sodium L-glutamate)s in water [97]. The formation of platelets was rationalized by the formation of leucine zippers in a three-dimensional array. Block copolymers comprising helical oligo(L-leucine) and charged poly(L-lysine) segments can produce hydrogels at polymer concentrations as low as 0.25 wt% [98, 99]. The scaffold of the gel is built of 5–10 nm wide twisted fibrillar tapes, with the oligo(L-leucine) helices packed perpendicular to the fibril axes (Fig. 15.4).

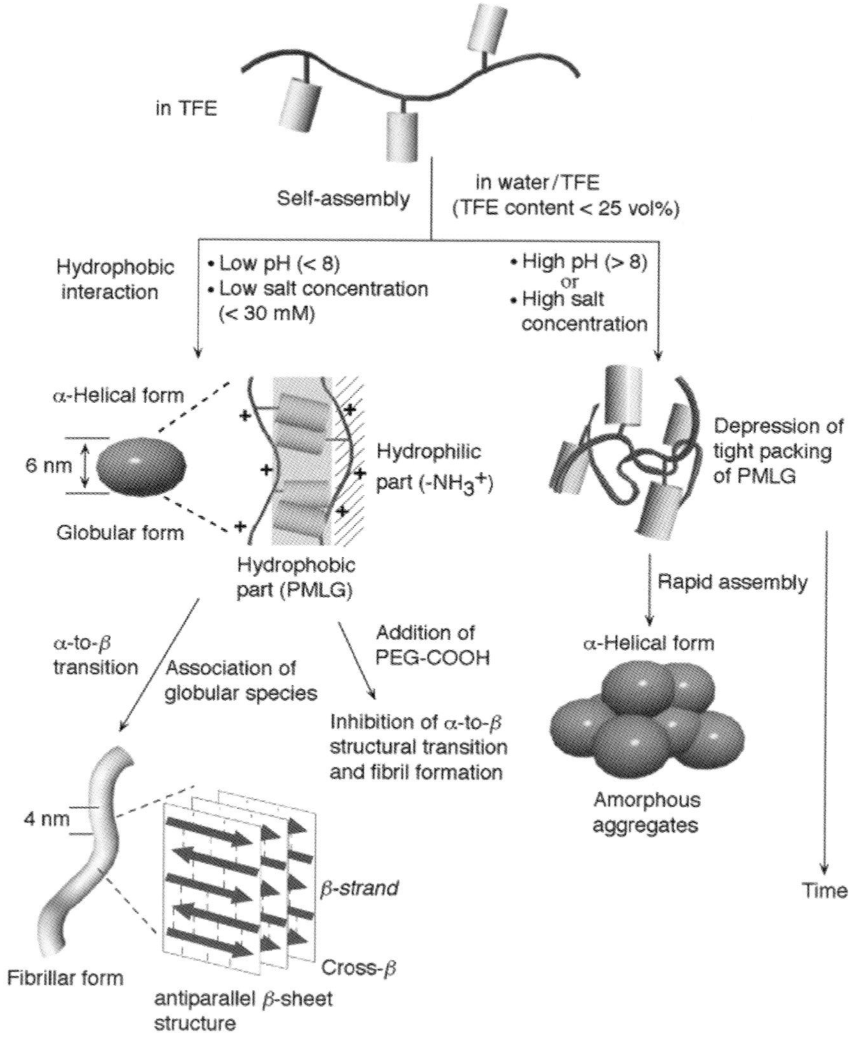

Fig. 15.3 Schematic illustration of the different morphologies and conformations of a polyallylamine-*graft*-poly(γ-methyl L-glutamate) in aqueous solution. Reprinted with permission from [94], copyright 2003 Wiley-VCH.

Asymmetric block copolypeptides made of L-leucine and ethylene glycol-modified L-lysine residues, which are all-helical molecules, self-assembled in aqueous solutions into vesicles as well as sheet-like membranes. The assembly into bilayers is an atypical phenomenon for asymmetric block copolymers and is related to a secondary structure effect. Accordingly, racemic samples with the

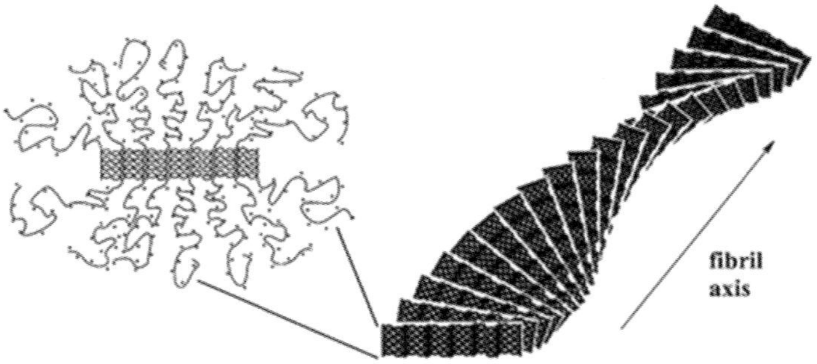

Fig. 15.4 Drawing showing the packing of block copolypeptide into twisted fibrillar tapes. Reprinted with permission from [98], copyright 2005 The Royal Society of Chemistry.

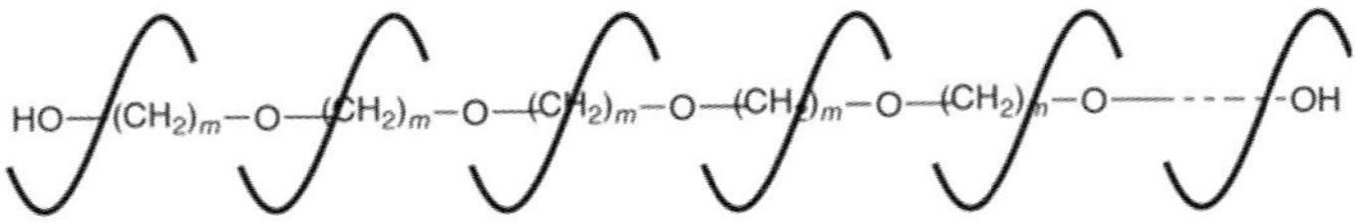

Fig. 15.5 Sketch of the structure of an amylose–polyether inclusion complex. Reprinted with permission from [103], copyright 2002 Wiley-VCH.

same composition but non-helical chain conformation were not found to form vesicles [100].

Polysaccharide-based copolymers appear to be rather difficult to handle, especially concerning equilibration of the samples. However, polystyrene–amylose block copolymers do not seem to be able to form other than spherical aggregates, in a selective solvent for polystyrene (tetrahydrofuran) or for amylose (water) [101]. A secondary structure effect due to the amylose α-helix could not be recognized.

Amylose can act a host molecule for hydrophobic polyesters or polyethers in aqueous solution, the amylose being wrapped around the synthetic polymer chain like a vine (Fig. 15.5) [102, 103]. Such inclusion complexes are formed when the enzymatic polymerization of α-D-glucose-1-phosphate is performed in the presence of the guest polymer. Complex formation fails, however, when amylose is just mixed with polymer.

Stimulus-responsive aggregates

Polybutadiene–poly(L-glutamate)s are considered to be stimulus-responsive materials in aqueous solution. Changing the pH has impact on the degree of ionization and hence the secondary structure of the poly(L-glutamate) segment. However, it was found that the morphology of aggregates (micelles or vesicles) does not change when the conformation of the polypeptide is switched from a random coil (pH > 6) to an α-helix (pH < 5) [84, 85]. Coiled and α-helical polypeptide chains seem to have similar spatial requirements at the core–corona interface [83]. On the other hand, the hydrodynamic size of aggregates was found to decrease by up to 20%, which was attributed to the different contour lengths of all-*trans* and α-helical poly(L-glutamate) chains [85]. An even greater change in size (up to 50%) was observed for polyisoprene–poly(L-lysine) micelles in water [104].

Block copolypeptides with short segments of L-glutamate and L-lysine (15 units each) can assemble into "schizophrenic" vesicles in aqueous solution [105]. At low pH (< 4), the oligo(L-glutamate) block is assumed to be in an α-helical conformation and thus be forming the core of the vesicle membrane; the corona is made of charged oligo(L-lysine). At pH > 10, the structure of the vesicle membrane is reversed. No aggregates are present in the region between pH 5 and 9.

Vesicles of block copolypeptides with L-leucine and ethylene glycol-modified L-lysine residues gained pH responsiveness when L-leucine residues in the hydrophobic part were statistically substituted by L-lysine. Acidification of the solution resulted in protonation of amine functionalities and the nearly instantaneous disruption of the membrane of vesicles [100].

15.3.1.2 **Solid-state Structures**

Bulk structures made of a polybutadiene, polystyrene or poly(ethylene oxide) synthetic block and α-helical poly(γ-benzyl L-glutamate) or poly[(*Z*)-L-lysine)] peptidic block(s) were described first by Gallot and coworkers [106, 107] and have subsequently been extensively examined for di- and triblock copolymers [108]. Almost exclusively observed were lamellar superstructures, irrespective of the composition of the copolymer, with dimensions of a few tens of nanometers. The preferential formation of lamellae is related to the stiffness of the polypeptide layer. Driven by dipole–dipole interactions, the polypeptide helices are usually arranged in an anti-parallel orientation and densely packed into a two-dimensional hexagonal array; the distance between helices is about 1.5 nm. Such hexagonal-in-lamellar structures can therefore be considered as a hierarchical structure. Owing to a non-uniform chain length distribution of helices, lamellar interfaces are not planar but undulated or zigzagged (Fig. 15.6) [109, 110]. Long helical chains are usually fold back to compensate for the internal dipole moment which obviously can only be built up to a critical value.

Structures with a higher level of hierarchical ordering could be seen for solvent-annealed thin films of polystyrene–poly(γ-benzyl L-glutamate) on a silicon substrate [111]. On the smallest length-scale, the structure was found to be built of short ribbons or lamellae of interdigitated polymer chains. Depending on the time of solvent annealing, different ordered structures on the micrometer length scale could be observed (Fig. 15.7). So far, a comprehensive picture of the processes involved in the formation of these structures is lacking.

Low molecular weight oligostyrene–poly(γ-benzyl L-glutamate)s can form various morphologies including hexagonal-in-hexagonal, hexagonal or lamellar

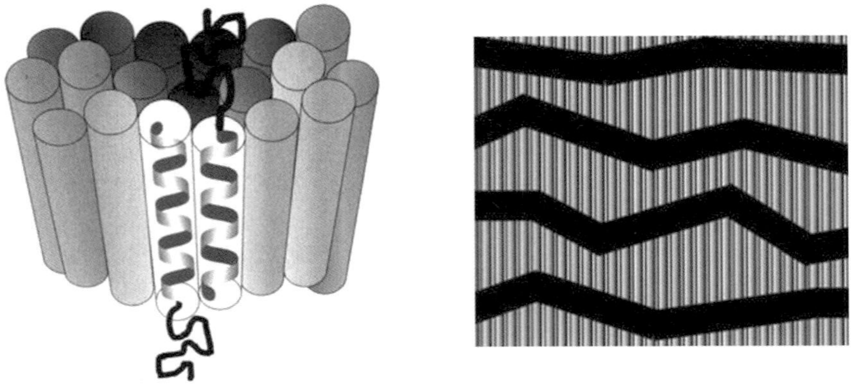

Fig. 15.6 Drawing of the hexagonal-in-lamellar morphology of polypeptide block copolymers. Reprinted with permission from [110], copyright 2004 American Chemical Society.

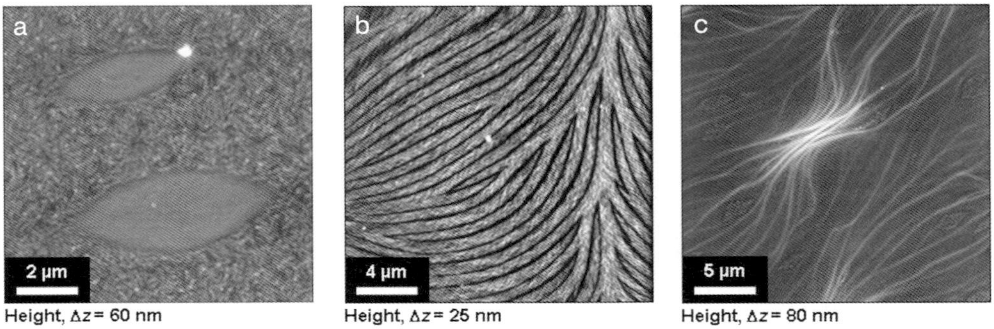

Fig. 15.7 Scanning force micrographs of a thin film of polystyrene–poly(γ-benzyl L-glutamate), solvent-annealed for (a) 3.5, (b) 22.5 and (c) 42 h. Reprinted with permission from [111], copyright 2004 American Chemical Society.

structures [112, 113]. Lamellar structures were also observed if the peptide block was too short to form a stable α-helix and formed a β-sheet instead, e.g. oligo(γ-benzyl L-glutamate) with an average number of 10 glutamate units.

15.3.2
Structure Formation Driven by Specific Interactions

15.3.2.1 Directed Peptide–Peptide Interactions

β-Sheet motif

An extension of the previously described concept is the use of specific interactions between polymer segments, for instance hydrogen bridging interactions

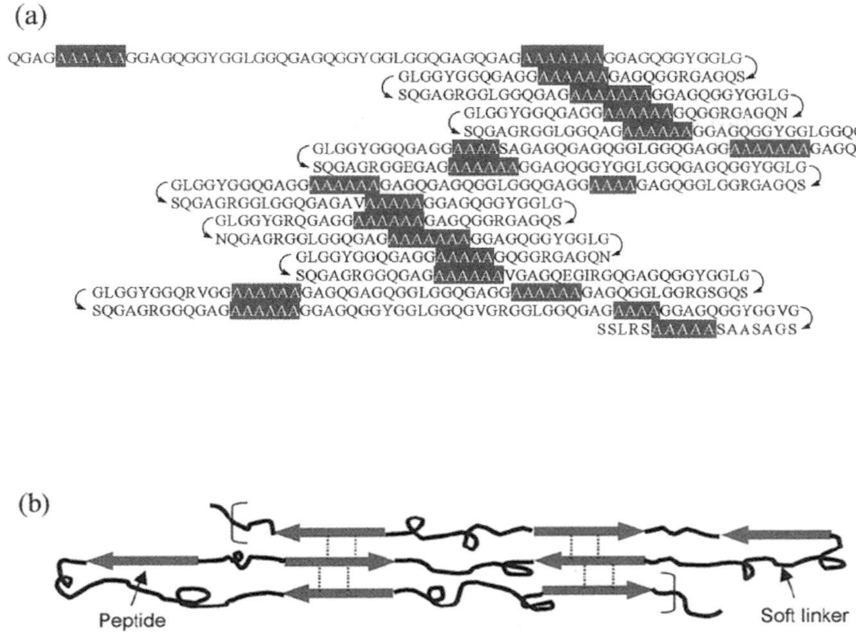

Fig. 15.8 Schematic illustration of natural silk mimics: *N. clavipes* major ampullate sequence showing the alanine-rich *β*-sheet domains (highlighted) and glycine-rich amorphous regions (a) and generic structure of silk-inspired multiblock copolymers in which the amorphous segments are replaced by PEG blocks (b). Reprinted with permission from [115], copyright 2001 American Chemical Society.

between *β*-strand-forming peptides. Inspired by the structure of natural silk, Sogah and coworkers introduced a synthetic oligoalanine–PEG multiblock copolymer as a silk mimic (Fig. 15.8) [114–116]. Due to the high propensity of alanine oligomers to adopt a *β*-sheet secondary structure, the peptide segments form physically cross-linked *β*-sheet domains. Since these are interconnected by flexible PEG segments, the polymer behaves like a thermoplastic elastomer. It was demonstrated by IR spectroscopy and X-ray diffraction that the function of the peptide domains was intact and in a *β*-sheet conformation.

In an early example, an aromatic hairpin turn was used to nucleate parallel *β*-sheet formation in silk-based multiblock copolymers composed of PEO and the peptide hairpins (Fig. 15.9). It was shown by Sogah and coworkers that preformed secondary structures (phenoxathin templated parallel *β*-sheets) within appropriate building blocks keep their structure in the final polymer, both in the solid state and in solution [114, 117]. The length of the PEO linker was found to be important, the longer the chain the higher the flexibility and the better the ordering.

However, even if the molecular structure of spider silk had been mimicked appropriately, the properties of the material were not comparable with the high-

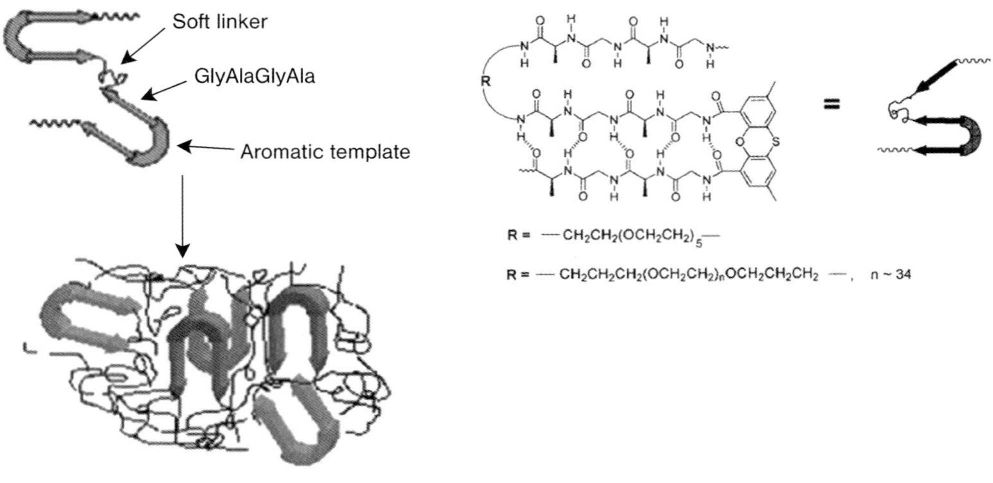

Fig. 15.9 Schematic illustration of a segmented block copolymer consisting of soft PEO and (Gly–Ala)$_2$ β-sheets, templated by phenoxathin. Reprinted with permission from [116], copyright 2001 American Chemical Society.

performance properties of drag line spider silk, etc. This is due to the elaborate processing of spider silk in the silk glands, which is an extrusion process in which sheer orientation is combined with controlled increase of peptide concentration. It is obvious that mimicking Nature is a complex task and multiple disciplines are required in order to achieve similar or even better properties or functions than existing native systems.

The β-sheet secondary structure motif might be one of the most suitable organization motifs for peptide-guided assembly of synthetic polymers. This is due to intense investigations, contributing to the understanding of this basic secondary structure element [118–124]. Unisometric tape structures can be accessed by directed self-organization of oligopeptides that are highly interesting for material science (Fig. 15.10c) [119]. The tape propagation frequently proceeds in an autocatalytic manner and could lead in principle to structures with indefinite length. Moreover, several distinct hierarchical structure levels exist, as outlined in Fig. 15.10. These levels are spanning structures from nanometers to several micrometers since β-strands can assemble to β-sheets, which can further stack to double sheets, denoted as ribbons (Fig. 15.10, b → c → d). Additional aggregation can lead to fibrils and fibers with a well-defined inner structure and, depending on the peptide sequence, a length of micrometers (Fig. 15.10, e → f).

Fibrils or fibers are important structural elements in native materials. Diverse properties such as anisotropic strength [125], structural stability [126], precise spacing of chemical functionalities in defined geometries [127, 128] and directed transport [126] can be realized with these structures. In materials science, nanofibers are of high interest due to their potential applications such as high-

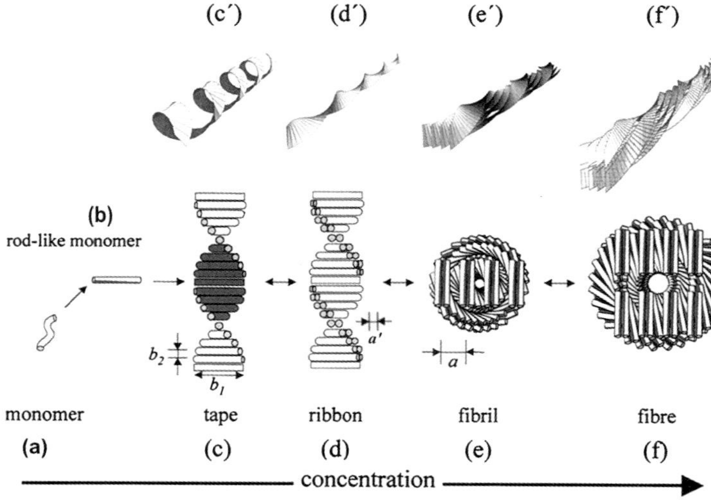

Fig. 15.10 Schematic illustration of hierarchical self-assembly of β-sheet forming peptides. Local arrangements (c)–(f) and the corresponding global equilibrium conformations (c′)–(f′): peptide unimer, statistical segment configuration (a); peptide unimer, β-strand configuration (b); peptide aligned to form helical β-sheet tapes (c, c′); twisted ribbons (d, d′); fibrils ($e_{front\ view}$, e′); and fiber ($f_{front\ view}$, f′). Geometric sizes: inter-strand spacing in a β-sheet (b_2=0.47 nm); tape width, equal to the length of a β-strand ($b_1 = n_{amino\ acids} \times 0.35$ nm); β-sheet-β-sheet spacing in a ribbon (a'=0.9–1.2 nm); inter ribbon distance in the fibril (a=1.6–2.4 nm). Reprinted with permission from [119], copyright 2001 National Academy of Sciences USA.

strength components in composite materials [129], nanowires [130], fibers for medical applications [131], or macromolecular actuators [132, 133]. Therefore, preparation of polymer nanofibers via peptide-guided routes is currently a field of high interest [134].

Lynn and coworkers pioneered this route to self-assembled polymer fibers exhibiting a peptide core and a polymer shell [135, 136]. A conjugate of PEO with the Aβ10–30 fragment of the amyloid sequence, associated with e.g. Alzheimer's disease, was assembled in a controlled manner to achieve soluble amyloidal aggregates. Although the work aimed to prevent the lateral aggregation of the amyloid fibrils, allowing the investigation of fibrillogenesis at dispersed structures in aqueous solution, the concept of peptide-guided assembly of polymers was already visible. It was shown that the association of the Aβ10–30 fragment into β-sheet structures results in a well-defined peptide core, forcing the PEO into the shell of this fibrillar aggregate.

Collier and Messersmith demonstrated that also shorter oligopeptides (e.g. 11-mer) have the inherent potential to self-organize PEO into soluble fibers [137]. The observed fibrillar structures (10–15 nm in width and several hundred nanometers in length) exhibit a strong reduction of the lateral association due to the PEO shell (cf. Koga et al. [94], Fig. 15.3). The applied peptide was devel-

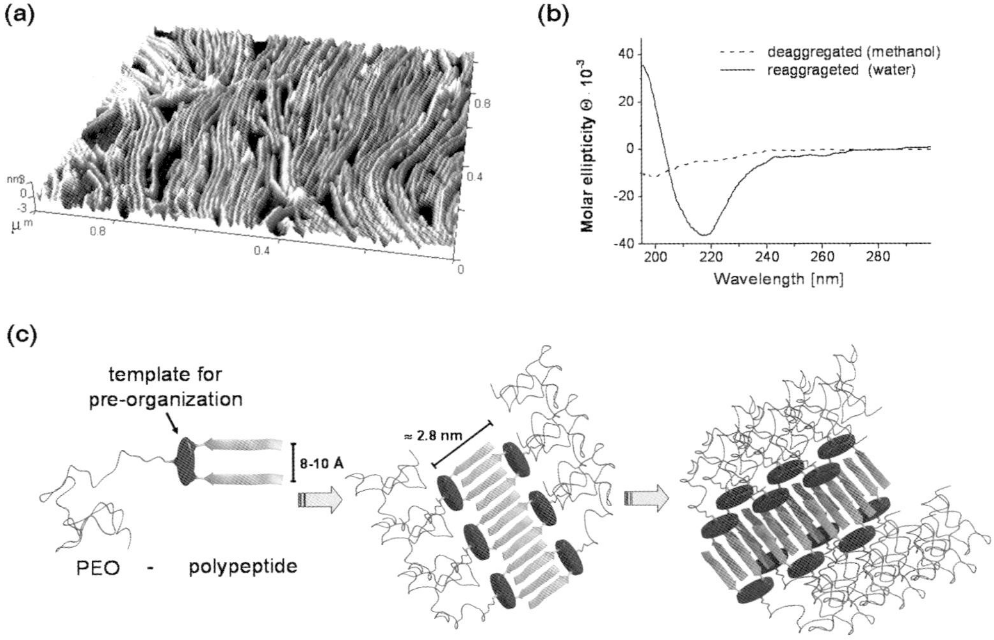

Fig. 15.11 (a) AFM height micrograph (tapping mode) of self-assembled PEO–fiber aggregates (0.4 mg mL^{-1} aqueous solution, spin-coated on mica) and (b) CD spectra of 1 mg mL^{-1} solutions before and after assembly. (c) Tentative mechanism of the formation of fiber-like aggregates exhibiting peptide core and PEO shell. Reprinted with permission from [139], copyright 2005 The Royal Society of Chemistry.

oped as a transglutaminase substrate for enzymatic functionalization of supramolecular structures and proved to have a tendency to assemble into β-sheet fibrils. This supports the hypothesis that fibrillar β-sheet structures can be potentially seen as energy minimal structures that appear to be more common than expected and not restricted to certain amino acid sequences [138]. However, assembly of the PEO–peptide conjugate required a defined pH and ionic strength to nucleate aggregation.

Börner and coworkers strongly increased the tendency of the oligopeptide segment in a PEO–peptide conjugate to form defined β-sheet structures, avoiding the synthesis of long peptides with high aggregation potential [139]. In SPPS such peptides are denoted as "difficult sequences" because of their high tendency towards aggregation [140, 141]. However, the synthesis of peptides with difficult sequences could be avoided by tethering two short tetrapeptides [$(Val\text{-}Thr)_2$] to a template. This results in the preorganization of the peptide strands, meeting the geometry of the β-sheet (Fig. 15.11 c) [118] and thus enhancing the aggregation tendency due to entropic reasons. It was shown that the linear $(Val–Thr)_4$ peptide, if conjugated to a PEO, could not induce aggregation, even though the sequence was two times longer than the preorganized $(Val–Thr)_2$ strands.

A controlled de-aggregation/re-aggregation approach was used to obtain tape-like structures showing a rather stiff appearance as indicated by the nematic-like order (Fig. 15.11 a) [139]. During the aggregation process, a structural transition of the peptide strands from a statistical configuration to the β-sheet could be observed by UV (CD) circular dichroism spectroscopy (Fig. 15.11 b). The analysis of the fibers by means of TEM and AFM revealed an average height of the fiber of 1.4 nm, a width of 14 nm and a maximum length of about 2 μm. As illustrated in Fig. 15.11 c, the fiber structure formed is stabilized by the formation of an anti-parallel β-sheet of the oligopeptide units. Thus, the aggregate exhibits a core–shell structure comprising an oligopeptide β-sheet core and a PEO shell. The latter contributes to solubility of the fiber and suppresses lateral aggregation. The fiber structures are composed of double β-sheets denoted ribbons. The ribbon formation is driven by the hydrophobic effect in order to minimize the contact energy of the hydrophobic valine face of the β-sheets with water. This assumption is supported by the average height of the fibers (~1.4 nm). Since a ribbon usually exhibits a 0.9–1.2-nm sheet–sheet spacing, it might be within the range of experimental error [139].

Van Hest and coworkers have extended the approach of peptide-guided assembly from peptides to genetically engineered proteins [142]. Two different $[(AG)_3EG]_n$ peptides with $n=10$ and 20 were synthesized by genetically modified *E. coli* hosts using recombinant methods. The peptide sequence consists of repetitive β-sheet-forming domains $(Ala–Gly)_3$ followed by a Glu–Gly sequence that provides flexibility to the strand and allows a reverse turn. The N- and C-terminal conjugation of PEO to this multi-block protein results in an ABA conjugate that exhibits a central β-sheet domain (Fig. 15.12 a). The PEO blocks prevent lateral aggregation of the peptides, directing the assembly process towards distinct needle-shaped lamellar crystals (Fig. 15.12 b). The formation of extended plate-like, multi-lamellar crystals, as was shown by Tirrell and coworkers, is prevented [143, 144]. Two-dimensional crystallization results in well-defined fibrils (Fig. 15.12 c, d) that possess an inner β-sheet core with a height defined by the $(AG)_3$ β-sheet domain and a width approximately correlating with the number of $[(AG)_3EG]_n$ repeats, as was shown for congener systems [144].

Hamley et al. [145] investigated the self-assembly of diverse PEO–peptide conjugates in aqueous solution. The conjugates were constructed from amphiphilic β-strand peptide sequences and PEO blocks of various molecular weights and positions. The peptide sequence utilized was a derivative of a peptide reported by Fukushima [146]. This stimuli-responsive peptide exhibits an amphiphilic α-amino acid sequence and thus a transition in secondary structure could be induced. This transition leads from isolated α-helices under acidic or basic conditions to extended β-sheets at neutral pH, making this peptide responsive to external stimuli, such as pH or sample concentration [146]. In this study, the Fukushima sequence was derivatized and the H-GELDLELDQQKLKLKLKG-OH and also the PEO conjugate H-GELDLELDQQKLKLKLKG-NH-PEO (molecular weight of PEO: 3 kDa) was investigated among other systems [145]. CD experiments indicated that conjugation of PEO stabilizes the β-sheet secondary struc-

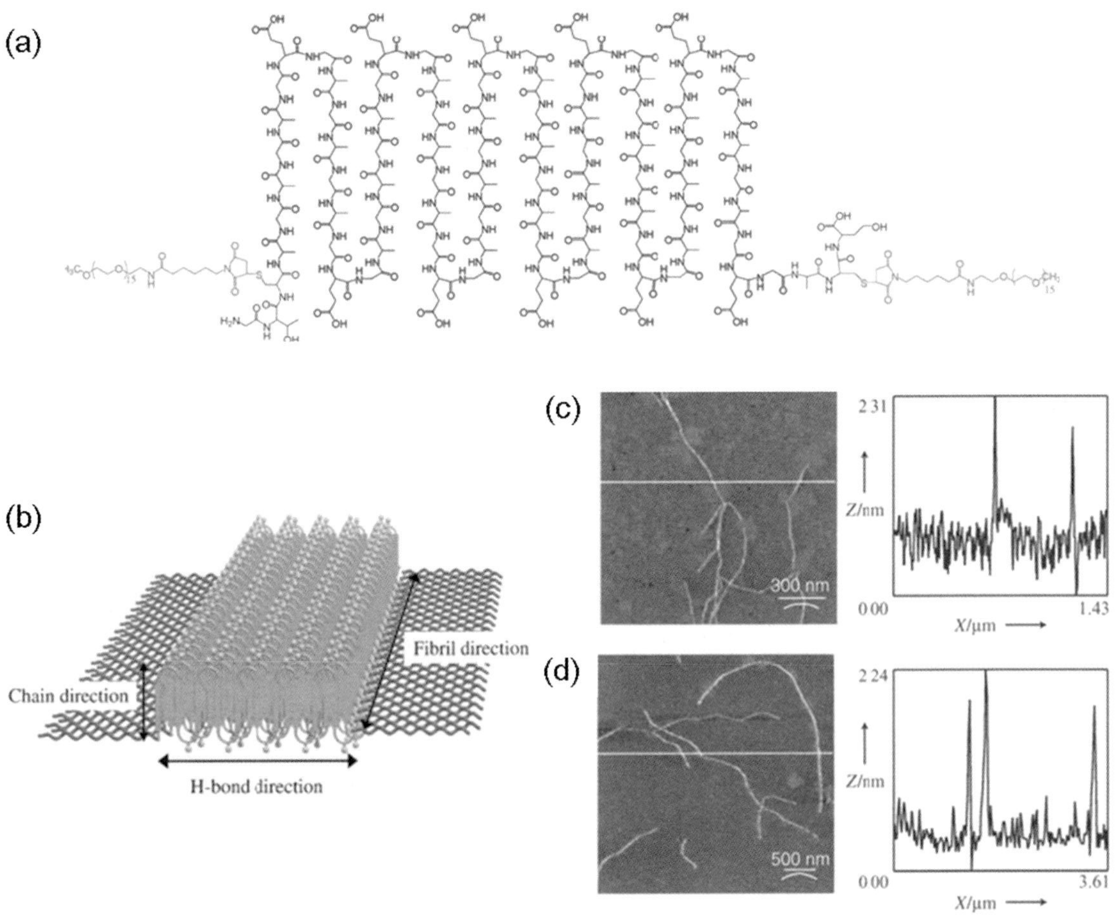

Fig. 15.12 (a) ABA-type block copolymer formed by reaction of maleimide-functionalized PEO with the cysteine-flanked $[(AG)_3EG]_n$ β-sheet domain. (b) Schematic illustration of the proposed assembly of the conjugate in the β-sheet fibrils. Tapping-mode AFM images of β-sheet fibrils on a mica: height image and height profiles of (c) PEO–$[(AG)_3EG]_{10}$–PEO and (d) PEO–$[(AG)_3EG]_{20}$–PEO conjugates. Reprinted with permission from [142], copyright 2005 Wiley-VCH.

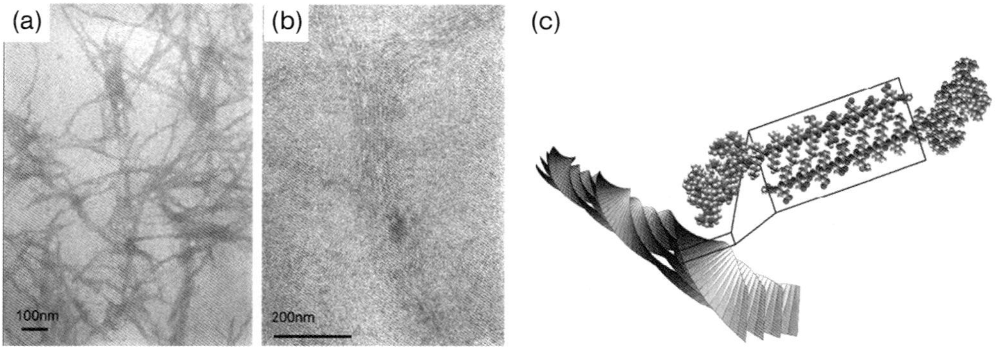

Fig. 15.13 TEM images of dried films of (a) non-conjugated peptide (pH 11) and (b) a PEO conjugate (pH 3). (c) Sketch of a fibril formed by aggregation of β-strands. Reprinted with permission from Ref. [145], copyright 2001 American Chemical Society.

ture in the conjugate as compared with the non-conjugated peptide sequence. Moreover, a reduced sensitivity of the peptide secondary structure transition toward pH variations was observed, making the structural transition difficult. It is noteworthy that SAXS experiments suggested the formation of rod-like aggregates for all of the investigated samples, irrespective of peptide secondary structure. This was confirmed by TEM showing fibrillar structures for both the conjugated and the non-conjugated peptides at different pH (Fig. 15.13 a, b). It has been suggested that these structures might be formed by self-assembly of the conjugates in twisted, helical tapes (Fig. 15.13 c). These structures might then stack into fibrillar rods, as was observed for non-conjugated peptides [119]. However, even though the concept was shown to work, it remains difficult to realize a distinct structural transition in bio-conjugate systems by applying moderate stimuli between pH 5 and 9.

Van Hest and coworkers investigated a peptide-polymer conjugate with a peptide comb structure, differing from the linear conjugates described above [50]. An ABA-triblock ("pom-pom") copolymer was synthesized by ATRP of a methacryloyl-terminated $(Ala\text{-}Gly)_2$ sequence using a bifunctional PEO-based macroinitiator. The oligopeptide side-chains have a tendency to form β-sheet structures, as verified by IR spectroscopy. However, the restriction in flexibility of the peptides by attaching the segments to a polymer backbone seemed to prevent the formation of extended β-sheets or larger ordered structures.

Elastin mimics

Elastin can be considered as a very important structural protein in Nature [147–149]. It is commonly found in the skin and ligaments and also arteries and lung tissue of mammals, expressing the function of an elastic material. Most frequently, tropoelastin was studied as the precursor protein of the mammalian elastin. The functional domain of elastin contains a repetitive sequence VPGVG (V = valine, P = proline, G = glycine). The aggregation, conformational and mechanical properties of both chemically synthesized and genetically engineered poly(VPGVG) were extensively investigated. Poly(VPGVG) exhibits lower critical solution behavior (LCST), originating from a conformational change in elastin from statistical chain segment configuration to a β-spiral. This completely reversible transition is driven by desolvation of the valine side-chains. The transition temperature can be fine-tuned by substitution of the second valine. Moreover, pH and temperature sensitivity can be introduced by the addition of either acidic or basic amino acids in the fourth position.

Van Hest and coworkers synthesized different branched polymers with oligoelastin side-chains [49]. These were obtained by ATRP of a methacryloyl-functionalized VPGVG segment. The polymerization of this oligopeptide macromonomer was initiated either by a low molecular weight initiator, resulting in a comb-shaped polymer, or by a bifunctional PEO macroinitiator, resulting in a "pom-pom"-shaped polymer (see above). CD measurements indicated that the VPGVG side-chain segments of the ABA block copolymer maintained their function. The characteristic LCST behavior could be observed upon heating a solution of the conjugate in water. The resulting turbidity change was correlated with the reversible assembly of the block copolymer, as verified by light scattering. Depending on the pH of the solution, the LCST varied between 40 and 70 °C.

Coiled-coil folding motif

The coiled-coil structure motif can be found in more than 200 native proteins. It results from a directed aggregation of right-handed α-helices into a left-handed superhelical structure [150]. The reason for the bundle formation of two or four or sometimes more α-helices is the amphiphilicity of these α-helices that exhibit hydrophobic patches along the helix being grouped on one side. Thus, the assembly in an aqueous environment is mainly driven by the hydrophobic or, being more exact, entropic effect. Further energy contributions frequently result from the formation of ion pairs, additionally stabilizing the superstructure. The primary structure of a coiled-coil domain is characterized by a heptad repeating sequence, $(abcdefg)_n$. Positions a and d are occupied by hydrophobic amino acid residues leading in the α-helical structure to a hydrophobic patch along the helix side. The formation kinetics and stability of coiled-coil structures depend on the amino acid sequence and the number of heptad repeats [151].

The distinctive features of the coiled coils with respect to specific spatial recognition, association and fully reversible dissociation of helices was used by two groups independently to induce order in synthetic polymers [53, 152, 153]. Pechar et al. designed a series of coiled-coil forming peptides and conjugated these to PEO with a molecular weight of 2 kDa [153]. The repetitive sequence $(VSSLESK)_n$ (S = serine, E = glutamic acid, K = lysine) was chosen with $n = 2$–6 as common peptide sequence, meeting the requirements for a coiled-coil domain [154]. It was shown that the helicity of peptides increased with increasing length

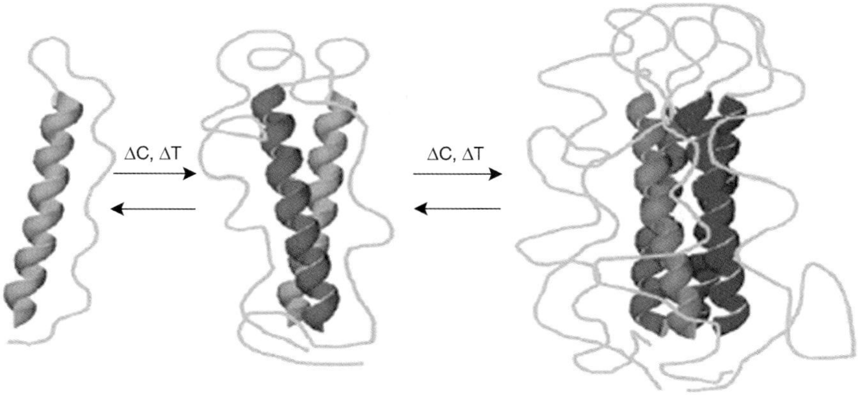

Fig. 15.14 Self-assembly of peptide–PEO conjugates into coiled-coil aggregates. Reprinted with permission from [53], copyright 2003 American Chemical Society.

in a cooperative manner and that the peptide–PEO conjugates with 35 and 42 amino acid residues adopted a two-stranded α-helical coiled-coil conformation in aqueous solution. The conjugation of the PEO block to the peptides was proven to have no influence on the conformation of the peptides and the formation kinetics of the coiled-coil superstructure. However, a stabilizing effect was observed with respect to the superstructure, showing higher thermal stability.

Klok and coworkers investigated the structure and dynamics of the self-assembly of a series of *de novo* peptides and compared them with those of peptide–PEO conjugates [53, 152]. It was evident that the conjugation of PEO to the peptide led to increased thermal stability of the coiled-coil aggregates. Moreover, the longer the PEO chain, the lower was the relative concentration of the coiled-coil aggregates, which could be attributed to steric hindrance. Electron spin resonance (ESR) measurements on spin-labeled conjugates indicated that the PEO chains are tightly wrapped around the aggregated coiled-coil domain and that the peptide helices are aligned in parallel (Fig. 15.14).

15.3.2.2 **Directed Oligonucleotide Hybridization (DNA–DNA Interactions)**

DNA provides a highly selective and easily predictable organization principle, associated with a hybridization event. Two single-stranded DNAs with a complementary sequence of nucleotide bases can form a helical dimer or duplex with well-defined structure. This hybridization is highly sequence specific and can be observed already with oligonucleotides made of 8–15 base pairs. Duplex formation is a fully reversible process that proceeds in a cooperative manner. The deaggregation can be triggered by temperature, pH and ionic strength. The transition temperature is tunable within a limited range, depending on the length of the oligonucleotide and, most important, on the ratio of adenine–thymine and

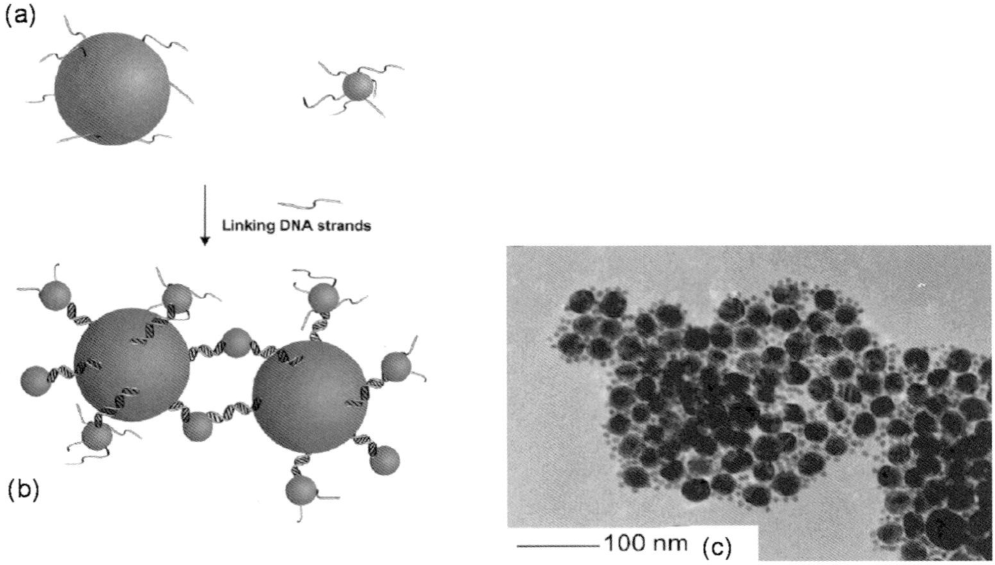

Fig. 15.15 Idealized mechanism for the DNA-mediated formation of binary nanoparticle networks (a → b). (c) TEM images of the binary DNA-linked network as an assembly formed from 8- and 30-nm gold colloids. Reprinted with permission from [70, 160], copyright 1998, 2000 American Chemical Society.

guanine–cytosine base pairs. The latter can be attributed to the increased stability of guanine–cytosine base pairs, since three hydrogen bonds are formed per pair, whereas adenine–thymine forms only two hydrogen bonds.

Mirkin and coworkers demonstrated the potential of DNA-controlled structure formation by the selective aggregation of gold colloids, mediated by hybridization of oligonucleotides with complementary sequences. It is noteworthy that the assembly process could be observed with the naked eye, due to the size dependence of the plasmon resonance band of colloidal gold. Moreover, by programming two different batches of gold colloids with selective single-strand oligonucleotide labels (Fig. 15.15 a) it was possible to form two- and three-dimensional structures (Fig. 15.15 b, c) [70, 155]. This approach was extended to other colloids, including diverse nanoparticles [156–159], in order to achieve functional architectures. However, the majority of work was done with gold colloids, which are easy to handle and straightforward to functionalize, but are not very interesting with respect to function.

The reversible cross-linking of polymer–DNA hybrids was demonstrated by Mirkin and coworkers using oligonorbornenes with multiple DNA segments grafted on to the backbone [72]. Two differently modified oligonorbornenes were prepared, each with complementary 12-mers of DNA. After mixing aqueous so-

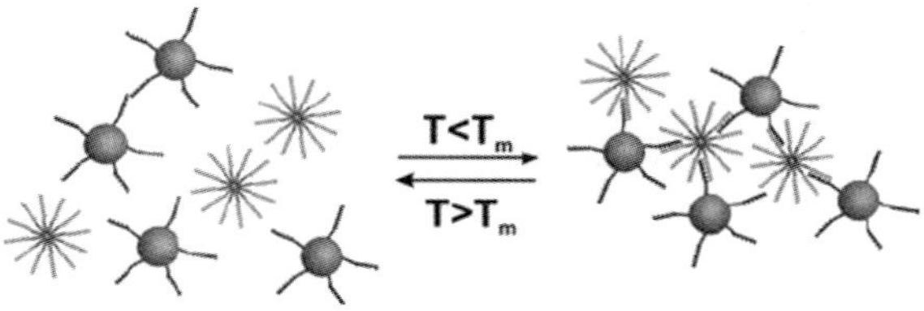

Fig. 15.16 DNA-directed sequence-specific assembly of DNA-block-polymer micelles and DNA-modified gold nanoparticles. Reprinted with permission from [71], copyright 2004 American Chemical Society.

lutions of both DNA–polymer conjugates, hybridization occurs instantaneously and triggers the precipitation of the polymer. The cooperative hybridization between the complementary side-chain DNA segments of the different polymers was thermally reversible and indicates that the attachment to a polymer does not hinder the recognition properties of the oligonucleotides.

This concept was expanded to realize probes for electrochemical detection of DNA binding events. Therefore, the reversible binding of polynorbornene block copolymers containing oligonucleotide and ferrocenyl or dibromoferrocenyl side-chains to a gold electrode could be triggered by an analyte DNA. The hybridization event leads to a readable redox signal that allows the detection of a single-strand DNA segment with a distinct sequence in the pM range and with one point-mutation selectivity [161].

The generation of block copolymer micelles by the aggregation of an amphiphilic polymer combining a hydrophobic polystyrene block and a hydrophilic 20-mer-oligonucleotide segment leads to colloidal building blocks with recognition properties [71]. These can be programmed via the DNA sequence. An organic–inorganic network was assembled through DNA-directed sequence-specific hybridization by mixing the micellar structures and DNA-modified gold nanoparticles (Fig. 15.16). The assembled aggregates can be reversibly de-aggregated by heating above the "melting temperature" of the duplex strand interconnects, as could be monitored by the surface plasmon band of the gold nanoparticles.

Comparable to the hybridization of two single-strand DNA segments with complementary base sequences, duplex formation also takes place with PNAs as synthetic analogues. This was demonstrated by Armitage and coworkers, who investigated the PNA-directed, sequence-specific assembly of a PNA–polymer hybrid with a trifunctional PNA-linker molecule (Fig. 15.17) [80]. The self-assembly occurs in a thermo-reversible fashion, mediated by the duplex formation between complementary PNA strands of the polymer and the trifunctional cross-linker. In this basic study, only trimer formation was observed by light scattering. However, the utilization of telechelic polymers with two terminal PNA segments would lead to gel formation. By extending this self-assembly strategy, stable hydrogels could be made that are more tolerant towards hydrolysis and enzymatic degradation and variations of pH and ionic strength as compared with DNA-mediated interlinks.

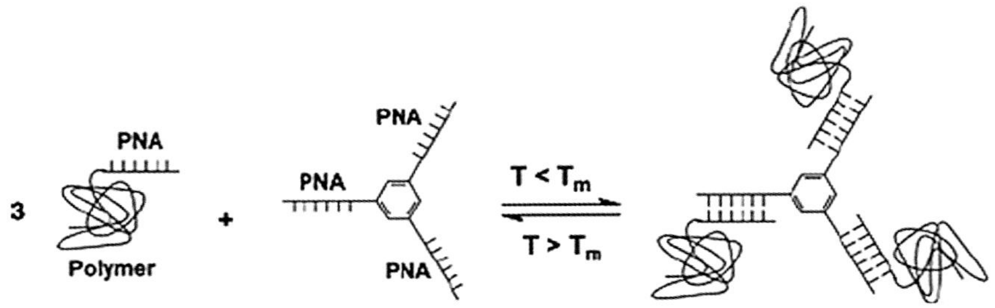

Fig. 15.17 Schematic illustration of thermally reversible coupling via PNA strands on a trifunctional cross-linker and the complementary sequence on a PNA–polymer hybrid. Reprinted with permission from [80], copyright 2005 American Chemical Society.

15.4 Future Perspectives

A large number of excellent studies have contributed to the field of bioinspired block copolymers with secondary interactions, especially by paving synthetic pathways, but the field is still in its infancy. Special effects and benefits are waiting to be harvested. Looking at the superior materials properties of natural polymer structures such as tissue, tendon, spider silk and the ultra-hard and ultra-thin shell of a virus cage, one can imagine the huge potential of biomimetic polymers.

In our opinion, self-reinforcing polymers where textures develop in film or bulk after processing (e.g. by depsipeptide switching) are just one step beyond current synthetic possibilities. If this assembly can be directed by an outer stimulus, human technology has made the first step towards responsive structures such as the supported growth of plant structures and plant motility.

Similarly nearby for technology are films with a biofunctional surface, made by classical extrusion or precipitation spinning steps, but without demanding post-functionalization. Here, minor amounts of the appropriate polymer–biopolymer conjugate are mixed with the corresponding classical bulk polymer and interface effects will drive the biofunctionality to enrich at the interface, for instance by spinning through a water-bath. Such fabrics and films could stimulate cell adherence (by a sufficiently exposed RGD motif), exhibit antibiotic function (for medical tissues) or simply improve colorability and touch. The task here is mainly clever engineering.

For applications in pharmacy, chimera blocks, bioconjugates and peptide-grafted polymers have already reached advanced clinical trial stages [162, 163]. Actual targets here are still tissue specificity by appropriate outer active or passive homing codons and more effective gene transfection. In order to translocate through both the outer cell membrane and the nuclear lipid bilayer barrier, multistage shuttle concepts with multiresponsive polymers are potentially needed.

Less demanding and much closer to classical polymer science is to apply controlled polymerization techniques not only to peptide systems but also to non-peptidic systems. The combination of a functional monomer library with a sequence-controlled arrangement of single entities has a potential which makes a much broader application very meaningful. The resulting multifunctional or "hyperfunctional" polymers are certainly very important in classical high-value applications of functional polymers, such as pigment and toner stabilization, mineralization and precipitation control or for galenical purposes.

With the possibility of having the aggregation behavior controlled by specific secondary interaction, aggregates are inherently functional in an operative sense. For instance, by modifying the secondary interactions, by either pH, salt or temperature, a small change is amplified into a larger structural change and this macroscopic change can be used to generate motility, release, adherence or flow, to mention just a few. Although structural features are certainly important for generating extreme polymer properties such as super-hardness or super-elasticity, it is also the functional side which makes biological polymers so promising for a number of key applications.

References

1 T. Thurn-Albrecht, J. Schotter, C.A. Kastle, N. Emley, T. Shibauchi, L. Krusin-Elbaum, K. Guarini, C.T. Black, M.T. Tuominen, T.P. Russell, *Science* **2000**, *290*, 2126.

2 C.J. Hawker, T.P. Russell, *MRS Bull.* **2005**, *30*, 952.

3 I.W. Hamley, *Nanotechnology* **2003**, *14*, R39.

4 I.W. Hamley, *Angew. Chem. Int. Ed.* **2003**, *42*, 1692.

5 M. Antonietti, C. Goltner, *Angew. Chem. Int. Ed.* **1997**, *36*, 910.

6 H. Leuchs, *Ber. Dtsch. Chem. Ges.* **1906**, *39*, 857.

7 B. Gallot, *Prog. Polym. Sci.* **1996**, *21*, 1035.

8 H. Schlaad, M. Antonietti, *Eur. Phys. J. E* **2003**, *10*, 17.

9 T.J. Deming, *Nature* **1997**, *390*, 386.

10 I. Dimitrov, H. Schlaad, *Chem. Commun.* **2003**, 2944.

11 W. Vayaboury, O. Giani, H. Cottet, A. Deratani, F. Schué, *Macromol. Rapid Commun.* **2004**, *25*, 1221.

12 T. Aliferis, H. Iatrou, N. Hadjichristidis, *Biomacromolecules* **2004**, *5*, 1653.

13 G. Ziegast, B. Pfannemüller, *Carbohydr. Res.* **1987**, *160*, 185.

14 K. Akiyoshi, M. Kohara, K. Ito, S. Kitamura, J. Sunamoto, *Macromol. Rapid Commun.* **1999**, *20*, 112.

15 K. Loos, A.H.E. Müller, *Biomacromolecules* **2002**, *3*, 368.

16 G. Ziegast, B. Pfannemüller, *Makromol. Chem.* **1984**, *185*, 1855.

17 K. Loos, R. Stadler, *Macromolecules* **1997**, *30*, 7641.

18 S. Kamiya, K. Kobayashi, *Macromol. Chem. Phys.* **1998**, *199*, 1589.

19 W.T.E. Bosker, K. Ágoston, M.A. Cohen Stuart, W. Norde, J.W. Timmermans, T.M. Slaghek, *Macromolecules* **2003**, *36*, 1982.

20 B. Pfannemüller, *Naturwissenschaften* **1975**, *62*, 231.

21 D.H. Haddleton, K. Ohno, *Biomacromolecules* **2000**, *1*, 152.

22 K. Kobayashi, S. Kamiya, N. Enomoto, *Macromolecules* **1996**, *29*, 8670.

23 T.J. Magliery, *Med. Chem. Rev. Online* **2005**, *2*, 303.

24 S.J. Anthony-Cahill, T.J. Magliery, *Curr. Pharm. Biotechnol.* **2002**, *3*, 299.

25 T.A. Cropp, P.G. Schultz, *Trends Genet.* **2004**, *20*, 625.
26 A.J. Link, M.L. Mock, D.A. Tirrell, *Curr. Opin. Biotechnol.* **2003**, *14*, 603.
27 J.M. Xie, P.G. Schultz, *Methods* **2005**, *36*, 227.
28 K.P. McGrath, M.J. Fournier, T.L. Mason, D.A. Tirrell, *J. Am. Chem. Soc.* **1992**, *114*, 727.
29 P.E. Dawson, S.B.H. Kent, *Annu. Rev. Biochem.* **2000**, *69*, 923.
30 L. Hartmann, E. Krause, M. Antonietti, H.G. Börner, *Biomacromolecules* **2006**, *7*, 1239.
31 P.E. Nielsen, in *Pseudo-peptides in Drug Discovery*, VCH-Wiley, **2004**, p. 256.
32 G.J. Cotton, M.C. Pietanza, T.W. Muir, *Drugs Pharm. Sci.* **2000**, *101*, 171.
33 J.P. Tam, Q. Yu, Z. Miao, *Biopolymers* **2000**, *51*, 311.
34 S. Kent, *J. Pept. Sci.* **2003**, *9*, 574.
35 P.E. Dawson, T.W. Muir, I. Clarklewis, S.B.H. Kent, *Science* **1994**, *266*, 776.
36 L.E. Canne, S.J. Bark, S.B.H. Kent, *J. Am. Chem. Soc.* **1996**, *118*, 5891.
37 P. Botti, M.R. Carrasco, S.B.H. Kent, *Tetrahedron Lett.* **2001**, *42*, 1831.
38 T.W. Muir, *Annu. Rev. Biochem.* **2003**, *72*, 249.
39 M.L. Becker, J.Q. Liu, K.L. Wooley, *Chem. Commun.* **2003**, 180.
40 M.L. Becker, J. Liu, K.L. Wooley, *Biomacromolecules* **2005**, *6*, 220.
41 S.-T. Yang, S. Yub Shin, Y.-C. Kim, Y. Kim, K.-S. Hahm, J.I. Kim, *Biochem. Biophys. Res. Commun.* **2002**, *296*, 1044.
42 Y. Mei, K.L. Beers, H.C.M. Byrd, D.L. Van der Hart, N.R. Washburn, *J. Am. Chem. Soc.* **2004**, *126*, 3472.
43 H. Rettig, E. Krause, H.G. Börner, *Macromol. Rapid Commun.* **2004**, *25*, 1251–1256.
44 J. Pyun, T. Kowalewski, K. Matyjaszewski, *Macromol. Rapid Commun.* **2003**, *24*, 1043.
45 N. Ayres, D.M. Haddleton, A.J. Shooter, D.A. Pears, *Macromolecules* **2002**, *35*, 3849.
46 S. Angot, N. Ayres, S.A.F. Bon, D.M. Haddleton, *Macromolecules* **2001**, *34*, 768.
47 M. Husseman, E.E. Malmstrom, M. McNamara, M. Mate, D. Mecerreyes, D.G. Benoit, J.L. Hedrick, P. Mansky, E. Huang, T.P. Russell, C.J. Hawker, *Macromolecules* **1999**, *32*, 1424.
48 M.G.J. ten Cate, H. Rettig, K. Bernhardt, E. Krause, H.G. Börner, *Macromolecules* **2005**, *38*, 10643.
49 L. Ayres, M.R.J. Vos, P. Adams, I.O. Shklyarevskiy, J.C.M. van Hest, *Macromolecules* **2003**, *36*, 5967.
50 L. Ayres, P.H.H.M. Adams, D.W.P.M. Loewik, J.C.M. Van Hest, *Biomacromolecules* **2005**, *6*, 825.
51 L. Ayres, G.M. Grotenbreg, G.A. van der Marel, H.S. Overkleeft, M. Overhand, J.C.M. van Hest, *Macromol. Rapid Commun.* **2005**, *26*, 1336.
52 A. Godwin, M. Hartenstein, A.H.E. Müller, S. Brocchini, *Angew. Chem. Int. Ed.* **2001**, *40*, 594.
53 G.W.M. Vandermeulen, C. Tziatzios, H.A. Klok, *Macromolecules* **2003**, *36*, 4107.
54 A. Rosler, H.A. Klok, I.W. Hamley, V. Castelletto, O.O. Mykhaylyk, *Biomacromolecules* **2003**, *4*, 859.
55 F. Albericio, *Curr. Opin. Chem. Biol.* **2004**, *8*, 211.
56 J.P. Tam, J. Xu, K.D. Eom, *Biopolymers* **2001**, *60*, 194.
57 K.E. Beatty, F. Xie, Q. Wang, D.A. Tirrell, *J. Am. Chem. Soc.* **2005**, *127*, 14150.
58 A.J. Link, D.A. Tirrell, *J. Am. Chem. Soc.* **2003**, *125*, 11164.
59 J.A. Opsteen, J.C.M. van Hest, *Chem. Commun.* **2005**, 57.
60 J.-F. Lutz, H.G. Börner, K. Weichenhan, *Macromol. Rapid Commun.* **2005**, *26*, 514.
61 A.J. Dirks, S.S. van Berkel, N.S. Hatzakis, J.A. Opsteen, F.L. Van Delft, J.J.L.M. Cornelissen, A.E. Rowan, J.C.M. van Hest, F.P.J.T. Rutjes, R.J.M. Nolte, *Chem. Commun.* **2005**, 4172.
62 M. Morpurgo, F.M. Veronese, *Methods Mol. Biol.* **2004**, *283*, 45.
63 A.S. Hoffman, P.S. Stayton, *Macromol. Symp.* **2004**, *207*, 139.
64 D. Lowik, J.C.M. van Hest, *Chem. Soc. Rev.* **2004**, *33*, 234.
65 K. Velonia, A.E. Rowan, R.J.M. Nolte, *J. Am. Chem. Soc.* **2002**, *124*, 4224.
66 H.A. Klok, *J. Polym. Sci., Part A: Polym. Chem.* **2005**, *43*, 1–17.

67 B. S. Lele, H. Murata, K. Matyjaszewski, A. J. Russell, *Biomacromolecules* **2005**, *6*, 3380.

68 J. J. Storhoff, C. A. Mirkin, *Chem. Rev.* **1999**, *99*, 1849.

69 C. M. Niemeyer, *Angew. Chem. Int. Ed.* **2001**, *40*, 4128.

70 C. A. Mirkin, *Inorg. Chem.* **2000**, *39*, 2258.

71 Z. Li, Y. Zhang, P. Fullhart, C. A. Mirkin, *Nano Lett.* **2004**, *4*, 1055.

72 K. J. Watson, S. J. Park, J. H. Im, S. T. Nguyen, C. A. Mirkin, *J. Am. Chem. Soc.* **2001**, *123*, 5592.

73 N. L. Rosi, C. A. Mirkin, *Chem. Rev.* **2005**, *105*, 1547.

74 B. Samori, G. Zuccheri, *Angew. Chem. Int. Ed.* **2005**, *44*, 1166.

75 P. E. Nielsen, M. Egholm, R. H. Berg, O. Buchardt, *Science* **1991**, *254*, 1497.

76 P. E. Nielsen, *Bioconj. Chem.* **1991**, *2*, 1.

77 E. Uhlmann, A. Peyman, *Chem. Rev.* **1990**, *90*, 543.

78 E. Uhlmann, A. Peyman, G. Breipohl, D. W. Will, *Angew. Chem. Int. Ed.* **1998**, *37*, 2797.

79 T. Ratilainen, A. Holmen, E. Tuite, G. Haaima, L. Christensen, P. E. Nielsen, B. Norden, *Biochemistry* **1998**, *37*, 12331.

80 Y. Wang, B. A. Armitage, G. C. Berry, *Macromolecules* **2005**, *38*, 5846.

81 F. S. Bates, G. H. Fredrickson, *Annu. Rev. Phys. Chem.* **1990**, *41*, 525.

82 S. Förster, M. Antonietti, *Adv. Mater.* **1998**, *10*, 195.

83 A. Nakajima, K. Kugo, T. Hayashi, *Macromolecules* **1979**, *12*, 844.

84 H. Kukula, H. Schlaad, M. Antonietti, S. Förster, *J. Am. Chem. Soc.* **2002**, *124*, 1658.

85 F. Chécot, A. Brûlet, J. Oberdisse, Y. Gnanou, O. Mondain-Monval, S. Lecommandoux, *Langmuir* **2005**, *21*, 4308.

86 H. Arimura, Y. Ohya, T. Ouchi, *Biomacromolecules* **2005**, *6*, 720.

87 A. Lübbert, V. Castelletto, I. W. Hamley, H. Nuhn, M. Scholl, L. Bourdillon, C. Wandrey, H.-A. Klok, *Langmuir* **2005**, *21*, 6582.

88 K. Naka, R. Yamashita, T. Nakamura, A. Ohki, S. Maeda, *Macromol. Chem. Phys.* **1997**, *198*, 89.

89 D. Tang, J. Lin, S. Lin, S. Zhang, T. Chen, X. Tian, *Macromol. Rapid Commun.* **2004**, *25*, 1241.

90 A. Toyotama, S.-i. Kugimiya, J. Yamanaka, M. Yonese, *Chem. Pharm. Bull.* **2001**, *49*, 169.

91 C.-M. Dong, K. M. Faucher, E. L. Chaikof, *J. Polym. Sci., Part A: Polym. Chem.* **2004**, *42*, 5754.

92 C.-M. Dong, X.-L. Sun, K. M. Faucher, R. P. Apkarian, E. L. Chaikof, *Biomacromolecules* **2004**, *5*, 224.

93 H. Schlaad, *Adv. Polym. Sci.* **2006**, *202*, 53.

94 T. Koga, K. Taguchi, Y. Kobuke, T. Kinoshita, M. Higuchi, *Chem. Eur. J.* **2003**, *9*, 1146.

95 T. Doi, T. Kinoshita, H. Kamiya, Y. Tsujita, H. Yoshimizu, *Chem. Lett.* **2000**, 262.

96 T. Doi, T. Kinoshita, H. Kamiya, S. Washizu, Y. Tsujita, H. Yoshimizu, *Polym. J.* **2001**, *33*, 160.

97 A. Constancis, R. Meyrueix, N. Bryson, S. Huille, J.-M. Grosselin, T. Gulik-Krzywicki, G. Soula, *J. Colloid Interf. Sci.* **1999**, *217*, 357.

98 T. J. Deming, *Soft Matter* **2005**, *1*, 28.

99 V. Breedveld, A. P. Nowak, J. Sato, T. J. Deming, D. J. Pine, *Macromolecules* **2004**, *37*, 3943.

100 E. G. Bellomo, M. D. Wyrsta, L. Pakstis, D. J. Pochan, T. J. Deming, *Nat. Mater.* **2004**, *3*, 244.

101 K. Loos, A. Böker, H. Zettl, M. Zhang, G. Krausch, A. H. E. Müller, *Macromolecules* **2005**, *38*, 873.

102 J.-i. Kadokawa, Y. Kaneko, A. Nakaya, H. Tagaya, *Macromolecules* **2001**, *34*, 6536.

103 J.-i. Kadokawa, Y. Kaneko, S.-i. Nagase, T. Takahashi, H. Tagaya, *Chem. Eur. J.* **2002**, *8*, 3322.

104 J. Babin, J. Rodríguez-Hernández, S. Lecommandoux, H.-A. Klok, M.-F. Achard, *Faraday Discuss.* **2005**, *128*, 179.

105 J. Rodríguez-Hernández, S. Lecommandoux, *J. Am. Chem. Soc.* **2005**, *127*, 2026.

106 A. Douy, B. Gallot, *Polymer* **1982**, *23*, 1039.

107 B. Perly, A. Douy, B. Gallot, *Makromol. Chem.* **1976**, *177*, 2569.
108 H.A. Klok, S. Lecommandoux, *Adv. Polym. Sci.* **2006**, *202*, xxx.
109 H. Schlaad, H. Kukula, B. Smarsly, M. Antonietti, T. Pakula, *Polymer* **2002**, *43*, 5321.
110 H. Schlaad, B. Smarsly, M. Losik, *Macromolecules* **2004**, *37*, 2210.
111 S. Ludwigs, G. Krausch, G. Reiter, M. Losik, M. Antonietti, H. Schlaad, *Macromolecules* **2005**, *38*, 7532.
112 H.-A. Klok, J.F. Langenwalter, S. Lecommandoux, *Macromolecules* **2000**, *33*, 7819.
113 S. Lecommandoux, M.-F. Achard, J.F. Langenwalter, H.-A. Klok, *Macromolecules* **2001**, *34*, 9100.
114 M.J. Winningham, D.Y. Sogah, *Macromolecules* **1997**, *30*, 862.
115 O. Rathore, D.Y. Sogah, *J. Am. Chem. Soc.* **2001**, *123*, 5231.
116 O. Rathore, D.Y. Sogah, *Macromolecules* **2001**, *34*, 1477.
117 O. Rathore, M.J. Winningham, D.Y. Sogah, *J. Polym. Sci., Part A: Polym. Chem.* **2000**, *38*, 352.
118 C.L. Nesloney, J.W. Kelly, *Bioorg. Med. Chem.* **1996**, *4*, 739.
119 A. Aggeli, I.A. Nyrkova, M. Bell, R. Harding, L. Carrick, T.C.B. McLeish, A.N. Semenov, N. Boden, *Proc. Natl. Acad. Sci. USA* **2001**, *98*, 11857.
120 W.A. Loughlin, J.D.A. Tyndall, M.P. Glenn, D.P. Fairlie, *Chem. Rev.* **2004**, *104*, 6085.
121 S.G. Zhang, *Biotechnol. Adv.* **2002**, *20*, 321.
122 A. Aggeli, M. Bell, N. Boden, J.N. Keen, T.C.B. McLeish, I. Nyrkova, S.E. Radford, A. Semenov, *J. Mater. Chem.* **1997**, *7*, 1135.
123 A. Aggeli, M. Bell, L.M. Carrick, C.W.G. Fishwick, R. Harding, P.J. Mawer, S.E. Radford, A.E. Strong, N. Boden, *J. Am. Chem. Soc.* **2003**, *125*, 9619.
124 H.A. Lashuel, S.R. LaBrenz, L. Woo, L.C. Serpell, J.W. Kelly, *J. Am. Chem. Soc.* **2000**, *122*, 5262.
125 M.S. Sacks, C.J. Chuong, R. More, *ASAIO J.* **1994**, *40*, 632.
126 J. Howard, *Mechanics of Motor Proteins and the Cytoskeleton*, Sinauer Press, Sunderland, MA, **2001**.
127 R.S. Farmer, L.M. Argust, J.D. Sharp, K.L. Kick, *Macromolecules* **2006**, *39*, 162.
128 E. Yoshikawa, M.J. Fournier, T.L. Mason, D. A. Tirrell, *Macromolecules* **1994**, *27*, 5471.
129 R.H. Baughman, A.A. Zakhidov, W.A. de Heer, *Science* **2002**, *297*, 787.
130 L. Dai, in *Encyclopedia of Nanoscience and Nanotechnology*, Vol. 8, Ed. H.S. Nalva, American Scientific Publishers, Stevenson Ranch, CA, **2004**, p. 763.
131 Z.M. Huang, Y.Z. Zhang, M. Kotaki, S. Ramakrishna, *Compos. Sci. Technol.* **2003**, *63*, 2223.
132 I.Z. Steinberg, A. Oplatka, A. Katchalsky, *Nature* **1966**, *210*, 568.
133 M.O. Gallyamov, B. Tartsch, A.R. Khokhlov, S.S. Sheiko, H.G. Börner, K. Matyjaszewski, M. Möller, *Chem. Eur. J.* **2004**, *10*, 4599.
134 A. Aggeli, N. Boden, S. Zhang (Eds) *Self-Assembling Peptide Systems in Biology, Medicine and Engineering*, Kluwer, Dordrecht, **2001**.
135 T.S. Burkoth, T.L.S. Benzinger, D.N.M. Jones, K. Hallenga, S.C. Meredith, D.G. Lynn, *J. Am. Chem. Soc.* **1998**, *120*, 7655.
136 T.S. Burkoth, T.L.S. Benzinger, V. Urban, D.G. Lynn, S.C. Meredith, P. Thiyagarajan, *J. Am. Chem. Soc.* **1999**, *121*, 7429.
137 J.H. Collier, P.B. Messersmith, *Adv. Mater.* **2004**, *16*, 907.
138 J.I. Guijarro, M. Sunde, J.A. Jones, I.D. Campbell, C.M. Dobson, *Proc. Natl. Acad. Sci. USA* **1998**, *95*, 4224.
139 D. Eckhardt, M. Groenewolt, E. Krause, H.G. Börner, *Chem. Commun.* **2005**, 2814.
140 L.A. Carpino, E. Krause, C.D. Sferdean, M. Schuemann, H. Fabian, M. Bienert, M. Beyermann, *Tetrahedron Lett.* **2004**, *45*, 7519.
141 M. Quibbel, T. Johnson, *Difficult Peptides. Fmoc Solid Phase Peptide Synthesis. A Practical Approach*, Oxford University Press, Oxford, **2000**.
142 J.M. Smeenk, M.B.J. Otten, J. Thies, D.A. Tirrell, H.G. Stunnenberg,

J. C. M. van Hest, *Angew. Chem. Int. Ed.* **2005**, *44*, 1968.
143 M. T. Krejchi, E. D. T. Atkins, A. J. Waddon, M. J. Fournier, T. L. Mason, D. A. Tirrell, *Science* **1994**, *265*, 1427.
144 M. T. Krejchi, S. J. Cooper, Y. Deguchi, E. D. T. Atkins, M. J. Fournier, T. L. Mason, D. A. Tirrell, *Macromolecules* **1997**, *30*, 5012.
145 I. W. Hamley, I. A. Ansari, V. Castelletto, H. Nuhn, A. Rosler, H. A. Klok, *Biomacromolecules* **2005**, *6*, 1310.
146 Y. Fukushima, *Polym. J.* **1995**, *27*, 819.
147 D. W. Urry, C.-H. Luan, C. M. Harris, T. Parker, in *Protein-based Materials*, Eds K. McGrath, D. Kaplan, Birkhäuser, Boston, **1997**. p. 133.
148 M. Manno, A. Emanuele, V. Martorana, P. L. San Biagio, D. Bulone, M. B. Palma Vittorelli, D. T. McPherson, J. Xu, T. M. Parker, D. W. Urry, *Biopolymers* **2001**, *59*, 51.
149 J. C. M. van Hest, D. A. Tirrell, *Chem. Commun.* **2001**, 1897.
150 A. Lupas, *Trends Biochem. Sci.* **1996**, *21*, 375.
151 O. D. Monera, C. M. Kay, R. S. Hodges, *Biochemistry* **1994**, *33*, 3862.
152 G. W. M. Vandermeulen, D. Hinderberger, H. Xu, S. S. Sheiko, G. Jeschke, H. A. Klok, *Chem. Phys. Chem.* **2004**, *5*, 488.
153 M. Pechar, P. Kopeckova, L. Joss, J. Kopecek, *Macromol. Biosci.* **2002**, *2*, 199.
154 T. J. Graddis, D. G. Myszka, I. M. Chaiken, *Biochemistry* **1993**, *32*, 12664.
155 T. A. Taton, R. C. Mucic, C. A. Mirkin, R. L. Letsinger, *J. Am. Chem. Soc.* **2000**, *122*, 6305.
156 Y. Sun, C.-H. Kiang, in *Handbook of Nanostructured Biomaterials and Their Applications in Nanobiotechnology*, Vol. 2, Ed. H. S. Nalwa, American Scientific Publishers, Stevenson Ranch, **2005**, p. 223.
157 E. Katz, I. Willner, *Angew. Chem. Int. Ed.* **2004**, *43*, 6042.
158 C. M. Niemeyer, in *Nanoparticle Assemblies and Superstructures*, Ed. N. A. Kotov, CRC Press, Boca Raton, FL, **2006**, p. 227.
159 C.-J. Zhong, L. Han, N. Kariuki, M. M. Maye, J. Luo, in *Nanoparticle Assemblies and Superstructures*, Ed. N. A. Kotov, CRC Press, Boca Raton, FL, **2006**, p. 551.
160 R. C. Mucic, J. J. Storhoff, C. A. Mirkin, R. L. Letsinger, *J. Am. Chem. Soc.* **1998**, *120*, 12674.
161 J. M. Gibbs, S. J. Park, D. R. Anderson, K. J. Watson, C. A. Mirkin, S. T. Nguyen, *J. Am. Chem. Soc.* **2005**, *127*, 1170.
162 R. Duncan, *Nat. Rev. Drug Discov.* **2003**, *2*, 347.
163 K. Kataoka, A. Harada, Y. Nagasaki, *Adv. Drug Deliver. Rev.* **2001**, *47*, 113.

16
Complex Functional Macromolecules

Zhiyun Chen, Chong Cheng, David S. Germack, Padma Gopalan, Brooke A. van Horn, Shrinivas Venkataraman, and Karen L. Wooley

16.1
Introduction

There are many scholarly arguments across diverse fields from philosophy to science that consider how natural or synthetic objects gain complexity. One currently contentious issue is the debate of intelligent design versus the scientific facts of evolution. The intelligent design postulate that biological complexities are beyond construction in the absence of intervention and, therefore, are irreducibly complex [1], is overcome by the recognition that hierarchical assembly of evolved sub-components produces increasing levels of complexity. Remarkably, much of this argument is presented by considering a macroscopic device, a mousetrap [2, 3]. However, for completely different reasons and based upon completely different motivations, this same recognition that the building of increasing complexity via stagewise assembly of increasingly complex components has also been applied to complex synthetic functional materials of nanoscopic dimensions; the dimensional evolution of synthetic organic chemistry [4]. Although still far behind the billions of years over which natural systems have been allowed to evolve, the preparation, assembly and manipulation of well-defined synthetic macromolecules and nanostructures has undergone an enormous growth of activities and advances over the past couple of decades. By connecting atoms into small molecules, using those small molecules to construct synthetic macromolecules and then assembling those macromolecules into nanoscale objects, the stepwise increase in the complexity of the building blocks allows for higher degrees of control and complexity for the final functional system than could be achieved by construction directly from the smallest entities, the atomic components. There is much investigation remaining to achieve complexities for synthetic materials that can compare with those of biological nanomachines.

This chapter emphasizes synthetic methodological approaches toward complex functional macromolecules, for which the term "complex" is defined as

Macromolecular Engineering. Precise Synthesis, Materials Properties, Applications.
Edited by K. Matyjaszewski, Y. Gnanou, and L. Leibler

ISBN: 978-3-527-31446-1

originating from several parts and not necessarily analogous with complicated macromolecules. Discussion begins with the supramolecularly based self- or directed assembly of linear polymers to afford complex nanoscale structures having internal morphological sophistication. Emphases are placed upon assembly processes both in solution and in the bulk state. Initially, the fundamental parameters under which block copolymer assemblies in solution can afford discrete nanoscale objects of unique shapes and sizes are highlighted. Examples are then presented that involve the incorporation of conjugated polymer segments as components, to transform passive nanostructures to complex electronically and optically active functional nanoscopically resolved assemblies. Synthetic methodologies that allow for the preparation of functional brush copolymers are then described from the perspective of providing for increased control over the three-dimensional composition and structure of the resulting nanoscopic objects, by combining covalent chemical bonds to achieve particular molecular architectures with supramolecular interactions to gain macromolecular conformational control. Hierarchical structures that include organic and inorganic components, with assembly processes and structural designs borrowed from Nature and including functional macromolecular and nanoscale components prepared synthetically, are discussed with outlooks towards the future expectations for such hybridized systems to advance further the evolution of synthetic and biological materials. Finally, the ultimate fate of the materials and the need to impose mortality to the functional materials by using degradable components are addressed. These examples represent development mainly over the past 5 years to transform what may have been considered previously to be common "plastics" into sophisticated and functional devices, designed to perform their function and then undergo a dismantling process.

16.2 Functional Self-assembled Materials

16.2.1 Overview

One of the most versatile and powerful methods to construct complex functional macromolecular structures of nanoscopic dimensions involves the reliance upon macromolecular self-assembly. The local organization of macromolecular structural components and the long-range periodicity within resulting aggregates, which are governed by the self-assembly processes, have large effects on the properties of the final materials. Excellent reviews on self-assembled materials already exist with focus on templating applications [5], stimuli-responsive behaviors [6] and many other future directions for further development of well-defined supramolecular assemblies [7]. Summarized here are recent studies of self-assembled materials with emphasis on the functionality imparted by the association of many individual components into final polyvalent and multi-functional nanoscopic objects.

Self-assembly of block copolymers is of major interest because it allows for organizational control on the nanometer to micrometer scale, which cannot be achieved easily by other means [7, 8]. In block copolymers, two or more polymeric chains are covalently attached together. These distinct chain segments, usually thermodynamically immiscible, undergo phase segregation. The great flexibility to tune the size and functionality of the resulting nanostructures in a controllable manner makes multiblock copolymers excellent candidates for tailor-made materials.

Two main themes have surfaced from recent studies of the self-assembly processes of block copolymers, both of which are discussed in this section. One research thrust explores the parameters that govern self-assemblies and the scope of nano-objects that can be achieved by self-organization of block copolymers. The second focus places emphasis on the functionalization of self-assembled structures and the production of "smart materials".

16.2.2
Nanostructures from Self-assembly of Block Copolymers

Great efforts have been exerted to provide precise compositional and morphological controls, which are critical to the application of nanotechnology. The functionality of nanostructures depends on the properties of the monomeric units incorporated into the final structures, the spatial distribution and periodicity and the structural complexity.

The scope of monomers that have been incorporated into block copolymers has been expanded beyond the traditional recipe of styrene and its derivatives, (methyl)acrylates, dienes, vinylpyridines, ethylene and propylene oxides [9, 10], to new types of monomers including peptides [11–15], lactides [16–18], dendrimers [19], fluorinated monomers [20], ionomers [21], organometallics [22] and siloxanes [23, 24], among others. This expansion of compositional design space provides different properties and functionalities to the assembled structures. For example, stimuli-responsive vesicles have been constructed from synthetic-peptidic block copolymers of poly{N_ε-2-[2-(2-methoxyethoxy)ethoxy]acetyl-L-lysine}-*b*-poly(L-leucine) ($K^P_xL_y$) (Fig. 16.1). The transportation rates of vesicle-encapsulated Fura-2 dye or external calcium ions across the membrane barrier increased from several days to a few seconds when the pH of the solution was adjusted from 10.6 to 3.0 by addition of hydrochloric acid [25]. Klok et al. also demonstrated that the folding and associative behaviors of glutamic acid peptides were not affected by incorporating them as polymeric segments within diblock copolymers [14], suggesting interesting prospects to combine the structure-forming ability of synthetic polymers with the specific biological response of peptide sequences to prepare novel self-assembled and biologically active materials. In fact, the antimicrobial activity of tritrypticin was maintained and even enhanced, when conjugated as the hydrophilic chain terminus of poly(acrylic acid)-*b*-polystyrene

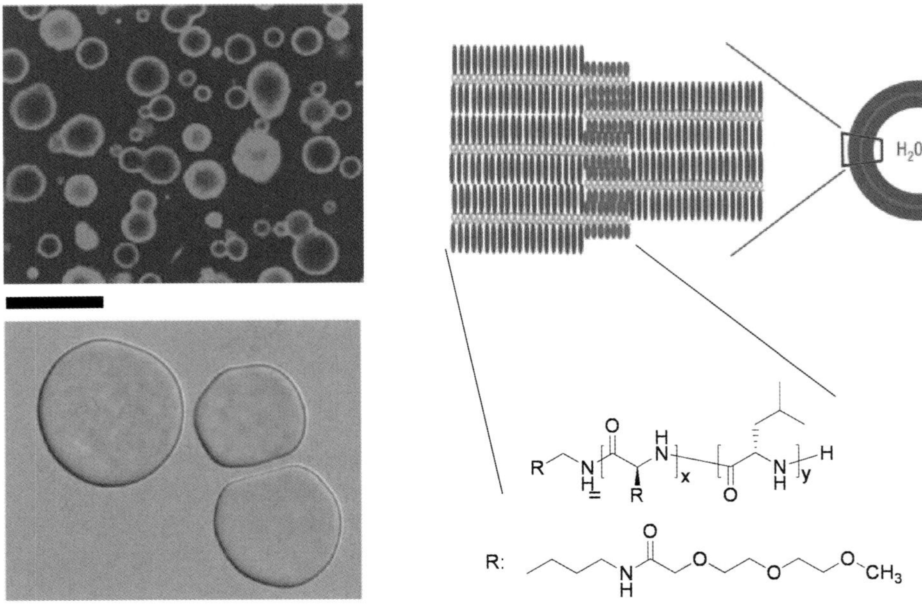

Fig. 16.1 Stimuli-responsive vesicles from self-assembly of $K^P_xL_y$: laser scanning confocal image of a $K^P_{100}L_{20}$ suspension visualized with $DiOC_{18}$ dye and a *z*-direction slice thickness of 490 nm (above) and differential interference contrast optical micrograph of the $K^P_{150}L_{40}$ sample (below). Scale bar: 20 μm. Reprinted with permission from [25].

(PAA-*b*-PSt) synthetic amphiphilic diblock copolymers and, thereby, presented polyvalently from the surface of polymer micelles in aqueous solution [26].

There is great expectation that building blocks of unique shapes will provide for variability and versatility for fabrication of hierarchical structures by "bottom-up" construction methods toward the development of nanotechnology. Spherical, cylindrical and layered structures are typical examples that are readily accessible from the self-assembly of diblock copolymers. It is also realized that nanostructures with novel morphologies can be obtained by using multiblock copolymers, because of the increased complexity of interactions that can occur between multiple blocks with thermodynamically distinct properties, varied according to the well understood Flory–Huggins interaction parameters. In addition to the increased number of types of interactions, their interplay, due to different spatial orientation, provides the fundamental basis for structural diversity and complexity and possibilities for higher order segregation. For example, toroidal micelles (Fig. 16.2) have been observed for the assembly of a linear triblock copolymer [27]. It was also demonstrated that no similar structures could be obtained from its diblock copolymer analogues or even triblock copolymer pseudo-isomer under identical assembly conditions [28]. Another novel self-as-

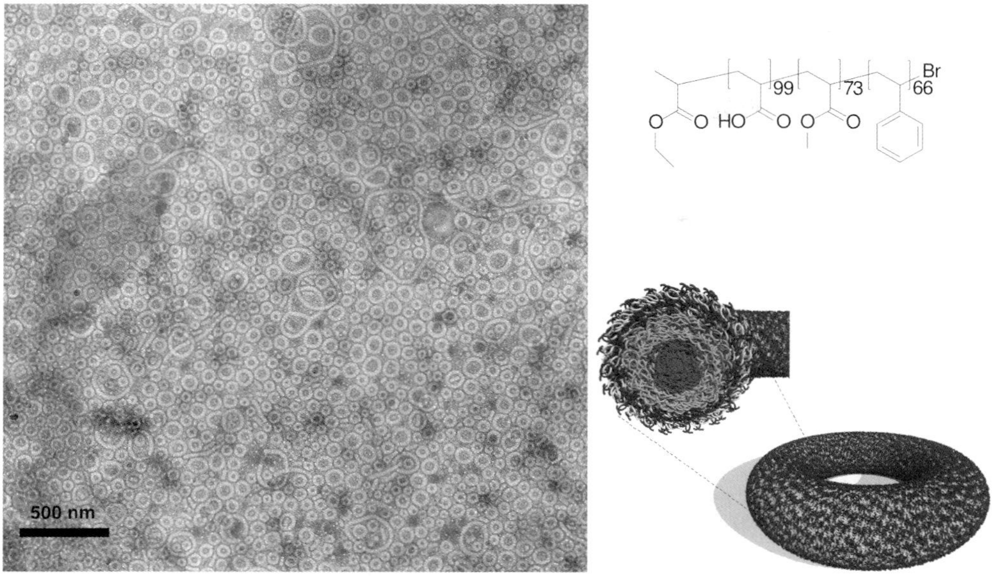

Fig. 16.2 Novel nanostructures: toroidal assembly from PAA_{99}-*b*-PMA_{73}-*b*-PSt_{66}. Reprinted with permission from [27].

sembled morphology, consisting of stacked discs, was reported from both the self-segregation of nonionic, three-arm-shaped triblock copolymers [29] and linear polyelectrolyte triblock copolymers [30], indicating the universality of this type of aggregation. Other novel nanostructures from self-assembly have included bowl-shaped micelles [31] and bamboo-like structures [32].

Another goal of investigating self-assembly of block copolymers is to achieve structural complexity or secondary phase segregation, which mimics the diverse structures of biological systems such as those offered by the virus and the bacterium. For a common core–shell forming protocol for block copolymers, secondary segregation can be made to take place within either the shell or the core region. Erhardt and coworkers demonstrated that Janus micelles were constructed by chemically fixing and resuspending a self-assembled triblock copolymer [33, 34]. These Janus micelles had a condensed hydrophobic core with the shell shared by two immiscible polymeric chains (Fig. 16.3 a). In another account, Liu and coworkers prepared micelles with segregated cores by mixing AB- and AC-type block copolymers [35]. Recently, the same group reported a clever mixing strategy to build a shell-segregated "bumpy" nanostructure sharing the same core component [36] (Fig. 16.3 b). Vesicles with differential inside and outside functionalities across the wall, which mimic the asymmetric membrane of a living cell, have been assembled from both block ionomers [37, 38] and ABC-type triblock copolymers [23]. Small molecules have also been used to co-assemble with block copolymers to form hybrid complex structures. These

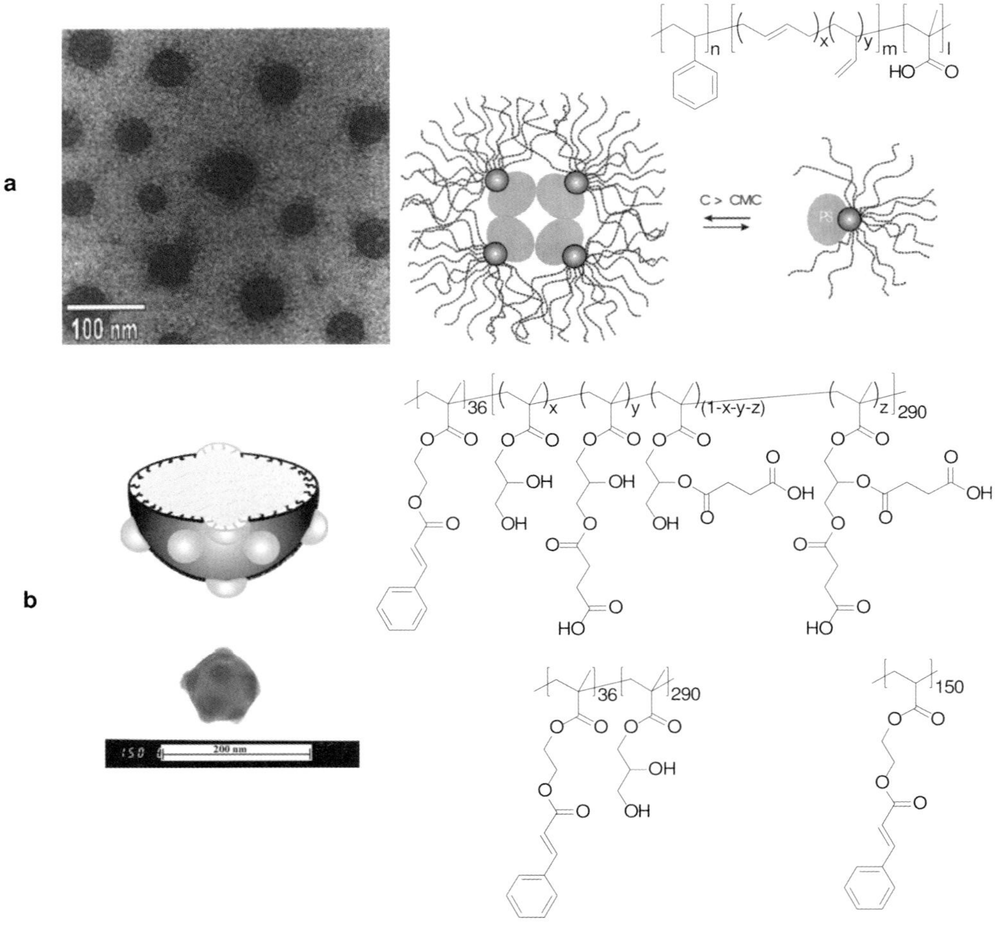

Fig. 16.3 Complex structures from high order self-assemblies: (a) Janus micelles [34]; (b) "bumpy" micelles [36]. Reprinted with permission from [34] and [36].

small molecules include surfactants [39], molecules that form hydrogen bonds [40–42] and ion pairs [21, 43]. As an example, Weck and coworkers synthesized one block copolymer as a universal scaffold, in which different functional groups could be bound to the specific block orthogonally by either hydrogen bonding or metal coordination [44, 45]. Following the same logic with increasing block number, the groups of Lodge [29] and Kubowicz [46] developed multicompartment micelles by utilizing the additional block of triblock copolymers to increase the structural complexity. It was also demonstrated that two different molecules could be sequestered into two different domains in the same multicomponent micelles due to differences in their partition ability [47].

16.2.3
Functionalized and Stimuli-responsive Materials from Self-assembly of Block Copolymers

Functionalization of self-assembled structures from block copolymers extends their significance beyond simple roles as building blocks. Wooley and coworkers have developed both premixing [48, 49] and post-micellization [50] functionalization strategies to synthesize nanostructures decorated with functional groups (Fig. 16.4). These "tagged" nano-objects showed promise for biological applications, for example in the development of targeted imaging [51] and delivery agents [50, 52].

Other significant studies have resulted in the self-assembly of block copolymers into "smart" nanostructures, which undergo dramatic changes in the physical and/or chemical properties under external stimuli. pH- and ion-responsive materials have been prepared from block copolymers with ionizable or hydrogen-binding monomer units [25, 53]. The incorporation of polymeric chains with lower critical solution temperature into block copolymers yields a temperature-responsive material. As an example, Sugihara et al. established that triblock copolymers having three different phase-transition temperature polymeric blocks underwent multi-stage association under temperature programming [54]. Similarly, responsive polymer-based morphological changes were shown to modify surface properties over macroscopic dimensions, which were induced to exhibit either hydrophilic or hydrophobic character under different environments [55, 56].

16.2.4
Applications

The periodicity of the self-assembly of block copolymers forms the basis of "bottom-up" nanofabrications, with recent literature indicating that large areas of patterned surfaces have been constructed free of defects [57] (Fig. 16.5). The ori-

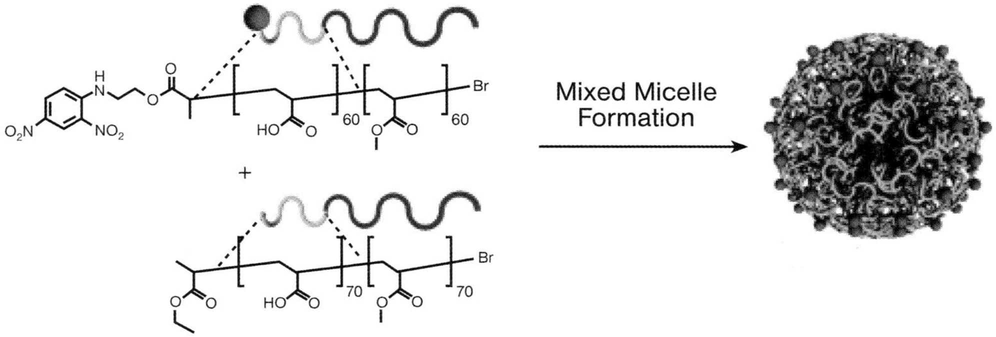

Fig. 16.4 Premixing strategy for functionalization of block copolymer self-assemblies. Reprinted with permission from [49].

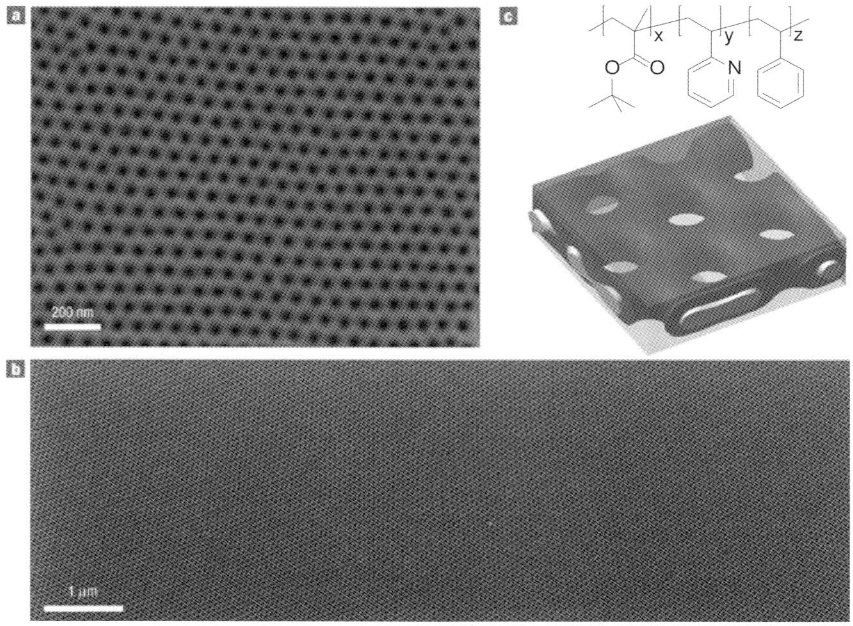

Fig. 16.5 Nanoscopically patterned film, free of defects across microscopic dimensions, self-assembled from PSt-*b*-P2VP-*b*-P*t*BMA. Reprinted with permission from [57].

entation of these patterns can be further aligned by the application of an external field [58]. Hawker and coworkers demonstrated a novel approach to create a universal surface by applying a cross-linkable random copolymer as an initial thin film upon various substrates. The subsequent self-assembly of block copolymers on to this film indicated no influence from the underlying substrates [56, 59]. Combined with other techniques, the error-proof dynamic organization processes and the size, size regularity and size control of the fabricated nanoscale morphologies make the self-assembly of block copolymers an ideal engineering tool for nanolithography.

The self-assembly of block copolymers incorporating inorganic precursors as one or more of their blocks, followed by post-assembly processing, allows for their use as template matrices for the preparation of organic–inorganic hybrids with regular spatial orientation [60, 61] and, furthermore, for their transformation into well-defined inorganic nanostructured materials [5]. Another application of block copolymers is as matrices for catalytic metals. For this purpose, functional block copolymers with coordinative binding ability have been constructed, loaded with metal precursors that after partitioning into specific phase-segregated domains were reduced. Such resulting metal particles were then shown to be stable and retain catalytic activity after many catalytic cycles [62, 63].

Block copolymers have also been used as structural materials for the fabrication of purposeful objects, such as nanocontainers or nanovessels. In addition to their role in providing structural integrity, block copolymers carry chemical functionality to determine the chemical and biological function of the containment devices. Recently, the chemical differentiation of the external and internal surfaces of nanoscale cage-like frameworks was demonstrated [52, 64]. In many cases, the core–shell structure of self-assembled block copolymers has been examined as drug delivery systems, in which hydrophobic drug molecules or polyelectrolyte complexes reside in the internal region, thereby being passivated or protected by the nanocontainer [65–68].

16.2.5 Outlook

Progress in the development of controlled polymerization techniques and related chemistries makes it possible to prepare multiblock copolymers having well-defined structures and architectures [69–72]. Various functional units can be incorporated into a self-assembled structure through the use of functional monomers, chain ends and other reactive handles as part of the block copolymers or after their assembly into complex functional macromolecular morphologies. Moreover, increased understanding of the parameters that govern the assembly process will illuminate future designs to be built by a programmed stepwise approach.

16.3 Functional Conjugated Polymer Assemblies

16.3.1 Overview

The fundamental understanding of polymer supramolecular assembly processes and the creation of interestingly shaped nanostructures from well-defined polymers are being extended to polymer components that contain degrees of function beyond customary chemical and biological activities, to include electrical, optical, magnetic and other physical properties. Common to all sections of this chapter is the synthesis of diverse block and/or graft copolymer architectures and their assembly into numerous interesting and useful nano- and mesoscale structures. These techniques and approaches are now being applied to the synthesis and organization of conjugated homo, block and graft (co)polymers and their unique supramolecular assemblies. These recent efforts have been driven by the desire to both exploit the optoelectronic properties of conjugated polymers for specific applications (e.g., as components of photovoltaic devices, field-effect transistors, etc.) and to gain further understanding of the fundamental aspects of their physical and photophysical properties. Recent reviews of func-

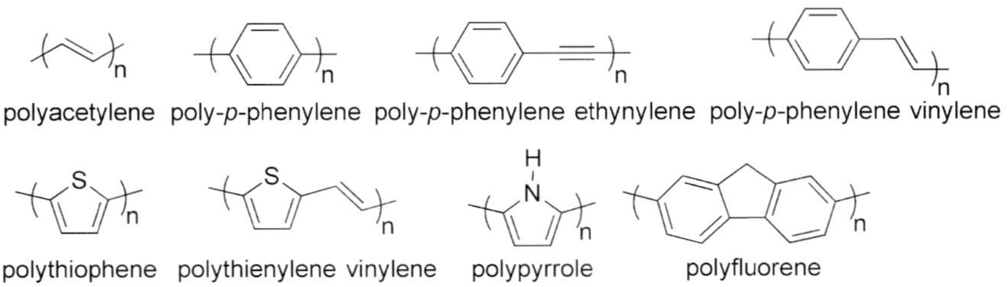

Fig. 16.6 A selection of conjugated polymer backbones that have been studied as components in functional block copolymer assemblies and others that represent future materials for incorporation into conjugated block copolymer-based nanostructured materials.

tional polyacetylenes [73] and polythiophenes [74], supramolecular assemblies of conjugated polymers [75–77], application of polythiophenes as light-emitting materials [78] and the application of poly(arylene ethynylenes) as bio- and chemosensing agents [79] are demonstrative of the degree of interest in these functional macromolecules.

Figure 16.6 illustrates some of the wide variety of conjugated polymer backbones that are currently under investigation, each with unique chemical, physical and electro-optical properties. Most conjugated polymers are synthesized by either Lewis acid-catalyzed oxidative polymerization, metal-catalyzed polymerization or electro-polymerization, which offer varying degrees of control over regiochemistry, molecular weight and polydispersity in addition to functional group tolerance. Advances in synthetic methodologies have included microwave-assisted, metal-catalyzed polymerization [80], enzymes or other biocatalysts for oxidative polymerization [81–85], click chemistry [86] or another mechanism by which to prepare the conjugated polymer backbone, with various side-chain substituents also being included.

The assembly of functional conjugated macromolecules can be driven by intermolecular hydrogen bonding, ionic interactions, π-stacking and phase segregation or, alternatively, through templating by biomacromolecules, micelles, porous silicon glass or other substrates, to produce morphologies ranging from discrete, solution-state particles to ordered, solid-state lamellae and crystalline phases. Once assembled, the inherent properties of these polymers impart functionality to the assembly, at times differing from the properties of the free polymer. Such supramolecularly facilitated association phenomena can affect the photophysics of a conjugated polymer, including changes in the intrinsic radiative and non-radiative decay rates of inter-chromophore coupling [87]. These and other effects resulting from assemblies of conjugated macromolecules relative to their polymeric precursors will be discussed throughout this section. Focus is upon complex functional polymeric systems recently reported in the literature, narrowed to linear conjugated polymer and multiblock copolymers.

Owing to space limitations, interesting areas of research into new conducting monomers, polymers and copolymers [88–92] and other types of conjugated materials, including oligomeric organic conducting molecules [93–97], conducting dendrimers [98] and polymers with pendant chromophoric molecules [99, 100], will be omitted from our discussion.

16.3.2
Solid-state Assemblies

The solid-state orientation of conjugated polymers is a crucial parameter for many of the applications for which they are being researched. For example, alignment of the plane of conjugation of polythiophenes perpendicular to the gate–insulator–semiconductor interface optimizes the efficiency of field-effect transistors by enhancing the on/off current ratio and the field-effect mobility. A potentially desirable consequence of aligning conjugated polymers is the creation of optical activity, which leads to the possibility of opto-electronic devices.

For polythiophenes, the confluence of two separate goals (improved processability and control of the regiochemistry of polymers) has resulted in polymers that assemble into lamellar sheets in the solid state. To enhance solubility of polythiophene in common organic solvents, alkyl groups have been introduced on the thiophene monomer in the 3-position. The development of new methods for the polymerization of thiophenes to improve the regioregularity of the polymer backbone, including the Grignard metathesis (GRIM) method [101] and the Rieke zinc method [102], has allowed the preparation of regioregular polythiophene. The solid-state assembly of poly(3-alkylthiophene) produces lamellae and is driven by interdigitation of crystalline side-chains. Collard and coworkers improved on this monolayer packing by preparing a substituted bithiophene monomer with an alkyl side-chain on one thiophene unit and a fluoroalkyl side-chain on the other. Preparation of the regioregular polymer by GRIM followed by casting films from chloroform or xylene resulted in a highly ordered film consisting of lamellar bilayers driven by both phase segregation of the side-chains and the crystallinity of the side-chains [103]. More recently, a regioregular polythiophene that incorporated electron-withdrawing perfluoroalkyl side-chains directly attached to the thiophene backbone was prepared, in order to manipulate the electronic properties of the polymer while maintaining control over the solid-state assembly [104].

An alternative method for controlling the solid-state assembly is the synthesis of conjugated block copolymers that contain non-conjugated blocks, which benefit from micro-phase segregation. McCullough and coworkers coupled existing methods for the controlled synthesis of regioregular poly(3-alkylthiophene)s with atom transfer radical polymerization (ATRP) or condensation polymerization to prepare block copolymers for the solid-state assembly, most notably of nanowires (Fig. 16.7) [105, 106]. In contrast, Meijer and coworkers prepared rod–crystalline coil diblock copolymers by synthesizing an alkenyl-terminated

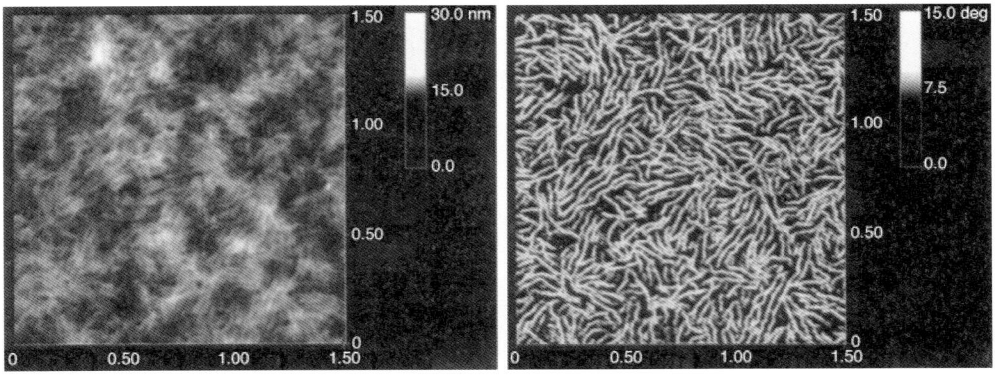

Fig. 16.7 AFM height (left) and phase (right) images of nanowires. Reprinted with permission from [105].

polythiophene, which served as a capping agent for the ring-opening metathesis polymerization (ROMP) of cyclooctene [107].

Conjugated oligomers and polymers when deposited from solution are especially promising in organic thin-film transistor applications. Most conjugated oligomers often exhibit a rich range of both lyotropic and thermotropic liquid crystalline (LC) phase transitions, which are advantageous as the deposition conditions can be controlled to target certain high-mobility LC phases. For example, formation of LC phases during solution deposition was identified as an important route to the formation of self-assembled interconnected film [108].

Polyfluorenes (PF) are another class of conjugated polymers which are used as blue light-emitting active layers in polymeric light-emitting diodes (PLEDs). PFs display very high photoluminescence efficiency in both the solution and solid states. The photophysics of PFs is affected by the film-forming conditions, which alter the microstructure of the polymer. A low-energy absorption band due to isolated chain segments, also referred to as the β phase, is attributed to increased intrachain fluorene–fluorene planarity. Typically, the choice of solvent or film-forming conditions controls the dominance of one conformational isomer over another. Typically, photoluminescence from PFs cast from toluene, *p*-xylene and chlorobenzene show the low-energy absorption band due to the β phase [109]. In addition, rod–coil diblock copolymers containing hole-transporting flexible segments, such as poly[2-(9-carbazolyl)ethyl methacrylate], have been studied [110] to address the poor hole injection properties of PF homopolymers. These copolymers were shown to self-assemble into nanoaggregates and worm-like morphologies from solution.

Utilization of the intrinsic properties of the polymer (or copolymer), solvent or deposition method is not the only tool for controlling the solid-state morphologies of block copolymers containing conjugated chain segments. DNA–polymer conjugates [111, 112] and also block copolymer host templates [113] have been employed recently to create arrays of aligned nanowires and polymer films

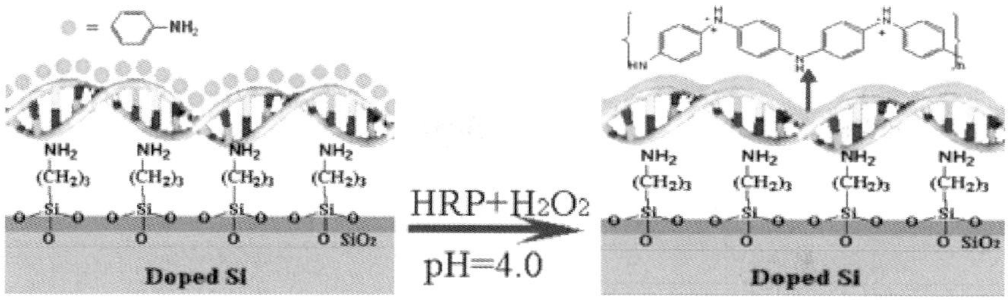

Fig. 16.8 The alignment of aniline monomers along a DNA template allows for the synthesis of nanowires under horseradish peroxidase (HRP) conditions. Reprinted with permission from [112].

with mechanically tunable polarized emission spectra. Additionally, porous silica glasses have been used to template the polymerization of conjugated polymers to create arrays of nanowires for chemosensing [114].

Nakao et al. utilized poly(phenazasilinetrimethylammonium iodide) to interact via intercalation of the repeat units in the base stack and ionic interactions with the negatively charged DNA backbone, which formed conjugated nanowires that were found to reduce $HAuCl_4$ to Au on their surface creating metallic Au nanowires [111]. Using soft templating, He and coworkers prepared polyaniline nanowires by incubating aniline monomer with stretched DNA and then initiating polymerization along the DNA template with horseradish peroxidase, which allowed the oxidative polymerization to occur without degrading the DNA template (Fig. 16.8) [112]. As an alternative approach using synthetic materials, Thomas and coworkers combined host–guest interactions and roll casting to prepare films of poly(phenylene ethylene)-*g*-polystyrene copolymers blended with PSt-*b*-PIp-*b*-PSt that exhibited optical anisotropy and enhanced mechanical properties [113].

16.3.3 Solution-state Assemblies

Assembly of macromolecules in the solution state continues to be one of the most interesting and diverse areas of current research in materials science. The drive to develop discrete solution-state objects with conjugated macromolecules stems from a desire both to explore the effect of nanoscale architecture on the electro-optical properties and to access certain structural motifs not easily available from solid-state assembly. A number of different methods, including reprecipitation, phase segregation self-assembly and soft and hard templating, have been exploited to create nano- and mesoscale structures from solution. Nanostructures of conjugated polymers can possess electro-optical properties that differ from those of the free chains in solution, most notably changes in absorption or emission, known as the solvatochromic effect.

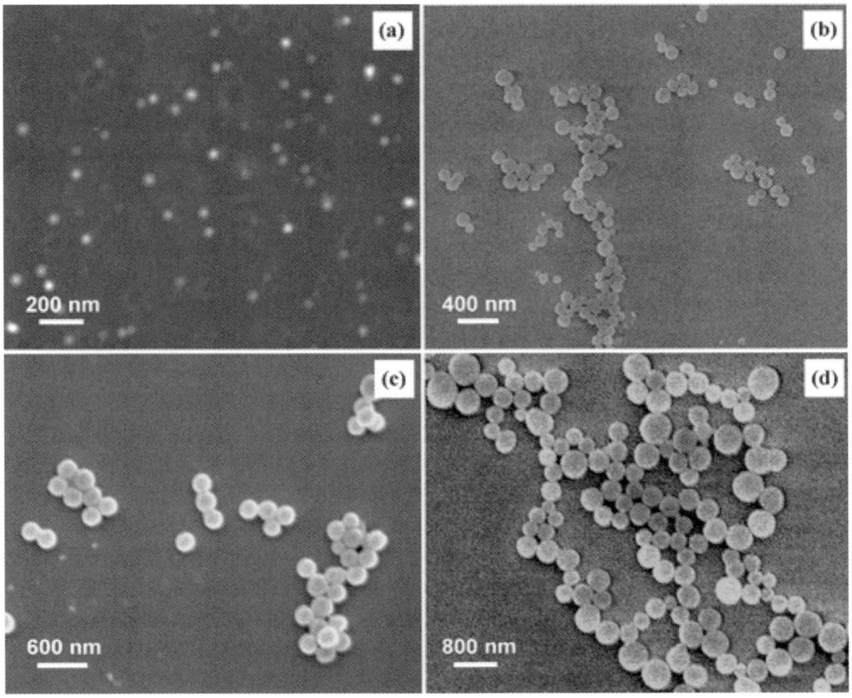

Fig. 16.9 Spherical nanoparticles prepared by reprecipitation of poly(thiophene-*alt*-azulene) copolymers. Reprinted with permission from [115].

Reprecipitation was employed by Lai and coworkers to create spherical nanoparticles comprised of multiple chains of poly(thiophene-*alt*-azulene) copolymers [115] (Fig. 16.9), with control over nanoparticle size dependent on the time in solution and the composition of the solubilizing side-chain. A red shift in the absorption spectra with increasing particle size (a linear relationship with particle size) was noted along with fluorescence behavior not observed for the free polymer in solution. Reprecipitation has also been used to prepare spherical nanoparticles from single polymer chains of poly{2-methoxy–5-[(2-ethylhexyl)oxy]-*p*-phenylenevinylene}, which exhibited a solvatochromic blue shift in the absorption spectrum that resulted from the formation of defects in previously long stretches of conjugation [116]. In a report from McCarley and coworkers [117], which will be discussed further in Section 16.4, a similar blue-shift solvatochromic effect was observed for a polythiophene-g-poly(*N*-isopropylacrylamide). When the graft copolymer was heated above the lower critical solution temperature of the acrylamide grafts, disruption of long conjugated segments of the backbone chain occurred, due to steric interference from the collapsed grafts.

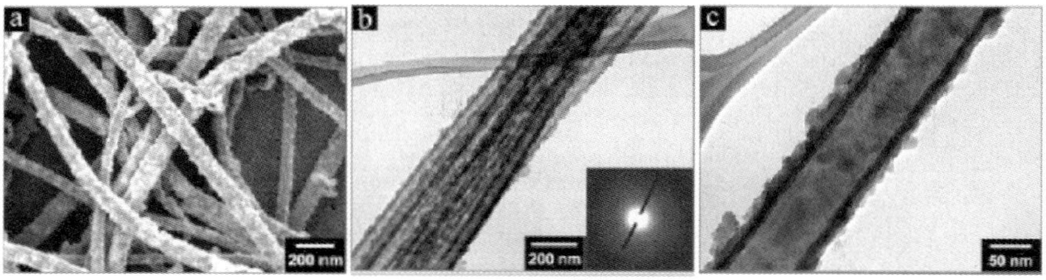

Fig. 16.10 TEM images of chiral polyaniline tubes demonstrating the smooth inner wall indicative of the templated polymerization. Reprinted with permission from [119].

Soft-templated assembly has also been employed to prepare hollow or solid nanotubes of conjugated functional polymers. Zhang and Wan prepared chiral polyaniline (PANI) nanotubes utilizing micelles of either the L- or D-isomer of camphorsulfonic acid as a template for *in situ* doping and polymerization with ammonium persulfate as the oxidant [118]. The resultant nanotubes could be made hollow or filled depending on the molar ratio of the aniline monomer to the chiral dopant, which also induced chirality, as evidenced by their circular dichroism spectra. Jang and Yoon proposed as a mechanism for the formation of hollow polypyrrole nanotubes from inverse micellar templates [119] that the tubular, micellar template arranged catalyst ions (in this case, $FeCl_3$) along the template surface that oxidized the monomer (pyrrole) and allowed polymerization to occur along the template surface and outwards from it. This mechanism was supported by the observed smooth inner wall and rough outer wall of the nanotubes (Fig. 16.10). The diameter of the tubes could be controlled by the weight ratio of the catalyst to surfactant and the reaction temperature.

A more common approach to create stable solution-state nano-objects from polymers is to employ phase segregation of block copolymers to prepare micelles or vesicles in solution by self-assembly. Meijer and coworkers produced shape-tunable vesicles from oligomeric ABA triblock amphiphiles with polythiophene as the B block and oligo-ethyleneoxy A blocks [120]. Cloutet and coworkers investigated the solution-state assembly of rod–coil and coil–rod–coil block copolymers composed of polythiophene rods and poly(*n*-butyl acrylate) (P*n*BA) coils by coupling acid-catalyzed oxidative polymerization and ATRP from a bifunctional initiator [121]. Rod–coil diblock copolymers with oligomeric polythiophene blocks assembled in cyclohexane to form stable, hollow vesicles, but copolymers containing longer blocks of polythiophene did not form stable structures in solution; this trend was also observed for the triblock copolymers. These initial reports demonstrate that the solution-state self-assembly of conjugated rod–coil multiblock copolymers presents a number of challenges, but that stable, discrete nano-objects are obtainable.

16.3.4
Outlook

The diversity and creativity applied to the assembly of conjugated polymers have resulted in theoretical and practical foundations that are being extended towards many applications, including several beyond the development of electronic devices, solar cells and optical devices. Among the most exciting alternative applications for assemblies of functional conjugated materials are biological and chemical sensing. The aggregation of conjugated polymers in solution (and the resulting solvatochromic shift) have been used to prepare colorimetric probes for biologically relevant cations and anions [101]. Fluorescence resonance energy transfer was employed with a conducting polymer–DNA conjugate to prepare a fluorescence detector for K^+ ions [122]. Poly(*p*-phenylene vinylene) (PPV) chains with pendant, acid-terminated side-chains have been prepared as intense dyes that could be employed as substitutes for small-molecule dyes [123]. Peptides bearing 2,4-dinitrophenyl groups have been coupled to PPVs to exploit fluorescence quenching by nitro-aromatics with a biologically relevant sensing moiety, in a similar manner as has been previously demonstrated with TNT and other explosive compounds [124]. Sensing microarrays of polydiacetylenes that have been shown to detect the presence of α-cyclodextrin or PAA colorimetrically were prepared by assembling vesicles of diacetylene monomer, immobilizing them on a surface and curing them through a mask [125]. As evidenced through the examples in this section, conjugated polymers and their assemblies are promising materials for a variety of applications; therefore, it is necessary that more research is conducted to continue to develop synthetic routes to prepare them in controlled manners and to explore their assembly to yield materials with improved performance in all areas.

16.4
Functional Brush Copolymers

16.4.1
Overview

Over the last decade, brush copolymers, i.e. graft copolymers with high grafting density, have attracted increasing attention due to their special properties and promising applications. Relative to linear and lightly branched polymers, brush copolymers possess a broad variety of unique molecular conformations, imparted by the intramolecular chain–chain interactions and steric effects, experience decreased attractive secondary inter-chain forces and also carry a large number of chain ends. Therefore, brush copolymers have significant application potential as single-molecule templates in the preparation of nanomaterials and as vehicles for drug delivery, gene therapy, catalysis and sensors. There are several detailed reviews related to brush copolymers available [126–129], so only recent advances in this area will be addressed here.

Because polymeric architectures and molecular conformations of brush copolymers have crucial influences on their properties and applications, synthetic design and conformational control are among the major aspects in the research of brush copolymers. Similarly to typical graft copolymers, brush copolymers have been prepared by a "grafting from" (polymerization from a polyfunctional macroinitiator), a "grafting through" (polymerization of macromonomers) or a "grafting onto" approach (coupling reactions using a polyfunctional coupling agent) [133]. In some cases, these methods can be combined to build up brush copolymers with complex structures. With both steric restriction and intramolecular self-assembly as the key contributing factors, brush copolymers can have a series of interesting molecular topologies and conformations (Fig. 16.11), providing unique features unavailable by self-assembly of block copolymers (see Section 16.2 for self-assembly of block copolymers).

The two primary advantages resulting from the exact preparation of brush copolymers, relative to supramolecular polymer assemblies, are control of the overall molecular dimensions (e.g. the lengths of cylinders) and accessibility to asymmetrically shaped nanostructures. In general, brush copolymers adopt a spherical conformation when the backbone remains short compared with the grafts and as the backbone becomes longer and comparable in length to the grafts, the steric hindrance among these grafts stiffens the backbone and forces the brush copolymers to adopt a cylindrical conformation (Fig. 16.11 a) [130, 134, 135]. Moreover, brush copolymers having a long backbone with gradient grafting density can present tadpole conformations (Fig. 16.11 b) [131] and brush copolymers with a star-like backbone can have multi-arm star conformations (Fig. 16.11 c) [132, 136]. At the same time, conformational transitions of brush copolymers can be induced by environmental changes [117, 130, 131, 136–140]. Combining these molecular conformations with polymer block segments having

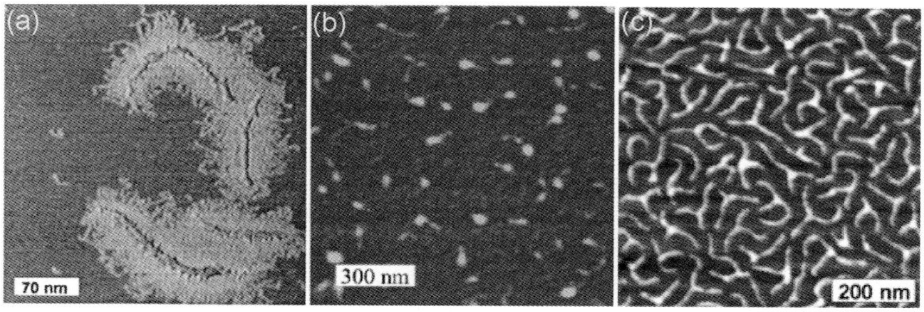

Fig. 16.11 SFM micrographs of single molecules of (a) brush copolymers with cylindrically uniform structure [130], (b) brush copolymers with a tadpole architecture [131] and (c) brush copolymers with a three-arm star topology [132]. Reprinted with permission from [130], [131] and [132].

different compositions or structures, a series of novel polymeric architectures, such as core–shell brush copolymers [122, 141–143], brush–linear copolymers [144–148], brush–brush copolymers [147] and linear–brush–brush copolymers [148] have been obtained.

16.4.2
"Grafting from"

Brush copolymers can be synthesized using polyfunctional macroinitiators via a "grafting from" polymerization process, affording precise structural control over backbone, grafts and grafting density of the resulting brush copolymers. Living polymerization techniques are preferred in both the synthesis of polyfunctional macroinitiators and the "grafting from" step.

ATRP has become a very powerful polymerization technique [149, 150], with a broad range of brush copolymers being synthesized via the "grafting from" method. Haloesters are universal ATRP initiation functionalities that can be prepared readily and a variety of haloester-based polyfunctional macroinitiators have been developed and employed in "grafting from" synthetic approaches [117, 132, 140–143, 145, 146, 151–157]. Matyjaszewski and coworkers reported the synthesis of a series of well-defined brush copolymers with different conformational characteristics using 2-(2′-bromopropionyloxy)ethyl methacrylate (BPEM)-based macroinitiators with various structural features (Scheme 16.1) [132, 141, 145, 151–154]. Brush copolymers having a uniform cylindrical topology (Scheme 16.1, route (a)) and adopting a cylindrical conformation

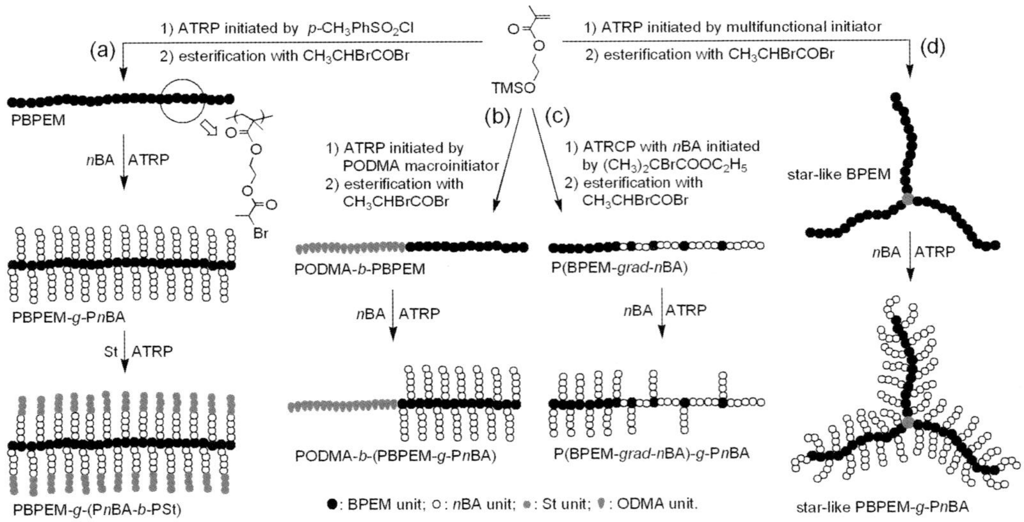

Scheme 16.1

(Fig. 16.11 a) were synthesized by ATRP "grafting from" using a linear PBPEM homopolymer as macroinitiator [141]. When *n*BA was used as the monomer, high initiation/grafting efficiency was achieved for the "grafting from" system at relatively low monomer conversion [151, 152]. By sequential ATRP using different monomers, core–shell cylindrical brush copolymers were also prepared from PBPEM [141]. Cylindrical brush–linear block copolymers were obtained by "grafting from" using diblock and triblock macroinitiators consisting of PBPEM blocks and crystallizable poly(octadecyl methacrylate) (PODMA) blocks (Scheme 16.1, route (b)) [145]. These copolymers exhibited intermolecular self-assembly behaviors different from their cylindrical brush analogues. By using statistical copolymers with gradient distribution of BPEM units as macroinitiators, brush copolymers with a gradient distribution of grafts along the backbone were prepared (Scheme 16.1, route (c)) [153, 154] and tadpole conformations of these brush copolymers were observed (Fig. 16.11 b) [131]. Brush copolymers with star-like architectures (Scheme 16.1, route (d)) and conformations (Fig. 16.11 c) were also synthesized from star-like PBPEM with three or four arms [132, 136].

Relative to the BPEM-based macroinitiators with α-bromopropionate functionalities, the macroinitiators with α-bromoisobutyrate functionalities have higher initiation capacity and, therefore, have a broader range of suitable monomers. Several groups have reported the syntheses of brush copolymers using poly[2-(2′-bromoisobutyryloxy)ethyl methacrylate] (PBIEM) to initiate ATRP of monomers including *n*BA [143, 155], *tert*-butyl acrylate (*t*BA) [142, 143], methyl methylacrylate (MMA) [151], 3-*O*-methacryloyl-1,2 : 5,6-di-*O*-isopropylidene-D-glucofuranose (a protected sugar-functionalized methacrylate, giving saccharide side-chain units after hydrolysis) [156], styrene [142, 155] and *N*-isopropylacrylamide (NIPAAm) [140]. Core–shell cylindrical brush copolymers were also prepared from PBIPM by sequential ATRP of *t*BA and *n*BA and further modified by selective hydrolysis of *t*BA units [143]. Subsequently, the resulting amphiphilic core–shell cylindrical brush copolymers were used as molecular templates for the preparation of magnetic/semiconducting nanoparticles [158, 159]. In addition, by using diblock PBIEM-*b*-PSt as macroinitiator, cylindrical brush–linear copolymers were synthesized by this ATRP "grafting from" strategy [146].

Besides BIEM units, other types of repeat units with α-bromoisobutyrate functionalities were also introduced into the structures of polyfunctional macroinitiators for the synthesis of brush copolymers by ATRP. A brush copolymer with a poly(thiophene)-based backbone and PNIPAAm grafts was prepared by the synthesis and oxidative polymerization of an α-bromoisobutyrate-functionalized thiophene, followed by ATRP of NIPAAm initiated by the resulting macroinitiator [117]. This brush copolymer not only was highly water-soluble and thermally responsive, undergoing reversible conformational transitions from an extended coil to a collapsed globule between 28 and 35 °C in water, but also exhibited interesting optical and electronic properties. As another example, a macroinitiator with four α-bromoisobutyrate functionalities per dendritic repeat unit was used in ATRP "grafting from" and the resulting brush copolymers might have unusually high grafting density [157].

Polyfunctional macroinitiators for anionic polymerization and ring-opening polymerization (ROP) were also used in the "grafting from" preparation of brush copolymers. A polyfunctional anionic macroinitiator with one 1,1-diaryl-alkyllithium initiator functionality per repeat unit was prepared from a polyfunctional 1,1-diphenylethylene agent and it was then used in the synthesis of core–shell brush copolymers with quantitative initiation/grafting efficiency [122]. As a simple protein mimic, a brush copolymer having polyleucine-based grafts was prepared by ROP "grafting from" and exhibited pH-regulated reversible conformational and morphological transitions [160]. Cylindrical brush copolymers with polypeptide-based [poly(L-lysine)- or poly(L-glutamate)-based] grafts and polylactide grafts were also synthesized by ROP "grafting from" [161, 162].

16.4.3 "Grafting through"

The "grafting through" method has been used to prepare brush copolymers by the polymerization of a broad variety of well-defined macromonomers, resulting in excellent control over graft structures of the brush copolymers formed. However, because the reactivities of macromonomers decrease with increase in their chain sizes, "grafting through" generally is not suitable for the preparation of brush copolymers with relatively long grafts, especially when vinyl-functionalized macromonomers are used. Moreover, compared with a linear polymer propagation center, the poly(macromonomer) propagating center typically has much lower reactivity due to steric hindrance and, therefore, it is somewhat difficult to prepare brush copolymers with a long backbone by "grafting through".

Nonetheless, complicated structures of brush copolymers have been accomplished based on copolymerization "grafting through" strategies, employing a variety of vinyl-ended macromonomers [127]. For example, amphiphilic cylindrical brush–linear block copolymers were prepared by sequential coordination polymerizations of methacryloyl-ended PSt macromonomer and *tert*-butyl acrylate (*t*BMA), followed by hydrolysis and neutralization with CsOH [144]. The resulting copolymer with many PSt grafts and a main-chain block having many $COO^{-}Cs^{+}$ functionalities formed giant spherical micelles by intermolecular self-assembly. In another study, brush copolymers with two types of grafts were prepared by ATRP "grafting through" using poly(dimethylsiloxane) (PDMS) and poly(ethylene oxide) (PEO) macromonomers, each with one methacryloyl chain end [163].

ROMP, an important ROP technique, has been used broadly in the synthesis of brush copolymers with different structural characteristics because norbornene-ended macromonomers have much higher reactivity than commonly used vinyl-ended macromonomers, for thermodynamic and steric reasons [162, 164]. As an interesting development, Fréchet and coworkers [165] reported the preparation of a fullerene C_{60}-based norbornene derivative and its application in the synthesis of fullerene-containing polymers by ROMP (Scheme 16.2). With the structural uniqueness, composed primarily of C_{60} and having a molecular

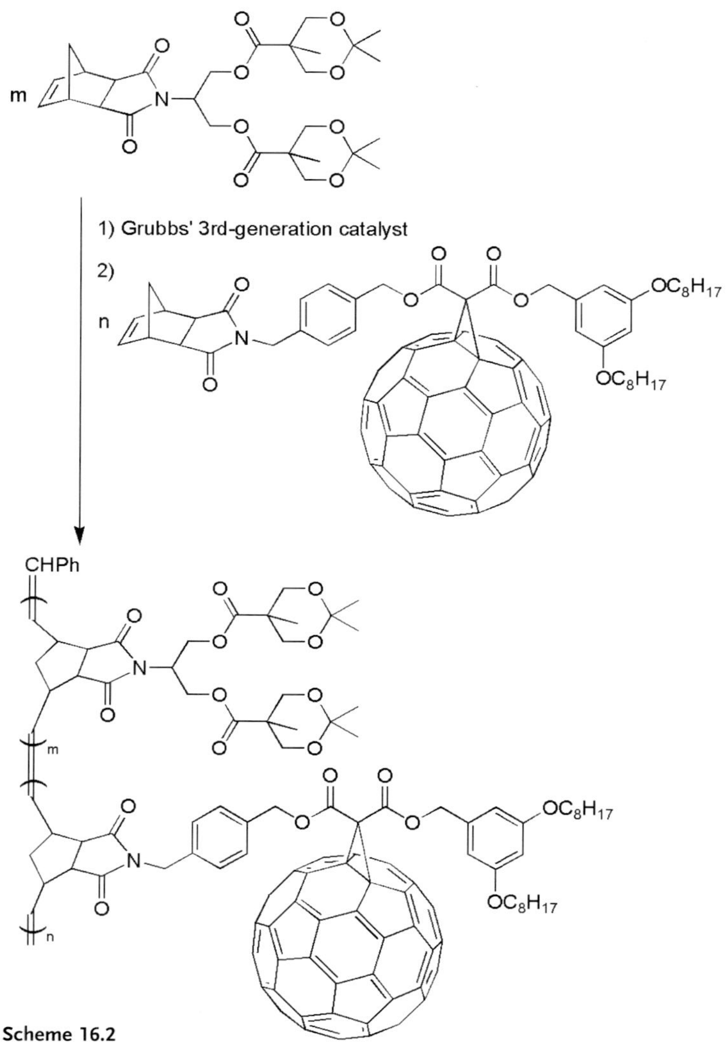

Scheme 16.2

weight of 1434 Da, this norbornene derivative can be considered as an atypical macromonomer. Diblock copolymers with high fullerene content were prepared by sequential ROMP of a small molecule norbornene derivative and this C_{60}-based macromonomer initiated by Grubbs' third-generation catalyst (Scheme 16.2) and they exhibited interesting self-assembly patterns.

16.4.4
"Grafting onto"

The "grafting onto" method involves the attachment of polymer species with one active chain end to polyfunctional polymers through coupling reactions. The brush copolymers prepared by "grafting onto" can have well-defined backbone and grafts by using well-defined polymeric reactants. However, because these polymeric reactants have much lower reactivities than their small molecule analogues, coupling efficiency is a major concern and only a few "grafting onto" systems have reached high coupling efficiency to yield brush copolymers. Living ionic polymers have been used in "grafting onto" due to their high reactivities, but a broad range of functional groups could not be directly introduced into the resulting macromolecular structures because of the poor functional group tolerance of ionic polymerization [128]. Recently, "click chemistry" has been used in "grafting onto" (Scheme 16.3) to prepare dendronized copolymers with a linear main chain by "click" reactions between Fréchet-type dendritic azides and the pendent alkynes of polyvinylacetylene with high efficiency [166].

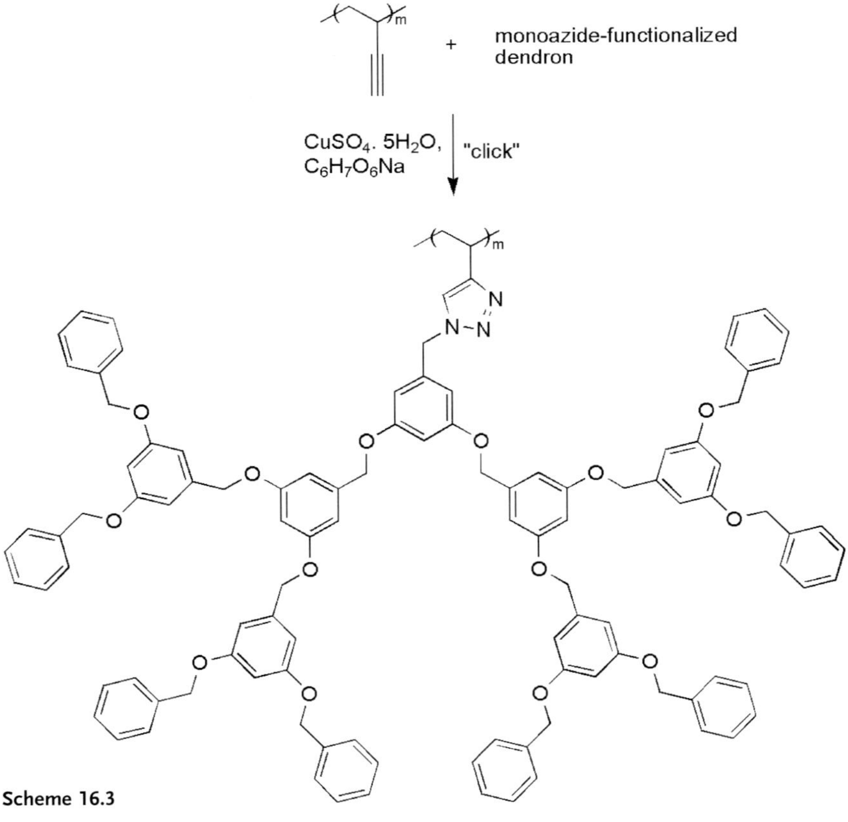

Scheme 16.3

The very bulky third-generation dendritic repeat units of the copolymer allowed for a rigid cylindrical conformation. Theoretically, linear grafts can also be introduced by "grafting onto" based on "click chemistry".

16.4.5
Combined "Grafting from" and "Grafting through"

Each method of "grafting from", "grafting through" and "grafting onto" has its own synthetic advantages and limitations. Therefore, the strategic combination of these methods is not only theoretically interesting, but also of practical significance for the preparation of brush copolymers with relatively complicated architectures. Recently, several types of brush copolymers synthesized by combined "grafting from" and "grafting through" have been reported.

As shown in Scheme 16.4, "double-grafted" brush copolymers were synthesized by using PBPEM as the polyfunctional "grafting from" macroinitiator to conduct "grafting through" a methacryloyl-ended PEO macromonomer (PEOMA, $DP_{\text{PEO}} = 5$) by ATRP [167]. The resulting copolymers possessed more branching than typical brush copolymers and exhibited soft rubbery properties, providing a potential application as a new type of elastomeric material. A type of brush–brush copolymer carrying PEO grafts and poly(2-hydroxyethyl methacrylate) (PHEMA) grafts was prepared by sequential ATRP of PEOMA ("grafting through") and HEMA, followed by transformation of a portion of the HEMA units to BIEM units and finally using the resulting brush–linear macroinitiators for ATRP of HEMA by "grafting from" [168]. Even more complicated grafting architectures can also be achieved through successive "grafting from" and "grafting through" techniques. For example, a linear–brush–brush copolymer carrying PSt grafts and PEO grafts was prepared by sequential reversible addition fragmentation chain transfer (RAFT) of glycidyl methacrylate (GMA), 4-vinylbenzyl chloride (VBC) and PEOMA ("grafting through"), followed by "grafting from" using the benzylic chloride functionalities on PVBC block of the copolymer to initiate ATRP of styrene [148].

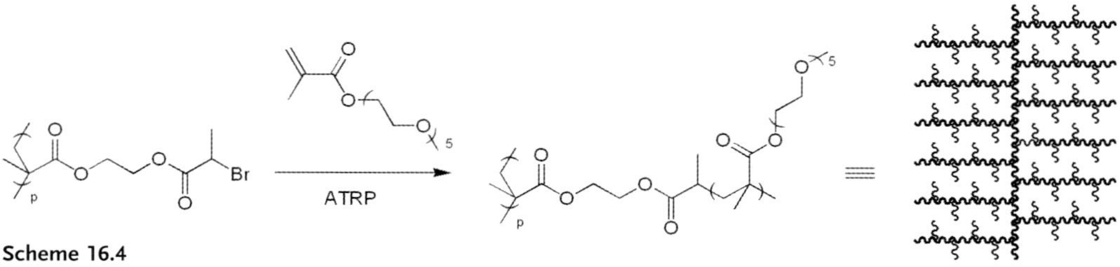

Scheme 16.4

16.4.6
Outlook

Along with the development of living polymerization techniques, more types of well-defined brush copolymers with precise composition and structure that exhibit conformational control will be synthesized. Although most examples have utilized ATRP, other types of living radical polymerization techniques, such as RAFT and nitroxide-mediated radical polymerization, can also be used in "grafting from" to extend the range of applicable monomers and the resulting brush copolymer compositions. To achieve conformational control, modified synthetic strategies, such as "grafting from" using polyfunctional macroinitiators with complicated non-linear structures and "grafting through" using multifunctional initiators to polymerize macromonomers, should be pursued. Coincidentally, combinations of "grafting from", "grafting through" and "grafting onto" methods are also expected to have increasing importance in the preparation of brush copolymers with complex compositions and architectures.

16.5
Functional Biomimetic Hybrid Nanostructures

16.5.1
Overview

As products of evolution, numerous functional hybrid nanostructures are available in Nature. The 'lotus effect' of lotus leaves [169], the mechanical behavior of mollusk shells [170] and tapping of solar energy by plants are some exemplary technologies that Nature has evolved, based upon hybrid hierarchical nanostructures. Ambient energy-efficient processing conditions [171] and the integration of multiple and biodegradable components via simple self-assembly are some of the interesting features of these systems that have intrigued researchers for years [172]. Interestingly, many hybrid structure syntheses are inspired by natural systems, requiring multidisciplinary approaches to model and study the individual components and the entire assembled systems [173]. Considerable effort has been directed to gain a fundamental understanding of how Nature engineers such systems and to translate this knowledge to commercially viable applications. As a summary of some of the latest pioneering work, we will focus on the preparation of functional hybrid materials, emphasizing biosynthetic and biomimetic synthetic pathways, which utilize biology versus chemistry, respectively, as the predominant means by which the materials are prepared.

16.5.2 Biosynthetic Hybrid Materials

Biological materials are the products of constant evolution. The time-line associated with evolution, however, can be a limitation [174], which has been overcome by recent advances made in molecular biology and genetic engineering. Access to combinatorial libraries has allowed for the rapid selection of materials that are not yet available in Nature [175]. In addition, recombinant synthesis has paved a facile route to unnatural peptides via biomachinery. As one example, by employing protein engineering approaches, significant advances have been made in unraveling the fundamental molecular and chemical bases for the interesting properties exhibited by the spider silks [176]. Biosynthesis has also been identified as a potential route to prepare functional hybrid macromolecules [177]. Natural–synthetic hybrid polymers have been biosynthesized by *Pseudomonas oleovorans* in the presence of diethylene glycol to produce hybrid polyhydroxyalkanoates [178], which are desired for their biodegradability, a property covered in greater detail in Section 16.6. Moreover, the strategies to design and engineer systems to incorporate unnatural amino acids have extended the range of accessible functionalities [179, 180]. Likewise, the ability to manipulate the genome of cells and viruses has been utilized to make materials through these biomachineries [174, 181, 182]. Clever manipulation of viral genetics allowed for selective placement of thiol functionalities at distinct surface sites, which was then followed by the growth of Au nanoparticles at those selective sites on the viral surface (Fig. 16.12 a) [183]. Virus-based scaffolds have, alternatively, been explored as a means to direct the synthesis of various magnetic and semiconducting materials of dimensions much larger than the virus itself, as illustrated in Fig. 16.12 b [184]. Recently, cowpea mosaic virus crystals have been demonstrated as porous scaffolds that can be used to template the production of robust nanocomposites as shown in Fig. 16.12 c–f [185]. Microbial organisms [186, 187], plant extracts [188] and other natural systems, including fungal strains of *Fusarium oxysporum*, have also been used for the biosynthesis of nanoparticles [189]. These examples of materials produced biologically represent the chemistry–biology interface and include broad extensions of the emerging field of synthetic biology.

16.5.3 Bioinspired Synthetic Materials

Nature's hierarchically organized hybrid materials provide opportunities for chemists to learn how to exert precise control over thermodynamic and kinetic components in the production of synthetic analogues [190]. Biomimetic materials chemistry involves identifying the key features of the natural system along with the associated individual components and their functions and then harnessing this knowledge to construct, via chemical processes, materials having similar structure and function [191].

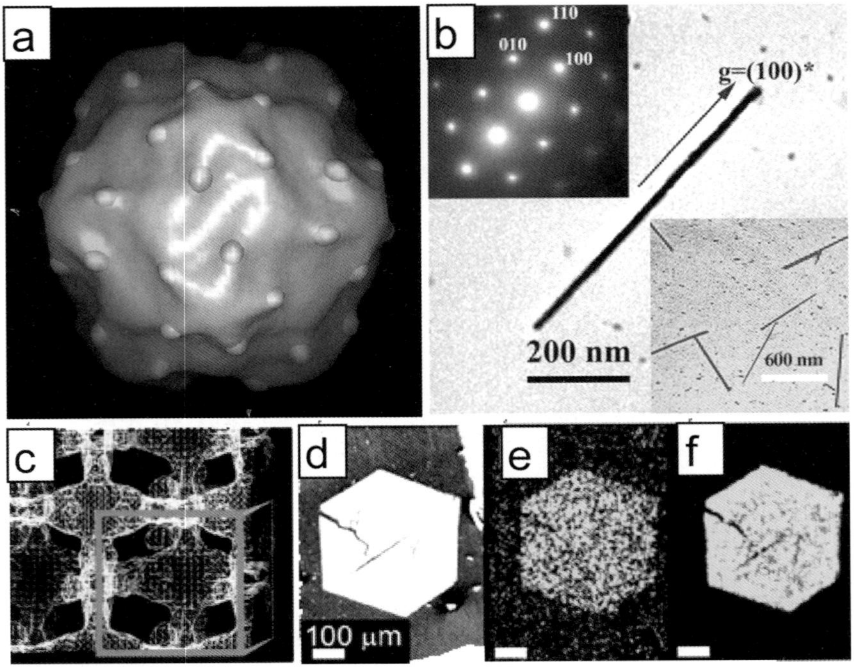

Fig. 16.12 Several interesting materials originating from genetic manipulation of viruses: (a) three-dimensional model derived from cryo electron microscopy images of virus decorated with gold nanoparticles [183]; (b) TEM image of a single crystal of ZnS nanowire formed after annealing, with inset showing electron diffraction pattern [184]; (c) porous spaces in the viral crystal structure, generated from x-ray diffraction data; (d) SEM image of a mineralized and cross-linked cowpea mosaic virus crystal; and (e) and (f) X-ray back-scattering of the palladium and platinum distribution in the crystal, respectively [185]. Reprinted with permission from [183], [184] and [185].

Cytoskeleton, cellular matrix and glycoproteins, in general, have received considerable attention in the development of synthetic structural materials for application as prosthetics or tissue regrowth. For example, hydrogels [192, 193] have been designed to mimic the function of elastin, by incorporating functional methacrylate and methacrylamide monomers to impart cell adhesion and mineral binding functionalities to nucleate the growth of 2-D calcium phosphate-based composite structures (Fig. 16.13a and b). As a novel strategy to improve interfacial adhesion, nanotextured poly(L-lysine)–calcium phosphate hybrid coatings have been shown to improve the bioactivity of titanium substrates, a widely used skeletal implant material, allowing for the preparation of better implants with enhanced cell proliferation and cell migration. Self-assembling interactions between peptides and apatites have been identified as playing a significant role in the formation of biominerals [194].

Calcium carbonate, one of the most common biominerals, is found to be sculpted by the action of proteins to have controlled shape, polymorph and me-

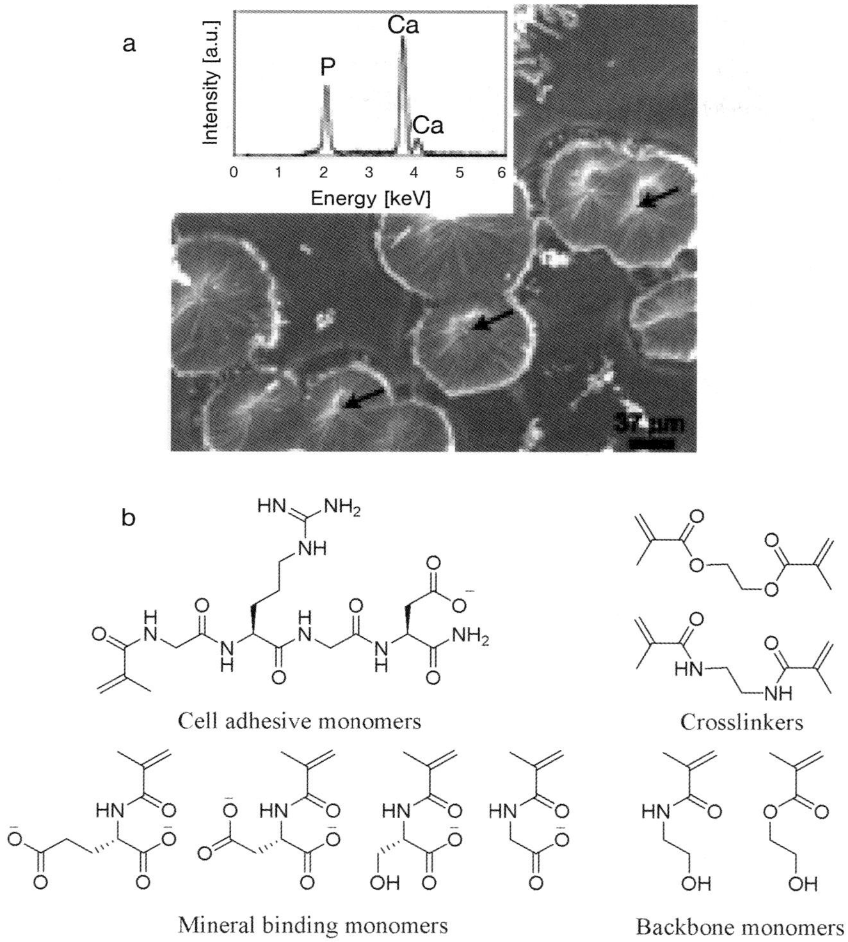

Fig. 16.13 (a) Calcium phosphate–hydrogel composite structure; (b) synthetic monomers used to prepare elastin mimetic hydrogels. Reprinted with permission from [192].

chanical properties [172]. Reports of acidic biomacromolecules [195, 196] isolated from natural systems and their *in vitro* calcium carbonate mineralization behavior have inspired the design of synthetic materials with acidic functionalities [197]. Literature examples demonstrate the importance of the nature of these additives as they have a profound influence on polymorph and morphology of the calcium carbonate crystals produced [198–200]. One of the key features of biogenic biomacromolecules is that they can stabilize amorphous calcium carbonate (ACC) and trigger oriented nucleation of large single crystals. This principle was used to achieve a controlled nucleation of calcite on a large

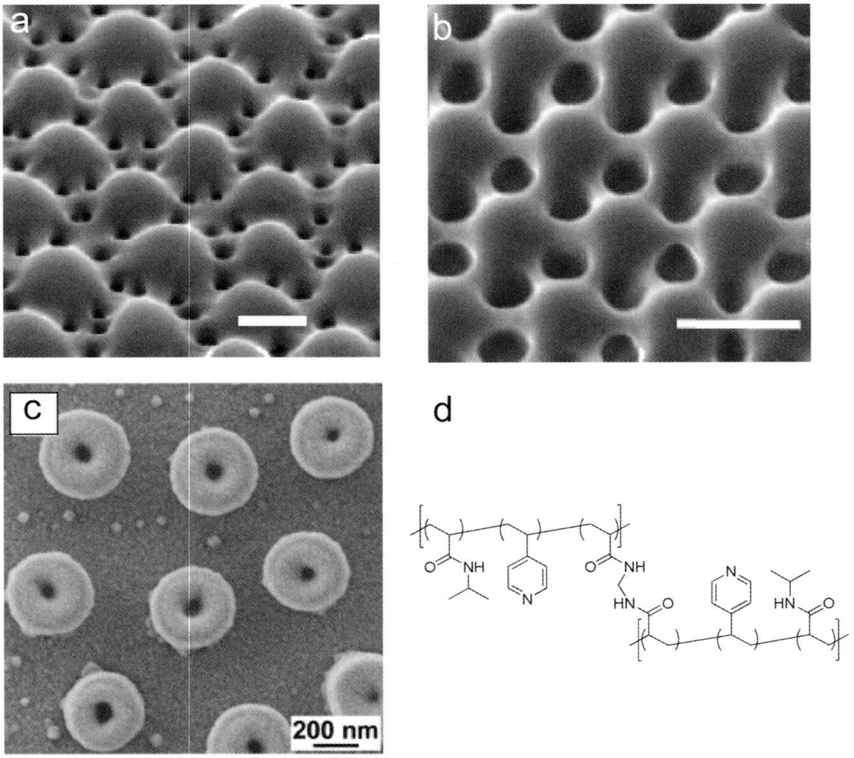

Fig. 16.14 (a) Porous microlens arrays formed by light-sensitive brittlestar, *Ophiocoma wendtii* (scale bar: 50 μm); (b) biomimetic microlens array (scale bar: 5 μm) [202]; (c) (PNIPAmVP)–$CaCO_3$–CdTe hybrid structure; (d) the chemical structure of PNIPAmVP [203]. Reprinted with permission from [202] and [203].

micro-patterned substrate with stabilized ACC through the deposition of specific functionalities that induced oriented nucleation, triggering the growth of single crystals across the entire region [201]. By combining this strategy with three-beam interference lithography, synthetic microlens arrays inspired from light-sensitive brittlestar have been produced with complex microstructure as shown in Fig. 16.14 a and b [202]. These lenses combine the functional aspects of lens and pores into a single structure, wherein the optical properties of the resultant photonic device could be tuned, and was also shown to have better focusing ability. These reports signify that complex functionalities can be incorporated by combining the bioinspired principles with existing technologies.

The design of materials that integrate numerous functionalities into a single system is an emerging trend. As an example, Kuang et al. have demonstrated the incorporation of nanocrystals of CdTe into hydrogel spheres of poly(*N*-isopropylacrylamide-*co*-4-vinylpyridine) (PNIPAm-*co*-VP) followed by the growth of $CaCO_3$ to produce hybrid nanostructures with novel morphologies [203]. Figure 16.14 c and d illustrate the formation of hybrid structures resembling the shape of red blood cells and possessing photoluminescent properties. Methodologies to integrate inorganic components with soft materials, such as antibodies, enzymes or proteins, to prepare conjugates that can be used in medical diagnostics and therapeutics have been developed [51, 204]. These techniques have led to elegant methods for the incorporation of multiple functionalities to produce probes that are useful in deconvoluting complex biological processes *in vitro* and *in vivo* [205, 206].

16.5.4
Outlook

Hybrid functional materials are produced by embracing biomimetic and biosynthetic strategies that can be applied for the production of novel materials. Bioinspired approaches have led to interdisciplinary programs where researchers produce materials with capabilities that often transcend the limits of Nature. By controlling, individually, the organic and inorganic components, synthetic methodologies are being developed to provide for the evolution of hybrid synthetic materials towards complex functional nanomaterials, while also taking into consideration the use of environmentally benign and energy-efficient processes to allow for the formation of functional bioresorbable materials.

16.6
Complex Functional Degradable Materials

16.6.1
Overview

Given the rising concern regarding the fate of synthetic materials and the diminishing resources available for small molecule precursors, chemical syntheses have begun to more substantially utilize renewable resources and natural materials [207–209]. Various modes of degradation have been explored to allow for responsible disposal of the substance and possible reuse of the products [210]. When the term "degradable" is used as an adjective to describe a material, it implies that an interaction or chemical linkage exists in the structure that can be broken to yield smaller constituent components. Degradation may occur through a number of mechanisms or under specific conditions and recent literature reports investigate hydrolysis [211], enzymatic action [212], thermal or photochemical exposure [213–216] and oxidation/reduction [217, 218] as methods for breakdown. Materials having multiple levels of degradability offer higher degrees of complexity in their performance and, therefore, application; some break down from large macromolecular architectures to individual polymer chains, whereas other designs allow the cleavage to proceed to oligomers or small molecules.

Synthetic materials having a particular function or a defined role, such as drug delivery [219–221], tissue engineering scaffolds [222–225] and sacrificial filler materials [226, 227], among others [228], may require inertness on a specific time-scale, followed by degradation, release of payload or removal from the system. In these cases, degradability of the functional macromolecule or structure is a necessary part of the material design and crucial for its successful application. A variety of synthetic polymeric materials, some demonstrating biocompatibility [229], have been prepared, but their applications are often limited by the incorporation of non-degradable synthetic polymer components. Broadening the composition of such macromolecular constructs to include degradable and biocompatible polymers will advance synthetic material usage in biomedical applications. This section of the chapter aims to highlight the creative syntheses and characterization of natural, unnatural and modified polysaccharides and polypeptides, and also bio-inspired synthetic polyesters, polyamides and polycarbonates, as degradable complex functional materials.

16.6.2
Biomaterials from Nature

Many of the interesting nanoscale materials available in Nature, such as those presented in Section 16.5, incorporate various forms and levels of degradability. Nature synthesizes polysaccharides, polypeptides and polynucleotides and assembles these macromolecules into complex architectures that are capable of performing multiple level functions on a given time-scale, dictated by the chemical form of the degradable linkages. One method to prepare materials with inherent degradability, along with the desired physical and chemical properties, is to utilize renewable, natural macromolecules. Although polypeptides [12, 230–233], polynucleotides [234] and other biomolecules [235–237] are receiving significant attention in the design of degradable materials, this section will focus upon the use of polysaccharides and their inherent functionality and degradability, as investigated in recent materials syntheses.

Polysaccharides represent an interesting and complex class of biomacromolecules, for which the sugar monomeric repeat units offer a high degree of functionality and the glycosidic linkages provide for degradability by enzymatic and chemical means. Zhong et al. [238] reported the formation of scaffolds for soft tissue engineering applications from electrospun solutions of collagen and chondroitin sulfate (a glycoaminoglycan, Fig. 16.15) yielding uniform fibers of nanoscale dimensions. These architectures were then cross-linked through reaction of pendant amines with glutaraldehyde vapor, producing a more biostable and collagenase-resistant scaffold material. As noted by the authors, however, polysaccharides can be difficult to process (i.e. by electrospinning), due to their relative insolubility in organic solvents. Hydrophobic modification of the pendant acid groups of alginate, depicted in Fig. 16.16, upon reaction with *n*-octyla-

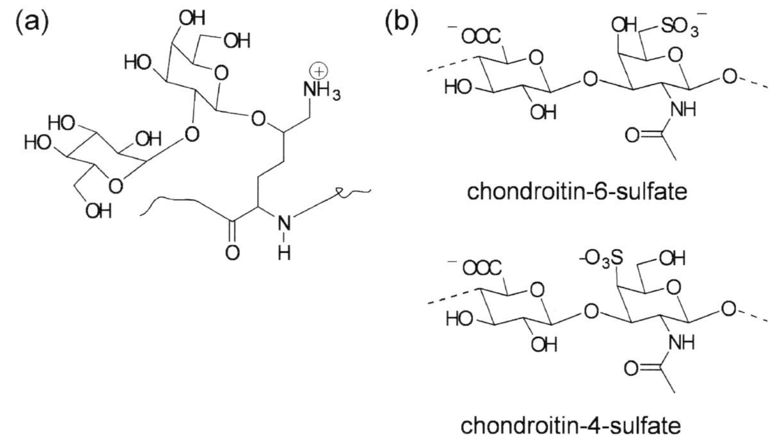

Fig. 16.15 (a) Glycosylated 5-hydroxylysine found in collagen; (b) representative saccharide repeat units of chondroitin sulfate [238].

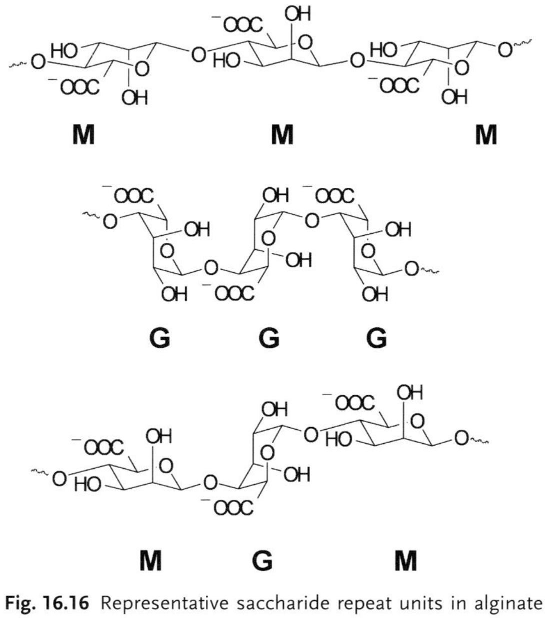

Fig. 16.16 Representative saccharide repeat units in alginate (**G** = guluronic acid and **M** = mannuronic acid) [239].

mine [239] illustrates an increase in interaction of the hydrophobically modified alginate molecules with sodium dodecyl sulfate (SDS), relative to the native, unmodified alginate, leading to stronger networks of biopolymer and micelles upon shearing. Biodegradable polymers, such as aliphatic polyesters, can also be used to form nanocomposites with polysaccharides, as demonstrated in a recent report [240] in which poly(lactic acid) (PLA) polymers were blended with acid-hydrolyzed microcrystalline cellulose to yield PLA-coated cellulose whisker composites.

The saccharide functionality and the overall shape of *β*-cyclodextrin, a macrocyclic polysaccharide containing seven glucose units linked through 1,4-glycosidic bonds and bearing 21 hydroxyl functionalities, have been exploited in the preparation of unique hybrid macromolecular architectures containing degradable components. Transformation of the hydroxyl groups into *α*-halo esters allowed for ATRP of *t*BA from the cyclodextrin surface to yield organosoluble star polymers with a degradable cyclodextrin core that may function as a polyelectrolyte after removal of the *tert*-butyl groups, as demonstrated by Karaky et al. [241]. The hexameric macrocycle of *α*-cyclodextrin and its host properties have been utilized in the formation of entirely hydrolytically degradable supramolecular inclusion complexes with polymers, such as star poly(*ε*-caprolactone) (PCL) [242, 243] (Scheme 16.5). The inclusion complex structure was confirmed through observation of reduced crystallinity of the PCL segments, as measured by differential scanning calorimetry and multi-step thermal degradation, as indicated by thermogravimetric analysis, along with characterization by nuclear magnetic resonance spectroscopy and X-ray diffraction studies.

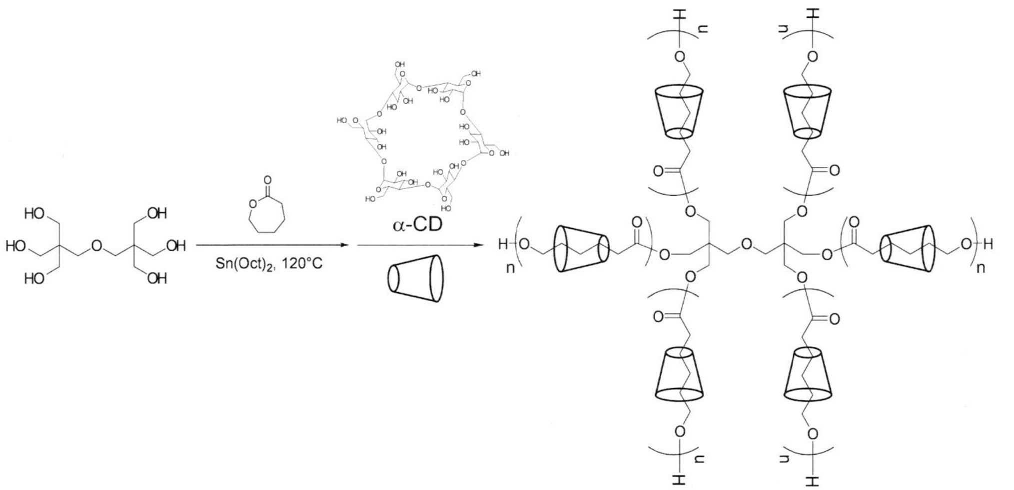

Scheme 16.5 Synthesis of six-arm PCL star and illustration of the supramolecular inclusion complex formed with *α*-cyclodextrin [243].

Polysaccharides, polypeptides and other polymeric biomolecules are assembled beautifully by organisms into intricate structures and offer inherent degradability with recyclability. Although these natural products can be utilized to prepare complex functional materials, their successful application may require additional control over the biomolecular composition and structure, which can be accomplished by enzymatic incorporation of functional monomeric units *in vitro* and *in vivo* (Section 16.6.3) or by purely synthetic means (Section 16.6.4).

16.6.3
Biosynthetic Degradable Materials with Synthetic Components

The synthesis of small molecules or macromolecules in Nature occurs through biosynthetic pathways that often utilize enzymes to perform specific chemical reactions, such as amide or ester bond formation or the reduction/oxidation of functional groups. For materials scientists and synthetic chemists, such chemical transformations are often conducted with the assistance of metallic catalysts, which can be expensive and may leave behind hard-to-remove metal contaminants. The ability to perform chemistry using Nature's enzymatic machinery allows for the production of synthetic biomaterials, without the expense or hazard of metal-mediated chemistries. It also allows for the incorporation of synthetic building blocks, including unnatural amino acids or nucleotides into the biomolecules, provided that the chemical modifications do not interfere with the ability of the enzymatic system to perform the synthesis. This section highlights some recent examples of enzymatic syntheses of biocompatible and biodegradable materials from both man-made and naturally occurring monomers.

The production of aliphatic polyesters, such as PCL, PLA and poly(glycolic acid) (PGA), is routinely accomplished through the ROP of cyclic monomers (i.e. a lactone or lactide), which is commonly mediated by metal-based catalysts including aluminum, tin, scandium or bismuth. Interestingly, aliphatic esters are also being polymerized through the use of commercially available Novozyme 435, a B lipase derived from *Candida antarctica* and bound to an acrylic macroporous resin. Specifically, homopolymers of 1,5-dioxepan-2-one (DXO) and copolymers of DXO and ε-CL were produced through ROP in the presence of Novozyme 435, demonstrating that mild reaction conditions could be employed to achieve high molecular weight polymers using the lipase biocatalyst, followed by its removal simply by filtration [244, 245]. The proposed mechanism for polymerization of DXO using Novozyme 435 was investigated and is included in Scheme 16.6. In addition, the transesterification ability of Novozyme 435 has also been employed in other functional materials syntheses [246, 247] and may be used in polymer recycling efforts [248]. Kallinteri et al. [249] prepared hydroxyl-functionalized polyesters through lipase-catalyzed polycondensation/transesterification of glycerol and divinyl adipate, followed by partial acylation of the hydroxyl groups, to yield functional polymers that were then assembled with surfactants into nanoparticles and studied as drug delivery vehicles (Scheme 16.7). Gross and coworkers have also contributed significantly to the implemen-

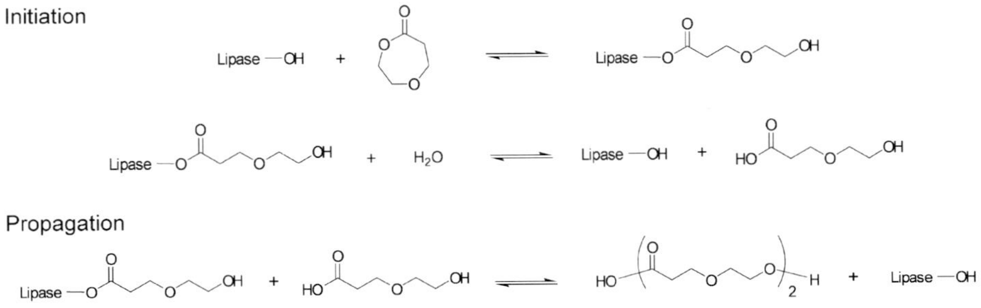

Scheme 16.6 Proposed mechanism for lipase-catalyzed ROP of DXO [244].

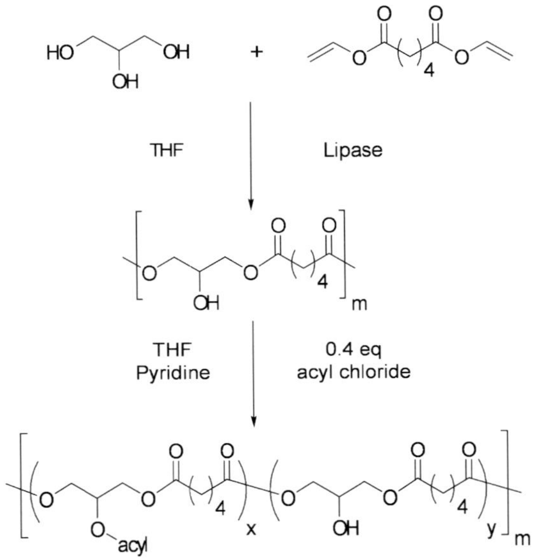

Scheme 16.7 Lipase-catalyzed polycondensation to form poly(glycerol adipate) acylated copolymers [249].

tation of the Novozyme 435 enzymatic system in the synthesis of functional degradable targets, ranging from modified starch particles [250] to linear polyesters bearing pendant carboxylic acid groups [251] and complex hyperbranched copolyesters [252].

The use of enzymatic machinery *in vivo* to produce biomolecules [218, 253, 254] offers a high degree of control over the polymer molecular weight, composition and sequence, as occurs in Nature, which has significant advantages over the *in vitro* polyester syntheses, for example. Tirrell and coworkers [255] extended this technique to include not only natural amino acids, but also the synthesis of artificial polypeptides containing unnatural amino acids. In one recent

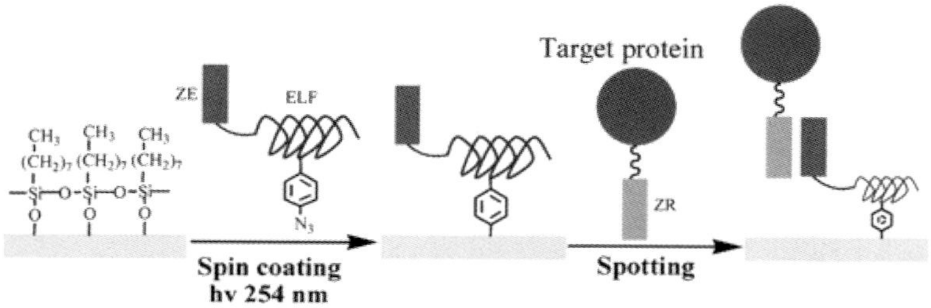

Fig. 16.17 Azide-containing ELF hydrophobic polypeptide for surface attachment of proteins. Reprinted with permission from [256].

case [256], by constructing an *E. coli* mutant phenylalanyl-tRNA synthetase (A294G) in a bacterial host, the photoreactive azidoamino acid *p*-azidophenylalanine could be incorporated in place of some of the phenylalanine residues. The resultant polypeptide contained photoreactive units within the hydrophobic elastin mimetic domain (ELF, Fig. 16.17) that is specifically designed for the attachment of a protein to a hydrophobic surface.

Manipulation of enzyme machinery has also allowed for optimization and expansion of the polymerization of saccharides to achieve unique polysaccharide syntheses by chemo-enzymatic means [257–260]. In one instance [259], a deletion enzyme missing the binding pocket responsible for degradation was utilized to produce polysaccharides of high molecular weight. Interestingly, this report by Nakamura et al. suggests that improvements in biopolymer synthesis through enzymatic means may be achieved through the identification of specific enzyme catalytic domains.

16.6.4 Purely Synthetic Degradable Materials

When Nature does not provide a suitable or processable material or a means to prepare one using modified, unnatural building blocks, purely synthetic methods are chosen to generate degradable complex functional macromolecules. Polyesters, with their well-studied degradation mechanisms, continue to attract significant attention as the number of methods for their preparation and functionalization increases. The literature for 2005 contains several clever functional polyester syntheses, including P(CL-*co*-PLLA)-perfluoropolyether composite materials [261], single-walled nanotubes grafted with PCL to increase PCL toughness [262], thiopropylsilyl-endcapped PCL and PLLA [263], mechanically strengthened PLA–PEO–PLA hydrogels [264] and P*t*BA-*b*-poly[(*R*)-3-hydroxybutyrate)]-*b*-P*t*BA triblock copolymers [265], illustrating the increased variety of synthetic approaches to and potential uses of polyester materials.

Novel polysaccharides, constructed through carbonate, glycoside, amide and other linkages, are also being reported to generate degradable materials. For ex-

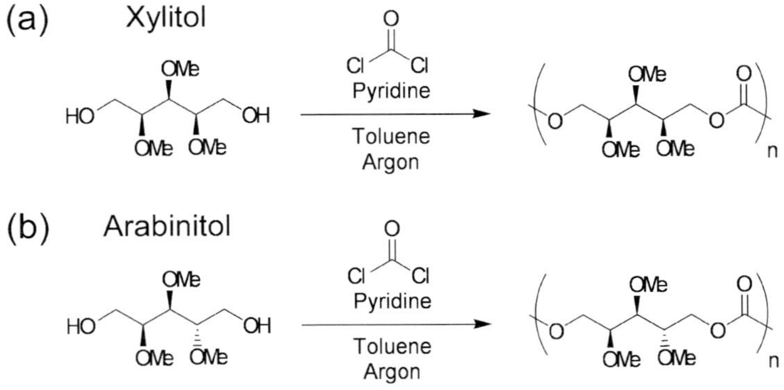

Scheme 16.8 (a) Xylitol and (b) arabinitol condensation polymerization [266].

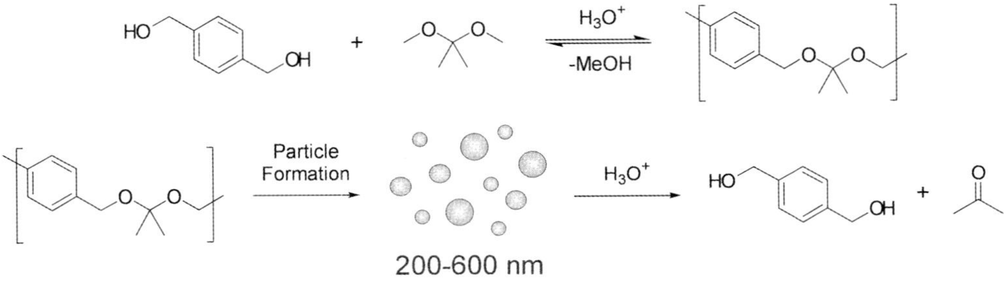

Scheme 16.9 Poly(ketal) nanoparticle synthesis from acetal exchange polymerization and oil-in-water emulsion particle formation followed by acid degradation [270].

ample, polycarbonates have been produced from arabinitol and xylitol sugar monomers by condensation polymerization, followed by study of their lipase-catalyzed degradation [266] (Scheme 16.8). The creation of various synthetic glycosyl donors and acceptors [267] exemplifies another interesting strategy for oligo- and polysaccharide preparation and significant progress has been made in controlling the stereoselectivity of the products [268]. An alternative route to purely synthetic polysaccharides employed amidation reactions between adipoyl chloride and amine-functionalized lactobionic acid, which yielded sugar-based gemini surfactants [269]. These surfactants, having pendant saccharides and constructed with amide backbone linkages, represent an interesting example beyond their composition, as they were also capable of assembling into micellar systems with tunable degradability.

Other innovative synthetic designs for degradable polymeric materials, extending beyond polysaccharides and polyesters, have been developed. Specifically, acetal and ketal units are being investigated as acid-labile linkages in the preparation of degradable nanoparticles for drug delivery through oil-in-water emulsion techniques [270] (Scheme 16.9) and through micellar assembly of linear-dendritic block copolymers [271]. Oligomeric spiro-orthocarbonates [272], other polycarbonates [273–277], synthetic polypeptides [278, 279], poly(ester amide) hydrogels [280] and hyperbranched poly(amino ester)s [281] typify the breadth of composition, degradable linkage and architecture that is being examined in laboratories to widen and improve degradable complex functional macromolecules.

16.6.5 Outlook

The recent polymer and materials science literature illustrates the diversity of chemical approaches being developed to create degradable macromolecules that are assembled into functional materials for greater applications. Within this section, examples of polyesters, polysaccharides and other unique degradable macromolecules have been presented. Materials syntheses are becoming more environmentally responsible and natural sources of monomers and macromolecules are being utilized. Key to the success of these Nature-based synthetic methodologies is the incorporation of useful chemical handles within the degradable materials to allow for further modification and conjugation chemistries. The recent literature indicates that this need is being addressed and already contains examples of polyester and polysaccharide syntheses that integrate specific latent functional groups and provide characterization data that demonstrate the presence and accessibility of these handles. In addition to functionalization, considerable emphasis is being placed on the orchestrated and controlled assembly of synthetic macromolecules (degradable or otherwise) into higher order materials, which are also discussed in previous sections of this chapter.

References

1 Behe, M. J., Discovery Institute. Irreducible complexity is an obstacle to Darwinism even if parts of a system have other functions, http://www.discovery.org/scripts/viewDB/index.php?program=CRSC%20Responses&command=view&id=1831, **2004**.

2 McDonald, J. H. A reducibly complex mousetrap, http://udel.edu/~mcdonald/mousetrap.html.

3 Miller, K. R. The mousetrap analogy or trapped by design, http://www.millerandlevine.com/km/evol/DI/Mousetrap.html.

4 Wooley, K. L. *J. Polym. Sci., Part A: Polym. Chem.* **2000**, *38*, 1397–1407.

5 Park, C., Yoon, J., Thomas, E. L. *Polymer* **2003**, *44*, 6725–6760.

6 Rodriguez-Hernandez, J., Checot, F., Gnanou, Y., Lecommandoux, S. *Prog. Polym. Sci.* **2005**, *30*, 691–724.

7 Hamley, I. W. *Nanotechnology* **2003**, *14*, R39–R54.

8 Förster, S., Plantenberg, T. *Angew. Chem. Int. Ed.* **2002**, *41*, 688–714.

9 Epps, T. H., Chatterjee, J., Bates, F. S. *Macromolecules* **2005**, *38*, 8775–8784.

10 Allen, C., Maysinger, D., Eisenberg, A. *Colloids Surf. B: Biointerfaces* **1999**, *16*, 3–27.

11 Cornelissen, J. J. L. M., Donners, J. J. J. M., de Gelder, R., Graswinckel, W. S., Metselaar, G. A., Rowan, A. E., Sommerdijk, N. A. J. M., Nolte, R. J. M. *Science* **2001**, *293*, 676–680.

12 Holowka, E. P., Pochan, D. J., Deming, T. J. *J. Am. Chem. Soc.* **2005**, *127*, 12423–12428.
13 Kukula, H., Schlaad, H., Antonietti, M., Forster, S. *J. Am. Chem. Soc.* **2002**, *124*, 1658–1663.
14 Klok, H.-A., Vandermeulen, G. W. M., Nuhn, H., Roesler, A., Hamley, I. W., Castelletto, V., Xu, H., Sheiko, S. S. *Faraday Discuss.* **2004**, *128*, 29–41.
15 Luebbert, A., Castelletto, V., Hamley, I. W., Nuhn, H., Scholl, M., Bourdillon, L., Wandrey, C., Klok, H.-A. *Langmuir* **2005**, *21*, 6582–6589.
16 Ho, R.-M., Chiang, Y.-W., Tsai, C.-C., Lin, C.-C., Ko, B.-T., Huang, B.-H. *J. Am. Chem. Soc.* **2004**, *126*, 2704–2705.
17 Rzayev, J., Hillmyer, M. A. *Macromolecules* **2005**, *38*, 3–5.
18 Salem, A. K., Rose, F. R. A. J., Oreffo, R. O. C., Yang, X., Davies, M. C., Mitchell, J. R., Roberts, C. J., Stolnik-Trenkic, S., Tendler, S. J. B., Williams, P. M., Shakesheff, K. M. *Adv. Mater.* **2003**, *15*, 210–213.
19 Cho, B. K., Jain, A., Gruner, S. M., Wiesner, U. *Science* **2004**, *305*, 1598–1601.
20 Bertolucci, M., Galli, G., Chiellini, E., Wynne, K. J. *Macromolecules* **2004**, *37*, 3666–3672.
21 Berret, J.-F., Vigolo, B., Eng, R., Herve, P., Grillo, I., Yang, L. *Macromolecules* **2004**, *37*, 4922–4930.
22 Rider, D. A., Cavicchi, K. A., Power-Billard, K. N., Russell, T. P., Manners, I. *Macromolecules* **2005**, *38*, 6931–6938.
23 Stoenescu, R., Meier, W. *Chem. Commun.* **2002**, 3016–3017.
24 Du, J., Chen, Y., Zhang, Y., Han, C. C., Fischer, K., Schmidt, M. *J. Am. Chem. Soc.* **2003**, *125*, 14710–14711.
25 Bellomo, E. G., Wyrsta, M. D., Pakstis, L., Pochan, D. J., Deming, T. J. *Nat. Mater.* **2004**, *3*, 244–248.
26 Becker, M. L., Liu, J., Wooley, K. L. *Biomacromolecules* **2005**, *6*, 220–228.
27 Pochan, D. J., Chen, Z., Cui, H., Hales, K., Qi, K., Wooley, K. L. *Science* **2004**, *306*, 94–97.
28 Chen, Z., Cui, H., Hales, K., Li, Z., Qi, K., Pochan, D. J., Wooley, K. L. *J. Am. Chem. Soc.* **2005**, *127*, 8592–8593.
29 Li, Z., Kesselman, E., Talmon, Y., Hillmyer, M. A., Lodge, T. P. *Science* **2004**, *306*, 98–101.
30 Li, Z., Chen, Z., Cui, H., Hales, K., Qi, K., Wooley, K. L., Pochan, D. J. *Langmuir* **2005**, *21*, 7533–7539.
31 Liu, X., Kim, J.-S., Wu, J., Eisenberg, A. *Macromolecules* **2005**, *38*, 6749–6751.
32 Chen, T., Wang, L., Jiang, G., Wang, J., Wang, X. J., Zhou, J., Wang, W., Gao, H. *Polymer* **2005**, *46*, 7585–7589.
33 Erhardt, R., Böker, A., Zettl, H., Kaya, H., Pyckhout-Hintzen, W., Krausch, G., Abetz, V., Müller, A. H. E. *Macromolecules* **2001**, *34*, 1069–1075.
34 Erhardt, R., Zhang, M., Boker, A., Zettl, H., Abetz, C., Frederik, P., Krausch, G., Abetz, V., Müller, A. H. E. *J. Am. Chem. Soc.* **2003**, *125*, 3260–3267.
35 Lu, Z., Liu, G., Liu, F. *Macromolecules* **2001**, *34*, 8814–8817.
36 Zheng, R., Liu, G., Yan, X. *J. Am. Chem. Soc.* **2005**, *127*, 15358–15359.
37 Luo, L., Eisenberg, A. *Angew. Chem. Int. Ed.* **2002**, *114*, 1043–1046.
38 Schrage, S., Sigel, R., Schlaad, H. *Macromolecules* **2003**, *36*, 1417–1420.
39 Vangeyte, P., Leyh, B., Auvray, L., Grandjean, J., Misselyn-Bauduin, A. M., Jérôme, R. *Langmuir* **2004**, *20*, 9019–9028.
40 Sivakova, S., Bohnsack, D. A., Mackay, M. E., Suwanmala, P., Rowan, S. J. *J. Am. Chem. Soc.* **2005**, *127*, 18202–18211.
41 Chen, D., Jiang, M. *Acc. Chem. Res.* **2005**, *38*, 494–502.
42 Tokarev, I., Krenek, R., Burkov, Y., Schmeisser, D., Sidorenko, A., Minko, S., Stamm, M. *Macromolecules* **2005**, *38*, 507–516.
43 Solomatin, S. V., Bronich, T. K., Bargar, T. W., Eisenberg, A., Kabanov, V. A., Kabanov, A. V. *Langmuir* **2003**, *19*, 8069–8076.
44 Stubbs, L. P., Weck, M. *Chem. Eur. J.* **2003**, *9*, 992–999.
45 Nair, K. P., Pollino, J. M., Weck, M. *Macromolecules* **2006**, *39*, 931–940.
46 Kubowicz, S., Baussard, J.-F., Lutz, J.-F., Thuenemann, A. F., von Berlepsch, H., Laschewsky, A. *Angew. Chem. Int. Ed.* **2005**, *44*, 5262–5265.

47 Lodge, T. P., Rasdal, A., Li, Z., Hillmyer, M. A. *J. Am. Chem. Soc.* **2005**, *127*, 17608–17609.
48 Qi, K., Ma, Q., Remsen, E. E., Clark, C. G., Jr., Wooley, K. L. *J. Am. Chem. Soc.* **2004**, *126*, 6599–6607.
49 Joralemon, M. J., Smith, N. L., Holowka, D., Baird, B., Wooley, K. L. *Bioconjug. Chem.* **2005**, *16*, 1246–1256.
50 Pan, D., Turner, J. L., Wooley, K. L. *Chem. Commun.* **2003**, 2400–2401.
51 Rossin, R., Pan, D., Qi, K., Turner, J. L., Sun, X., Wooley, K. L., Welch, M. J. *J. Nucl. Med.* **2005**, *46*, 1210–1218.
52 Turner, J. L., Chen, Z., Wooley, K. L. *J. Control. Release* **2005**, *109*, 189–202.
53 Weaver, J. V. M., Armes, S. P., Liu, S. *Macromolecules* **2003**, *36*, 9994–9998.
54 Sugihara, S., Kanaoka, S., Aoshima, S. *J. Polym. Sci., Part A: Polym. Chem.* **2004**, *42*, 2601–2611.
55 Crowe, J. A., Genzer, J. *J. Am. Chem. Soc.* **2005**, *127*, 17610–17611.
56 Xu, C., Fu, X., Fryd, M., Xu, S., Wayland, B. B., Winey, K. I., Composto, R. J. *Nano Lett.* **2006**, *6*, 282–287.
57 Ludwigs, S., Böker, A., Voronov, A., Rehse, N., Magerle, R., Krausch, G. *Nat. Mater.* **2003**, *2*, 744–747.
58 Stoykovich, M. P., Mueller, M., Kim, S. O., Solak, H. H., Edwards, E. W., de Pablo, J. J., Nealey, P. F. *Science* **2005**, *308*, 1442–1446.
59 Ryu, D. Y., Shin, K., Drockenmuller, E., Hawker, C. J., Russell, T. P. *Science* **2005**, *308*, 236–239.
60 Wang, X. S., Arsenault, A., Ozin, G. A., Winnik, M. A., Manners, I. *J. Am. Chem. Soc.* **2003**, *125*, 12686–12687.
61 Du, J., Tang, Y., Lewis, A. L., Armes, S. P. *J. Am. Chem. Soc.* **2005**, *127*, 17982–17983.
62 Fahmi, A. W., Stamm, M. *Langmuir* **2005**, *21*, 1062–1066.
63 Horiuchi, S., Fujita, T., Hayakawa, T., Nakao, Y. *Langmuir* **2003**, *19*, 2963–2973.
64 Turner, J. L., Wooley, K. L. *Nano Lett.* **2004**, *4*, 683–688.
65 Savic, R., Luo, L., Eisenberg, A., Maysinger, D. *Science* **2004**, *303*, 627–628.
66 Kakizawa, Y., Kataoka, K. *Adv. Drug Deliv. Rev.* **2002**, *54*, 203–222.
67 Sant, V. P., Smith, D., Leroux, J.-C. *J. Control. Release* **2005**, *104*, 289–300.
68 Minko, T., Batrakova, E. V., Li, S., Li, Y., Pakunlu, R. I., Alakhov, V. Y., Kabanov, A. V. *J. Control. Release* **2005**, *105*, 269–278.
69 Hawker, C. J., Wooley, K. L. *Science* **2005**, *309*, 1200–1205.
70 Matyjaszewski, K., Spanswick, J. *Mater. Today* **2005**, *8*, 26–34.
71 Monteiro, M. J. *J. Polym. Sci., Part A: Polym. Chem.* **2005**, *43*, 3189–3204.
72 Runge, M. B., Dutta, S., Bowden, N. B. *Macromolecules* **2006**, *39*, 498–508.
73 Lam, J. W. Y., Tang, B. Z. *Acc. Chem. Res.* **2005**, *38*, 745–754.
74 Barbarella, G., Melucci, M., Sotigu, G. *Adv. Mater.* **2005**, *17*, 1581–1593.
75 Leclère, P., Surin, M., Viville, P., Lazzaroni, R., Kilbinger, A. F. M., Henze, O., Feast, W. J., Cavallini, M., Biscarini, F., Schenning, A. P. H. J., Meijer, E. W. *Chem. Mater.* **2004**, *16*, 4452–4466.
76 Hoeben, F. J. M., Jonkheijm, P., Meijer, E. W., Schenning, A. P. H. J. *Chem. Rev.* **2005**, *105*, 1491–1546.
77 Schenning, A. P. H. J., Meijer, E. W. *Chem. Commun.* **2005**, 3245–3258.
78 Perepichka, I. F., Perepichka, D. F., Meng, H., Wudl, F. *Adv. Mater.* **2005**, *17*, 2281–2305.
79 Zheng, J., Swager, T. M. *Adv. Polym. Sci.* **2005**, *177*, 151–179.
80 Carter, K. R. *Macromolecules* **2002**, *35*, 6757–6759.
81 Akkara, J. A., Senecal, K. J., Kaplan, D. L. *J. Polym. Sci., Part A: Polym. Sci.* **1991**, *29*, 1561–1574.
82 Bruno, F. F., Fossey, S. A., Nagarajan, S., Nagarajan, R., Kumar, J., Samuelson, L. A. *Biomacromolecules* **2006**, *7*, 586–589.
83 Hu, X., Zhang, Y.-Y., Tang, K., Zou, G.-L. *Synth. Met.* **2005**, *150*, 1–7.
84 Mazhugo, Y. M., Caramyshev, A. V., Shleev, S. V., Sakharov, I. Y., Yaropolov, A. I. *Appl. Biochem. Microbiol.* **2005**, *41*, 283–287.
85 Song, H.-K., Palmore, G. T. R. *J. Phys. Chem. B* **2005**, *109*, 19278–19287.
86 van Steenis, D. J. V. C., David, O. R. P., van Strijdonck, G. P. F., van Maarseveen, J. H., Reek, J. N. H. *Chem. Commun.* **2005**, 4333–4335.

87 Clément, D., Makereel, F., Herz, L. M., Hoeben, F. J. M., Jonkheijm, P., Schenning, A. P. H. J., Meijer, E. W., Friend, R. H., Silva, C. *J. Chem. Phys.* **2005**, *123*, 084902-1-7.

88 Beinhoff, M., Bozano, L. D., Scott, J. C., Carter, K. R. *Macromolecules* **2005**, *38*, 4147–4156.

89 Wu, F.-I., Shih, P.-I., Tseng, Y.-H., Chen, G.-Y., Chien, C.-H., Shu, C.-F., Tung, Y.-L., Chi, Y., Jen, A. K.-Y. *J. Phys. Chem. B* **2005**, *109*, 14000–14005.

90 Li, Z., Dong, Y., Qin, A., Lam, J. W. Y., Dong, Y., Yuan, W., Sun, J., Hua, J., Wong, K. S., Tang, B. Z. *Macromolecules* **2006**, *39*, 467–469.

91 Lee, B., Seshadri, V., Sotzing, G. A. *Langmuir* **2005**, *21*, 10797–10802.

92 Yasuda, T., Sakai, Y., Aramaki, S., Yamamato, T. *Chem. Mater.* **2005**, *17*, 6060–6068.

93 Murphy, A R., Chang, P. C., VanDyke, P., Liu, J., Fréchet, J. M. J., Subramanian, V., DeLongchamp, D. M., Sambasivan, S., Fischer, D. A., Lin, E. K. *Chem. Mater.* **2005**, *17*, 6033–6041.

94 Chen, J., Ratera, I., Ogletree, D. F., Salmeron, M., Murphy, A. R., Fréchet, J. M. J. *Langmuir* **2005**, *21*, 1080–1085.

95 Letizia, J. A., Facchetti, A., Stern, C. L., Ratner, M. A., Marks, T. J. *J. Am. Chem. Soc.* **2005**, *127*, 13476–13477.

96 Beckers, E. H. A., Meskers, S. C. J., Schenning, A. P. H. J., Chen, Z., Würthner, F., Marsal, P., Beljonne, D., Cornil, J., Janssen, R. A. J. *J. Am. Chem. Soc.* **2006**, *128*, 649–657.

97 Thomas S. W. III, Long, T. M., Pate, B. D., Kline, S. R., Thomas, E. L., Swager, T. M. *J. Am. Chem. Soc.* **2005**, *127*, 17976–17977.

98 Xia, C., Fan, X., Locklin, J., Advincula, R. C., Gies, A., Nonidez, W. *J. Am. Chem. Soc.* **2004**, *126*, 8735–8743.

99 Mitsumori, T., Craig, I. M., Martini, I. B., Schwartz, B. J., Wudl, F. *Macromolecules* **2005**, *38*, 4698–4704.

100 Whiting, G. L., Snaith, H. J., Khodabakhsh, S., Andreasen, J. W., Breiby, D. W., Nielsen, M. M., Greenham, N. C., Friend, R. H., Huck, W. T. S. *Nano Lett.* **2006**, *6*, 573–578.

101 Ewbank, P. C., Loewe, R. S., Zhai, L., Reddinger, J., Sauvé, G., McCullough, R. D. *Tetrahedron* **2004**, *60*, 11269–11275.

102 Chen, T.-A., Wu, X., Rieke, R. D. *J. Am. Chem. Soc.* **1995**, *117*, 233–244.

103 Hong, X. M., Tyson, J. C., Collard, D. M. *Macromolecules* **2000**, *33*, 3502–3505.

104 Li, L., Collard, D. M. *Macromolecules* **2005**, *38*, 372–378.

105 Liu, J., Sheina, E., Kowaleski, T., McCullough, R. D. *Angew. Chem. Int. Ed.* **2002**, *41*, 329–332.

106 Iovu, M. C., Jeffries E. L. M., Sheina, E., Cooper, J. R., McCullough, R., D. *Polymer* **2005**, *46*, 8582–8586.

107 Radano, C. P., Scherman, O. A., Stingelin-Stutzmann, N., Müller, C., Breiby, D. W., Smith, P., Janssen, R. A. J., Meijer, E. W. *J. Am. Chem. Soc.* **2005**, *127*, 12502–12503.

108 Katz, H. E., Siegrist, T., Lefenfeld, M., Gopalan, P., Mushrush, M., Ocko, B., Gang, O., Jisrawl, N. *J. Phys. Chem. B* **2004**, *108*, 8567–8571.

109 Cheun, H., Tanto, B., Chunwaschirasiri, W., Larson, B., Winokur, M. J. *Appl. Phys. Lett.* **2004**, *84*, 22–24.

110 Lu, S., Liu, T., Ke, L., Ma, D.-G., Chua, S.-J., Huang, W. *Macromolecules* **2005**, *38*, 8494–8502.

111 Nakao, H., Hayashi, H., Iwata, F., Karasawa, H., Hirano, K., Sugiyama, S., Ohtani, T. *Langmuir* **2005**, *21*, 7945–7950.

112 Ma, Y., Zhang, J., Zhang, G., He, H. *J. Am. Chem. Soc.* **2004**, *126*, 7097–7101.

113 Deng, T., Breen, C., Breiner, T., Swager, T. M., Thomas, E. L. *Polymer* **2005**, *46*, 10113–10118.

114 Doherty III, W. J., Armstrong, N. R., Saavedra, S. S. *Chem. Mater.* **2005**, *17*, 3652–3660.

115 Wang, F., Han, M.-Y., Mya, K. Y., Wang, Y., Lai, Y.-H. *J. Am. Chem. Soc.* **2005**, *127*, 10350–10355.

116 Szymanski, C., Wu, C., Hooper, J., Salazar, M. A., Perdomo, A., Dukes, A., McNeill, J. *J. Phys. Chem. B* **2005**, *109*, 8543–8546.

117 Balamurugan, S. S., Bantchev, G. B., Yang, Y., McCarley, R. L. *Angew. Chem. Int. Ed.* **2005**, *44*, 4872–4876.

118 Zhang, L., Wan, M. *Thin Solid Films* **2005**, *477*, 24–31.
119 Jang, J., Yoon, H. *Langmuir* **2005**, *21*, 11484–11489.
120 Shklyarevsky, I. O., Jonkheijm, P., Christianen, P. C. M., Schenning, A. P. H. J., Meijer, E. W., Henze, O., Kilbinger, A. F. M., Feast, W. J., Del Guerzo, A., Desvergne, J.-P., Maan, J. C. *J. Am. Chem. Soc.* **2005**, *127*, 1112–1113.
121 de Cuendias, A., Le Hellaye, M., Lecommandaux, S., Cloutet, E., Cramail, H. *J. Mater. Chem.* **2005**, *15*, 3264–3267.
122 Cheng, C., Yang, N.-L. *Macromol. Rapid Commun.* **2005**, *26*, 1395–1399.
123 Amara, J. P., Swager, T. M. *Macromolecules* **2005**, *38*, 9091–9094.
124 Wosnick, J. H., Mello, C. M., Swager, T. M. *J. Am. Chem. Soc.* **2005**, *127*, 3400–3405.
125 Kim, J.-M., Lee, Y. B., Yang, D. H., Lee, J.-S., Lee, G. S., Ahn, D. J. *J. Am. Chem. Soc.* **2005**, *127*, 17580–17581.
126 Zhang, M., Müller, A. H. E. *J. Polym. Sci., Part A: Polym. Chem.* **2005**, *43*, 3461–3481.
127 Hadjichristidis, N., Pitsikalis, M., Iatrou, H., Pispas, S. *Macromol. Rapid Commun.* **2003**, *24*, 979–1013.
128 Hadjichristidis, N., Pitsikalis, M., Pispas, S., Iatrou, H. *Chem. Rev.* **2001**, *101*, 3747–3792.
129 Pyun, J., Kowalewski, T., Matyjaszewski, K. *Macromol. Rapid Commun.* **2003**, *24*, 1043–1059.
130 Sheiko, S. S., Prokhorova, S. A., Beers, K. L., Matyjaszewski, K., Potemkin, I. I., Khokhlov, A. R., Möller, M. *Macromolecules* **2001**, 8354–8360.
131 Lord, S. J., Sheiko, S. S., LaRue, I., Lee, H.-I., Matyjaszewski, K. *Macromolecules* **2004**, *37*, 4235–4240.
132 Matyjaszewski, K., Qin, S., Boyce, J. R., Shirvanyants, D., Sheiko, S. S. *Macromolecules* **2003**, *36*, 1843–1849.
133 Hsieh, H. L., Quirk, R. P. *Anionic Polymerization: Principles and Practical Applications.* Marcel Dekker, New York, **1996**.
134 Denesyuk, N. A. *Phys. Rev. E* **2003**, *68*, 031803.
135 Gerle, M., Fischer, K., Roos, S., Müller, A. H. E., Schmidt, M., Sheiko, S. S., Prokhorova, S., Möller, M. *Macromolecules* **1999**, *32*, 2629–2637.
136 Boyce, J. R., Shirvanyants, D., Sheiko, S. S., Ivanov, D. A., Qin, S., Börner, H., Matyjaszewski, K. *Langmuir* **2004**, *20*, 6005–6011.
137 Sun, F., Sheiko, S. S., Möller, M., Beers, K., Matyjaszewski, K. *J. Phys. Chem. A* **2004**, *108*, 9682–9686.
138 Gallyamov, M. O., Tartsch, B., Khokhlov, A. R., Sheiko, S. S., Börner, H. G., Matyjaszewski, K., Möller, M. *Chem. Eur. J.* **2004**, *10*, 4599–4605.
139 Stephan, T., Muth, S., Schmidt, M. *Macromolecules* **2002**, *35*, 9857–9860.
140 Li, C., Gunari, N., Fischer, K., Janshoff, A., Schmidt, M. *Angew. Chem. Int. Ed.* **2004**, *43*, 1101–1104.
141 Börner, H. G., Beers, K., Matyjaszewski, K., Sheiko, S. S., Möller, M. *Macromolecules* **2001**, *34*, 4375–4383.
142 Cheng, G., Böker, A., Zhang, M., Krausch, G., Müller, A. H. E. *Macromolecules* **2001**, *34*, 6883–6888.
143 Zhang, M., Breiner, T., Mori, H., Müller, A. H. E. *Polymer* **2003**, *44*, 1449–1458.
144 Neiser, M. W., Muth, S., Kolb, U., Harris, J. R., Okuda, J., Schmidt, M. *Angew. Chem. Int. Ed.* **2004**, *43*, 3192–3195.
145 Qin, S., Matyjaszewski, K., Xu, H., Sheiko, S. S. *Macromolecules* **2003**, *36*, 605–612.
146 Khelfallah, N., Gunari, N., Fischer, K., Gkogkas, G., Hadjichristidis, N., Schmidt, M. *Macromol. Rapid Commun.* **2005**, *26*, 1693–1697.
147 Ishizu, K., Satoh, J., Sogabe, A. *J. Colloid Interface Sci.* **2004**, *274*, 472–479.
148 Cheng, Z., Zhu, X., Fu, G. D., Kang, E. T., Neoh, K. G. *Macromolecules* **2005**, *38*, 7187–7192.
149 Matyjaszewski, K., Xia, J. *Chem. Rev.* **2001**, *101*, 2921–2990.
150 Kamigaito, M., Ando, T., Sawamoto, M. *Chem. Rev.* **2001**, *101*, 3689–3745.
151 Neugebauer, D., Sumerlin, B. S., Matyjaszewski, K., Goodhart, B., Sheiko, S. S. *Polymer* **2004**, *45*, 8173–8179.
152 Sumerlin, B. S., Neugebauer, D., Matyjaszewski, K. *Macromolecules* **2005**, *38*, 702–708.

153 Börner, H.G., Duran, D., Matyjaszewski, K., Silva, M.d., Sheiko, S.S. *Macromolecules* **2002**, *35*, 3387–3394.
154 Lee, H.-I., Matyjaszewski, K., Yu, S., Sheiko, S.S. *Macromolecules* **2005**, *38*, 8264–8271.
155 Beers, K.L., Gaynor, S.G., Matyjaszewski, K. *Macromolecules* **1998**, *31*, 9413–9415.
156 Muthukrishnan, S., Zhang, M., Burkhardt, M., Drechsler, M., Mori, H., Müller, A.H.E. *Macromolecules* **2005**, *38*, 7926–7934.
157 Zhang, A., Barner, J., Gössl, I., Rabe, J.P., Schlüter, A.D. *Angew. Chem. Int. Ed.* **2004**, *43*, 5185–5188.
158 Zhang, M., Drechsler, M., Müller, A.H.E. *Chem. Mater.* **2004**, *16*, 537–543.
159 Zhang, M., Estournes, C., Bietsch, W., Müller, A.H.E. *Adv. Funct. Mater.* **2004**, *14*, 871–882.
160 Higuchi, M., Inoue, T., Miyoshi, H., Kawaguchi, M. *Langmuir* **2005**, *21*, 11462–11467.
161 Zhang, B., Fischer, K., Schmidt, M. *Macromol. Chem. Phys.* **2005**, *206*, 157–162.
162 Jha, S., Dutta, S., Bowden, N.B. *Macromolecules* **2004**, *37*, 4365–4374.
163 Neugebauer, D., Zhang, Y., Pakula, T., Matyjaszewski, K. *Macromolecules* **2005**, *38*, 8687–8693.
164 Grande, D., Six, J.-L., Breunig, S., Héroguez, V., Fontanille, M., Gnanou, Y. *Polym. Adv. Technol.* **1998**, *9*, 601–612.
165 Ball, Z.T., Sivula, K., Fréchet, J.M.J. *Macromolecules* **2006**, *39*, 70–72.
166 Helms, B., Mynar, J.L., Hawker, C.J., Fréchet, J.M.J. *J. Am. Chem. Soc.* **2004**, *126*, 15020–15021.
167 Neugebauer, D., Zhang, Y., Pakula, T., Sergei, S.S., Matyjaszewski, K. *Macromolecules* **2003**, *36*, 6746–6755.
168 Ishizu, K., Kakinuma, H. *J. Polym. Sci., Part A: Polym. Chem.* **2004**, *43*, 63–70.
169 Barthlott, W., Nienhuis, C. *Planta* **1997**, *202*, 1–8.
170 Mayer, G. *Science* **2005**, *310*, 1144–1147.
171 Kuhn, L.T., Fink, D.J., Heuer, A.H. In *Biomimetic Materials Chemistry*, Mann, S. (Ed.). Wiley-VCH, New York, **1996**, pp. 41–68.
172 Mann, S. *Biomineralization*. Oxford University Press, New York, **2001**.
173 Aizenberg, J., Weaver, J.C., Thanawala, M.S., Sundar, V.C., Morse, D.E., Fratzl, P. *Science* **2005**, *309*, 275–278.
174 Flynn, C.E., Lee, S.-W., Peelle, B.R., Belcher, A.M. *Acta Mater.* **2003**, *51*, 5867–5880.
175 Sarikaya, M., Tamerler, C., Schwartz, D.T., Baneyx, F. *Annu. Rev. Mater. Res.* **2004**, *34*, 373–408.
176 Wong Po Foo, C., Kaplan, D.L. *Adv. Drug Deliv. Rev.* **2002**, *54*, 1131–1143.
177 Asakura, T., Nitta, K., Yang, M., Yao, J., Nakazawa, Y., Kaplan, D.L. *Biomacromolecules* **2003**, *4*, 815–820.
178 Sanguanchaipaiwong, V., Gabelish, C. L., Hook, J., Scholz, C., Foster, L.J.R. *Biomacromolecules* **2004**, *5*, 643–649.
179 Zhang, K., Diehl, M.R., Tirrell, D.A. *J. Am. Chem. Soc.* **2005**, *127*, 10136–10137.
180 Wang, L., Schultz, P.G. *Angew. Chem. Int. Ed.* **2005**, *44*, 34–66.
181 Douglas, T., Young, M. *Nature* **1998**, *393*, 152–155.
182 Whaley, S.R., English, D.S., Hu, E.L., Barbara, P.F., Belcher, A.M. *Nature* **2000**, *405*, 665–668.
183 Wang, Q., Lin, T., Tang, L., Johnson, J.E., Finn, M.G. *Angew. Chem. Int. Ed.* **2002**, *41*, 459–462.
184 Mao, C., Solis, D.J., Reiss, B.D., Kottmann, S.T., Sweeney, R.Y., Hayhurst, A., Georgiou, G., Iverson, B., Belcher, A.M. *Science* **2004**, *303*, 213–217.
185 Falkner, J.C., Turner, M.E., Bosworth, J.K., Trentler, T.J., Johnson, J.E., Lin, T., Colvin, V.L. *J. Am. Chem. Soc.* **2005**, *127*, 5274–5275.
186 Pum, D., Sleytr, U.B. *Trends Biotechnol.* **1999**, *17*, 8–12.
187 Sastry, M., Ahmad, A., Khan, I.M., Kumar, R. *Curr. Sci.* **2003**, *85*, 162–170.
188 Shankar, S.S., Rai, A., Ankamwar, B., Singh, A., Ahmad, A., Sastry, M. *Nat. Mater.* **2004**, *3*, 482–488.
189 Senapati, S., Ahmad, A., Khan, M.I., Sastry, M., Kumar, R. *Small* **2005**, *1*, 517–519.
190 Navrotsky, A. *Proc. Natl. Acad. Sci. USA* **2004**, *101*, 12096–12101.
191 Mann, S. in *Biomimetic Materials Chemistry*, Mann, S. (Ed.). Wiley-VCH, New York, **1996**.

192 Song, J., Malathong, V., Bertozzi, C. R. *J. Am. Chem. Soc.* **2005**, *127*, 3366–3372.
193 Song, J., Saiz, E., Bertozzi, C. R. *J. Am. Chem. Soc.* **2003**, *125*, 1236–1243.
194 Spoerke, E. D., Stupp, S. I. *Biomaterials* **2005**, *26*, 5120–5129.
195 Fu, G., Valiyaveettil, S., Wopenka, B., Morse, D. E. *Biomacromolecules* **2005**, *6*, 1289–1298.
196 Gotliv, B.-A., Addadi, L., Weiner, S. *ChemBioChem* **2003**, *4*, 522–529.
197 Meldrum, F. C. *Int. Mater. Rev.* **2003**, *48*, 187–224.
198 Deng, S. G., Cao, J. M., Feng, J., Guo, J., Fang, B. Q., Zheng, M. B., Tao, J. *J. Phys. Chem. B* **2005**, *109*, 11473–11477.
199 Mukkamala, S. B., Powell, A. K. *Chem. Commun.* **2004**, 918–919.
200 Estroff, L. A., Incarvito, C. D., Hamilton, A. D. *J. Am. Chem. Soc.* **2004**, *126*, 2–3.
201 Aizenberg, J., Muller, D. A., Grazul, J. L., Hamann, D. R. *Science* **2003**, *299*, 1205–1208.
202 Yang, S., Chen, G., Megens, M., Ullal, C. K., Han, Y.-J., Rapaport, R., Thomas, E. L., Aizenberg, J. *Adv. Mater.* **2005**, *17*, 435–438.
203 Kuang, M., Wang, D., Gao, M., Hartmann, J., Möhwald, H. *Chem. Mater.* **2005**, *17*, 656–660.
204 Michalet, X., Pinaud, F. F., Bentolila, L. A., Tsay, J. M., Doose, S., Li, J. J., Sundaresan, A., Wu, A. M., Gambhir, S. S., Weiss, S. *Science* **2005**, *307*, 538–544.
205 Gould, P. *Mater. Today* **2004**, *7*, 36–43.
206 Medintz, I. L., Uyeda, T. H., Goldman, E. R., Mattoussi, H. *Nat. Mater.* **2005**, *4*, 435–446.
207 Steinbuchel, A. *Curr. Opin. Biotechnol.* **2005**, *16*, 607–613.
208 Mohanty, A. K., Wibowo, A., Misra, M., Drzal, L. T. *Polym. Eng. Sci.* **2003**, *43*, 1151–1161.
209 Mohanty, A. K., Misra, M., Drzal, L. T. *J. Polym. Environ.* **2002**, *10*, 19–26.
210 Jayasekara, R., Harding, I., Bowater, I., Lonergan, G. *J. Polym. Environ.* **2005**, *13*, 231–251.
211 Wang, M., Gan, D., Wooley, K. L. *Macromolecules* **2001**, *34*, 3215–3223.
212 Cai, Q., Shi, G. X., Bei, J. Z., Wang, S. G. *Biomaterials* **2003**, *24*, 629–638.
213 Pandey, J. K., Reddy, K. R., Kumar, A. P., Singh, R. P. *Polym. Degrad. Stability* **2005**, *88*, 234–250.
214 Tyler, D. R. *J. Macromol. Sci. Polym. Rev.* **2004**, *C44*, 351–388.
215 Nagai, N., Matsunobe, T., Imai, T. *Polym. Degrad. Stability* **2005**, *88*, 224–233.
216 Chiellini, E., Corti, A., Swift, G. *Polym. Degrad. Stability* **2003**, *81*, 341–351.
217 Tsarevsky, N. V., Matyjaszewski, K. *Macromolecules* **2005**, *38*, 3087–3092.
218 Gao, H., Tsarevsky, N. V., Matyjaszewski, K. *Macromolecules* **2005**, *38*, 5995–6004.
219 Dailey, L. A., Wittmar, M., Kissel, T. *J. Control. Release* **2005**, *101*, 137–149.
220 Qiu, L. Y., Bae, Y. H. *Pharm. Res.* **2006**, *23*, 1–30.
221 Svenson, S., Tomalia, D. A. *Adv. Drug Deliv. Rev.* **2005**, *57*, 2106–2129.
222 Boccaccini, A. R., Blaker, J. J. *Exp. Rev. Med. Dev.* **2005**, *2*, 303–317.
223 Randolph, M. A., Anseth, K., Yaremchuk, M. J. *Clin. Plast. Surg.* **2003**, *30*, 519–524.
224 Mahoney, M. J., Anseth, K. S. *Biomaterials* **2006**, *27*, 2265–2274.
225 Shin, H. J., Lee, C. H., Cho, I. H., Kim, Y. J., Lee, Y. J., Kim, I. A., Park, K. D., Yui, N., Shin, J. W. *J. Biomater. Sci. Polym. Ed.* **2006**, *17*, 103–119.
226 Fantner, G. E., Oroudjev, E., Schitter, G., Golde, L S., Thurner, P., Finch, M. M., Turner, P., Gutsmann, T., Morse, D. E., Hansma, H., Hansma, P. K. *Biophys. J.* **2006**, *90*, 1411–1418.
227 Mecerreyes, D., Lee, V., Hawker, C. J., Hedrick, J. L., Wursch, A., Volksen, W., Magbitang, T., Huang, E., Miller, R. D. *Adv. Mater.* **2001**, *13*, 204–208.
228 Langer, R., Tirrell, D. A. *Nature* **2004**, *428*, 487–492.
229 Batrakova, E. V., Vinogradov, S. V., Robinson, S. M., Niehoff, M. L., Banks, W. A., Kabanov, A. V. *Bioconjug. Chem.* **2005**, *16*, 793–802.
230 Haynie, D. T., Zhang, L., Rudra, J. S., Zhao, W., Zhong, Y., Palath, N. *Biomacromolecules* **2005**, *6*, 2895–2913.
231 Brzezinska, K. R., Deming, T. J. *Macromol. Biosci.* **2004**, *5*, 566–569.
232 Lele, B. S., Murata, H., Matyjaszewski, K., Russell, A. J. *Biomacromolecules* **2005**, *6*, 3380–3387.

233 Parras, P., Castelletto, V., Hamley, I.W., Klok, H.-A. *Soft Matter* **2005**, *1*, 284–291.
234 Turner, J.L., Becker, M.L., Li, X., Taylor, J.-S.A., Wooley, K.L. *Soft Matter* **2005**, *1*, 69–78.
235 Stine, R., Pishko, M.V., Hampton, J.R., Dameron, A.A., Weiss, P.S. *Langmuir* **2005**, *21*, 11352–11356.
236 Shikanov, A., Domb, A.J. *Biomacromolecules* **2006**, *7*, 288–296.
237 Martin, B., Sainlos, M., Aissaoui, A., Oudrhiri, N., Hauchecorne, M., Vigneron, J.P., Lehn, J.M., Lehn, P. *Curr. Pharm. Des.* **2005**, *11*, 375–394.
238 Zhong, S., Teo, W.E., Zhu, X., Beuerman, R., Ramakrishna, S., Yung, L.Y.L. *Biomacromolecules* **2005**, *6*, 2998–3004.
239 Bu, H., Kjøniksen, A.-L., Knudsen, K. D., Nyström, B. *Langmuir* **2005**, *21*, 10923–10930.
240 Kvien, I., Tanem, B.S., Oksman, K. *Biomacromolecules* **2005**, *6*, 3160–3165.
241 Karaky, K., Reynaud, S., Billon, L., François, J., Chreim, Y. *J. Polym. Sci., Part A: Polym. Chem.* **2005**, *43*, 5186–5194.
242 Wang, J.-L., Wang, L., Dong, C.-M. *J. Polym. Sci., Part A: Polym. Chem.* **2005**, *43*, 5449–5457.
243 Wang, L., Wang, J.-L., Dong, C.-M. *J. Polym. Sci., Part A: Polym. Chem.* **2005**, *43*, 4721–4730.
244 Srivastava, R.K., Albertsson, A.-C. *J. Polym. Sci., Part A: Polym. Chem.* **2005**, *43*, 4206–4216.
245 Srivastava, R.K., Albertsson, A.-C. *Macromolecules* **2006**, *39*, 46–54.
246 Kumar, R., Sharma, A.K., Parmar, V.S., Watterson, A.C., Chittibabu, K.G., Kumar, J., Samuelson, L. A. *Chem. Mater.* **2004**, *16*, 4841–4846.
247 Peeters, J.W., Palmans, A.R.A., Meijer, E.W., Koning, C.E., Heise, A. *Macromol. Rapid Commun.* **2005**, *26*, 684–689.
248 Kaihara, S., Osanai, Y., Nishikawa, K., Toshima, K., Doi, Y., Matsumara, S. *Macromol. Biosci.* **2005**, *5*, 644–652.
249 Kallinteri, P., Higgins, S., Hutcheon, G.A., St. Pourçain, C.B., Garnett, M.C. *Biomacromolecules* **2005**, *6*, 1885–1894.
250 Chakraborty, S., Sahoo, B., Teraoka, I., Miller, L.M., Gross, R.A. *Macromolecules* **2005**, *38*, 61–68.
251 Kulshrestha, A.S., Sahoo, B., Gao, W., Fu, H., Gross, R.A. *Macromolecules* **2005**, *38*, 3205–3213.
252 Kulshrestha, A.S., Gao, W., Gross, R.A. *Macromolecules* **2005**, *38*, 3193–3204.
253 Maier, T.H.P. *Nat. Biotechnol.* **2003**, *21*, 422–427.
254 Maskarinec, S.A., Tirrell, D.A. *Curr. Opin. Biotechnol.* **2005**, *16*, 422–426.
255 Kirshenbaum, K., Carrico, I.S., Tirrell, D.A. *ChemBioChem* **2002**, *3*, 235–237.
256 Zhang, K., Diehl, M.R., Tirrell, D.A. *J. Am. Chem. Soc.* **2005**, *127*, 10136–10137.
257 Fujikawa, S.-I., Ohmae, M., Kobayashi, S. *Biomacromolecules* **2005**, *6*, 2935–2942.
258 Wang, X., Wu, Q., Wang, N., Lin, X.-F. *Carbohydr. Polym.* **2005**, *60*, 357–362.
259 Nakamura, I., Yoneda, H., Maeda, T., Makino, A., Ohmae, M., Sugiyama, J., Ueda, M., Kobayashi, S., Kimura, S. *Macromol. Biosci.* **2005**, *5*, 623–628.
260 Ochiai, H., Ohmae, M., Mori, T., Kobayashi, S. *Biomacromolecules* **2005**, *6*, 1068–1084.
261 Bongiovanni, R., Malucelli, G., Messori, M., Pilati, F., Priola, A., Tonelli, C., Toselli, M. *J. Polym. Sci., Part A: Polym. Chem.* **2005**, *43*, 3588–3599.
262 Buffa, F., Hu, H., Resasco, D.E. *Macromolecules* **2005**, *38*, 8258–8263.
263 Kricheldorf, H.R., Hachmann-Thiessen, H., Schwarz, G. *J. Polym. Sci., Part A: Polym. Chem.* **2005**, *43*, 3667–3674.
264 Tew, G.N., Sanabria-De Long, N., Agrawal, S.K., Bhatia, S.R. *Soft Matter* **2005**, *1*, 253–258.
265 Zhang, X., Yang, H., Liu, Q., Zheng, Y., Xie, H., Wang, Z., Cheng, R. *J. Polym. Sci., Part A: Polym. Chem.* **2005**, *43*, 4857–4869.
266 García-Martín, M.G., Pérez, R.R., Hernández, E.B., Espartero, J.L., Muñoz-Guerra, S., Galbis, J. A. *Macromolecules* **2005**, *38*, 8664–8670.
267 Pornsuriyasak, P., Gangadharmath, U.B., Rath, N.P., Demchenko, A.V. *Org. Lett.* **2004**, *6*, 4515–4520.
268 Ramakrishnan, A., Pornsuriyasak, P., Demchenko, A.V. *J. Carbohydr. Chem.* **2005**, 24649–24663.

269 Yoshimura, T., Ishihara, K., Esumi, K. *Langmuir* **2005**, *21*, 10409–10415.

270 Heffernan, M. J., Murthy, N. *Bioconjug. Chem.* **2005**, *16*, 1340–1342.

271 Gillies, E. R., Jonsson, T. B., Fréchet, J. M. J. *J. Am. Chem. Soc.* **2004**, *126*, 11936–11943.

272 Hino, T., Endo, T. *J. Polym. Sci., Part A: Polym. Chem.* **2005**, *43*, 5323–5327.

273 Yokoe, M., Aoi, K., Okada, M. *J. Polym. Sci., Part A: Polym. Chem.* **2005**, *43*, 3909–3919.

274 Feng, Y., Zhang, S. *J. Polym. Sci., Part A: Polym. Chem.* **2005**, *43*, 4819–4827.

275 Guan, H.-L., Xie, Z.-G., Zhang, P.-B., Wang, X., Chen, X.-S., Wang, X.-H., Jing, X.-B. *J. Polym. Sci., Part A: Polym. Chem.* **2005**, *43*, 4771–4780.

276 Kricheldorf, H. R., Rost, S. *Macromolecules* **2005**, *38*, 8220–8226.

277 Endo, T., Kakimoto, K., Ochiai, B., Nagai, D. *Macromolecules* **2005**, *38*, 8177–8182.

278 Aliferis, T., Iatrou, H., Hadjichristidis, N. *J. Polym. Sci., Part A: Polym. Chem.* **2005**, *43*, 4670–4673.

279 Shen, W., Lammertink, R. G. H., Sakata, J. K., Kornfield, J. A., Tirrell, D. A. *Macromolecules* **2005**, *38*, 3909–3916.

280 Guo, K., Chu, C. C. *J. Polym. Sci., Part A: Polym. Chem.* **2005**, *43*, 3932–3944.

281 Wu, D., Liu, Y., Jiang, X., Chen, L., He, C., Goh, S. H., Leong, K. W. *Biomacromolecules* **2005**, *6*, 3166–3173.